Planning Production and Inventories in the Extended Enterprise

International Series in Operations Research & Management Science

Volume 151

For other volumes:
http://www.springer.com/series/6161

Karl G. Kempf • Pinar Keskinocak • Reha Uzsoy
Editors

Planning Production and Inventories in the Extended Enterprise

A State of the Art Handbook, Volume 1

Springer

Editors
Karl G. Kempf
Intel Corporation
5000 W. Chandler Blvd.
MS CH3-10
Chandler, Arizona 85226, USA
karl.g.kempf@intel.com

Pinar Keskinocak
Georgia Institute of Technology
School of Industrial
& Systems Engineering
Ferst Drive NW., 765
30332-0205 Atlanta Georgia
USA
pinar@isye.gatech.edu

Reha Uzsoy
Edward P. Fitts Department
of Industrial and Systems Engineering
North Carolina State University
Raleigh, NC 27695-7906
USA
ruzsoy@ncsu.edu

ISBN 978-1-4614-2609-7 ISBN 978-1-4419-6485-4 (eBook)
DOI 10.1007/978-1-4419-6485-4
Springer New York Dordrecht Heidelberg London

Printed on acid-free paper

Springer is part of Springer Science+Business Media (www.springer.com)

Contents

Acknowledgements

The process by which these volumes came to fruition has been a long one, and we are grateful to many different people for their support, advice and contributions. First among these are the contributing authors – without them, there would be no book, and we are deeply grateful to them for bearing with our repeated requests for materials and revisions while providing the high-quality contributions worthy of these volumes. The many reviewers who gave their time and effort to help improve the chapters are also gratefully acknowledged.

Thanks are also due to Professor F. Hillier, the Editor of this Series, for supporting the project, and to Gary Folven, under whose guidance the project took shape, and who sustained us through the first 4 years until his well-deserved retirement. His many contributions to operations research over a long and distinguished publishing career will stand for a long time. Neil Levine and Matthew Amboy of Springer saw the volumes through production, nudging the editorial team when necessary and exhibiting resourcefulness and patience above and beyond the call of duty. Thanks are also due to Carrie Brooks of North Carolina State University for preparing the list of authors for indexing at very short notice.

An undertaking of this magnitude and duration would simply not be possible without a collegial, mutually supporting editorial team. It has been a privilege to work together on this project; we have all learnt from each other and from the authors who have contributed, and are grateful to have had this opportunity to work with some of the best in our field.

Finally, we would like to thank our families, who have had to put up with late nights and grumpy mornings over the duration of the project, and without whose support and understanding our lives would be much poorer in many ways.

Contributors

Bruce E. Ankenman Department of Industrial Engineering and Management Sciences, Northwestern University, 2145 Sheridan Rd., Evanston, IL 60208, USA, ankenman@northwestern.edu

Dieter Armbruster Department of Mathematics, Arizona State University, Tempe, AZ 85287-1804, USA, armbruster@asu.edu

Tolga Aydinliyim Decision Sciences Department, Charles H. Lundquist College of Business 1208, University of Oregon Eugene, OR 97403-1208, USA, tolga@uoregon.edu

Jennifer M. Bekki Department of Engineering, Arizona State University at the Polytechnic Campus, 7231 E. Sonoran Arroyo Mall, Room 230, Mesa, AZ 85212, USA, jennifer.bekki@asu.edu

Ebru K. Bish Grado Department of Industrial and Systems Engineering (0118), 250 Durham Hall, Virginia Tech, Blacksburg, VA 24061, USA, ebru@vt.edu

Robert C. Carlson Management Science and Engineering Department, Stanford University, Stanford, CA 94305-4026, USA

Carlos Daganzo Department of Civil and Environmental Engineering, Institute of Transportation Studies, University of California, Berkeley, CA 94720, USA

Borga Deniz The Joseph W. Luter, III School of Business Christopher Newport University Newport News, VA 23606, USA, borga.deniz@cnu.edu

Laura Dionne Intel Corporation, Hillsboro, OR, USA

Salah E. Elmaghraby North Carolina State University, Raleigh, NC 27695-7906, USA, elmaghra@eos.ncsu.edu

Feryal Erhun Department of Management Science and Engineering, Stanford University Stanford, CA 94305-4026, USA, ferhun@stanford.edu

John Fowler Department of Industrial Engineering, Arizona State University, PO Box 875906, Tempe AZ 85287-5906, USA, john.fowler@asu.edu

Stephen C. Graves MIT, 77 Massachusetts Avenue, E62-579, Cambridge MA 02139-4307, USA, sgraves@mit.edu

Aliza Heching IBM Research Division, T. J. Watson Research Center, 1101 Kitchawan Road, Yorktown Heights, NY 10598, USA, ahechi@us.ibm.com

Jonathan R. M. Hosking IBM Research Division, T. J. Watson Research Center, 1101 Kitchawan Road, Yorktown Heights, NY 10598, USA, hosking@watson.ibm.com

Itir Z. Karaesmen Kogod School of Business, American University, 4400 Massachusetts Avenue NW, Washington DC 20016, ikaraesme@american.edu

Karl G. Kempf Intel Corporation, 5000 W. Chandler Blvd., MS CH3-10, Chandler, Arizona 85226, USA, karl.g.kempf@intel.com

Pinar Keskinocak School of Industrial and Systems Engineering, Georgia Institute of Technology Atlanta, GA 30332-0205, USA, pinar@isye.gatech.edu

Alan King IBM Research Division, T. J. Watson Research Center, 1101 Kitchawan Road, Yorktown Heights, NY 10598, USA, kingaj@us.ibm.com

Erjen Lefeber Department of Mechanical Engineering, Eindhoven University of Technology, PO Box 513, Eindhoven, The Netherlands, A.A.J.Lefeber@tue.nl

Gerald T. Mackulak Department of Industrial Engineering, Arizona State University, PO Box 875906, Tempe AZ 85287-5906, USA, mackulak@asu.edu

Bacel Maddah Engineering Management Program, American University of Beirut, Beirut 1107 2020, Lebanon, bacel.maddah@aub.edu.lb

Kenneth N. McKay Department of Management Sciences, University of Waterloo, Waterloo, ON, Canada, kmckay@uwaterloo.ca

Hubert Missbauer Department of Information Systems, Production and Logistics Management, University of Innsbruck, Universitätstrasse 1,5 Innsbruck A-6020, Austria, hubert.missbauer@uibk.ac.at

Brenda Munroe Hannaford Bros., 145 Pleasant Hill Rd, Scarborough, ME 04074, USA, BMunroe@hannaford.com

Barry L. Nelson Department of Industrial Engineering & Management Sciences, Northwestern University, 2145 Sheridan Road, Room C250, Evanston, IL 60208-3119, USA, nelsonb@northwestern.edu

Yanfeng Ouyang Department of Civil and Environmental Engineering, University of Illinois at Urbana-Champaign, Urbana, IL 61801, USA, yfouyang@illinois.edu

Irfan M. Ovacik Intel Corporation, 6505 W. Chandler Blvd., Chandler, AZ 85226, USA, irfan.m.ovacik@intel.com

Özalp Özer The University of Texas at Dallas, School of Management, Office: SOM 3.412, 800 West Campbell Road, Richardson, Texas 75080-3021, oozer@utdallas.edu

Joesph Pekny College of Engineering, Purdue University, West Lafayette, IN 47907, USA, pekny@purdue.edu

Dariush Rafinejad Management Science and Engineering Department, Stanford University, Stanford, CA 94305-4026, USA, rafineja@stanford.edu

Gintaras V. Reklaitis Burton and Kathryn Gedge Distinguished Professor of Chemical Engineering, Purdue University, West Lafayette, IN 47907-1283, USA, reklaiti@ecn.purdue.edu

Alan Scheller-Wolf Tepper School of Business Carnegie Mellon University Pittsburgh, PA 15215, USA, awolf@andrew.cmu.edu

Reha Uzsoy Edward P. Fitts Department of Industrial and Systems Engineering, 300 Daniels Hall, Campus Box 7906, North Carolina State University, Raleigh, NC 27695-7906, USA, ruzsoy@ncsu.edu

George L. Vairaktarakis Department of Operations, Weatherhead School of Management Case Western Reserve University 10900 Euclid Avenue Cleveland, OH 44106-7235, USA, gxv5@case.edu

Feng Yang Industrial and Management Systems Engineering, West Virginia University, PO Box 6070, Morgantown, WV 26506-6107, USA, feng.yang@mail.wvu.edu

Juan Camilo Zapata Eli Lilly & Co., Indianaplois, IN, zapata_juan_camilo@lilly.com

Chapter 1
Preface

Karl G. Kempf, Pinar Keskinocak, and Reha Uzsoy

1.1 Overview of the Problem

Production planning has been an integral part of industry since the beginnings of craft production. Basic quantitative research in this area is at least 50 years old (Modigliani and Hohn 1955). However, it is widely recognized that there are significant gaps between the research problems addressed, the state of industrial practice, and the needs of the industrial community. It is our experience that production planning is viewed as a narrow function at the interface of production and sales, and the basic modeling and solution approaches associated with this view have remained largely unchanged for several decades.

We first discuss the traditional, narrow view of production planning as the allocation of production capacity to different products and customers to optimize some measure of the firm's financial performance. This view has been widely studied in the academic and practice literatures. To place the traditional view of production planning in context, we present a set of realistic production planning problems that are minimal in the sense that the removal of any particular feature significantly reduces the fidelity of the problem to industrial practice. We discuss some basic approaches to these problems taken in industry, relate them to the quantitative research to date and conclude that a complete, general, and repeatable solution to this minimal but realistic problem set does not yet exist.

Having made a case that even the traditional, narrowly defined production planning problem has not yet been well solved, we suggest that a variety of economic and technological trends, such as the globalization of supply chains and the accessibility of real-time data, have significantly altered the environment within which production planning now takes place. To illustrate the possibilities, we present brief discussions of three axes along which we feel the scope of this function in the firm has been expanded in recent years: (a) looking upstream, explicit consideration of subcontractors' and suppliers' capabilities and risk profiles,

K.G. Kempf (✉)
Intel Corporation, 5000 W. Chandler Blvd., MS CH3-10, Chandler, Arizona 85226, USA
e-mail: karl.g.kempf@intel.com

K.G. Kempf et al. (eds.), *Planning Production and Inventories in the Extended Enterprise*, International Series in Operations Research & Management Science 151,
DOI 10.1007/978-1-4419-6485-4_1,

(b) looking downstream, active integration of demand management with production planning, and (c) looking internally, the role and structure of production planning models in an environment with multiple interacting organizations with different incentives and objectives.

1.2 Realistic Production Planning Problems

The traditional view of production planning as espoused in both the academic and practice literatures since the late nineteenth century has focused on the efficient and effective use of production resources to further the firm's financial goals. This involves the allocation of capacity to products in support of customer satisfaction and the firm's profitability.

To illustrate the core decision complexity of the production planning challenge as it exists in practice, we will use a concrete instance of a simple manufacturing system. On the one hand, from a modeling perspective, the instance can be completely described by a handful of entities and parameters. It is minimal in the sense that removing any entity from the structure or making simplifying assumptions diminishes the mapping of the sample problem to reality. On the other hand, from a production planning perspective, it contains much of the decision complexity managed on a daily basis by practicing production planners. Of course, any number of additional entities and complicating features can be included, some of which we will discuss later. But the core decision complexity of production planning as encountered in industrial practice lies in this simple example.

On the supply side of the problem (bottom of Fig. 1.1), there is one existing factory with equipment and a production process in place. The equipment set is composed of two machines, one that batches (i.e., Ma processes a number of parts

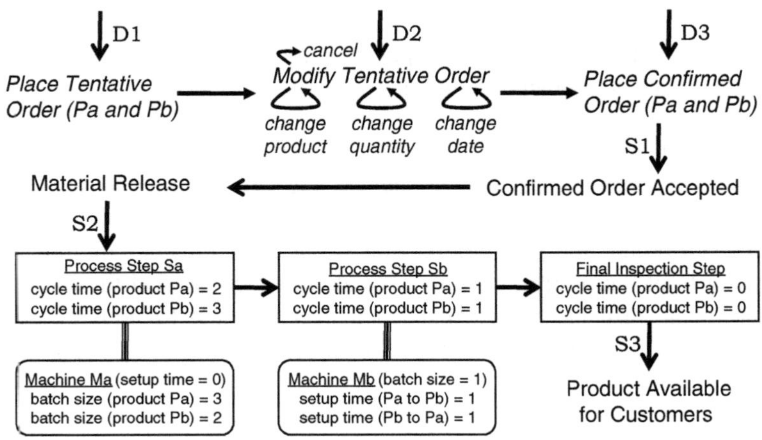

Fig. 1.1 Demand and supply in the realistic minimal problem set

simultaneously) and one that processes one part at a time (Mb). The production process has only one route with two strictly ordered processing steps (Sa before Sb). Ma is qualified to run only Sa, and Mb to run only Sb. Two products can be manufactured in the factory (Pa and Pb) on the full production route (Sa, Sb).

Machine Ma batches three items of product Pa with a processing time of 2 units. Machine Mb runs one item of product Pa at a time with a cycle time of 1 units. At steady state achieving maximum throughput while producing only product Pa, the average output of the factory will be one item every 1 time unit. The utilization of machine Ma will be roughly 67%, while the utilization of machine Mb will approach 100%.

Machine Ma processes batches consisting of two items of product Pb with a processing time of 3 units. Machine Mb runs one item of product Pb at a time with a cycle time of 1 units. At steady state, achieving maximum throughput while producing only product Pb, the average output of the factory will be one item every 1.5 time unit. The utilization of machine Ma will approach 100%, while the utilization of machine Mb will be roughly 67%.

Machine Mb can be setup to run in two configurations, one required for running product Pa and the other required for running product Pb. It takes 1 time unit to change setups.

Factory throughput time (TPT) is deterministic and distributed. In the case of production dedicated to Pa, from the release of raw material to Sa through the completion of Sb, TPT would exhibit a uniform distribution with 1/3 of the items exiting the factory at each of 3, 4, and 5 time units. In the case of dedicated Pb production, 1/2 of the items would exit at each of 4 and 5 time units. Mean TPT when focused on producing Pa is 4 time units and $4^1/_2$ time units when focused on Pb (or worse in both cases depending on setup changes incurred). The factory throughput rises with rising factory work in process inventory (WIP) until the maximum throughput of the constraining machine is reached. Beyond this point, if release of raw materials continues, throughput remains constant with increasing WIP and factory TPT.

Finally, an inspection is performed after the completion of step Sb before the product leaves the factory that requires negligible time and capacity to perform. This inspection will reject a random number of units drawn from a uniform distribution spanning from 1 to 5 for every 100 instances of either product Pa or Pb produced.

Full data are available on past system performance. There are no other constraints on the system: there is an adequate supply of raw materials, machines are available 100% of the time, there is adequate WIP storage inside the factory, and so on. However, any planning algorithm for this system will be required to manage the limited capacity including throughput and TPT nonlinearity with load, the dependence of throughput, TPT, batching and setup changes on product mix, and the stochasticity of product yield.

On the demand side of the problem (top of Fig. 1.1), there are two existing customers (Ca and Cb), each ordering both products Pa and Pb. An order consists of *product name: desired delivery quantity: desired delivery date*. Product substitutions and partial delivery quantities are not acceptable. However, the customer is willing to allow a tolerance around the delivery date prenegotiated as an earliest and latest acceptable. The delivery will not be accepted if it arrives before the earliest acceptable time and the order will be automatically cancelled if it has not been filled by the latest acceptable time.

At any given point in time, there are a variety of data available for production planning purposes.

(a) Forecasted or tentative orders due in the future placed as an indication of what the customer might order. These orders can be cancelled or changed in product, quantity, and due date until order confirmation at the discretion of the customer.
(b) Confirmed orders due currently or in the future that cannot be changed or cancelled except for late delivery.
(c) A historical record of the dates that tentative orders were placed, changed, cancelled, confirmed and that confirmed orders were delivered or automatically cancelled due to tardiness.

Note that the demand from either customer for each product may change over time in unpredictable ways. Any planning algorithm will be required to manage these nonstationary, uncertain demand signals.

This is an ongoing supply-demand system and so a production plan must be generated repeatedly on a regular basis many periods into the future. The decisions that can be made by the customers in each time interval include:

D1) Placing a tentative order,
D2) Changing a tentative order including cancellation, and
D3) Confirming a tentative order.

The decisions that must be made by the producer in every time interval include:

S1) What tentative orders to accept on confirmation,
S2) How much material to release into the factory, and
S3) What orders to fill with products exiting the factory.

The goal of the production planning system is to maximize profits. There is a cost per unit to manufacture (different for Pa and Pb), and a selling price per unit that has been prenegotiated (different for Pa and Pb). Transportation costs are assumed to be included in the selling price.

In the example presented thus far, all of the supply side parameters have been specified, but none of the demand side parameters are given. This allows the specification of a rich set of subspaces based on the relationship between the supply and demand parameters (assuming both customers are ordering both products).

Subspace 1: There is a set of production planning problems in a subspace defined by the value of the mean demand with respect to mean supply.

1. If the mean demand is less than the mean supply, the supplier is more focused on decision S2. Given extra supply capacity, deciding which tentative orders to accept (S1) becomes simpler, as does deciding what to do when production material exits the factory (S3).

2. If the mean demand is greater than the mean supply, the problem for the supplier becomes more difficult. All orders cannot be taken because of capacity limits (S1). Given a heavily loaded factory, nonlinearities in throughput and TPT become more important, stressing effective management of material release into the factory (S2), and matching factory output to orders becomes more of a challenge (S3).
3. If the mean demand is approximately equal to the mean supply averaged over many periods, but may be randomly less or greater than the mean supply in individual periods, the material release decision (S2) for the supplier involves "build-ahead" to use excess capacity in low-demand periods to manufacture products to (hopefully) cover orders in high-demand periods.

Subspace 2: There is a set of production planning problems in a subspace defined by the value of the mean of the demand lead time (when the tentative order is confirmed relative to the requested due date) with respect to the mean of the supply lead time (when the items exit the factory relative to when the raw materials to produce them was released into the factory).

1. If the mean of the demand lead time is greater than the mean of the supply lead time, the supplier is more focused on decision S2. Given extra supply lead time, deciding which tentative orders to accept (S1) becomes simpler, as does deciding what to do when production material exits the factory (S3).
2. If the mean of the demand lead time is less than the mean of the supply lead time, the producer must consider the interaction and integration of decisions S1, S2, and S3, since the material needed to fill tentative orders will have to be released into the manufacturing facility before there is a confirmed demand signal.

Other features of the demand parameterization can make the production planning problem easier or harder to formulate and/or solve through their impact on the inherent uncertainty that must be managed. Zero tolerance for delivery timing makes the problem more difficult in every time period. This means that the factory must make the right amount of product at the right time (or risk holding inventory) in spite of the nonlinearity and stochasticity in the production system. As these tolerances loosen, it becomes easier to accommodate these supply planning difficulties. Increased frequency and/or increased magnitude of placing and/or changing tentative orders makes the problem harder over time. Solving at any particular point in time is not impacted, but solving from period to period with a set of tentative orders that is changing frequently in large steps means that the resulting plan may contain large changes from period to period causing undesirable thrash in the factory.

Clearly a great deal of our ability to plan production in the system above depends on the amount, timing and quality of information we have available. For the sake of simplicity, we shall follow the majority of the literature in assuming that the firm involved in this minimal example has complete access to all its internal data for planning purposes and that its operations are governed by a single well-defined objective, in this case maximizing profits. These assumptions are all questionable in practice. Few firms own their entire supply chain, and competing objectives are common even among different divisions of the same firm. We will return to these issues later.

The realistic minimal problems outlined above combine two fundamental aspects that render the production planning problem both difficult and interesting. The first of these is the presence of limited capacity with nonnegligible TPT. The effect of this constraint is that the system cannot respond to demands instantaneously resulting in significant production lead times. Limited capacity leads to a number of nonlinear relationships between critical system variables such as utilization and lead time (e.g., Hopp and Spearman 2001), and its allocation between competing products or customers is a significant source of complexity. While a firm may eliminate these difficulties by maintaining significant levels of excess capacity, this is generally not an economical proposition, especially in capital-intensive industries.

The second aspect of the minimal problem is the prevalence of uncertainty. Historically, uncertainty due to supply side considerations generally originates within the firm and is thus at least in theory amenable to analysis and improvement by the firm's personnel. Uncertainties associated with the demand side tend to originate outside the firm and are often associated with factors such as the general economic climate and the actions of competitors about which the firm often has limited information and control. We will discuss these two aspects separately in the following sections.

1.3 Quantitative Research in Production Planning

The realistic minimal problems described above provide an interesting lens through which to view the extensive body of literature on quantitative models developed to support some aspect of production planning. Several such streams of literature have developed, often largely independent of each other and increasingly specialized around the particular mathematical tools they require, such as mathematical programming, queuing, stochastic inventory theory, or dynamic programming. In general, each stream of literature has developed its own set of broadly accepted assumptions and problem formulations that are often divergent to some degree from industrial reality even in the form of the realistic minimal problems above.

Consider the extensive literature on mathematical programming models for production planning (e.g., Johnson and Montgomery 1974; Hax and Candea 1984; Hackman and Leachman 1989; Voss and Woodruff 2003). This body of research generally takes a completely deterministic view of demand and ignores the nonlinearities due to congestion in the production process. The basic problem formulation focuses on the supply side issues of assigning work to resources in discrete planning periods that has not substantially changed since the first papers in this area in the 1950s (Modigliani and Hohn 1955). The prevalent representation of a production facility includes a fixed lead time with an aggregate capacity limit over the basic planning period of interest.

It should be noted that there are relatively few references on production planning problems in the leading management science journals in the last few years and many of these are focused not only on formulations but also on solution techniques (for example, the application of cutting plane techniques to lot sizing problems). With a few exceptions, these models take the perspective of a centralized planning department with full access to all relevant internal information and complete control of all production capacity of interest. There is little explicit consideration of the stochastic aspects of demand in the production plan, although some models consider stochastic process yields in some detail. Despite broad recognition that most production systems are governed by the dynamics of queuing, there has been very little effort to apply the extensive body of insights derived from the queuing literature to optimization models of production planning problems. Relationships with partners outside the firm, such as suppliers or subcontractors, are represented at best as hard constraints on the amount of material available or the amount of production that can be subcontracted. The techniques of stochastic programming (Birge and Louveaux 1997), which have received extensive application in areas such as energy planning and financial services, have not been applied to production planning problems to any comparable extent.

While there is an extensive literature on forecasting, examples of research linking forecasting to production planning are again relatively few (e.g., Graves et al. 1998; Heath and Jackson 1994; Kaminsky and Swaminathan 2001) although they are more common in the inventory literature (e.g., Eppen and Martin 1988). The vast majority of the forecasting literature treats the problem of demand forecasting as an exercise in econometrics or time series analysis with no consideration of any supply side issues or what will be done with the forecast once it has been obtained. In the area of facility location models, the vast majority of models focus on the tradeoff between transportation costs and the fixed costs of facility location, ignoring operational aspects such as lead times and the ensuing safety stock requirements that define a significant portion of the operating cost of the supply chain.

The extensive body of literature that has developed around the management of inventories (e.g., Hadley and Whitin 1963; Zipkin 1997) appears at first glance to bring together the demand and supply aspects of the problem that have been considered separately in the work mentioned above. However, this body of work also has its set of accepted assumptions that limit its applicability to our difficult minimal problem set. The vast majority of stochastic inventory models view the inventory system independently of the manufacturing system from which it is replenished, generally representing this system by a replenishment lead time that is usually either a known constant or random with a time-stationary probability distribution. The nonlinear relationships between workload and lead time in a capacitated manufacturing system are generally not considered, although there are some notable exceptions (e.g., Zipkin 1986; Liu et al. 2004). The emphasis is heavily on infinite-horizon models and performance measures that are defined as long-run steady-state expectations. In recent years, the advent of supply chain management has led to the extension of the basic models to incorporate some microeconomic aspects, such as multiple independent actors, information asymmetries between supply chain participants or models of customer behavior, but these are generally highly stylized models of small systems that limit their practical value.

Deterministic inventory models (e.g., Kimms 1997; Bahl et al. 1987; Elmaghraby 1978) are generally direct lineal descendants of the economic lot sizing model of (Harris 1915) focusing on the tradeoff between fixed setup costs and inventory holding costs in the presence of deterministic demand and, at best, very limiting assumptions on capacity. Much of this work has been driven by applications in material requirements planning (MRP) systems (Orlicky 1975; Vollmann et al. 2005), but it is important to note that MRP is one particular solution to a limited aspect of production planning and does not by any means define the problem in full.

Inventories have become such a prevalent part of the production landscape that the problem of production or supply chain management is often presented as an inventory management problem. We would argue that while inventories represent a widespread solution to the problem of economically matching supply to demand in the face of limited production capacity and supply and demand uncertainties, they should not be confused with the problem itself. In recent years, a number of alternative means of managing uncertainties and limited capacity, such as creative contractual arrangements with customers and suppliers, use of financial instruments such as options, and active demand management through mechanisms such as dynamic pricing, have begun to be considered. These constitute a major expansion of the scope of the production planning problem and will be discussed more extensively below.

In general, we conclude that the existing body of quantitative models addresses at best only a subset of even our realistic minimal problem, failing to provide an integrated solution that considers the basic aspects of limited capacity and widespread uncertainty in a generalizable, scalable manner.

1.4 Industrial Approaches to Production Planning

In contrast to the relatively recent contributions to the academic literature on quantitative decision support models, industry has a long history of working to address the effects of limited capacity and widespread uncertainty that are encountered daily. On the supply side, these take the form of improving the production process to increase capacity and reduce the variability of the production process. On the demand side, a variety of approaches have been pursued to try to obtain better demand forecasts and employ them effectively.

Ever since the beginning of industrial production firms have recognized the importance of understanding and effectively managing their shop-floor operations. Early examples of this approach were often based on the application of scientific principles for better understanding of the core manufacturing processes. The work at companies such as the Carnegie Steel Corporation and Standard Oil in the second half of the nineteenth century culminated in the scientific management movement led by Frederick Taylor and his colleagues, most prominently in the machining industries (Kanigel 1997). The theme of these efforts, mostly undertaken by engineers, is a focus on data-driven continuous improvement involving extensive data collection on operations and its evaluation and monitoring. The basic idea is to understand the nature of both the capacity of the production process and the uncertainties inherent in the environment by measuring them, identifying opportunities to improve them where possible, and developing strategies to manage the system in the face of the uncertainty and capacity limitations that are either not economical or not possible to remove. The direct descendants of these early efforts today are the Just in Time/Lean Manufacturing approaches (e.g., Liker 2004), the Six Sigma movement (e.g., Pande et al. 2000), and Theory of Constraints (Goldratt and Fox 1986). Each has a focus on detailed understanding of shop-floor operations and systematic processes for continuous improvement, manifested as kaizen in JIT/Lean Manufacturing, the Define–Measure–Analyze–Implement–Control (DMAIC) cycle in Six Sigma, and the Identify–Exploit–Elevate–Repeat cycle in Theory of Constraints.

The prevalent approach to managing the effects of widespread uncertainty and limited capacity is the use of inventories. When different parts of the production system operate with different lead times, it is often desirable to hold inventory between the two stages so that the production and replenishment processes of each segment can be planned and managed independently. For example, a seller holds inventory in a make to stock or retail environment where the replenishment lead time is longer than the lead time acceptable to the customer. Similarly, a firm can build inventories to meet demand during peak periods where demand is predicted to exceed capacity.

This use of inventory to manage the effects of uncertainty and limited capacity is not applied only to the deployment of inventories within the factory, although it probably originated there. A great deal of effort in supply chain management is aimed at determining the appropriate location and quantity of inventories at different points in the supply chain, such as raw materials, semifinished goods, and finished goods inventories in the distribution system.

It is probably safe to say that in most firms, production organizations are relatively passive observers of demand uncertainty. They can observe it and try to prepare for it, but they cannot affect it. The business of altering demand, presumably in a manner favorable to the firm, is generally the domain of the sales organization in the short term and the marketing organization in the long term. The vast majority of conventional production planning efforts try to quantify demand uncertainty as well as possible, and then manage its effects mainly through the use of safety stocks at strategic points in the system.

1.5 Planning in the Expanding Supply Chain

As discussed above, the basic problem of production planning involves, at a minimum, the coordination of production through a capacitated manufacturing facility in the face of uncertain demand for multiple products. External agents such as customers and suppliers are represented largely by exogenous inputs typified by demand forecasts. We now consider broader supply chain planning problems that we believe illustrate the expanded scope of the planning activity in today's global supply chains.

The minimal problem we discussed earlier involved a single production facility, and the subproblems dealt with cases where the limited capacity of this factory was related to the volume of demand (subspace 1) and where the supply lead time was related to the demand lead time (subspace 2). In either case, an agonizing situation for a production planner is to be constrained by the company's manufacturing capacity (the one thing directly under the planner's control). In industrial practice, there are other firms that can potentially manufacture products Pa and Pb. While it is not the responsibility of production planners to deal with these other firms directly, when there is not sufficient capacity to meet demand, temporary or permanent outsourcing is a possibility. The production planner plays a vital role in these decisions including which products to outsource to whom in what volumes for what periods of time. The planner brings a detailed knowledge of capacitated stochastic manufacturing facilities that will be necessary to understand what commitments the other firm can make. A rudimentary understanding of contracting including options is vital for the production planner to play this role. However, for most products, finding an adequate capacity subcontractor is not possible from a tactical perspective and is left for discussion in the next section on proactive risk management. The best the planner can do is consult with the sales and marketing organization for a clear prioritization among products and customers to maximize revenue, while minimizing damage to the company's reputation and relationships.

The example problem assumed that there is always an adequate supply of raw materials. This luxury is not often afforded to industrial production planners. Material suppliers also have capacity restrictions and sell to a variety of customers. When materials cannot be acquired to support a production plan, the production planner must be included in replanning considering the availability of materials, a deep understanding of the needs of the firm, and a broad understanding of manufacturing physics including the material supplier's factory. It could be that some materials are more constraining than others or that rearrangement of the sequencing of orders to the material supplier can remedy the problem. Another possibility that the production planner can explore is revising the production plan to match the delivery of materials while minimizing the impact on satisfying the demand for the company's products. Once again, a rudimentary understanding of contracting including options is vital for the production planner to play this role.

In the initial problem, production planners received a demand signal that they had very little control over. In industrial practice, planners should be involved in decision making on capacity allocation with the sales and marketing personnel who are directly engaged with the customers. Idle capacity and unused inventory are incurring cost but not contributing to revenue, while unmet demand is a missed opportunity for revenue. Poor collaboration between planners and marketers can result in all three of these conditions simultaneously. When demand is the constraint, the production planner can again be in a precarious position. Since many factories are measured on equipment utilization, and no factory manager wants an idle factory, there will be the risk of building extra product. This will add manufacturing and inventory holding costs to build products that have at least a possibility of being scrapped. Working with sales and marketing to find ways to stimulate demand through pricing moves and special offers becomes part of the production planner's task. Care must be taken in these efforts and production planning expertise must be included to be sure that the effect that the sales and marketing team envisions is actually possible for the manufacturing team to execute in a timely fashion. For example, special sales and marketing activities such as price promotions intended to influence demand that cannot be supported by manufacturing will not only fail but might harm the reputation of the company. A basic understanding of the market allows the production planner to be more creative in using the firm's manufacturing capability to minimize cost and maximize revenue.

Potentially the worst situation for a production planner is a constraint that moves between capacity and materials and demand periodically over time. Proactively managing these possibilities is far preferable to attempting to manage them dynamically. Consider the initial example problem extended to include one capacity subcontractor, one material vendor, the possibility to influence demand, and a suite of financial instruments with which to manage these extensions. As with the initial problem, this extended problem includes interesting and tactically important subspaces.

1.5.1 Supply Side: Inventory vs. Other Proactive Supply Management Techniques

In this expanded view of production planning, the planner is the primary agent to allocate the resources of the company to satisfy demand and the consultative agent to support acquiring materials and extra capacity. It is certainly possible to stockpile raw materials and outfit factories with extra capacity to mitigate the risk of these items constraining demand satisfaction. However, there are alternative techniques that may be more effective at mitigating risk as well as less financially burdensome.

On the topic of material acquisition, an array of financial instruments and techniques may be required to manage risk over different time periods. Looking a few weeks into the future, it may be necessary to resort to the spot market to manage material shortages or to divest materials that are surplus to requirements (although the more specialized the materials the less useful this mechanism becomes). Whether our firm is the seller or the buyer, grounding in modern auction theory is essential for this activity. Looking a little farther into the future, where inventory is still a part of the solution, various types of contacts can be employed. Assuming that historical practice was fixed price procurement contracts including materials to be held as safety stocks and/or consigned inventory held at the company's site but owned by the supplier until used, options contracts can provide useful augmentation. Where applicable, options can reduce the amount of safety stock needed wherever it is located and whomever is its owner. The interesting challenges are negotiating whether the option is for a quantity at standard lead time or for a quantity with a reduced lead time and of course the option and strike prices. The production planner is a consultant to these negotiations given knowledge of the actual needs of the company including the demand the company is trying to satisfy into the future as well as having practical insight into the nonlinear stochastic manufacturing engine of the supplier. Since the supplier operates a factory with all of the difficulties described above, a default penalty might be a useful addition to the terms and conditions.

When considering capacity subcontracting, a different set of issues are encountered. The most obvious is the potential lack of a broad spot market. For another company to make your product to be sold into the market under your name, there have to be a number of preparatory activities. At least the subcontractor has to participate in some kind of qualification activity to assure capability and quality. Once this is accomplished, however, a mixture of conventional contracts and options is feasible. In pursuing options, the interesting negotiations are again exactly what the option is for in terms of quantities and lead times and how much the option is worth. Penalties are a topic of concern again since the subcontractor is operating a nonlinear and stochastic factory. A new concern arises here since the subcontractor may purchase materials from the same vendor as our company. Uncoordinated options written with the materials vendor and the capacity subcontractor may lead to an unsatisfiable situation. In any case, the production planner serves as a consultant in these negotiations and grounding in modern options theory is essential for this activity. Equally important is communication and collaboration between production planners and purchasing agents.

1.5.2 Demand Side: Forecasting vs. Other Proactive Demand Management Techniques

As discussed previously, much of the conventional view of production planning takes a somewhat passive view of demand as an exogenous process, where the best we can hope for is to predict it with better accuracy. Forecasting demand is a complex process that requires a carefully selected and analyzed subset from the abundance of information available today, including past and current demand data, "tentative" orders, the size of the customer base, seasonality of demand, the effects of product life-cycle, price and promotions, information about competitors, information about industry trends and new products, nonprice attributes such as warranties, service guarantees, etc., and other external factors such as economic trends and indicators. Rapid developments in information technology are making it possible to collect, organize and mine this type of data in a manner that was unimaginable even a few years ago.

While most forecasting efforts in the past have assumed the existence of adequate historical data, this assumption is increasingly questionable in today's rapidly changing markets. In our expanded view of the production planning problem in the supply chain, it is important to realize that our demand may depend on the recent actions of other actors several degrees removed from us in the supply chain. In the case of complex products or services, a customer may place an order after some discussions and negotiations with the sales force, and possibly after testing the product for some time. Such "tentative" orders may be very useful in forecasting demand, in particular, if the company also keeps information about how long it takes for the customer to place an order after the initial discussions start, what percentage of these orders are eventually confirmed, with what kind of modifications, at the individual customer level. However, tentative orders can sometimes lead to overly optimistic forecasts. For example, in 2001 Solectron's big customers, including Cisco, Ericsson, and Lucent were expecting explosive growth for wireless phones and networking gear. Basing its own projections on these expectations, Solectron ordered materials from its suppliers, and when the demand did not realize, ended up with $4.7 billion in inventory.

Since the final price a customer pays for a product impacts the customer's decision of purchasing, it is important to have information about prices, promotions, and other incentives offered to customers, even if the firm does not directly control such incentives. For manufacturers who do not sell directly to the end consumers of their products, an added challenge in planning and forecasting is the "middleman" between the manufacturer and its end customers. For example, auto manufacturers sell their products through dealers, consumer package goods manufacturers sell through retailers, and the demand observed by electronics contract manufacturers is dictated by the demand of their customers' customers. In such cases, the manufacturer usually receives the sales information about end customers with a significant delay or sometimes not at all. The demand of a product might depend on the other complement and substitute products (and their attributes such as price and quality) available in the market. Are new products coming to the market soon that will impact the sales of this product? These questions suggest the need for all-around awareness of what is happening in the larger supply chain. That requires immense amounts of data to maintain and techniques of analysis on large data sets that are the focus of research efforts. There is also a strong need to identify information that can help predict the demand for new products more accurately, and allowing for idiosyncrasies of the data collection method, such as arises in the retail sector when no record is kept of lost sales, rendering the tracking of demand (as distinct from sales) difficult.

Moving beyond the use of information on pricing, promotions, and other environmental factors to develop better demand forecasts, at the next level of demand management, companies not only try to respond to demand, but also to influence it in a manner favorable to them. For example, telephone sales personnel can steer customers towards available configurations. This moves beyond the functional/distributed approach where sales and production operate largely independently, requiring regular communication and coordination between the production planner and the sales and marketing functions.

Probably the most important tool used by companies to influence demand is pricing. For most manufacturers, pricing decisions are currently done in a reactive rather than a proactive way. That is, most manufacturers currently turn to discounts and promotions to generate additional demand for absorbing excess inventory. For example, manufacturers can offer dealers up to 30 days of free credit and encourage them to sell below list prices if they are overstocked. Similarly, customer-class oriented promotional pricing such as cash-back and various financing plans are additional tools used by manufacturers to influence demand. For the most effective use of pricing and promotions, it is again essential to have collaboration between the manufacturing organization and the sales and marketing organization since the careless use of promotions without considering their impact on the supply side can lead to increased costs of labor, materials, capacity, and transportation, reducing the overall profitability of the firm.

1.5.3 Centralized vs. Distributed Planning Environments

While it would be ideal from a systems point of view to manage all these expanded activities in a coordinated manner between purchasing, production planning, sales, and marketing, such an integrated approach is seldom the case in either research or practice. Even in cases where the firm owns significant portions of its supply chain, different segments are likely to be assigned to different organizations. The managers of these organizations are given different forms of financial and other incentives to ensure that their actions support the overall success of the firm. In the narrow, conventional view of the production planning problem this is not as much of an issue. The management of the particular unit for whom the planning is being done formulates the characteristics of a good plan (explicitly or implicitly) and plans evolve to reflect these characteristics. However, when multiple organizational units are involved, a solution to the planning problem will necessarily involve negotiations and compromises between the different players. Setting the incentives is a major challenge, as attested by the extensive body of management literature on performance measurement both in the supply chain and in the firm. In any case, execution of the resulting plan requires interaction from several different organizational units and functional areas, and as such is necessarily distributed.

We can conjecture a number of reasons for this situation. In all but the smallest firms, these different activities require different skill sets that are often difficult to find in the same individual. As manufacturing technology has become more complex, for instance, it has become more difficult for an engineer or manager to have a working knowledge of the entire firm's manufacturing operations. This progressive specialization was also very much in line with the reductionist approach of the scientific management philosophy that was influential in the latter part of the nineteenth century when the modern industrial enterprise had its origins (e.g., Chandler 1980). This led quite early on to the organization of firms into specialized functional groups each of which considered one particular aspect of the firm's operations, such as production, procurement, or marketing. By necessity, each of these functional groups developed its own performance measures, such as minimizing costs or maximizing revenues. This lead to a distributed decision making environment within the firm including a mechanism to coordinate decisions through incentives offered to each group. For example, production operations are generally viewed as a cost center, while demand-related functions such as sales and marketing are considered as profit centers.

A number of quantitative models addressing these issues have begun to emerge in recent years. In the research community, some use the tools of economics such as game theory and principal-agent theory to explore the effects of information sharing and different types of collaboration on the behavior and performance of the supply chain. Many of these models explore very simple, stylized systems with the aim of obtaining insight into how to incentivize different parties to obtain behavior approximating that of the centralized system. Some researchers have explored using the tools of mathematical programming and price-based decomposition approaches that seek to coordinate the behavior of a number of decision entities by setting prices for internal transactions. A number of others have explored these issues computationally using tools such as system dynamics and agent-based models. In industry, there is a strong movement towards the development of collaborative supply chain planning tools, although the precise nature of these appears to be in flux. Several industry-specific efforts aimed at structuring the interactions between members of the supply chain have arisen in a number of industrial sectors, such as Collaborative Planning, Forecasting and Replenishment in the consumer goods sector and Quick Response in the textile sector.

The role of the planning model, whatever mathematical form it takes, is likely to be quite different in this collaborative environment than in a centralized one. Rather than being prescriptive in nature, its main function may be to facilitate negotiations, providing all parties with a better view of the consequences of their decisions and those of others. Individual models and planners must now have an idea of how the other stakeholders in the system with whom they interact are incentivized, a clear understanding of how the costs and revenues of their organization are affected by the decisions of others, and how external parties are likely to behave in various circumstances. An understanding of contracts and financial options enters the picture here as well as the need to maintain relationships and sharing information with multiple parties without breaching confidentiality requirements. An example of such an alternative view of the problem is that taken by researchers in the artificial intelligence community (e.g., Zweben and Fox 1994) who have viewed problems of production scheduling and planning as requiring the identification of a solution satisfying a set of constraints, generally resulting from the needs of different individuals or groups, rather than as an optimization problem with a single, well-defined objective function.

It is interesting to contrast these ideas with the nature of current Enterprise Resource Planning (ERP) systems whose domain spans wide areas of the firm, at least from a data management perspective. The results of this functional focus are clearly manifested in the assumptions made by these modeling efforts. Several of these are pervasive, especially the modeling of exogenous information in relatively simple ways such as point estimates for demand forecasts, fixed replenishment lead times independent of order quantities in inventory models, and access to all relevant information for the planning problem under consideration. Note that these are, upon reflection, quite consistent with the functional view of the world. Demand is beyond our control, so the best we can do is try to predict it. Inventories in the distribution channel belong to the distribution organization, not the production organization so production quotes a fixed lead time for replenishments. The scope of the planning decision is functional, so the functional group has access to all information arising within its own organization, but must either estimate or make simplifying assumptions about the ill-understood exogenous parameters, providing complete information about a very narrow segment of the problem space under the control of this particular functional group.

1.6 Production Planning in Industrial Practice: Data and Scalability

Our discussion until now has used the basic and expanded realistic problems presented above to discuss the state of the research literature, outline the essential characteristics of the production planning problem and suggest the need for an expansion of its scope relative to the conventional definition of the problem to achieve its potential. However, no-one would pretend that these realistic minimal problems are complete reflections of industrial practice. It is worthwhile, therefore, exploring how industrial problems differ from the somewhat stylized problems we have presented until now.

One of the most important dimensions of difference is the sheer size of the problems. While we have presented minimal problems with two products, two capacity types and two customers, industrial problems will routinely involve dozens of capacity types distributed over tens of geographic locations around the globe producing hundreds (if not thousands) of products for thousands (if not millions) of customers. One implication of this is that the scalability of any solution procedure proposed for the problem, or even a part of the problem, is crucial to its being implementable.

An important corollary is that the extraction, organization, and maintenance of the data to support planning models on this scale is a very significant effort in its own right, often dwarfing the actual modeling and solution methodology in its complexity. Since no planning solution of any kind is ever implemented in a vacuum, it must be integrated into a business process that fits the organization in which it will be used, interface with existing information systems and other business processes, and be intelligible to its users.

Another aspect of industrial practice is the emphasis placed by the firm on different aspects of the production planning problem will vary significantly from industry to industry. For example, for companies in the aerospace and automotive industries that make a complex product with hundreds of subsystems manufactured by other firms, the effective management of suppliers is the crux of their production planning. For firms in capital intensive industries such as semiconductor manufacturing, managing the factory to maintain high output and reliable cycle times at high utilization is essential to success. For fashion apparel companies, early testing of markets with small batches of a variety of styles and colors is crucial for refining demand forecasts.

Evolving corporate information systems render possible the collection of vast amounts of data to support these activities, but these data must be supported by planning models and processes that exploit them to their best advantage. It is interesting to note that in contrast to production planning, ERP systems at the majority of large corporations are specifically intended to be implemented across organizational groups within the firm, and increasingly with key suppliers and customers. However, in many of these systems, the basic planning technology is based on the MRP paradigm developed more than 50 years ago, whose ability to manage even our realistic minimal problem effectively is highly limited. Indeed, the lack of effective production planning tools in at least the first generation of these systems led to the emergence of an entire software sector of "bolt-on" Advanced Planning and Scheduling solutions, which used the ERP system for transactional data and execution, but used their own algorithms to provide improved planning capability over and above that inherent in the ERP system. Developments in this area have generally been driven exclusively by advances in information technology, most importantly relational databases, client–server computing and the Internet. The systems tend to be developed, deployed, and managed by Information Technology personnel, who have their own set of priorities (e.g., ease of maintaining a homogeneous platform as opposed to homegrown systems tailored to the firm's peculiar planning needs) and often lack an understanding of the firm's supply chain and production operations.

There remains today a large segment of the industrial community in which production planning is viewed essentially as a software problem that can be resolved by applying faster computers and better data collection. However, there is growing evidence in industry today that our ability to collect, transmit, and organize data on supply chain operations is outstripping our ability to use it to more efficiently and effectively plan and run our supply chains. A better understanding of the expanded production planning problem is essential to developing a "science base," a generalizable, extendable body of knowledge, to support this aspect of the firm's activities.

1.7 Motivation for this Volume

Over 50 years ago, Simon and Holt (1954) stated in the introduction to their survey paper on control of inventories and production rates:

> The problem of discovering appropriate decision rules is greatly complicated by the fact that decisions... need to be made at several stages in a procurement-manufacturing-warehousing-selling sequence. In even a relatively simple situation we might have the following stages: (a) procurement of raw materials and purchased parts, (b) raw material inventory, (c) parts manufacture, (d) semi-finished inventory, (e) finished-goods assembly, (f) finished-goods inventory, (g) shipment to district warehouses, (h) warehouse inventory, (i) sales to customers.

They then summarize research in ordering, production rate, and scheduling decision making considering static and dynamic cases with and without uncertainty and include a brief discussion of forecasting. Towards the end of their description of the then state-of-the-art they conclude:

> It should be clear from this summary that substantial progress has been made in several directions in devising procedures for making rational decisions about production and inventory control. . . . Ideally, one would like to think that some combination of these approaches would handle simultaneously all classes of decisions including scheduling. They would embrace the interactions in the many-commodity case, and include the interactions in a whole series of 'cascades'. What would appear to be needed is some kind of 'dynamic nonlinear programming with built-in forecasting.'

Our purpose in this essay has been to make a case that two important concepts persist from the Simon and Holt survey. On the one hand, the same core problem is still faced by production planners on a daily basis. On the other hand, while substantial progress has been made, there still exist significant gaps between the needs of industrial practice and the decision support tools provided by the existing body of quantitative research addressing production planning. We began by outlining a conventional, narrow view of the production planning function and presenting some historical perspective on why this view of the problem has prevailed. We have outlined a set of minimal but difficult problems of this type, characterized by the prevalence of uncertainty in both supply and demand as well as nonlinear behavior of the supply process due to limited production capacity and nonnegligible TPTs. Examining the major streams of existing literature suggests that a complete, scalable solution to even this difficult minimal set of problems is currently not available. The existing literature has separated into a number of increasingly independent streams, largely based on the mathematical tools they employ, and each of these is capable of addressing at best a subset of even the difficult minimal problem. In industrial practice, the deployment of inventories at different locations in the factory and the supply chain has been the prevalent approach to dealing with this problem, to the extent that the overall problem is perceived as an inventory management problem by a broad section of both the industrial and practitioner communities.

While this finding would be disturbing enough by itself, we believe it comes at a time when the scope of the production planning problem that industry needs solved is expanding in response to the growing recognition of the need to consider the broader supply chain to maintain an ongoing competitive advantage. This suggests our expanded problem set, where the scope of the original problem is expanded to include management of capacity subcontracting and material supply through contracts and other financial instruments; management of demand through tools such as dynamic pricing; and the recognition that in the vast majority of industrial environments the production planning process will take place among a number of different organizational entities with different constraints and objectives, leading to a distributed view of the problem in contrast to the prevalent centralized, prescriptive approach of most models.

Along with this expansion of the *scope* of the problem, reflected in the scope of the decisions it involves, is the realization that the *scale* of industrial problems requiring solution is so large as to constitute a significant challenge in its own right. To be viable in practice, a solution technology must be scalable across global supply chains with highly diverse product portfolios, multiple production locations involving capacity of different types, as well as many different customers. While developments in computing and information technology are increasingly converging to provide data on all aspects of supply chain operation, this rapid increase in the scale of the problem poses significant challenges to designing algorithms that can be used in effective solutions.

Our objectives in these volumes have been twofold: to assemble a clear picture of the state of the art in production planning, defined broadly, and how we got to it, and then to suggest directions for future work in this field that will be required to bring decision support technologies, again broadly defined, to a level where they can contribute to effective solution of industrial problems. Rather than emphasizing focused contributions of an analytical nature to a specific aspect of the problem, we include chapters that suggest promising directions for future work, presenting sufficient analysis, mathematical or computational, to support the case for the proposed direction.

References

Bahl HC, Ritzman LP, Gupta JND (1987) Determining lot sizes and resource requirements: a review. Oper Res 35:329–345

Birge JR, Louveaux F (1997) Introduction to stochastic programming. Springer, New York, NY

Chandler AD (1980) The visible hand: the managerial revolution in American Business. Belknap Press, Cambridge, MA

Elmaghraby SE (1978) The Economic Lot Scheduling Problem (ELSP): review and extensions. Manag Sci 24:587–598

Eppen G, Martin RK (1988) Determining safety stock in the presence of stochastic lead times. Manag Sci 34:1380–1390

Goldratt E, Fox RE (1986) The race.. North River Press, New York, NY

Graves SC, Kletter DB, Hetzel WB (1998) Dynamic model for requirements planning with application to supply chain optimization. Oper Res 46(3):35–49

Hackman ST, Leachman RC (1989) A general framework for modeling production. Manag Sci 35:478–495

Hadley G, Whitin TM (1963) Analysis of inventory systems. Prentice-Hall, Englewood Cliffs, NJ

Harris FW (1915) Operations and cost. Factory management series. A. W. Shaw Co, Chicago, IL

Hax AC, Candea D (1984) Production and inventory management. Prentice-Hall, Englewood Cliffs, NJ

Heath DC, Jackson PL (1994) Modeling the evolution of demand forecasts with application to safety stock analysis in production/distribution systems. IIE Trans 26(3): 17–30

Hopp WJ, Spearman ML (2001) Factory physics: foundations of manufacturing management. Irwin/McGraw-Hill, Boston, MA

Johnson LA, Montgomery DC (1974) Operations research in production planning, scheduling and inventory control. Wiley, New York, NY

Kaminsky P, Swaminathan JM (2001) Utilizing forecast band refinement for capacitated production planning. Manuf Serv Oper Manag 3(1):68–81

Kanigel R (1997) The one best way. Viking, New York, NY

Kimms A (1997) Multi-level lot sizing and scheduling. Springer, Berlin

Liker J (2004) The Toyota way: 14 management principles from the world's greatest manufacturer. McGraw-Hill, New York, NY

Liu L, Liu X, Yao DD (2004) Analysis and optimization of multi-stage inventory queues. Manag Sci 50:365–380

Modigliani F, Hohn FE (1955) Production planning over time and the nature of the expectation and planning horizon. Econometrica 23(1):46–66

Orlicky J (1975) Material requirements planning: the new way of life in production and inventory management. McGraw-Hill, New York, NY

Pande P, Neuman RP, Cavanagh RR (2000) The Six Sigma way: how GE, Motorola and other top companies are honing their performance. McGraw-Hill, New York, NY

Simon HA, Holt CC (1954) The control of inventories and production rates – a survey. Oper Res 2(3):289–301

Vollmann TE, Berry WL, Whybark DC, Jacobs FR (2005) Manufacturing planning and control for supply chain management. McGraw-Hill, New York, NY

Voss S, Woodruff DL (2003) Introduction to computational optimization models for production planning in a supply chain. Springer, Berlin; New York, NY

Zipkin PH (1986) Models for design and control of stochastic, multi-item batch production systems. Oper Res 34(1):91–104

Zipkin PH (1997) Foundations of inventory management. Irwin, Burr Ridge, IL

Zweben M, Fox M (eds) (1994). Intelligent scheduling systems. Morgan Kaufman, San Francisco, CA

Chapter 2
The Historical Foundations of Manufacturing Planning and Control Systems

Kenneth N. McKay

2.1 Introduction

Before probing the history of manufacturing control systems, we have to ask a question. What is a manufacturing planning and control system? In theory, there is a simple answer – it is a system of "things" that work together to control what is done where, by who, using what set of resources (machines, tools, and materials), at what time, and in what quantity. Controlling also implies setting expectations or knowing what the expectations are, measuring the realized manufacturing performance and outcomes, knowing what is where, what has been done, providing feedback, and making necessary corrections so that the expectations and goals can be met, or at least be better met in the future. So, what are the elements of a manufacturing planning and control system?

First, these systems include, or at a minimum are influenced by, the harsh, physical reality of the plant–machine layout (functional or process), machine and tooling capability, inventory locations, space around machines, and material movement options. The physical or technological aspects provide potential options for, and constraints on, what can and cannot be physically done in the plant, and thus provide a form of *physical manufacturing control.* Flexibility can exist in the physical components and can range from different fixtures to physically altering machines (combining or separating). Exploiting any potential flexibility in the physical system may have associated costs, lead-time issues, and risks.

Second, superimposed over the physical reality are the logical components – policies and procedures. The policies and procedures, if at all accurate, can never overstate or violate the physical reality and usually act as constraints on the physical flexibility. For example, if the floor loading will support 10,000 kg/m^2, a logical policy allowing inventory to exceed 10,000 is nonsensical. However, a policy restricting the inventory to 8,000 kg might make a good inventory control policy independent of any economic order quantity. Policies and procedures may also be influenced by

K.N. McKay (✉)
Department of Management Sciences, University of Waterloo, Waterloo, ON, Canada
e-mail: kmckay@uwaterloo.ca

K.G. Kempf et al. (eds.), *Planning Production and Inventories in the Extended Enterprise*,
International Series in Operations Research & Management Science 151,
DOI 10.1007/978-1-4419-6485-4_2,

outside forces. For example, governmental rulings regarding health and safety might also constrain the degrees of freedom. The set of applicable policies and procedures used by a firm form the *procedural manufacturing control*. Depending on the nature of the issues they address, some policies and procedures have latent elasticity and thus constitute relatively soft constraints which can be negotiated based on the business context.

Third, between the procedural and physical manufacturing levels, there are the actual tools and instruments used for control – *manufacturing control technology*. This includes *monitoring and informing* techniques for knowing what has happened, where things are, the status of work orders, as well as *activation and notification* techniques for indicating when work should start, how much should be produced, or where inventory should be moved to. For example, barcode readers and sensors can determine the production results at a machine, and a Kanban card or signal can indicate that production can start on a particular item. The manufacturing control technology also includes the *logic and analysis* behind all of the production control process for determining what is needed when and what is the best way of doing it. Included in this subcategory is the MRP logic, planning, scheduling and dispatching tasks, and so forth.

To summarize, a manufacturing planning and control system will combine the following elements:

- Physical manufacturing control
- Procedural manufacturing control
- Manufacturing control technology – monitoring and informing, activation and notification, and logic and analysis

While it is possible to think of these elements as being reasonably independent, they become interrelated and interdependent during manufacturing execution. For example, in the 1920s a form of mechanical scheduling using conveyor systems as a controlling mechanism was described (Coes 1928). Everything was moved by conveyor – if there was something on the incoming conveyor and your machine was not blocked by the outgoing conveyor, you performed your task. If you were starved or blocked, you did no work. This controlled the total inventory in the plant and forced machines and operators to an idle status. There was no inventory outside the conveyor system. This is very similar to the effects of Kanban and illustrates a combination of the physical, procedural, and technology components.

If all three elements are appropriately matched to the manufacturing situation, it is likely that manufacturing will be efficient and effective. If one or more of the components is significantly inappropriate to the context, then manufacturing is more likely to be chaotic and definitely not efficient or effective. For example, expecting a job shop to respond effectively and efficiently to a Kanban pull from multiple assembly areas without adequate bins and quantities to cover the variable lead time that will result is an example of a mismatch. Another mismatch arises when inventory control techniques such as order point are used when the majority of work is composed of dependent demand and the demand is highly variable. This also does not work well. Thus, a goal of manufacturing system design is to know what form of manufacturing planning and control is suitable for a given level of quality, volume, speed of manufacturing, and mix. The design challenge is compounded by the fact that requirements will undoubtedly change over time – the three axes represented by the procedural, physical, and control elements must be periodically reviewed to ensure that balance is retained. A common example of requirements evolving over time is an increase in product variety with a decrease in volume for each product type. Depending on the magnitude of each change, the situation may flip from a high volume, low mix situation to a low volume, high mix situation, which might warrant layout changes accompanied by changes to manufacturing control.

As we review the history of manufacturing planning and control systems, these elements and their interplay will be used to anchor the discussion. Section 2.2 will provide a brief history of manufacturing planning and control systems prior to the start of the twentieth century. Section 2.3 will focus on the early systemization and developments during the first three decades 1900–1930. Section 2.4 will look at the situation in the 1950s, 1960s, and early 1970s. In this chapter, we will not focus on the physical shop floor execution systems that sense and track the actual production, but on the logical controls using this information.

2.2 Pre-1900

Before 1900, there were isolated examples of relatively large-scale, systemized manufacturing concerns which involved various types of manufacturing control. For example, Lane (1934) describes the fifteenth century Venetian Arsenal. The arsenal had a moving assembly line for the rapid completion of galleys. In the early 1700s, the Lombe brothers constructed a large-scale silk mill in Derby (Cooke-Taylor 1891) and by the end of the 1700s, Boulton and Watt had collaborated at the famous Soho Manufactory (Doty 1998, Immer 1954). The Portsmouth Block Mill of the early 1800s (Wilkins 1999) and the American armouries of the same time period (Smith 1977) also introduced manufacturing control innovations. In each of these examples, high volumes of products were manufactured with little logical manufacturing control per se; the majority of control was imposed by the layout and physical nature of the plant. There was little competition and in the case of the military units, the focus was on basic production and not on the efficiency and effectiveness of the manufacturing process. Lombe and Boulton/Watt also had business structures similar to that of a monopoly and while manufacturing was controlled, extreme levels of transactional analysis was not needed. By the 1830s, the concepts of quantity and economic production were being discussed (Babbage 1832).

Starting in the mid-1800s, manufacturing activity expanded rapidly (Hounshell 1984) and by the late 1800s manufacturers were looking for ways of reducing the chaos. The American Society of Mechanical Engineers (ASME) started to publish a number of papers and discussions of manufacturing management and techniques in the late 1800s. As the most thorough and thoughtful treatise on the matter,

Lewis (1896) might be the first manufacturing management text. Lewis provides a very thorough discussion and mapping of a factory's decision making and information flow, details the cost accounting, and describes how manufacturing was orchestrated from order taking to shipping. While others such as Taylor (Kanigel 1997) focused on a subset of manufacturing management, Lewis described the whole organization. This was a British publication, and no equivalent North American publication appeared for at least two decades. We have been unable to find a reference to Lewis' work in any of the early Industrial Engineering texts. Generally, any systematic form of control and organization was an exception and as noted by Emerson (1909), the general state of manufacturing was mass confusion and the first order of business was to systemize and bring some order to the process. The systemization of manufacturing would prove to be challenge.

2.3 1900–1930

During these three decades, the manufacturing sector saw the mass systemization of production and associated activities (McKay 2003). There were many articles and publications written during this period covering an immense range of topics (Cannon 1920). Subjects included what individual operators were doing and how to supervise them (Taylor 1903, 1911), incentive schemes (Gantt 1910), how to organize resources in functional and process arrangements (Diemer 1910, Kimball 1913), mass production and continuous improvement (Ford 1926), focused factory organizations (Robb 1910), economic analysis of inventory (Erlenkotter 1989), supply chain management and hand-to-mouth inventory (Nash 1928, Alford 1934), statistical process control (Shewhart 1931), and reconfigurable production lines (Barnes 1931, Koepke 1941). As noted by Alford (1934), the majority of manufacturing activity was relatively basic and did not require or use many of these ideas. There were, however, occasional exceptions. The idea of just-in-time inventory (hand-to-mouth) was widely used in the 1920s. For example, Nash (1928) had daily turns of inventory at his automotive assembly plant and there were many examples of lean supply chains and controlled inventory. Throughout industry there were many assembly lines and material handling devices deployed, but there was little evidence of large numbers of companies actually using statistical process control, economic analysis of inventory, or other algorithmic types of procedural or logistical manufacturing control. Knoeppel and Seybold (1937) pointed out that in general the prescriptions, tools, and concepts created by the early Industrial Engineers were rarely used.

It appears that the majority of manufacturing management effort was focused on tracking and basic control. Almost a century later, this reasoning still appears to be the primary reason for firms implementing ERP systems and this will be discussed in later sections. Even in the early 1900s, the physical and control technology was rudimentary and was composed of the typical functional, product and process flows combined with basic book-keeping. There were many self-help style publications on how to codify parts, organize tools, and have smoother flow of goods through a factory. These were still popular topics in the early 1930s even though they started to be discussed in the early 1900s. One of the most common topics throughout this era was that of tracking and planning boards. Gantt's 1919 work is perhaps the best known of the graphical display methods, but there were others such as Knoeppel (1915, 1920) who was a consultant like Gantt and proposed a variety of ways to present, plan, and track work through factories. In this period, many mechanical planning boards and production control tools were developed to help plan, track, and predict factory operations (Simons and Dutton 1940). There were also many card and filing systems created to account for and to track production, some using IBM's punched cards (Simons and Dutton 1940).

In terms of procedural or logistical control, very little of a mathematical nature can be found before mid-century. There was the work on economic order quantity (Erlenkotter 1989) which was adopted by some portion of industry, but it is hard to ascertain how popular the method was or who actually benefited from it. Other mathematical techniques were applied to quality control, especially sampling (Robertson 1928a, b), and control limits (Shewhart 1931). Based on descriptions found in American Statistical Association (1950), it would appear that statistical tools such as sampling methods were used extensively in the World War II production effort but not before. This is supported by the minimal reference to SQC/SPC in popular texts of the time (e.g., Koepke 1941). There was substantial work on forecasting, of course (White 1926), but this was not really of an operational nature as it focused on the longer term. For actual planning and scheduling, very little mathematical analysis was proposed to help with sequencing decisions. This is not in itself surprising since the manual effort to mathematically analyze any reasonably realistic industrial scheduling problem would have been prohibitive. Knoeppel (1915) recommended using a numerical ratio to determine the best time to start a subsequent operation when overlapping tasks, which is the only example of a scheduling or dispatching heuristic that can be found from this era. Note that this dealt with the operations within a job and not really sequencing multiple operations on the same machine. There is no solid evidence that anyone used his suggestion. In texts dedicated to routing and scheduling (e.g., Younger 1930), there is no mention of advanced reasoning beyond that of looking at overlapping operations for better sequencing and the body of the texts are focused on creating departments, systemizing paperwork, setting up basic plans, and tracking. The only other "advanced" algorithmic suggestion found in the literature related to work flow through a factory. Knoeppel (1920) suggested that work should be pulled through factories using an approach not unlike MRP and this message was still appearing in 1940 publications (Simons and Dutton 1940).

2.4 The 1950s, 1960s, and early 1970s

In the era before the wide-spread use of computerized MRP, a factory with even a modest amount of product variety and process complexity found that they had a very difficult planning and scheduling task. Koepke (1941) describes one factory using approximately thirty people to generate a 2-week plan. It took them 2 weeks with overtime to prepare the next 2-week plan. The amount of paperwork and manual effort was just daunting. It is little wonder that the majority of manufacturing used statistical, order point or quantity reordering schemes which were independent of actual orders or demand even as late as the early 1970s (Orlicky 1975). As Orlicky comments, given the information systems they had, it was the best they could do. Orlicky estimated that there were approximately 150 firms using MRP-type systems by 1970 and that this number would grow to 700 by 1975. Part of this increase was expected by Orlicky because of the MRP Crusade mounted by APICS in the early 1970s. This slow evolution from the first business computer usage for inventory management and control in the early 1960s to this awareness in the early 1970s was considered a major development.

The elements of MRP are now well known. There are bills of material, time-phasing concepts, calculating component item demand, lot sizing, estimation of coverage, priority control, and load projections. A key concept in MRP was clearly distinguishing dependent demand based on end item forecasts and orders from independent demand such as that arising from component sales and service parts. Plossl (1973) identifies an article by Orlicky in 1965 that describes the independent and dependent demand principle as being the breakthrough development. Hopp and Spearman (2001) provide a very good description of how MRP works and the subsequent migration to MRP II. In MRP II, the integrated computer system pulled together the additional functions of "demand management, forecasting, capacity planning, master production scheduling, rough-cut capacity planning, dispatching, and input/output control" (Hopp and Spearman 2001).

Interestingly, the basics of an integrated MRP were described and actually implemented in the early part of the twentieth century (Knoeppel 1915) and the number of clerical staff required can only be imagined. This early MRP system had almost all the key features of the later computerized systems: bill of material, routings, backward loading of work with lead times from shipping (or required) dates, finite capacity loading and analysis and *daily* updates, and was also used to provide availability information to sales. Unfortunately, the issues later commented upon by Orlicky became the reality for the people who recognized the benefits arising from a more integrated planning approach. To do any systemized and thorough manufacturing control process required a great deal of dedication and effort. This might be the reason why no other integrated MRP-type systems with a short horizon can be found in the literature for the next 50 years – examples or concepts. Knoeppel's factory example and system description appears as an anomaly.

How were the 1970s reached? The first computer was sold for business purposes in approximately 1954 (Orlicky 1969) and the traditional areas were pay-roll, sales analysis, accounting, and cost analysis (Reinfeld 1959). Reinfeld points out that the adoption of computers into production control was very slow with only a few companies doing it (e.g., large aircraft companies, General Electric, and a few others) and that the majority of applications were mostly listing and posting. Computers were still very expensive in this period and it is not surprising that few were used and that production control was performed with little black books and what could be found in the head of the foreman (Reinfeld 1959). Even by 1973, Plossl (1973) was estimating that less than 1% of the computer base was being used for manufacturing and he speculated that there were two main reasons: first, suitable applications did not exist that would actually do the job, and second, production and inventory control personnel did not see the need for such a system in the first place.

Before continuing the MRP discussion, it is useful to step back and consider the general order point approach and what was happening. The statistical order quantity or order point was being used for almost every type of manufacturing by 1970. This was not always the case. Plossl (1973) describes a version of material requirements planning that had been done quarterly in many firms (e.g., tool building, ship building, aircraft, locomotives, and other heavy products). Unfortunately, "As product models proliferated and as the product complexity increased, however, it became increasingly difficult to develop a practical, workable production schedule for finished products far enough in advance, explode all of the bills of material, net out available stocks, and trigger replenishment orders. The work of making the calculations and record comparisons was too time-consuming, the technique was impractical for products of even moderate complexity without the use of a computer." (Plossl 1973, p. 69). This led to over reliance on the independent demand techniques. We of course know that independent demand techniques can and do work quite well in the right situation. Manufacturers in the 1920s (McKay 2003) and Japanese firms later in the century figured this out. If you can stabilize the forecast for a reasonable period of time (e.g., 6 weeks $\pm$ 10%) and balance your production capacity accordingly, you can use very simple production control methods, the most famous being Kanbans. The Just-In-Time approach is, in essence, an independent demand order point system. Unfortunately, this does not work well in many other situations.

Plossl's comment is interesting in that the 1950s started off as a very simple production environment. The Anglo-American Council on Productivity 1949) noted the few models and minimal options provided by American manufacturers and the British writers compared this to the proliferation of parts and high-mix facing the British counterparts. Koepke (1941) describes a variety of different master schedules, each depending on the type of situation. For example, in mass production where almost everything was dedicated and connected by moving conveyors, chains, and other mechanical devices, the master schedule was considered the schedule for the final part as everything else just flowed together and did not need to be scheduled, At the start of the second half of the century, the situation was relatively simple for the large manufacturers operating in a mass production mode. For projects or intermittent production, Koepke described master schedules that synchronized components and sub-assemblies using a bill of material and there were other master schedules as well. The focus on safety-stock between final assembly and the customer, and the general dismissal of upstream uncertainty fits the idea of dedicated lines with low variety operating at high volumes. Simple order points would suffice for materials and small parts in mass production settings and the quarterly ordering method described above by Plossl would satisfy the project-oriented production. If a functional approach, and not a product layout, was used for mass production the result might have been different. If the mix was low, resource conflict almost non-existent, and there was not much re-entry, the functional layout would behave like a virtual flow line and not require sophisticated planning. However, the functional layout would not have been a difficult physical control concept to plan and orchestrate for mass production when high-mix existed and resource and material conflicts arose.

By the 1970s, production was no longer simple. There were many models, many options, and manufacturing was getting messy. It was no longer possible to have dedicated and duplicated equipment for each end product, and it appears that many firms adopted a functional factory style. Automotive plants were the exception as they continued to use drag-chains and main assembly lines. Functionally organized factories found that the order point approach was not successful at controlling inventory (avoiding stock outs or minimizing inventory throughout the plant). If the existing functional layout and basic manufacturing execution control was to be retained, the basic MRP approach was the perceived answer.

The various assumptions and difficulties associated with the traditional MRP approaches are eloquently explained in Hopp and Spearman (2001). The key issue relates to the way that lead time is used independent of the plant status. Because of this assumption, MRP can model and plan a dedicated, automated line with high reliability – the bottlenecks cannot float, work cannot enter or exit the line, and everything is predictable (within limits). The large, drag-chain automotive assembly lines once common in the industry are examples where MRP was relatively successful. If you have a job shop, or free-flowing virtual assembly lines with highly variable loading, MRP cannot estimate when anything will be used or produced. Another assumption is that the model of the process and product is static and stable over the planning horizon. There are usually some records and some fields in MRP which have effective date controls, but the majority of the production model is static and if the factory is rapidly or constantly changing, the model is inaccurate. An inaccurate model results in infeasible planning and expectations. MRP systems can also exhibit nervousness and sensitivity to changes in the forecast or production results. The generated plan can ask for different quantities and dates every time it is created. This is not a problem on a resource that can react quickly. However, if there are infrastructure issues (e.g., the resources needed to set up the machine, additional helpers, etc.) or supply issues (e.g., cannot easily change the schedule at a steel mill) the nervousness can be annoying. Ongoing nervousness can also result in increased errors and problems as the plant tries to do what the plan asks for.

In its approach to inventories, MRP is reactive – it will try to maintain certain levels of finished goods, work-in-process, and raw. MRP logic does not typically include strategic use of inventory that will allow the level to go below the level specified. For example, it might make sense to make more of one product and less of another during a short horizon because of situational factors and then recover to the desired levels a bit later. This type of reasoning and commonsense is absent. The final limitation of MRP we will note is that of inventory bank health. By this, we mean the degree to which the right parts are in the bank in the right quantities to minimize conflicts on machines and to space out (e.g., cycle) production. If there is a significant disturbance in the factory that upsets this balance, how does the factory recover to the desired levels of inventory for each part that avoids conflict? Often the human is expected to review the plans each day and make the necessary adjustments – increasing the quantity of one part, decreasing the quantity of another, locking in dates etc. MRP does not have the tools or the ability to determine whether the inventory is not in balance and how to recover, and it is not possible to do it manually in a realistic industrial environment.

Taking all this into account, it is sufficient to say that MRP was not the answer to all manufacturing situations and was unwisely applied to a number of situations. The MRP and MRP II approach has always relied upon a number of assumptions and unless they are satisfied, grossly infeasible schedules and plans will result. There are assumptions about the type of manufacturing system being applied to, and there are assumptions about how the MRP system is actually used. If there are too many resource conflicts and manufacturing execution is highly variable, the amount of manual intervention to resolve exceptions and conflicts is immense. Since sustained attention to such issues has proven to be difficult for almost all firms, MRP- and MRP II-based systems (e.g., ERP systems) have remained problematic whenever the manufacturing environment is less than ideal. The efforts to combine advanced planning systems with MRP logic is an attempt at bridging some of the gap and associated difficulties. By including finite modeling, more feasible schedules can be attained and if things go well, the results will be better than guessing.

2.5 Conclusion

It has become a historical and sometimes hysterical truth – if the various components of manufacturing planning and control are well matched to the environment, things will go well, and if they do not make sense, chaos and mayhem will dominate the milieu. Use MRP type concepts in a situation either not suitable or without the discipline and matching philosophy, and you will get what others have got before you. Use Just-In-Time concepts incorrectly and you are unlikely to get the Toyota halo-effect. As the mix and volumes change in products, so must the manufacturing system. This includes the physical, logical, and technological. Historically, it can be seen that the basic ideas for good production control were known and advocated in the early 1900s, but the technology required for effective, sustainable implementation did not then exist.

In the second half of the twentieth century, manufacturing was focused on what they could do with the information systems they had and this rapidly changed when computer-based systems became widespread. This change coincided with the problem changing from one of simple mass production to that of complex mass production and more functional layouts. Once MRP-based systems started to be popular in the early 1970s, manufacturing was then focused on MRP as the solution for all manufacturing problems and became momentarily blinded to other concepts. Periods became locked in, inventory checked in and out of stores, and expediting and chaos became routine. The models and varieties continued to increase and then competition appeared that make simpler products with very few options. This competition was able to flow material through factories, with many inventory turns a year and without complicated planning and scheduling. Toyota was able to do in the later stages of the twentieth century what the North American manufacturers once were capable of doing. Nash and others might have been amused.

References

Alford LP (ed) (1934) Cost and production handbook. Ronald
American Statistical Association (1950) Acceptance sampling – a symposium, Washington
Anglo-American Council On Productivity (1949) Simplification in industry, Anglo-American Council On Productivity
Babbage C (1832) Economies of manufactures, 2nd edn., Knight
Barnes RM (1931) Industrial engineering and management, McGraw-Hill
Cannon HGT (1920) Bibliography of industrial efficiency and factory management. George Routledge, London
Coes HV (1928) Mechanical scheduling. In: Dutton HP (ed) 110 Tested plans that increased factory profits. McGraw-Shaw, New York, pp. 69–74
Cooke-Taylor RW (1891) The modern factory system. Kegan Paul Trench Trubner
Diemer H (1910) Factory organization and administration. McGraw-Hill
Doty R (1998) The soho mint & the industrialization of money. Smithsonian Institution, British Numismatic Society. Spink
Emerson H (1909) Efficiency as a basis for operation and wages, The Engineering Magazine
Erlenkotter D (1989) An early classic misplaced: Ford W. Harris's economic order quantity model of 1915. Manage Sci 35(7):898–900
Ford H, Crowther S (1926) Today and tomorrow. Doubleday, Page
Gantt HL (1910) Work, wages, and profits. The Engineering Magazine
Gantt HL (1919) Organizing for work. Harcourt, Brace and Howe, New York.
Hopp WJ, Spearman ML (2001) Factory physics, 2nd edn. McGraw-Hill
Hounshell DA (1984) From the American system to mass production, 1800–1932. Johns Hopkins
Immer JR (1954) The development of production methods in Birmingham 1760–1851. PhD Dissertation, Oxford
Kanigel R (1997) The one best way, Viking
Kimball DS (1913) Principles of industrial organizations. McGraw-Hill
Knoeppel CE (1915) Installing efficiency methods. The Engineering Magazine, New York
Knoeppel CE (1920) Graphic production control. The Engineering Magazine, New York
Knoeppel CE, Seybold EG (1937) Managing for profit. McGraw-Hill
Koepke CA (1941) Plant production control. Wiley
Lane FC (1934) Venetian ships and shipbuilders of the renaissance. 1992 Reprint. Johns Hopkins
Lewis JS (1896) The commercial organization of factories. Arno Press Reprint 1978. Spon & Chamberlain
McKay KN (2003) Historical survey of production control practices. Int J Prod Res 41(3):411–426

Nash CW (1928) Purchasing for a fast rate of turnover. In: Dutton HP (ed) 110 Tested plans that increased factory profits. McGraw-Shaw, pp. 169–173
Orlicky J (1969) The successful computer system. McGraw-Hill, New York
Orlicky J (1975) Material requirements planning. McGraw-Hill, New York
Plossl GW (1973) Manufacturing control – the last frontier for profits. Reston, Reston
Reinfeld NV (1959) Production control. Prentice-Hall, New York
Robb R (1910) Lectures on organization. Private printing
Robertson WL (1928a) Quality control by sampling. Factory Ind Manage 76(3):503–505
Robertson WL (1928b) Quality control by sampling. Factory Ind Manage 76(4):724–726
Shewhart WA (1931) Economic control of quality of manufactured product. Van Nostrand
Simons AM, Dutton HP (1940) Production management, original 1926, revised 1939. American Technical Society, Chicago
Smith RM (1977) Harpers ferry armory and the new technology. Cornell University Press
Taylor FW (1903) Shop management. Trans ASME 24:1337–1480
Taylor FW (1911) Principles of scientific management. Harper
Wilkins FS (1999) The application of emerging new technologies by Portsmouth dockyards 1790–1815. PhD Dissertation, Open University
Younger J (1930) Work routing in production including scheduling and dispatching. Ronald, New York

Chapter 3
Advanced Planning and Scheduling Systems: The Quest to Leverage ERP for Better Planning

Irfan M. Ovacik

3.1 Introduction

In this chapter, we focus on advanced planning and scheduling (APS) systems: their emergence as a replacement of materials requirements planning (MRP) systems and their eventual evolution into supply chain management (SCM) systems. This chapter provides a view from a software application provider's perspective and is based on the author's 10+ years of experience in various roles in one of the software application providers. As it represents one point of view, it is not meant to be comprehensive, nor to include all activities and developments that took place in academia or in industry in the time period covered by this chapter.

The 1990s began with a number of relatively large software companies dominating the production planning software market with solutions based on MRP. These companies typically competed with internal, home-grown planning systems, also based on MRP logic, which had been established a few decades before. Also present in the market were a number of small, agile companies who competed with their larger counterparts by exploiting the weaknesses of MRP, and carving themselves a slice of the market under the umbrella of the term APS.

After the emergence of APS solutions, the planning software industry went through a significant transformation, its initial growth fueled by a series of technology-enabled innovations. This growth continued as software companies first exploited the fears around Y2K, and then rode the Internet wave. With the burst of the Internet bubble, the market started shrinking as businesses shied away from new technology investments and software companies focused on survival by cutting their research and development budgets. As the overall state of the economy improved, companies started investing in off-shore custom development projects and integrating the supply chain solutions in which they had already invested, instead of continuing to invest in new software solutions.

I.M. Ovacik (✉)
Intel Corporation, 6505 W. Chandler Blvd., Chandler, AZ 85226, USA
e-mail: irfan.m.ovacik@intel.com

K.G. Kempf et al. (eds.), *Planning Production and Inventories in the Extended Enterprise*, International Series in Operations Research & Management Science 151,
DOI 10.1007/978-1-4419-6485-4_3, © Springer Science+Business Media, LLC 2011

Interestingly enough, after a decade and a half of turmoil, the market is now dominated by a number of large software companies providing a set of dull solutions. Also present are a number of small, agile companies who are flourishing by focusing on niches they have carved for themselves by exploiting the gaps in the solutions provided by the larger companies. While the future remains mostly unknown for the planning software companies, it appears that history has once again proved to repeat itself.

3.2 How It All Started

The previous chapter by McKay provides an excellent overview of the historical foundation of planning systems leading up to the establishment of MRP as the key component of the planning processes in almost all major manufacturing companies.

As MRP was implemented across a large number of companies, the assumptions it makes, such as fixed leads time and infinite capacity, were well documented. These are discussed in more detail in Hopp and Spearman's book *Factory Physics* (McGraw-Hill 1996). In the early 1990s, a number of software companies started to distinguish their solutions by exploiting the documented weaknesses of MRP solutions, using advanced algorithms leveraging new hardware and software technologies to develop production planning solutions. Most of the emerging companies focused on developing algorithms that addressed the factory planning problem of satisfying demand subject to the material and factory constraints. These solutions led to the emergence of the term APS, which broadly classified this new generation of planning solutions. While the established companies burdened themselves by supporting legacy mainframe solutions, the newly emerging companies took advantage of UNIX and advances in graphical user interface (GUI) technologies to develop sleek applications providing better quality solutions much faster than MRP. The new technologies allowed factory planning to become a daily activity where plans could be generated in a matter of hours and minutes instead of waiting for an MRP to run that could only run on a weekend when all resources could be dedicated to it. Reduced planning cycle times enabled factories to work with more timely information and react to changes in demand much more effectively while reducing finished goods and work in process inventories.

These technology advances that allowed fast planning cycles and enabled sleek GUIs came at a time when the physicist Eliyahu Goldratt introduced business management to his concept of the Theory of Constraints, a body of knowledge on the effective management of (mainly business) organizations. Goldratt's ideas were disseminated through his book *The Goal: A Process of Ongoing Improvement* (North River Press, May 1992), which received wide industry acceptance. Emerging software companies used these ideas in combination with the advances in technology to create solutions that addressed the known deficiencies of MRP while carving out a significant market share for themselves.

The algorithms made extensive use of in-memory object models that allowed representation of a rich set of factory and material constraints in the same model, as well as their interactions including pegging of material and manufacturing orders to customer demand. This characteristic allowed the propagation of decisions made in the factory both upstream, to understand their impact on purchasing requirements, and downstream, to understand their impact on demand satisfaction.

Material and capacity constraints were considered simultaneously (or at least in the same model), which allowed planners to understand how material purchasing decisions impacted factory loading, and how factory loading impacted material purchasing decisions. For example, if an order was pushed out due to a capacity constraint, then purchasing requirements due to that order were either pushed out or the material released for an order that needed it earlier. Using the same mechanisms, it was also possible to minimize WIP inventory and use material effectively by making sure that manufacturing was planned only when all incoming assembly materials were available. If production was delayed due to a single item not being available out of many within an assembly operation, then timing of purchasing requirements or the upstream production for the other items could be synchronized with the new schedules, minimizing inventory on the factory floor. This was a critical capability for electronics original equipment manufacturers (OEMs), where the majority of cost is tied to the materials that are assembled into the end product. It was also critical for producers of large products (e.g., furniture and heavy machinery) where there is usually not enough room on the factory floor for a lot of WIP. The same principle also benefited companies which took orders for a combinations of products that all need to come out of various manufacturing lines simultaneously for loading into trucks.

Only a few of the companies who led the way in the earlier days of advanced production planning systems remain in their original form. Intellection was founded by Sanjiv Sidhu, then a researcher at Texas Instruments, and Ken Sharma, a former colleague of Goldratt. They were joined by Cyrus Hadavi who had been working on developing planning systems at Siemens. Intellection later became i2 Technologies. After 2 years, Hadavi left i2 and founded Paragon, which later became Adexa. Another key player in the market was Red Pepper Software, founded by former NASA scientists. Red Pepper's solution evolved from the ground processing scheduling system developed for NASA to refurbish space shuttles between missions and was based on artificial intelligence techniques. Red Pepper was later acquired by Peoplesoft (1997) which was later acquired by Oracle (2004). Also in the market was Numetrix, based in Toronto, which focused on the consumer packaged goods (CPG) industry. Numetrix was later acquired by JD Edwards (1999) which, in turn, eventually was bought by Peoplesoft (2003).

3.3 Growth Phase

The companies providing advanced planning solutions saw considerable growth in parallel with technology advancements. Technology allowed the companies to model larger systems and to develop more complex algorithms. As the factory planning space was addressed by a number of software providers, the natural extension was to start considering the larger problem of matching supply and demand across the entire enterprise. This led to the emergence of the terms supply chain planning (SCP) and SCM, as well as the consolidation of a wide range of efforts in production planning and scheduling, logistics, and inventory control under one term. This consolidation of terms also allowed a number of other software vendors who had been providing transportation planning, distribution planning, and logistics solutions to enter the SCM market. Soon, through the power of marketing, it became very hard to distinguish between the truly innovative solutions and the solutions that repackaged existing solutions under a new marketing message.

The SCM market was defined by a number of leading solution providers and a number of early adopters who invested in these solutions with the vision that SCM would make them better competitors in their respective markets. The grand vision of a seamless planning process across the entire supply chain is still the holy grail of enterprise planning. It encompasses the entire enterprise, all of its divisions (purchasing, manufacturing, distribution, and sales) and all planning horizons (from short-term scheduling of production, distribution, and transportation, all the way to the long-term design of the supply chain).

This vision requires that short-term decisions are made using very granular, detailed models and fast, agile algorithms that are able to react to last minute changes in the system. Longer-term decisions are made with more aggregated models to mitigate the risks around uncertainty and the accuracy of available information. As time moves forward, information becomes more accurate and more information becomes available. As a result, more aggregate decisions seamlessly roll forward to become more detailed decisions.

The planning algorithms rely on a combination of exact methods and powerful search algorithms that provide near-optimal results. This ensures that longer-term decisions that often require more investment (such as capacity expansions) are optimal and provide good guidance to the shorter term, more detailed decisions. As the vision calls for a seamless planning process across the enterprise, it also recognizes the need to extend beyond the four walls of the enterprise and include partners through the use of intelligent request-promise mechanisms. The idea allows a particular enterprise to extend the planning process to the entire supply chain by collaborating with its customers and suppliers, enabling the entire supply chain to operate with more information and, more importantly, with *better* information. Collaboration with customers provides up-to-date information on the demand forecast, ensuring the company is planning the right products at the right time and in the right quantities to meet customers' needs. Similarly, collaboration with suppliers allows the company to send the latest and most accurate demand signal to the suppliers while making sure that requested quantities are within each supplier's capabilities.

Unfortunately, realizing this grand vision became elusive as providers started building the solutions and early adopters started implementing them. Both parties ran into a number of problems:

- Availability of powerful hardware enough to handle the computation requirements of such a solution.
- Availability of both heuristic and exact algorithms to optimize across the entire enterprise.
- Magnitude of the organizational changes needed to adopt the business processes implied by these solutions.
- Availability of enterprise data to support such a solution.

All these problems led both the solution vendors and adopters to take a more practical look at the problem, while the marketing message continued to push the grand vision. As optimizing the entire supply chain in a single, complete, seamless model proved to be elusive, the vendors retreated to familiar methods for solving large complex problems. The problem of optimizing the supply chain was decomposed into a hierarchical set of problems where a long-term aggregate plan across the company provided guidance to a set of mid-range master plans, typically for each division or product line. The master plans, in turn, fed the short-term production scheduling and distribution/transportation planning functions.

The decomposition of the larger planning problem into smaller components helped those solution providers who had been in business for a while, but were late to the SCM market. Since most of these companies had solutions on hand for one or more of the components of the bigger problem and the associated customer base, it became a matter of adopting the marketing message and expanding the solution footprint by building new components and integrating them with the older pieces.

The decomposition also allowed the vendors to tailor solutions to the needs of specific industries. For example, for manufacturing-intensive companies, the master production scheduling (master planning) component allowed supply and demand matching across divisions and factories, whereas production scheduling focused on short-term manufacturing facility planning. In distribution-intensive companies, on the other hand, master planning focused on distributing the planned production to the distribution network in order to best meet the forecasted demand, with transportation planning and scheduling handling the short-term detailed planning needs.

The master planning problem worked with more aggregate data to mitigate the risk of inaccuracies in data. This negated the need to model detailed constraints such as lot sizing and setups. As a result, the master planning problem lent itself to exact methods such as linear programming with proven and scaleable algorithms and guaranteed optimality. This was all the better since it was important for the higher level planning processes to come up with solutions as close to optimal as possible in order to provide the right guidance to the other processes that used master planning results as a starting point.

The factory planning problem required the modeling of a number of constraints that were discrete and nonlinear. The algorithms developed were heuristically based on the theory of constraints and had the goal of producing plans that were considered good and feasible, but not necessarily optimal. These algorithms addressed the known deficiencies of MRP and allowed the planners to consider quantity-dependent lead times, take into account material and resource constraints, and coordinate assemblies based on material and resource availability.

The latter half of the 1990s brought expansion of the market as solution providers, some old and some new, some large and some small, focused on components of the SCP problem. The adoption rate of these solutions accelerated as the new planning systems already addressed Y2K issues and many companies opted to buy new systems instead of trying to reengineer legacy systems to be Y2K compliant.

3.4 Stagnation Phase

During the "growth phase" while the companies offering SCP solutions were doing well, so were the larger enterprise resource planning (ERP) companies such as SAP, Peoplesoft, and Oracle who were riding the tails of the Y2K wave. The success of the SCP vendors did not go unnoticed when they started getting larger and larger chunks of the total investment dollars in deals involving SCP and ERP solutions. Therefore, the ERP companies who already had MRP modules (but whose main competency up to that point was building solutions around transaction management systems such as order management, purchasing, and human resources) entered the advanced planning world either through acquiring smaller companies already in the market or by launching their own development efforts. The entry of ERP companies into the advanced planning market was one of the key contributors leading to the "stagnation phase" covered in this section.

The bursting of the Internet bubble and the subsequent economic downturn had a profound effect on the advanced planning solutions market. Customers' attention turned toward reducing costs rather than new investments in technology and software, and hence to getting the most out of already existing investments. These decisions put a considerable financial burden on solution providers who had invested heavily in research and development during the boom days and relied heavily on new revenue to survive. The result was a major consolidation in the market space where smaller companies either went out of business or were acquired by the larger ERP companies. The ERP companies took advantage of changes in the market by pushing their "integrated" solutions with the premise that solutions from a single company would reduce the integration effort and cost. As the market shrunk, ERP companies leveraged their installed bases to increase their market share. Advanced planning became one of the many solutions that the ERP vendors offered. These companies did not have the accumulated knowledge needed to enhance the current planning solutions, nor did they have the motivation to do so since planning was just one more "add-in" as opposed to the primary driver for new deals. As a result, the research and development investment, and therefore the innovation, that went into planning solutions dwindled to a bare minimum.

Another change that occurred around this time was the association of selection and implementation of planning systems with the information technology (IT) or information systems (IS) departments. In the earlier days of advanced planning systems, the selection was primarily done by the business units themselves, often led by people who had a good understanding of the company's business and planning processes and appreciated the benefits of implementing advanced planning systems in order to be able to reduce planning cycle times, reduce inventories, and improve customer satisfaction.

There were a number of factors that contributed to the rise of the IT/IS departments as the key decision makers for the selection of planning systems. IT/IS departments had been growing in size and influence as the personal computer became an integral part of daily life and software became available to replace the paper-based, manual business processes which were the standard of the preceding 20 years. As year 2000 approached and companies invested heavily in new information systems to replace older systems, their influence expanded and, naturally, IT/IS departments took a bigger role in the selection of any software – including the planning solutions.

As mentioned in the previous section, there were a number of practical problems related to the availability of hardware and software to realize the concept of a seamless, integrated SCM process. The result was a set of smaller components that were linked to each other through a hierarchy. From a technical perspective, this also meant that all these components had to be integrated with each other, as well as the enterprise transaction systems (ERP or legacy) to get the necessary data. This often was not a trivial activity, requiring complex transformations and in most cases requiring companies to build new systems to stage the data coming from ERP and legacy systems and to maintain data that did not exist elsewhere in the enterprise. The efforts required for integration increased to the level where the cost to build systems to house and move data among the components of the overall system became a major component of the cost of implementing the planning systems. All these integration activities fell into the domain of the IT/IS departments.

Another factor that led to the increase of IT/IS influence was the general hype about the benefits of Internet technologies which made the implementation of planning systems orders of magnitude more complex than in the earlier days of APS, as they now required increasing numbers of technical resources. In the earlier days, a particular factory planning software could be shipped in four files: One executable for the engine, one executable for the GUI, and one text file each for the configuration of the engine and UI. Building a factory model using real data took as little as a week and was mainly constrained by the efficiency of the one or two technical resources who extracted the data from the legacy systems. There was no need for any additional software, and the only hardware needed was a UNIX box. In the new era of database servers, application servers, Web servers, middleware and Web-based user interfaces, the same effort required a number of people with different skills, a number of middleware and other supporting software systems, and weeks, sometimes months, to just set up the environment for the planning system. Naturally, all this hardware, software, and resources were within the domain of IT/IS departments.

While it seems natural for the IT/IS departments to influence the selection of planning systems, it is also easy to see how the same influence eventually led to the "stagnation phase" that we cover in this section.

Traditionally, IT departments had been the primary customers of the ERP companies whose value offering often revolved around cost of ownership. As the ERP companies expanded their offerings to planning, their primary contact remained the IT departments and their value offering remained the same. Thus planning systems became just one more software solution that was to be put in place to automate existing processes and reduce cost of ownership. The real value of advanced planning systems in terms of reducing of planning cycle times, reducing inventories, and improving responsiveness to customer needs was forgotten or conveniently ignored. As the focus shifted to the cost of ownership, even the traditional SCM companies shifted their messages and investment strategies. Realizing that companies were spending a large portion of their budgets on building custom solutions to provide the data needed by the planning solutions, the solution providers shifted their investments to building tools for master data management (MDM) and product information management (PIM) systems. All of this came at the expense of planning solutions which received little or no share of the available research investments.

The other impact of IT influence and focus on cost of ownership was on the technology used to deliver planning solutions. The perceived benefits of Web-based systems in terms of cost of ownership through central control led the IT departments to put a lot of pressure on solution vendors to convert their solutions to the new technology. Supporting the new Web-based technology became a prerequisite to entry into new markets. As a result, solution vendors directed valuable research and development funds into converting their architectures from client–server technologies to three-tier Web technologies. These investments came at the expense of any innovation in planning processes and planning algorithms. During this time, two important considerations were forgotten: (1) the difference between planning and transaction management and (2) the needs of planners who required a lot more analysis and a lot less forms. Unfortunately, the new technologies were much better at supporting the latter than the former.

Another blow to the traditional SCM companies was the emergence of off-shore development centers that offered their resources at a fraction of the cost of United States- or Europe-based companies. Suddenly, at least on paper, it became cost efficient to develop custom planning systems that included low-cost integration services instead of having to buy packaged software and investing a lot of time and effort into configuring and integrating it to the enterprise systems. Missing from the picture were the millions of hours of intellectual property embedded in the packaged software.

Also absent was the valuable learning from the past implementations. Buyers of planning solutions had traditionally been the actual business users who understood the benefits of using advanced methods for running their factories. This gave the solution providers the opportunity to work directly with the end users, allowing them to configure the planning solutions to meet the business requirements. This also allowed the providers to learn from the interactions and build their learning into future releases of their software. Solutions that were implemented with business buy-in and involvement were often extremely successful since the business owned the solution. With the emergence of IT as the key decision maker in investment decisions, this direct connection between solution providers and the business users was largely lost. The IT departments took on the responsibility of understanding business needs, translating them into requirements (and often interpreting them), and interacting with the solution providers to select the "best" solutions. Similarly during deployment, IT departments continued to act as intermediary between the solution providers and business users, often changing the message to accommodate the goals and objectives of the IT departments. Naturally, this led to a lot of supply chain implementations that were declared successful by the IT departments, but were soon abandoned by the end users who did not feel like they owned the solutions.

All the factors above resulted in the reduction of investment by both solution vendors and their customers on the core of their planning systems. As the focus shifted into other areas, innovation slowed down to a trickle and stagnation was the result. But this was just part of it. Another result of the stagnation phase was the reemergence of spreadsheet applications as the primary medium for planning. In the next section, we look at the spreadsheet phenomenon and its impact on planning.

3.5 The Spreadsheet Phenomenon

Any discussion of the evolution of advanced planning solutions would not be complete without understanding the impact of spreadsheet applications, specifically Microsoft Excel which has become the de facto standard in industry when it comes to planning. Today, every organization makes use of Excel in one form or the other, some running their entire organizations with it, and most using it as a stealth backup to cover the deficiencies of "official" enterprise planning systems.

The emergence of Excel as the primary planning tool across a very wide range of industries and planning functions can be viewed as a reaction to the increased influence of IT organizations in the selection and implementation of planning applications although the use of spreadsheets for planning goes as far back as the early days of spreadsheets.

The author's earliest experience goes back 1986 when the spreadsheet was used to assist the master production planning process for a major appliance manufacturer in Turkey. The spreadsheet was used to replace the hand calculations and hand-built tables used in creating the master plan. Another example is from the mid-1990s where a division of a major United States-based electronics OEM was using around 200 Excel worksheets that were connected to each other with an elaborate logic to support master production scheduling activities. The planning involved painstakingly preparing the data in each spreadsheet, then activating the calculations which sometimes took as long as 2 days to complete. Since planning required a number of iterations and sometimes creation of new spreadsheets to work around the tool's limitations, the monthly planning cycle sometime took as long as 5 weeks.

The success of Excel as a planning tool actually has a very obvious explanation. One only has to look at what planners do; interestingly, this is something that solution providers and IT department have overlooked for years. Their focus is usually on how to create an "optimal" plan, and not on what planners do with that plan. Also often overlooked is the fact that the plan is with respect to a model that is an abstraction of the real system being planned and the data is with respect to a snapshot of the system that may be hours or days old.

The planners' primary task is to analyze the plans, identify issues, and look for ways to resolve the problems. Some of these problems may be due to the infeasibility of the plan with respect to the real world, or they may be real problems that need to be resolved to meet business goals. This process requires the planner to understand the current situation, develop potential alternatives, and understand the impact of these alternatives on the rest of the system. The analysis often requires dealing with a lot of data – filtering, transforming, comparing it to get to results – all functions that can be done very intuitively and quickly using a spreadsheet.

As a result, even if an enterprise system or other planning application is used to create the initial plan, the accepted best practice has become to first extract the plans from the enterprise systems into Excel, perform the necessary analysis, and then feed the information back into the enterprise systems (often manually). Note that since the plans are already captured in the spreadsheets, and since the spreadsheets used often have all the necessary constructs for analysis and replanning, it becomes easy to abandon the IT-supported planning systems and go back to all Excel solutions.

The planners' need to access a lot of data for analysis was also overlooked during the emergence of Web-based applications. The ability to deliver applications through a browser was a great innovation from an IT perspective because it allowed control of applications and data from a central location which could be distributed to large communities of users through the use of a browser without having to install anything on the user's computer. This worked really well for applications such as purchasing and human resources where transactions are the key components and these transactions can be delivered to the users one at a time, using screens similar to the paper forms that were used before.

Unfortunately, what worked for purchasing or human resources did not work so well for planning. Planners needed access to a lot of data and the ability to analyze that data, and the Web-based architectures were not designed to serve that purpose. One option was to serve up the data in small quantities and do all the filtering and manipulation on the server. This meant that, at any time, the planner had access to only a subset of the data needed and had to rely on multiple trips between the server and the browser to get anything done. This did not work very well since a lot of time was spent negotiating the network traffic and the server which was shared among many users. The other option was to serve up the necessary data all at once, and provide the analysis capability within the browser. This also did not work since the browser technology could never match the capabilities provided by Excel. Later on, the browser just became a tool to bring the data to the user, so that it could be transferred into Excel. As Excel later acquired the capability to directly access databases, it became just as easy to get the data directly into Excel, eliminating the browser all together.

Recently, a number of solution providers have started adopting Excel as the front end to the planning applications instead of developing their own. The data and plans still reside in an enterprise system which can be centrally controlled, but the users interact with the data only through Excel. This seems like a good compromise as it gives the IT departments the control that they need over data and applications and the users the analysis capability that they need to do their jobs successfully.

All in all, Excel still continues to be a primary planning medium for a lot of planning processes even if the initial plans are created by IT-supported enterprise applications. As of this time, there does not seem to be a good alternative provided either by the application vendors or the IT departments.

3.6 Future

Interestingly enough, the current market for planning solutions looks pretty similar to the market in the earlier days of the advanced planning systems. Back then, the market was dominated by a few companies proving solution based on MRP logic, where as the smaller, more agile companies were presenting solutions based on new ideas and exploiting the weaknesses of the larger companies. Today a number of ERP companies, such as SAP and Oracle, dominate the planning solution market with fairly pedestrian solutions. While only few of the early pioneers of advanced planning solutions (such as Adexa) still survive, there are a number of small companies who are doing very well by focusing on specific planning problems. Examples are Servigistics for service parts planning and Optiant for inventory optimization. Unfortunately, due to the close relationships of the ERP companies with the IT departments, barriers to entry are very high – much higher than back in the early 1990s when advanced planning solutions were just emerging. So the ability of these small companies to survive depends on being able to reach the real users of the planning solutions and on being able to distinguish their solutions from those of the larger companies.

Chapter 4
The Transition to Sustainable Product Development and Manufacturing

Robert C. Carlson and Dariush Rafinejad

4.1 Introduction

In this chapter, we provide an overview of the state-of-the-art in sustainable product development and manufacturing and of the challenges in ubiquitous adoption of sustainable development practices in business. Environmental and business sustainability are examined in a holistic framework underscoring their interdependence on both spatial and temporal scales. We review the evolutionary rise in sustainability awareness including the development of methodologies for the assessment and development of sustainable products/manufacturing.

The major global corporations and manufacturers such as IBM, HP, 3M, Toyota, Shell, Nestle, Monsanto, GE, and others have gone through multiple evolutionary phases in their outlooks toward environmental sustainability. These and other companies have generally evolved through stages of no concern, to pollution control, to pollution prevention, and to resource efficiency maximization in lockstep with governmental regulations. And in all these stages, firms have sought opportunities for product differentiation/branding, tried to influence governmental regulations, developed relationships with environmentalists, worked with their suppliers, and adopted environmental and social responsibility metrics for internal audit and marketing purposes. However, the adoption of sustainable development as an imperative strategic vision is often lacking in the industrial enterprises (World Watch Institute 2006–2007).

According to the UN Commission led by Gro Harlem Brundtland in 1987, sustainable development is "*development that meets the needs of the present without compromising the ability of future generations to meet their own needs.*" This definition has led to much discussion in the fields of (ecological) economics, public policy, and environmental ethics. The prevalent interpretation of (economic) sustainability in business is growing (or at least nondiminishing) economic output (that is generally measured in terms of gross national product or GNP). This interpretation of

D. Rafinejad (✉)
Management Science and Engineering Department, Stanford University,
Stanford, CA 94305-4026, USA
e-mail: rafineja@stanford.edu

K.G. Kempf et al. (eds.), *Planning Production and Inventories in the Extended Enterprise*,
International Series in Operations Research & Management Science 151,
DOI 10.1007/978-1-4419-6485-4_4,

sustainability is referred to as *weak sustainability*. On the other hand, sustainability that is interpreted as nondiminishing life opportunities is branded as *strong sustainability*. In the latter, human focus is extended beyond economic (manufactured) capital to ecological and social capital, and development is considered as human development in quality of life. An extension of strong sustainability is coined "deep economy" where humans, as integral part of the environment, seek development in harmony with nature. An excellent overview of the concepts of weak and strong sustainability is provided by Ayres et al. (1998).

In this chapter, we primarily focus on issues that relate to firm-level business sustainability. The definition of business sustainability (and corresponding strategies and practices of a firm for sustainable product development and manufacturing) is strongly related to and must be congruent with the above sustainable development concepts, which encompass the macroeconomic context within which the firm operates. In other words, an enterprise can sustain achievement of its business objectives, if its operation is aligned with the broader sustainable development framework. Similarly, sustainable products and manufacturing practices are aligned with and serve the broader societal sustainable development goals. These ideas are explored in an associated paper by the authors (Carlson and Rafinejad 2010).

The initiatives for minimizing environmental impact beyond pollution control are frequently referred to as environmentally sustainable development by industry leaders. Minimizing environmental damage, although a prerequisite for sustainable development, merely slows down the environmental degradation process. Furthermore, the initiatives of individual firms are inherently competitive and un-integrated and hence insufficient in addressing the environmental sustainability issues, which are often interrelated and have global impact.

While weighing the impact of economic development on natural and human capitals are (critically) necessary, the resulting impact assessment and mitigation measures are not sufficient conditions for a sustainable economy.

Most academic and popular literature tend to state the looming sustainability problem, emphasize the need for the so-called triple bottom line objectives (profitable growth, environmental friendliness, and social responsibility) and propose initiatives that would mitigate environmental harm caused by human economic activities. In spite of the broad recognition of the need for sustainable products and manufacturing, progress in fulfilling this need is grossly inadequate. There is a general belief that the development of commercially successful "sustainable products" faces an insurmountable challenge in the current economic context.

In spite of humankind's fantastic technological ingenuity and accomplishments, the current economic system has not been a panacea and has led us to severely undesirable consequences. We are "trapped" within the confines of material possession, although it is merely the satisfaction of the most fundamental human need for survival and is at the bottom of the hierarchy of needs for the actualization of human potential. We have forsaken the pursuit of esthetics, art, spirituality, community bonding, the relationship to people/nature and hence *happiness* and have framed them as "monetized" commodities. We have failed to achieve a minimal degree of equity among the majority of humanity, even in the satisfaction of the basic needs of food, shelter, and health. And lastly, our current mode of economic activities has led us to the brink of environmental and resource calamity (e.g., global warming) that could result in the irreversible loss of humanity's collective accomplishments.

Perhaps we need to examine critically our unquestioned assumptions of the following: growth as the overarching objective of business, the win-lose approach toward nature, survival-of-the-fittest as the desirable (and unavoidable) mechanism for development, the focus on short-term gains, globalization for exploitation of resources (of the earth and labor) without globalization of equity, detachment from local and community issues, and consolidation/centralization for the control of resources in a zero-sum game.

Sustainable product development and manufacturing might require a new economic model that supports the *sustainable growth of human development* (vs. sustained growth of shareholder value). A new economic structure that fundamentally changes the "basis of competition," shifts the concept of "development" away from raw materialism, and changes "purchasing power" as the sole indicator of success, happiness, and self-actualization. The United Nations (UN) has developed the Human Development Index (HDI) as a representative metric to integrate the impact of economic, human, and natural capitals on the standard of living. HDI is calculated from the three factors of life expectancy (health), adult literacy (education/skills), and gross domestic product (GDP) per capita at purchasing power parity (PPP).

There is an urgent need for research in multiple areas, including the following: the development and implementation of a sustainable economic framework in which sustainable products and manufacturing could flourish and become the norm in industrial activities; development of methodologies to characterize sustainable economic activities, products and manufacturing through measurable indicators, and the education of the business (and societal) leaders of tomorrow.

In what follows, we shall examine the causal factors underlying the current heightened concerns for environmental and business sustainability and review methodologies for assessing products and manufacturing and the best practices for sustainable product development and manufacturing.

4.2 Driving Forces that Threaten Business Sustainability in the Twenty-First Century

The rapid economic development of the twenty-first century has resulted in a significant impact on the natural resources both in the depletion of nonrenewable energy and material resources and in exceeding the earth's capacity to absorb and reprocess the life cycle effluents of human-made products (including the end-of-life disposal). Nature's endowments of resources and reprocessing services that are often referred to as its sources and sinks establish the critical input and output fluxes of economic activities and their depletion jeopardize business sustainability.

Depletion of petroleum reserves and global warming to a perilous state are only two of the major manifestations of the adverse impact of human economic activities on the sources and sinks of the earth. It is important to note that oil consumption (and hence the depletion of the reserves) and global warming (caused in a large part by the carbon dioxide generated from oil consumption) are strongly interrelated. Similarly, many other impacts on the earth's sources and sinks are interrelated and reinforce each other through positive-feedback loops.

Former US Vice-President Al Gore's well-publicized film, "The Inconvenient Truth" presents a convincing argument for rapidly accelerating global warming and urges an immediate and drastic action by all countries of the world, particularly by the highly industrialized economies who contribute the most to CO_2 generation. The recognition of global warming as a threat to human civilization and call for urgent action for reduction in CO_2 generation (up to 80% by 2050) have been underscored through numerous publications by international organizations and governments (see Notes 4–6).

The interactions among the stock of nonrenewable sources, processes of the natural sinks (for regeneration of renewable resources and other reprocessing services; see Note 7) and human economic activities occur in local, global, and temporal dimensions. For example, global warming is caused by green house gases (GHG) irrespective of where on earth they are generated. Also the dangerous level of CO_2 in the earth atmosphere is reached by a time-integrated accumulation of CO_2 from human activities beyond the earth's reprocessing capability (through the photosynthesis process in trees). The examples of local and short-term effects of human activities include deforestation, defacing of the landscape by strip-mining and air, water, and land pollution. Such strong interdependence calls for ecosystem level thinking in product development and manufacturing where product life cycle and manufacturing closed loop interactions with the environment are integrated into design decisions.

The depletion of nonrenewable resources is not limited to petroleum. The example in Fig. 4.1 shows the concentration of copper (Cu) in remaining mines declining by almost eightfold in the twentieth century. As resources are depleted, the cost of extracting and processing raw material increases as does the extent of associated environmental damage (larger area and the earth's mass must be disturbed for a given

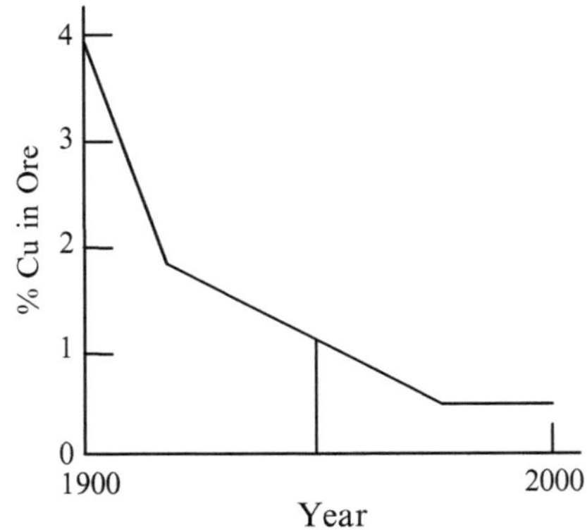

Fig. 4.1 Change in copper concentration in remaining mines (Manahan 1999; Graedel and Allenby 2003)

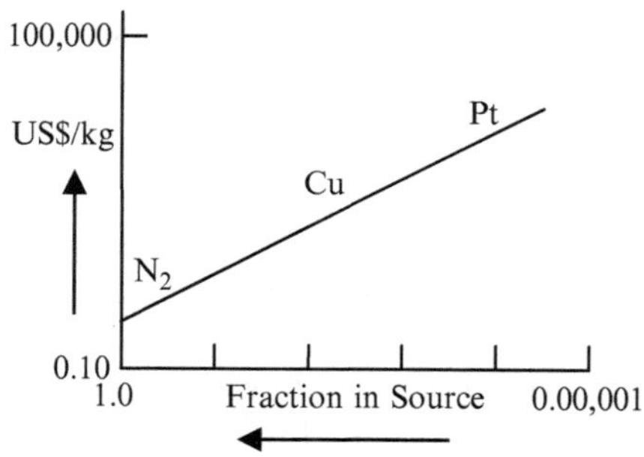

Fig. 4.2 Raw material price vs. concentration in the source (Manahan 1999; Allen and Behmanesh 1994; Ayers and Ayers 1996)

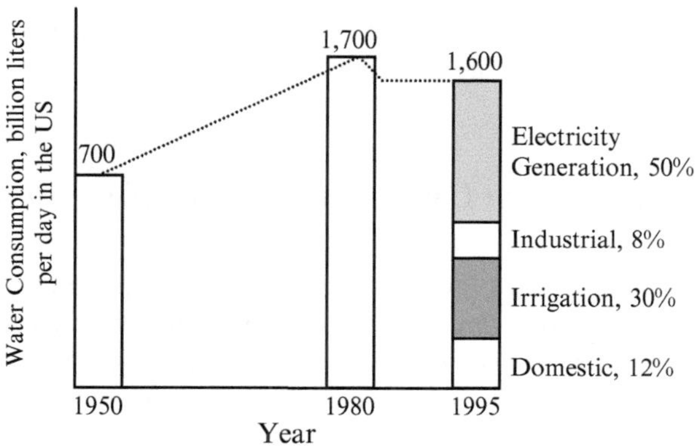

Fig. 4.3 Interdependence of energy efficiency and water efficiency (Manahan 1999)

gain in raw material). The cost (price) of raw material used in industry generally correlates with the fraction of the material in the source, as shown in Fig. 4.2 for nitrogen, copper, and platinum. While increasing extraction cost of nonrenewable resources could make alternative environmentally friendly resources viable (e.g., fossil fuels vs. solar energy), it might also encourage more destructive extraction practices (e.g., oil exploration vs. tar sands).

Another example of interdependence among various factors of economic activities on resources and sinks of the earth is shown in Fig. 4.3. This chart depicts water consumption level in the second half of the twentieth century in various sectors of the US economy. The overall consumption increased dramatically from 1950 to 1980 and decreased over the following 15 years as a result of improvement in energy efficiency in various sectors.

The World Wide Fund for Nature (WWF) provides semiannual data on the *ecological footprint* of more than 150 nations in its Living Planet Report. Ecological footprint is defined as the land area required for providing the resources and absorbing the emissions in support of human production and consumption activities (Wackernagel et al. 2002). Wackernagel et al. show that human ecological footprint exceeded the earth's carrying capacity in the 1980s.

The impact of human economic activities on the ecological footprint can be estimated according to the following formula (Hart 1997):

$$I = P \times A \times T,$$

where I is the impact (the ecological footprint) of any population upon the earth's sources and sinks, P is population, A is affluence (proportional to consumption), and T is technology that supports the affluence (the degree by which it increases the footprint). According to this model, "sustainability means constant or declining impact (I) or ecological footprint". Hart (1997) argues that A must increase tenfold (for the majority of the world) to stabilize the population at the ten billion level and (sustainability) technology must improve 16-fold to stabilize I. Meeting these challenges offer a great business opportunity. For further discussion on the IPAT equation, see Chertow (2000) and Graedel and Allenby (2003).

4.2.1 Impact of Human Consumption on the Environment

In the following paragraphs, a few recent examples – many from China – are cited to underscore the staggering impact of human consumption on the earth's sources and sinks. It is important to note that the present impact is primarily caused by only 30% of human population who have achieved a reasonable living standard and per capita GDP of more than US$5,000 per year (see Note 15).

Battery recycling: Deleterious substances in a single button cell battery can pollute 600 m^3 of water and one deteriorated D-sized battery may render 1 km^2 of soil useless. Heavy metals leaked from discarded batteries on the land contaminate not only the soil but the underground water supplies as well (see Note 16).

Each year more than 70 billion button cell batteries alone are consumed in China, with a total weight exceeding 1.4 million tons. In the Guangdong province, 200,000 tons of general batteries are disposed annually. If they are recycled, they could be turned into 100,000 tons of lead, 23,000 tons of zinc, 1,000 tons of nickel, with a total value of US$180 million (according to Chen Hongyu, dean of the environment department at South China Normal University).

The average European household uses 21 batteries a year, according to EU figures (see Note 17). In 2002, that added up to more than 158,000 metric tons of batteries. For industrial use, Europe went through 190,000 metric tons of lead acid batteries.

A new law approved by the EU Parliament requires distributors to take used batteries back at no charge. Battery producers and distributors will foot most of the bill for implementing the recycling program mandated by the law, including educating the public where to turn in the batteries. The EU Commission estimates the program to cost industry between 200 and 400 million euro. The law will also ban some portable cadmium batteries and prohibit dumping in landfills or burning of automotive and industrial batteries. By 2012, a quarter of all batteries sold must be collected at end-of-life (EOL). By 2016, the target will rise to 45%.

Washing jeans can be costly to the earth: Washing, tumble drying, and ironing jeans can be costly, to the earth, concluded a study commissioned by France's environment agency on the ecological impact of a pair of denims (see Note 18). An average pair of jeans is made with 600 g of denim, lined with 38 g of polyester, with six rivets, and a button, is worn 1 day a week for 4 years, washed every third time in a high-energy machine at 40 °C, and, in a singularly French twist, ironed before wear. The study by the research firm Bio Intelligence Service, looked at the jeans' life cycle, from material production to daily wear. It concluded that machine washing, tumble drying, and ironing were responsible for 47% of the eco damage the jeans caused – 240 kW of energy a year. Dry cleaning, furthermore, was "an environmental disaster."

Extrapolating the data to the USA and assuming that 50 million jeans are worn by Americans, the energy usage would amount to 12 billion kW whose production would result in eight million tons of CO_2 emission into the atmosphere.

Toxic chemicals harming arctic animals: Toxic chemicals are harming Arctic animals including polar bears, beluga whales, seals, and seabirds, the environmental group World Wild Fund (WWF) in Switzerland said in a June, 2006 report (see Note 19).

The report said pollutants such as flame retardants, pesticides and fluorinated chemicals made Arctic wildlife vulnerable to health problems including immune suppression and hormone disturbances. The chemical contamination of the Arctic threatened the survival of many of the regions' animal species, who also faced possible habitat and food supply loss due to climate change.

"The bodies of some belugas from the St. Lawrence estuary in Canada are so contaminated that their carcasses are treated as toxic waste," said the WWF. The Arctic is far from industrial centers but many long-lasting chemicals get swept north by winds and ocean currents and build to damage levels in fatty tissues of creatures in the region.

WWF appealed for the "urgent and significant strengthening" of European Union (EU) legislation designed to protect people and the environment from the adverse effects of chemicals found in paint, detergents, cars, and computers. The bill, known by the acronym REACH, has drawn criticism from the USA and other countries who say its provisions hurt trade and would be difficult to implement.

Weihe River in china faces long-term cleanup: Just 15 years ago, fishing boats used to ply the Weihe River but now locals call this section of the River the "sewer" (see Note 20). "If it did not rain the water was red and it stunk," said 53-year-old grandmother Lao Wong, a villager in Qishan County in northwestern China's Shaanxi Province. The Weihe River is one of the most polluted in China and a massive cleanup has been underway the past couple of years by the central government.

The shrinking river has left fishing boats rotting on high ground, and the wide banks, which used to be a submerged fish habitat, are now planted with vegetables. The human waste and factory discharge dumped into the Small Weihe River joins more pollution in the 800-km long Weihe River, which finally meets up with the Yellow River before spilling into the Bohai Sea.

The Shaanxi environmental protection bureau says in 2004 more than 600 million tons of waste was discharged into the Weihe River (from hundreds of paper mills along the river among other sources). The water quality of nine of 13 sections of the Weihe was found to be below class V, meaning it is undesirable even for irrigation, said Tian Xijun, vice director of environmental pollution control department of the State environmental protection agency (EPA).

The Shaanxi government has ordered all paper mills with an annual production capacity of <20,000 tons to close by the end of this year. Major pollutants have been cut from 600 in 2001 to 149. In the next 5 years, Shaanxi plans to build a sewage and waste treatment plant in each county along the river basin.

Environmental damage to the Chinese economy: "The World Bank says the environmental damage costs us about 7% of our GDP. The highest figure I have seen is 18%. That wipes out our economic growth. The growth we have is inflated – it is not sustainable", says Mr. Xue Ye, executive director of Friends of Nature in Beijing (see Note 21).

China's e-waste capital chokes on old computers (see Note 22): "Guiyu, China is a modern day gold rush town. Workers sift through piles of broken old computer parts in acrid smelling shacks, smelting down parts with crude equipment to extract valuable metals such as gold and copper."

According to a 2005 UN report, up to 50 million metric tons of e-waste is generated annually, as people upgrade laptops and PCs and throw out old models. The China Quality News estimates that about 72% of that e-waste is smuggled into China by sea. Much ends up in Guiyu, a rough town on the southern Chinese coast, not far from Hong Kong. E-waste is not supposed to be exported without the consent of the importing country. To bypass it, e-waste is labeled as "used PCs" or "mixed metals" according to Greenpeace and smuggled in from Hong Kong.

During the disposal process, workers, including women and sometimes children, are exposed to a toxic cocktail of chemicals and are injured by exploding computer parts or burns from the furnaces. There is little regard for safety – no masks, little ventilation, and few signs of government officials enforcing what safety rules do exist in China. State media estimated almost nine of ten people in Guiyu suffered from problems with their skin, nervous, respiratory, or digestive systems.

After the useful metals are taken out, leftover parts are often dumped into landfills or rivers or simply burnt. Piles of old computers even block the traffic in some parts of Guiyu. Reporters and green activists are not welcome.

The state-run newspaper the People's Daily said last year that Guiyu's more than 5,500 e-waste businesses employed over 30,000 people. This business is estimated to be worth one billion yuan (US$130.9 million) in Guiyu alone. Yet many of the workers, who come from all parts of China, are paid as little as three US dollars a day.

Even small things make a difference (Meadows et al. 2004): The invention in 1976 of the pop-up opener on the aluminum soda cans meant that the tab stayed with the can, therefore passing back through the recycling process, rather than being thrown away. Around the year 2000, Americans used 105 billion aluminum cans per year, of which some 55% were recycled. The recycling of those tiny tabs saved 16,000 tons of aluminum and around 200 million kW of electricity per year. It also prevented 136,000 tons of CO_2 emission.

Copper cables: 2000 lb. of copper (Cu) cable can be replaced by 65 lb. of fiber optic cable. Cu production generates toxic heavy metal wastes such as arsenic and requires destruction of large land areas by strip mining. The fiber production consumes only 5% of the energy required for the copper (Billatos and Basaly 1997).

Development of biodegradable plastics: According to an EPA projection, 25.7 million tons of plastic waste per year will be generated in the USA by 2010 (9% by weight and 20% by volume of all landfill wastes). Many plastics take hundreds of years to decompose. Biodegradable plastic is a preferred alternative and usually refers to the combination of 6% cornstarch with plastic polymers. Cornstarch is the bonding material and disintegrates, leaving a fine polymer dust (Billatos and Basaly 1997). Biodegradable plastics are unsuitable for food and retail packaging and they are currently only used for grocery bags. There is also a class of photodegradable plastics that decompose when exposed to sunlight.

4.2.2 Factors Driving Utilization and Scarcity of (Nonrenewable) Natural Resources

The accumulated human consumption over time has depleted many nonrenewable resources to an alarmingly low level and has rendered the long held implicit assumption of the inexhaustibility of earth's resources void. Such is also true about the assumption that environmental sink and reprocessing capacity of nature are insensitive to human economic activities. Furthermore, global population and per capita production and consumption are growing at an accelerated rate in regions of the world that have not previously been significant players in the world market. The factor that perhaps has the highest impact on the depletion of nonrenewable resources is the prevalent framework of economic activities over the centuries of industrial development. These factors are further explored in the following sections.

4.2.2.1 Global Population Growth

Population growth is a function of fertility and mortality rates which in turn depends (in a nonlinear fashion) on the populace standard of living and quality of life as schematically shown in Fig. 4.4.

Table 4.1 lists the mortality, fertility, and population growth rates for selected countries illustrating the trend in Fig. 4.4. Cohen (1995) observes a historical pattern in demographic transition and population growth that corresponds to the population growth trend in Fig. 4.4. Cohen reports that a population undergoes an idealized demographic transition in four stages: (1) high birth and death rates; (2) high

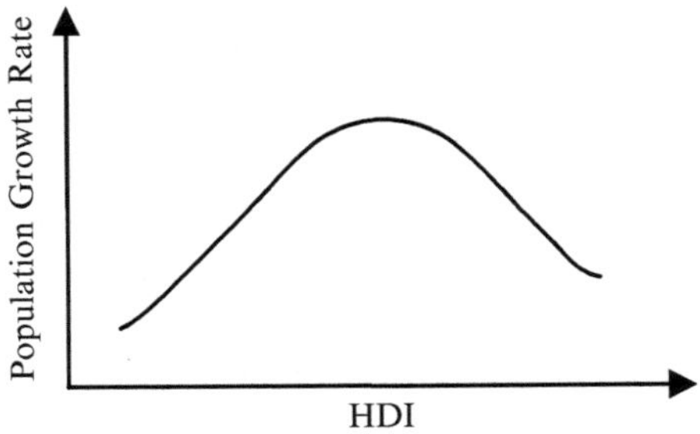

Fig. 4.4 Trend in population growth rate as a function human development index (HDI)

Table 4.1 Population mortality, fertility, and growth rate vs. economic development

Countries and GDP	Mortality rate	Fertility rate	Population growth rate (fertility minus mortality)
Poor Countries (Africa), GDP < US$1,000	High (poor hygiene and medical care); Swaziland = 3.0%	High (economic needs); Swaziland = 2.7%	Low; Swaziland = −0.3%
Developing Country: China, GDP = US$4,000	Moderate to low; China = 0.7%	Low (legal mandate); China = 1.3%	Moderate to low; China = 0.6%
Developing Countries; GDP< US$10,000	Moderate to low Niger = 2%, Brazil = 0.6%	High (economic needs); Niger = 4.9%, Brazil = 1.6	Moderate to high; Niger = 2.9%, Brazil = 1%
Developed Countries, GDP > US$30,000	Low; Italy = 1%, Japan = 0.9%	Low Italy = 0.9%, Japan = 0.8%	Low; Italy = 0.01%, Japan = −0.1%

Source: US Central Intelligence Agency, 2007

birth rate and low death rate; (3) lower birth rate and low death rate; and (4) low birth and death rates. The countries listed in Table 4.1 follow similar demographic transitions.

As more countries develop their economies and a larger segment of the world population enjoys higher HDI, population growth will accelerate in the near future. The world population grew by 4.3 billion during the twentieth century and is estimated to grow by more than 3 billion (to the total of nine billion) in just the first half of the twenty-first century (see Note 15). The emerging growth in HDI, GDP, and consumption for an increasingly larger population means an even more arduous strain on the natural sources and sinks.

4.2.2.2 Accelerated Growth of GDP and Standard of Living Globally

Ubiquitous growth of GDP at PPP across the globe, particularly in the densely populated developing countries such as China and India, is increasing the rate of depletion of nonrenewable resources. Even renewable resources are being consumed at a significantly higher rate than nature is capable of replenishing (e.g., the deforestation of the Amazon, the near extinction of beluga sturgeon in the Caspian Sea, the drying-out of two-thirds of the Aral Sea in Central Asia). Nevertheless, these changes have not yet reached the threshold of immediate impact on the global industry that would necessitate a strategic inflection. Hence in the development of new products, sustainability provisions and attributes (such as reusability, recyclability, and efficiency in manufacturing), beyond the mandatory requirements, are usually treated as low priorities compared to the requirements for function and time-to-market.

How would the state of the world economy (i.e., availability and price of resources, economics of sustainability, and resource-value efficiency) look if the whole world were developed to the current level of the US, EU, and Japan? In other words, if the GDP per capita of 6 billion people were US$40,000, how would the sustainability factors be weighed in the new product development process? Would the basis of comparative advantage shift radically?

One answer is offered by Wackernagel and Rees (1995) who noted that "... Current appropriations of natural resources and services already exceed earth's long-term carrying capacity... If everyone on earth enjoyed the same ecological standards as North Americans, we would require three earths to satisfy aggregate material demands, using prevailing technology... To accommodate sustainably the anticipated increase in population and economic output of the next four decades, we would need six to twelve additional planets." This assessment implies that the imminent rise of China, India, and Brazil requires a fundamental and radical shift away from the current economic model and practices to the economics of sustainability, if a major conflict over the resources and sinks is to be averted.

4.2.2.3 Prevalent Economic Framework

The third and perhaps the most important factor influencing utilization and scarcity of natural resources is our economic framework.

Human economic activities are aimed at improving quality of life, which starts with satisfying the existential needs including physiological (food, water, and shelter) and safety/security. Transcending these basic needs, Maslow in his theory of hierarchy of needs (Maslow 1970), conceived three higher levels including the needs for belonging/love, esteem, and self actualization which become motivators of human behavior once the most basic needs are satisfied.

The basic needs are primarily satisfied by consumption of goods which directly contribute to material wealth generation through industrial production. The higher level needs, however, can be satisfied through significantly less material-intensive means. Community, socialization, creativity, art, spirituality, esthetics, and nature have been instrumental in fulfilling the needs of belonging and love and creating the opportunity for realization of one's potential and self actualization.

In the modern capitalist economic system, the desire for maximization of profit has led to the inescapable requirement for continuous growth in economic activities of production and consumption. In other words, the GDP and GNP have to grow continuously. In order to achieve this objective when the basic needs of consumers in the society have been satisfied, marketers have cleverly created ingenious associations between the consumers' higher level needs (in Maslow's model) and material consumption. For example, consumption of an expensive facial cream is positioned not as mere application of a chemical compound to enhance skin health but as a means of enhancing self-esteem and self-expression. The consumption of multiple vitamin pills a day is positioned as essential for feeling good and enjoying life. Material wealth has increasingly become the sole metric of success and social status. Preventive methods of solving health and social ills (e.g., heart ailment or crime) have given way to corrective means which contribute to more production and consumption. Furthermore, investment in technology and product development tends to be focused on the needs of affluent consumers who have the purchasing power. This leads to a rise in inequity and a widening gap between the haves and have-nots and to over production and consumption by affluent societies, while the basic needs of the societies with unattractive demand conditions (low-purchasing power) are left unmet (electricity, clean water, cure for AIDS.)

In our prevalent economic system, per capita GDP has become the most important indicator of the living standard (purchasing power) and the quality of life (HDI and happiness) of a nation. Consumption level is synonymous with economic health and social wellbeing, and it is tracked diligently by economists and governments. Two-thirds of all economic activities in the USA is in the consumer sector – which amounted to US$2 trillion per quarter in 2003 (see Note 28).

Another key feature of the prevailing economic paradigm is how natural resources and societal commonwealth (the common property) are treated. The value of a "common" property (e.g., air, fresh water, oil, minerals, and most of the ecosystem services) is determined by the cost of access to the resource rather than its contributed value in comparison with alternate means. Only the creation and depreciation of private and corporate wealth are accounted for in the economic models and the commonwealth is left out of business decision metrics. In other words, natural resources and sinks, which have been available at relatively low cost, are not treated as assets and more importantly they are assumed to be boundless.

These assumptions about and the treatment of common goods encourage: development of products that have a short life, disposal instead of repair, and reuse of products, focus on short-term market needs (vs. long-term consequences) and steady accumulation of waste and unused possessions. Furthermore, as mentioned above, only products and technologies will be developed, which fulfill the needs of consumers in the regions with attractive market potential.

While many consumer products are made in low-cost regions, they are predominantly consumed in the developed countries where labor cost is higher. As a result, manufacturing processes that use low virgin material/extraction cost (from undeveloped regions) and deploy low-cost manufacturing labor (also from "low-cost regions") make the cost-of-goods-sold (COGS) of new products low in comparison with the cost of (local) maintenance/repair, refurbishment, and recycling in developed regions.

Operational strategies and supply chain decisions are often cost based and aimed at circumventing compliance to environmental regulations, which are viewed as tactical necessities rather than strategic opportunities. For example, General Motors Corporation developed its electric car only to raise the average fuel efficiency of its aggregate product line in compliance with US-EPA standards. The company decided to discontinue the product line later in spite of its enthusiastic acceptance by a few thousand early adopters who had leased the car (see Note 29).

The above behavior has resulted in excessive utilization of (nonrenewable) natural resources and the looming scarcity that could threaten business sustainability on a broad scale.

4.2.3 The Case for Sustainable Product/Process Design and Manufacturing

Development, production and delivery of products, and services are central to most business enterprises. Hence sustainable product design and manufacturing is integral to the core strategy of a corporation that adopts sustainability as its principle operational imperative. In other words, the case for sustainable products and processes is also the case for sustainable business strategy which can be supported through the following arguments.

Social and environmental responsibility: More than ever before, large multinational corporations feel the need for integrating social and environmental responsibility into their espoused core strategy. They express support for the overarching goal of sustainability to achieve economic prosperity, while protecting the natural systems of the planet and to ensure a high quality of life for current and future generations across the globe. The drivers for the current trend include changing preferences of customers and investors, government regulations, and the pressure that corporations feel from the media and nongovernmental (advocacy) organizations (NGOs). More than 30 years ago, the famed economist Milton Friedman offered a counter argument absolving corporations of social responsibility beyond legal compliance (Friedman 1970). Friedman argued that social welfare is the business of governments who safeguard the commons through enactment of laws and regulations. This minimalist view is workable if governments were not subjugated to corporations and assumed ownership of the commonwealth and the responsibility for its protection.

Maximizing resource-value chain efficiency for short-term business benefits through reduction of operating cost: Improving energy efficiency, reducing manufacturing waste, and other resource efficiency measures translate directly to reduction in operating cost and increased profitability. At the same time, these efficiency improvement measures are good for the environment and for reduction in the depletion of nonrenewable resources.

Business opportunities in meeting the market demand/preference for sustainable products and services: The Prius hybrid car is a good example in this category.

Toyota Motor Corporation envisioned the basis of competitive advantage in the future to be fuel efficiency and invested in the hybrid technology against the conventional marketing wisdom in early to mid-1990s.

Entrepreneurial business opportunities in environmental protection which has been referred to as "enviro-capitalism" (Anderson and Leal 1997) *and as "doing well by doing good"*: For example, entrepreneurial opportunities might arise in promoting tourism through environmental protection, in turning byproducts (waste) of one industrial process into marketable products in another sector and thereby reducing the damage to the environment, and in real estate development and improvement of property value through environmental revitalization and enhanced esthetics.

Sustained business leadership for long-term shareholder value growth: The motivation in this category is to circumvent medium to long-term threat of resource scarcity and environmental management cost, and the resulting jeopardy to business sustainability. There are several counter arguments against this motivation: First, one firm (no matter how large) can only have a limited impact on the global scale, and the long-term measures can only be effective if adopted by all producers. Second, voluntary strategies might be self-discriminatory in the fiercely competitive global landscape. And third, the overarching business objective of sustained growth in return-on-equity is achieved through growth in market opportunity and maximization of people/capital productivity, i.e., maximization of output per unit of labor cost and the deployed capital (see Note 32). This objective encourages continuous: (a) growth in production and corresponding consumption, (b) reduction in worker value (lower wages/benefits and longer working hours), and (c) increased production output capability (through automation, economy of scale, and centralization of production), all of which tend to be counter to sustainable utilization of resources. In other words, the goal for continuous business growth appears irreconcilable with the environmental and resource sustainability objective.

Our fantastic technological advancement and material wealth generation have also resulted in excessive individualism, alienation, and breakdown of interdependence. In many affluent societies, material possession is abundant but people feel insecure of the future and frightened as being "on their own" without a social support network from neighbors, communities, and the society as a whole.

As we will discuss later, the apparent dichotomy between business and environmental sustainability might be reconciled with a change in the underlying conceptual frameworks. For example, an alternative to the business strategy of "sustained shareholder value growth" might be sustained *stakeholder* value growth where stakeholders are shareholders, employees, and the communities and environment where the firm's economic activities take place across the entire supply chain. In this revised strategic viewpoint, value is also redefined and takes a broader meaning beyond the mere financial metrics. The revised definition of value might be *quality of life*. While the definition of "quality of life" is appropriately vague and subjective, it has many universally shared dimensions such as: physical need satisfaction, security, happiness, liberty, equity, opportunity, and community.

In the remainder of this chapter, our discussion is focused on product/process development and operational strategies that serve business sustainability within the current economic framework. However, the proposed concepts and methodologies are not in conflict with the evolutionary process of transition to a new economic framework where strong sustainability is materialized.

4.2.4 Incentives for Sustainable Product/Process Development

Table 4.2 presents prevailing incentives which drive firms to develop sustainable products and processes and to adopt sustainable manufacturing practices in today's economic framework. The relative weight that is assigned to these incentives in managerial tradeoff analysis and decision making is also listed in the table. While the Table 4.2 incentives are rooted in the factors in Sect. 4.2.3, they are constrained by the overarching business objective of shareholder value growth. Furthermore, sustaining the growth over short-term reporting periods forms the

Table 4.2 Driving factors in strategy for sustainable product development and manufacturing

Driver	Weight High, Medium, Low
Laws and government regulations (local, national, international, and global treaties); Risk avoidance	M–H (see the following section on prioritizing compliance)
Operating cost benefits of efficiency in the resource value chain for reducing the operating cost, COGS, and cost-of-ownership (COO). Note that COGS and COO can be reduced through: (a) Product and process design and (b) outsourcing of operational processes (manufacturing, HR, IT, and financial transactions) to low-cost regions	L–M
Marketing 1. Users' preference for eco-friendly products (varies by socio-cultural factors) (see Note 33) 2. Competitive differentiation through: (a) Lower COO by reducing consumption of resources (e.g., energy efficient appliances, hybrid vehicles) and (b) Aesthetics value (through designs that minimize material content) 3. Pricing flexibility through lower cost base: Manufacturing waste reduction and efficiency in energy and water use 4. Branding Social responsibility and eco-friendly growth embedded in the firm's mission statement and espoused strategy; also referred to as "Triple Bottom Line Strategy" (see Note 34)	L–M
Entrepreneurial opportunities in fulfilling market needs where business opportunity intersects environmental protection/improvement: • Products that reduce resource utilization and improve operating cost; e.g., energy-saving devices such as motion-sensing light switches (see Note 35)	M–H (if aligned with the firm's business strategy)

basis for prioritization of the drivers in Table 4.2. Hence, in product development, time-to-market, and competitive differentiators in the functional attributes of a product are commonly treated as high priority and trump other considerations in the design.

4.2.5 *Prioritizing Compliance to Laws and Government Regulations in Product and Process Design Decisions*

As noted in Table 4.2, compliance to government regulations might be treated as medium-to-high priority in product development strategy. Different regulations impact different stages of a product life cycle (manufacturing, use, and end-of-life), and deciding how to weigh-in the requirements for compliance in design depends on the firm's operational choices and marketing strategy. Firms can optimize their strategy for where the product is manufactured, where it is marketed (used), and where it is disposed according to the environmental requirements of the location. For example, if the firm decided to manufacture the product in a region that has less-stringent regulations than where the product would be marketed for use, it can design the product according to the less-stringent environmental requirements. Similarly, the firm's obligations and strategy in disposing of the product at its end-of-life can be optimized. That is why so many electronic products end up in China at the end of their use in North America and the EU.

In the regions where environmental regulations are stringent and cause a significant increase in COGS, the firm might decide not to comply with regulations and to forego the market opportunity if it did not produce an attractive return on investment. For example, a stringent restriction on the use of hazardous materials in products (including lead and cadmium) was enacted in the EU as of July 2006 (see Note 36). A manufacturer of photovoltaic solar modules that uses cadmium might decide not to market the product in the EU rather than embarking on a major R&D effort to find a substitute for cadmium.

Another consideration is compliance to international treaty regulations under auspices of the UN or regional trade agreements. These regulations usually become enforceable on the local and national level only if they were adopted by the host country where the product is manufactured, used, or disposed of. For example, the Kyoto protocol for GHG emission is not adopted by the USA and does not impact the products manufactured in the USA.

Table 4.3 is a matrix of governmental regulation class vs. life cycle phases of a product. This table illustrates the relative impact of regulations on various stages of a product life cycle. The numbers in Table 4.3 are illustrative examples in comparing the impact of regulations on design and manufacturing strategy depending where the product is made, marketed, and disposed of.

Note that in Table 4.3 example (as in the other examples discussed above), the firm's strategy for sustainable design is constrained by shareholder value growth and short-term return on investment optimization. The long-term factors such

Table 4.3 Product strategy matrix in compliance to government regulations

Class of laws and regulations	Product life cycle phase					
	Make		Use		Dispose	
Local	2 Shanghai	3 Malaysia	5 Germany	4 California	4 California	1 Xian
National	1 China	3 Malaysia	5 EU	3 USA	3 USA	2 China
Global Treaty Guideline (if adopted as local or national laws); e.g., Kyoto GHG protocol	3 China	3 Malaysia	5 EU	1 USA	1 USA	3 China

The numbers in the table represent the impact of regulations on Design/Manufacturing Practices: 1 = minimal, 5 = strong

Table 4.4 Potential adverse and favorable impacts of design for sustainability

Potential adverse impacts of design for sustainability	Potential favorable impacts of design for sustainability
(a) Limits designer's choice and compromises product performance.	(a) Improves efficiency and reduces cost of product manufacturing, installation, after sale service.
(b) Lengthens the development cycle time and delays time-to-market.	(b) Reduces cost of consumables in use hence improving competitive differentiation.
(c) Increases cost of goods sold and lowers product profitability.	
(d) Increases cost of product development project.	

as minimizing utilization of nonrenewable resources and environmental impacts beyond compliance to regulations are assumed to be low priority and only of public relation value.

4.2.6 Potential Impact of Sustainability Requirements on Product/Process Development

Sustainability requirements constrain product design and manufacturing in many respects and while some constraints might create an opportunity for operational efficiency improvement and cost reduction, many are perceived as undesirable limitations on speed to market and cause for increase in the product's manufacturing cost. Table 4.4 lists potential adverse and favorable impacts of sustainability requirements on commercialization success of a product. Note that the discussion in this section is focused on sustainability requirements which are above and beyond those imposed by mandatory laws and government regulations. The latter are not sources of competitive differentiation except in the way that the supplier implements them in design and manufacturing similar to all other product requirements.

The benefits that are not listed above, as we discussed earlier, are resource availability and environmental benefits which impact long-term business sustainability and are often beyond the life cycle horizon of a product and do not impact its commercialization success. Because of this tension between the short- and long-term considerations, farsighted firms are increasingly faced with an ethical challenge. Should sustainability requirements be subordinated to imperative short-term business results?

4.3 Product and Process Assessment

Product and manufacturing sustainability requires vigilance and proactive action throughout the product and process life cycle. Defining sustainability metrics, setting product and process targets, and executing appropriate action plans are imperatives of sustainable development. Mindfulness (knowledge) of environmental sustainability factors plus creativity and product innovation can significantly enhance sustainability of products and processes without adverse impact on business results.

Many of management actions to increase operational efficiency for profitability and competitive advantage also help environmental sustainability, including short cycle time, short time-to-market, lean manufacturing, low materials and manufacturing-labor costs, high inventory turns, low installation cost, coordination within the supply chain, and low cost-of-ownership. Through proactive identification of sustainability metrics, assessment of their impact and setting targets, firms can articulate the firm's contribution to sustainability for marketing promotion, increase workforce/management awareness and sensitivity to sustainability issues, identify further opportunities for sustainability improvement in products, services, manufacturing, operations, and the supply chain, and implement targeted and high-impact sustainability improvement action plans.

Sustainability can be assessed in several dimensions including: (a) performance of individual products and manufacturing processes; (b) overall corporate (or business unit) performance aggregating all products, processes, and the enterprise operation; and (c) the quality (maturity) of the firm's strategy and business processes pertaining to leadership in sustainability.

The following sections present the sustainability metrics in each of the above categories. These metrics are intended not only to assess a firm's sustainability performance but also to drive its operational strategies through the desired values that are specified for the metrics as design and process requirements. The proposed metrics are based on authors' experience and on the best practices at leading manufacturers and are not intended to be exhaustive of all product and process types. Nevertheless, the concepts can readily be adjusted to new situations.

We start with the metrics measuring the overall sustainability performance of a firm (or a business unit). Metrics for assessing sustainability of suppliers are discussed next as firms should assume full ownership of sustainability across their global supply chain. We will then present the sustainability metrics for products, manufacturing processes, and distribution channels.

4.3.1 *Overall Corporate and Business Unit Level Metrics*

The following is a list of metrics that measure a firm's overall sustainability performance. Many corporations, particularly large multinationals, have adopted these metrics (to some extent) and publish annual reports based on an internal audit of progress against annual improvement targets.

1. ISO 14001 certification for having an Environmental Management System Standard.
2. Fines and penalties for lack of compliance to external regulatory requirements (number of violations, dollar value of fines per year, and percent change from prior years).
3. Percent of products ranked stellar in energy efficiency according to the US EPA Energy Star Tag and EU Energy Label.
4. Compliance to the leading global sustainability standards and regulations for products, manufacturing, and transportation, including:
   (a) Percent of products in the corporate portfolio which comply with EU directives, RoHS, WEEE, and EuP.
   (b) Compliance with Chemical Ban List (Montreal and Kyoto protocols)
   (c) Compliance with hazardous material transportation regulations (local, national, international)
   (d) EU directive on End-of-Life Vehicles (ELVs)
   (e) Japan's Law on Recycling of ELVs.
   (f) Rating according to the EU Ecolabel.
5. Recycled material use in manufacturing of products in units of kilogram per unit of product or revenue, and percent change from prior year.
6. Landfill use rate by end-of-life (EOL) products: in percent of EOL products disposed in landfills (kg/y).
7. Total gaseous, liquid, and solid discharge into the air, water, and land (vs. the zero-emission goal) from production, enterprise operation and use/EOL disposal of all products.
8. Investment in sustainability, associated cost savings, and the saving to investment ratio. For example in 2006, IBM reported the saving/investment ratio of 2 over a 7-year period.
9. Innovation and investment in *disruptive technology for sustainability* including annual budget and the number of patents filed. For example, investment in products and technologies that substitutes for nonrenewable materials and energy sources at reduced cost.
10. Cost-of-Sustainability (COS) as defined in Sect. 4.3.1.1
11. Education: training employees, suppliers, distributors, subsuppliers, and disseminating best known practices. Workforce education is further discussed below.
12. Promotion: Sustainability Innovation Award (SIA) to employees including sustainability-kaizen in product design, manufacturing, and enterprise operation.

13. Societal outreach (community, government, NGOs) and stakeholder engagement encompassing the firm's global supply chain and served markets.
14. Sustainability score across the supply chain. Measure and aggregate the above 13 metrics for all suppliers, including upstream/downstream logistics/transportation and distributors to assess the firm's total sustainability impact.
15. Summarize firm's performance in the above 14 categories in an Overall Score Card showing the actual performance rating vs. target, in each category and the company's total ecological footprint.

As a result of assessment and monitoring of corporate sustainability metrics, some corporations have come across entrepreneurial opportunities or innovative ideas for reducing the firm's ecological footprint. Dupont measures SVA (Shareholder Value Add) per pound of production to maximize weight reduction in products (Holliday et al. 2002). To achieve the SVA goal, Dupont added service to its portfolio of business offerings. In the carpet business for example, Dupont started offering installation, maintenance, and end-of-life (EOL) service in addition to just selling the carpets. It is not clear how strongly the SVA-per-pound metric correlates with reducing the ecological footprint of carpet production and use. For example, the installation service does not reduce environmental impact although it improves the SVA/lb. There is, however, an opportunity for improving sustainability if the maintenance service increases carpet life and EOL service includes material recycling.

STMicrolectronics' CEO, Pasquale Pistorio has set a company goal of "zero-equivalent CO_2 emission by 2010". ST's plan for achieving this goal is twofold: (a) increase energy efficiency and use of renewable sources in production and (b) plant 35,000 ha of trees to compensate for CO_2 emission that cannot be achieved through efficiency enhancing actions. Planting trees to make up for CO_2 emissions is not yet regarded as a way of offsetting emissions, because the trees need to be maintained in perpetuity (or at least for the lifetime of the greenhouse gas emissions, typically on the order of 500 years or so). However, new carbon neutral standards are being developed including carbon neutral certification, and many firms are participating in worldwide emission trading markets that provide economic incentives for achieving reductions in the emissions through a cap and trade mechanism. Furthermore, trade associations such as the International Emissions Trading Association (IETA) are established. IETA aims to promote an integrated view of the emissions trading system; participate in the design and implementation of national and international rules and guidelines; and provide information on emissions trading and greenhouse gas market activity.

4.3.1.1 Cost-of-Sustainability

The concept for COS is modeled after cost-of-quality (COQ) proposed by Joseph Juran in 1951. He defined COQ as unavoidable cost minus avoidable cost of quality, where avoidable cost includes cost of rework of defects in manufacturing, cost of repairing failures in the product warranty period, and cost of customer dissatisfaction.

The unavoidable cost of quality is the cost of preventive measures including increase in product cost of manufacturing because of design quality improvements and of added manufacturing inspection and statistical process control.

We define COS for a product as:

$$\text{COS} = \text{unavoidable cost} - \text{avoidable cost} - \text{opportunity cost},$$

where unavoidable cost is the incremental cost of design-for-sustainability (DfS) + increase in product manufacturing cost caused by DfS constraints + cost of being late to market due to incremental R&D; avoidable cost is the savings through operational efficiency improvement (e.g., using less water and nonrenewable energy) + lowering product manufacturing cost by reduction in weight and in use of (expensive) restricted materials + savings associated with reduction in nonrenewable consumables (in the cost-of-consumables or COC), and opportunity cost is the market share gain through sustainability differentiation + reduction in ecological footprint of the product.

Note that the terms in the COS equation reflect only the cost associated with actions which improve product sustainability beyond the satisfaction of market requirements for competitive parity and compliance with environmental regulations. COS might, however, depend on the cost-of-compliance (to regulations) because the global regulatory landscape is nonhomogeneous. This situation often shifts product development and manufacturing decisions away from the strategy for a sustainable solution to a global sourcing strategy where managers opt to manufacture products in "low-cost" regions with lax environmental restrictions.

In the fiercely competitive high-tech industries and particularly when a business is on a steep growth curve, managers tend to focus on market share and revenue growth strategies rather than on cost reduction. Therefore, minimizing the COS will for the most part be a strategic priority if it delivers enhanced short-term customer value. That is why, it is important to develop a life cycle business model for products, where resource value chain efficiency maximization and COS improvement are integral parts of a product strategy from the onset and the benefits are passed on to customers (in the form of lower initial price, lower COC, and improved life cycle experience – ease of service, EOL takeback, and replacement.)

4.3.2 Supplier Metrics

Today's globalized supply chain extends a product's impact on environmental and resource sustainability far beyond the operational control volume inside a firm. Therefore, firms should assume full ownership of sustainability across their global supply chain and the product life cycle, starting with the supplier selection process. Contractual demand, information sharing (of best known methods), training, and audit should be deployed across the supplier base to ensure a high level of product and manufacturing sustainability on par with the firm's operational strategy. The firm and its suppliers should monitor the following metrics and implement action plans for achieving established targets.

1. Monitor the metrics listed in Sect. 4.3.1 for all suppliers, upstream/downstream and assess the total sustainability impact of a product.
2. Environmental Product Declaration (EPD) (Michelsen et al. 2006) identifying the impact of supplier's embedded products.
3. Content declaration on supplier's products and subassemblies.
4. Reduction in packaging material and solid waste disposal, as percent per year.
5. SMI (Sustainability Maturity Index) for the supplier.

4.3.3 Product Sustainability Metrics

The metrics in this section are aimed to drive product design. The desired value set for each metric must be treated as a product design requirement in conjunction with other customary requirements for product performance in satisfying customer needs. Note that the following is not intended to be an exhaustive list and it must be tailored to firm's unique product characteristics.

1. Compliance to the requirements of ISO 14001 and other global environmental regulations such as RoHS and WEEE.
2. Product energy efficiency and Ecolabel ratings; e.g., EPA Energy Star and EU Ecolabel.
3. Product Sustainability Index (PSI) calculated as the aggregate numerical rating of the product's sustainability impact in manufacturing, transportation, use, and end-of-life (EOL) as listed in items 4–7 below.
4. Manufacturing: fraction by weight or number of parts of the following classes of material by design specification:
   (a) Recycled material
   (b) Restricted, scarce, or nonrenewable material
   (c) Materials whose extraction and delivery are energy intensive
   (d) Materials whose extraction and delivery have a significant environmental impact
   (e) Remanufactured parts
5. Transportation:
   (a) Reusability and recyclability of the packaging material (see Note 39).
   (b) Nonrenewable energy consumption and other environmental impacts of transportation (perhaps the total distance traveled by a product to reach the end use is a good indicator – the distance traveled from the manufacturing plant to various distribution centers and finally to the consumer). Since the energy consumption and environmental impacts are dependent on the mode of transportation, appropriate weight factors must be applied to this metric. US EPA has initiated the Smartway Transport Partnership program to improve freight transport efficiency.

6. Product use
   (a) Biodegradable and nonrenewable materials in the consumables (e.g., tires of a car)
   (b) Energy efficiency, percent
   (c) Water efficiency, percent recycled, and retreated
   (d) Climate control impact (e.g., product lifetime GHG emission in kg)
7. EOL manageability
   (a) Product life
   (b) Repairability
   (c) Upgradability.

   Note that EOL manageability is different than EOL management (in Sect. 4.3.5). Manageability is a property of product design as it is reflected in the above metrics, while EOL management is a property of product life cycle management.
8. Recyclability
   - Recyclable material content, percent (e.g., Toyota cars are 85% recyclable)
   - Dissimilar material content that cannot be readily separated
   - Reusability: percent of material/components used back in product manufacturing and used in another product

Graedel and Allenby (1998) rank products according to the choice of material, energy use, and effluents into the environment over the product life cycle. They propose the following "Environmentally Responsible Product Rating" system:

$$\text{Environmentally responsible product rating, } R_{\text{ERPT}} = \sum_i \sum_j Mij,$$

where $M_{i,j}$ is the (i, j) element of the assessment matrix in Table 4.5 and is summed over all i and j entries. The integer value assigned to $M_{i,j}$ ranges from 0 (for greatest impact) to 4 (for lowest impact), respectively. The range of R_{ERPT} is from 0 to 100.

Table 4.5 The environmentally responsible product-assessment matrix

	Environmental concern				
Life stage	Materials choice	Energy use	Solid residues	Liquid residues	Gaseous residues
Premanufacture					
Product manufacture					
Product delivery					
Product use					
Refurbishment, recycling, disposal					

Note that in Table 4.5, entries in the left two columns and in the right three columns represent the product impact on environmental "sources" and "sinks", respectively.

Several other researchers have also proposed metrics for assessing eco-efficiency of products (Schmidheiny 1992; Henrik 2000; Frankl and Rubik 2000; Billatos and Basaly 1997; Fabio Giudice et al. 2000) (Schmidheiny 1992; Henrik 2000; Frankl and Rubik 2000; Billatos 1997; Giudice 2000). Manufacturing Science and Technology Center of Japan and Clean Japan Center have also developed metrics for product eco-efficiency.

4.3.4 Process Sustainability Metrics

The metrics in this section are aimed to drive process design and operational practices at the factory and throughout the supply chain. The desired value that is set for each metric must be treated as process design requirement and as operational guideline in conjunction with other customary requirements for factory design, operational plans, and process control. Note that the following is not intended to be an exhaustive list and it must be tailored to firm's product and process characteristics.

1. Energy efficiency:
   (a) (Energy saving + renewable energy use)/total energy consumption; percent.
   (b) Total energy use/sales.
   (c) Energy consumption per unit of production; percent reduction per year.
2. Water efficiency:
   (a) Water consumption reduction per year; percent
   (b) Water consumption per unit of production; percent reduction per year
3. Climate change impact:
   Total production of the following effluents (in kg per constant-dollar sales, and percent reduction per year)
   - PFC (perfluorocarbon) emission
   - CO_2 emission
   - Other greenhouse gas (GHG) emission – Kyoto & non-Kyoto gases
4. Hazardous emissions:
   - Volatile organic compound (VOC) emission
   - TRI (toxic release inventory) chemical
5. Waste management:
   (a) Overall Waste Generation Index (hazardous and nonhazardous); kg/sales; kg/unit of production, percent reduction per year

Table 4.6 The environmentally responsible process-assessment matrix

	Environmental concern				
Life stage	Materials choice	Energy use	Solid residues	Liquid residues	Gaseous residues
Resource extraction					
Process implementation (process equipment)					
Process operation					
Complimentary process implications					
Refurbishment, recycling, and disposal (of process equipment)					

(b) Fraction of process byproducts and effluents in each of the following categories; percent:
- Reduced
- Reused
- Recycled (e.g., 99% of scrap metal at Toyota car manufacturing plants is recycled)
- Treated (chemical, physical)
- Disposed (incinerate, landfill, release to air or water)

6. Process equipment: in a life cycle assessment technique, the impact of process equipment (that is often purchased from third party suppliers) must also be considered.

Graedel and Allenby (1998) rank processes according to the choice of material, energy use, and effluents into the environment similar to the methodology for assessing a product. They propose the following "Environmentally Responsible *Process* Rating" system:

$$\text{Environmentally responsible process rating, } R_{\text{ERPrR}} = \sum_i \sum_j Mij,$$

where $M_{i,j}$ is the (i, j) element of the assessment matrix in Table 4.6 and is summed over all i and j entries. The integer value assigned to $M_{i,j}$ ranges from 0 (for greatest impact) to 4 (for lowest impact), respectively. The range of R_{ERPrR} is from 0 to 100.

Note that in Table 4.6 "process implementation" accounts for the environmental impact of process equipment which is often designed and manufactured by an independent supplier.

4.3.5 End-of-Life Management Metrics

The EOL management metrics impact a firm's product life cycle management and the supply chain strategy. The desired values set for these metrics affect the customary optimization tradeoffs such as collection and refurbishment rate of used products in servicing product warranty. In this category, the percent of products that are manufactured and sold is monitored for each of the following categories at the end of product life (EOL).

1. Resold
2. Refurbished and reused
3. Recycled
4. Disposed in landfill
5. Incinerated
6. Ratio of the number of products recycled plus reused and resold to the number of new products manufactured. For IBM PC/Workstations in 2003, this ratio was 37% in the USA and 17% worldwide.

4.3.6 Sustainability in Distribution, Logistics, and Sales

A Sustainability Management System (SMS) should be implemented at distribution, logistics, and sales centers to include the firm's performance metrics for:

1. ISO 14001 compliance
2. Energy and water efficiency
3. Recycling and reuse
4. Waste minimization
5. Disposal
6. Shipping methods (and hazardous material or HAZMAT compliance)
7. Logistics: metrics for transportation of products and for the distributors (warehousing, storage, value-added resellers, etc.) must be established. As mentioned before, transportation metrics should discern sustainability impact of local vs. distant manufacturing.
8. Information dissemination and training in SMS best practices

4.3.7 Sustainability at Service Centers

The SMS should be established at service centers to include performance metrics for:

1. ISO 14001 compliance
2. Compliance to regulations (HAZMAT)
3. Zero emissions (GHG, liquid discharge)
4. Energy and water efficiency
5. Use of remanufactured parts
6. Recycling and reuse
7. Disposal
8. Information dissemination and training in SMS best practices

4.3.8 *Assessment Audit and Reporting*

The firm should perform periodic audits of its sustainability performance based on the above metrics and report the results to its stakeholders. The following reports and audits are recommended:

1. Corporate Sustainability Report to Stakeholders (annually)
2. Audit for compliance to external and internal requirements:
   (a) Annual self assessment by every manufacturing, product business unit, and R&D center.
   (b) Annual assessment by Corporate Internal Audit Organization
      i. Report to management
      ii. Follow up for accountability and action status
   (c) ISO 14001 audit (annual)

4.4 Best Practices in Product and Manufacturing Sustainability

The impact of a product on natural resources and the environment extends in both space and time dimensions and the associated causal factors are interrelated. For example, the factors influencing the environmental impact of a discharge from a process might include: weight of pollutants, toxicity (short-term/long-term impacts on humans, animals, and plants), impact regime (air, land, and water), interactive nature of pollutants (changing to something else, having a long-life, upsetting the ecological balance), localization, and transport (atmospheric transport, land seepage, sewer > sea > fish > human transport). Management and engineering decisions at the product and process design stage profoundly impact the product's sustainability metrics throughout its life cycle and must be made from a global and holistic systems viewpoint.

In our discussion, nonrenewable resources such as oil, coal, and elemental materials refer to the natural resources whose economic reproduction is not feasible in the time scale of a human generation or an economic cycle. And environment refers to the earth's ecological systems whose balance impacts human life and in turn is impacted by human economic activities.

In the following sections, we present a general guideline and qualitative specification for the development of sustainable products and processes, integrating the practices of several leading global corporations such as IBM, Toyota, 3M, and Intel. We will also discuss the life cycle analysis (LCA) methodology and considerations for designing a "closed loop product/process cycle" for sustainability.

4.4.1 Sustainable Product Design Guidelines

Schmidheiny posits that a sustainable product has the highest eco-efficiency defined as the ratio of product value to its environmental impact (Schmidheiny 1992). The following is a list of attributes that a sustainable product must possess and they apply to the "whole product" including packaging and ancillary devices that make the product *useful* to the end user: (i) Efficiently (see Note 42) incorporates environmentally preferred materials and finishes. (ii) Requires minimal consumption of resources in various stages of the product life cycle, including: manufacturing (energy, water, etc.), use (consumables), and end-of-life (land for disposal). (iii) Causes minimal solid, liquid, and gaseous discharge into the air, water, and land in use and in manufacturing. (iv) Has a long life. (v) Can be upgraded to extend the product life. (vi) Has high reliability and maintainability with low cost of repair. (vii) At the end of life, its components can be reused and its material-of-construction can separated and recycled. Note that many of these attributes, such as long product life and low consumable use, are contrary to the business model for many consumer products which encourage short product life (with frequent technology/feature change – such as in consumer electronics) and count on high consumable usage to compensate for the low initial price (e.g., the computer printer and its cartridge or razor and razor blade). Sustainable product development processes can succeed only if they are in alignment with the firm's business model.

During the product design process, tradeoff considerations to abide by the above guidelines should take a holistic system view spanning the entire resource cycle (including raw material extraction/processing, transportation, and delivery) and the entire product life cycle (including manufacturing, logistics, use, and end-of-life management). For example, designers should avoid specifying a material whose extraction and delivery are energy intensive and have a significant environmental impact through harmful discharges. The following examples highlight the need for a holistic approach to sustainability.

- The Chinese government's recent green push imposes a tax on disposable chopsticks (see Note 43). This regulation is designed to safeguard the 1.3 million cubic meters of Chinese timber lost to chopstick production every year. Is the washable (plastic or metal) chopstick a more sustainable solution? In order to arrive at an optimal strategy, one must weigh the resource consumption and environmental impact of the two options.
- How superior is an electric car (in sustainability) to an efficient (hybrid) gasoline combustion engine car if electricity for the electric car comes from a fossil fuel power plant? Again, one must consider the entire resource cycle in comparing the overall nonrenewable energy consumption and environmental impact of the two options, as noted below:

    1. Fossil fuel consumption tradeoff based on the overall well-to-wheel efficiency:

        (a) Electric car overall efficiency = combined efficiency of fossil fuel to electric power conversion at a stationary power plant, electricity delivery, and battery charge/discharge operation.

(b) Hybrid combustion engine overall efficiency = combined efficiency of fossil fuel to gasoline conversion, delivery, and combustion process in a hybrid cycle.

2. Environmental tradeoff: impact on air quality and climate change as the result of stationary source pollution (power plants) vs. distributed source pollution (cars). If fuel for the hybrid gasoline car or electricity for the electric car was from sources other than fossil fuel, the entire tradeoff study would change. For example, the tradeoff might be between a biofuel hybrid car and an electric car that is charged by electricity from a solar/wind power plant.

- In assessing alternate automobile design technologies, Toyota Motor Corporation considers energy efficiency and environmental impact of these technologies over the entire resource cycle (see Note 44). Toyota defines *"well-to-wheel"* efficiency as the product of *"well-to-tank"* and *"tank-to-wheel"* efficiencies. For a combustion-engine car, well-to-tank efficiency is the combined efficiency of the cycle from extracting the fuel from the ground, refining it into usable fuel, and getting it to the vehicle. The tank-to-wheel efficiency is the familiar term for vehicle efficiency and is usually expressed as miles per gallon. Energy efficiency of alternate vehicle designs is shown in Table 4.7.

A similar table can be constructed to compare the environmental impact of alternate vehicle technologies in terms of "well-to-wheel" CO_2 discharge, as shown in Table 4.8.

- Intel's power supplies had been inefficient because of the lack of an integrated system design approach. Many desktop power supplies are only 50% efficient during normal operating conditions. That is, half of the supplied energy is dissipated in heat. Power supplies are often designed to deliver optimal performance (maximum efficiency) at or close to full load. However, most desktop systems operate at <50% of full load under normal operating

Table 4.7 Energy-efficiency tradeoff study of alternate vehicle technologies by Toyota

Alternate vehicle design	Well-to-tank efficiency (%)	Tank-to-wheel efficiency (%)	Overall, well-to-wheel efficiency (%)
Gasoline	79	16	18
Diesel	82	23	19
Hybrid (Gasoline)	79	37	29
Hydrogen fuel cell (compressed H2)	58	38	22
Hydrogen fuel cell – hybrid (compressed H2)	58	50	29
Hydrogen fuel cell target	70	60	42

Note that because of the low efficiency of hydrogen production and delivery (58%), the overall efficiency of hydrogen fuel cells (22%) is not as high one expects (vs. hybrid or even diesel)

Table 4.8 Emission tradeoff of alternate vehicle technologies by Toyota

Alternate vehicle design	Well-to-tank Co_2 emission[a]	Tank-to-wheel Co_2 emission[a]	Overall, well-to-wheel Co_2 emission[a]
Gasoline	15	85	100
Diesel	7	68	75
Hybrid (Gasoline)	5	39	44
Hydrogen fuel cell – hybrid (compressed H2)	37	0	37
Hydrogen fuel cell with solar energy	5	0	5

[a]Relative to gasoline = 100

conditions. Working with the Natural Resource Defense Council (NRDC), Intel made changes to its power supply design guidelines to encourage the development and adoption of more energy efficient power supplies (see Note 45). The US EPA estimates that the environmental impacts of achieving the recommended targets established in the design guideline would result in the following savings in the USA alone: (a) Electricity savings of over 16 billion kilowatt hours per year. (b) Reduced CO_2 emissions of over 10 million tons annually. (c) Cost savings to the end user of US$1.25 billion annually. (d) Reduced cost of ownership of a typical desktop PC of US$50 over 3 years.

4.4.2 Sustainable Manufacturing and Process Design Guidelines

A manufacturing operation that is at the "externally supportive" (see Note 46) stage of maturity enables the corporation to lead in sustainability. The following is a list of attributes in a sustainable process and manufacturing operation.

- Minimal utilization of nonrenewable resources.
- Zero use of restricted resources (e.g., RoHS materials).
- Energy efficiency, defined as the ratio of energy needed to the energy consumed and can be maximized through conservation and efficient equipment and processes.
- Use energy from a renewable source. For example, Google is powering its campus in Mountain View, California with solar energy and so does Macy's at many of its retail stores nationwide.
- Maximize water utilization efficiency:
  - Recycle the process water.
  - Reduce water use through innovative process design.
- Maximize material consumption efficiency, defined as the ratio of quantity needed to quantity consumed, through process innovation and recycling.

For example, an IC manufacturing process is generally quite inefficient; and in many process steps, more than 80% of (largely toxic) chemicals are discharged as effluents into the environment or captured in scrubbers for subsequent disposal in landfills.

- Zero discharge of pollutants in sold, liquid, and gaseous forms.
  - As a high-priority subset of this requirement, manufacturing processes must have zero toxic release inventory (TRI) including copper compounds, nitrate compounds, xylene, etc.
  - No climate change impact through zero emission of greenhouse gases and use solvent-less processes.
- Eliminate material waste – including hazardous and nonhazardous wastes. Waste is primarily, the unused raw material from an inefficient process. The generally prescribed rule is the 3Rs of: reduce, reuse, and recycle (Jawahir 2007). Recycling usually involves a treatment process. Disposal of nonbiodegradable materials through incineration and landfill must be treated as the last resort. In case of biodegradable materials, landfill approaches such as composting can be beneficial and necessary for soil health. Similarly, combusting biomass waste for generating process heat and energy can make these processes carbon neutral or even carbon positive.

Increasingly, leading manufacturers are deploying sustainability guidelines in their operation. For example, at Toyota's car manufacturing plants, antichip paint is now applied with a roller rather than a sprayer. This saves paint, reduces emissions, eliminates the need for plastic masking, and holds down cleaning costs. The redesigned process has reduced wastes by 40%. Another ongoing project at Toyota is the development of water-based paint instead of solvent-based paint.

The above process guidelines must be implemented through the full manufacturing cycle across the supply chain, from material extraction and delivery to the final packaged product. Successful implementation of these guidelines in process design requires domain expertise and managerial commitment to process sustainability. Note that the above guidelines should be tailored to a manufacturer's unique process conditions.

4.4.3 Product Packaging and End-of-Life Management Guidelines

- Apply the principles of reduce, reuse, and recycle.
- Design for "long"-life, reuse, repair, and end-of-life recycling.
- Use recyclable and recycled content. For example, no PVC or plastic "peanuts" should be used in packaging a product.

IBM has a packaging requirements manual that can be downloaded from the company website.

4.4.4 Supply Chain Partnership for Sustainability

It is crucial to assume ownership of the product sustainability and to partner with customers and suppliers for product/process design, packaging, logistics, distribution, reuse, and end-of-life recycling. The above guidelines for sustainable product design and manufacturing should be implemented across the supply chain including the supplies and subsuppliers. Sustainability considerations must be integral to contractual, operational, and value-distribution issues in dealing with the supply chain.

4.4.5 Life Cycle Management and Closed Cycle Design

According to ISO 14040, LCA is a technique for assessing the environmental aspects and potential impacts associated with a product over its life cycle, by defining goal and scope of LCA, compiling an inventory of inputs and outputs of a product system, evaluating potential impacts, and interpreting the results in relation to objectives of the study.

Klostermann (1998) describes the LCA methodology for identifying the environmental impact of a product and discuss mitigation options through various examples. Economic feasibility of a sustainable design can be assessed by performing a total cost assessment (TCA) in parallel with the LCA.

United Nations Economic Development (UNED) organization has initiated a Life Cycle Management (LCM) program to develop "concepts, techniques, and procedures with the goal of creating sustainable development." LCM includes the use of tools such as LCA and TCA. UNED has also defined the concept of Life Cycle Thinking (LCT), or Cradle-to-Grave thinking, as: "an approach where analysis is done while considering the impacts of a product or process from its genesis through to its disposal."

The output of a product development process (PDP) is generally a specification for the product assembly, materials of construction, manufacturing steps/flow/tolerances, and finally the user's operation and maintenance instructions. The input to the PDP is information about the necessary material, components, and subassemblies which could be acquired as purchased parts or made-to-print for the purpose of manufacturing the product according to the specifications. The traditional PDP optimizes the output of the process within the constraints of the input to achieve the desired return on investment, time-to-market, and market-share goals for the product. In other words, the traditional PDP adopts a very narrow view of the PDP "system" and limits the consideration of the product impact to the interactions with its immediate stakeholders upstream (suppliers) and downstream (users). The LCM/LCA methodology, in contrast, takes a holistic view of the PDP system and seeks to optimize the dynamic interactions of the product with all upstream and downstream stakeholders including the natural resources and ecosystems sinks.

Figure 4.5 depicts the life cycle stages of a product with its manufacturing and useful life by the users at the center. In the LCM methodology, all upstream and

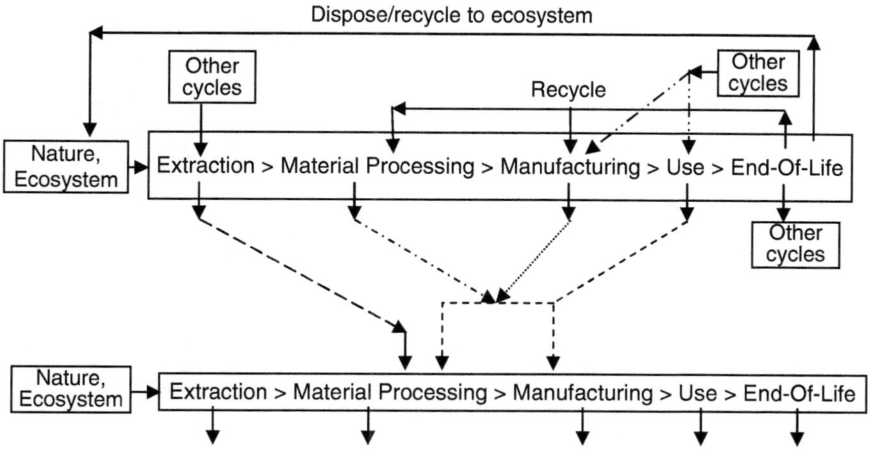

Fig. 4.5 Closed loop product and process life cycle

downstream processes that are impacted by a product (as shown in Fig. 4.5) are considered as integral stages of the product life cycle. Raw material extraction that draws resources from the nature and its ecosystem and the subsequent material processing steps precede manufacturing of the product. On the downstream side of its useful life, the product interacts with nature as a sink for absorption and reprocessing of its wastes.

In a closed loop LCM system, potential recycle- and reuse loops between various stages of a product life cycle are included where material recovered from the product at the end-of-life substitutes for virgin material in manufacturing. Furthermore, life cycles of different products are linked where effluents from the manufacturing stage of one product are fed into the manufacturing stage of another product; and similarly, the end-of-life reused and recovered materials cross alternate product cycles as shown in Fig. 4.5.

Examples of the closed loop approach are abundant in nature where waste from one process becomes an input to another process. Whole sectors of ecosystems, particularly in the soil, work to take nature's waste materials apart, separate them into usable pieces, and send them back into living creatures again (see Note 7). Figure 4.6 illustrates an example of a closed cycle processing design by humans. Figure 4.6 is the schematic of an ecosystem in Denmark demonstrating how by-products and effluents of one system are creatively used as input to another system (Manahan 1999; Ayers and Ayers 2002).

Applied sustainability LLC (ASLCC) created a new business based on "By-Product Synergy (BPS)" by converting waste into saleable commodities (used in other products; see Note 50). Applied sustainability brought diverse companies together and identified BPS opportunities among them. ASLLC faced numerous challenges, including a workable business model, ROI justification for clients, long-term nature of the projects, and contractual/regulatory restrictions that led to their demise. To succeed, the BPS concept requires cooperation and coordinated operational planning.

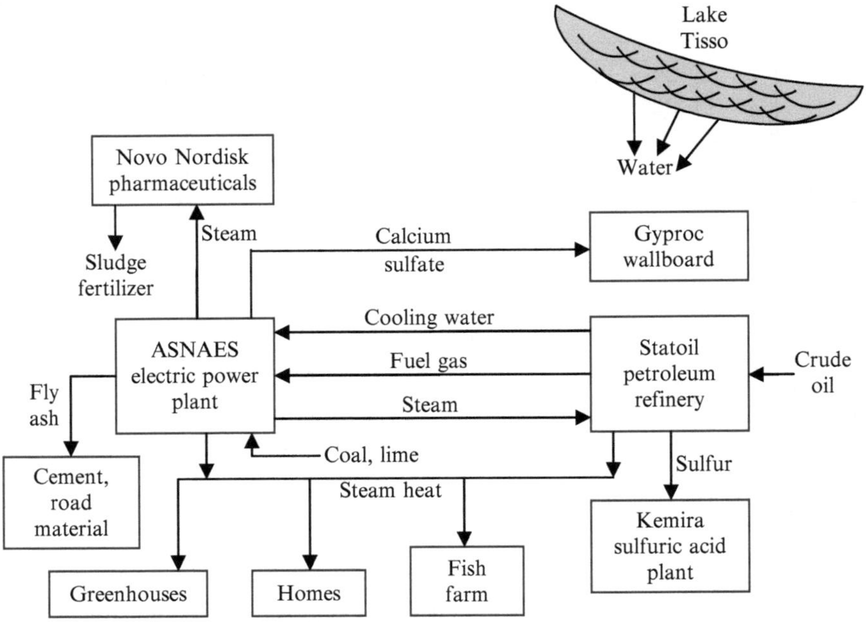

Fig. 4.6 Schematic of industrial ecosystem in an industrial park in Kalunborg, Denmark

Another example of closed cycle processing can be found in Sweden. The city of Vaxjo in Sweden is seeking a fossil fuel-free future, and it is almost halfway there without having sacrificed lifestyle, comfort, or economic growth (see Note 51). "The starting point on the way towards Vaxjo's – and Sweden's – success was the city power plant. Today its giant smokestack towers over the pristine lakes, parks and cycleways, barely emitting a puff of steam. Inside the plant there's a huge furnace, similar to those that burn coal. Woodchips, sawdust and other wood waste discarded by local forestry industries are burning at extremely high temperatures to produce electricity. Instead of the cooling water being dumped, as in most power stations, it is pumped out to the city's taps and into another network of insulated pipes, which runs hot water through heaters in homes and offices. The water leaves the plant at close to boiling point, travels as far as 10 kilometers and comes back warm to be reheated, over and over. An enormous municipal hot water tank acts as a back-up, so showers never go cold."

In addition to identifying the recycling, reuse, and closed cycle opportunities, LCA enables product development to minimize waste generation. Figure 4.7 illustrates multiplication of waste upstream of the product manufacturing cycle (Wenzel et al. 2000). Producing 1 ton of consumer goods often requires processing 5 tons of material in manufacturing and 20 tons of waste at the extraction stage of the product life cycle. Therefore, reduction in consumption and design-for-long life and repairability significantly reduce waste in the product life cycle.

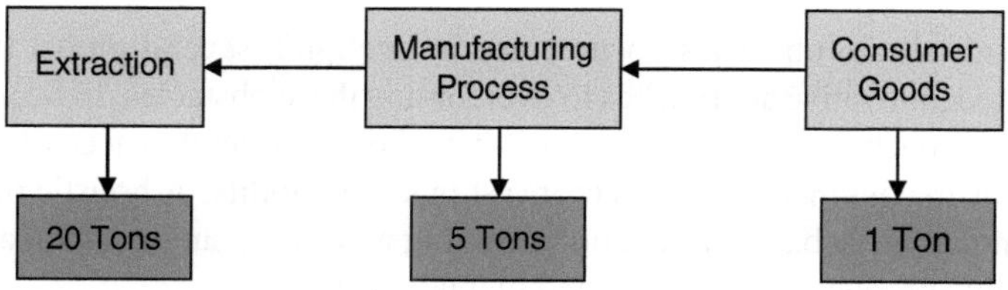

Fig. 4.7 Waste multiplies upstream of the product manufacturing cycle

It is important to note that often a formal and comprehensive LCA becomes quite cumbersome and loses its utility as a design tool. LCA should be an element of Life Cycle Thinking where eco-efficiency and sustainability are encouraged throughout a product design and manufacturing process as Kaizen and are institutionalized as organizational culture influencing strategic and tactical decision making. For example, Intel has found sustainable manufacturing in developing countries challenging because of cultural differences, lack of infrastructure (including procedures and enforcement), and a relative shortage of employees with relevant EHS experience.

Norm Thompson Outfitters (NTO), a consumer catalog company in Portland Oregon set out to achieve "triple bottom line of sustainability" as being economically viable, environmentally responsible, and socially active. Specifically the company had the following two objectives: (1) no negative forestry impact by using recycled paper for NTO catalogs and (2) no toxins in products (and associated processes) merchandised by NTO. To achieve these objectives, NTO had to change the behavior of its buyers and suppliers through education and the use of a scorecard reward system. Marshall and Brown have analyzed the implementation of NTO's sustainability action plan by applying the system dynamics methodology (Marshall and Brown 2003). Their paper comprises a list of proactive actions that the company took in achieving its sustainability objectives.

4.5 Summary

In Sect. 4.2, we examined the obstacles that current economic models and activities place in the path of global sustainability. Developed and developing economies present different barriers; and it is likely that changes required to overcome these barriers will also differ. The examples cited show just how large the negative effects have become. To this point, a "hands off" approach with regard to economic entities has resulted in only limited success.

The need for product and process assessment was addressed in Sect. 4.3. Assessment cannot occur without metrics; and we suggested a comprehensive set of metrics for the firm (or business unit), the product, the process, and all elements of the supply chain. We also define a measure (equation) for determining the life cycle cost of sustainability. Such a measure will be critical for managers deciding whether to implement sustainability-enhancing recommendations.

Even with all the obstacles documented in Sect. 4.2, several global firms have implemented innovative approaches to overcoming these obstacles. In Sect. 4.4, several such approaches are reviewed. Some have been extremely successful. It often takes a LCA to convince managers to invest in sustainability. A holistic operational strategy and an LCA-based production planning across the supply chain are critical to sustainable product development and manufacturing.

Acknowledgement We would like to thank Professor Feryal Erhun of the Management Science and Engineering Department at Stanford University for her valuable review comments that helped us improve the content and readability of the paper.

References/Notes

According to World Watch Institute, Vital Signs Report, 2006–2007, pp. 122: In 2004, nearly 1,800 transnational corporations or their affiliates filed corporate responsibility reports, up from virtually none in the early 1990s. While this reflects growing transparency and commitment to social and environmental principles, 97.5% of the nearly 70,000 TNCs worldwide still do not file such reports

Ayres RU, van den Bergh JCJM, Gowdy JM (1998) View point: weak versus strong sustainability. Tinbergen Institute Discussion Papers, # 98–103/3. Available online at: http://www. Tinbergen.nl/discussionpapers/98103.pdf. Sept 1998

Carlson RC, Rafinejad D (2010) Economic models for environmental and business sustainability in product development and manufacturing. Stanford University Paper, MS&E June 19, 2010

Climate Change (2007) The physical science basis, UNEP Report by The Intergovernmental Panel on Climate Change

The Stern Review (2006) The economics of climate change. A UK Government Publication, October

Harvard business review on business & the environment. (1999) HBS Press Book, Product No. 2336; 17 December 1999

Ecosystem services include: air and water purification, water absorption and storage, sequestering and decomposition of wastes, regeneration of soil nutrients, pollination, seed, and nutrient dispersal, climate stabilization, etc. See Hassan R et al (eds) (2005) The millennium ecosystem assessment series & synthesis reports. Island; Available at the following link: http://www.millenniumassessment.org/en/Condition.aspx#download

Manahan S (1999) Industrial ecology. CRC, Boca Raton

Graedel TE, Allenby BR (2003) Industrial ecology, 2nd edn. Prentice Hall, New Jersey

Allen DT, Behmanesh N (1994) Wastes as raw materials. In: Allenby BR, Richards DJ (eds) The greening of industrial ecosystems. National Academy Press, Washington, pp. 69–89

Ayers RU, Ayers LW (1996) Industrial ecology: towards closing the materials cycle. Edward Elgar Publishing, Cheltenham, UK, (Chapter 1)

Wackernagel M et al (2002) Tracking the ecological overshoot of human economy. Proc Acad Sci 99(14):9266–9271

Hart SL (1997) Beyond greening, strategies for a sustainable world, HBR Jan–Feb 1997, Reprint No. 97105

Chertow M (2000) The IPAT equation and its variants. J Ind Ecol 4(4):13–29

United Nations Statistics Division – National Accounts, http://www.unstats.un.org

China Daily, 12 June 2006

International Herald Tribune, 4 May 2006

The Guardian, 13 June 2006

World Wild Fund (WWF) http://www.worldwildlife.org

Shanghai Daily, 17–18 June 2006
BBC, 16 April 2006
Reuters, 11 June 2007
Meadows DH, Randers J, Meadows DL (2004) Limits to growth: the 30-year update. Chelsea Green, Vermont
Billatos S, Basaly N (1997) Green technology and design for the environment. Taylor & Francis, New York
Cohen JE (1995) How many people can the earth support. Norton, New York
Wackernagel M, Rees W (1995) Our ecological footprint: reducing human impact on the earth. New Society, Philadelphia
Maslow AH (1970) Motivation and personality, 2nd edn. Harper & Row, New York
US Department of Commerce. "Who killed the electric car?," a film by Chris Paine, DVD by Sony Pictures Classics, 2006
Friedman M (1970) *The social responsibility of business is to increase its profits*. NY Times Magazine 33:122–126
Anderson TL, Leal DR (1997) Enviro-capitalists, doing good while doing well. Rowman & Littlefield, Maryland
Cobb–Douglas production function, where output $Y = AL^{\alpha}K^{\beta}$; $L =$ labor input, $K =$ capital input and A, α, and β are constants determined by technology ($\alpha+\beta = 1$ and $\alpha,\ \beta \geq 0$). Also, the individual utility function is modeled a logarithmic function of his/her consumption (that is, the contribution of an individual to production is proportional to his/her consumption (Kempf H, Rossignol S (2007) Is inequality harmful for the environment in a growing economy? Econ Polit 19:53–71)
The following is a 2006 excerpt from Toyota–North America web site: "Our research continues to show that the majority of customers will not compromise performance for environmental benefits and are not inclined to pay a premium for an environmentally sensitive vehicle. However, with rising fuel prices, an increasing number of customers are considering the vehicle's fuel economy as well as its price and performance, when they make purchase decisions. The onus is on the manufacturer to design and build environmental products with as few compromises as possible and sometimes in advance of market signals or regulatory requirements."
GE promotes an image of environmental leadership. "GE seeks leadership in environmental sustainability for shareholder value and not for charity", said Jeff Immelt, the CEO in a 2006 interview with the Talk Show Host Charlie Rose. (b) BP undertakes a renewable energy ad campaign to overcome the negative public perception against oil companies. (c) Corporations communicate their triple-bottom-line programs to investors and customers in their annual sustainability reports.
Wind turbines marketed by GE for power generation. (b) IBM created a new market opportunity as an element of its Product End-of-Life Management (PELM) program. IBM's service is called Asset Recovery Solution and targets commercial customers to manage their computers and workstations at end-of-life. A limited service is also provided to individual consumers through an "advance recycling fee" program. Asset Recovery Solution includes: data security management, disk overwrites, resale of products (remarketing), refurbish, and recycling service. In 2003, 69,000 metric tons were processed and 830,000 PCs were recycled, reused, or resold. (c) 3M has developed biodegradable paint remover without methylene chloride and nonsolvent-based ink without volatile organic compounds
Directive 2002/05/EC of The European Parliament and of the Council of 27 January 2003: on the restriction of the use of certain hazardous substances (RoHS) in electrical and electronic equipment
Holliday C, Schmidheiny S, Watts P (2002) Walking the talk: the business case for sustainable development. Brett-Koehler Publishers, San Francisco
Michelsen O et al (2006) Eco-efficiency in extended supply chain: a case study of furniture production. J Environ Manag 79:290–297

For example, six or more packages are used in transporting and delivering tooth paste to a consumer, including: tooth paste plastic tube/plastic security seal wrap/paper box/plastic wrap for a dozen tubes/cardboard box/shipping crate
Graedel TE, Allenby BR (1998) Design for environment. Prentice Hall, New Jersey
Schmidheiny S, Business Council for Sustainable Development (1992) Changing course: a global business perspective on development and the environment. MIT, Cambridge
Efficiency means achievement of the intended performance with minimal utilization of resources
BBC, April 16, 2006
Toyota Motor Corporation: http://www.toyota.com/about/environment-2006
The Power Supply Design Guide is available on the web at: http://www.formfactors.org
Stanford MS&E 268 Course – There are four stages in the strategic role of manufacturing in support of a firm's business success. These stages in increasingly degree of contribution are: internally neural, externally neutral, internally supportive, and externally supportive. In the latter stage, manufacturing is a significant contributor to the firm's competitive advantage
Jawahir IS (2007) Machining process and other case studies in sustainable manufacturing. University of Kentucky, Kentucky
Klostermann J (ed) (1998) Product innovation and eco-efficiency. Academic, The Netherlands
Ayers RU, Ayers LW (2002) A handbook of industrial ecology. Edward Elgar, UK
Applied Sustainability LLC, Business case for by-product synergy, Stanford, Prod No E118, 2–2002
Sydney Morning Herald/Australia, Saturday, 23 June 2007
Wenzel H et al (2000) Environmental assessment of products. Methodology, tools and case studies in product development, vol. 1. Springer, New York
Marshall SR, Brown D (2003) The strategy of sustainability: a systems perspective on environmental initiatives. California Manage Rev 46(1) Reprint CMR 271, 10/1/03
Henrik W, Michael ZH, Alting L (2000) Environmental assessment of products. Methodology, tools and case studies in product development, vol. 1, 1st edn. Springer, New York
Frankl P, Rubik F, Institute for Ecological Economy, Germany (2000) Life cycle assessment (LCA) in industry and business (Adoption Patterns, Applications & Implications) – in Germany, Italy, Sweden, Switzerland, Springer, Hiedelberg
Billatos S, Basaly N, Tayler & Francis (1997) Green technology and design for the environment. University of Connecticut, Storrs, CT
Giudice F, Rosa GL, Risitano A (2000) Product design for the environment. CRC, Boca Raton

Chapter 5
Uncertainty and Production Planning

Stephen C. Graves

The intent of this chapter is to review and discuss how uncertainty is handled in production planning. We describe and critique current practices and then prescribe possible improvements to these practices. In particular, we argue that there are a set of tactical decisions that are critical to the proper handling of uncertainty in production planning. We observe that current planning systems do not provide adequate decision support for these tactical decisions; we regard this shortcoming as an opportunity for new research and development, which could significantly improve the practice of production planning.

This chapter is based primarily on personal observations of production planning practices in a variety of industrial contexts. As such it is written more as an essay than as a scientific research article. We make no effort to review or survey the research literature on production planning under uncertainty; we do provide a commentary on the research literature, and identify a few points of entry for the interested reader. We also cite a few illustrative references, albeit primarily from our research. Similarly, our observations on practice are not the outcome of a carefully designed field study, but rather are derived from a potpourri of projects over many years. Again, our intent is to provide a framework and a set of observations on current practice and to provoke some new thinking on how we might do better.

We organize the chapter into five sections. In the first section, we briefly describe and discuss major sources and types of uncertainty, how these uncertainties are realized and how they affect the production plan.

In the second section, we comment on existing research in production planning as it relates to the theme of this chapter. We note that much of the research literature is based on deterministic models, and does not explicitly account for uncertainty. With regard to the research that does include uncertainties, we discuss its applicability and the challenges to its transfer to practice.

S.C. Graves (✉)
MIT, 77 Massachusetts Avenue, E62-579, Cambridge MA 02139-4307, USA
e-mail: sgraves@mit.edu

K.G. Kempf et al. (eds.), *Planning Production and Inventories in the Extended Enterprise*, International Series in Operations Research & Management Science 151,
DOI 10.1007/978-1-4419-6485-4_5,

In the third section, we introduce a stylized framework for describing current production planning systems to provide a basis for the subsequent discussion and critique of production planning under uncertainty.

In the fourth section, we make a series of observations on the generic treatment of uncertainty in production planning in practice, and note that most systems for production planning do not recognize or account for uncertainty. Yet these systems are implemented in uncertain contexts; thus the planning organizations need to develop coping strategies. We describe and critique the most common coping strategies.

In the final section, we identify a set of tactical decisions that we view as critical for handling uncertainty in production planning. We describe how these tactics can be incorporated into production planning systems as proactive countermeasures to address various forms of uncertainty. We provide a perspective on the key trade-offs in making these decisions and identify both examples of relevant work from the research literature and opportunities for new research on developing effective decision support for these tactics.

5.1 Types of Uncertainty

In this section, we identify and discuss the three major types of uncertainty that arise in manufacturing contexts and that can affect a production plan. We contend that the production plan needs to account in some ways for these uncertainties. In subsequent sections, we discuss and critique the research literature and current practice, and then will propose possible tactics for handling these uncertainties.

5.1.1 Uncertainty in Demand Forecast

This is usually the largest single source of uncertainty. All production plans rely on a demand forecast or a demand plan as an input. A demand forecast extends over a multiperiod planning horizon and represents the firm's best guess at the future demand. The forecast is based on a combination of inputs, the specifics of which depend on the context. These include a projection of historical demand data, as might be done by a statistical forecasting package; advanced orders in contexts where at least some of the production is make-to-order; a corporate demand plan for firms that operate with a sales and operations planning (S&OP) process; any customer forecasts that the customer is willing to share with the firm; and market intelligence, often in the form of expert judgments.

As a firm gets more and better information about future demand, it updates the forecast. Indeed, in most planning systems, there is a regular cycle in which the forecasts are moved forward and revised; for instance, at the start of each week, a new forecast for demand over the next 13 weeks is released. The new forecast reflects information inferred from the observed demand since the last forecast update, any changes to customer orders or forecasts, as well as any changes on the market outlook.

Forecasts are never perfect, and the actual demand realization will differ from the forecast, resulting in a forecast error. To address the uncertainty due to forecasts, we need to characterize the forecast errors. Typically, we view the forecast errors as random variables for which we will want to know (at least) the first two moments. It is important to recognize here that the forecast for a particular product is usually a vector of forecasts, which cover the planning horizon. That is, at any time t, for each product we have a forecast for future time periods $t + i$, for $i = 1, \ldots, H$, where H is the length of the planning horizon. Thus, we have H forecasts; for instance with weekly time periods, we have a 1-week forecast, a 2-week forecast, and so on. We then need to characterize the errors for each type of forecast, as each forecast has a different impact on the production plan.

5.1.2 *Uncertainty in External Supply Process*

A second type of uncertainty is associated with the external supply process. A production plan results in orders placed on outside suppliers and has expectations on the fulfillment of these orders. That is, a plan might initiate an order for ten steel plates of a certain dimension and grade, and then expect that these plates will arrive and be available for processing according to a stated lead time of, say, 8 weeks. Nevertheless, there can be uncertainty in the delivery date due to uncertainty and capacity constraints in the supplier's manufacturing and distribution processes; for instance, the order might take 10 weeks to arrive due to a work stoppage or delays attributable to the weather.

Furthermore, in many contexts, there can be uncertainty in the amount of the delivery. For instance, a supplier might be permitted by contract to deliver $\pm 10\%$ of the amount ordered; in other contexts, the buyer might reject some portion of the delivery due to quality considerations. To model this uncertainty, one needs to characterize the uncertainty in the replenishment lead times and in the replenishment quantities.

5.1.3 *Uncertainty in Internal Supply Process*

A production plan also needs to account for uncertainty in the internal supply process, which is similar to the uncertainty in the external supply process. A production plan results in work or job orders placed on the internal manufacturing, transportation, and supply processes. Furthermore, the production plan has expectations on how these processes will perform. That is, a plan might set the number of wafer starts into a semiconductor fabrication (fab) facility with expectations on both the yield from these wafers and the flow time or process duration for these wafers within the fab. Again, there is uncertainty on both accounts. The actual flow or completion time will deviate from the expectation depending upon the work-in-process (WIP) in the shop, the equipment availability and the dispatch rules; the wafer yield is inherently random and depends on numerous process factors and conditions. Again, one needs to characterize the uncertainty in the flow or process lead times and in the yield quantities for each process step.

5.2 Observations on Production Planning Research

In this section, I provide general comments on the operations research-based research in production planning; I do not attempt a survey of the literature, but try to provide relevant references for a few entry points into this vast literature.

5.2.1 Deterministic Models

The dominant thrust of the research literature has been the formulation of deterministic models for production planning, and the development of solution procedures, both optimal and heuristic, for these models. The primary intent of these models has been to specify the requirements for a feasible production plan and to capture the key cost tradeoffs that depend on the production plan. Typically, a feasible production plan is one that satisfies the given demand over the planning horizon with no backorders or lost sales, abides by specified production recipes for each final product, and does not violate any capacity constraints. The models attempt to optimize total costs, which often include: sourcing and production costs, including setup-related costs; holding costs for pipeline inventory and cycle stock; costs for adjusting production capacity, such as hiring and overtime costs; and logistics costs for transportation and warehousing. For reviews of this research, we suggest Thomas and McClain (1993), Shapiro (1993), and Graves (2002).

This literature is largely oblivious to uncertainty. Much like research on the economic-order-quantity (EOQ) model, the contention is that the value of these models is in optimizing critical cost tradeoffs, often in the context of tight constraints. The research perspective is that dealing with uncertainty is of secondary importance to getting the tradeoffs right; furthermore, there is the assumption that the uncertainties can be handled by other measures, which are independent of the determination of the production plan. Nevertheless, there is also the recognition that the deterministic assumptions are a shortcoming of this research, but are necessary in order to keep the models tractable.

5.2.2 *Hierarchical Production Planning*

Hax and Meal (1975) introduced hierarchical production planning (HPP) as a framework for production planning and scheduling, motivated by the desire to create an applicable structure for developing effective planning systems. A hierarchical approach partitions the production planning problem into a hierarchy of subproblems, often corresponding to the organizational hierarchy of the planning organization. In any planning period, the subproblems are solved sequentially, with solutions from the upper-hierarchy subproblem(s) imposing constraints on the lower-hierarchy subproblem(s). The planning system implements only the solutions for the immediate period and resolves the subproblems each period in a rolling horizon fashion. See Bitran and Tirupati (1993) for a review of research literature on HPP, and Fleischmann and Meyr (2003) for a review of HPP and advanced planning systems.

The literature identifies three advantages for HPP relative to the alternative of solving a monolithic problem: it is computationally simpler; depending on the formulation, it can have less onerous data requirements; and it has implementation advantages to the extent that the subproblems are aligned with the hierarchy of decision makers.

Within the research literature, the HPP approach has primarily been applied to deterministic models for planning and scheduling problems. As such, it is subject to the same criticisms raised in the prior subsection. Yet, the approach would seem to have an advantage in the consideration of uncertainties, in that it might be possible to tailor the lower-hierarchy subproblems to account for uncertainty, e.g., short-term demand uncertainty. Indeed, there is some research along this premise: Bitran et al. (1986), Lasserre and Merce (1990) and Gfrerer and Zapfel (1995). In each case, the research attempts to develop within an HPP framework a deterministic aggregate plan that is robust to item-level demand uncertainty; that is, as the item-level demand uncertainty is realized, there is some assurance that the lower-hierarchy subproblem can disaggregate the aggregate plan into a good detailed schedule. This is an interesting line of research, but so far has been limited to fairly specific, single-stage production contexts.

5.2.3 *Production Planning with Quadratic Costs*

One of the earliest production-planning modeling efforts was that of Holt, Modigliani, Muth and Simon (HMMS) (Holt et al. 1960), who developed a production-planning model for the Pittsburgh Paint Company. They assume a single-aggregate product and have three sets of decision variables for production, inventory, and work force level in each period. More notable are their assumptions on the cost function, entailing four components. The regular payroll cost is a linear function of the workforce level. The second component is the hiring and layoff costs, which are assumed to be a quadratic function in the change in work force from one period to the next. The production cost is also modeled by a quadratic function.

HMMS assume for a given workforce level that there is an ideal production target and that the incremental cost of deviating from this target (representing either overtime or idle time) is symmetric and quadratic around this target. Finally, they model inventory and backorder costs in a similar way. The inventory and backorder cost is a single quadratic function of the deviation between the inventory and an inventory target that depends on the demand level.

With these cost assumptions, HMMS minimize the expected costs over a fixed horizon, where the expectation is over the demand random variables. The analysis of this optimization yields two key and noteworthy results. First, the optimal solution can be characterized as a *linear decision rule*, whereby the aggregate production rate in each period is a linear function of the future demand forecasts, as well as the work force and inventory level in the prior period. Second, the optimal decision rule is derived for the case of uncertain demand, but only depends on the mean of the demand random variables. That is, we only need to know (or assume) that the demand forecasts are unbiased in order to apply the linear decision rule. Both results depend crucially on the assumptions of a quadratic objective function.

This research stands out from other production planning research in that it explicitly allows for uncertain demand, and it develops an easy-to-implement plan, namely the linear production rule. However, this line of research has fallen out of favor for a couple of reasons. One is discomfort with the assumptions on the cost functions. A second reason is that the simplicity of the result depends on the restriction to one aggregate product with a single capacity; the form of the production plan gets more complex with more products or resource types.

5.2.4 Stochastic Programming

Over the past 10–15 years, there has been an increasing interest in explicitly adding uncertainty to production planning models. Mula et al. (2006) provide an extensive review of this research. A good portion of this research examines the applicability of stochastic optimization, particularly stochastic programming methods, to production planning models. Stochastic programming is notoriously computation intensive for many problem contexts; production planning is no exception. Yet, with the ever-increasing computational power and the improvements in our optimization algorithms, there has been more exploration of the feasibility of using stochastic programming for production planning. Escudero et al. (1993) show how to formulate a multistage stochastic program for production planning and explore its computational implications. Graves et al. (1996) report on the application of two-stage stochastic programming for the optimization of production and supply chain planning for the Monsanto Crop Protection business.

Although stochastic programming methodology has promise as a methodology for capturing uncertainty, it also has significant limitations. In many contexts, it remains computationally prohibitive when there are many periods in the planning horizon and frequent replanning. Also, the resulting scenario-based production plans can be difficult to communicate and hence difficult to implement.

5.3 A Generic Framework for Production Planning

In this section, I present a highly simplified generic framework to describe current production planning practices. My intent in introducing this framework is to create a "straw man" on which to comment and present some observations about planning practices. In the following section, I will use this framework to characterize and critique how many planning systems address uncertainty.

Almost all text books that discuss production planning provide a framework that lays out the various elements of a planning system (e.g., Hopp and Spearman 2007; Nahmias 2008; Silver et al. 1998). In Fig. 5.1, I provide a stylized version, given my intent as discussed above.

A planning system starts with a forecast of future demand over some forecast horizon of length H periods. The long-term portion of this forecast is an input into an aggregate planning module that assesses whether there is sufficient capacity to satisfy the demand forecast. This is the aggregate production plan as it is usually done by using aggregate products and large time buckets, and must account for the key capacity considerations within the manufacturing system and/or supply chain. To the extent that there is a mismatch between the available capacity and the long-term demand forecast, the module needs to examine and decide how to rectify this gap.

In general, there are four common ways that this may be done. First, in contexts with seasonal demand, one might develop an aggregate production plan that builds inventory during the off season in anticipation of a seasonal demand peak. Second, the mismatch might be addressed by adding to or augmenting current capacity, for

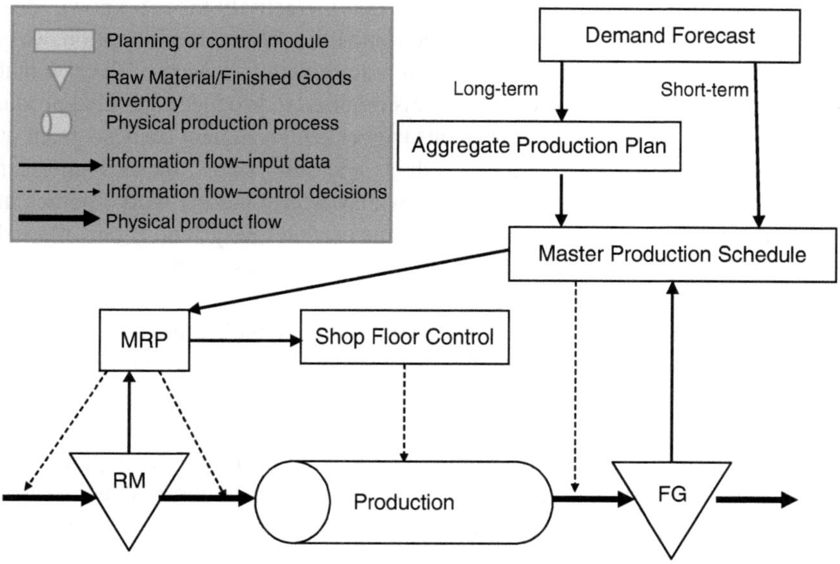

Fig. 5.1 Framework for classical production planning

instance through overtime or subcontracting. Third, when it is not possible to build anticipation inventory or add capacity, one would delay meeting demand. This is usually done in terms of extending the backlog by quoting longer and longer delivery lead times to customers. Finally, there may be some downward revision of the forecast so as to eliminate the gap between the firm's supply capacity and the anticipated demand; this might be an outcome from an S&OP process that equates the forecast to a sales plan and then aligns the sales plan to the production plan. In practice, a firm would rely on a mixture of these tactics in developing its aggregate production plan.

The next step in Fig. 5.1 is to convert the forecast over the shorter term into a detailed master production schedule (MPS), subject to the guidelines or constraints from the aggregate production plan. The aggregate production plan determines, at a gross level, how and when customer demand is met. Given the aggregate production plan and the current finished goods (FG) inventory, the MPS determines the necessary production output to meet the demand forecast over the short term. Relative to the aggregate production plan, the MPS is at a much more detailed level, both in terms of products and time periods. In some contexts, production lot or batch sizing is done as part of the master scheduling, so as to account for economies of scale in the production process. There is also sometimes additional processing to check the feasibility of the master schedule relative to the available capacity.

To determine the inputs into the production system, we need to convert the MPS into a plan for the raw materials (RM) and intermediate products. This is traditionally done with the logic of materials requirement planning (MRP), based on a bill of materials and process recipes for each final product (Vollman et al. 2004). The key assumption is that we have planned lead times for each required activity to produce the product; that is, for the procurement of each raw material, as well as for each process and transportation step, we assume that the activity takes a known, deterministic amount of time, termed the planned lead time. With this assumption, we can readily translate the MPS into time-phased requirements for each of the raw materials, intermediate products, and subsystems required to produce each final product. For the raw materials, these requirements trigger replenishment requests from outside suppliers. For the intermediate products and subsystems, these requirements are input to the shop floor control system, which determines the work releases into production and the job priorities throughout the production system. Again, these decisions are based on and guided by the planned lead times for the process steps in the production operation.

The MRP step might also account for lot sizing considerations, whereby a lot sizing heuristic or algorithm is applied for each process step or component in the product bill of material. In some planning systems, this step of the planning process might consider some capacity constraints on production, usually by means of heuristics such as a forward loading scheme. However, even in these cases, the system must rely to some extent on planned lead times in order to coordinate the replenishments for multiple components and subsystems whose requirements derive from the demand for an assembly product or multiproduct order.

5.4 Observations on Common Approaches for Handling Uncertainty

In this section, I provide a series of observations on how uncertainty is addressed in practice by many planning systems, where we will use the framework from the prior section. Admittedly, these observations are largely anecdotal but are based on a large set of industry-based projects and other interactions.

Most systems for production planning do not explicitly account for uncertainty: The planning system operates as if its world were deterministic. That is, plans are created based on assumptions that the forecast is perfect, the internal production processes are perfect and the outside suppliers are perfect. The plans presume that the supply and production processes perform exactly as prescribed by their planning parameters and that customer demand occurs as predicted by the demand forecast.

Most implementations for production planning systems make no effort to recognize the uncertainty in their environment: Indeed, in many planning systems that I have observed, there is limited, if any, attempt to track and measure the uncertainty in the demand forecast or the replenishment processes. For instance, we find that most planning systems do not retain and measure forecast errors. Once a forecast has been revised, we observe that the old forecast is written over by the new forecast, and the old forecast is lost. As a consequence, there is no record from which to measure the forecast errors. Similarly, we still find implementations of planning systems that do not track the actual replenishment times from suppliers, although this is slowly changing as lead-time performance becomes a more common metric in supply contracts. The execution component of most planning systems now has the capability to track the actual flow times within the production operations; however, again we seldom see these data used to understand the uncertainty in these processes. On the other hand, yield data do get recorded more routinely, and at least the statistics on average yield seem to be more routinely used in the planning systems.

Even though these planning systems do not explicitly recognize or account for uncertainty, they do operate in an uncertain world. The planning organization thus develops various strategies and tactics for coping with this reality. We describe our observations on the major tactics as follows.

Safety stocks: Most planning systems do allow the users to set safety stocks for finished goods (FG) and for raw materials (RM), as depicted in Fig. 5.1. Safety stocks for raw materials can protect against supplier uncertainty, both in lead times and in yield. In a make-to-stock environment, it is possible to install a finished goods safety stock that can buffer against uncertainty from the production processes as well as uncertainty due to forecast errors.

Replanning: A second common practice is replanning. That is, at some regular frequency, the planning system is rerun to create a new plan. Often the replanning frequency corresponds to the frequency with which the demand forecast gets updated; in other cases, replanning is done even more frequently in order to capture the dynamics from the internal and external supply processes.

This replanning might be done each week or month, even though the plan might extend for 3–12 months into the future. The revised plan would account for the changes in the demand forecast, as well as the realizations from both the external supply and internal production processes. In this way, the system reacts to the uncertainty as it occurs, by revising its plans and schedules based on the updated information on demand and the supply processes.

Time fences and frozen schedules: One consequence of this replanning is that it induces additional uncertainty in the form of schedule churn. That is, with each forecast revision, we generate a new MPS, which then can result in new detailed schedules for all raw materials, components, and subsystems. The due dates for some replenishment orders and production jobs are accelerated, while others are delayed. But a change in priority or due date in one revision might often be reversed by the next revision in the next time period. This churn or schedule nervousness leads to additional costs as changing priorities inevitably lead to inefficiencies in any production operation, as lower priority work gets put aside in order to expedite the higher priority work.

The schedule churn also leads to dissatisfaction and distrust of the planning system. Indeed, we find that suppliers (both internal and external) will often develop their own forecast of requirements, rather than accept and follow the requirements schedule that gets passed to them; in effect, they "second guess" the requirements schedule that is given to them. The suppliers know that the requirements schedule from the customer's planning system will keep changing, and they think they can do a better job with their own forecast. Nevertheless, it is not at all clear whether this second guessing helps or hurts.

One tactic to protect against the induced uncertainty from replanning is the freezing of the MPS. A firm might decide that no changes are permitted for part of the master schedule, say for the next 4 weeks. The frozen schedule provides some stability in the short term, as any short-term changes in the demand forecast get accumulated and then deferred until beyond the frozen period. A related tactic is to use time fences to establish varying limits on the amount of change permitted to the MPS. For instance, a firm might set a time fence at week 4, 8, and 13, and then specify that the MPS is frozen within the first time fence of 4 weeks, but permitted to change by at most (say) $\pm 10\%$ for the weeks 5–8 and (say) $\pm 25\%$ for the weeks 9–13. Beyond the last time fence, there might be no restrictions on how much the MPS could change. Again, this type of policy can help to mollify and dampen the dynamics introduced by production replanning. However, these time fences and frozen schedules act as a constraint on the replanning; thus, they necessarily limit the effectiveness of replanning as a tactic for handling and responding to uncertainties in the demand forecast and in the supply and production replenishment processes.

Flexible capacity: In some contexts, firms maintain a capacity buffer to respond to uncertainty. In effect, there is a reserve capability that permits the manufacturing system to recover from supply disruptions and/or to respond to unanticipated changes in demand volume or mix. A common example is overtime work, which is employed on an as needed basis to handle contingencies. However, even though many firms will rely on flexible capacity as a way to cope with uncertainties, their planning systems do not formally recognize or plan for any capacity buffer; that is, there is no means within the planning system to record or track utilization of the capacity buffer, let alone provide guidance on its size and deployment.

Backlog management: Another tactic for dealing with uncertainty is how a firm manages its order backlog. In particular, a firm might vary the size of the backlog or correspondingly vary the delivery or service times quoted to customers. Thus, the backlog grows (falls) if demand is greater (lesser) than forecast and/or if the supply process is slower (faster) than planned. This is only possible in contexts in which the manufacturer has sufficient market power to do this.

Inflated planned lead times: The last tactic that I have observed is the use of planned lead times to indirectly create additional safety stock throughout the production system. As noted above, most planning systems rely extensively on planned lead times. There are several good reasons for this to be the case.

- Planned lead times greatly simplify any complex planning problem by permitting decomposition. The assignment of a planned lead time to each process step permits the scheduling of a multistep serial production activity to be separated into a series of single-step activities, whereby each process step has a specific time window within which to accomplish its task. In effect, the planned lead times convert the final due date for a multistep activity into intermediate due dates, one for each process step.
- Planned lead times facilitate coordination whenever multiple components or subsystems need to be joined or assembled together into a final product or assembly. As above, the planned lead times permit a decomposition by which we can establish intermediate due dates for each component or subsystem and then can manage each replenishment process independently.
- Planned lead times often serve as a proxy for dealing with capacity constraints (see Graves, 1986 for a discussion and analysis). Many planning systems do not explicitly account for capacity constraints; instead, they rely on the planned lead times to compensate for this oversight. The planned lead times are set to reflect the impact of limited capacity. A constrained work center that is heavily loaded will have a longer planned lead time than one with a lighter load. A highly utilized work center requires more smoothing of its work arrivals in order to level its load to match its capacity. A longer planned lead time results in a larger queue at the work center, which permits more smoothing.

Yet, beyond these reasons, we also observe that firms use their planned lead times as a way to create another buffer to protect against uncertainty. As noted earlier, the planning systems use the planned lead times to determine work and order releases; as a consequence, a planned lead time translates directly to a level of WIP. If a work center has a planned lead time of 3 days, then we expect that it will have 3 days of WIP on average. (This follows from Little's law, if we assume that the actual realized lead time is on average equal to the planned lead time.) As noted above, for capacity-constrained work centers, this level of WIP might be dictated by the need to smooth the work load through the work center. However, we often observe that the WIP (and planned lead times) exceeds that which is needed for work smoothing. In these cases, we find that the WIP is actually acting as a safety stock for the production system; its purpose is to provide an additional buffer to protect against uncertainties in demand and/or in the supply processes.

This WIP safety stock differs from the raw material (RM) and finished good (FG) safety stocks in a couple of important ways. First, the RM and FG safety stocks are established by setting their planning parameters in the planning system. In contrast, there are no planning parameters to directly set a WIP safety stock; rather, the WIP safety stock is a byproduct of the planned lead times, and as such is not the result of any deliberate planning decisions. Second, the WIP safety stock typically does not reside in a warehouse, but sits on the shop floor in the form of WIP. One consequence of these two observations is that this buffer is usually under the radar of the "inventory police" and is not recognized as a planned safety stock. Indeed, we have seen operations eliminate their safety stocks and close their warehouses as part of an inventory reduction initiative, only to have that inventory recur on the shop floor in WIP, as the inherent uncertainties of demand and supply remain unchanged and thus a (unrecognized) safety stock is still required.

The inflated lead times also result in a control phenomenon known as *launch and expedite*. Based on a demand forecast, a firm releases work orders to initiate production to meet the demand forecast; that is, the firm pushes or *launches* the work into the shop based on the shop lead time. This creates a large WIP in various stages of completion, depending on how inflated are the planned lead times. The actual demand deviates from the forecast. As actual orders come in, the firm matches the orders with the available WIP and pulls or *expedites* the work out of the shop to meet the true demand. (see Sahney 2005 for a case study)

This type of operation is often subject to an unhealthy dynamic. A longer planned lead time results in more work getting pushed into the shop, creating a larger WIP safety stock. This is attractive to the planner as there is more work from which to select to expedite to meet demand, once it has been realized. However, a longer planned lead time results in more uncertainty in the demand forecast; with more uncertainty, the shop needs even more safety stock, which results in pressure to increase further the planned lead times. This can lead to a so-called vicious cycle.

In summary, we have argued in this section that most production-planning systems do not recognize or account for uncertainty. Plans are typically created based on assumptions that the forecast is perfect, and the production and supply processes are perfect. Yet these systems are implemented in uncertain contexts; as a result, the planning organization needs to develop coping strategies. These coping strategies take the form of rapid and regular replanning (subject to time fences and frozen short-term schedules), with the resulting churn; explicit FG and RM safety stocks as well as hidden safety stock in the form of WIP; and a mixture of fluctuating service or delivery times along with flexible capacity. We find substantial inefficiency in the deployment of these tactics. This is not surprising, as these responses are generally reactive ad hoc measures. In the next section, we review the key tactics for dealing with uncertainty and discuss how these might be explicitly incorporated into the planning system.

5.5 A Proactive Approach to Uncertainty in Production Planning

In this section, I discuss possible counter measures and practices for handling uncertainty. We take as given that the current practices and systems for production planning are not likely to change radically in the immediate future. There is a huge installed base of planning systems and the accompanying IT support systems; as described above, these planning systems are largely oblivious to uncertainties. The question here is what might help. What can be done to help these systems be more proactive with respect to the uncertainties in their environments?

The first step is certainly to do *more routine tracking and measurement of the uncertainty*, in whatever form it occurs. This includes the measurement of forecast errors, lead-time variability, and yield variability. This by itself is not overly challenging, as it primarily entails keeping track of deviations from the norm or a target. But to use this data in planning requires some characterization of each type of variability so as to be able to model its occurrence. For instance, should the random deviations from a target yield be modeled as an additive or multiplicative process? Are these yield deviations correlated over time? Alternatively, the yield process might be better modeled by a Bernoulli process, for which there is some probability of having a yield bust, namely a zero yield. Systems are needed to not just capture the data but also to help in building models to characterize the uncertainty.

Similarly, modeling the lead-time variability for the purposes of inventory planning can have some subtlety. Consider an example based on a project to size a finished goods inventory for a semiconductor wafer fabrication facility (Johnson 2005). We measured the lead times for production lots in the facility and found the coefficient of variation (ratio of standard deviation to mean) to be on the order of 0.20. Based on this, we made inventory recommendations, which were rejected by the factory as being excessive.

Upon closer inspection, we saw why. There was substantial amount of "order-crossing" within the wafer fab, as the production lots for the same product were not consistently processed in a first-in, first-out sequence for a variety of reasons. As a consequence, the completion order of the lots differed significantly from the order in which the lots were released into the shop. But from the standpoint of the finished goods inventory, the order of the output did not matter; what did matter was the cumulative output process, indicating how much had been completed by any point in time.

Based on this observation, we redefined the lead times: we ordered the start times and completion times for all of the production lots, from earliest to latest; we then defined the nth lead time to be the difference between the nth completion time and the nth start time. We found that the coefficient of variation was now less than half of the original number. This resulted in a more reasonable inventory recommendation. Muharremoglu and Yang (2010) have shown that ordering lead times in this way is optimal for single-stage base-stock systems and near-optimal for serial systems.

We need also to understand the forecast evolution process so as to decide how best to characterize and model it. That is, given a forecast at one point in time, how should we think about the update process in the next time period? We know that with new market information and advanced orders the forecast will change, but does the forecast improve? How does a forecast change or evolve over a number of update cycles? And ultimately, does the forecast improve and by how much, as it is updated from period to period? Graves et al. (1986), Heath and Jackson (1994), and Gallego and Özer (2001) provide some examples and approaches for the characterization of the forecast evolution process.

Under the assumption that we can characterize the uncertainties, the second step is to develop *a more explicit consideration of the tactical decisions that provide counter measures to the uncertainty*. We identify and discus five categories of tactical decisions. We contend that most planning systems address these decisions in an ad hoc way and that there is a great opportunity to do better. Indeed, we think the key to improving our existing planning systems is to devise more systematic, proactive ways to find the right mix of tactics that match the uncertainty.

We will describe how these tactical decisions interface with the planning system depicted in Fig. 5.1; for the most part, the connection is by means of setting planning parameters. Also, as will be clear, the tactical decisions are highly interdependent and their deployment will depend very much on the context.

Customer Service Times: One tactic is to decide the service or delivery times to quote to customers, and how to adapt this in light of the uncertainty in demand volume and mix. This tactical decision would be incorporated into the aggregate production planning module in Fig. 5.1, as it is key to deciding how to match demand and capacity.

Allowing the customer backlog to vary with demand permits more efficient utilization of the production and supply system and/or less inventory buffers. However, varying the customer service times can result in market-related costs, such as lost sales; indeed, in some contexts, this option is not economically feasible. To set these service times require an examination of these trade-offs. Whereas in theory this is not difficult, this is not the case in practice. The trade-off requires an understanding of the customer sensitivity to the service times, as well as the cost inefficiencies from varying the production and supply processes.

MPS smoothing: A second tactical decision is how to convert the short-term forecast into a detailed MPS in a way that is consistent with the aggregate production plan and cognizant of the demand forecast uncertainty. There are two sets of questions to consider. First, should the MPS smooth the demand forecast and by how much? Second, when the forecast changes, how should the MPS be updated to accommodate these changes?

The MPS acts as a gatekeeper between the demand forecast process and the upstream production system and supply chain. The MPS determines how much of the uncertainty in the forecast and in the forecast updates gets seen by the production and supply system. Hence, a key tactical decision is to decide how wide this gate should be. The more smoothing of the demand forecast by the MPS, the less uncertainty gets sent upstream. The same is true with the response by the MPS to forecast updates; the revision to the MPS can dampen the forecast updates and reduce the schedule churn, and thus provides more stable signals to the production and supply system.

There is a cost when the MPS buffers the uncertainty in this way. The replenishment schedules for production and supply are, by design, less responsive to actual changes in demand, and thus a larger finished good inventory is required to assure some level of customer service. Again, we have a trade-off between the cost impact from passing the forecast uncertainty to the production and supply system and the cost of buffering the uncertainty with the MPS. We noted earlier that time fences and freezing of the MPS are current practices for smoothing the MPS. However, these approaches tend to be deployed in an ad hoc fashion, with limited consideration of the trade-offs and the alternatives. I think there is an opportunity to develop decision support tools for master scheduling, which would account for the uncertainty in the demand forecast process and provide a more complete treatment of the trade-offs; one example of such an approach is Graves et al. (1998).

Inventory buffers: An important tactical decision is where to position inventory buffers within the production and supply system. The stylized model in Fig. 5.1 assumes an inventory buffer of raw material (RM) and of finished goods (FG), but with no other buffers between them. This seems to be nonoptimal in many settings that have any level of complexity. We noted earlier that some planning organizations circumvent this shortcoming in their planning systems by using inflated lead times to create a WIP safety stock. In other cases, we suspect that excessive finished goods inventory and/or underutilized capacity is required for the systems to function.

We contend that a better approach is to designate several inventory buffers, strategically located across a production and supply system. These buffers could act as decoupling buffers; that is, each buffer is sized to protect the upstream supply processes from the noise or uncertainty in the downstream demand and to protect the downstream supply processes from any uncertainty in the upstream replenishment times or quantities. In effect, these buffers create a safety stock that allows the downstream to operate independently from any hiccups in the upstream process and vice versa.

Thus, the placement of these buffers can define relatively independent operating units within the production and supply system. Depending on the context, this can be an important consideration in deciding how many buffers and where to locate. Beyond this, one would of course need to account for the inventory holding cost; there is a cost for each buffer that depends on the size of the buffer and the value of the contents. The buffer size depends on the magnitude of both the downstream demand uncertainty and the variability in the upstream replenishment process over its lead time. The value of the inventory depends on where it is in the process. Graves and Willems (2000) provide one framework for determining the location and size of these buffers. Schoenmeyr and Graves (2008) show how to extend the framework and analyses to account for a forecast evolution process. This remains a fruitful area for developing decision support tools for guiding these tactical decisions.

Capacity buffers: In some contexts, it may be more economical to employ a capacity buffer, rather than build an inventory buffer. For instance, in a make-to-order assembly operation, it may not be feasible to have a finished goods inventory buffer, due to all of the possible combinations that can be built. Instead, the daily variability in demand might be handled by varying the production capacity. To do this requires that there be some reserve capacity to respond to upswings in demand; this reserve is often the capability to expand or lengthen the work day, by working a longer shift.

A capacity buffer can be more flexible than an inventory buffer, as it can be used to create multiple types of inventory. A capacity buffer can take several forms. One, as noted above, is the ability of a production unit to expand or flex its capacity by working longer hours. In a one- or two-shift operation, this might occur by extending the length of each eight-hour shift to (say) ten hours. Alternatively, in a 5-day operation, there might be an option to work a sixth or seventh day. A second way of creating a capacity buffer is by explicitly scheduling "idle" time on a work center. That is, we reserve time in a schedule where the actual use of the time will be decided later. In this way, we postpone the decision of what product will be produced until we have a better resolution of the demand or process uncertainties. A capacity buffer might also take the form of an option. We might contract with a supplier or contract manufacturer to reserve a certain amount of capacity, which can be exercised at a later date at some exercise price.

One common context for a capacity buffer is when there is yield uncertainty in a manufacturing process. Miller (1997) provides one example based on a project examining the production planning practices for film manufacturing at Kodak. The bottleneck operation is the film sensitizing operation, which at the time was subject to significant process variability. A single capital-intensive machine performs the film sensitizing operation for multiple-film products. The machine is highly utilized and changeovers are expensive. The machine is operated with a cyclic schedule that sequences through each type of film; the size of the batch run for each film is set to meet its short-term requirements.

One element of the process variability was the occurrence of incident failures, whereby there is a major discrepancy between the actual output of good film and the planned output. Over the course of the year, there were about one incident failure per week, with 95% of the weeks having zero, one or two incident failures. As all film types are vulnerable to these incident failures, using inventory buffers to protect against this uncertainty was deemed unreasonable. Instead, Miller (1997) implemented a capacity buffer that could be used for any film type; in each production cycle, a certain amount of capacity was reserved at the end of the cycle. If there were one or two incident failures during the cycle, then the reserve capacity would be used to run a second batch of each of the affected films. If there were no incident failures, then the reserve capacity would not be used and either the machine would be idled or the next production cycle would be moved forward in time.

This example is indicative of how one might deploy a capacity buffer. However, I am not aware of any systematic approach to thinking about this tactic, especially for a complex multistage production system. In particular, one would want to identify which process steps are good candidates for a capacity buffer, and how best to create and size these buffers. There is also the question of how to use capacity buffers in conjunction with inventory buffers: where and how should they be positioned and for what types of uncertainties would each buffer be deployed. This seems like a good opportunity for research and the development of decision support tools.

Planned lead times: Planned lead times are critical planning parameters in current planning systems. As discussed earlier, a planned lead time establishes the level of WIP in a manufacturing process step or the pipeline stock in a supply step. This inventory serves to dampen or absorb variability; in particular, it permits the smoothing of time-varying requirements. The longer is the planned lead time, the more smoothing is possible.

Typically, the planned lead times are set in correspondence to the actual lead times. Sometimes they are set to equal the average or median observed lead time. In other instances, we have seen the planned lead times set "conservatively" so as to cover the actual lead time with high probability; for instance, the planned lead time might be set to match the 80th percentile for the actual lead times.

There does not seem to be a standard practice for setting these parameters. Furthermore, I question the validity of setting a planned lead time based on observations of the actual lead time, as there should be a strong interdependence between the planned and actual lead times. If the planned lead time sets the WIP at a work center, then Little's law would say that the actual lead time should equal the planned lead time on average. That is, if the planned lead time at a work center were 3 days, then the planning system pushes work to the work center to create 3-days of WIP. If the work center is staffed according to its work load, then it will process roughly 1 day of work each day. One then expects the actual lead time to match, at least on average, the planned lead time. Indeed, we find that the planned lead times can often become self-fulfilling prophecies.

We regard the determination of the planned lead times to be a critical tactical decision as these parameters dictate the local tactics for dealing with uncertainty within a series of process steps. At each process step, a longer planned lead time translates into using more WIP for damping the effects from demand and process uncertainty; in contrast, a shorter planned lead time requires more flexible capacity as the means to handle the uncertainty. We think there is a great opportunity for developing decision support to help planners in understanding the trade-offs and in setting these parameters in a more scientific way. Hollywood (2000) and Teo (2006) provide model developments of one line of approach to finding the planned lead times, based on the framework from Graves (1986). However, this work requires assumptions that might not apply to every setting. We expect that there would be great value to practice from a more concerted effort on this problem domain.

An alternative approach is to replace the planned lead times with load-dependent lead times. This would entail a significant modification to current planning systems; yet this could result in a more accurate formulation of the planning problem. One challenge here is to model the relationship between the work load at a production step and its lead time, capturing the congestion effects, and supply uncertainties.

Another challenge is then to incorporate this relationship into a planning model. Asmundsson et al. (2006) and Ravindran et al. (2008) provide viable approaches to these challenges and establish this as a promising avenue for future research and development.

In summary, we have identified five tactical decisions for handling uncertainty in the context of production planning. We have discussed each of these in terms of the trade-offs and considerations and pointed out opportunities for developing more explicit approaches for making these decisions. We contend that getting these decisions right presents a huge opportunity for improvement to the current practice of production planning.

Acknowledgment The preparation of this chapter has been supported in part by the MIT Leaders for Manufacturing Program, a partnership between MIT and global manufacturing firms; and by the Singapore-MIT Alliance, an engineering education and research collaboration among the National University of Singapore, Nanyang Technological University, and MIT. The author acknowledges and thanks the anonymous reviewer and the editors for their helpful comments that have improved the chapter.

References

Asmundsson JM, Rardin RL, Uzsoy R (2006) Tractable nonlinear production planning models for semiconductor wafer fabrication facilities. IEEE Trans Semicond Manufact 19:95–111

Bitran GR, Haas EA, Matsuo H (1986) Production planning of style goods with high set-up costs and forecast revisions. Oper Res 34:226–236

Bitran GR, Tirupati D (1993) Hierarchical production planning. In: Graves SC, Rinnooy Kan AHG, Zipkin PH (eds) Logistics of production and inventory. Handbooks in operations research and management science, vol. 4. Elsevier, Amsterdam, pp. 523–568

Escudero LF, Kamesam PV, King AJ, et al (1993) Production planning via scenario modeling. Ann Oper Res 43(1–4):311–335

Fleischmann B, Meyr H (2003) Planning hierarchy, modeling and advanced planning systems. In: de Kok AG, Graves SC (eds) Handbooks in operations research and management science. Supply chain management: design, coordination and operation, vol. 11, Elsevier, Amsterdam, pp. 457–523

Gallego G, Özer Ö (2001) Integrating replenishment decisions with advance demand information. Manage Sci 47(10):1344–1360

Graves SC (1986) A tactical planning model for a job shop. Oper Res 34:522–533

Graves SC (2002) Manufacturing planning and control. In: Pardalos P, Resende M (eds) Handbook of applied optimization. Oxford University Press, New York, pp. 728–746

Gfrerer H, Zapfel G (1995) Hierarchical model for production planning in the case of uncertain demand. Eur J Oper Res 86:142–161

Graves, SC, Gutierrez C, Pulwer M, Sidhu H, Weihs G (1996) Optimizing Monsanto's supply chain under uncertain demand. Annual conference proceedings – council of logistics management, Orlando FL, pp. 501–516

Graves SC, Meal HC, Dasu S, Qiu Y (1986) Two-stage production planning in a dynamic environment. In: Axsater S, Schneeweiss Ch, Silver E (eds) Lecture notes in economics and mathematical systems, multi-stage production planning and inventory control, vol. 266. Springer, Berlin, pp 9–43

Graves SC, Kletter DB, Hetzel WB (1998) A dynamic model for requirements planning with application to supply chain optimization. Oper Res 46:S35–S49

Graves SC, Willems SP (2000) Optimizing strategic safety stock placement in supply chains. Manufact Serv Oper Manage 2 68–83

Hax AC, Meal HC (1975) Hierarchical integration of production planning and scheduling. In Geisler MA (ed) Studies in management sciences. Logistics, vol. 1, Elsevier, New York, pp. 53–69

Heath DC, Jackson PL (1994) Modeling the evolution of demand forecasts with application to safety stock analysis in production/distribution systems. IIE Trans 26(3):17–30

Hollywood JS (2000) Performance evaluation and optimization models for processing networks with queue-dependent production quantities. PhD Thesis, MIT Operations Research Center, Cambridge, MA

Holt CC, Modigliani F, Muth JF, Simon HA (1960) Planning production, inventories and work force. Prentice-Hall, Englewood Cliffs, NJ

Hopp WJ, Spearman ML (2007) Factory physics: foundations of manufacturing management, 3rd edn. Irwin/McGraw-Hill, Burr Ridge, IL

Johnson JD (2005) Managing variability in the semiconductor supply chain, MS Thesis, MIT Engineering Systems Division, Cambridge, MA

Lasserre JB, Merce C (1990) Robust hierarchical production planning under uncertainty. Ann Oper Res 26:73–87

Miller MP (1997) Business system improvement through recognition of process variability, MS Thesis, MIT Leaders for Manufacturing Program, Cambridge, MA

Muharremoglu A, Yang N (2010) Inventory management with and exogenous supply process. Oper Res 58(1):111–129

Mula J, Poler R, García-Sabater JP, Lario FC (2006) Models for production planning under uncertainty: a review. Int J Prod Econ 103:271–285

Nahmias S (2008) Production and operations analysis, 6th edn. Irwin/McGraw-Hill, Boston

Ravindran A, Kempf KG, Uzsoy R (2008) Production planning with load-dependent lead times and safety stocks. Research report. In: Fitts EP (ed) Department of industrial and systems engineering, North Carolina State University

Sahney MK (2005) Building operational excellence in a multi-node supply chain, MS Thesis, MIT Leaders for Manufacturing Program, Cambridge, MA

Schoenmeyr T, Graves SC (2008) Strategic safety stocks in supply chains with evolving forecasts. Manufact Serv Oper Manage 11(4):657–673

Shapiro JF (1993) Mathematical programming models and methods for production planning and scheduling. In: Graves SC, Rinnooy Kan AHG, Zipkin PH (eds) Handbooks in operations research and management science. Logistics of production and inventory, vol. 4, Elsevier, Amsterdam, pp. 371–443

Silver EA, Pyke DF, Peterson R (1998) Inventory management and production planning and scheduling, 3rd edn. Wiley, USA

Teo CC (2006) A tactical planning model for make-to-order environment under demand uncertainty PhD Thesis, NTU

Thomas LJ, McClain JO (1993) An overview of production planning. In: Graves SC, Rinnooy Kan AHG, Zipkin PH (eds) Handbooks in operations research and management science. Logistics of production and inventory, vol. 4. Elsevier, Amsterdam, pp. 333–370

Vollman TE Berry WL Whybark DC Jacobs FR (2004) Manufacturing planning and control systems 5th edn. McGraw-Hill, New York

Chapter 6
Demand Forecasting Problems in Production Planning

Jonathan R.M. Hosking

6.1 Introduction

A recent survey of 247 senior finance executives (CFO Research Services, 2003) found that "accurately forecasting demand" was the most commonly occurring problem in their companies' supply chain management. Forecasting is recognized as a hard problem. "It is difficult to predict, especially the future," according to a quotation attributed to Niels Bohr (among many others). To forecast demand, a quantity that can be difficult to measure and even to define, is particularly challenging, even in the simplest case of forecasting demand for a single product. Yet demand forecasts are essential for production planning. Any manager who makes a decision to produce a particular quantity of a product is using, explicitly or implicitly, a forecast of the demand for the product.

The general problem that we consider here is common in large industrial enterprises. To make production planning decisions, demand forecasts are required for both individual products and groups of products, perhaps hundreds or thousands of products in all. Demand forecasts are principally derived from data on product sales, in units of products or as revenue amounts, which are typically available in weekly or monthly time buckets (though other time granularities are possible). Forecasts are required some time in advance of the time period whose demand is being forecast: this "lead time" may vary from a few days to a few months or even years. The shorter lead times are relevant for setting manufacturing schedules or planning sales efforts; longer lead times for planning product lifecycles and making decisions to close production facilities or open new ones. Forecasts are made or revised within a planning cycle that involves regular meetings of executives or operations managers with responsibilities for production and marketing. The planning cycle attempts to reach consensus estimates of future demand, taking into account all relevant information: extrapolations of historical demand and sales; historical effects of pricing and promotion actions, and prospects of future such actions; introductions of new

J.R.M. Hosking (✉)
IBM Research Division, T. J. Watson Research Center, 1101 Kitchawan Road, Yorktown Heights, NY 10598, USA
e-mail: hosking@watson.ibm.com

K.G. Kempf et al. (eds.), *Planning Production and Inventories in the Extended Enterprise*, International Series in Operations Research & Management Science 151,
DOI 10.1007/978-1-4419-6485-4_6,

products and phasing out of old products; identification of product and customer groups with perceived changes in level of demand; production capacity and supply constraints; and any other relevant data or judgements. As part of this cycle, forecasts must be updated at regular intervals, typically weekly or monthly. Forecasts may be made as single (point) values or as ranges of plausible values, depending on the requirements of subsequent stages of the production planning process.

Within this general forecasting process many problems can arise. The following sections each describe a class of problems that may need to be considered. The problems can be classified into three main groups.

First there are basic questions of definition. What is demand, the elusive quantity that we are trying to forecast? What forecasts are needed to support the production planning process? How should we measure the accuracy of the forecasts, and the overall success of the entire forecasting system?

Some fundamental problems arise from the nature of the demand data itself. The data structure, often a hierarchy reflecting the relations between different products, must be accounted for when making forecasts. Demand or sales data may have particular features that make forecasting difficult, such as intermittent demand or short product lifecycles.

Additional information beyond the raw demand data is often available, and another class of problems concerns how to make effective use of it. What are the effects of sales campaigns and price changes? Can the behavior of individual customers be modeled and used to improve forecasts? Are outside influences, such as the overall health of the economy, relevant, and can knowledge of them be used to improve forecasts?

Finally, we shall briefly consider how the nature of the overall production planning process may itself reflect the accuracy with which demand forecasts can be made.

6.2 Definition of Demand

Demand can be a difficult concept to define. A customer's buying decision is influenced by the practical suitability, aesthetic attractiveness, price of a product relative to its competitors, and the product's availability. In some markets fashion can also play a role, with a product's current popularity serving either to increase or to reduce future demand.

In practice, an appropriate proxy for demand must be found in data that can be reliably measured and regularly reported. Unit sales data are usually available, and certainly have a close relation to demand. But sales can underestimate demand when supply is constrained, and need to be interpreted together with the price at which the product was sold. Forecasting demand by extrapolating past sales also fails to account for changes in demand caused by the actions of competitors, such as the introduction, upgrading or withdrawal of a competitive product. Considerable caution is therefore required when treating sales as a proxy for demand.

Customer orders are available for some types of products, and can overcome some of the supply-related issues that affect sales data. If an order is made with a request for delivery in January but the product was not be delivered until March, it is reasonable to assert that the demand occurred in January, rather than on the March date that would be inferred from the sales data.

In some industries data on how long a product is offered for sale can be used as a proxy for demand. In automotive sales, for example, the number of days that a vehicle spends on the dealer's lot before being sold is arguably an indicator of the strength of demand for that vehicle's particular configuration (model type, engine size, and attached options).

A further question is whether demand can be treated simply as an observed quantity, or whether the producer's own actions need to be allowed for. Most forecasting algorithms treat the quantity being forecast as a natural phenomenon that evolves independently of outside influences. Demand, in contrast, is something that producers actively try to influence, for example by changing prices or by advertising. The effects of these actions are often poorly understood in quantitative terms. Allowing for them in an objective forecasting procedure is therefore problematical, but can be very necessary: sales promotions, and customers' expectations of future sales promotions, can have a major impact on the pattern of demand for a product.

Forecasts based on extrapolation of past behavior require a measure of demand that is uniform over time, after allowing for outside influences such as sales promotions, other price changes, and the actions of competitors. At least four approaches can be used. First, observed demand, or a simple proxy for it such as sales, can be forecast in isolation, with outside influences regarded as part of a "business as usual" environment that will be similar in the future. This approach can be effective if outside interventions are small or follow a regular pattern, but is vulnerable to the occurrence of large infrequent events, which can exert undue influence on the forecasts. Second, the forecaster may try to estimate some kind of "baseline" demand from which external influences have been removed, for example, peaks attributed to sales campaigns have been smoothed away. This may not be straightforward to do, and the concept of "baseline demand" may not be meaningful in an environment in which demand is actively managed. Third, industry experts might be able to specify the direction of future sales. This can be incorporated into a formal system such as the "rule-based forecasting" of Collopy and Armstrong (1992). Finally, the forecaster can use a model that explicitly models the effects of interventions. This requires that the effects be well understood, that the timing and nature of future interventions can be predicted, and that the models of their effects will remain valid through the forecasting horizon.

6.3 Requirements for the Forecasts

Forecasts are made to support business decisions and may be used for many purposes. Major uses include production scheduling, planning the ordering of parts from suppliers, inventory management, marketing, and plant and equipment decisions. Each of these applications places different requirements on the forecasts, in terms of the level of accuracy, degree of detail, and lead time that may be needed. Devising a forecasting system that can satisfy all these requirements is challenging. Yet the alternative of having forecasts made independently by each business unit for its own purposes is likely to be worse, risking duplication of effort, inconsistency, and inefficiency.

What kind of forecasts should be provided? A single number, or point forecast, is still probably the most common kind of forecast. An interval forecast provides more information, in the form of upper and lower bounds on the likely future values of the quantity being forecast. To be useful in quantitative analyses, these bounds should have formal probability interpretations, as values that will be exceeded with specified probability. For example, the upper and lower bounds might correspond to exceedance probabilities of 25% and 75%, or 10% and 90%. The interval between the bounds is then a prediction interval for the quantity being forecast (the difference between prediction intervals and other statistical intervals such as confidence intervals is discussed by Vardeman (1992) and Hahn and Meeker (1991, Sect.2.3). More generally, forecasts can be made for several exceedance probabilities. For example, exceedance probabilities of 10%, 50%, and 90%, corresponding respectively to optimistic, typical, and pessimistic outlooks, may adequately summarize the forecaster's knowledge. Finally, a predictive distribution is a complete specification of the forecaster's overall belief. It takes the form of a probability distribution that specifies for each level of demand the probability that it will reached; it can also be regarded as specifying a set of forecasts for every exceedance probability between 0 and 1. Figure 6.1 illustrates a point forecast, an interval forecast, and a predictive distribution in a particular forecasting situation.

As inputs to human decisions, point and interval forecasts may be as much information as can be understood and effectively used. In more automated systems, use of predictive distributions provides the most information and enables the most

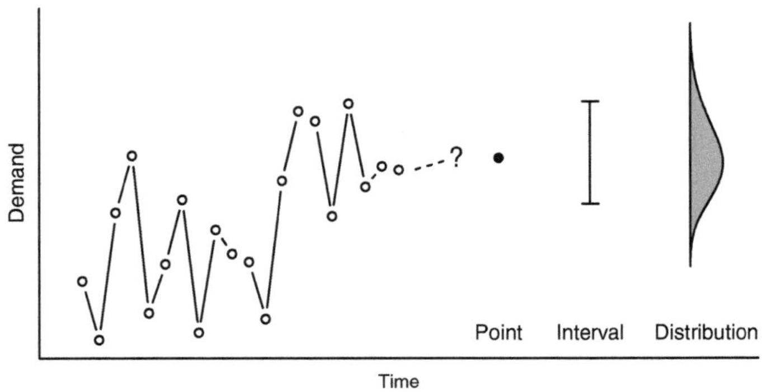

Fig. 6.1 Point and interval forecasts, and a predictive distribution, in a particular forecasting situation. For the predictive distribution, the width of the shaded area at any value y on the vertical axis is proportional to the forecast probability that demand will be in a narrow interval containing y

efficient decisions. But predictive distributions bring additional issues. It seems reasonable to require that forecast accuracy should be measured in such a way that a forecaster who seeks to achieve maximal accuracy must give an honest representation of his or her belief: in other words, the accuracy measure must be a "proper scoring rule," as discussed by Gneiting and Raftery (2007). Specifying predictive distributions for multiple forecasts simultaneously brings further problems of computational effort. For n products one must forecast not just n demands but also the $\frac{1}{2}n(n-1)$ correlations between them. To reduce the problem to a manageable size, simplification of the correlation structure may be essential. Graphical models such as Markov networks can be used for this purpose (Gebhardt et al. 2003).

A further consideration is whether a single forecast is adequate; the possibility of qualitatively different alternative future states may be entertained (the price of oil jumps, a large customer is gained or lost, etc.) with separate forecasts provided for each alternative. It may be important to subdivide the possible patterns of future events in this fashion, particularly if some scenarios can affect demand for many products simultaneously.

In general forecasts are needed at a range of lead times. Recall that lead time is the length of the interval between the time at which the forecast is made and the end of the time period for which demand is being forecast. When the range is wide, which is the typical case, further problems arise. It may be necessary to use fundamentally different approaches for forecasting at different lead times. As a simple example, extrapolation of a linear trend may be adequate for a short-term forecast, but its suitability for long-term forecasting requires an assessment of how long the trend might continue. Separate forecasts may be made for short- and long-range forecasting, with the overall forecast being a weighted combination of the two where the short-term (long-term) forecast is given more weight at short (long) lead times (Armstrong et al. 2001).

An important factor when making business decisions is the stability of forecasts over time. Changes to a production schedule can be disruptive, and it is undesirable to have these changes driven by fluctuations in forecasts that reflect no more than random noise. There may therefore be a need to control fluctuations in forecasts, particularly the variation in forecasts for a fixed target time point in the future as the lead time decreases. The pattern of such fluctuations can often be modeled by the "martingale model of forecast evolution" (Heath and Jackson 1994); Toktay and Wein (2001) used this approach to derive a production policy that minimizes inventory-holding and back-order costs. This is particularly a problem with point forecasts. If interval forecasts or predictive distributions are provided and are properly calibrated, the fluctuations between successive forecasts can be compared with the width of the interval forecast or the dispersion of the predictive distribution to give an indication of the business significance of the fluctuations.

6.4 Complex Data Structures

Demand forecasts are typically required for many products simultaneously. Demands for different products can be correlated, and the correlation can be positive (similar products have similar demand patterns) or negative (substitution of nearly interchangeable products). A key question is whether to forecast demand for separate products independently, or to allow for dependence.

In a large organization, products are typically arranged in a hierarchy. The hierarchy may be based on the nature of the product (e.g., materials, technology used), the purpose of the product (e.g., intended customer set), marketing considerations (e.g., customers classified by industry group), or administrative considerations (e.g., management responsibilities). Several hierarchies may be active simultaneously,

As an example, Heching et al. (2004) describe a system for forecasting demand for IBM semiconductor products. In this application the products are arranged in a hierarchy with each node in the hierarchy corresponding to the business responsibility of a manager or executive. Products are also grouped into families by technology or by the purpose for which the products are intended: there are several dozen of these product families. Forecasts are provided within a set of hierarchies, one for each product family. In each hierarchy, the top level is the product family, intermediate levels correspond to the business line managers' responsibilities, the penultimate level contains the individual products (part numbers), and the lowest level corresponds to combinations of part number and customer. A typical hierarchy has 8 levels and 500 nodes.

Forecasts may be required at several levels of the hierarchy. There is usually a requirement that forecasts at different levels be consistent. For point forecasts this means that the forecast at a parent node of the hierarchy must be equal to the sum of the forecasts at its child nodes. For predictive distributions the requirement is that the distributions of forecast demand for any combination of products must be mutually consistent.

Consistency can be achieved by making forecasts at the lowest level of the hierarchy and aggregating them to higher levels (the "bottom-up" approach). The author's experience is that directly forecasting demand at some higher level of the hierarchy typically gives, at that level, more accurate forecasts than those obtained from aggregating low-level forecasts; however, some researchers have reached the opposite conclusion (Armstrong 2006). Whether direct or aggregated forecasts are more accurate is likely to depend on the pattern of variation within the data. Forecasts made directly at high levels of aggregation are likely to be more accurate when the patterns in the data that can be used as the basis for forecasting are more apparent at higher levels of aggregation, for example, variation over time is smaller and seasonal variations are more regular. In addition, the effects of external influences such as the overall strength of the economy may be more easily modeled at higher levels of aggregation. When this is required, forecasts are most naturally made at high levels of aggregation and must be disaggregated to give forecasts at lower levels of the hierarchy (the "top-down" approach). In another approach, forecasts are made independently at different levels of the hierarchy and are then modified to achieve consistency, for example, by multiplying all of the forecasts at a lower level of the hierarchy to make their sum equal to the forecast at a higher level. The choice of the best level or levels at which to make the forecasts, and of an appropriate method of imposing consistency on the forecasts, are difficult issues for which few general guidelines exist. One possibility is to make forecasts at different levels, aggregate or disaggregate them to the level of interest, and combine them using the methods discussed in Sect. 6.6.

Hierarchies can be dynamic: new products or technologies are introduced, old products are retired, products are moved from one branch of the hierarchy to another, and customers merge or spin off. Through all of these activities, forecasts must be consistent over time. This can be a problem, particularly for forecasts at higher levels of aggregation, where the meaning of the higher-level data can change over time. It may be necessary to restate historical data in terms of a new alignment of products, in order to achieve consistency of forecasts.

6.5 Measurement of Forecast Quality

The aim of a forecasting system is to provide accurate forecasts, but deciding on how to measure accuracy, or judging whether one forecasting method is better than another, is already a difficult problem. Many forecast accuracy measures have been proposed. Surveys have been given by Armstrong (1985, Chap. 13), Armstrong and Collopy (1992), and Hyndman and Koehler (2006).

For a simple case consider a set of point forecasts F_i, $i = 1, \ldots, n$, and the corresponding actual values A_i, $i = 1, \ldots, n$. Commonly used measures of forecast accuracy include the mean absolute error (MAE),

$$\text{MAE} = n^{-1} \sum_{i=1}^{n} |F_i - A_i| \,, \tag{6.1}$$

and the mean absolute relative error (MARE),

$$\text{MARE} = n^{-1} \sum_{i=1}^{n} |F_i - A_i| / A_i \,, \tag{6.2}$$

often expressed as a percentage and called the mean absolute percentage error (MAPE). These measures can be problematical in practice. Forecast errors tend to be larger for products whose overall level of demand is large, and when combining data for a range of products the MAE value is often dominated by contributions from a few high-volume products. In extreme cases use of MAE as the accuracy criterion can imply that the best approach to forecasting is to concentrate on making accurate forecasts for the highest-volume products and set the demand forecasts to zero for all other products. This is a counterintuitive conclusion that is unlikely to be popular with production managers or successful in practice.

MARE, in contrast, gives high weight to the forecast performance for low-volume products, and can be unduly influenced by poor forecasts of products with minimal economic significance. Because of its asymmetry – a high forecast when the actual is low is more heavily penalized than a low forecast when the actual is high – a policy of minimizing MARE can lead to systematic underforecasting.

More elaborate accuracy measures can be useful. A symmetrized form of MARE,

$$\text{SMARE} = n^{-1} \sum_{i=1}^{n} L(F, A) \tag{6.3}$$

where

$$L(F, A) = \begin{cases} 0 & \text{if } F = A = 0, \\ |F - A| / \{\tfrac{1}{2}(A + F)\} & \text{otherwise}, \end{cases} \tag{6.4}$$

aims to correct the different weighting given to high and low forecast errors by MAE and MARE. SMARE, called the adjusted MAPE by Armstrong (1985), gives equal weight to underforecasting and overforecasting by the same factor – for example, the penalty is the same for a forecast that is half the actual value and one that is twice the actual value – and has an upper bound on the contribution from any single forecast. Heching et al. (2004) found that SMARE worked well in practice in a complex demand forecasting problem with hierarchical data. However, explaining SMARE, or any other accuracy measure that involves nonlinear transformations, to nontechnical executives can be a challenge.

The foregoing measures assess forecasting accuracy for a single data series. Other measures may be more useful when comparing different forecasting methods across multiple data series with the aim of choosing an overall "best" method. Here it can be important to allow for the different degrees of difficulty in forecasting the different series. The relative absolute error

$$\text{RAE} = |F - A| \,/\, |F^{\text{B}} - A|$$

measures the accuracy of a forecast F compared with a baseline forecast F^{B} that might for example be a random-walk forecast (forecast for all lead times is the most recent observed value). RAE is a positive quantity, with values less than 1 indicating that forecast F is more accurate than the baseline forecast. Because RAE can take extreme values, it is often appropriate to truncate or "Winsorize" its values, replacing values larger than some value C (say $C = 10$ or $C = 100$) by C itself and values smaller than $1/C$ by $1/C$ (e.g., Armstrong et al. 2001, p. 471). To combine RAE values across different series, it is appropriate to use the geometric mean of the Winsorized RAE values,

$$\text{GMRAE} = \left(\prod_{m=1}^{M} \text{RAE}_m \right)^{1/M},$$

where RAE_m is the Winsorized RAE of the mth series, $m = 1, \ldots, M$.

Forecast accuracy measures for interval forecasts or predictive distributions have not been much studied. In principle, given a loss function $L(F, A)$ for specific values F and A of forecast and actual demand respectively, a loss measure for a predictive distribution is given by the expected value of $L(F, A)$ over the distribution of F. This and other approaches are discussed in the survey of scoring rules by Gneiting and Raftery (2007).

The choice of forecast accuracy measure is affected by the use to which the forecast is put (as in Sect. 6.3). The accuracy measure may need to accommodate multiple objectives. Choice of an accuracy measure may also be affected by the nature of the product. For example, one would expect to use different criteria for perishable as opposed to long-lived products, reflecting the speed at which the value of unsold product depreciates.

Further issues arise when forecasting is considered as part of the business process. Forecasts may become self-realizing: once a demand forecast has been shown to high-level executives, it may become a sales target and a performance criterion for managers, and the firm will do whatever it takes to meet the forecast, even by shrinking or eliminating profit margins. By evaluating forecast performance as part of a model of the business, the forecast accuracy measure can be chosen explicitly to reflect business objectives. This approach has been used to choose from among competing forecast methods so as to maximize profitability (Price and Sharp 1986) or to minimize inventory costs for a given service level (Gardner 1990).

Forecasts are typically made sequentially, and the importance of the patterns of successive forecast errors should not be overlooked. Table 6.1 gives an example, showing actual demand for a product in 2 consecutive months, and two sets of forecasts. Suppose that in two consecutive periods the actual demands for a product are 10 and 50 units. Forecast A has the correct overall amount of demand, but places most of it in the wrong time period. Forecast B has lower MAE, but underestimates the total demand. Which forecast is more useful may well depend on the business requirements (e.g., whether items produced in one month can be profitably kept in inventory for sale in a later month) or on the market environment (e.g., whether demand that is unsatisfied in one month will still be present the following month or whether consumers will have bought a competitor's product instead).

Although ideally the forecast accuracy measure should be attuned to the business requirements, it can be difficult to establish exactly what is required of a good forecast. Perhaps the best approach is, so far as is possible, to set up the forecast as an unbiased estimate of actual demand and let other stages of the decision-making procedure account for business considerations such as whether overforecasting is more costly than underforecasting or vice versa. There is still the issue of spreading

Table 6.1 Illustrative example of sequential forecast errors. Which forecast is more accurate?

	January	February
Actual demand	10	50
Forecast A	50	10
Forecast B	10	10

awareness throughout the business process of what the forecasts can and cannot be expected to do. Acknowledgement of the limitations of the accuracy of the forecasts, by provision of interval forecasts or predictive distributions, should be helpful in this respect.

6.6 Combining Multiple Forecasts

For demand of any product or group of products over some time period, several forecasts may be available. These may include automatic forecasts made by different methods (exponential smoothing, time series models, and regression on explanatory variables), forecasts that use different assumptions (no trend, linear trend, and damped trend), judgemental forecasts by different people, etc. When a single overall forecast is needed, the question arises of how best to combine the individual forecasts.

One possibility is to select the best forecast method based on its historical performance according to some criterion, such as GMRAE defined in Section 6.5, and use this as the final forecast. Alternatively, forecasts can be combined, typically by taking a weighted average of the individual forecasts. This approach is popular and has been surveyed by Clemen (1989) and Armstrong (2001, pp. 417–435). Armstrong (2001, p.428) summarized 30 studies of forecast combination and found that combining forecasts reduced the forecast error in each case, by an average of 12.5%. Even a simple average of all of the forecasts often performs better than the best individual forecast. When there is good evidence that one method has been more accurate than another in the past it is better to give it more weight. Schmittlein et al. (1990) quantify this effect in some simple cases (forecast errors with identical normal distributions in each time period), showing for example that weights estimated from the accuracy of ten historical forecasts are better than using equal weights when the forecast accuracies differ by more than 20%.

6.7 Irregularities in Historical Sales Data

Demand forecasts often based on extrapolation of historical sales data. But sales are a censored version of demand: observed sales are the smaller of demand and supply. This is clearly a problem for the production of demand forecasts, and its resolution to some extent depends on what happens to unsatisfied demand. There are several possibilities: it may be transferred to competing products, either the producer's own or a competitor's; it may be carried over, as a whole or in part, to next time period; or it may vanish entirely. In each case the requirements that this lays on demand forecasts, and the consequent requirements on production planning, are not well understood.

One possible response to the censored-demand problem is to set the level of supply in such a way as to obtain information about unmet demand. Demand forecasting and supply setting then become combined in a dynamic decision-making problem. Ding et al. (2002) showed that when demand is censored the optimal inventory level is larger than would be the case if demand were known even when it cannot be met, and explain this as the cost of acquiring information about the distribution of demand. Tan and Karabati (2004) derived a policy for updating the optimal inventory level as observations of censored demand arrive sequentially.

Even without censoring, data issues may make forecasting difficult. Simple forecasting methods are based on extrapolation of historical data, and require some regularity in the data. Some of the problems encountered in time series of sales or estimated demand include high variability, intermittent demand, short product lifecycles, and irregularities in trend or seasonality. Figure 6.2 gives some typical examples from a demand forecasting project in the semiconductor industry (Heching et al. 2004). These problems are particularly acute for products with few customers, for which demand may be dependent on the customers' particular buying patterns.

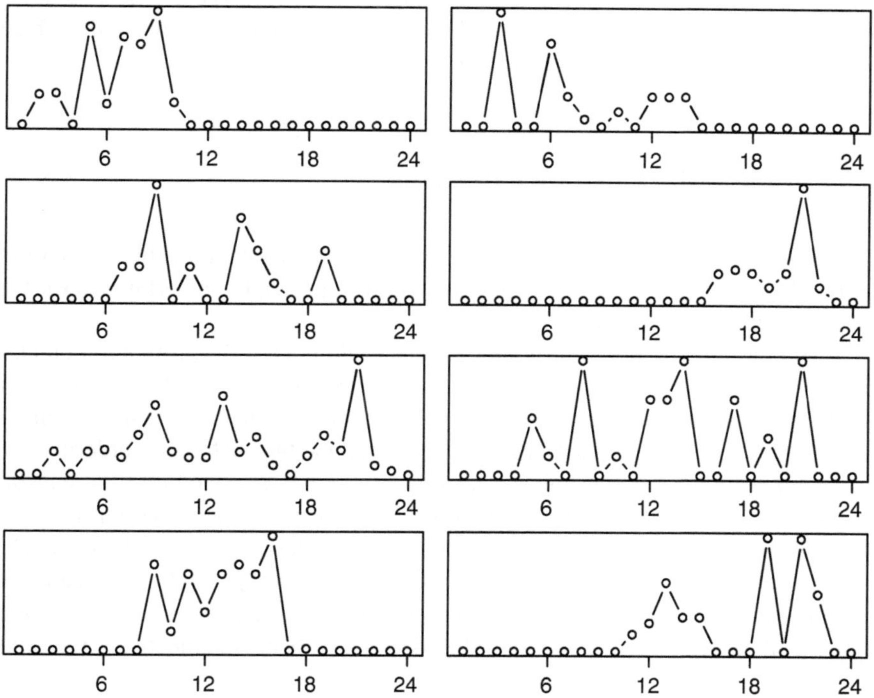

Fig. 6.2 Monthly demand for some typical IBM semiconductor products

6.8 Interventions in the Sales Process

As noted in Sect. 6.2, most forecasting algorithms treat the quantity being forecast as a natural phenomenon that evolves independently of outside influences. Producers, however, actively try to influence demand for their products, and allowing for these influences is another problem in demand forecasting.

The basic problem is irregularity in the time series of demand resulting from nonuniformity of marketing effort. Specific sources of nonuniformity include sales campaigns, other price changes, and attempts to achieve sales targets (corporate or individual) by designated dates. Supply constraints could be considered here too. Ideally, these effects should be removed from historical data and allowed for when making forecasts. Blattberg and Levin (1987) developed a regression-based model of the effectiveness and profitability of trade promotions, incorporating reactions of both retailers and consumers to temporary price reductions offered by producers to retailers.

Dynamic effects are also important. The effect of a temporary price reduction of 10% will have a different effect on demand if it is a one-time event as opposed to one in a regular sequence of promotions. Kopalle et al. (1999) describe a model of the dynamic effect of discounting on sales, which can be used to estimate the effects of past promotions and to optimize future promotions.

6.9 Forecasting Customer Behavior

Demand for a product reflects the intentions of the product's potential purchasers, and understanding their behavior offers, at least in principle, opportunities for improving demand forecasts.

When the identity of the customer for each sale is known, information on customer behavior can potentially be used for demand forecasting. This information can be particularly valuable if a few customers account for majority of sales. General information about a customer's behavior may include the customer's ordering patterns: for example, the customer may place orders at regular intervals, causing seasonality in demand for particular products. Specific information is often available through marketing channels, when members of the sales force are in direct contact with the customer and may be able to assess, at least subjectively, the likelihood that the customer will make a particular purchase. How reliable these estimates are, and how they can be effectively introduced into an otherwise objective forecasting procedure, are challenging questions.

The spread of the internet has made large-scale data on customer purchasing intentions and histories much easier to obtain. Retail locations can now collect details of customer purchases, for example, as part of customer loyalty programs, and this information may potentially be available to producers. Some information may be directly obtained by producers from customer visits to their web sites. In the automotive industry, for example, producers can obtain information about vehicle configurations inquired about by web site visitors, and estimates are available of the customer's propensity to buy, based on the sequence of web page downloads. How best to make use of this kind of information to predict future demand is still an open question.

6.10 Use of External Information

Demand for a product can be influenced by many things other than the product itself and the producer's actions. Some of these include the general state of the economy, the actions, observed or anticipated, of customers, and the actions, observed or anticipated, of competitors. In some cases, data can be obtained that measures, directly or indirectly, these external influences. These data sources are potentially useful for demand forecasting, but making effective use of them is rarely straightforward. Many sources may be potentially available. Their relevance and reliability must be judged. Qualitative judgements about external data must be converted into quantitative changes in forecasts.

As an example, a common belief is that demand for products is affected by the strength of the overall economy, or of particular industries, or of particular customers. Making use of this information can be difficult in practice. The ideal is to find a leading indicator, such that changes in the indicator translate into changes in demand at some future time that is useful for forecasting. Otherwise, even if a relationship between an external indicator and demand can be found, there is the need to forecast the indicator itself in order to obtain forecasts of demand. Another problem can be nonstationarity in the relationships: for example, a relationship between an economic variable and demand for a product may have a markedly different pattern when the economy is in a period of recession rather than growth.

6.11 Concluding Remarks

We have seen that many problems arise in demand forecasting. Some of these are well known, but still difficult to deal with; others are frequently encountered in practice but are rarely considered in the academic literature, even in such established texts as Makridakis et al. (1998) and Armstrong (2001). Issues that in the author's opinion fall in this latter category are controlling fluctuations in forecasts, allowing for timing effects in accuracy measurement, and allowing for the censoring of demand by available supply.

Finally, we note that there are circumstances in which demand forecasting can be reduced in importance. This can be advantageous when forecast accuracy is hard to achieve. One case is when the producer can actively manage demand. When the producer's actions, by controlling supply, changing prices, and changing the level of marketing, have a measurable and well-understood effect on demand, the problem of demand can be as much one of control as of forecasting. Another case arises when considering how the accuracy with which demand can be forecast can affect other parts of the production planning system. If demand cannot be forecast accurately, an effective response is to design a production planning process that is robust to inaccurate demand forecasts. Aviv and Federgruen (2001) recommend delayed product differentiation and quick response as production strategies when demand forecasts are inaccurate. Reduction in manufacturing lead times, just-in-time manufacturing, and the use of configure-to-order rather than build-to-plan production schedules can also reduce the dependence on accurate demand forecasts. To achieve a sufficiently robust production-planning environment may be the most effective way of dealing with demand forecasting problems.

References

Armstrong JS (1985) Long-range forecasting, 2nd edn. Wiley, New York.
Armstrong JS (ed.) (2001) Principles of forecasting: a handbook for researchers and practitioners. Kluwer Academic Publishers, Boston, Mass.
Armstrong JS (2006) Findings from evidence-based forecasting methods for reducing forecast error. Int J Forecast 22:583–598.
Armstrong JS, Adya M, Collopy F (2001) Rule-based forecasting: using judgment in time-series extrapolation. In Armstrong (2001), pp. 259–282.
Armstrong JS, Collopy F (1992) Error measures for generalizing about forecasting methods: empirical comparisons. Int J Forecast 8:69–80.
Aviv Y, Federgruen A (2001) Design for postponement: a comprehensive characterization of its benefits under unknown demand distributions. Oper Res 49:578–598.
Blattberg RC, Levin A (1987) Modelling the effectiveness and profitability of trade promotions. Mark Sci 6:124–146.
CFO Research Services (2003) CFOs and the supply chain. CFO Publishing Corp., Boston, Mass. http://www.ups-scs.com/solutions/documents/research_CFO.pdf.
Clemen RT (1989) Combining forecasts: a review and annotated bibliography. Int J Forecast 5:559–583.
Collopy F, Armstrong JS (1992) Rule-based forecasting: development and validation of an expert systems approach to combining time series extrapolations. Manage Sci 38:1394–1414.
Ding X, Puterman ML, Bisi A (2002) The censored newsvendor and the optimal acquisition of information. Oper Res 50: 517–527.
Gardner ES, Jr (1990) Evaluating forecast performance in an inventory control system. Manage Sci 36:490–499.
Gebhardt J, Detmer H, Madsen AL (2003) Predicting parts demand in the automotive industry: an application of probabilistic graphical models. In Uncertainty in Artificial Intelligence: Proceedings of the 19th Conference (UAI-2003). Morgan-Kaufman, San Francisco.
Gneiting T, Raftery AE (2007) Strictly proper scoring rules, prediction, and estimation. J Am Stat Assoc 102:359–378.
Hahn G, Meeker W (1991) Statistical intervals: a guide for practitioners. Wiley, New York.
Heath DC, Jackson PL. (1994). Modeling the evolution of demand forecasts with application to safety stock analysis in production/distribution systems. IIE Trans 26:17–30.
Heching AR, Hosking JRM, Leung YT (2004) Forecasting demand for IBM semiconductor products. Research Report RC23074, IBM Research Division, Yorktown Heights, New York.
Hyndman RJ, Koehler AB. (2006) Another look at measures of forecast accuracy. Int J Forecast 22:679–688.

Kopalle PK, Mela CF, Marsh L (1999) The dynamic effect of discounting on sales: empirical analysis and normative pricing implications. Mark Sci 18:317–332.

Makridakis SG, Wheelwright SC, Hyndman RJ. (1998) Forecasting: Methods and Applications 3rd edn. Wiley, New York.

Price DHR, Sharp JA (1986) A comparison of the performance of different univariate forecasting methods in a model of capacity acquisition in UK electricity supply. Int J Forecast 2:333–348.

Schmittlein DC, Kim J, Morrison DG. (1990) Combining forecasts: operational adjustments to theoretically optimal rules. Manage Sci 36:1044–1056.

Tan B, Karabati S (2004) Can the desired service level be achieved when the demand and lost sales are unobserved? IIE Trans 36:345–358.

Toktay LB, Wein LM (2001) Analysis of a forecasting-production-inventory system with stationary demand. Manage Sci 47:1268–1281.

Vardeman SB (1992) What about the other intervals? Am Stat 46:193–197.

Chapter 7
Production Capacity: Its Bases, Functions and Measurement

Salah E. Elmaghraby

7.1 What is "Capacity"?

The word "capacity" invokes in the minds of most people the notion of "capability" – this is what people usually understand when one states that "the capacity of the school bus is 40 students" or "the capacity of the water dam is 2.5 million cubic meters". Matters become more fuzzy when the discussion relates to *productive* entities, mainly because now "capability" is intertwined with "performance" to the point where the two become one and the same in the minds of many. In fact, concern with the productive "capacity" of a worker in a factory created the whole field of *Motion and Time study* with its own vocabulary of *time standards, normal output, allowances, levelling, and rating.* Clearly, the concern here is not only so much with what a worker is *capable* of doing, but also with what the worker's *output* should be. "Capacity in the sense of capability" is transformed into "capacity in the sense of output". For a lucid historical perspective and critical evaluation of the very fundamentals of this important field of industrial engineering, you are directed to the illuminating monograph of Davidson (1956).

The same subtle metamorphosis of the meaning of capacity can be observed in many books, papers, and conference proceedings by distinguished researchers in the field of production. For instance, a recent (2005) survey by Pahl et al. (2005) of the literature on the relationship between "lead time" and the "work-in-process" (WIP) ahead of a facility; Fig. 7.1 is presented, which is a direct quotation from a paper by Karmarkar (1989). Observe that the vertical axis is labeled "capacity" while in reality it measures *productivity*. This can be easily gleaned from the fact that at very low WIP the productivity is also low, and rises steadily until the system is saturated (i.e., as the WIP grows without bounds). Evidently, the capacity of the facility, in the sense of its capability, is the same throughout the range of the WIP independently of its value and is represented by the horizontal line

S.E. Elmaghraby (✉)
North Carolina State University, Raleigh, NC, USA
e-mail: elmaghra@eos.ncsu.edu

K.G. Kempf et al. (eds.), *Planning Production and Inventories in the Extended Enterprise*, International Series in Operations Research & Management Science 151,
DOI 10.1007/978-1-4419-6485-4_7,

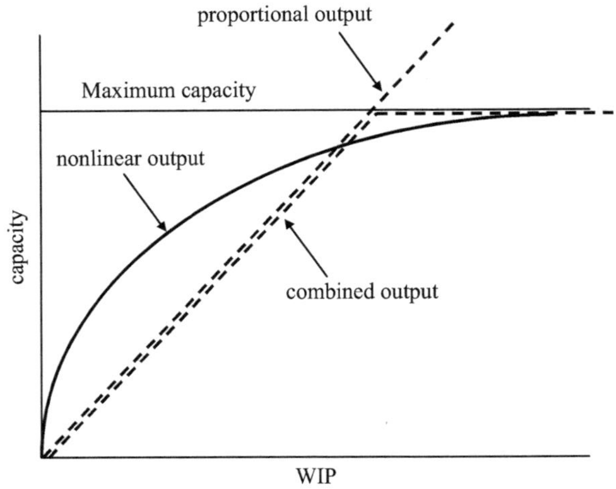

Fig. 7.1 Diverse forms of clearing functions (Karmarkar, Srinivasan et al.)

labeled "maximum capacity" – it is *productivity* that varies with WIP. The same misinterpretation of the notion of "capacity" permeates almost all the references cited by Pahl et al. that treat capacity or its related topics.

The confusion relative to the concept of capacity is the more perplexing because of its ubiquitous presence of in all operations research/industrial engineering/production engineering treatises on production or operations planning and control; see, as a small sample, the books by (arranged chronologically): Nahmias (2006), Stevenson (2005), Chase and Aquilano (1985), Manipaz (1984), Buffa (1983), Dilworth (1979), Johnson, Newell and Vergin (1974), and Johnson and Montgomery (1974). Consulting any one of these books or any of the hundreds of published papers on the subject, reveals elaborate models on "capacity planning" under a variety of assumptions on the demand pattern, the cost factors, the reliability of the facility, the length of the planning horizon, the probability distribution of any of its parameters and so forth. All researchers in the field attempt to construct a verbal description of capacity, a daunting undertaking in itself, in order to be able to use the "value" secured into the various mathematical models that are constructed for the purpose of planning and control. The paper by Elmaghraby (1991) gives direct quotations from three of the books cited above which result in the following crop of "capacities" mentioned in just these three books: "*true capacity*", "*time dimension of capacity*", "*design capacity*", "*theoretical maximum capacity*", "*effective capacity*", "*actual capacity*", and "*potential capacity*". Undoubtedly, the list would be longer if more books were quoted. The message conveyed by this collection of "capacities" is that there is indeed need to *define* what is meant by capacity, as a prerequisite for its *measurement*. As it turns out, the understanding of what is meant by "capacity" and the determination of its value are no minor feats.

Capacity is too vital an element of the production system to be ignored. Productive capacity, which has long been relegated to the obvious with little thought to its meaning or its measurement, is in need of careful study since a great deal depends on its proper understanding and correct measurement as detailed in the next section. There is no escape from a serious attempt at clarifying what it is that we are talking about, and at measuring it. As was stated by Elmaghraby (1991) "what is really important is not exactitude in an absolute sense, but rather agreement on entities that are *measurable*, and a demonstration of the operational utility of these measures". It is the objective of this chapter to do just that.

This brief review of the "state of the art" at the time of this writing would be incomplete without mention of two reports that postdate the paper by Elmaghraby (1991) cited above. The comments made in the previous two paragraphs are equally applicable to these reports.

The first is a paper by Patterson et al. (2002) in which they introduce the concept of "*protective capacity*", defined as "a given amount of extra capacity at nonconstraints above the system constraint's capacity, used to protect against statistical fluctuations." In their research, they conducted a full factorial experiment with a simulation model to explore issues associated with the quantity and location of processing variance in a five-station manufacturing cell.

The second is the final report on the Measurement and Improvement of Manufacturing Capacity project (MIMAC) issued by SEMATECH; Inc., and authored by Fowler and Robinson (1995). There are several research documents on this project, which was set up "to identify and measure the effects and interactions of the major factors that cause the loss in semiconductor manufacturing capacity". Interestingly enough, and we quote,

> "... (several among the industrial) participants in the study defined capacity in terms of throughput, or the amount that a facility could produce in a given time. Sometimes the definition was narrowed to refer to the maximum amount that the bottleneck or bottlenecks could produce. Assumptions about equipment availability, line yield, setups, and product mix were recommended as part of some definitions, as were assumptions on budget, quality, operators, "hot lots", lot sizes, and batch sizes. In a few cases, cycle time was cited as important to the definition of capacity. For example, capacity "while achieving the cycle time the customer asks for" was defined, as was the "output rate that will sustain and not exceed that (given) cycle time." The standard deviation of cycle time was also mentioned occasionally as a parameter."

Most importantly, the MIMAC Team later defined capacity as

> "... the maximum output rate (or start rate) sustainable for a particular factory with a given product mix and a constraint on the average cycle time. This capacity can be measured by generating the characteristic curve of cycle time versus output rate (or start rate) for the factory and by finding the output rate that corresponds to the specified cycle time constraint. The characteristic curve is obtained using either simulation or queuing models. And it focused on understanding the impact of the factors on the capacity planning process."

The authors also found the following 18 factors that influence the manufacturing capacity, they are: tool dedication, batching, breakdowns, dispatching/sequencing, end-of-shift effect, factory shutdown, hot lots, inspection/yield, operator cross-training, lot sizes, mix, operator availability, order release/WIP limits, redundant tools, rework, setup, and time bound sequences.

7.2 The Uses of Capacity

A measure of capacity of a plant, or a shop in a plant, or a machine within a shop, is required to perform any of the following functions: plant expansion or contraction that may be accompanied with manpower hiring and layoff– alternatively, for decisions related to investment in additional resources or divestment of existing ones, commitment to delivery of products and the timing of such delivery (see the discussion of the relationship between capacity and lead time below), short-term job scheduling/sequencing to meet any of several objectives, such as the minimization of the makespan, make-or-buy decisions relative to assemblies and subassemblies, the evaluation of suggestions on diversification or contraction of the products offered by the firm, the identification of idle capacity and its causes to render available to the decision makers the facts concerning suggested changes in resource utilization, etc.

It is our contention that there is not one but four forms of capacity, which we label as *nominal, operational, planned* (over a specified horizon), and *utilized* capacities. The next section presents their definitions and the approaches to measure them.

7.3 Proposed Definition of Capacity

We suggest that there is not one but several capacities that are used in practice and that it is high time to identify them individually for the various purposes of production planning and control outlined above. The basic premise of our thesis is simply the following. A production facility is installed with a given "nominal" (or theoretical, or maximal) capacity. When a facility is run in day-by-day operation, one does not achieve the rated nominal capacity since different factors enter into play that inhibit maintaining the nominal output except for a very short time. A more detailed discussion of some of these factors is offered below. Over a planning horizon of 1 year, say, the facility must plan according to some realistic available capacity that takes these factors into account. Assuming that the original facility was correctly planned, its "available" capacity should be fully utilized in order to realize the maximal return on the investment made. But when production is in fact realized, the utilization of the productive facility may fall below the available capacity. This underutilization may be intentional, in which case we speak of "planned" capacity. On the other hand, underutilization may not be intentional because of the many reasons detailed below. The difference between the available capacity and the "actual" utilization represents idle capacity, which would have been utilized if the causes of output decrement are eliminated. It is natural for the firm to wish to identify the idle capacity and its causes to make available to the decision makers the facts concerning future corrective actions and the degree of their impact on possible improvements.

In the following, we assume that the different capacities are defined relative to a specific "workday" or "workweek", which may vary from one enterprise to another or within the same enterprise from "shop" to "shop" or from season to season.

7.3.1 The Nominal Capacity

The *nominal capacity* is the productive capability assuming continuous availability of process (or machine) and all its support facilities, such as labor, maintenance, material, tools, jigs and fixtures, electric power, warehousing facilities, materials handling and transportation, etc., when the process is devoted to the production of a single "standard" product or the execution of a "standard" activity. Our nominal capacity is alternatively labeled by others as "theoretical" or "maximal" capacities. We prefer the label "nominal" because of its positive connotations that avoid the stigma of either "theory" or "maximization".

It is worth noting that there exist production facilities that produce only one product – which is then the "standard" product referred to in the above definition – such as electric power generators and water management systems, in which case the nominal capacity is quite meaningful. Naturally, the problem arises when the productive facility does *not* produce a single product or perform a single activity; how does one then measure the process nominal capacity? This issue shall occupy a good portion of our subsequent analysis.

The definition of nominal capacity insists on assuming perfect support activities. This may not be palatable to some scholars in the field who prefer to exclude the required maintenance and overhauls from the definition of the nominal capacity on the basis that they consume time that is known to be not available for production from the outset. We reject this argument because while the *execution* of these activities is mandated from the outset, their *timing* is not. For instance, maintenance is scheduled by some firms to be performed during the nonproductive periods such as the second shift in one-shift operation shops or during the weekends in 5- or 6- days operations, and overhauls are scheduled during the annual shutdowns for vacation. This is a managerial decision, taken after carefully weighing the pros and cons (and the costs) of such actions. Consequently, if we are to have a base datum for all firms in the same industry, the definition cannot provide a ready-made loophole that allows one firm to gain an advantage over another by showing a more favorable nominal capacity.

There is one other advantage to maintaining the definition of nominal capacity to be the maximum attainable under the best of circumstances, namely, that now the determination of the "operational" capacity, defined next, will assist the analyst who is interested in the measurement of productivity to distinguish among firms (or among shops within the same firm) more easily, consistently, and more rationally. In other words, if two firms have exactly the same nominal capacity, but one plans its maintenance activities during weekends and the other does not, then the operational capacity of the latter will clearly be smaller than that of the former. Such cause of discrepancy can then be more easily isolated, and its cost-to-benefit ratio is evaluated. If favorable, the management of the latter firm may be directed to improve its performance by rescheduling its maintenance activities, if that is feasible. Such comparison would not have been possible if the nominal capacities of the two firms were reported to be different in the first place.

7.3.2 The Operational Capacity

The *operational capacity* is the productive capability after subtracting from the nominal capacity the *anticipated* and *unavoidable* loss in productivity due to the age of the facility, the cumulative use of the facility, the required maintenance and overhauls, the *optimal* change-over loss in capability as dictated by the product mix, the standard reject allowance; etc., but still assuming that all support facilities mentioned in Sect. 7.3.1 are present. It is sometimes referred to by some authors as the *realizable* or *disposable* capacity.

Several of these factors that detract from the nominal capacity are *random* in nature and are often *correlated*. For instance, the required maintenance and overhauls are functions of the "condition" of the facility which, in turn, is determined by the age of the facility and its use – an old truck may be in excellent condition because it has not been used a lot, and conversely, a not-so-old truck may be in poor condition because it was used a lot in adverse environment. The same can be said about the rejects: they vary randomly from day to day with the same equipment and personnel, and often from shift to shift with different personnel.

As will become amply clear as the discussion progresses, while the nominal capacity may be stated as a single (crisp) number, it is often the case that the operational capacity can be stated only as an *interval* with attached probability distribution. This will be exemplified later.

The determination of the operational capacity includes two elements that require care in their determination. The first is relative to the "optimal setup (or change-over) time between products" and the second is relative to the "standard rate of rejects". We discuss these two items in more detail.

The optimal setup time for the quantities actually produced in a period of time (say a week) may be determined, or closely approximated, by a mathematical programming model that takes the demands for the products and the storeroom capacity into account. The issue is not *how* it is done, which we leave to the operations research experts on this problem, but *why* it should be done. The absence of a benchmark in the form of what setups *should be* for a particular production plan results in the blind acceptance of the time lost in setups, no matter how large it is, as a legitimate consequence of having to produce a mix of items. This is wrong and should not be acceptable. In fact, when such a benchmark is available, then the performance of the production scheduling function of the firm can be objectively evaluated. To seek optimality in the setup times taking account all other factors of production, such as inventory accumulation in both in-process and finished goods, is a prerequisite to being able to judge the efficacy of the production planning function across shops or across firms, or within the same shop or firm over time. It also helps in pinpointing internal or external factors that impact adversely on the operational capacity, such as the size of the warehouse or the efficacy of the materials handling system.

We now come to the issue of the "standard rate of rejects". Why should it be taken into account? Because "output" should be measured in terms of *good* product – the churning out of inferior or unusable product is no "production" at all; it is waste and should not be counted as "production". We propose to use the industry standard as the datum, until the firm establishes its own standard that is lower than the industry average. In other words, if the industry experiences an average rate of 0.5% reject (or one in 200), then that should be the initial allowance in all resources used in the determination of the operational capacity. Actually, the picture is more complicated than that, which shall be illustrated in more detail in the examples cited below.

It is evident that some of the difference between the nominal and operational capacities is a measure of the *idle* capacity – in other words, capacity that should have been used but was not. An example would be the time devoted to setup; a factor that is amenable to managerial decision and rational analysis. But there are other factors that contribute to idle capacity, which can be detected only through careful study of that difference. Here are some possible causes for it: labor absenteeism, shortage of equipment, shortage of raw material, electric current brownouts or blackouts; etc. Each of these causes reflects poorly on the performance of some support activity or on the industrial infrastructure available to the firm. Correction can be achieved only after the diagnosis of the cause has been completed; hence the need for the distinction between the nominal and operational capacities.

7.3.3 The Planned Capacity Utilization

The *planned capacity utilization* is the portion of the operational capacity that is planned to be used over the planning horizon. The planned capacity utilization may be less than the operational capacity for several reasons, not the least significant of which is the lack of external demand for the product(s), or it may be *more* than the nominal (i.e., realizable) capacity when it is deemed beneficial to exert some "pressure" on the facility to excel – see the above discussion of the relation between the WIP and the lead time and their impact on productivity.

One important reason for the presence of "over-planning" and "under-planning" is that it is almost impossible to achieve perfectly balanced loads on all shops, or all facilities in the same shop, simultaneously and at all time. This is due to the varied processing requirements of the different products on the individual shops or machines. Consequently, "bottlenecks" shall exist, which shift among the facilities is dependent on the product. This will become apparent in the examples discussed below. If this is the nature of the products and the production facility, then the planned capacity will necessarily leave some shops underutilized, while others are fully loaded and possibly overloaded. Naturally, the underutilized resources cannot be penalized for sitting idle – a point that escaped the "efficiency experts" of the 20s and 30s, with concomitant labor strife and social unrest.

7.3.4 *The (Actual) Utilized Capacity*

The *actual utilized capacity* is the capacity, in whatever measure it is done, actually utilized in the realization of the products delivered over a given planning horizon. Remember that even the best laid plans can go awry, and for several reasons, many of which are beyond the control of either management or personnel. In the final accounting, it may turn out that the utilized capacity is less – and sometimes much less – than the planned capacity. Normally, the utilized capacity is less than the planned capacity due to uncontrollable external causes – including acts of God – which could not be foreseen or avoided. A short list of such causes include: labor strike, material shortage, breakdown in the transportation system, etc.

Note that planned capacity refers to the *future* utilization of projected operational (or available) capacity, while the actual capacity usage measures *past* utilization, which may be more, or less, than what was planned. The difference between the "plan" and the "actual" may serve to measure the accuracy of the firm's forecast of its sales and its productive capabilities. Alternatively, it may measure the effort of the sales department in increasing the sales.

7.4 Capacity and Sales / Marketing

Consider the following scenario. A machine shop is composed of four sections, each specializes in a particular operation. Although there is machinery in each section, production is actually controlled by the availability of workers, not the machines. The interesting aspect of this shop is that labor is easily trained in all operations so that, for all practical purposes, the labor in the shop is available to work on any section. The time of shifting labor from one section to another is minimal (of the order of 5 min) and therefore negligible.

Suppose the question asked is: What is the shop capacity per week in hours? The obvious response is: the number of people multiplied by the net available hours per week, or $h \cdot W$, where h is the net working hours per person per week and W is the number of workers in the shop.

But suppose the question asked is: What is the shop capacity per week in *weight* or in *value*? Now the answer is not so obvious because different products require different times in the shop but have different weights and values. Analysis has shifted from simple arithmetic to the statistical domain where the answer depends on the *distribution* of the orders received. An unexpected consequence of this shift in the mode of analysis is to raise the following question, which has little to do with *capacity* but a lot to do with *sales and marketing*: suppose that a subset of the manufacturing orders received exhibit a high ratio of value to (shop) load, can the shop promote these products at the expense of the orders that ask for products with the reverse characteristics (low ratio of value to load)? Here, concern with capacity spills over to concern with other disciplines of the enterprise, and the whole production system and its support activities are intertwined as an indivisible entity.

Orders Received				
			Nominal	
Order #	**Time**	**Weight**	**lbs/day**	**ratio**
1	7006	4208	3634	0.601
2	3953	4464	6832	1.129
3	6564	6210	5724	0.946
4	6070	6964	6941	1.147
5	2329	2269	5896	0.974
6	5171	6391	7478	1.236
7	11146	4916	2668	0.441
8	3591	6172	10399	1.719
9	2453	2899	7151	1.182
10	1822	3740	12417	2.052
11	8596	8331	5863	0.969
12	2745	6791	14967	2.474
13	5770	5903	6189	1.023
14	2791	3068	6649	1.099
15	3526	3314	5687	0.940
16	3943	2185	3353	0.554
17	7190	9527	8016	1.325
18	8423	9265	6655	1.100
19	12146	10810	5385	0.890
20	10015	6209	3751	0.620
total time, min =	115251	**min =**	2668	0.441
nominal occupancy, days =	19.0497	**max =**	14967	2.474

Statistical Analysis	
Range *1000	**frequency**
2.0 -2.999	1
3.0 - 3.999	3
4.0 - 4.999	0
5.0 - 5.999	5
6.0 - 6.999	5
7.0 - 7.999	2
8.0 - 8.999	1
9.0 - 9.999	0
10 - 10.999	1
11 - 11.999	0
12 - 12.999	1
13 - 13.999	0
14 -14.999	1

Average =	6783
Variance =	8845739
Std. Dev. =	2974
95% =	11675
0.05% =	1891

Shop Nominal Capacity = 6050 min/day

days req'd to produced demand = 19.04968days

Fig. 7.2 A sample of orders received and their parameters (processing time and weight)

The point made here is best explained by an example.

The data shown in Fig. 7.2 is real-life data (scaled to respect privacy) from a plant over a short period of a few days. The nominal shop capacity is 6,050 min/day. Despite the paucity of the information available, intentionally made so to conserve space, it is sufficient to illustrate the analysis to be conducted if more extensive data were available.

The table on the left of Fig. 7.2, labeled "Orders Received", lists the customer orders in the sequence in which they were received. The second column gives the standard total processing time for each order. The third column gives the weight of the product, in lbs. The fourth column calculates the "nominal weight per day" based on the shop nominal capacity in minutes and the order standard time requirements. For instance, order #1 is characterized by long-processing time (it occupies 7,006 min) but rather light weight (produces 4,208 lb). Hence the weight produced per day of this order based on the shop's nominal capacity is $(4{,}208/7{,}006) \times 6{,}050 = 3{,}634$ lb. In other words, if production is confined to orders identical to this order the plant would produce 3,634 lb/day. However, if production is confined to orders similar to #8, which is characterized by short-processing time (it occupies only 3,591 min) and rather heavy weight (produces 6,172 lb), the daily production would be 10,399 lb/day. The table on the right, labeled "Statistical Analysis", gives the frequency of orders received in this period categorized into slots of 1,0001b, from 2,000 to 15,000 lb.

What is the nominal capacity of this plant *in pounds*?

Let us assume that the nominal capacity of 6,050 min/day is fully loaded, which we know is not always possible because of the discrete nature of the units produced. Still, the assumption can be forgiven for the sake of driving home the point we wish to make. (A job may be partially "advanced" or "retarded" to fill the available capacity.) The result of such manipulation is the production of 6,050 min each day, with the result in pounds as shown in the production profile shown in Fig. 7.3, also depicted graphically in Fig. 7.4. (The table is cut-off at day 19 because loading day 20 to the available capacity requires knowledge of the orders received in day 21, which is not available.)

Day	1	2	3	4	5	6	7	8	9	10
Production/day, ton	3634	6448	6043	6594	7201	2677	3873	9399	6809	9821

Day	11	12	13	14	15	16	17	18	19
Production/day, ton	6352	6105	4938	7821	6655	5701	5385	4171	3422

Fig. 7.3 Production to fill nominal capacity

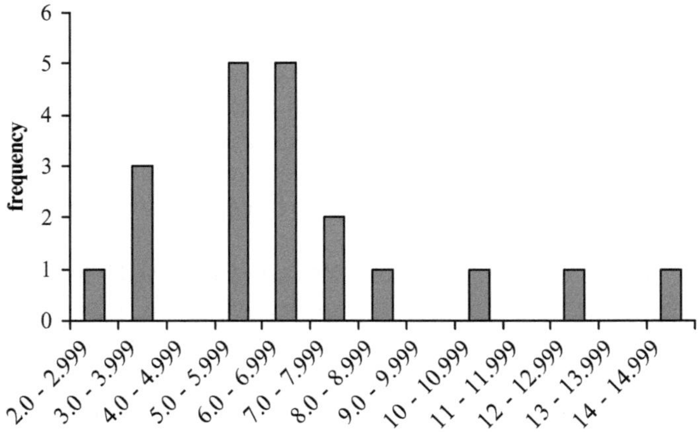

Fig. 7.4 Frequency of order size, Nominal lbs

Analysis of these (admittedly meager) data results in the following statistics:

Average production/wk = 5950 lb/day,
Std. Dev.= 1906 lb
minimum = 2677 lb/day
maximum = 9821 lb/day
mode = 6000–6499 and 6500–6999 lb/day
The 90% interval $\approx [2815, 9085]$.

When presented with this analysis management retorted that it always knew that the plant can produce some 10,000 lb/day. In reality, such an output was (almost) achieved only once in 20 days.

This illustration demonstrates two important insights. First, variability in the product specifications vitiates the possibility of stating a single number as *the* capacity of the plant; the answer has to be couched in probabilistic terms. Secondly, management almost always remembers the most favorable result and sets it as the standard, to the detriment of morale in the plant and the continued disappointment of management.

Finally, this real-life situation exemplifies the point made earlier in this section. Close scrutiny of the 20 orders received by the plant reveals that some of them had a rather small ratio of weight to time. The minimum ratio is ≈ 0.44 and the maximum is ≈ 2.47, or some 5.6 *times* as much. Put differently, if weight of the product translates into profit to the firm, the plant would realize some five times the profit if it can devote its time to the products that have the higher ratios. Can the marketing/sales department realize such target? To repeat, at the risk of being redundant, concern with capacity spilled over to concern with other activities of the enterprise, and the whole production system and its support activities must be viewed as an indivisible entity.

7.5 Capacity and Lead Time: A Vicious Circularity

There is little argument that available capacity determines the production lead time (LT), where the latter is (narrowly) defined as the time between release of a production order to the shop and the delivery of the finished product to the finished goods inventory (FGI). In the words of Asmundsson et al. (2002),

> "... lead times increase nonlinearly in both mean and variance as system utilization approaches 100%. Hence, deterministic production planning models have suffered from a fundamental circularity. In order to plan production in the face of time-varying demands, they often use fixed estimates of lead times in their planning calculations. However, the decisions made by these models determine the amount of work released into the facility in a given time period, which determines the utilization and, in turn, the lead times that will be realized!"

It is interesting to note that studies of the variation in the lead time concentrated on the relationship between the LT and the shop "load", where the load, as well as the LT itself, are measured in units of time (hours, say); see the recent review article by Pahl et al. (2005) which is titled "Production Planning With Load-Dependent Lead Times".

Critical reading of the extensive literature on the subject quoted by these and other authors reveals several references to "capacity" as a determining factor in the study of both "load" and "lead time" – but little is said about what is meant by capacity, how it was determined in the first place, how it is measured, or *its* influence on either the "load" or the LT!

A fundamental law in queueing theory states that (WIP) $= \lambda\overline{p}$, in which λ is the arrival rate of jobs, measured in units per unit time, and $\overline{p}$ is the average residence time of a job including its own processing. This "law" should be intuitively appealing and may be reasoned as follows. In the "steady state" the input must equal the output. The processing of a job consumes, on average, $\overline{p}$ time units. During that time, the number of arrivals is $\lambda\overline{p}$, leaving the length of the queue intact at WIP. Translated into our language, this law may be written as WIP = (job arrival rate) $\times$ (average job residence time). Where does *capacity* figure in this formulation? It must reside in $\overline{p}$, the residence time of a job; a larger capacity must *reduce* $\overline{p}$ and thereby also reduce WIP for the same rate of job arrivals λ, which implies a reduction in the LT.

Consider the following simplified view of a plant that may be considered as "one facility" – say a chocolate factory – which produces a variety of products, each with its processing time p_i, $i = 1, \ldots, N$, assumed to be deterministically known. The shop itself is continuously available (any maintenance or repair is done off-production time). Assume that the production orders for all N products were released to the shop at the same time (now), and sequenced[1] in the order $1, \ldots, N$. A new order, $N + 1$, would have its LT equal to $\sum_{i=1}^{N} p_i + p_{N+1}$ if it is released to the shop immediately. Any delay in releasing the new job would just add to its LT. In view of these considerations, it is not be difficult to see why the LT is intimately related to the work in process (WIP), which is related to capacity. Clearly, if the sum of the processing times ahead of job $N + 1$, $\sum_{i=1}^{N} p_i$, is small, the job shall be finished early and the opposite shall happen if $\sum_{i=1}^{N} p_i$ is large. In the steady state, the length of the WIP is more-or-less constant, which translates into an average LT of $\overline{LT} = \overline{p} \cdot (\text{WIP})/2$.

The above discussion assumes that the production facility, which is composed of people and machines, is not "flustered" by the "load" ahead of it; i.e., the WIP. In other words, an implicit assumption in the above analysis is that, on average, the production facility outputs $1/\overline{p}$ jobs per unit time, provided there are enough jobs in the WIP to keep the facility "busy". But empirical evidence shows that production facilities are indeed "flustered" by the load as measured by the WIP ahead of them and that such adverse effect on the productivity occurs well before they reach their "nominal capacity". They cease to produce at the rate $1/\overline{p}$ but produce only at a

[1] We are not concerned here with the methodology of sequencing the jobs, but take the sequence as given.

fraction thereof. The interesting observation is that *the fraction decreases as the WIP increases*! In other words, *beyond a certain load*, the productivity of the facility decreases as a function of the load. This gave birth to the concept of "*clearing function*", first coined by Graves (1986). In the words of Asmundsson et al. (2003):

> "The basic idea of *clearing functions* is to express the expected throughput of a capacitated resource over a given period of time as a function of some measure of the system workload over that period, which, in turn, will define the average utilization of the production resources over that period."

To phrase this concept mathematically, let $h(W)$ denote the *throughput* per unit time when the WIP is W. Many functional forms have been proposed, some based on stochastic arguments and some based on purely deterministic considerations. As a sample of the former we quote Medhi's (1991) result that is based purely on queueing arguments and relates the expected length of the queue W (the WIP) to the "utilization ratio" ρ (which is the ratio of the input rate to the output rate) and the coefficients of variation (defined as the standard deviation divided by the mean) of the arrival and production processes, denoted by ν_s and ν_a, respectively,

$$\begin{aligned} W &= \frac{\nu_a^2 + \nu_s^2}{2}\frac{\rho^2}{1-\rho} + \rho \\ &= \frac{\nu^2\rho^2}{1-\rho} + \rho; \quad \text{where} \quad \nu^2 = \frac{\nu_a^2 + \nu_s^2}{2} > 1. \end{aligned} \tag{7.1}$$

This expression can be inverted to yield ρ as a function of W,

$$\begin{aligned} \rho(W) &= \frac{\left[\sqrt{(W+1)^2 + 4W(\nu^2 - 1)} - (W+1)\right]}{2(\nu^2 - 1)} \\ &= \frac{\left[\sqrt{W^2 + 2W(2\nu^2 - 1) + 1} - (W+1)\right]}{2(\nu^2 - 1)}. \end{aligned} \tag{7.2}$$

At $W = 0$ the utilization is $\rho(0) = 0$, as to be expected; and as $W \to \infty$ the utilization $\rho(W) \to 1$ (the denominator of $\nu^2\rho^2/1-\rho$ approaches 0). Figure 7.5 shows ρ as a function of W for two values of ν. As expected, the larger ν impacts adversely on utilization.

The message conveyed by this analysis is the following. Assuming that the input demand (i.e., arrival) is random with coefficient of variation ν_a and that the production facility service time is also random with coefficient of variation ν_s, then the full capacity of the production facility shall *never* be fully utilized for any finite "load" (or WIP).

Depressing as this conclusion may be, it is the inevitable consequence of the random variation in both the input and the output mechanisms. Increasing the capacity of the production facility will succeed in increasing the *volume* of the output proportionately; but it will not alter the fundamental relation between the two; the capacity is never fully occupied.

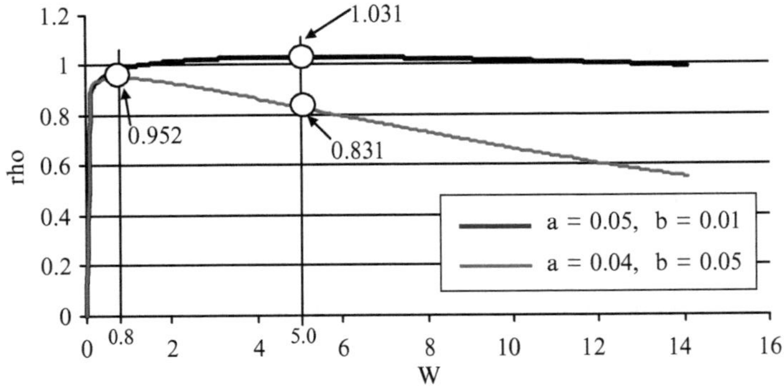

Fig. 7.5 ρ versus WIP W in deterministic model for two different values of υ

Turning now to the deterministic case in which output is a function of the "load" as represented by the WIP, several researchers tried to emulate the stochastic behavior by assuming a variety of functional relations. For instance, Karmarkar (1989) and Srinivasan et al. (1988) proposed functions of the form

$$h(W) = \frac{k_1 W}{k_2 + W} \tag{7.3}$$

$$h(W) = k_1 \left(1 - \mathrm{e}^{-k_2 W}\right), \tag{7.4}$$

where k_1 is the "maximum capacity" (presumably equivalent to our nominal capacity) and k_2 is the estimate of the decay in productivity (the "*clearing function*"). Both functions in (7.3) and (7.4) have value 0 at $W = 0$, and both approach k_1 as $W \to \infty$, rising in a concave fashion between these two limits. Again, full utilization of the available capacity is not achievable except for an infinitely large WIP.

It is thus seen that in these studies emphasis has gradually shifted from correcting the *clearing function* proposed by Graves (1986), which assumes *infinite* capacity (the straight line in Fig.7.1, labeled "proportional output"), to the development of a *production function* that recognizes the finiteness of the production capacity and gives the throughput $h(W)$ as a function of the WIP W. This function was subsequently used by manys researchers ([see, e.g., Asmundsson et al. (2002)]) to develop mathematical programs that purport to optimize the release of jobs to the production facility in order to optimize its throughput. We shall not pursue this development any further since it would take us far away from the main theme of this chapter.

Pending the availability of compelling empirical evidence to substantiate the above conclusions, it may be unpalatable to many to accept the premise that the capacity of a production facility is not fully utilized unless the WIP is choking the production system, a highly unacceptable scenario for practical people. To such people the following model may be more acceptable:

$$h(W) = k_1 W^a e^{-bW}, \quad a, b > 0, \quad 0 \le W \le \overline{W} < \infty, \tag{7.5}$$

in which k_1 still represents the "maximum capacity" and $\overline{W}$ is a *finite* upper bound on the WIP. The advantage of such a model is that $h(W)$ starts at 0 when $W = 0$, which stands to reason, but rises to a maximum (secured by differentiation and equating the derivative to zero) equal to

$$W_{\text{opt}} = \frac{a}{b} \tag{7.6}$$

well before the WIP grows to infinity. At this value of WIP the throughput $h_{\max}(a/b)$ is

$$h_{\max}\left(\frac{a}{b}\right) = k_1 \left(\frac{a}{b}\right)^a e^{-a}, \tag{7.7}$$

which can be larger or smaller than k_1 depending on the value of the parameters a and b. For instance, for $a = 0.05$, $b = 0.01$, $h_{\max}(5) = 1.031$; but for $a = 0.04$, $b = 0.05$, $h_{\max}(0.8) = 0.952$. The performance of the production process for these two sets of parameters is depicted in Fig. 7.5. The model of (7.7) with the first set of the parameters a and b represents the case in which a "little pressure" on the production facility, as manifested in the magnitude of the WIP, motivates it to perform well – even to (slightly) exceed it "maximal capacity".

7.6 In What Unit Should Capacity Be Measured?

A perennial question is: in what unit should one measure capacity (any of the four capacities)? Our answer may appear evasive, but it is truthful: it depends on the nature of the product. Sometimes it is best to measure capacity in terms of weight (tons of steel produced per year); sometimes in terms of distance (thousands of kilometers of road track); sometimes in terms of area or volume (floor space or storage volume); and sometimes it is best to convert both capacity and output into a common measure such as value (dollars) or time (hours). Often, when the product is more-or-less uniform, it is meaningful to measure capacity in terms of units produced per year. Since the product is frequently the same but with variations in size (small-to-large refrigerators) or weight (light sheet metal to heavy sheets) or "optional parts" (cars with varying musical systems), a measure in units typically refers to an "average" unit – which may not exist. An enterprise that manufactures different kinds of custom-ordered scientific equipment may opt to measure its capacity in terms of man-hours, while another in terms of dollar value.

In the final analysis, it is immaterial the units in which capacity and output are measured; what is important is that they are in fact measured correctly in units that are meaningful for decision making within the firm, and for meaningful comparative analysis across the whole industry.

7.7 Difficulties in Measuring Capacity

Granted that knowledge of the requisite capacity is vital for the design and initiation of the productive activity before its existence, and knowledge of the current capacity is mandatory for managing its healthy performance after it has been installed and running; and granted that there are these different classes of capacity in the daily operation of the facility, why is it so difficult to estimate them prior to the realization of the facility or achieve their measurements after its installation?

The following analysis details the reasons for the difficulty and delineates the pitfalls that will surely result in errors, minor or gross, intentional or otherwise, in the firm's estimation of the various capacities available to it, to the point of vitiating the utility of the data at all levels of management. The origin of these errors resides in the very definitions of the various capacities. And avoiding the errors requires more sophisticated analysis than the usual chatter on "averages" – one must be willing, and capable, of dealing with statistical variation as a fact of life.

We shall list six reasons. The first five are technical in nature and therefore are amenable to analytical rationalization. The sixth is only partially technical, being anchored in the social and cultural environment of the firm. We hope that its inclusion shall alert the managers of production systems to these aspects of the processes under their purveyance, aspects which are typically ignored or glossed over by the more technically oriented analysts.

In the following discussion, we shall use "capacity" to mean any of the four capacities defined above. The reader is free to adapt the discussion to her/his favorite one.

7.7.1 The Problem of Product Mix

The most prevalent reason for the inability to measure the capacity of a facility is that it is clearly dependent on the product mix that happens to be produced in any period or over any finite horizon. But capacity is usually defined externally (to the firm) in terms of some hypothetical "standard" (or "average") product which does not exist, at least does not exist all the time nor does the firm even produce the same mix of products period after period so that the said "standard" product is realized. The capacity, so the argument goes, is not a fixed entity. Its measurement is like aiming at a moving target; therefore it cannot be accurately determined!

We begin by explaining why the existence of a product mix plays havoc with the measurement of capacity. The problem resides in the existence of a "bottleneck" stage of production that depends on the product produced; change the product and the bottleneck shifts from one facility (or machine) to another. In fact, if all products possess the same bottleneck operation, then the capacity is identical to that operation's capacity, which is relatively easy to measure. Consequently, when the firm (or plant) is considered as a whole, one has several measures of capacity that depend on the items produced in any period, say a week. And since the product mix varies from week to week (or month to month or, in general, from period to period), one really has 50 product mixes per year (allowing for 2 weeks vacation), and the determination of the capacity appears to be an insurmountable problem which solution defies analysis!

The problem is a real one, and its resolution rests on the concepts of statistical distribution of capacity. The concept is best illustrated by an example.

Consider a shop that produces three different items, call them A, B, C. For simplicity of exposition, assume that in any week all items are produced, but at any time one and only one product is produced by the firm and all shops are set up to accommodate its production. Changing from the production of one product to another involves considerable setup time, which will be the subject of the next section. We shall use the *nominal* capacity as the vehicle of illustration; later sections explain how the numbers can be changed to represent other capacities. The firm's production facilities are composed of four "shops", which capacities are as given in Table 7.1.

The meaning of these figures is as follows. Consider product A: the capacity of shop 1 is 20 tons/week, of shop 2 is 15 tons/week; etc. The bottleneck capacity is clearly the capacity of shop 2 of 15 tons/week. If the shop was devoted to the production of product A alone, that would be the nominal amount that can be produced in any week. It would also mean that shops 1, 3, 4 will be underutilized (i.e., will have idle capacity) of, respectively, 5, 6, and 17 tons/week, which represent 25, $33\frac{1}{3}$, and 53.125% of the respective capacities. Similar comment can be made if the shop was devoted to the production of product B or C alone.

What is the ideal product mix for this plant?

The last section of the table gives the answer to this question in the column "ideal mix". It is secured as the solution of the linear program (writing 1 for A, 2 for B, and 3 for C):

LP:

$$\max\ z = \sum_{i=1}^{3} v_i x_i \tag{7.8}$$

subject to :

$$\sum_{i=1}^{3} \frac{x_i}{c_{i,j}} \le 1, \quad j = 1, \ldots, 4$$

$$x_i \ge 0. \tag{7.9}$$

Table 7.1 Nominal capacity (in tons/week)

	Shop capacity					Bottleneck		
	Shop:	1	2	3	4	BNC†	Shop‡	Ideal mix
	A	20	15*	21	32	15	2	6.00
Product	B	14	16	12*	21	12	3	4.66
	C	32*	38	36	43	32	1	11.73

*Identifies the bottleneck shop
†Bottleneck capacity
‡The identity of the bottleneck shop

The objective function of this LP is to maximize the total value produced by the plant in which 1 ton of product i generates v_i units of value. The constraint set in (7.9) limits the sum of the fractions of "load" in each shop to 1 (full capacity), where $c_{i,j}$ is the nominal production capacity of shop j of product i. The inverse $1/c_{i,j}$ measures the fraction of capacity of shop j used by 1 ton of product i. For example, in reference to Table 7.1, shop 1 can produce 20 tons of product A; hence 1 ton of A would occupy $1/20 = 0.05$ of the shop's capacity. The fractions for products B and C in shop 1 are $1/14$ and $1/32$; respectively. Hence, for shop 1 constraint (7.9) would read:

$$\frac{1}{20}x_1 + \frac{1}{14}x_2 + \frac{1}{32}x_3 \leq 1. \tag{7.10}$$

There are four equations of the genre of (7.10). If all the items are of equal value, $v_i = v$, then the solution of this LP is

$$x_1 = 0 = x_2,\ x_3 = 32.$$

This solution would load shop 1 fully, but leave shops 2, 3, 4 greatly underutilized:

Shop	Utilization(%)
1	100
2	84.21
3	88.89
4	74.42

Clearly such a solution is unacceptable (only product C is produced!). However, knowing that shop 4 has excess capacity, we may insist that the other three shops be fully utilized. This would result in the solution of three equations in three unknowns:

$$\text{shop 1}: \frac{1}{20}x_1 + \frac{1}{14}x_2 + \frac{1}{32}x_3 = 1$$

$$\text{shop 2}: \frac{1}{15}x_1 + \frac{1}{16}x_2 + \frac{1}{38}x_3 = 1$$

$$\text{shop 3}: \frac{1}{21}x_1 + \frac{1}{12}x_2 + \frac{1}{36}x_3 = 1$$

yielding,

$$x_1^* \approx 6.0, \quad x_2^* \approx 4.66, \quad x_3^* \approx 11.73 \tag{7.11}$$

$$\text{and } z^* \approx 22.39 \text{ tons.} \tag{7.12}$$

This is the vector that appears in the last column of Table 7.1 under "ideal mix". The interpretation of this result is that, ideally, the plant should produce (approximately) 6 tons of product A, 4.66 tons of product B, and 11.73 tons of product C each week in order to utilize the plant's capacity in the first three shops fully. The fractional occupancy of each product in each shop is shown in (7.13), from which is seen that shops 1, 2, 3 are fully utilized but shop 4 is underutilized by some 32% of its (nominal) capacity.

Shop	1	2	3	4
A	0.30	0.40	0.29	0.19
B	0.33	0.29	0.39	0.22
C	0.37	0.31	0.33	0.27
Total occupancy	1.00	1.00	1.00	0.68

(7.13)

To sum up the analysis thus far, one may conclude that *with the specified shop capacities* for the production of each product there is an *ideal mix* of the products which, if produced all the time, will fully utilize the capacity of shops 1, 2, 3 but not shop 4 which capacity should be modified to achieve *its* full utilization.

But the ideal mix of the products is *not* realized week after week since the demand for the various items varies over time. Table 7.2a gives the amounts *actually* produced of the various items over a quarter (13 weeks), and Table 7.2b gives the same date in percentages of the plant capacity. For instance, product A was produced in quantities $0.81, 1.98, 4.90, 2.61, \ldots, 2.73$ tons in weeks 1 through 13; respectively. Other rows of Table 7.2a are interpreted similarly. Assume, for the moment, that the actual production coincides with the planned production. The bottom row of Table 7.2a gives the total tonnage (planned and) produced in each of the 13 weeks.

How well is the plant doing in face of varied demand profile for the various items?

Observe that the total output over the 13 weeks varied from a low of 11.49 tons to a high of 24.77 tons per week, or by $(24.77 - 11.49)/11.49 \approx 115.47\%$, which may raise many eyebrows and be cause for concern. Should it be?

We start by translating the actual production in tons into the proportion of the nominal plant capacity used by each product in each week. This translation leads to

Table 7.2 Simulated actual demand over one quarter

Product	wk 1	wk 2	wk 3	wk 4	wk5	wk6	wk7	wk8	wk9	wk10	wk11	wk12	wk13
A	0.81	1.98	4.90	2.61	1.43	1.86	3.53	0.68	0.99	2.38	1.67	4.11	2.73
B	4.16	0.84	3.37	0.72	3.42	3.74	7.99	5.47	4.17	1.25	5.09	3.22	2.26
C	11.72	13.98	5.97	16.55	14.55	5.89	1.33	15.68	14.38	15.18	16.30	11.94	19.78
	16.69	16.79	14.23	19.87	19.40	11.49	12.85	21.83	19.55	18.80	23.06	19.27	24.77
											total production =		238.60

(a) Simulated actual production; ton.

Product	wk 1	wk 2	wk 3	wk 4	wk5	wk6	wk7	wk8	wk9	wk10	wk11	wk12	wk13
A	0.0538	0.1317	0.3264	0.1740	0.0956	0.1240	0.2354	0.0454	0.0660	0.1584	0.1115	0.2737	0.1818
B	0.3465	0.0697	0.2806	0.0598	0.2850	0.3117	0.6656	0.4562	0.3478	0.1038	0.4241	0.2686	0.1881
C	0.3664	0.4368	0.1864	0.5171	0.4546	0.1842	0.0416	0.4900	0.4494	0.4744	0.5094	0.3732	0.6182
Utilization:	0.7667	0.6381	0.7935	0.7508	0.8351	0.6199	0.9426	0.9916	0.8633	0.7366	1.0451	0.9155	0.9881

(b) Capacity utilization

the data given in Table 7.2b. The calculation of this table is simple. For example, the production of 0.81 tons of A in week 1 represents $0.81/15 = 0.054$ (the apparent error is due to truncation for a more pleasant appearance; the figures are actually, to six significant figures, 0.807077 and 0.0538051). It can be seen from Table 7.2b that, almost all the time, the ratios of the actual production is far from the "ideal" mix specified in the last column of Table 7.1. Similarly, for the other figures; each evaluated relative to the bottleneck shop that determines the plant capacity for each product. The sum of the percent utilization of the plant capacity is shown in the row opposite the "utilization" in Table 7.2b, from which it is seen that the plant's best performance was in week 11 with 104.5% utilization (slightly over full utilization) and its worst was in week 6 with 62% utilization of its nominal capacity.

Variation in production was in response to variation in demand, truncated at (slightly over) plant capacity. This demand pattern was randomly created assuming a normal distributions for all three products with means and variances as given in Table 7.3. The fourth column of this table labeled "90% range" gives the expected range of demand for the product, with confidence of 0.90.
Assuming independence of the demand for the three products, these values imply that, with probability 90%, the total demand on the plant shall be in the interval $[8.82, 35.18]$ tons/week. Table 7.2a verifies these bounds since the total demand was well within this interval all the time.

In the absence of knowledge on how the demand was generated, management would reach the conclusion that each of the products has mean and variance as shown in Table 7.4. The reason for the smaller means and variances is due to truncation by the firm to fit within capacity.

The message conveyed by this elementary analysis is that *for purposes of estimating demand one must use the actual demand data rather than the actual production data,* because the latter is the filtered demand data to fit the plant bottleneck capacity. This is important in responding to questions related to the adequacy, or otherwise, of the existing capacity. Interestingly enough, this simple rule is often ignored in real life!

We shall assume that the plant has estimated the correct mean and variance of the demand of each product on the basis of *demand* data, not the *production* data,

Table 7.3 Demand parameters, assuming normal distribution

Product	Mean	Variance	90% Range
A	4	6.25	$[0, 8.11]$
B	3	9	$[0, 7.93]$
C	15	49	$[3.49, 26.51]$
Total	22	64.25	$[8.82, 35.18]$

Table 7.4 Demand parameters as derived from actual production

Product	Mean	Variance
A	2.28	1.65
B	3.51	4.09
C	12.56	26.92
Total	18.35	32.66

and hence it knows Table 7.3. Then it should be aware of the analysis shown in the right-most column of the table, which gives the range of demand for each product, and for the plant as a whole. Two remarks are immediately present.

1. The "ideal" product mix would generate 22.39 tons/week and fill the capacity of three of the four shops. And yet demand not exceeding that magnitude will occur about 52% of the time! In other words, over 48% of the time the demand will exceed the shop capacity *even if it conformed to the "ideal" product mix*, which it will almost surely not do. In fact, with probability of almost 25% the demand is no more than 16.5 tons/week. (In Table 7.2a the number of weeks in which demand did not exceed 16.5 tons/week is 3, yielding $(3/13) = 23.08\%$.)
2. The *expected* demand on the plant of 22 tons/week is very close to the demand of the "ideal" product mix (of 22.39 tons/week). Basing the judgement on *averages* would indicate that all is well, which we know is not true. Evidently, it is the *variation* in the product mix that causes the variation in capacity utilization.

These two remarks translate into the following. Under the best of conditions one would expect this shop to be underutilized some 50% of the time and would be overloaded the other 50%. But even when it is overloaded, the "imbalance" in the product mix (i.e., its deviation from the "ideal") would cause some shops to be severely underloaded. The reader can verify this from Table 7.5 which is abstracted from Table 7.2a and deals only with the 3 weeks in which the total demand was "close" to the average. We have selected shop 2 to illustrate our contention; tables for the three other shops can be constructed in a similar fashion.

Table 7.5 gives the actual (simulated) demand realized in these 3 weeks, while Table 7.6 gives the proportion of the capacity of shop 2 that was occupied by each product in each of the 3 weeks. The absolute values of the proportions are of little significance by themselves; only in comparison with the utilization under the "ideal" product mix do they gain significance. As can be easily verified, the proportion of the shop used by each product varies wildly from the "ideal". For instance, product A varies from 11.35% $(= 0.045/0.4)$ to 45.44% $(= 0.182/0.4)$ of the "ideal", product B varies from 48.43%

Table 7.5 Actual demand in three weeks, tons

	wk 8	wk 11	wk 13
A	0.68	1.67	2.73
B	5.47	5.09	2.26
C	15.68	16.30	19.78
Total	21.83	23.06	24.77

Table 7.6 Capacity utilization, Shop 2

	wk 8	wk 11	wk 13	'Ideal'
A	0.045	0.112	0.182	0.400
B	0.342	0.318	0.141	0.291
C	0.413	0.429	0.521	0.309
Total	0.800	0.859	0.843	1.000

Table 7.7 Limits of shop occupancy, based on 90% confidence

	Shop 1	Shop 2	Shop 3	Shop 4
A	[0, 0.407]	[0, 0.541]	[0, 0.386]	[0, 0.253]
B	[0. 0.567]	[0, 0.496]	[0, 0.661]	[0, 0.378]
C	[0.109, 0.828]	[0.092, 0.698]	[0.097, 0.736]	[0.081, 0.617]

$(= 0.141/0.291)$ to 117.49% $(= 0.342/0.291)$ of the "ideal", and product C varies from 133.66% $(= 0.413/0.309)$ to 168.66% $(= 0.521/0.3091)$ of the "ideal". Similar analysis can be done for other shops, with the same conclusion: variation in the product mix plays havoc with the utilization of the available capacities of individual shops. Table 7.7 gives the bounds, at the 90% confidence level, on individual shop occupancy by each of the three products. It is determined based on the 90% ranges of Table 7.3 divided, respectively, by the nominal capacity of each shop. For instance, shop 2 has capacity of 15 tons/week of product A. The range of demand for product A is given in Table 7.3 as $[0, 8.11]$ which, when divided by 15 gives the range $[0, 0.541]$; etc. As the reader can see, all ranges are rather wide, varying by several orders of magnitude from their lower to their upper limits.

7.7.2 The Problem of the Setup Time

Another reason for the inability to provide an accurate measure of the capacity of a plant, or a shop in that plant, or even a machine within a shop, is that the presence of a product mix causes a *nonmeasurable* loss in productivity due to the need for frequent changeovers (note the emphasis on "nonmeasurability"). In particular, since in any long enough period different items are produced to satisfy market demands, this necessitates setting up the process when production changes from product to product. This setup (or changeover) time reduces the available time for production and, consequently, the operational capacity. And since the production mix varies from period to period, it is not possible, so the argument goes, to determine the available capacity of the facility even if the nominal capacity were known.

We readily concede that setups consume machine time that could have been utilized for production, if demand is there. And we concur with the contention that production planning to satisfy a specified demand *and* minimize setups and inventory build-up is indeed a very difficult problem to resolve. Industrial engineers are taught the principles and techniques of "economic production lot sizes" that achieve the optimal compromise between minimizing setup times and the piling up of inventory, which should include the *sequence* in which the products are produced and the impact of that sequence on the *dynamic* piling up and depletion of stocks, as well as on the changeover time between products. Recall that the minimization of setup times through off-line stand-by jigs and fixtures constituted one cornerstone in the triumph of the Japanese production practice over the rest of the industrialized world so that at least that portion of the total problem can be ignored.

But there is a pitfall in the above argument that must be avoided; to wit, that excessive setup times may be self-inflicted since a vicious cycle can easily develop due to poor production planning. In particular, if the "economic quantity" produced of a product in any period is large enough to cause the starvation of some other product (or products), then its "optimal" size is ignored and a *smaller* lot is produced to allow other products to be produced to satisfy *their* demand. This necessitates new setups which decrease the time available for production which, in turn, require smaller lots which require further additional setup times, which further reduce the available capacity, and so on until the plant appears to be in the absurd position of always being set up for smaller and smaller quantities of the different products!

If one compares the actual expenditure in time in setups with the *optimal* setups for the given product mix demanded in each period, determined, for instance, by any of the mathematical programming models presented in the literature, then one would gain a clearer indication of the adequacy, or the lack of it, of the production planning function in the plant. Adding over the periods should give the total time spent on setups throughout a quarter or a year, and a measure of the overall efficacy of the production planning function.

The reporting of setup information is thus seen to be important from at least two points of view.

1. It permits the evaluation of the production control function in the firm relative to a datum based on its demand pattern.
2. It permits the determination of the "opportunity cost" due to restrictions imposed on inventory, or on the satisfaction of the customer's demand, if such restrictions are in fact present, which necessitate frequent changeovers in the production lines.

To render our discussion more concrete, consider the three items mentioned in Sect. 7.1, and suppose the items were produced in the sequence A–B–C. Assume, further, that after the completion of the weekly production a maintenance/cleanup operation is undertaken that sets all the machinery in all four shops back to their "neutral" position so that a setup is required when production is initiated in the following week. We label the "neutral" state as "state Ø". The setup times are given in (7.14). The production time/ton is the reciprocal of the production quantities shown in Table 7.1, assuming nominal capacity of 40 h/week, presented for convenience in Table 7.8. The setup times under the sequence A–B–C occupies a total of $2 + 0.8 + 2.7 + 0.3 = 5.80$ h.

Table 7.8 Production times, h/ton

Product	Shop 1	Shop 2	Shop 3	Shop 4
A	2.000	2.667	1.905	1.250
B	2.857	2.500	3.333	1.905
C	1.250	1.053	1.111	0.930

Setup times, h

Product	Ø	A	B	C
Ø	0	2.0	1.5	0.2
A	0.3	0	0.8	1.4
B	0.4	3.0	0	2.7
C	0.3	0.28	2.8	0

(7.14)

For these setup times, the optimal sequence is C – A – B, which occupies a total of $0.2+0.28+0.8+0.4=1.68$ h, resulting in a saving of 4.12 h/week, or 10.30% of the normal capacity of each shop (of 40 h/week). The immediate impact of this improved setup sequence reveals that the *actual* shop utilization for production is even worse than that reported in Table 7.6. This rather shocking result is detailed in Table 7.9.

The significance of this table is instructive. Had the plant adopted the optimal sequence it would have saved some 214 h each quarter, or slightly over five shop-weeks of nominal capacity, which translates into over 20 shop-weeks' of nominal capacity per year! This is not an insignificant amount. Put differently, the plant could have produced at least 20.00 tons of A ($=4.12/2.67\times 13$) or 16.00 tons of B ($=4.12/3.33\times 13$), or 42.85 tons of C ($=4.12/1.25\times 13$) *more* each quarter, which translate into 80, 64, and 171.40 tons per year, respectively. Since the total production of the three items in the quarter were, respectively 29.67, 45.69, and 163.25 tons, for items A, B, and C (secured by adding the production in the simulated actual demand of Table 7.2) the difference in setup times represent, respectively, $20/29.67=67.41\%$, $16/45.69=35.02\%$, and $42.85/163.25=26.25\%$ *increase* over current production (assuming the available time in the plant is devoted to the production of each item).

Table 7.9 Comparing the actual utilization of each shop under current (A–B–C) and optimal setup (C–A–B) sequences

SUMMARY DATA													
SHOP 1	Sequence A,B,C												
	Total hrs:	33.95	29.61	32.67	33.75	36.62	27.57	35.24	33.83	37.68	33.09	37.53	37.04
	Sequence C,A,B												
	Total hrs:	29.83	25.49	28.55	29.63	32.50	23.45	31.12	29.71	33.56	28.97	33.41	32.92
SHOP 2	Sequence A,B,C												
	Total hrs:	30.69	27.87	33.55	31.97	33.48	26.32	34.74	30.41	34.01	31.23	34.55	35.89
	Sequence C,A,B												
	Total hrs:	26.57	23.75	29.43	27.85	29.36	22.20	30.62	26.29	29.89	27.11	30.43	31.77
SHOP 3	Sequence A,B,C												
	Total hrs:	33.92	27.58	32.68	31.24	35.79	28.06	37.87	33.33	37.28	31.04	37.16	36.27
	Sequence C,A,B												
	Total hrs:	29.70	23.36	28.46	27.02	31.57	23.84	33.65	29.11	33.06	26.82	32.94	32.05
SHOP 4	Sequence A,B,C												
	Total hrs:	25.34	22.56	23.58	25.52	27.34	20.43	24.96	25.41	28.07	24.96	27.81	27.19
	Sequence C,A,B												
	Total hrs:	21.12	18.34	19.36	21.30	23.12	16.21	20.74	21.19	23.85	20.74	23.59	22.97
	Total Annual Loss due to setup, hr =				824								
	Percentage of annual capacity =			10.30%	of total plant capacity								

7.7.3 The Problem of Varying Efficiency

A third reason given for the inability to measure capacity is that the variation in product mix from period to period has a *subtle and nonmeasurable* detrimental effect on the ability to report on productive capacity: to wit, the *efficiency* of the production facility itself varies with the product produced, or the mix thereof, with the same facility having different production rates for different items, and therefore it is not possible to report on its capacity. Improved efficiency due to "learning", for instance, is absent under such circumstances. Worse still, the loss in efficiency is indeterminable. If only the plant can produce the same product for "long enough" time, things would be much better and capacity measurement and its utilization would be greatly improved, so the argument goes!

Granted that the efficiency of a facility changes with the product produced, due to different reasons not the least significant of which is varying labor efficiency with different products, it should still be possible to determine the productivity of the facility for each product, if it was produced *all the time*. This, then, represents the correct capacity of the facility, any of the four capacities listed above, and it is the figure used on the evaluation of the plant's capacity. This will have the salutary effect of identifying the loss, if any, due to the presence of product mix, a loss which may be corrected by any one of several avenues open to management. One such avenue, which is certainly not the only one or even the preferred one, is to produce the same spectrum of output *sequentially* instead of *simultaneously*.

To illustrate what is intended here, consider the capacity figures listed in Table 7.1. The above analysis was based on the *simultaneous* production of all three products. By that is meant that in any week all three products are produced – the products share the available capacity of the plant, leading to loss in productivity (or so it is claimed). It is also claimed that, if the plant is permitted to concentrate on one product at a time, its productivity would increase by at least 10%. Can such a mode of operation be adopted, and would it be more efficient (from the capacity utilization point of view) if the items were produced *sequentially*, one after the other, say in the order A , B , C?

The total tonnage produced under the simultaneous mode of operation of the three products is 29.67, 45.69, and 163.25 tons for products A, B, C when produced in that sequence; respectively, which we now augment by 10% to result in 32.63, 50.26, and 179.58 tons. Given the bottleneck capacities one can see that they requires 2.175,4.188, and 5.611 weeks, respectively, to satisfy all the (augmented) demand for the three products. Thus the plant shall produce according to the following schedule, in which we allowed week 3 to produce B after completing A and week 7 to produce C immediately after completing B, assuming with little effect on efficiency (*!*).

Week:	1	2	3	4	5	6	7	8	9	10	11	12	Total
A	15	15	2.63										32.63
B			9.89	12	12	12	4.36						50.26
C							20.36	32	32	32	32	31.21	179.58

Table 7.10 Fractional occupancy of the four shops under sequential production

Shop Capacity Utilization Under Sequential Production												
	wk 1	wk 2	wk 3	wk 4	wk5	wk6	wk7	wk8	wk9	wk10	wk11	wk12
shop 1	0.750	0.750	0.838	0.857	0.857	0.857	0.948	1.000	1.000	1.000	1.000	0.975
shop 2	1.000	1.000	0.794	0.750	0.750	0.750	0.809	0.842	0.842	0.842	0.842	0.821
shop 3	0.714	0.714	0.950	1.000	1.000	1.000	0.929	0.889	0.889	0.889	0.889	0.867
shop 4	0.469	0.469	0.553	0.571	0.571	0.571	0.681	0.744	0.744	0.744	0.744	0.726

This plan of operation appears to have, in addition to producing 10% more output, the salutary effect of shortening the manufacturing interval by 1 week, from 13 to 12 weeks!

Surprisingly, from the *capacity utilization point of view* this schedule, while better due to increased production, is still disappointing, as can be seen from the fractional occupancy of each shop shown in Table 7.10. To be sure, each bottleneck shop is fully utilized when producing its "favorite" product; otherwise it is underutilized, and sometimes grossly so. Over the 12 weeks, the imbalance in shop utilization varies from a low of 71.40% to a high of 100% in the "busy" shops 1, 2, and 3.

One is driven to conclude that the *design of the shop capacity* is faulty from the outset and should be corrected taking into account the mode of operation (simultaneous or sequential) as well as the variation in the individual product processing requirements.

7.7.4 The Problem of Scrap/Dropout

A problem that interferes with the correct estimation of operational capacity stems from a different dimension, namely, that of a *variable proportion of scrap/dropout* which, naturally enough, causes a drop in the output – the plant is rewarded for good products; bad products are waste of time and resources. The difficulty arising from this consideration is sometimes subtle, especially when it is concerned with the estimation of future performance.

Suppose that the industry standards for the different products are known. Then it should be a simple matter indeed to secure an estimate of the expected dropout for each product and/or of each shop (if the standards vary from shop to shop), as well as for the plant as a whole, dependent on the proportion of capacity devoted to the production of each product. The industry standard helps in defining the expected *operational* capacity, at least the part of it that concerns dropouts. This, in turn, would give a more precise picture of the performance of the shop. The subtle difficulty in estimating capacity stems from the fact that, by definition, the operational (i.e., available) capacity is based on industry standards, while actual scrap/dropout may be more, or less, than that standard and usually varies over time. If the actual scrap/dropout is more than the standard in any period, then there is little possibility of error or misinterpretation since the plant is performing worse than what is anticipated, which is easily identified. But if the converse is true, i.e., if the firm performs better than the industry average, hence its scrap/dropout is less than the standard, then one may end up with compensating errors that hide poor performance in some other criterion or, worse still, with actual capacity utilization that exceeds the operational capacity, which we have ruled out from the outset, by definition of the various capacities. This would play havoc with the concept of capacity which we have been trying so hard to define.

To render these considerations more concrete, consider again the example given in Tables 7.1 and 7.2. Assume that the industry standard of scrap/dropout for product A is 1.4%, for B is 0.78%, and for C is 1.85, and that the firm's actual performance is respectively, 3.9, 2, and 4.20%. The difference is 2.5% for A, 1.22% for B, and 2.35% for C, which translate into 0.77[2] (=30.44 – 29.67), 0.57 (=46.26 – 45.69), and 4.00 (=167.25 – 163.25) tons of the products in the quarter, respectively, for a total of ≈5.35 ton/quarter, which represents some 2.24% of the plant production (which would increase from 238.60, which is the actual quantity produced in the quarter, to 243.95 tons/quarter (2.24 ≈ (243.95 – 238.60)/243.95). This analysis is summarized in Table 7.11. Of course, the industry "average" is just that – an average. By the definition of the word "average", there must be plants with dropout less, and other plants with dropout more, than the average. The plant's ambition should be on the lower side of the industry average, which means that the increased production determined above constitutes a *lower bound* on what should be produced.

The scenario presented above was relatively easy to analyze because the plant's performance was worse than the industry average. But suppose the plant's performance was better than the industry average, then the plant would have produced more than was anticipated. Unless data on the scrap/dropout of the products are estimated independently from the actual production data, there is the possibility of the

Table 7.11 Dropout Analysis (relative to industry standard)

		Summary			
				tons	
	Dropout		Actual	Indstry avge	
	Actual%	Ind.Avge%	production	production	Increase
A	3.90	1.40	29.67	30.44	0.77
B	2	0.78	45.69	46.26	0.57
C	4.20	1.85	163.25	167.25	4.00
		total / qrtr =	238.60	243.95	5.35

[2] This is the difference between the sums. Individual entries for each product for each week are evaluated separately. For instance, from Table 7.2 the actual production of A in week 1 was 0.81 tons. Therefore production under industry standards is $(0.81/1 - 0.039)(1 - 0.014) =$ 0.83 ton; etc.

Table 7.12 Estimates of operational capacities

Estimate of plant operational capacity, Current				
19.22	14.42	20.18	30.75	
13.72	15.68	11.76	20.58	
30.66	36.40	34.49	41.19	
Estimate of plant capacity, Industry Average				
19.72	14.79	20.71	31.55	
13.89	15.88	11.91	20.84	
31.41	37.30	35.33	42.20	

good performance relative to quality camouflaging poor performance on some other dimension such as the sequence of production relative to setup considerations.

A discussion of scrap/dropout is not complete without consideration of the *input* to the process in the form of materiel and other support activities such as maintenance and replacement. But such considerations would take us far afield from our main concern with capacity.

To sum up this discussion on scrap/dropout with respect to capacity and its measurement, it is clear that scrap/dropout should be measured independently of the other measures available, such as the total (or individual) production. Information on scrap/dropout is invaluable in determining the operational (i.e., available) capacity. The operational capacities of the plant under current scrap/dropout conditions and under industry standard are shown in Table 7.12. Though the differences may appear small, you should remember that this is the difference due solely to scrap/dropout. It should be considered in conjunction with the other factors that determine the operational capacity. Its availability over time is also invaluable not only relative to the plant's practice and progress (or lack of it), but also concerning the plant's performance relative to the industry in general.

7.7.5 The Problem of Semifinished Items or Subassemblies

A remarkable phenomenon occurs when the reported production exceeds the operational, and sometimes even the nominal, capacity! We say "remarkable" because we have insisted throughout that the operational capacity forms an upper bound on availability; and more so for the nominal capacity. Then how is it possible that actual production exceeds its bound?

The answer lies in the insertion of semifinished products at some point in the process so that the bottleneck operation(s) is circumvented. Production supervisors would gleefully cite this device as one more reason for their inability to report "correct" figures on capacities. Evidently, subcontracting part of the work is another device used to boost the actual production beyond available capacity, and thus make a mockery of the figures accumulated on either the operational or nominal capacities.

To see the import of these capacity augmenting devices, consider once more Table 7.1. Suppose that the plant can secure partially processed product A that requires only shops 3 and 4. Then the old bottleneck, operation 2, has been circumvented and the plant's capacity of product A has now jumped to 21 ton/week, an increase of 40% (= (21 − 15) ÷ 15) over the old capacity. Put differently, the plant is now capable of producing close to 21 tons/week which makes it appear to be performing 40% above its nominal capacity, a magnificent feat indeed for which the workers would deserve a hefty bonus!

The fallacy lies in the manner in which data are collected. There is no inherent problem in determining the correct capacity in the presence of semifinished products or subassemblies. Of course, the data collection method should recognize the presence of semiprocessed items, which are to be treated as new items with their own bottlenecks and their own share of the shop time. One more example of the interdependence of *all* the activities in the production process, this time with the plant's information system.

The picture gets more complicated if not all the planned production of a product is supplied as semiprocessed parts which require a partial set of the total processes. In this instance, one must be careful lest the total capacity of the plant (or shop) be underestimated, because of the possibility of simultaneously performing production on two items that are the same in terms of the end-product, but are different in their processing requirements.

To exemplify the calculation under such circumstances, consider once more product A and its production in week 3 of 4.9 tons as shown in Table 7.2. Suppose that in fact some of that tonnage was secured as a partially processed product A that requires only shops 3 and 4. To be able to distinguish this lot, call it product A1. Then its bottleneck is shop 3 with nominal capacity of 21 ton/week. If some 32.64% (abbreviated in Table 7.2 to 33%) of the plant's capacity is still devoted to the production of product A, of which 10% was for product A and the remaining 22.64% was for product A1, then the plant would produce $0.1 \times 15 + 0.2264 \times 21 \approx 6.25$ tons and not the mere 4.90 tons previously reported. The difference of $6.25 - 4.90 = 1.35$ tons represents the difference between producing A "from scratch" and injecting A1 in the plant at shop 3. On the basis of the old nominal capacity of the plant, the 6.25 tons of A, if reported as such, would account for $6.25 \div 15 = 0.4167$ of the plant capacity. Adding this figure to the production of items B and C would result in total usage equal to $0.417 + 0.281 + 0.186 = 0.884$, a hefty increase over the original value of 0.793. In fact, items A and A1 would have consumed, respectively, 0.1 and 0.2264 of the shop capacity, for a total of 0.3264 as originally envisaged. The jump in productivity is illusory.

The problem of the use of semifinished products and/or subassemblies to augment plant of shop capacity poses no particularly difficult problems *provided that the data on which the calculations are based reflect faithfully the processing requirements of the items produced*. If such capacity-augmenting devices are possible, and are in fact used, even on an ad hoc basis, then the data collection system should include the possibility of reporting such capacity-augmenting devices and distinguishing among inputs in view of the fact that outputs are indistinguishable. Here we encounter once more the need to view the production system in its totality, which necessarily includes its information system.

7.7.6 Some Sociological/Cultural/Economic Factors

Strange as it may seem, inability to accurately determine the capacity of a productive facility may have its roots in nontechnical aspects of the issue, namely, the social/cultural/economic elements of the enterprise. In particular, in an environment where the faking of data to hide the truth from management is endemic, it seems odd to insist on correct data for this particular purpose. A plant within a company, or a shop within a plant, or even a machine operator within a shop, may possess the correct data and may know how to use those data correctly to measure its capacity (any of the four), but may not be willing to divulge these data to its management lest it be deprived of certain benefits or advantages, such as bonus payments when the planned performance is raised by the firm's planning office to match its true capabilities. The literature of industrial engineering is replete with examples of reaction to "time standards" which vary from "soldiering" to outright destruction of product and equipment to prevent management from having access to such information.

It is well known that information is power, possessed by those who have access to it. For information to be shared by all, an atmosphere of trust, understanding, and working for the common good must exist and nurtured by management. The course of action to create such environment varies with the particular circumstances of the firm, and there is a vast literature on it, which is well outside the scope of this chapter. But one can safely state that unless, and until, such environment is fostered there is no escape from being trapped in an atmosphere of uncertainty of unknown magnitude.

7.8 A Recipe for the Determination of the Operational Capacity

Section 7.7 detailed why it is difficult to measure capacity (any of the four defined in Sect. 7.3), and it is time to address the question of: How does one go about determining its magnitude, assuming the requisite data, uncertain as it may be, is available?

The most important concept forwarded here is that you should not expect "a number", but a "distribution" with probabilities attached to it. The concept of a *random* estimate of capacity may be new to industry, but it is inescapable if one wishes to come up with meaningful results. We use the example data presented in Tables 7.1 and 7.2 to illustrate the concepts presented.

Our take-off point is the nominal capacity. It is the easiest to determine in case of machine-controlled production because it is the "rated" output of the machine, or facility, declared by the manufacturer of the equipment. In case of labor intensive operations, matters are not that "crisp" and one must resort to the "official and declared net working hours". The translation of these hours into the units of output, such as pounds, yards, dollars, etc., should be statistical in nature.

Consider product A in Table 7.1. The plant's nominal capacity is given as 15 ton/week based on the bottleneck shop 2. Then, we had made the implicit assumption that production is machine controlled on the basis of 40 h/week of operation, and that is how we secured the "crisp" number 15. But suppose it is labor intensive; then it would be more meaningful to speak of the nominal capacity as varying between 12 and 18 ton/week, say, with mean 15. In the absence of any information about the variability of the nominal capacity, one can proceed assuming the *uniform distribution* between these two limits; otherwise, the known distribution would be used. For the moment, we shall assume the following values for product A in all four shops, with all values uniformly distributed between the specified limits. (In a real-life situation, the analyst should substitute the appropriate distributions.) Observe that in all cases the average of the distribution is the "crisp" estimate given in Table 7.1.

Shops nominal capacity of product A, ton/week

Shop:	1	2	3	4
A	$\sim U\,[18, 22]$	$\sim U\,[12, 18]$	$\sim U\,[18, 24]$	$\sim U\,[28, 36]$

(7.15)

Further, the production engineer should have some information on the various factors that distinguish the *nominal* capacity from the *operational* (or available) capacity. For purpose of illustration, assume the following information is available on the product:

quality information: as specified in Table 7.12 under column "Ind. Avge".
setup information: as specified in the tabulation (7.14)

For simplicity of exposition, suppose that all four shops have identical parameters for which the following statistics were available from past records, still on the basis of 40 h/week:

maintenance: regular maintenance = 0.50 h/week.
electric power blackout: there are two factors that behave independently, both uniformly distributed:

number of occurrences per week $N_{\mathrm{B}} = 2.0 \pm 0.36$,
duration of blackout per occurrence $T_{\mathrm{B}} = 1.5 \pm 0.25$ h.

Then one would proceed as follows to calculate the operational (i.e., available) capacity in the plant. For the sake of brevity, we shall detail the calculations only for product A in shop 1, product A in other shops as well as product B and C in all shops can be analyzed similarly. The duration (or length) of the "working week" is denoted by L, which shall be subscripted by the factor that reduces its value from the nominal.

As a footnote to our analysis, we would be remiss if we do not mention that considerations of the decrement in productivity due to maintenance, machine breakdowns, and quality yield are also the concern of the area of industrial engineering labeled "total productivity maintenance" (TPM); seeVenkatesh (2005)and the references cited therein. The main distinction between the concerns of TPM and our concerns in this chapter are three: (1) TPM is focused only on machine maintenance activities in a productive system, while we are concerned with the capacity of the productive system; (2) TPM is a "business-management" oriented view (as opposed to a "technically" oriented view) of the machine maintenance activities in a productive system. The focus on the managerial aspects (as opposed to the engineering aspects) is evident in their talk about the "5 S" principle. These are the five Japanese terms for the so-called "pillars" of TPM: Seiri, Seiton, Seiso, Seiketsu, Shitsuke; (3) TPM is "deterministic" in its analysis, dealing mostly with averages, while we argue for stochastic considerations.

7.8.1 Accounting for Maintenance Requirements

The *maintenance* requirements, if performed during the working hours, are actually random variables (r.v.'s) depending on the condition of the plant facilities at the time of maintenance. For simplicity of the subsequent analysis, we shall assume them to be fixed constants. In particular, for illustrative purposes, we assume that maintenance would subtract 0.50 h each week, thus reducing the nominal working week to $L_{\mathrm{M}} = 39.5$ h, in which the subscript stands for having accounted for maintenance.

7.8.2 Accounting for Blackouts

The reader is advised that we use "blackout" as a stand-in for all the unforeseen and uncontrollable disruptions that interrupt the use of the production facility, such as, in addition to electric current blackout, machine failure, shortage of tools and fixtures, labor absenteeism, etc.

In a given time interval, say a week, which we take as our "loading time unit" the total time lost due to blackouts, denoted by B, is impacted by two factors: the *number* of occurrences in a week N_B, a r.v., and the *duration* of each occurrence $T_B(n)$, $n = 0, 1, \ldots,$ assumed to be independent and identically distributed r.v's. N_B and $\{T_B(n)\}$ are assumed to be two independent occurrences; hence blackout duration (= process unavailability) in a week is a random variable equal to the sum

$$B = \sum_{n=1}^{N_B} T_B(n).$$

It is desired to derive the mean and variance of B. Using the relations given in the Appendix, we know that

$$\mathcal{E}[B] = \mathcal{E}[N] \cdot \mathcal{E}[T_B],$$

and,

$$\begin{aligned}\mathrm{var}[B] &= \mathcal{E}\left[N_B \sigma_{T_B}^2\right] + \mathrm{var}[N_B \cdot \mathcal{E}[T_B]] \\ &= \sigma_{T_B}^2 \cdot \mathcal{E}[N_B] + (\mathcal{E}[T_B])^2 \cdot \sigma_{N_B}^2\end{aligned}$$

We have that the number of blackouts/week is $N_B \sim U[1.64, 2.36]$, with $\mathcal{E}[N_B] = 2$, and $\mathrm{var}(N_B) = 0.72^2/12 = 0.0432$; and that the duration of a single blackout is $T_B \sim U[1.25, 1.75]$ h/occurrence, with $\mathcal{E}[T_B] = 1.5$ h and $\mathrm{var}(T_B) = 0.5^2/12 = 0.02083$. We have that

$$\mathcal{E}[B] = 2 \times 1.5 = 3 \text{ h/week},$$

and,

$$\begin{aligned}\mathrm{var}(B) &= 0.02083 \times 2 + 1.5^2 \times 0.0432 = 0.13887 \\ \Rightarrow \sigma_B &= \sqrt{0.13887} = 0.37265.\end{aligned}$$

Then, assuming that B is approximately normally distributed, one can assert that with probability 0.99 the loss in production time due to blackout B lies in the interval

$$\mathcal{E}[B] - z_{0.99}\sigma_B \leq B \leq \mathcal{E}[B] + z_{0.99}\sigma_B,$$

in which $z = 2.5758$, to yield,

$$2.040 \leq B \leq 3.960.$$

Therefore the net available hours per week lie in the interval

$$[39.5 - (3.960, 2.040)] = [35.540, 37.460].$$

Thus far the net available time per week is a r.v., call it $L_{M,B}$ (for *maintenance* and *blackouts*), which lies in the interval [35.540, 37.460]; i.e.,

$$L_{M,B} \sim U[35.540, 37.460], \quad \text{with mean } = 36.5 \text{ h/week.} \tag{7.16}$$

7.8.3 Accounting for Personal Time

Of the 40 h in the *nominal* work week, allowance is usually made for meals and personal needs which amount to some 50 min/day ($0.8\dot{3}$ h), or 4.167 h/week, which when subtracted from the result in (7.16) yields,

$$L_{M,B,P} \sim U[35.540 - 4.167,\ 37.460 - 4.167] = U[31.373, 33.293], \tag{7.17}$$

in which $L_{M,B,P}$ stands for the length of available time in a week after accounting for *maintenance, blackouts,* and *personal* needs.

Observe that, so far, the "*crisp*" value of 40 h/week have been modified into the *interval* $[31.373, 33.293]$.

7.8.4 Estimation of Hourly Output

Let $X_{A,\mathrm{nom}}(1)$ denote the nominal *hourly* production of product A in shop 1. Then

$$X^{(1)}_{A,\mathrm{nom}} \sim \frac{\mathcal{A}^{(1)}}{L_{M,B,P}},$$

in which $\mathcal{A}^{(1)}$ denotes the nominal capacity of shop #1 of product A, see the plant nominal capacity table in (7.15), a random variable given as $\mathcal{A}^{(1)} \sim U[18, 22]$ and measured in ton/week. Taking note of the net available capacity per week $L_{M,B,P}$ of (7.17) after recognizing the loss due to maintenance, blackouts, and personal needs, one can deduce the distribution of the nominal hourly production. The theoretical derivation of the exact distribution of $X^{(1)}_{A,\mathrm{nom}}$, which is the ratio between two random variables, is presented in the Appendix to this chapter.

We know from the data in (7.15) and (7.17) that it lies between $18/33.293 \approx 0.541$ and $22/31.373 \approx 0.701$, and we learn from the Appendix that it is *not* uniformly distributed between these two bounds. With abuse of probability and for the sake of simplicity of exposition, we shall continue to assume that it is approximately uniformly distributed between these two limits (a quick glance at Fig. 7.A.3 in the Appendix would convince you that this is really not a bad approximation), then we have

$$X^{(1)}_{A,\mathrm{nom}} \sim U[0.541, 0.701], \quad \text{ton/h}.$$

7.8.5 Accounting for Scrap/Dropout

We are given that the industry average of the *scrap/dropout* rate is $p_s = 0.014$. Therefore, the limits of the expected production rate would be further reduced to $0.537(1 - 0.014) \approx 0.533$ and $0.7065(1 - 0.014) \approx 0.691$, and we end up with

$$X^{(1)}_{A,\mathrm{nom}} \sim U[0.533, 0.691], \quad \text{ton/h}. \tag{7.18}$$

We therefore conclude that the operational capacity of shop 1 of product A is a random variable that is distributed between $31.373 \times 0.533 = 16.725$ ton and $33.526 \times 0.696 = 23.019$ ton, with an average of 19.872 ton. In the new Table 7.13, the interval $[16.725, 23.019]$ should replace the single entry of 20 of

Table 7.13 Operational Capacity/wk, ton for product A

Shop	$[l.b., u.b]$	Average	Estimate†	Support‡
1	[16.725, 23.019]	19.872	20	6.295
2	[11.150, 18.834]	14.992	15	7.684
3	[16.725, 25.112]	20.918	21	8.387
4	[26.016, 37.668]	31.842	32	11.652

† Secured from Table 7.1
‡ Support =u.b.− $l.b.$ Any discrepancy between the stated value and the arithmetic operation is due to round-off

Table 7.1 in cell (A, shop1). A reader who insists on exactitude should replace the uniform distribution in Table 7.13 suggested here with the exact distribution given in the Appendix.

Without cluttering this chapter with further calculations, the operational capacity for product A in the other three shops is determined in a similar fashion and the result for all four shops is as shown in Table 7.13. Recall that these values are *approximations* of the exact values, which are not uniformly distributed between their respective limits. Finally, observe the closeness of the average values to their estimated deterministic equivalents of Table 7.1.

When similar analysis is conducted for the other two products B and C, we shall have in hand the operational capacity of all products in all shops after accounting for the four factors: maintenance, blackouts, personal, and scrap/dropout.

7.8.6 The Determination of the Bottleneck Capacity

The bottleneck shop for each product will have a distribution that is given by the minimum of the four random variables along its row across the shops. This distribution can be evaluated from the four distributions, which we assumed to be uniform. To be more precise, Let $Y_{A,\text{oper}}$ denote the operational capacity of the bottleneck shop for product A. Then,

$$\overline{F}_{Y_{A,\text{oper}}}(y) = \prod_{s=1}^{4} \overline{F}_{X_{A,\text{oper}(s)}}(y), \quad \text{ton/week}$$

in which $\overline{F}_{Y_{A,\text{oper}}}(y)$ is the complementary cumulative distribution function (ccdf), $\overline{F}_{Y_{A,\text{oper}}}(y) = 1 - F_{Y_{A,\text{oper}}}(y)$, and similarly for $\overline{F}_{X_{A,\text{oper}}(s)}(y) = 1 - F_{X_{A,\text{oper}}(s)}(y)$. The expected (bottleneck) production is given by

$$\mathcal{E}\left(Y_{A,\text{oper}}\right) = \int_{y=0}^{\infty} \overline{F}_{Y_{A,\text{oper}}}(y)\, dy, \qquad \text{ton/week.}$$

To illustrate these calculations, assume, for simplicity, that as far as product A is concerned the same parameters (maintenance, blackout, and personal allowance) apply to the other three shops. Then,

$$\overline{F}_{Y_{A,\text{oper}}}(y) = \left(1 - \max\left\{0, \frac{y - 16.725}{6.295}\right\}\right)\left(1 - \max\left\{0, \frac{y - 11.150}{7.684}\right\}\right)$$
$$\times\left(1 - \max\left\{0, \frac{y - 16.725}{8.387}\right\}\right)\left(1 - \max\left\{0, \frac{y - 26.016}{11.652}\right\}\right),$$

with $11.150 \le y \le 37.668$.

It is clear that shop 4 plays no role in the determination of the bottleneck capacity, since its distribution starts after the completion of the other three distributions. Further, since the distribution of shops 1 and 3 start at 16.725, the probability distribution in the interval [11.055, 16.725] is identical to the distribution of the operational capacity of shop 2, which we assumed to be uniform. In the interval [16.725, 18.834], shops 1 and 3 may be the bottleneck shop and the distribution is no longer the simple uniform. The exact cumulative distribution is given by

$$F_{Y_{A,\text{oper}}}(y) = \frac{y - 11.150}{7.684}, \quad \text{for } 11.150 \le y \le 16.725,$$

and for $16.725 \le y \le 18.834$ we have,

$$F_{Y_{A,\text{oper}}}(y) = 1 - \left(1 - \max\left\{0, \frac{y - 16.725}{6.295}\right\}\right)\left(1 - \max\left\{0, \frac{y - 11.150}{7.684}\right\}\right)$$
$$\times\left(1 - \max\left\{0, \frac{y - 16.725}{8.387}\right\}\right).$$

The cumulative distribution of the bottleneck is shown in Fig.7.6 from which it is seen that one can indeed take the density function to be approximately uniform between the two limits; i.e., if we denote the bottleneck capacity by $Y_{A,\text{oper}}$ then it is approximately uniformly distributed between 11.150 and 18.834.

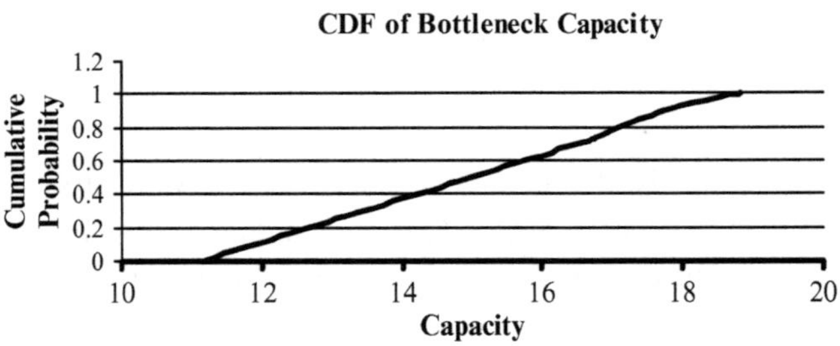

Fig. 7.6 The cumulative distribution $F_{A,\text{oper}}(y)$ of the bottleneck operation for product A

The upshot of this analysis is twofold: first, the bottleneck operational capacity of the plant for product A varies *randomly* from as low as 11.150 ton/week to as high as 18.834 ton/week, and second, the average capacity is (approximately) $11.150 + 18.834/2 \approx 15$ ton/week, as previously suggested (see Table 7.1 under column "BNC").

7.9 Conclusions

The concept of *capacity* in humans and processes is fuzzy and elusive, and is, unfortunately, often confused with other notions such as *performance, output, throughput, etc.* As a consequence, the measurement of "capacity", which is often taken as "well known", has remained uncharted ground.

The folklore of industrial engineering contains the following anecdote. Four railroad repairmen were on a "platform" on the track fixing some of its joints when they heard the express train approaching them. They realized that if they stay on the platform they shall certainly die, and if they jump off the platform and leave it on the track the train may be derailed and hundreds of passengers shall certainly be hurt. They had only one option, to remove the platform off the track; which they did just in time for the express train to pass safely. Afterwards they tried to put the platform back on the track but could not – it was too heavy for them. It took 12 men to lift the platform and place it back on the track! Question: what is human lifting capacity?

It is our contention that there are not one but *four* kinds of "capacity", and whoever talks about "capacity" should clearly identify which one is referred to; they are: *nominal, operational, planned,* and *realized* capacities. It is also our contention that in the majority of cases one cannot measure capacity, any one of the four types, by a crisp number, but that its proper identification requires distribution of values and a probability statement attached to its values. This view necessitates a different type of analysis (of existing processes) and synthesis (of proposed ones) that is different in methodology as well as conclusions.

Knowledge of a plant's capacity is required for the myriad of uses to which such knowledge is put, particularly the whole gamut of activities that are related to the field of production planning and control, from plant location and sizing to the daily scheduling of operations.

We trust that this chapter will motivate research in at least three important and related issues.

First, the establishment of a taxonomy of "applicable capacity concepts" in the various fields of production and logistics. One would expect that the needs of a "flowshop" in a chemical process are different from those of a "jobshop" in a general engineering firm.

Second, not all the six factors listed in Sect. 7.7 are at play at any one time, or are present in the same firm. It is possible that management, if made cognizant of the role these factors play in determining its capacity, will identify the factors that are relevant to its particular environment. Management may even be able to go one step further and determine, with some degree of confidence, the variability of these factors and the degree of their impact on capacity. In which case, the construction and validation of models for the accurate and precise measurement of the *combined* impact of these factors, taking stochastic behavior into account and incorporating the relevant concepts of production engineering relative to the degradation of performance with use and age would be a profitable undertaking. This would certainly be much preferred to trying to produce a set of *separable* numbers which represent the capacity of the facility for each item produced. We gave rather rudimentary examples of how one can proceed in this arena, but a great deal more needs to be done.

Third, the implementation of such paradigms on select prototype enterprises that would demonstrate the subtleties and degree of applicability of the proposed procedures to real-life situations, and perhaps point out the need for other classes of capacity.

Appendix: Mathematical Background and Derivations

The Variance of Compound Random Variables

We are interested in the variance of a compound random variable (r.v.). In particular, we wish to establish a relation between the variance of the r.v. X and its variance when conditioned on another r.v. Y. We start with a lemma which we use to derive the desired result.

Lemma 7.1.

$$\operatorname{var}(X) = \mathcal{E}\left[\operatorname{var}(X\,|\,Y)\right] + \operatorname{var}\left(\mathcal{E}\left[X\,|\,Y\right]\right). \tag{7.A.1}$$

Proof.

$$\begin{aligned}
\mathcal{E}\left[\operatorname{var}(X\,|\,Y)\right] &= \mathcal{E}\left[\mathcal{E}\left[X^2|Y\right] - (\mathcal{E}\left[X\,|\,Y\right])^2\right], \text{ by definition}\\
&= \mathcal{E}\left[\mathcal{E}\left[X^2|Y\right]\right] - \mathcal{E}\left[(\mathcal{E}\left[X\,|\,Y\right])^2\right],\\
&= \mathcal{E}\left[X^2\right] - \mathcal{E}\left[(\mathcal{E}\left[X\,|\,Y\right])^2\right].
\end{aligned} \tag{7.A.2}$$

The second equality is secured by pushing the expectation through the square bracket, and the third equality follows from the fact that the expectation of the conditioned X^2 is the unconditioned expected value of X^2 itself.

We also have by the definition of variance,

$$\begin{aligned}
\operatorname{var}(\mathcal{E}\left[X\,|\,Y\right]) &= \mathcal{E}\left[(\mathcal{E}\left[X\,|\,Y\right])^2\right] - (\mathcal{E}\left[\mathcal{E}\left[X\,|\,Y\right]\right])^2, \text{ by definition}\\
&= \mathcal{E}\left[(\mathcal{E}\left[X\,|\,Y\right])^2\right] - (\mathcal{E}\left[X\right])^2.
\end{aligned} \tag{7.A.3}$$

Adding these two expressions yields

$$\mathcal{E}\left[\mathrm{var}(X|Y)\right] + \mathrm{var}(\mathcal{E}\left[X|Y\right]) = \mathcal{E}\left[X^2\right] - (\mathcal{E}\left[X\right])^2,$$

in which the right side is the variance of X, and the result in (7.A.1) follows. □

Let $X_1, X_2, \ldots$ be independent and identically distributed r.v.'s (each representing the duration of an interruption) with cumulative distribution function (*cdf*) F having mean $\mathcal{E}\left[X\right] = \mu_X$ and variance $\mathrm{var}(X) = \sigma_X^2$. Assume that the X_t's are independent of the nonnegative r.v. N (the number of occurrences). The r.v. $S = \sum_{i=1}^{N} X_i$ is called a "compound r.v." in which the variance is secured by the following argument that uses Lemma 7.1. We have that

$$\mathcal{E}\left[S|N = n\right] = n\mu_X, \tag{7.A.4}$$

which

$$\Rightarrow \mathcal{E}\left[S|N\right] = N\mu_X. \tag{7.A.5}$$

Upon removing the conditioning, we get

$$\mathcal{E}\left[S\right] = \mu_X \mathcal{E}\left[N\right] = \mathcal{E}\left(X\right)\mathcal{E}\left[N\right]. \tag{7.A.6}$$

To determine the variance of S, we first condition on N,

$$\begin{aligned}\mathrm{var}(S|N = n) &= \mathrm{var}\left(\sum_{i=1}^{n} X_i | N = n\right), \\ &= \mathrm{var}\left(\sum_{i=1}^{n} X_i\right) = n\sigma_X^2.\end{aligned}$$

Therefore,

$$\mathrm{var}(S|N) = N\sigma_X^2,$$

in which N is a r.v. Hence,

$$\mathcal{E}\left(\mathrm{var}(S|N)\right) = \sigma_X^2 \mathcal{E}\left[N\right] = \mathrm{var}(X)\mathcal{E}\left[N\right]. \tag{7.A.7}$$

To utilize the result in Lemma 7.1, we need to evaluate $\mathrm{var}\left(\mathcal{E}\left[S|N\right]\right)$, which is given as follows:

$$\begin{aligned}\mathrm{var}\left(\mathcal{E}\left[S|N\right]\right) &= \mathrm{var}\left(N\mathcal{E}\left[X\right]\right), \quad \text{by (7.A.5)} \\ &= \left(\mathcal{E}\left[X\right]\right)^2 \mathrm{var}(N), \quad \text{since } \mathcal{E}\left[S\right] \text{ is a constant.}\end{aligned} \tag{7.A.8}$$

Using the result in (7.A.1) and substituting from the last two results, we have that

$$\begin{aligned} \text{var}(S) &= \mathcal{E}\left[\text{var}\left(S\,|\,N\right)\right] + \text{var}\left(\mathcal{E}\left[S\,|\,N\right]\right) \\ &= \text{var}\left(X\right)\mathcal{E}\left[N\right] + \left(\mathcal{E}\left[X\right]\right)^2 \text{var}(N). \end{aligned} \tag{7.A.9}$$

The Compound Poisson Process

An interesting special case which has many applications in practice has N as a Poisson r.v. Then $S = \sum_{i=1}^{N} X_i$ is called a "compound Poisson r.v." Because the variance of a Poisson r.v. is equal to its mean, it follows that for a compound Poisson r.v. with $\mathcal{E}\left[N\right] = \lambda$, we have,

$$\text{var}(S) = \lambda\sigma^2 + \lambda\mu^2 = \lambda\mathcal{E}\left[X^2\right],$$

where X has the cdf F.

The Distribution of a Ratio: A Geometric Argument

Suppose we have two r.v.'s X and Y that are uniformly distributed in the intervals $[a, b]$ and $[c, d]$, respectively, with $0 < a < b$ and $0 < c < d$. Let $W = X/Y$. The problem is to find the cdf of W, $F_W(w)$. Since X and Y are strictly positive r.v.'s, it is clear that

$$F_W(w) = \Pr\left(W \le w\right) = 0 \quad \text{if } w < \frac{a}{d} \text{ or } w > \frac{b}{c}.$$

Thus, we may take $b/c \ge w \ge a/d$.

Now,

$$\begin{aligned} \Pr\left(W \le w\right) &= \Pr\left(\frac{X}{Y} \le w\right) = \Pr\left(X \le wY\right), \\ &= \Pr\left(Y \ge \frac{X}{w}\right), \\ &= \Pr\left[(X, Y) \in S_w\right], \end{aligned}$$

where

$$S_w = \left\{(x, y) : a < x < b,\ c < y < d,\ y \ge \frac{x}{w}\right\}.$$

The ratio space as shown in Fig. 7.A.1 [assuming that $(da/c) < (cb/d)$] is divided into three regions.

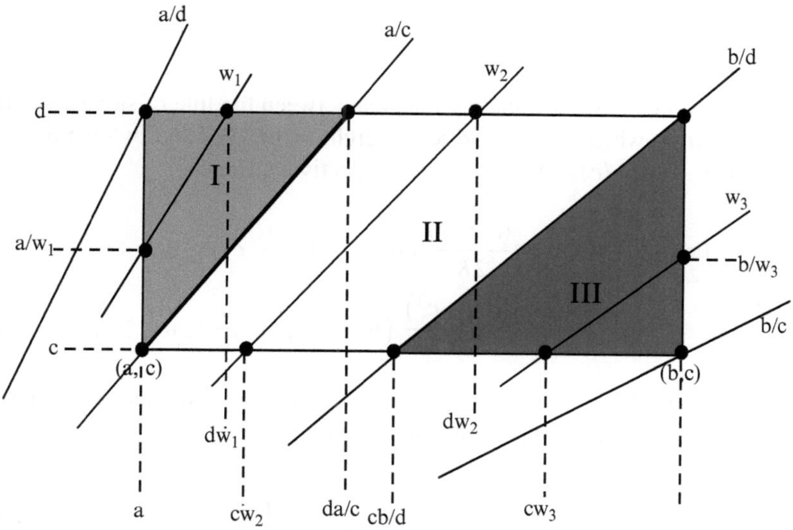

Fig. 7.A.1 Ratio space for geometric development of the distribution of a radio of random variables

Region I

We seek the value of

$$\Pr\left[(X,Y)\in S_w\right]=\frac{1}{K}\int_{x=a}^{da/c}\left(\int_{y=x/w}^{d}f_{X,Y}\left(x,y\right)\mathrm{d}y\right)\mathrm{d}x, \tag{7.A.10}$$

in which

$$K=(b-a)\,(d-c)\,.$$

The probability in (7.A.10) is given by the area of the triangle above the line $x/y = w_1$, given by (the base) × (the height of the triangle), where the base is $dw_1 - a$ and the height is $d - (a/w_1)$; hence

$$\Pr\left[\frac{X}{Y}\le w_1\right]\Rightarrow\frac{1}{2K}\left(dw_1-a\right)\left(d-\frac{a}{w_1}\right),$$
$$=\frac{\left(dw_1-a\right)^2}{2Kw_1};\quad \frac{a}{d}\le w_1\le\frac{a}{c}. \tag{7.A.11}$$

As a check, $\Pr\left[X/Y\le w_1\right]_{w_1=a/d}=0$, as it should be. We shall need the next value for addition to the results of the other regions,

$$\Pr\left[\frac{X}{Y}\le w_1\right]_{w_1=\frac{a}{c}}=\frac{a}{2Kc}\left(d-c\right)^2. \tag{7.A.12}$$

Region II

Similar reasoning leads to evaluating the area between the line of slope w_2 and the line of slope a/c, which is a trapezoid of height $= (d-c)$ and two parallel sides given by $(dw_2 - (da/c))$ and $(cw_2 - a)$, respectively. Hence,

$$\Pr\left[\frac{X}{Y} \leq w_2\right] \Rightarrow \frac{(c+d)}{2K}\left(w_2 - \frac{a}{c}\right)(d-c),$$
$$= \frac{(d^2-c^2)}{2K}\left(w_2 - \frac{a}{c}\right); \quad \frac{a}{c} \leq w_2 \leq \frac{b}{d}. \tag{7.A.13}$$

As a check, $\Pr[X/Y \leq w_2]_{w_2=a/c} = 0$, as it should be. We shall need the next value for addition to the results of the other regions,

$$\Pr\left[\frac{X}{Y} \leq w_2\right]_{w_1=\frac{b}{d}} = \frac{(d^2-c^2)}{2K}\left(\frac{b}{d} - \frac{a}{c}\right). \tag{7.A.14}$$

Region III

Similar reasoning leads to evaluating the area between the line of slope w_3 and the line of slope b/d, which is easier to compute as it is equal to the area of the whole triangle of region III less the lower triangle below the line of slope w_3. The area of region III is,

$$\frac{1}{2K}\left(b - \frac{cb}{d}\right)(d-c). \tag{7.A.15}$$

The triangle below the line of slope w_3 has area equal to

$$\frac{1}{2K}(b - cw_3)\left(\frac{b}{w_3} - c\right). \tag{7.A.16}$$

The desired probability is the difference between these two expressions,

$$\Pr\left[\frac{X}{Y} \leq w_3\right] = \frac{1}{2K}\left[\left(b - \frac{cb}{d}\right)(d-c) - (b-cw_3)\left(\frac{b}{w_3} - c\right)\right],$$
$$= \frac{1}{2K}\left[\frac{b(d-c)^2}{d} - (b-cw_3)\left(\frac{b}{w_3} - c\right)\right], \quad \text{for } \frac{b}{d} \leq w_3 \leq \frac{b}{c}. \tag{7.A.17}$$

As a check, $\Pr[X/Y \leq w_3]_{w_3=b/d} = 0$, as it should be. We shall need the next value for addition to the results of the other regions,

$$\Pr\left[\frac{X}{Y} \leq w_3\right]_{w_3=\frac{b}{c}} = \frac{b(d-c)^2}{2Kd} \tag{7.A.18}$$

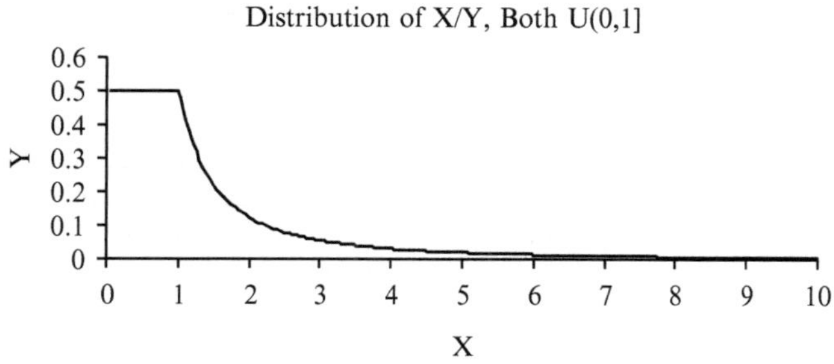

Fig. 7.A.2 Cumulative distribution function of X/Y with both U(0,1]

The final check is the sum of the three regions,

$$\begin{aligned}&(7.A.12) + (7.A.14) + (7.A.18),\\ &= \frac{a}{2Kc}(d-c)^2 + \frac{(d^2-c^2)}{2K}\left(\frac{b}{d}-\frac{a}{c}\right) + \frac{b\,(d-c)^2}{2Kd},\\ &= \frac{(d-c)}{2K} \times 2\,(b-a) = 1, \text{ by the definition of } K.\end{aligned}$$

If both X and Y are uniformly distributed over the interval (0, 1], then the cumulative distribution would appear as shown in Fig. 7.A.2.

Application of this theory to our case, the variable X is the capacity and the variable Y is the operational time per week (referred to in the text as L). In shop #1, we have the following parameters:

$$\begin{aligned} a &= 18, & \tfrac{a}{d} &= 0.541\\ b &= 22, & \tfrac{a}{c} &= 0.574\\ c &= 31.373 & \tfrac{b}{d} &= 0.661\\ d &= 33.293 & \tfrac{b}{c} &= 0.701 \end{aligned}$$

Substituting these parameters in expressions (7.A.11),(7.A.13) and (7.A.17) results in the complete specification of the probability distribution function over the full range of w,

$$F_W(w) = \begin{cases} 0.057 \times \frac{(33.293w-18)^2}{w}, & 0.541 \le w \le 0.574,\\ 0.121 + (w - 0.574), & 0.574 \le w \le 0.661,\\ 0.618 - 0.073\,(22 - 31.373w)\left(\frac{22}{w} - 31.373\right), & 0.661 \le w \le 0.701. \end{cases}$$

The plot of $F_W(w)$ is given in Fig. 7.A.3, which is almost linear except at the two exterminates (especially between 0.582 and 0.660 where it is = 0.0142).

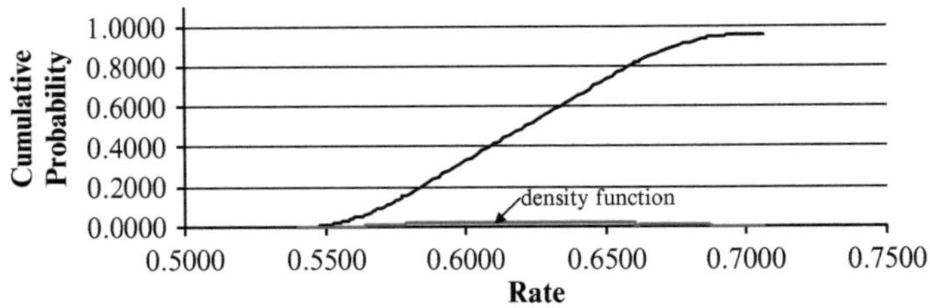

Fig. 7.A.3 Cumulative distribution function (cdf) of hourly rate

The Distribution of a Ratio: An Algebraic Argument (Contributed by J.R. Wilson)

Suppose that $X \sim U[a, b]$ and $Y \sim U[c, d]$ are independent r.v.'s with $0 < a < b$ and $0 < c < d$.[3] Thus, X has probability density function (pdf)

$$f_X(x) = \begin{cases} \frac{1}{b-a}, & \text{for } a \le x \le b, \\ 0, & \text{otherwise;} \end{cases} \tag{7.A.19}$$

and the cdf of X is

$$F_X(x) = \begin{cases} 0, & \text{if } x < a, \\ \frac{(x-a)}{(b-a)}, & \text{if } a \le x \le b, \\ 1, & \text{if } x > b. \end{cases} \tag{7.A.20}$$

Similarly, the pdf of Y is given by

$$f_Y(y) = \begin{cases} \frac{1}{d-c}, & \text{for } c \le y \le d, \\ 0, & \text{otherwise;} \end{cases} \tag{7.A.21}$$

and the cdf of Y is

$$F_Y(y) = \begin{cases} 0, & \text{if } y < c, \\ \frac{(y-c)}{(d-c)}, & \text{if } c \le y \le d, \\ 1, & \text{if } y > d. \end{cases} \tag{7.A.22}$$

We seek the cdf of $W = X/Y$. Let

$$F_W(w) = \Pr\{W \le w\}, \quad \text{for } -\infty < w < \infty, \tag{7.A.23}$$

[3] Professor James R. Wilson, Head of the Industrial and Systems Engineering Department, North Carolina State University, Raleigh, NC 27695-7906, USA.

denote the cdf in question. We see that

$$F_W(w) = 0 \quad \text{for } w \leq \frac{a}{d} \text{ and } F_W(w) = 1 \quad \text{for } w \geq \frac{b}{c}. \tag{7.A.24}$$

For $a/d \leq w \leq b/c$, we compute $F_W(w)$ by conditioning on the value of Y and applying the law of total probability:

$$\begin{aligned}
F_W(w) &= \Pr\left\{\frac{X}{Y} \leq w\right\}, \\
&= \int_c^d \Pr\left\{\frac{X}{Y} \leq w \mid Y = y\right\} f_Y(y)\mathrm{d}y, \\
&= \int_c^d \Pr\{X \leq wy\}\frac{dy}{d-c} \text{ (since } X \text{ and} Y \text{ are independent)}, \\
&= \frac{1}{d-c}\int_c^d F_X(wy)\mathrm{d}y, \\
&= \frac{1}{d-c}\left[\int_{\max\{c,a/w\}}^{\min\{b/w,d\}} \frac{wy-a}{b-a} + \int_{\min\{b/w,d\}}^{d} 1\,\mathrm{d}y\right].
\end{aligned} \tag{7.A.25}$$

The limits of integration in (7.A.25) are derived as follows. Observe that since $w > 0$, we have

$$wy \leq a \text{ if and only if } y \leq \frac{a}{w} \tag{7.A.26}$$

and

$$wy \leq b \text{ if and only if } y \leq \frac{b}{w}; \tag{7.A.27}$$

and combining (7.A.26) and (7.A.27) with (7.A.20), we obtain (7.A.25). It follows from (7.A.25) that

$$\begin{aligned}
F_W(w) &= \frac{1}{2(b-a)(d-c)w}\left[\left(w\min\left\{\frac{b}{w},d\right\}-a\right)^2 - \left(w\max\left\{c,\frac{a}{w}\right\}-a\right)^2\right] \\
&\quad + \frac{d-\min\{b/w,d\}}{(d-c)}, \\
&= \frac{w}{2(b-a)(d-c)}\left[\left(\min\left\{\frac{b}{w},d\right\}-\frac{a}{w}\right)^2 - \left(\max\left\{c,\frac{a}{w}\right\}-\frac{a}{w}\right)^2\right] \\
&\quad + \frac{d-\min\{b/w,d\}}{(d-c)},
\end{aligned} \tag{7.A.28}$$

for $a/d \le w \le b/c$. To get differentiable expressions for $F_W(w)$ on nonempty open intervals, we have to consider two distinct cases:

Case I: $b/a \le d/c$. In this case, (7.A.28) can be re-expressed as follows:

$$F_W(w) = \begin{cases} \frac{(dw-a)^2}{2w(b-a)(d-c)}, & \text{for } \frac{a}{d} \le w \le \frac{b}{d}, \\ \frac{2dw-(a+b)}{2w(d-c)}, & \text{for } \frac{b}{d} \le w \le \frac{a}{c}, \\ \frac{-b^2+2(bd-ad+ac)w-c^2w^2}{2w(b-a)(d-c)}, & \text{for } \frac{a}{c} \le w \le \frac{b}{c}. \end{cases} \tag{7.A.29}$$

Case II: $b/a > d/c$. In this case, (7.A.28) can be re-expressed as follows:

$$F_W(w) = \begin{cases} \frac{(dw-a)^2}{2w(b-a)(d-c)}, & \text{for } \frac{a}{d} \le w \le \frac{a}{c}, \\ \frac{w(c+d)-2a}{2(b-a)}, & \text{for } \frac{b}{c} \le w \le \frac{b}{d}, \\ \frac{-b^2+2(bd-ad+ac)w-c^2w^2}{2w(b-a)(d-c)}, & \text{for } \frac{b}{d} \le w \le \frac{b}{c}. \end{cases} \tag{7.A.30}$$

Expressions for the pdf are now easily obtained on each of the subintervals indicated in (7.A.29) and (7.A.30).

Although all this is somewhat tedious, the main point is that both the pdf and cdf are piecewise rational functions in w and thus are fairly easy to manipulate numerically so as to compute the moments of W, for example.

Remark 7.A.1. Notice that in both (7.A.29) and (7.A.30) we have equality of the right-side limit

$$F_W(z^+) \equiv \lim_{\substack{w \to z \\ w > z}} F_W(w) \tag{7.A.31}$$

and the left-side limit

$$F_W(z^-) \equiv \lim_{\substack{w \to z \\ w < z}} F_W(w), \tag{7.A.32}$$

for all real z; in particular, we have,

$$F_W(z^-) = F_W(z^+) \text{ for } z \in \left\{ \frac{a}{d}, \frac{b}{d}, \frac{a}{c}, \frac{b}{c} \right\}, \tag{7.A.33}$$

since W is a continuous r.v. and thus $F_W(z)$ must be continuous at every real z. Moreover, note that $F_W(a/d) = 0$ and $F_W(b/c) = 1$ in both (7.A.29) and (7.A.30) as required.

The noncentral moments of W are easily computed because X and Y are independent, so we have

$$\begin{aligned}
\mu_W &= E[W] = E[X]E\left[\frac{1}{Y}\right], \\
&= \left[\int_a^b x f_X(x)\mathrm{d}x\right]\left[\int_c^d y^{-1} f_Y(y)\mathrm{d}y\right], \\
&= \left(\frac{a+b}{2}\right)\left(\frac{\ln(d)-\ln(c)}{d-c}\right), \\
&= \frac{(a+b)\left[\ln(d)-\ln(c)\right]}{2\,(d-c)}
\end{aligned} \tag{7.A.34}$$

and for $k = 2, 3, \ldots$, the kth noncentral moment of W is given by

$$\begin{aligned}
\mu'_k &= E\left[W^k\right] = E\left[X^k\right]E\left[Y^{-k}\right], \\
&= \left[\int_a^b x^k f_X(x)\mathrm{d}x\right]\left[\int_c^d y^{-k} f_Y(y)\mathrm{d}y\right], \\
&= \left[\frac{b^{k+1}-a^{k+1}}{(k+1)(b-a)}\right]\left[\frac{c^{1-k}-d^{1-k}}{(k-1)(d-c)}\right], \\
&= \frac{\left(b^{k+1}-a^{k+1}\right)\left(c^{1-k}-d^{1-k}\right)}{(k^2-1)(b-a)(d-c)}.
\end{aligned} \tag{7.A.35}$$

Then the variance and third and fourth central moments of W are obtained in the usual way,

$$\begin{aligned}
\sigma_W^2 = \mathrm{var}[W] = E\left[(W-\mu_W)^2\right] &= \mu'^2 - \mu_W^2, \\
E\left[(W-\mu_W)^3\right] &= \mu'_3 - 3\mu'_2\mu_W + 2{\mu_W}^3, \\
E\left[(W-\mu_W)^4\right] &= \mu'_4 - 4\mu'_3\mu_W + 6\mu'_2{\mu_W}^2 - 3{\mu_W}^4;
\end{aligned}$$

and then the skewness and kurtosis of W can be computed from

$$\frac{E\left[(W-\mu W)^3\right]}{\sigma_W^3} \quad \text{and} \quad \frac{E\left[(W-\mu_W)^4\right]}{\sigma_W^4}, \tag{7.A.36}$$

respectively.

References

Asmundsson J, Rardin RL, Uzsoy R (2002) Tractable nonlinear capacity models for aggregate production planning. Working paper, School of Industrial Engineering, Purdue University, West Lafayette, Indiana, USA.

Asmundsson J, Rardin RL, Uzsoy R (2003) An experimental comparison of linear programming models for production planning utilizing fixed lead time and clearing functions. Working paper, School of Industrial Engineering, Purdue University, West Lafayette, Indiana, USA.

Buffa ES (1983) Modern production/operations management. Wiley, New York, NY.

Chase RB, Aquilano NJ (1985) Production and operations management. Irwin, Homewood, Il.

Davidson HO (1956) Functions and Bases of Time Standards. the American Inst. of Ind. Eng., Inc.

Dilworth JB (1979) Production and operations management : manufacturing and nonmanufacturing. Random House, New York, NY.

Elmaghraby SE (1991) Manufacturing capacity and its measurement: A critical evaluation. Computers Oper Res 18:615–627.

Fowler JW, Robinson JK (1995) Measurement and improvement of manufacturing capacity (MIMAC) project final report. SEMATECH Technology Transfer #95062861A-TR. Web URL: http://www.fabtime.com/files/MIMFINL.PDF

Graves SC (1986) A tactical planning model for job shops. Oper. Res. 34:522–533.

Johnson RA, Newell WT, Vergin RC (1974) Production and operations Management; a systems Concept. Houghton Mifflin, Boston.

Johnson LA, Montgomery DC (1974) Operations research in production planning, scheduling and inventory control. Wiley, New York, NY.

Karmarkar US (1989) Capacity loading and release planning with work-in-progress (WIP) and lead-times. J. Manufact Oper Manag 2:105–123.

Manipaz E (1984) Essentials of production and operations management. Prentice-Hall, Englewood Cliffs, N.J.

Medhi J (1991) Stochastic models in queuing theory. Academic Press.

Nahmias S (2006) Production and operations Analysis, 5th edn.

Pahl J, Voss S, Woodruff DL (2005) Production planning with load dependent lead times. 4OR 3:257–302.

Patterson JW, Fredendall LD, Craighead CW (2002) The impact of non-bottleneck variation in a manufacturing cell. Production Planning and Control 13:76–85.

Srinivasan A, Carey M, Morton TE (1988) Resource pricing and aggregate scheduling in manufacturing systems. Unpublished paper, GSIA, Carnegie-Mellon University.

Stevenson WJ (2005) Operations management. 8th edn. The McGraw-Hill/Irwin Series on Operations and Decision Sciences, Boston.

Venkatesh J (2005) An introduction to total productive maintenance (TPM). *Article on the web*, URL: http://www.plant-maintenance.com/articles/tpm_intro.shtml.

Chapter 8
Data in Production and Supply Chain Planning

Laura Dionne and Karl G. Kempf

8.1 Introduction

Charles Babbage, one of the inventors of mechanical engines capable of calculation, commented (Babbage 1864): "On two occasions I have been asked, – 'Pray, Mr. Babbage, if you put into the machine wrong figures, will the right answers come out?' ... I am not able rightly to apprehend the kind of confusion of ideas that could provoke such a question." Roughly 100 years later in the age of electronic engines capable of calculation, an IBM instructor in New York named George Fuechsel captured this idea more succinctly when he used "garbage in, garbage out" as a training mantra.

For the computer programs that form the core of all but the smallest production and supply chain planning systems, the "garbage in" side of the mantra refers to data and the "garbage out" to the results of the analysis carried out using the data. This chapter focuses on issues around data as inputs while most of the rest of this volume deals with algorithms that use this data to supply analysis leading to decisions. Large production and supply chain systems are so complex that there is no practical way to plan them without using computer systems. Unless the users of these planning systems understand in depth the ramifications of "garbage in," there is a tendency to assume "gospel out" by placing far too much confidence in the computer programs providing the analysis.

To a large degree the practical discussion of data and issues related to data that follows applies to any computer program, particularly the decision support systems that are the focus of this book.

Issue A: One must comprehend what data is required to perform the desired analyses to the point of being able to provide precise, unambiguous definitions for each data element. In a large business organization it is sometimes surprising to find how many data items must be included in a particular computation to satisfy all

K.G. Kempf (✉)
Intel Corporation, 5000 W. Chandler Blvd., MS CH3-10, Chandler, Arizona 85226, USA
e-mail: karl.g.kempf@intel.com

K.G. Kempf et al. (eds.), *Planning Production and Inventories in the Extended Enterprise*, International Series in Operations Research & Management Science 151,
DOI 10.1007/978-1-4419-6485-4_8, © Springer Science+Business Media, LLC 2011

the personnel involved. It is also surprising how many different, and sometimes conflicting, definitions may emerge for what would initially appear to be the same data element.

Issue B: Given a set of data elements with definitions, one must address the completeness, correctness, consistency, and currency of any data set that is used as input to any analysis process. Planners involved in day to day practice know that even with extreme diligence it is impossible to have a perfect data set for any discussion or decision. The goal is to have as good a data set as possible and a planning system that can tolerate ever-changing data issues.

Issue C: However good (or bad) the data system, attention must be given to service, maintenance, performance monitoring, and continuous improvement. Even in the best of cases, data problems move around the system, the business evolves over time, and the analysis requirements change as the company, products, customers, suppliers, and markets grow and mature.

Issue D: While these "technical" issues are engaging and very important, a major deciding factor influencing the quality and integrity of the corporate data systems is the corporate culture. It is a simple matter to determine whether a company is truly data driven by inspection of its data systems. Understanding of and respect for the concepts of "defined, complete, correct, consistent, and current data" is reflected in the corporate business processes that supply input data and/or consume output data.

Our discussion will not include the layers of hardware and software needed to implement high quality data systems. It is our experience that this technology is easily accessible in today's marketplace and is not the constraint in building a world class data system. That dubious distinction belongs to the points listed above and it is these that we will elaborate. In the spirit of this volume, we will focus on planning and decision support systems rather than transactional systems, noting that transactional data is equally important to the smooth and successful operation of the business, and often overlaps with data needed for planning. Nor will our discussion include the concept of information, since turning raw data into information through various forms and layers of analysis is the focus of the majority of chapters in this volume. We are interested in the foundational data on which analysis, decision making, and planning are constructed.

8.2 What Data Is Needed

The data required for planning depends on what questions will be asked in the planning process for local production planning and global planning in the supply-demand network. Since there are a practically infinite set of possible questions, the discussion here will provide illuminating examples without attempting to be encyclopedic. For the simplest planning problems outlined in the introductory chapter of this volume, the key questions for the firm included (1) what orders to accept, (2) what material to release into the factory, and (3) how to allocate the finished product among customers. The more sophisticated problems described in the introduction included additional questions about suppliers and customers across the distributed organization of the firm.

In practice, asking planning questions and devising answers frequently reduces to projecting the current state into the future utilizing past experiences. Consider a production planner addressing the simplest planning problem and trying to answer the "material release" question. The tactical answer would require at least data concerning the following:

– Orders to be filled over the factory throughput time including volumes and due dates.
– Finished units in inventory.
– Work in progress (WIP) in the factory along with estimates of:
  • How many finished units are expected to result?
  • When will the finished units be available for shipment?
  • How many more units can the production system accept before becoming unacceptably congested and negatively impacting factory throughput time?
– Raw material in inventory.

This data would allow the planner to do a time-staged forward projection, balancing supply that is in inventory and will be leaving the factory against demand shipping to customers to determine the appropriate "material release." Feasibility checks would complete the process considering the inventory of raw materials and the level of congestion that might result from implementing the desired releases.

However, even this simplest data is problematic. Depending on the firm's policy on customers changing orders, in volume or due date, data on the amount and timing of finished goods needed is not necessarily constant. Data on finished goods and raw material inventories can have multiple sources of error including human error in placement, retrieval or counting, and loss through aging, spoilage, or pilfering. Locating and counting the units distributed across the manufacturing process as WIP is usually accurate, but estimation of remaining throughput time (Backus et al. 2006), yield through final inspection (Nurani et al. 1998), and the impact of raw material release are difficult (Asmundsson et al. 2006). These estimates required for the current problem rely on analysis of historical data about factory performance and may include data on recent trends as well as long-term data.

Consider a more strategic production planner addressing the simple "material release" question who desires to look forward multiple factory throughput times to avoid surprises like excess inventory and stock outs. This planner will want access to historical ordering data from customers to better estimate future demand including seasonal effects. There will be similar data requests for past performance under various conditions for raw material suppliers as well as material orders that have been placed but not yet filled.

Expanding from the academic "material release" problem in the introduction to the practical "material release" problems solved daily by the firm's supply chain planners requires even more data. The inventory data must now include many products in many sites, perhaps scattered around the world, including units consigned to customers and distributors as well as units in transit, magnifying the difficulties mentioned earlier. The useful historical data on demand for even a medium-sized firm must cover multiple years over thousands of products and customers (sometimes tens or hundreds of thousands). The data set would include both forecasted and realized demand as well as important market events that influenced (or did not influence) demand. This event history would include pricing changes, sales and marketing campaigns, and other such data for both the firm and its competitors with their impact on the firm's demand. Parameters from past and present sales contracts may be required. In data-driven firms this very large data set is maintained and used by the forecasting department, usually within the sales and marketing group, to provide a demand projection for supply chain and production planners.

On the supply side of the "material release" question, the useful historical data again probably covers multiple years over thousands of materials and suppliers (or tens or hundreds of thousands). Forecasts the firm sent to the supplier as well as actual quantities purchased with prices would be included in the data set. Responsiveness to orders as well as pricing changes is important historical data in the descriptor set for each supplier. Current materials inventory with all of its attendant data issues is again an important component. The same is true for current and historical purchase contract parameters and current orders in the pipeline for many different materials from many different suppliers. The materials purchasing department usually maintains and uses this very large data set, calling upon it to respond to requests from supply chain and production planners.

The firm's own capacity data completes the set discussed here. On the one hand is data about the firm's current equipment set and its short and long-term performance history. This might be extended to include forward-looking capacity plans for rearranging and adjusting capacity based on future product plans and market growth. On the other hand is data about the firm's current WIP across multiple factories for multiple products. It is difficult enough to keep track of the position of each unit in each factory. Even more difficult is estimating the future yield of saleable final products from the WIP in multiple factories with varying local conditions as it nears the end of the production line. Most difficult of all is estimating the timing of specific units of output across multiple products in multiple factories.

In some cases, product identity can be added to yield and timing. In a fan-in supply chain where many parts are assembled to get to a few final products, individual parts may be used in multiple products and WIP that was started as a subassembly for one product may be routed into another product if demand changes during the production lead time. Identity data is even more difficult to disambiguate in a fan-out supply chain where a common set of starting materials is progressively differentiated in the production process since material released to support a broad family of products can be routed to specific individual products during manufacture. The production engineering department usually maintains this complex data set that provides capacity estimates for supply chain and production planners.

The planners' job is still basically the same as initially described above – considering the time-phased estimates of customer demand, current and future inventories of raw materials and finished products, and a deep understanding of the firm's capacity to decide the material release question. However, in practice this task requires a data set large enough to require the support of three different departments in the distributed organization of the firm. Of course there are other basic questions such as "order acceptance" and "finished product allocation" that require additional data. But a very large set of questions related to planning can be addressed with the appropriate data sets on customer demand, raw material supply, and the firm's capacity including inventory. The data sets need to incorporate historical and current (that tomorrow becomes historical) data and as well as forecasts for the future.

8.3 The Necessity and Difficulty of Data Definitions

However large or small the data set, however many or few organizations own the data, however many uses for the same data element, precise definition is required to avoid confusion when collecting, storing, retrieving, and using the data (Wang et al. 1995, Madnick and Zhu 2006). This definitional perspective on data has many aspects that can be demonstrated on the data used in the simplest planning problems outlined in the introductory chapter of this volume.

The most rudimentary is the base definition of the atomic data elements or "master data." These are the elements that are directly measured (rather than derived) and widely used across the enterprise and are the foundation for all analysis and planning. For example, what is the definition of "work in progress" or WIP data? Is it all the units in the factory between raw material input and final product output? Should the engineering units in the factory that are there for some process diagnostic or improvement experiment be counted? Is the WIP only those units that can be sold, a definition that would make sense to the sales and marketing organization? Or is the WIP any unit that will use the capacity of a piece of equipment, a definition that would make sense to the manufacturing organization? Or is it both, a definition that would make sense to finance since any unit that uses capacity may be thought of as generating a cost and any unit that can be sold as generating revenue. Or should there be engineering WIP and production WIP as separate data categories, both using capacity, generating cost, and influencing congestion, but only the later directly generating revenue? Within these categories how should units that have been placed on hold for some processing irregularity be included? These units may be scraped where they sit (pure cost) or released for further processing and exit as final product (contributing to cost and revenue).

As another example, how does one define "capacity" data? When considering an individual machine, the most obvious definition is the number of units that can be produced during a fixed period of time. But the machines in the introductory chapter produced different products at different rates. If an average is to be taken, should the machine that requires some setup change when switching from one product to another have that setup time included? In practice, machines require scheduled maintenance and suffer unplanned breakdowns. Machines are sometimes used for engineering exercises such as process improvement experiments or functional upgrades. How long a time period needs to be considered to properly account for the effect of these activities on production capacity? Considering that some of these events are stochastic, it would follow that the capacity of a tool can only be defined as a distribution relying on historical data about detailed machine performance over a long period of time (Dalvi and Suciu 2007). If capacity data for each machine in a factory needs to be defined as a distribution, how is the capacity of a factory defined?

This last question introduces the concept of aggregation in the definition of data. The example of historical sales data can be used to expand on this concept. Sales vary by product and by customer over time, and would be considered to be the atomic data. Manufacturing might find it useful to define an aggregation by product over all customers over time for production planning purposes. The detail of sales quantity and timing by product is useful to manufacturing but by customer is not. Logistics might define an aggregation over all products for all customers by geographic region. For transportation planning, geographic detail is useful but customer and product may not be (especially if the products are of uniform size and weight, like semiconductor chips, for example). A definition of an aggregation by time period to expose seasonality effects might be useful for the finance department for revenue planning. An aggregation defined over all products by customer might be useful to sales and marketing, allowing them to rank customers by overall revenue for prioritization in production and supply chain planning. A crucial concept here is that data held in its atomic form can be aggregated in a large number of ways, but data held in aggregate form is usually impossible to disaggregate. This must be taken into account when building data definitions to insure that the lowest levels are included as well as the various levels of aggregation on various axes (product, customer, geography, season, etc.).

The reality of practice is that all of the above definitions are useful, each to a different department for a different reason (Brackett 1994). This means that there is a need for definition administration. Who is empowered to set new definitions as need arises and retire old definitions that have passed from usefulness? How are the definitions documented, stored, distributed, and accessed? To avoid confusion, the definitions must be enforced. Who is responsible for this enforcement and what is the enforcement process? It is relatively easy to imagine how to use software tools to enforce definitions on any agent (human or electronic) that enters or manipulates data, although it may be argued that any such system can be compromised. Enforcing definitions on agents that access data and use it for various purposes is a much more difficult problem to solve.

Perhaps the most difficult data definition problem involves "demand." Since the definition of demand is such a central concept to production and supply chain planning, an extended thought experiment is warranted to describe this definitional conundrum. Consider a firm in a specific situation at the beginning of month M. The enterprise manufactures two products P1 and P2, each with a manufacturing lead time of 1 month that are used by its customers in their products. Six months ago in M-6, well outside the production lead time, Corporate Sales and Marketing (CSM) requested that the Corporate Manufacturing Organization (CMO) prepare itself to supply 150K units of P1 and 150K units of P2 per month. CMO responded by preparing to supply 160K units of each product. Although CMO committed only 300K total units to CSM, they held a 10K capacity buffer on each product since in the past (a) CSM had underestimated demand and CMO had been forced to stretch to cover the shortfall, and (b) CMO occasionally experienced production problems leading to lost output. CMO has an additional degree of freedom since its capacity is partially flexible. In the 320K total units, it could actually produce as many as 200K units of P1 (but then only 120K units of P2) or 190K units of P2 (but then only 130K units of P1). CSM has been given a basic model of this flexibility at the 300K total unit level that CMO has committed.

Based on their tactical econometric models for months M-6 through M-2, CSM has determined that the total available market (TAM) for P1 is 300K units and they believe that they should be able to capture a 60% share of market (SOM) against their weaker competitor. Therefore the CSM goal for P1 is set at 180K units for month M. Similarly the TAM for P2 has been determined to be 300K units with a possible SOM of 40% against a stronger competitor for a goal of 120K units for month M. The setting of these goals was heavily influenced by the CMO committed production capacity of 300K total units and the flexibility model between P1 and P2. This plan for month M was passed to CMO at the end of month M-2 to be executed in month M-1 to supply products in the market in month M.

As depicted in Fig. 8.1, the enterprise sells its products in geographies G1 and G2. Customers Ca and Cb are located in G1 and place their orders with the enterprise's G1 CSM representative. The enterprise also has a CSM representative in G2 who takes orders from customers Cx and Cy. The CSM representatives are in close contact with their customers communicating on a weekly basis.

In geography G1, the CSM representative took orders early in month M-2 for the coming month M. Ca placed orders for 40K units of P1s and 40K units of P2s

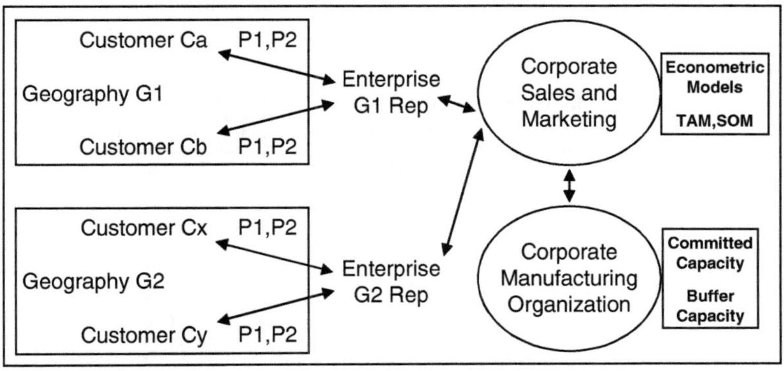

Fig. 8.1 Participants in the demand thought experiment

while Cb placed orders for 80K units of P1s and 60K units of P2s. Thus the G1 representative had total orders for 120K units of P1s and 100K units of P2s to pass on to headquarters. But this representative had seen the CSM global estimates of 180K units for P1 and 120K units for P2 and did not feel comfortable passing along such large orders. Since he felt that Cb was being overly optimistic, as they had consistently been in the past, he passed 100K P1 and 80K P2 for month M based on Cb being judged down by 20K units on each of its orders.

The orders that the CSM representative in G2 took early in month M-2 included Cx orders for 70K units of P1 and 30K units of P2 and Cy orders for 60K units of P1 and 80K units of P2. The total G2 orders were 130K units of P1 and 110K units of P2. With knowledge of the CSM global estimates for the two products, and with sales commissions clearly in mind, the G2 representative passed the customers' order quantities directly up to headquarters.

The CSM personnel at headquarters combine the judged geography data late in month M-2. Total orders for month M were 230K units for P1 and 190K units for P2, well over the committed capacity of CMO as well as the econometric predictions of CSM. Senior personnel decide to proportionally commit the planned use of CMO capacity to the orders to maintain a level playing field among the customers. This meant that early in month M-1 the month M planned production was committed to the customers through the CSM Geography representatives as shown in Table 8.1.

On the first day of month M as the committed production quantities began shipping, customer Ca contacted the CSM representative in G1 to ask for an additional 10K units of product P1. This request was passed to headquarters. A few days later, customer Cy contacted the CSM representative in G2 to ask to decrease its order for product P1 by 25K units. As this was within the terms of Cy's contract, the representative agreed but delayed passing this message to headquarters in case customer Cx asked for an increased quantity of product P1. This was based on the fact that Cx initially ordered 70K units of P1, but only 55K units were committed by the enterprise. In the middle of month M, customer Cb contacted the G1 representative asking to cancel its entire committed 25K units of product P2 based on the surprise retraction by one of its customers of a very large order that required the P2s as a component. This was a relief to CMO who encountered production problems and would be unavoidably short of P2s by 10K units by the end of month M. Finally at

Table 8.1 Initial orders in the demand thought experiment

Customer	Product	Requested	Judged	Committed
Ca	P1	40 K	40 K	30 K
	P2	40 K	40 K	25 K
Cb	P1	80 K	60 K	45 K
	P2	60 K	40 K	25 K
Cx	P1	70 K	70 K	55 K
	P2	30 K	30 K	20 K
Cy	P1	60 K	60 K	50 K
	P2	80 K	80 K	50 K
Total		460 K	420 K	300 K

the end of month M, customer Cx contacted the G2 representative to ship back 10K unused units of P2, and based on the terms of its contract, the appropriate arrangements were made for the return.

This is a simplified but realistic version of a typical month for a firm under conditions of initial orders exceeding capacity. (Note that an equally interesting version could be constructed for conditions of capacity exceeding initial orders.) This scenario raises many questions. There was enough burst capacity to satisfy more demand, but as it was hidden from CSM by CMO, commitments were lower than they could have been. The cancelation in one geography could have been used to satisfy demand in the other geography but was not.

These are useful questions concerning commitment and execution policies and dynamic demand fulfillment, but the focus of this thought experiment is demand identification. In the process of historical data collection for use in future demand management computations, the important question is "what was the demand in month M?" A number of candidates must be considered as shown in Table 8.2.

Candidate 1 is simply the initial orders placed by the customers. Candidate 2 includes the judgment of the enterprise's representatives in the field closest to the customers. But it would be difficult to consider these an accurate demand signal since the customers knew that the enterprise was limited on capacity from discussions with the field representatives and understood from past experiences that there was a "proportion limited capacity according to customer orders" rule in place. This knowledge induced them to inflate their initial orders. It might be appropriate to consider the original orders as an upper bound, although in this case it would be a high bound.

Candidates 3 through 5 represent CMO's view of the situation and are based on a doubly pessimistic approach: pessimism about CSM's ability to predict the market, and pessimism about CMO's ability to tightly control its production process.

Candidates 6 and 7 represent CSM's best efforts to manage the uncertainties in the market based on their models, understanding of CMO's capacity, and efforts to maintain equality in the treatment of their customers. The econometric models of CSM and the committed capacity of CMO (based on CSM's econometric models over a longer time horizon) clearly influence the enterprise's view of demand.

Table 8.2 Candidates for recording historical demand

	Source	P1	P2	P1 ± P2
1	Initial customer orders	250 K	210 K	460 K
2	Geography judged orders	230 K	190 K	420 K
3	CMO actual capacity	180 K	140 K	320 K
4	CMO realized capacity	180 K	130 K	310 K
5	CMO committed capacity	180 K	120 K	300 K
6	CSM econometric models	180 K	120 K	300 K
7	CSM initial commitments	180 K	120 K	300 K
8	CSM adjusted commitments	165 K	85 K	250 K
9	Shipped product	155 K	95 K	250 K
10	Shipped product less returns	155 K	85 K	240 K

Candidate 8 represents the firm's dynamic responses to market activity during month M (10K units of P1 requested by Ca, 25K units of P1 canceled by Cx, 25K units of P2 canceled by Cb, 10K units of P2 returned by Cy). Candidates 9 and 10 are slightly different ways to account for the same adjustments. It might be appropriate to consider these results as a lower bound on demand.

Although recording all these options is advisable for future forecasting efforts, assume that the personnel recording demand data select 155K units of P1, 85K units of P2, and 240K units overall as the official "demand" for month M. Note that these are 62%, 40%, and 52%, respectively, of what the customers initially ordered and 86%, 71%, and 80%, respectively, of what the econometric models of the CSM organization originally forecast.

Now consider the enterprise at time M+18. The "demand" data base has the data entered in month M along with data from a number of months before and all months after M. A few things have changed from the situation portrayed in Fig. 8.1. Customer Cy has acquired customer Cx in geography G2 and the resulting Cxy requires a different sales and marketing approach because of its size. A new enterprise Cz has started up in geography G2 and has become a customer, but is difficult to deal with since it aims to compete on low price. Product P1 has been phased out to be replaced by product P3 with expanded functionality. Product P2, which had only been released into the market in month M-4, is now a best seller. The enterprise representative in G1 has taken a different job and has just finished training his replacement. Based on strong enterprise results over the past six quarters, CMO has expanded their capacity aggressively and currently have roughly 20% more than the SOM estimated by CSM.

The forecasting cycle for months M+19 through M+24 is beginning. What data should be used to best support the forecasting process? On the one hand everything has changed: the customers, the products, the CSM field personnel, and the CMO capabilities. On the other hand nothing has changed: CSM is running their econometric models, CMO is buffering their capacity, the customers are ordering, and a tactical commitment and a strategic forecast are needed. Simple projection from the historical data is risky. The connection between the data in the "demand" data base and the actual demand at the time it was collected is tenuous on at least two levels. First, the data recorded in fact concerned what was shipped and has only a loose relationship to actual demand. Second, little if any of the contextual data associated with the time period was captured. It is debatable how much of that data it is possible to capture and how useful it would be even if it was.

All data from the most atomic to the most aggregated needs a crisp and clear definition agreed upon by all suppliers and customers of that data. Lack of definition assures confusion across the corporate planning systems. Few definitions are as easy to develop as might initially appear, and this difficulty increases with the number of users and uses. Definitions of WIP and capacity are examples of common atomic elements that are surprisingly difficult to specify. It is not uncommon to have to define more aggregations of atomic data than there are atomic elements (and higher level aggregations of lower level aggregations). Sales data provides an excellent example with multiple useful aggregations constructed from the atomic data.

Finally, some data types are inherently problematic, with demand as an outstanding example. Consensus on these items is especially difficult across the diverse organizations in the corporation, but especially important to achieve (Brackett 1994).

8.4 Completeness, Correctness, Consistency, and Currency of Data

Assuming that we have correctly identified all of the data elements needed to support all production and supply chain planning questions, and we have adequately defined each element and aggregation, we next apply the concept of "garbage in" to the broad topic of data quality. We will discuss completeness, correctness, consistency, and currency although the distinction is contrived considering the broad overlap between these topics (Wang et al. 2001).

By completeness we simply mean acquiring all of the data required for the next analysis. From a process perspective, completeness means the data collection system is working for each and every data element. The probability of this being true is inversely proportional to the size and complexity of the firm. Some of the data collection will be automated, and the chances of all sensors and systems working flawlessly all the time is vanishingly small. For example, tracking the dynamic location of a few million individual units in a few hundred warehouses distributed across the globe 24 hours a day 365 days per year with bar code readers will inherently involve data incompleteness. Bar code readers have various failure modes, as do the links connecting them to communication devices, the communication devices themselves, and so on. Some of the data collection will be performed by humans, some employees of the firm, some not. Again the probability of realizing complete data is vanishingly small since humans also have various failure modes, some physical, some psychological.

Completeness issues can be improved over time but probably not completely eliminated. The question therefore becomes that of how to minimize the impact of incompleteness. Since most analysis algorithms manipulate the data set mathematically, data fields with no data or zeros usually lead to execution difficulties (i.e., divide by zero error messages) or "garbage out." Program failure is arguably the best case since there is a clear signal that something is amiss in the data set. Apparently trouble-free program execution producing apparently sound results is the worst case since the garbage out that results may be treated as gospel in the absence of any indication of problems.

Data system design can help address this problem by at a minimum detecting, and at best repairing, incomplete data. Filling up a data structure that initially contains a nonsense character with respect to the expected data followed by a search for nonsense can be effective. In the simplest case a report can be generated for personnel who are responsible for completing the data set (failing to define such responsibility clearly and unambiguously is a major source of problems for obvious reasons). With some overhead, default values or moving averages can be maintained for use in auto-repair, again with notification to the appropriate personnel. In some cases, such as slowly changing data, for example, a simple form of auto-repair can be implemented by making a copy of the old data set and generating the new data set by overwriting the old. Previous values persist in positions where new data is incomplete. Checking for and repairing incompleteness is necessary before initiating analysis using the data. In some cases this may involve a meta-data definition specifying what data is needed for a particular analysis and what repair approaches are appropriate.

Detecting and repairing inaccurate data is a much more difficult problem (Dasu and Johnson 2003). In this case data is present but is wrong in some meaningful sense. Some type of error has occurred during data collection, perhaps due to sensor malfunction or human error. For example, a physical count of inventory in a warehouse could have missed or double counted some units, or the count was correct but the data entry was flawed, or both.

We have employed various approaches to this problem. One is to exercise human pattern recognition capabilities by displaying the differences between the current data set and some reference data set. The reference can be from the previous time frame or can represent a moving average. The differences can be displayed directly in a bar chart or can be sorted by size prior to display. Experienced humans can detect changes that are out of the ordinary and investigate. With some overhead, upper bounds for changes can be stored and applied with violations displayed to experts for investigation. With even more overhead, upper and lower limits of acceptable values for data items can be stored and used to check for plausibility (Petrovskiy 2003). This is easy for some elements and can be included in the data definition. For example total number of units that can be stored in a warehouse must be non-negative and has a spatially determined upper limit. For other elements this can be rather difficult. Upper and lower bounds on the weekly output of a factory, for example, might have to be specified by product mix. Other elements cannot usefully be treated in this manner. Bounds on the amount of finished product or raw material in transit at any point in time may have to be so broad as to be practically useless in detecting inaccurate data.

Checking consistency in the data set is another way to detect inaccurate data (Fan et al. 2001). Individual data elements may each pass all completeness and correctness tests but be inconsistent and hence contain some error. Using another warehouse example, the quantities of Product A and Product B in a particular warehouse may each be within bounds, but the sum may be over the storage capacity of the warehouse as stated in its capacity definition. The error may be in the quantity of Product A or Product B or the warehouse definition, or some combination of all these components.

Another root cause of consistency issues can be the currency or timeliness of data collection and entry (Ballou and Pazer 1995). In the example above, assume that warehouse data is collected at 9 AM and 9 PM each data and transmitted to the central data repository. At 9 AM, the quantities of Product A and Product B were individually and collectively accurate and acceptable. In the daily 1 PM delivery to the warehouse a large number of Product B arrived, was offloaded onto the dock and logged into the data system. At 6 PM a large amount of Product A was moved from storage onto the dock in preparation for shipment and the space vacated was used to store the newly arrived Product B. At 10 PM transport for Product A arrived 2 hours later than scheduled and Product A was loaded and logged out of the data system. Clearly the 9 PM inventory update would shown a plausible amount or Products A and B, but an implausible total inventory. Extending this simple example to a large company doing business across multiple time zones, it is obvious that a wide variety of data inconsistencies can having timing as their root cause.

The point of this extended (and perhaps tedious) set of issues and examples is to emphasize the vanishingly small probability of having a complete, correct, consistent, and current data set for production and supply chain planning. A large data set is susceptible to a very large number of individual errors and errors in various permutations and combinations. Detection and correction of each and every error is neither feasible nor economically practical, but planning activities must proceed in spite of this inherent difficulty.

The best defense is a thorough understanding of the system from the perspective of data providers, custodians, and consumers. The design and operation of the system must be based on a clear understanding of (a) the inherent limitations of the data providers, so that the system can account for detecting and correcting quality issues and (b) the basic requirements of the data consumers for accessing and utilizing the data (Chengalur-Smith et al. 1999, Jung et al. 2005). This is of course more easily declared than accomplished. On the one hand, there are a practically infinite number of data errors possible and the providers need to be aware of them. On the other hand, there are a practically infinite number of analyses that can be done on the data. Data consumers need to appreciate the system's limitations in the design of their analyses (Wang and Strong 1996, Raghunathan 1999). Of course different groups in the firm may have very different requirements for the same data element(s). The system designers need to consider both views as well as those of the custodians who will operate and maintain the system (Jarke et al. 1999, Ballou and Tayi 1999).

The master data at the bottom of the data hierarchy has the highest quality requirements. Since all aggregations and many calculations are directly based on it, it deserves the most attention (Loshin 2008). How frequently can it and should it be sampled? Should it be sampled at high frequency or low frequency, only on demand or triggered by particular events? With what accuracy can it and should it be sampled? Perhaps most importantly, how can it be captured and entered into the system with the highest quality? A data system that can avoid taking "garbage in" as much as possible becomes considerably easier to manage and tremendously more useful (English 1999). As usual, prevention is superior to detection and correction.

8.5 Service, Maintenance, and Improvement Issues

Identifying and defining data elements and achieving quality of those elements are necessary but not sufficient. The data is the product, and there are associated services (Kahn et al. 2002). The custodians of the data system have to provide a variety of functions to both the providers and the consumers of the data. The basic services include ease of use, responsiveness to change, and appropriate security. In our experience, the more cumbersome the data capture system, the lower the quality of the input data. Concerning manual entry (that we strive to keep to a minimum), we expend a great deal of effort with the data providers to understand their environment and concerns when developing data entry interfaces that focus on ease of use. Consumers of the data are focused on the analysis that they are developing. The last thing that they want to spend time on is finding or accessing the data. The data system cannot be too easy or too fast for the consumers. The offsetting consideration to ease and speed of use is security. Allowing indiscriminate access to the data entry interface risks accidental or malicious corruption of the data system. Clearly, some of the contents of the system are the most confidential data about the enterprise and its operations. In a system supporting production and supply chain planning, data about capacity, supplier contacts, future products, and so on must be protected. But such protection can slow or stop critical analysis tasks. The balance between ease of use and protection against misuse is a fundamental system design consideration. So is responsiveness. The only constant about data systems is change. The basic goal of the system is to reflect the ever changing world. However, the world changes not only in the values of the data elements but also in the data elements themselves. The product portfolio evolves over time, as does the roster of customers. The equipment set changes, and new warehouse locations arise. New data elements are added and old ones eliminated. New features are requested for data entry. New analyses must be supported with evolving data queries. Other changes are driven by suppliers, customers, and competitors. The data system has to be responsive to changing elements, definitions, and use cases while protecting quality.

Maintenance is a related but larger concern. Maintaining completeness, correctness, consistency, and currency is an ongoing daily concern. This involves monitoring data quality and since quality issues tend to move around over time this is a particularly difficult problem (Lee et al. 2002, Pipino et al. 2002, Pierce 2004). A practical solution is to include specific monitoring procedures in the definition of data elements. Another is to maintain records tracking elements relative to the bounds on values included in the definitions for purposes of correction. Injection and detailed tracking of test data can be useful. The custodians of the data must be creative in this endeavor considering the wide variety of origins, trajectories, and uses of the elements in the enterprise system. Maintenance also involves the spectrum of software transitions from installing the latest release of the current system to switching vendors usually impacting the quality of both the data and the service. This topic is beyond the scope of this chapter but is an issue with all of the software systems used by any firm.

Continuous improvement is another facet in this set of issues. On the one hand, providers and consumers of the data can be a source of improvement ideas for quality of data and service based on their use of the system. Implementing a process to collect their feedback is a relatively simple exercise that can yield very insightful suggestions. In our experience, such a process gains momentum if constructive criticism results in rapid and noticeable system improvements, especially if the person offering the suggestion is publicly recognized. On the other hand, quality monitors give a complementary insight. The historical record of identifying and correcting quality problems can be considered metadata for the continuous improvement effort. A recurring problem provides an improvement opportunity assuming the root cause can be identified.

8.6 The Importance of Developing a Data Culture

There are a number of inherent and inescapable technical problems associated with high quality in data and data services. The solutions to each and every one of these technical problems are dramatically influenced by the corporate culture. There are additional problems that have little or no technical component and are solely due to the corporate culture. The symptoms of these latter problems include lack of alignment on definitions, missing data, missing quality monitors, erratic performance of existing monitors, lack of trust in the data system, long delays in fixing known quality issues, lack of continuous improvement of the data system, and so on.

Development of a corporate culture around data quality eliminates the latter problems and supports developing solutions to the inherent technical problems (English 1999, Brackett 2000, Batini and Scannapieco 2006). There is recognition and commitment throughout the corporate hierarchy that data quality is a never-ending concern of every employee. A focus on data quality is embedded in every process at every level (Lee et al. 2004). Ownership of data definitions including quality monitors is clear, as are processes for review and improvement. Adherence to definitions is expected and deviations are not tolerated. Shoddy practices are exposed and addressed. Monitors are exercised and results are analyzed with appropriate actions following quickly. Critical success indicators are set for quality of the data and the services involving collection and use of the data. Expectations are set for continuous improvement. Reward structures are aligned with high data quality standards.

Constancy of purpose is an especially important component of developing and maintaining the corporate data quality culture. Progress will not always be steady but persistence is crucial. Developing a high quality data system is a daunting task. Developing the processes to maintain a high quality data system in the face of rapid and unpredictable changes in all facets of the business is the ultimate goal.

Over the past decade progress has been made on these topics using a convenient analogy. For many years various industries have been forced through competition to improve the quality of their products. This has resulted in a branch of manufacturing engineering dealing with quality assurance (QA) including quality control (QC). QA attempts to improve and stabilize production and associated processes to avoid (or at least minimize) issues that led to quality problems in the first place. In QA it is important to realize that quality is determined by the intended users or customers, not by some absolute standard. QA includes regulation of the quality of raw materials, components and sub-assemblies, production and inspection processes, services related to production, and management. QC as a component of QA emphasizes testing of products to uncover defects. There are various specific instantiations of QA, but all rely on three basic ideas. (1) Manufacturing processes have characteristics that can be measured, analyzed, controlled, and improved. (2) Efforts to achieve stable and predictable process results are of vital importance to quality. (3) Achieving sustained product quality, including continuous improvement, requires commitment from the entire organization, particularly from top-level management.

Since many firms have been forced by the marketplace to adopt this approach to product quality, including the necessary but extremely painful organizational changes, some researchers in data quality have advocated a similar approach (Wang et al. 1998, Ballou et al. 1998, Shankaranarayan et al. 2003). Considering data as a "product" makes accessible powerful tools from such well-developed and scientifically sound areas as reliability engineering and statistical process control. There are, however, difficulties to be overcome by continuing research and lessons from practical application. Manufacturing processes are based on physics and chemistry and can be characterized and repeated with precision. While the processes described in this chapter concerning data in production and supply chain planning involve computer programs that have similar properties, the processes also involve human that do not. Manufacturing processes result in physical products that can be measured and characterized using international standards of long standing. The same cannot be claimed for data.

Summary

During the past few decades developed countries have been expanding from industrial economies to include information management as a critical capability. Companies doing business in this environment must not only compete on design, manufacturing, and distribution but also on their ability to assimilate new information and respond appropriately. The information that fuels this new level of competition is built on a foundation of data. The quality of the tactical and strategic results obtained by the company reflects in large part the quality of the foundation data and the derived information. Production and supply chain planning provide a concrete example.

The danger is that a company will overestimate the quality of its data and underestimate the impact of poor data quality on its success (Redman 1998). From the perspective of production and supply chain planning, this double misestimation can directly impact financial success. Inside the company, low data quality will lead to operational inefficiencies (wasted time, materials, and capacity, inappropriate inventory, inefficient distribution, with higher than necessary costs) and poor decisions (wrong products in the wrong quantities in the wrong places at the wrong times, with lower than expected revenues). Outside the company, the results will be frustrated and alienated suppliers and customers. Dissemination of poor quality data and information up and down the supply chain will result in a loss of credibility for the company in its ecosystem, putting it at a serious (perhaps terminal) disadvantage relative to its competition.

From a different vantage point, building a corporate culture that deeply values measuring and continuously improving its data quality can supply a formidable competitive advantage. For production and supply chain planning this includes (a) comprehending what data is required to perform the desired analyses including precise definitions for each data element with actionable quality monitors, (b) focusing on the completeness, correctness, consistency, and currency of every data element across the all data capture, storage, manipulation, and analysis processes, and (c) establishing service, maintenance, and improvement processes that are robust in the face of changing requirements as the company, customers, suppliers, competitors, and markets morph and grow.

Recent advances in computer hardware, software, and networking provide the basic power tools to realize these data quality goals. Steady progress on measuring and improving physical product quality in manufacturing provides an analogical basis for data quality methods. But there are some crucial differences and therein lies the motivation for research in this area on technical topics as well as organizational dynamics.

References

Asmundsson J, Rardin RL, Uzsoy R (2006) Tractable non-linear production planning models for semiconductor wafer fabrication facilities. IEEE Trans Semicond Manufact 19(1):95–111

Babbage C (1864) Passages from the Life of a Philosopher. Longman and Co., pp 67

Backus P, Janakiram M, Mowzoon S, Runger G, Bhargava A (2006) Factory cycle-time prediction with a data-mining approach. IEEE Trans Semicond Manufact 19(2):252–258

Ballou D, Pazer H (1995) Designing information systems to optimize accuracy-timeliness tradeoff. Inf Syst Res 6(1):51–72

Ballou D, Tayi GK (1999) Enhancing data quality in data warehouse environments. Commun ACM 42(1):73–78

Ballou D, Wang R, Pazer H, Tayi GK (1998) Modeling information manufacturing systems to determine information product quality. Manage Sci 44(4):462–484

Batini C, Scannapieco M (2006) Data quality: concepts, methodologies and techniques (data-centric systems and applications). Springer-Verlag New York, Inc., Secaucus, NJ. ISBN 978-3450331728

Brackett MH (1994) Data sharing using a common data architecture. Wiley. ISBN 978-04713 09932

Brackett MH (2000) Data resource quality: turning bad habits into good practices, Addison-Wesley. ISBN 978-0201713060

Chengalur-Smith IN, Ballou D, Pazer H (1999) The impact of data quality information on decision making: an exploratory analysis. IEEE Trans Knowl Data Eng 11(6):853–864

Dalvi N, Suciu D (2007) Management of probabilistic data: foundations and challenges. In: Proceedings of 26th ACM SIGMOD-SIGACT-SIGART Symposium on Principles of Database Systems (Beijing), June 11–13, 2007, pp 1–12

Dasu T, Johnson T (2003) Exploratory data mining and data cleaning. John Wiley & Sons, Inc., New York, NY. ISBN 978-0471268512

English LP (1999) Improving data warehouse and business information quality: methods for reducing costs and increasing profits. Wiley. ISBN 978-0471253839

Fan W, Lu H, Madnick SE, Cheung D (2001) Discovering and reconciling value conflicts for numerical data integration. Inform Syst 26(8):635–656

Jarke M, Jeusfeld MA, Quix C, Vassiliadis P, Architecture and quality in data warehouse: an extended repository approach. Inform Syst 24(3):229–253

Jung W, Olfman L, Ryan T, Park Y (2005) An experimental study of the effects of contextual data quality and task complexity on decision performance. In: Proceedings of IEEE International Conference on Information Reuse and Integration, pp 149–154

Kahn B, Strong D, Wang R (2002) Information quality benchmarks: product and service performance. Commun ACM 45(4):184–192

Lee YW, Pipino L, Strong D, Wang R (2004) Process embedded data integrity. J Database Manage 15(1):87–103

Lee YW, Strong DM, Kahn BK, Wang RY (2002) AIMQ: a methodology for information quality assessment. Inform Manage 40(2):133–146

Loshin D (2008) Master data management. Morgan Kaufmann. ISBN 978-0123742254

Madnick S, Zhu H (2006) Improving data quality through effective use of data semantics. Data Knowl Eng 59(2):460–475

Nurani RK, Strojwas AJ, Maly WP, Ouyang C, Shindo W, Akella R, McIntyre MG, Derrett J (1998) In-line yield prediction methodologies using patterned wafer inspection information. IEEE Trans Semicond Manufact 11(1):40–47

Petrovskiy MI (2003) Outlier detection algorithms in data mining systems. Program Comput Soft 29(4):228–237

Pierce EM (2004) Assessing data quality with control matrices. Commun ACM 47(2):82–86

Pipino LL, Lee YW, Wang RY (2002) Data quality assessment. Commun ACM 45(4):211–218

Raghunathan S (1999) Impact of information quality and decision-maker quality on decision quality: a theoretical model and simulation analysis. Decis Support Syst 26(4):275–286

Redman TC (1998) The impact of poor data quality on the typical enterprise. Commun ACM 41(2):79–82

Shankaranarayan G, Ziad M, Wang RY (2003) Managing data quality in dynamic decision environment: an information product approach. J Database Manag 14(4):14–32

Wang RY, Lee Y, Pipino L, Strong D (1998) Managing your information as a product. Sloan Manag Rev Summer:95–106

Wang RY, Reddy MP, Kon HB (1995) Toward quality data: an attribute-based approach. Decis Support Syst 13(3–4):349–372

Wang RY, Strong DM (1996) Beyond accuracy: what data quality means to data consumers. J Manage Inform Syst 12(4):5–33

Wang RY, Ziad M, Lee YW (2001) Data quality. Springer. ISBN 978-0792372158

Chapter 9
Financial Uncertainty in Supply Chain Models

Aliza Heching and Alan King

9.1 Introduction

In this chapter, we discuss approaches to applying financial markets theory to address risk and uncertainty, such as uncertainty in pricing or production costs, in the formulation of supply chain models. Financial parameters such as revenues and costs are key factors that drive optimal decisions in production planning and supply chain problems. Yet typically, these factors are the most difficult to capture and quantify. The values of these parameters may depend on some future "state of the market," such as exchange rates, interest rates, or consumer prices, or they may depend on the behaviors of suppliers, customers, or competitors. Supply chain contracts generally contain provisions that determine revenues and costs based upon observable indices such as the consumer price index or interest rates. Even when contracts specify all dependencies, supply chain contracts often include features that allow either or both parties the flexibility to modify their commitments at a future date prior to delivery. Thus, significant uncertainty remains.

Financial markets theory addresses the problem of managing future uncertainty in returns from portfolios of financial securities. The key idea in financial markets theory is that risks may be hedged by trading securities with similar offsetting payoffs. Continuous trading activity acts to provide liquidity and to keep market prices in a dynamic equilibrium, from which mathematical relationships between security prices can be derived. Supply chain problems, however, are often of a hybrid type in which some risks – such as exchange rates or commodities prices – can be hedged through trading, but other risks – such as demand or production uncertainty – cannot be hedged away. The application of financial markets theory to supply chain problems is therefore an exercise in combining two very different application realms.

In this chapter, we present a basic linear programming formulation of trading that is suitable for adaptation to supply chain problems. We discuss the existing literature on financial modeling in the supply chain and show how it can be restated

A. Heching (✉)
IBM TJ Watson Research Center, 1101 Kitchawan Road, Route 134, Yorktown Heights, NY 10598, USA
e-mail: ahechi@us.ibm.com

K.G. Kempf et al. (eds.), *Planning Production and Inventories in the Extended Enterprise*, International Series in Operations Research & Management Science 151,
DOI 10.1007/978-1-4419-6485-4_9,

using this simple framework. We illustrate an application of financial markets theory to the newsvendor model. Finally, we discuss the significant differences between the assumptions required to address uncertainty in financial markets and uncertainty in the supply chain.

The outline of the chapter is as follows. Section 9.2 describes a realistic supply chain where the decision maker enters into supply chain contracts with various provisions for supply and demand exigencies. We review the significant literature on supply chain contracts in Sect. 9.3 from the perspective of financial risk modeling. In Sect. 9.4, we focus on the information that can be derived from financial markets. We consider the role of risk assessment using extreme risk measures such as value at risk and conditional value at risk, which are the focus of financial risk regulation. Next, we present a simple discrete time and space model of market valuation using the dual formulations of finding a replicating portfolio and identifying a stochastic discount factor. We discuss calibration of this model and also present the dual role of utility and penalization.

Section 9.5 applies the basic model of trading to illustrate how it can be used to unify recent contributions to the literature on financial modeling in the supply chain. Section 9.6 presents an example of how to incorporate the financial theory-based models into the traditional newsvendor model. Finally, in Sect. 9.7, we discuss why the ideas and theory developed in the finance literature may not be so easily transferable to the production planning setting.

Our exercise in extending supply chain modeling using financial markets theory highlights some key issues in bringing financial markets practices into the supply chain, namely: timeliness of market data, correlations, standardization, and transparency. Thus, while the financial markets theory approach may yield some insight into how the production planning and supply chain problems can be addressed, it does not readily yield solutions to these challenging problems.

9.2 A Sample Supply Chain Contract

Consider a manufacturer of multiple products. The manufacturing process is capital intensive. Thus, any investment in manufacturing equipment is a long-term investment. The manufacturer also has the option to outsource production. The manufacturer produces custom and standard "off-the-shelf" products. Products are grouped along categories such as geography, business line, and product type.

The manufacturer has production facilities around the world. Products produced at each of the facilities are designated to service specific geographic regions. The manufacturer enters into two types of contracts: (1) long-term procurement contracts with his supplier to procure the components required to build the products and (2) contracts with his customers to whom he sells his products.

The procurement contracts are in place over extended periods of time, typically 5 years. When the manufacturer enters into the procurement contract with his supplier, the procurement price for the current period is known. However, the procurement price in future periods is not specified and is determined by the supplier in the future, as it is a function of prevailing market conditions. Further, volume discounts may apply and the manufacturer does not know upfront the quantity of component that he will purchase in the future. Both of these sources of uncertainty present significant risk for the manufacturer. Similarly, future component availability is unknown.

Over the lifetime of the contract, the manufacturer places firm orders with the supplier. When a firm order is placed, the price per component is fixed based upon prevailing market conditions (such as interest rates, exchange rates, etc., as specified by the terms of the contract), order volume, and component availability. If the supplier does not have sufficient product available, the manufacturer can turn to outside suppliers (or to the spot market) to meet the unmet demand. The manufacturer does not know what price per component the outside supplier will offer, as he has no long-standing relationship with the outside supplier.

In the manufacturer's downstream relationship, the manufacturer signs contractual agreements with the customers to whom he sells his products. These contracts are also typically in place over extended periods of time.

Each customer provides the manufacturer with 6-month rolling horizon demand forecasts. Over time, the customer also requests product from the manufacturer which, after negotiation between the customer and the manufacturer (regarding request quantities and ship dates), are converted into firm order quantities (and equivalent ship quantities) with committed ship dates. Both the customer and manufacturer are bound by these order quantities and ship dates; the manufacturer is penalized if he does not have sufficient supply available by the committed ship date and the customer is penalized if he does not purchase enough product. Within a given time frame, say within each quarter, the manufacturer allows the customer some flexibility with his actual orders. Namely, at the end of a quarter, the manufacturer reviews the customer's total orders within the quarter. If the customer's total actual order quantity within that quarter does not equal the firm orders for that quarter, then the manufacturer ships the difference between the firm orders and the total actual orders during that quarter. Thus, the manufacturer grants the customer some flexibility during each quarter with respect to the timing of orders, but by the end of each quarter the total actual order quantity must equal the total firm order quantity.

The contract contains another element of flexibility, both for the customer and for the manufacturer: Although the customer order quantities after negotiation are considered firm, the contract allows for change within a contractually specified percentage. More specifically, the contracts are structured to allow for a $\gamma\%$ increase or decrease above or below the firm order quantity. The manufacturer is often allowed similar flexibility with respect to his committed ship quantities. This flexibility is sometimes referred to as *embedded optionality* within supply contracts. Embedded optionality poses significant risk to both the customer and the manufacturer.

Thus, the manufacturer's contracts with his suppliers and with his customers contain various financial parameters whose values are unknown and require estimation at the initial contract date. The uncertainty is introduced largely due to quantity flexibility and/or time flexibility that is built into the supply contracts. For example, contractual terms that permit customers to vary their order quantities can be classified as risk or uncertainty attributable to quantity flexibility. Similarly, uncertainty associated with contract terms such as per unit price at a future date, where price is related to prevailing market conditions, can be classified as risk or uncertainty attributable to time flexibility.

9.3 Financial Modeling in Supply Chain Literature Review

Traditional supply chain literature does not consider using financial markets to mitigate risk associated with uncertainties; instead, the literature analyzes risk mitigation and risk transfer within a supply chain framework (see Cachon 2002; Lariviere 1999; Tsay et al. 1999). There is also substantial literature on real options theory as a technique for analyzing optionality in business contracts (see Dixit and Pindyck 1994; Trigeorgis 1996). In this chapter, we focus on the formulation and analysis of supply chain problems where some part of the risk is related to securities traded in financial markets.

Kleindorfer and Wu [2003] provide a framework to enhance supplier–manufacturer interactions using financial markets. In addition, they provide an excellent literature review, which we do not attempt to replicate here. Birge [2000] explores the relationships between real and financial options in a stochastic programming capacity planning framework. Birge's basic setup is very close to the approach described in this chapter. Our contribution can be viewed in part as extending Birge's stochastic programming example into a more general framework that incorporates the dual concepts of calibration and utility. Kamrad and Ritchken [1991] lay out a basic framework for valuing fixed-price supply contracts under demand uncertainty that is correlated to financial markets. Smith and Nau [1995] adopt a discrete-time decision analysis framework for a project investment model. They first develop results under the assumption of a complete financial market and then consider the case where financial markets are incomplete. Kouvelis et al. [2001] discuss the application of foreign exchange market information to ownership structures. Gaur and Seshadri (2005) analyze a newsvendor problem: first assuming demand is perfectly correlated with market securities and then assuming that demand is partially correlated with market securities. Caldentey and Haugh [2006] present a fundamental model of a nonfinancial firm that trades continuously in financial markets.

The literature on financial modeling in the supply chain is quite daunting for readers with a typical background in the mathematics of operations research. There is a huge variation in notation, in the level of mathematical sophistication, and even in the basic conceptualization of the underlying hedging problems. Readers are required to posess sophisticated knowledge of mathematical financial modeling concepts to understand how these papers can be related one to another. Our extension of the Birge [2000] model can be used as a basis to discuss these models. In Sect. 9.5, we develop the models of Smith and Nau [1995], Gaur and Seshadri [2005], and Caldentey and Haugh [2006] within this simple framework.

9.4 Basic Concepts of Financial Modeling

The starting point for discussions of financial modeling is the "market," which is a mechanism by which buyers and sellers are brought together to exchange securities, for example, IBM shares. Options are the right to buy or sell a specified number of IBM shares, for example, at a fixed (strike) price on some future date. The "market for IBM securities" refers to the collection of all securities whose value can be derived from the price of a share of IBM equity ("derivative securities"). The market "state" is the collection of bid and ask prices in the market for these securities.

The main guiding principles of financial modeling derive from the assumption that prices of securities move in response to continuous buying and selling activity. Thus, more activity ("liquidity") in a given financial market increases the likelihood that market prices for related securities can be explained by a market equilibrium model. Market equilibrium implies a condition where supply and demand for securities are relatively equal. In the absence of equilibrium, for example, if demand exceeds the number of sellers, the price will rise, sellers attracted to this higher price will enter the market while buyers disinterested in this higher price will exit. This activity will result in restoration of market equilibrim.

There are two types of calculations that derive from a market equilibrium assumption (1) assessment of risk of a given position, or portfolio, of financial securities and (2) market valuation of related securities – securities such as options – whose prices depend on underlying securities such as futures, equities, and bonds.

9.4.1 Risk Assessment

The objective of risk assessment in a financial institution is to determine the required amount of capital to be held in reserve against losses in its portfolios. Risk assessment identifies a forecasting methodology for the tail behavior of losses. The basic risk measurements for portfolio losses are value at risk (VaR) and conditional value at risk (CVaR). VaR measures the quantile of a portfolio loss distribution over a specified time horizon and at a given probability level. For example, a portfolio with a 1-day 95% VaR of \$1M has a 5% likelihood that the value of the portfolio will decline by more than \$1M over the next day. CVaR measures the conditional expectation of losses, given that losses exceed the VaR level. Interest is building in CVaR because it is a coherent risk measure and because it has convexity properties that render it more compatible with optimization techniques (see Rockafellar and Uryasev 2002).

The data for risk measurements such as VaR and CVaR are usually derived from variances in historical price returns. *Portfolio stress testing* is a means for identifying unanticipated circumstances that may lead to larger than anticipated portfolio losses. Stress testing involves information that is not necessarily consistent with the observed histories. For example, banks model some crises by scenarios in which correlations between asset classes trend near unity – meaning that no diversification is possible. Part of their risk exposure calculation might include correlation scenarios that have been observed in exceptional market situations, such as crashes. Stress testing would then calculate VaR and CVaR given a market correlation scenario, and average them over some probability weight assigned to these scenarios. Monte Carlo simulation is a popular method for performing stress testing.

Financial institutions also measure risks such as *operational risk* and *counterparty risk*. Operational risk is defined as the risk of loss from failed internal processes, people, or systems, or from external events. Counterparty risk is the risk of loss associated with failure of the other parties to meet their contractual obligations.

Supply chain organizations are now attempting to adopt similar methods for stress testing the supply chain to understand how different events or market scenarios may impact supply chain performance. Similarly, they will analyze worst-case events that may impact the supply chain and measure the impact of these events on the supply chain performance. For further discussion and for examples, see Elkins [2005].

9.4.2 Market Valuation

The theory of pricing contingent claims begins with replicating the cash payoffs of a contingent claim using a porfolio of traded assets whose prices are known. This is known as a *replicating portfolio.* Assuming an efficient market (i.e., prices of traded assets reflect all known information), the law of one price prevails and the price of the contingent claim should equal the price of its replicating portfolio. This approach to pricing is also known as "arbitrage pricing." Arbitrage means incurring a positive reward for zero risk, that is, making something for nothing. For example, if the market value of the replicating portfolio were less than the cost of the asset that produced equivalent cash flows, then a trader could buy the replicating portfolio, sell the asset, and pocket the difference. In practice, the constant search for arbitrage trading opportunities forces the market to remain in equilibrium, thus enabling market valuation pricing.

A simple linear programming version of market valuation can be developed (see Edirisinghe et al. 1993; the approach we follow here appeared in King 2002). To avoid unnecessary technical complications, we measure time in discrete intervals $t = 0, 1, \ldots, T$ and require the stochastic process of security prices $\{S_t : t = 0, \ldots, T\}$ to be supported over finitely many states. Suppose the trader wishes to replicate a security F with cash flows $\{F_t\}$ that are completely dependent on the same underlying risks as the security prices S_t, that is, F is a contingent claim. Denote θ_t to be the trader's portfolio allocation at time t. A linear program for market valuation of F can then be written as

$$\begin{aligned} \min_{\theta} \quad & S_0\theta_0 \\ \text{s.t.} \quad & S_t\theta_t = S_t\theta_{t-1} - F_t, \quad t \geq 1 \\ & S_T\theta_T \geq 0. \end{aligned} \tag{9.1}$$

This system describes a *self-financing* trading policy $\{\theta_t\}$, meaning that purchases of assets are financed by sales of others so that new funds are not invested. Specifically, the proceeds from the previous period's investments, $S_t\theta_{t-1}$, minus the required payment, F_t, are reinvested into the current portfolio holdings, θ_t, at current prices, S_t. (The equations and inequalities are interpreted as holding with probability 1.) At the final period, the amount remaining in the trader's account is required to be non-negative. So the trading policy is said to "super-replicate" the risky cash flows $\{F_t\}$. A super-replicating trading policy generates the required payouts, and in addition may generate some nonzero positive balances at the end of the time horizon.

Interpreting the major result in financial mathematics (see Harrison and Pliska 1981) to the simple context of the self-financing linear program for contingent claim F (9.1), we note that by linear programming duality, an optimal market valuation exists for every dependent instrument F if and only if there is a nonempty set of *risk-neutral discount factors*, Π, satisfying the dual linear system

$$\Pi := \left\{ \pi \geq 0 \;\middle|\; \pi_t S_t = E\left[\pi_{t+1} S_{t+1} \mid S_t\right] \right\}, \tag{9.2}$$

in which case, the market valuation is equal to the optimization

$$\max_{\pi \in \Pi} \sum_{t=1}^{T} E\left[\pi_t F_t\right]. \tag{9.3}$$

The risk-neutral discount factors π_t are the multipliers for the self-financing equalities in the linear program (9.1), and represent the marginal value of an additional unit of value (a dollar, say) in each future state of the stochastic process. This is a well-known result; the particular argument presented here can be found in King [2002].

9.4.3 Complete Markets

The equality constraints in (9.2) are sometimes called "martingale equalities." A securities market is referred to as *complete* if all contingent claims can be perfectly hedged (i.e., perfectly replicated by self-financing trading policies with no surplus remaining under any future scenarios). Dually, a securities market is complete if there is exactly one risk-neutral discount factor that satisfies the martingale equalities (9.2). The complete markets assumption, while clearly an approximation, is an important step in the mathematical algorithms used to price contingent claims in the continuous time settings of security valuation. However, conclusions based on this property would not be generally applicable outside of extremely liquid financial markets.

The basic concern with the complete markets assumption is that it eliminates incentives to trade. If there were a single risk-neutral measure, then all risk-neutral participants would use it and they would arrive at identical prices.

There are three real-world market characteristics that tend to contradict the complete markets assumption. First, the martingale property is fragile: imposing realistic technological restrictions on trading activity (such as transaction costs or execution delays) weakens the martingale equalities for the dual feasible stochastic discount factors. Second, market security prices can be driven by many factors. It is highly unlikely that there are enough liquid options in the market that are sensitive to all possible factors. It follows that it is quite natural to suppose that there are many possible risk-neutral discount factors that are consistent with a market equilibrium, and that even risk-neutral market participants might select different discount factors. See King [2002] for additional discussion on this point. Finally, securities that are not frequently traded may have prices that are not consistent with the existence of a risk-neutral discount factor. This does mean technically that the market is not in equilibrium at these prices, but such arbitrage opportunities are usually due to the fact that the securities in question have stale prices. What this means is that these securities are not traded often and their offered bid or ask prices may be out of date with the current market equilibrium. When a trader expresses interest in these prices, the offer will be updated to reflect current market conditions.

These concerns arise particularly when attempting to calibrate a risk-neutral discount factor to market prices, as we now describe.

9.4.4 *Calibration*

Using the self-financing linear program (9.1), one can price derivative instruments by a trading program, provided the stochastic process for the underlying asset is known. When the stochastic process for the underlying asset is unknown, *model calibration* is a method for finding the stochastic process, given observations of related derivative security prices. The dual model (9.3) can be used to calibrate using observed market data. The calibration problem is the inverse of the contingent claim pricing problem. However, in the absence of complete markets, this problem is ill-posed: there may be multiple underlying processes that yield the same prices as those observed in the market or there may be none.

When the dimensionality of the uncertainty attributed to market securities is low, a linear programming approach (see King et al. 2005) can be developed to calibrate the pricing model. Suppose the market has M listed options $\{C^i, i = 1, \ldots, M\}$ with observed bid–ask prices $C^i_{\text{b}} < C^i_{\text{a}}$, and future payouts C^i_t. Then it would be natural to require that all feasible risk-neutral discount factors satisfy

$$C^i_{\text{b}} \leq \sum_{t=1}^{T} E\left[\pi_t F_t\right] \leq C^i_{\text{a}}, \quad i = 1, \ldots, M.$$

It turns out that this requirement is too strong. Sometimes, there are no discount factors that satisfy all these constraints, perhaps because some prices are stale. A better modeling practice is to transform these inequalities into soft constraints and penalize the constraint violations in the dual objective function, as follows:

$$\max_{\pi \in \Pi,\, c_{\mathrm{a}} \geq 0,\, c_{\mathrm{b}} \geq 0} \sum_{t=1}^{T} E\left[\pi_t F_t\right] - v(c_{\mathrm{a}} + c_{\mathrm{b}})$$
$$\text{s.t. } C_{\mathrm{b}}^i - c_{\mathrm{b}}^i \leq \pi C^i \leq C_{\mathrm{a}}^i + c_{\mathrm{a}}^i, \quad i = 1, \ldots, M. \tag{9.4}$$

The penalty parameter v would be related to some measure of the quality of the market prices for the listed options – for example, the volume traded during the current session. The program (9.4) describes the situation of a trader who is *selling* the security F. An exactly similar program that *minimizes* the objective describes the situation of a trader who wishes to compare the cost of *buying* the dependent security as compared to super-replicating it herself.

Let $\pi^{\mathrm{b}} F$ denote the minimum calibrated price and $\pi^{\mathrm{a}} F$ denote the maximum calibrated price for the dependent security F. The interval

$$\left[\pi^{\mathrm{b}} F\, ,\, \pi^{\mathrm{a}} F\right] \tag{9.5}$$

is called the *arbitrage interval*. The lower bound of the interval is interpreted as the maximum price that a risk-neutral trader could extract from the market with F as collateral, and the upper bound is the minimum price that a risk-neutral trader would accept in order to replicate the cash flows of F. When the bounds of the arbitrage interval are relatively tight, we can be confident that market prices indeed do specify a risk-neutral discount factor for any risks that depend on price movements S_t of the underlying security.

However, in many circumstances, these bounds (9.5) may not be tight. One explanation is that the driving stochastic factors for the cash flows come from the tails of the price movements S_t. It is a fact that listed option prices do a poor job of estimating the influence of tails, since the options that are influenced by tails are usually very lightly traded. A second explanation is that the cash flows may depend on uncertainties that are independent of the market. This phenomenon is highly relevant to supply chain settings, where market prices may not capture all the relevant factors that contribute to uncertainty. When the arbitrage bounds are wide, the only alternative for model calibration is to develop a nonlinear formulation that provides a capability to refine the optimal choice of risk-neutral discount factor. From a technical standpoint, it is easier to develop the nonlinear formulation as a primal problem. Toward this end, we introduce primal variables ξ_{a} and ξ_{b} that correspond to the calibration error constraints. These can be interpreted as initial positions taken by the trader in the listed options themselves that will be held until the end of the time horizon. The nonlinear aspect is modeled by a utility function $u(\cdot)$ and a probability measure P (which until this point has not played a role). The primal version of the nonlinear calibration problem maximizes the utility of the terminal surplus value of the hedging portfolio:

$$\begin{aligned} \max_{\theta,\xi,V} \quad & E^P\left[u(S_T\theta_T - V)\right] \\ \text{s.t.} \quad & V = S_0\theta_0 + (C_a\xi_0^a - C_b\xi_0^b), \\ & 0 = F_t + S_t(\theta_t - \theta_{t-1}) + (C_t\xi_0^a - C_t\xi_0^b), \\ & \xi_0^a, \xi_0^b \geq 0, \end{aligned} \tag{9.6}$$

The term V on the left-hand side of the period 0 equation (the wealth balance equation) is now a variable of optimization. To keep wealth from growing without bound, we subtract its value from the terminal wealth when calculating utility. The dual problem turns out to be identical to the calibration problem (9.4) except that the objective now contains a nonlinear term

$$-E^P\left[u^*(\pi/P)\right], \tag{9.7}$$

where $u^*(\cdot)$ is the *concave conjugate* function:

$$u^*(x) = \inf_w \{wx - u(w)\}. \tag{9.8}$$

This conjugate term plays the role of a *calibration penalty* on the risk-neutral discount factor π relative to the prior P. For example, when $u(w) = -e^{-w}$, the conjugate function is

$$u^*(x) = x - x\log x, \tag{9.9}$$

for $x > 0$, so the exponential utility creates a dual problem with a calibration penalty that is the *entropy of π relative to P*

$$-E^P\left[u^*(\pi/P)\right] = E^\pi\left[\log(\pi/P)\right]. \tag{9.10}$$

Other utility functions yield different calibration penalties. A utility that we will refer to below in discussing the supply chain literature is the quadratic utility $u(w) = w - lw^2$, for which the calibration penalty is $u^*(x) = -x/2l - x^2/4l$. Additional discussion and complete proofs of this duality can be found in King et al. (2009).

In summary, adding static positions in listed options to the linear program (9.1) allows us to calibrate to observed prices of related derivative securities where the stochastic process of the underlying asset is unknown. If the arbitrage bounds are too wide, then a utility formulation (9.6) may be required to narrow the gap, in which case the obtained valuation will depend upon the prior probability P as well as the utility/calibration penalty. This approach can apply to incomplete models in finance, as well as to models in the supply chain with uncertainty not captured in market prices.

9.5 Financial Modeling Applications in the Supply Chain

In this section, we consider in greater detail some of the main contributions in the literature integrating supply chain decision making with financial modeling techniques. We show how the basic trading models outlined in the previous section can be used to apply financial pricing theory to supply chain problems and can be used to unify the variety in notation and presentation that is found among these different models.

Smith and Nau [1995] discuss a project investment model where each investment policy α_t produces stochastic cash flows $F_t(\alpha_t)$. The policy decisions are discrete choices such as "invest," "defer," or "decline." Project cash flows depend on both market states, $\mathcal{M}_t$, and non-market states, $\mathcal{N}_t$, and it is possible to trade in market securities. The firm maximizes an additive exponential utility function of the hedged cash flows:

$$\begin{aligned} \max_{\alpha,\theta,W} \quad & E\left[-\sum_{t=0}^{T} k_t e^{-W_t/\rho_t} \right] \\ \text{s.t.} \quad & W_t = F_t(\alpha_t) - S_t(\theta_t - \theta_{t-1}), \\ & \theta_T = 0. \end{aligned} \tag{9.11}$$

The formulation permits the firm to use market instruments to borrow project investment funds and to reinvest project earnings.

The terminal condition $\theta_T = 0$ prevents "borrowing from over the horizon." It is common in market models, as a first approximation, to allow the trader to take negative positions without limit. But if a trader is permitted to borrow unlimited money in the final period, then it may be possible to hedge all sorts of irresponsible risk-taking in prior periods. Thus, the choice of T in such a trading model corresponds to some accounting interval, after which it is no longer necessary to model the borrowing of money.

Smith and Nau develop an integrated rollback algorithm based on two key assumptions. One key assumption, quite common in the financial literature, is that the martingale conditions (9.2) admit *exactly one solution*. This is called "the complete market" assumption. The second key assumption is that the utility is an additive exponential. The integrated rollback algorithm applies dynamic programming. At each backward step, risk-neutral pricing is performed at market states $\mathcal{M}_t$, using the unique risk-neutral density. Effective certainty equivalent pricing is then used at nonmarket states $\mathcal{N}_t$, using the firm's private probabilities. Under assumptions of completeness and exponential utility, Smith and Nau show that the algorithm produces an optimal investment policy that is decomposed into a hedging part, an investment part, and a part consisting of unhedgeable cash flows. An important benefit of the separation achieved by integrated rollback is that nonmarket probability distributions are specified conditional on market events – there is no need to develop a complete joint probability distribution. Smith and Nau further argue that the additive exponential utility is necessary since it evaluates the residual risk R_t independently of interperiod and wealth effects.

In the case of incomplete markets, not all risks can be hedged by trading securities. This situation is relevant in supply chain settings, where often the many risks faced by members of a supply chain cannot be fully hedged by market securities. Smith and Nau consider the situation of incomplete markets and show that the basic options-pricing value approach can be extended to produce bounds for the valuations of the investment policy decisions. These bounds will be the arbitrage bounds, as we discussed earlier.

Gaur and Seshadri [2005] analyze a single-period single-item newsvendor model where demand at time T is correlated with the price at time T of a market security that is actively traded, such as the S&P 500 index. In our notation, the cash flows of the newsvendor are $F_t(I)$ for $t \in \{0, T\}$ (and $F_t(I) = 0$ otherwise), where I is the initial inventory decision. The objective is to maximize utility at time T:

$$\begin{aligned} \max_{I,\theta} \quad & E\left[u(S_T\theta_T)\right] \\ \text{s.t.} \quad & F_t(I) = S_t(\theta_t - \theta_{t-1}). \end{aligned} \tag{9.12}$$

Gaur and Seshadri's formulation is comparable to that of Smith and Nau (9.11), except that here the utility is applied only to the terminal wealth, which is captured in the terminal portfolio $S_T\theta_T$. Gaur and Seshadri examine the impact of static hedging for general classes of utility functions in order to determine whether hedging increases or decreases initial inventory decisions and utility. The baseline static hedge is the *minimum variance* hedge θ^* that minimizes the squared error of newsvendor payoffs relative to the span of wealth achievable by trading only on market information:

$$\begin{aligned} \min_{\theta} \quad & E\left[|F_T(I) - S_T\theta_T|^2\right] \\ \text{s.t.} \quad & 0 = S_t(\theta_t - \theta_{t-1}). \end{aligned} \tag{9.13}$$

An equivalent expression for $S_T\theta_T$ for self-financing trades is to use the gains process formulation

$$S_T\theta_T = G_T(\theta) := \sum_{t=0}^{T-1} \theta_t(S_{t+1} - S_t). \tag{9.14}$$

It follows that the hedging process θ^* that optimizes (9.13) is that which generates the projection of $F_T(I)$ onto the span of the gains processes

$$G_T(\theta^*) = E\left[F_T(I) \,|\, \mathcal{M}_T\right]. \tag{9.15}$$

This is a calculation that can be performed by conditional integration; however, implementing a minimum variance hedge is complex and requires many trades. Gaur and Seshadri analyze the impact of two simpler hedges comprised of positions in the security and in call options. The first is a static hedge that cannot be revised during the horizon, and the second is a dynamic hedge with one portfolio rebalancing period. They perform numerical experiments with demand data and derive conclusions showing the impact of hedging on the optimal inventory. In general, their conclusions are that hedging reduces substantially the investment cost of the initial inventory and that optimal inventory levels are higher.

Caldentey and Haugh [2006] discuss a model of a nonfinancial firm that can trade *continuously* in financial markets. The uncertainty in returns is modeled by two independent geometric Brownian motions: one models market observables and the other models nonfinancial, firm-specific uncertainty. As in Gaur and Seshadri, the terminal wealth is equal to the payoff from operations plus gains generated by a trading policy. Caldentey and Haugh focus their analysis entirely on quadratic utilities, such as that in the minimum variance hedge model used as a baseline in Gaur and Seshadri. The quadratic utility function $u(\cdot)$, defined over terminal wealth, is of the form $u(w) = w - lw^2$. The analysis takes place under two alternative assumptions concerning the information process. The financial market is assumed complete and the two assumptions of Caldentey and Haugh apply to the nonfinancial uncertainty. The first alternative is "incomplete information," under which uncertain inputs to the operations are not observed, so the trading process cannot act on them; the second is a "complete information" setting in which uncertain inputs to the operations can be observed in addition to security prices.

Caldentey and Haugh apply the approach of Schweitzer [2001], which first identifies a minimal risk-neutral measure. The minimal risk-neutral measure, in our notation, is the risk-neutral discount factor that has smallest quadratic variation relative to the probability distribution P:

$$\min_{\pi \in \Pi} E^P \left[(d\pi/dP)^2 \right]. \tag{9.16}$$

This problem is dual to the minimum variance problem (9.13). A decomposition theorem of Schweitzer then provides a stochastic differential equation that can be solved for the optimal hedging policy. This optimal hedging policy turns out to be decomposed into the same three parts as in Smith and Nau, namely: a hedging part, an investment part, and a part consisting of unhedgeable cash flows. In addition, the decomposition indicates how to integrate the expression for the optimal value – which expression can then be optimized to find the optimal strategy. Caldentey and Haugh apply this method to a newsvendor and a production–inventory problem, as illustrations of the technique.

The approach of Caldentey and Haugh is representative of one of the current research frontiers in financial mathematics with respect to the treatment of market incompleteness. The application of these methods in a practical setting requires a specification of the operational payoffs sufficiently tractable to permit the application of Îto's lemma and to solve the resulting stochastic differential equations. For problems where operational risk may be modeled by diffusions or Poisson jump processes, this type of approach can be very useful, provided the differential equations can be solved.

9.6 Application to Newsvendor

The supply chain contract described in Sect. 9.2 encompasses a number of realistic sources of uncertainties that exist in a supply–demand contracting relationship. For example, the supplier is faced with cost uncertainty, unknown discount factors, current period and future period demand uncertainty, future pricing uncertainty, and supply uncertainty. Due to these uncertainties, both parties to the contract must contend with considerable amounts of risk. In this section, we consider how the traditional linear programming framework for newsvendor models can be extended to address these sources of uncertainty. Our approach here can be viewed as an extension of Birge [2000]. We utilize the model formulation of Gaur and Seshadri [2005] as a starting point for this discussion.

We define the following problem variables and parameters:

$$\begin{aligned} I &: \quad \text{initial inventory to be ordered,} \\ D &: \quad \text{future demand,} \\ p &: \quad \text{unit selling price,} \\ c &: \quad \text{unit cost,} \\ s &: \quad \text{salvage price,} \\ r &: \quad \text{risk-free interest rate.} \end{aligned} \tag{9.17}$$

The newsvendor has a time 0 cash flow that consists of the cost paid for the initial inventory of newspapers:

$$F_0(I) = -cI \tag{9.18}$$

and a time T cash flow that consists of the income from the sold newspapers plus a salvage term for the unsold inventory:

$$F_T(I) = p \min[D, I] + s \max[0, I - D]. \tag{9.19}$$

At all other intervening times, $t \in (0, T)$, there are no newsvendor cash flows, $F_t(I) := 0$. It is assumed that a simple statistical model relates future demand D to the price at time T of a market-traded security, S:

$$D = bS_T + \epsilon, \tag{9.20}$$

where ϵ is a random noise term that is independent of S_T. As in Smith and Nau above, we suppose that $\mathcal{M}_t$ describes the observable states for market prices S_t and $\mathcal{N}_t$ describes the observable states for ϵ. Such a model (9.20) could be constructed by regression of demand data against the security prices. Substituting for D in (9.19) and collecting terms yields

$$F_T(I) = pI - (p-s)(I-\epsilon) + (p-s)b\big(S_T - \max[0, S_T - (I-\epsilon)/b]\big). \tag{9.21}$$

This equation describes the newsvendor income in terms of $(p-s)b$ units of a long position in the security S and $(p-s)b$ units of a short position in a "call option" with strike $(I-\epsilon)/b$. Gaur and Seshadri then proceed to show how a portfolio consisting of a short position in the security and a long position in an option can partly hedge this payout. Deviating slightly from Gaur and Seshadri, we can obtain an equivalent but slightly more concise expression for the newsvendor income:

$$F_T(I) = pI - (p-s)b\big[(I-\epsilon)/b - S_T + \max[0, S_T - (I-\epsilon)/b]\big], \quad (9.22)$$

$$F_T(I) = pI - (p-s)b\max[0, (I-\epsilon)/b - S_T]. \quad (9.23)$$

This shows that the newsvendor time T cash flow equals the nominal income pI from selling all the inventory minus $(p-s)b$ units with cash flows that resemble the payoffs of "put options" with strike price $(I-\epsilon)/b$. This suggests that the optimal hedge is for the newsvendor to purchase such a put option, were it to exist.

The newsvendor is in the position of a *buyer* of securities that will be paid for out of the newsvendor profits, so one can express the risk-neutral newsvendor problem as follows:

$$\begin{aligned} \max_{I,\theta,\xi}\ & S_0\theta_0 + (C_a\xi_0^a - C_b\xi_0^b) \\ \text{s.t.}\ & F_t(I) = S_t(\theta_t - \theta_{t-1}) + (C_t\xi_0^a - C_t\xi_0^b), \\ & S_T\theta_T \ge 0, \\ & \xi_0^a, \xi_0^b \ge 0, \end{aligned} \quad (9.24)$$

where I^* denotes the optimal initial inventory. The optimal value V^* is the maximum value of a current portfolio in the securities S and the corresponding listed options C that can be purchased with the payments $F_t(I^*)$ as collateral. The initial portfolio holdings $(\theta_0, \xi_0^a, \xi_0^b)$ will be as large as possible such that when the portfolio is unwound, the uncertainty from the sale of the portfolio will overmatch the uncertainties in the newsvendor profits. For this reason, this formulation is called *superhedging*.

Superhedging models are generally criticized for being too conservative. In this setting, conservative means that the optimization (9.24) may generate a valuation V^* that is far too low. To check the effect of the conservatism of the superhedging solution, one computes the upper bound (9.4) with the optimal newsvendor cash flows $F_t(I^*)$. If these values are reasonably close, then the superhedging solution will work as well as any other alternative proposal.

If the comparison of the bounds indicates that the superhedging model leaves too much on the table, then the only alternative is, by analogy with (9.6), to recast (9.24) as a utility maximization problem:

$$\begin{aligned} \max_{I,\theta,\xi,V}\ & E^P\left[u(S_T\theta_T - V)\right] \\ \text{s.t.}\ & V = S_0\theta_0 + (C_a\xi_0^a - C_b\xi_0^b), \\ & F_t(I) = S_t(\theta_t - \theta_{t-1}) + (C_t\xi_0^a - C_t\xi_0^b), \\ & \xi_0^a, \xi_0^b \ge 0. \end{aligned} \quad (9.25)$$

In summary, calibration to externally observed option prices is the way in which information in the market equilibrium can be used to model financial uncertainty in the supply chain such as in the newsvendor problem. In models that use the partial completeness assumptions of market prices, such as those used by Smith–Nau, Gaur–Seshadri, or Caldentey–Haugh, one must estimate the model parameters as a separate step in the solution process. More recent models used to treat market incompleteness in finance, where calibration error is part of the problem specification, are quite naturally adapted to supply chain problems using the linear programming formulation.

9.7 Difficulties in Adapting Finance Approaches to Supply Chain

Similar to their financial counterparts, realistic supply chain models need to incorporate uncertainties into the basic parameters that drive revenue and costs. There is a growing literature that adapts market valuation techniques from finance to derive appropriate discounting factors for supply chain problems. However, many challenges remain. Fundamental to any market-based approach is the availability of a liquid market of relevant securities that can be used to hedge the uncertainty in the problem.

The major difficulty in hedging risk in supply chain problems is the difficulty in obtaining information that is relevant for this setting. We close with a brief discussion of some significant issues.

9.7.1 Timeliness of Market Information

A fundamental assumption in many financial models is the efficient market hypothesis. In its strong form, the efficient market hypothesis implies that all public and private information is already reflected in securities prices. However, in supply chain settings, there is often a lag between the time that information appears in the supply chain, such as changes in order or demand forecast quantities, and the time that this information is reflected in the financial markets. This is in part due to the nature of the long-term purchasing contracts that are common to supply chain participants. These contracts distort the view on the nature of true end consumer demand and hence on the health of the industry and supply chain participants as a whole. Consequently, attempts by one member of the supply chain to hedge exposures to orders or supplies from another member of the supply chain using that member's corporate securities market information may not yield desired results.

Recent academic literature (see, e.g., Mendelson and Tunca 2007) studies the balance between spot market trading and long-term contracting for supply chain members. As the relative use of spot markets increases, the increased participation by supply chain participants will enable more timely market information.

9.7.2 Correlation Between Markets

Risks from multiple sources are the norm in the supply chain. Calibration is a powerful technique, but trying to calibrate joint risk measures can quickly lead to intractable problems. Most articles in this literature in effect ignore correlations.

Froot and Stein [1998] provide an argument for ignoring the correlations. They argue that it is sufficient to hedge those risks that can be hedged, and to assess the remaining risk – even if it is correlated – using the implied risk adjustment in the market for the own-corporation securities. The argument hinges on the interests of the equity holders. The returns to the equity holders are affected by the various risks to which the company is exposed. Those risks (such as jet fuel for an airline) that are hedgeable in financial markets should probably be hedged by the company – since it is likely cheaper for the company to do this than an individual shareholder. However, the shareholders do want to be exposed to the nonhedgeable risks, since this is why they hold the stock in the first place. In setting corporate risk management policy, then, it seems reasonable to apply the implied risk reflected in the marketplace, for the equity in the company itself.

9.7.3 Standardization and Transparency

Contracts that are traded on financial exchanges contain standardized terms defining what is to be traded as well as consequences in the case that a counterparty fails to deliver on one of the attributes of the contract. This high level of standardization combined with the limited number of standardized contracts ensures market liquidity. In the supply chain context, contract terms exhibit high levels of customization that render these contracts difficult to introduce for third-party trade. Further, counterparties to supply chain contracts do not have incentives to reveal information that is critical for third parties to assess the future value of the contracts. Thus, there is no real market information that can be derived. This lack of standardization and transparency results in limited market data for supply chain contracts and is a key differentiator between risk in supply chain settings and financial settings.

Absent the existence of liquid markets for supply chain information, there is no possibility for successful calibration or risk hedging of price uncertainty. Moreover, if the counterparties do not have incentives to reveal information that is critical for third parties to assess the future risks, then no real market information can be derived. This lack of standardization and transparency is perhaps the major reason why market information is difficult to obtain in the supply chain.

This underlines the importance of creating markets in the supply chain, in which participants have an incentive to collect information and obtain rewards for being correct. Guo et al. (2006) suggest the creation of macroindices to enable trading in supply chain information. Some firms (Google, for one) have even set up betting pools among employees to obtain a virtual market of views on future technology developments. New Internet technologies can be used to facilitate this development; however, it will remain very difficult to enforce the transparency required for third parties to have information that is good enough for risk management in the supply chain.

References

Birge JR (2001). Option methods for incorporating risk into linear capacity planning models, Manufact Serv Oper Manage, 2:19–31

Cachon C (2002). Supply chain coordination with contracts. In: Supply chain management, handbook in operations res manage sci North-Holland

Caldentey R, Haugh M (2006). Optimal control and hedging of operations in the presence of financial markets. Math Oper Res 31:285–304

Dixit AK, Pindyck RS (1994). Investment under uncertainty. Princeton University Press, Woodstock, Oxfordshire, UK

Edirisinghe NCP, Naik V, Uppal R (1993). Optimal replication of options with transactions costs and trading restrictions. J Financ Quant Ana 28:117–138

Elkins D (2005) Enterprise risk management - cross-cutting applied mathematics research in finance, insurance, manufacturing, and supply chains. In: SIAM Conference on Mathematics and Industry, November

Froot K, Stein J (1998). Risk management, capital budgeting and capital structure policy for financial institutions: An integrated approach, J Financ Econ 47:55–82

Gaur V, Seshadri S (2005). Hedging inventory risk through market instruments. Manufac Serv Oper Manage 7:103–120

Guo Z, Fang F, Whinston AB (2006). Supply chain information sharing in a macro prediction market. Decis Support Sys 42:1944–1958

Harrison JM, Pliska SR (1981). Martingales and stochastic integrals in the theory of continuous time trading. Stoch Proc Appl 11:215–260

Kamrad B, Ritchken P (1991). Valuing fixed price supply contracts. Eur J Oper Res 74:50–60

King AJ (2002). Duality and martingales: A stochastic programming perspective on contingent claims. Math Program B 91:543–562

King AJ, Pennanen T, Koivu M (2005). Calibrated option bounds. Int J. Theor Appl Finance 8:141–159

King AJ, Streltchenko O, Yesha Y (2009). Private valuation of contingent claims: Discrete time/state model. In: Guerard J (eds) Handbook of Portfolio Construction: Contemporary Applications of Markowitz Techniques. Springer, New York, NY, USA

Kleindorfer PR, Wu DJ, (2003). Integrating long- and short-term contracting via business-to- business exchanges for capital intensive industries. Manage Sci 49:1597–1615

Kouvelis P, Kostas A, Sinha V (2001). Exchange rates and the choice of ownership structure of production facilities. Manage Sci 1063–1080

Lariviere M (1999). Supply chain contracting and coordination with stochastic demand. In: Quantitative Models of Supply Chain Management. Kluwer Academic Publishers, Boston, pp. 233–268

Mendelson H, Tunca TI (2007). Strategic spot trading in supply chains. Manage Sci 53:742–759

Rockafellar RT, Uryasev S (2002). Conditional value-at-risk for general distributions. J Bank Finance 26:1443–1471

Schweizer M (2001). A guided tour through quadratic hedging approaches. In: Jouini E, Cvitanic J, Musiela M (eds) Option Pricing, Interest Rates, and Risk Management. Cambridge University Press, Cambridge, MA, pp. 538–574

Smith JE, Nau RF, (1995). Valuing risky projects:Options pricing theory and decision analysis. Manage Sci 41:795–816

Trigeorgis L, (1996). Real Options. MIT Press, Cambridge, MA, USA

Tsay A, Nahmias S, Agrawal N (1999). Modeling supply contracts: A review. In: Quantitative models for supply chain management. Kluwer Academic Publishers, Norwell, MA, USA, pp. 299–336

Chapter 10
Field-Based Research on Production Control

Kenneth N. McKay

10.1 Introduction

Field-based research is a form of empirical research and involves actual experiments or observations. The experiments and observations are often used to support or test scientific claims. They are also often used to generate insights about a phenomenon or possible research topics for further analysis (e.g., find a suitable topic for a graduate student). Although we perform field-based research for a number of reasons, we hope that the results we obtain are sound, scientifically valid, and provide value to both the field and science. Through field research, we endeavor to understand and model the business process, or capture the important and salient characteristics of the problem so that we can include them in modeling and analysis. We might hope to incorporate findings from field research in automated tools and advanced algorithms, making them more realistic and useful. In the best situation we can hope for scientific results that can predict or guide processes to the best possible outcome. This might be the lowest inventory levels possible, highest quality, least scrap, the most nimble and responsive reaction to a change in demand, or the quickest completion of an order. Strong, rigorous science is often associated with characteristics such as awareness and minimization of bias, inclusion of the necessary and sufficient aspects of the problem, ability to be replicated, evidence-based reasoning, careful and supported lines of causality, consistency in the use of terms and definitions, and the ability to be generalized in different ways. The science can take different forms: descriptive, prescriptive, predictive, or normative. Each type of science has assumptions and limitations. Each type also has recognized methods and tests of scientific quality. This chapter discusses various types of field research and presents ideas and methods for the sound undertaking of each type.

Field-based research requires that a number of methods and concepts used for the results are considered to be strong and rigorous. If this is not consciously done, the results might not make any sense whatsoever. To illustrate this point, consider two

K.N. McKay (✉)
Department of Management Sciences, University of Waterloo, Waterloo, Ontario, Canada
e-mail: kmckay@uwaterloo.ca

K.G. Kempf et al. (eds.), *Planning Production and Inventories in the Extended Enterprise*, International Series in Operations Research & Management Science 151,
DOI 10.1007/978-1-4419-6485-4_10,

analogies based on real examples in which the analyses were not sound. In the first case, manufacturing engineers introducing just-in-time and kanbans decided that 25 parts per bin was a good number and that they would use two bins. There was no mathematical analysis or rationalization for the units per bin except that 25 sounded good and 25 parts would fit in a bin. There was no justification for the two bins except that everyone did it that way. These engineers clearly did not understand the operational concepts behind the use of bins, did not know that methods existed to derive the number of bins, and did not understand what might happen as a result of their actions and decisions. They knew machines, welding, and metal fabrication but did not know the science of inventory management. They were working outside of their field. It would be fair to say that these engineers were lacking in the appropriate education and were naively applying methods and ideas that they thought were best.

In the second case, financial personnel in a large factory mandated that the finished goods level for all parts be set at 3 days instead of 4 because of the cost – as it appeared on the financial statements. Again, this was a case of people well schooled in their own discipline working outside of it and naively making decisions with the best of intentions. They lacked the education and understanding to realize that some parts might require less or more than any targeted number. They also apparently did not understand what the finished goods inventory was used for – buffering against uncertainties in production and demand. They also did not realize that there were methods which could be used to determine suitable levels for various categories of parts. Individuals well versed in industrial engineering or operations management would never dream of making such decisions in these ways. Would they?

Hopefully not. Then why do so many industrial engineering and operations management researchers and practitioners make similar mistakes when they perform field-based research? They make the same types of assumptions with the best of intentions and unfortunately obtain the same quality of results – poor quality at best. They do not realize how bad the science that they are using is, and do not recognize nor acknowledge the weaknesses in their approach. Before delving into the specifics and details of field research, there are some general steps to keep in mind:

1. Clearly state and define what the intent or purpose of the field research is. What type of science or result is sought after? What level of research is striven for? For example, descriptive or prescriptive?
2. Consider what implicit and explicit assumptions you are making about what you will observe or gather and what it means. For example, what is your operational definition of a good schedule? While such assumptions might appear obvious, the issues are often more complicated than they appear (Kempf et al. 2000).
3. Ensure that you know what the problem that you are trying to study is. If you do not know, defining this is the first activity to explicitly undertake. Do you know the necessary and sufficient aspects to capture?
4. Before the study starts, make sure you know when, how, and what to capture and verify. What is the data and information you need to support your claims or level of science? Do you know how much data to collect and under what situations?

5. Reflect on the boundary and scope of the modifying or context-sensitive factors which could affect what you are observing and what the observations might mean. For example, is it a special time of year at the factory? Is something else happening concurrently (e.g., a new product launch)?
6. Critically examine any field instruments and methods – phrasing of questions, techniques for gathering the on-site field observations. Do you know how to avoid bias? Do you know how to build understanding from observations?
7. Make your study testable. Do you know how to make your field research replicable, thorough, and supportable? Could someone else independently repeat your study and obtain the same insights and results? In a different factory?
8. Establish a baseline measurement for all key factors you will be studying or making claims about. You need to know what the current situation is if you are going to make statements about performance or behavior staying the same or changing during the field study. For example, in a given situation, what was the theoretical optimal or minimum number of setups, how many were actually taking place at the beginning of the study, and how many were occurring at the end of the study? More than one study has on the surface disappointed the researchers because of less than optimal numbers, but pleased the practitioners because the results were better than if the science had not been applied. Having a comparison point is critical for all field-based analyses.
9. During the study, clearly document the status quo and any exceptions to the status quo – any special decisions which could not be predicted using basic manufacturing data. Special routings, batch sizes, machine assignments, or task assignments are clues to context-sensitive situations. Document the normal policies and then be sensitive to variance.
10. Capture the official plans given to management as well as any verbal or unofficial instructions supplied by the production control staff to the actual production workers. Also document what the scheduler expects to happen and what the scheduler would do if left to his or her own devices. These four observation points can provide valuable insights into the real constraints and objectives, the degrees of freedom existing in the system, and how any official schedule is to be interpreted.
11. Capture and document any other changes introduced just before or during the field study. For example, were any new policies introduced about inventory levels or part quality testing just prior to the study? Were any new products introduced, engineering changes, new vendors, different materials, new crews hired, etc.
12. Support any documented plans and schedules with actual execution data. That is, how well did the factory adhere to the plan? During the day? During the next day or two? During the next week?

While there are many other things to consider, these dozen points constitute an initial point of departure for any field study. In thinking through these issues and requirements for strong field research, it is important to critically examine what proof you might need to exhibit in supporting your claims. Like our misguided engineers and accountants described above, what education, training, and skill can be demonstrated? Is your credibility in doing field research above question? The following sections present taxonomy for classifying field research and discuss various issues the researcher must address within each classification.

10.2 Taxonomy for Conducting Field Research on Production Control

In production control, there are perhaps a number of ways to categorize or group such research, but for the purpose of this chapter, we will use the following schema.

First, there are two main segments of field research – those *involving humans in some way as subjects* and those *not involving humans as subjects*. This distinction is used as one dimension in the schema. Once a human participant is involved as a subject – learning what they do, understanding how or why they do something, comparing computational results to their results, prescribing what tasks they should do or how they should do their tasks, or assessing their capability or skill – the line is crossed and an entirely new set of research methodologies are involved. The human can be involved as the individual decision maker or as part of a larger organization, depending on the research focus. Thus, the field research may be focused on individual tasks and decisions or on a larger business process involving many people. In any research involving humans, the issues begin with how data is acquired and continue through how it is analyzed, and what can be said about the results. If the research is going to discuss how and why someone or a group of people did something, or use a line of reasoning provided by a human, or discuss how good a human-generated decision was, there are elements of research that must be incorporated if the conclusions are to be rigorous and scientifically sound. A human as subject or as an active part of the research does not include those situations where a human is merely operating a machine, moving material, or assembling a part – unless a human factors analysis is being performed. One of the purposes of this dimension is to clarify situations where certain methodologies are appropriate because of human subjects and where they are not.

The purpose of the field research serves as the second dimension in the schema. Why does someone do field-based research? The purposes can be quite varied and can include: *deriving* insights about the problem structure and relationships; *gathering* data such as demand patterns and machine behaviors to be used in experiments; *testing* theories and concepts in order to determine the applicability and quality of results; *applying* theories and concepts in order to improve a practical situation; and *training or education* of researchers.

The schema for categorizing field research is documented in Table 10.1.

The chapter will use this schema to discuss various types of field research situations and what might need to be considered. We will briefly deal with the simpler situations in which humans are not directly involved in the research. The majority of the chapter will focus on the human or sociotechnical situations as these latter are the most complex to study.

Table 10.1 Field-based research categorization

	Human subjects Involved	Human subjects not involved
Deriving		
Gathering		
Testing		
Applying		
Training		

10.3 Technological Situations

In this section, we will focus on production control situations which are not confounded by the presence of human subject as an active part of the research. There are still complications, but issues relating to human judgment, decisions, rationale, etc. are absent. Thus, the first question is whether human study methods are involved?

There are indeed some clear cases where they are not. For instance, automated assembly lines, gated or driven by mechanical means with little or no variability, are examples of what we would easily call technological situations. A computerized flexible manufacturing cell or the study of a single automated machine is others. If the situation is buffered from external influences such as weather, culture, and every day looks like any other day, then the human element is likely absent. In these types of cases, the production control problem comes down to the basics: demand, bills of material, operations, routings for the operations, resources for the setup and operation, sequence-dependent constraints for setup, time estimates per unit of manufacture, inventory controls, movement controls, and possible governance mechanisms between suppliers and customers. This type of data is sufficient for effective production control within a plant or a supply chain and the problem is largely context free. In technologically focused research activities, the various pieces of data and perhaps their statistical distributions are the major focus of any *acquisition* activity. For example, what is the basic business process, what are the average setup and operation processing times, what is the variance in the processing times, and what is the sufficient amount of data possible to determine a suitable data distribution.

If the human element is added in the way of operators, then the problem is still likely the same unless there are special and unusual characteristics about some (or all) of the operators that are used to determine significant aspects of the resource allocation and sequencing, or inventory levels. If the differences are simple skills, this can be viewed as quantitative data as well and the situation is basically the same as a technological one. However, if the differences are significant and relate to factors such as personalities, social situations, social behavior, and attitude shown during training, then the situation turns into a sociotechnical situation and cannot be dealt with as a purely technological situation. For example, we have seen examples of where operator attitude during training was used to select one crew over another for the initial production run of a new part.

Thus, a major concern with field research in technological situations is, *Is it really a technological situation?* It is important to take a little bit of time and effort to ensure that it is indeed free of any significant amounts of cultural, personnel, or environmental constraints or context. There will always be some level of context sensitivity, but the major factors that could affect the problem structure should be considered. For example, do the objectives change based on the time of the month? Does the factory concentrate on inventory levels more at the end of the month because of a corporate measurement metric? Are special or priority parts flowed through in nonstandard ways? Perhaps certain machines or cells are used for priority work associated with certain customers who have complained recently. These types of context-sensitive production control have been observed as standard routines in several factories. Issues such as these are important to understand if the researcher is going to observe a situation and comment upon it, or generalize from it. It is not sufficient to assume that a situation is not context sensitive, it is necessary to do sufficient investigation and data gathering to confirm that it is not.

Exploratory or preliminary research into *deriving the problem formulation* or basic phenomenon typically does not result in general, prescriptive, or normative claims. Researchers in the early phase of research are looking for the basic elements and existence proofs. Subsequent research will explore the power and general nature of the science. If general claims are the intent of the research, then a longitudinal approach provides added visibility and insights – useful in all phases of deriving, gathering, testing, applying, and training. A longitudinal study is one that is performed over an extended period of time and is not a single visit or a single snapshot of the situation. If it is possible to observe or experiment at different times (e.g., shifts, time of day, days of the week, weeks, months, and seasons), and when different activities are happening in the factory (e.g., line upgrades, tours, plant shutdowns, vacation, line introduction, new product launch, product phase out, and outsourcing) and *NOTHING* is different about production control, then the situation is probably safe to be declared *CONTEXT FREE.*

The word *context* is itself a problem. In this chapter, the larger or macrosense of context is used. Does it matter why the task is done, when the task is done, by whom, where, and how? An extreme view suggests that everything about production planning is context sensitive – there is a due date and this creates a context for prioritizing, sequence-dependent setup is context, what operations a machine can perform is also context. However, the minor aspects of context are often easy to capture, encode, and accommodate within logic while the major aspects of context involve information and issues beyond those normally represented in computer systems.

Why is it important to worry about context-free research and context-sensitive research? If the researcher is going to make a claim about why something has happened, is happening, will happen, or should happen, the researcher should be aware of when such a claim is valid or invalid. This relates to the power and generalization of the research. The more situations thoroughly tested, usually the more rigorous the study.

Consider the printed circuit board assembly factory studied by McKay (1992). The lines were highly automated, state-of-the-art, and in what was considered one of the top factories in a top corporation. On the surface, this factory situation should have been considered context free. In fact, one researcher associated with the study, but not deeply involved, predicted that factory would be a situation where context-sensitive issues would not be found. Contrary to this prediction, there were many examples of unexpected context found in the study, and one stands out as a nice example. Every spring for 1 month, yield would drop by 10%. This occurred because of the change from central heat to central air conditioning, and there were processes sensitive to humidity. Each year it took time to restabilize the processes and scrap was higher than the other times of the year. Not all factories have processes sensitive to humidity in the air, but this one did. If this fact was not noted, consider what errors might have occurred. If the study period did not include this month, then expectations and predictions for this month would have been invalid and could have led management astray. If the study period was only during this odd month, nonstandard production control would have been observed as the factory tried to deal with the lower yield and any generalization without additional data would have been erroneous. If the scheduler was closely observed the month before the expected drop in yield, the scheduler's decision process was again unique and would not be seen during the other 11 months. Interestingly, the noted researcher was not alone in assuming that this factory was context free. The factory's information systems department was constructing an automated scheduling tool that assumed 100% context-free performance on the factory floor, as dictated by a senior executive. As a result of the study, that documented roughly 10% of the daily decisions were actually context sensitive, the project was canceled. Consider what would have happened if the automated scheduling tool had been finalized and deployed!

It is also possible that context dependencies will sometimes enter into a situation that was once context free. Consider an automated manufacturing cell with several machining centers. When the cell was new, it was likely that any of the machines could work on any part assigned to the cell with equal performance. As the cell ages, it is possible that unique characteristics will develop for one or more of the machines and that certain work will have a machine preference. For example, a certain part might push the tolerances on a machine that has had mechanical problems or wear points, thus making a different machine a better choice. Observing nonrandom assignments made by human planners in such cases is useful in detecting these types of context-sensitive issues. This type of issue has been observed at many factories and is one of the first things to look for. A visit was made several years ago to a very high-tech manufacturer producing close tolerance parts who was trying to improve cell efficiency through improved flow and layout. Approximately six machining centers were in the cell, all with the same specifications and the same tooling. With automated material handling as well, it certainly looked like a context-free situation. Unfortunately, it was not. The centers were now 10 years old and had developed individual personalities – some centers performed better on some parts than other parts. This had to be taken into account in the analysis.

A field study that has a single data acquisition may have many stated and unstated assumptions which limit the descriptive or prescriptive nature of the research. As long as the claims match the rigor, there is no difficulty. However, if the claims extend beyond what can be reasonably claimed, there is a problem. For a field study that uses a single set of data, based on a single observation or experiment, it is probably safer to discuss everything relative to the single case and not make any suggestions about other situations. The evidence may suggest that certain relationships may hold elsewhere, but there is clearly no solid evidence to this effect. This is a general caution. However, if it is possible to state with confidence that the situation is context free, then a single data point might be sufficient for some powerful statements. In these cases, good experimentation and sensitivity analysis can provide insights for what might be and what should be. Any assumption of a situation being context free should be supported with evidence and solid reasoning.

The following sections will discuss various issues related to the deriving, gathering, testing, applying, and training in context-free situations.

10.3.1 Deriving

In the early phases of scientific inquiry, the researcher is attempting to discover what to include in the science. What is the essence of the problem? For example, in modeling a surface mounted printed circuit board assembly line for general flow, it might not be relevant to know if a machine is a pick and place insertion device; but it is important to know if the machine works on one or two boards at the same time and is at the start or end of the assembly line (if loading or unloading is performed). The challenge to the researcher is to include what is necessary – no more, no less. Why one relationship is included and others are not? This type of analysis is needed if field situations are going to be used for insight – model building, constraint identification, and objective specification. It is never possible to include all factors, so it is important to identify the dominant factors. Applying key critical thinking processes (McKay 2000; Browne and Keeley 2004) are usually sufficient to deal with context free or possibly context-free situations.

10.3.2 Gathering

In a context-free situation, data gathering is more likely to be done with computer files rather than with pencil and paper or a stop watch. Manufacturing execution systems (MES) and ERP (sales, sale forecasts, bill of material, material masters, routing tables, build history, shipped history, inventory history, and yield history) will be the sources for such data in studies of business processes or specific operational areas. In some cases, corporate information system support will be required to gather and extract certain data, but most data should be available at the plant level.

Questions to consider are:

- What types of data are necessary to gather
- Degree of granularity of the data
- Time period to consider
- Sources of error, types of error, magnitude of error, and implication of error

Completeness and accuracy of the data should also be tested and validated if strong claims are to be made – it should not be assumed. Validation of data should always be done, of course, but compromises are often made in exploratory studies – they should not be made in studies designed for normative or prescriptive results. The exploratory study might be looking for a basic existence proof and encouragement for further research, and an extensive study may not be wise or possible initially. The normative or prescriptive study will use the data for more definitive purposes, which warrants additional care.

10.3.3 Testing

In this type of field research, the researcher has a theoretical model or theory to test in an actual situation, or has empirically inspired science to test. This type of activity also includes test-of-concept implementations or demonstrations of a more practical nature (e.g., new scheduling software system to deploy).

All field tests, for the strongest claims, should have baselines or benchmarks for comparative purposes. The researcher needs to present the case for baseline or benchmark equivalence and also provide the appropriate reasoning for why the tested science resulted in any difference. It must be clear that the introduced science or technology provided the benefit and not something else in the factory. For example, a company introduced a sophisticated scheduling tool with powerful mathematical algorithms and obtained an improvement of 10–15% as a result. It would have been simple to claim that the mathematical algorithms were responsible for the gain. The scientists probed deeper and upon investigation, they discovered that the science did not provide the gain. By introducing the tool, consistency was obtained between shifts with less confusion, and the downtime associated with the shift changeover was dramatically reduced, producing the 10–15% gain. While it is not known what this company did about subsequent gains, it is possible to speculate about what might have been done (or could have been done). For example, after the system was installed and the environment stabilized, could the mathematical engine be downgraded to simple heuristics and compared with situations where the engine was allowed to operate at full power? While not perfect, such ideas can create some comparison baselines.

Credible experiments in the field are very difficult. In a real factory there are often many changes introduced and it is almost impossible to create a pure situation where the science can be guaranteed to be the cause for the effect. Continuous improvement programs imply that most factories face the challenge of managing during change and not management of change. It is a constant variable. Thus, it is important to investigate, document, and discuss any and all significant differences between the baseline and the experimental conditions. For example, are the operators the same? Are the parts the same in terms of complexity, maturity, learning curves? Have the operators been given the same training on the different equipment? Has the maintenance schedule been the same for the equipment? Is the demand profile the same? Objectives the same? Any factor that could introduce substantial variability in processing time and yield should be thought about, and should not be dismissed without consideration. For example, tests should be done at the same time of the month in most factories to control for demand patterns and production objectives. Tests should also avoid the time periods around other substantial or significant changes to the factory's profile (processes, methods, and products).

For stronger claims, replication of the tests is important – under different conditions if possible. For the strongest claims, it is also recommended that the field data be re-examined in the lab and computational experiments be performed to ensure that results derived from the laboratory experiments match those observed and predicted. If the results match, there is still not a clear or perfect linkage, but there is support. If the results do not match, further investigation is warranted.

10.3.4 Applying

There are many case studies in the literature where academic theories and models are applied in an industrial setting, some interesting results noted, and the results written up. The purpose is to show possible benefits and discuss underlying factors associated with the hope for improvement in the situation. There are also case studies or situations where existing or newly developed production planning aids are deployed by practitioners. The practitioner activities are in many ways similar. The introduced technology or new methods are intended to improve the situation and in many cases proof is needed to support the claims. The client or vendor might have performance clauses in the contract or claims want to be made about improvements to inventory levels, flow times, and fill rates.

The same care taken for the basic *testing* in context-free situations is required. If the application and analysis is of a single instance, then the same issues noted above exist. Evidence must be provided for any claims for generality and prescription. If the application has replication and a longitudinal research element, then it is possible to consider stronger claims. First, the situation must be shown to be context free and then, secondly, the results should be replicated to avoid the initial shock or introduction of change effect.

For production planning, it is possible to gather and support claims. For example, it is recommended that any tool or process be developed with a placebo capability. If a sophisticated scheduling algorithm is part of the package being applied, ensure that the system also has a very simple loading or sequencing algorithm. Then it is possible to run the system with and without the possible contributing factor. Just having a system (any system) might be a benefit, so if claims are to be made about the latest and greatest mathematical logic, the burden is placed upon the implementer to ensure that claims can be studied and supported. It is also possible to track the accuracy of planning and the use of the schedule. For several time horizons (e.g., shift, day, 2 days, week, 2 weeks, etc.), it is possible to keep track of what was planned, what was changed in the plan, and what the factory actually did. In one factory studied in the mid-1990s, it was observed that 75% of what was planned 1 day for the next 2 days of production was changed by the next day. In such a situation, improvements and values can be obtained from various factors, but not likely from well-tuned sequences.

In order to clarify claims and benefits associated with applied and practical technology such as advanced planning systems (APS), McKay and Wiers (2004) introduced a taxonomy. While the full description will not be replicated here, the basic ideas will be summarized. The possible areas to look for benefit depend on who might be viewing the situation and include:

- Saving time in the task
- Improving accuracy in task
- Improve quality in task
- Reduce effort in doing task
- Improve gathering of information
- Improve dissemination of information
- Reduce wastage, setups
- Reduce expediting, pre-emptions
- Reduce inventory levels
- Improve work flow through the plant
- Have better sequences

Although this list was specifically for planning and scheduling technology, many of the same points apply to general business process improvements. The APS focus will be used to illustrate the general idea. At the start of an implementation or introduction, it is important to document expectations associated with the various benefits sought. If possible, it is also important to document the current state of affairs. If this information is prepared in advance of the system introduction, it can be used in verifying claims and supporting any analysis. The McKay and Wiers taxonomy establishes levels of success which can match the expectations. The levels allow comparisons between implementations and also statements to be made within the project itself. The five levels of success are:

- Class A – better, more accurate sequences, and exploiting the full potential of the APS concept and technology
- Class B – determining feasible amounts of work
- Class C – identifying resource conflicts and loading issues
- Class D – improving visibility, communication, and consistency in the plan
- Class E – improving system integrity, cleaning up processes, and addressing data flows

Classes A and B have scales associated with the time horizon. For example, an A-1 implementation has a longer horizon (e.g., 2 weeks) during which time optimized sequences can be followed almost exactly and activities will take place when and where they were planned. An A-8 rating would provide optimized sequences that could actually be used for the next hour or two. If optimized logic is not used but the target is feasible loading, the B rating can be used. A B-1 rating is associated with good modeling of capacity – understanding what can be done. Thus, it is possible to talk about an A-8/B-1 situation in which the tool is very good but the ability to generate and follow the plan is dependent upon other factors, such as demand changing in the immediate time period.

While this framework is specifically designed for APS, it is possible for researchers to consider additional ways to clarify success and degrees of success in other applied production control situations. Every attempt should be made to avoid loose claims of success (or failure) which are not bounded.

10.3.5 Training

Context-free situations are often good examples of advanced manufacturing techniques and concepts. The level of automation is often high and there are few factors interfering with predicted or normative behaviors. If the situation is context free, the student or learner does not need to be concerned with field observation techniques involving human subjects. However, the student should be sensitive to possible sources of variability in supply, demand, and production. Some such sources might be machine age, numbers of engineering changes, prototype parts, and the possible implications associated with each. The student should also be aware that they are learning in a context-free situation and how that differs from a context-sensitive situation. They should understand that simple extrapolation and reusing assumptions can be harmful in a different situation.

The training and learning should include how things start (initiating triggers), how things move, how things terminate, and how things go wrong and are recovered (e.g., yield and rework). They should learn how and why production processes merge and diverge, and what orchestrates any merge point – why, when, and how many. Parallel, sequential, and transfer-in-batch flow controls are also important to observe and understand. Part of the learning should include probing any situation that does not match the usual – e.g., Is it always done this way?

New product introduction, product phase out, machine repairs, machine upgrades, machine introduction, machine decommissioning, process changes, material changes, product changes, and other such events should be observed if possible and documented. In a context-free situation, these types of events will have limited and isolated impact – only on those resources and processes directly impacted and will not impact other resources or processes in the plant. The student should be aware of how to identify and respond to such situations when performing other forms of field research: deriving, gathering, testing, and applying. If the student has not been trained in advance, a risk develops that a confounding situation is not identified or controlled for.

10.4 Sociotechnical Situations

It sounds very easy to do. Go out to the factory and ask the scheduler how they schedule (their objectives, constraints, and policies) and ask for some sample data. How hard can that be? All that is needed is a set of simple questions (e.g., How do you plan? Can you give us a copy of your schedule and input data?). It also sounds easy when a larger business process is being studied. Hold some focus groups and document the flow of data denoting who does what where. Unfortunately, it is not easy to do these types of studies in a scientifically rigorous fashion. How the field component is viewed will affect what data is collected, the accuracy and quality of the data, and what can be deduced or induced from the data.

The latter point is very important – what does the data say and what claims can be stated? In a business process situation, how confident can the researcher be that the appropriate steps were studied and accurately depicted? Furthermore, what data was missed in either type of field study? If the field methodology is not valid, the field results are no more scientifically valid than fireside stories. For example, if aggregate data is gathered on final plans and schedules without tracking transactions, sequences of transactions, and timing of transactions, it is difficult, if not impossible, to infer anything detailed about what the human did in reaching the final sequence, or why they did it. It is certainly difficult to infer anything about skill, knowledge, intent, or reasoning. It might be possible to analyze the final sequence but it is not clear what conclusions could be reached. To delve deep into the scheduler's reasoning, it is important to capture all special decisions or trade-offs – the sequence of decisions, constraints, and any special reasoning that resulted in changes to setup or processing estimates or resource allocation. The final sequence shows a static entity, but hides the process.

To study a business process, it is important to take similar precautions. Would you trust one or two people to explain how a complete and complex works? Is it safe to assume that a supervisor of an area knows what the individual workers do and how they do it? The supervisor might claim to know and might even say that they taught the workers everything they know. This claim has been made in almost every field study conducted by the author, and rarely has this been true. Thus, it is important in business process studies to think through the same issues and contemplate information sources, and think about what information might be hidden from view and how you might be able to access it.

To illustrate another aspect of the problem when focusing on schedule generation, consider the schedule itself. Ask the scheduler for the schedule that is being used to guide production. It will clearly document work allocation, timing, and sequences. At least this will not be a problem. Or, is it? It depends on what will be done with the schedule. We would not recommend using this schedule for quantitative comparisons due to the following reasons. As documented in McKay et al. (1992), there can be multiple schedules being used by the scheduler simultaneously. For example, it is possible that this schedule is political in nature; call this schedule number one. We have observed *political* schedules which did not correspond to what the scheduler was directing the factory to do; call this *directing* schedule number two. Furthermore, the schedule the scheduler was conveying to the factory floor did not correspond to what the scheduler actually thought would happen; call this *predictive* schedule number three. To top that off, none of the schedules or expected plans corresponded to what the scheduler wanted to do if left to their own decision making; not sacrificing quality or cost; call this *ideal* schedule number four. If the scheduler's skill in sequence creation is going to be compared, which schedule will be used?

If all the scheduler's constraints used to create the third plan (i.e., what they expect to happen) are included, then it might be possible or reasonable to compare schedules. If all the constraints guiding the sequencing are not included, then it would be best to ask the scheduler to create their dream sequence and not what is officially known as the schedule. The purpose of the schedule will affect its construction as the information will be used to craft the resource allocation. The information used by the scheduler in crafting a sequence is an interesting subtopic.

McKay et al. (1992) noted the existence of *enriched* data, data not normally included in scheduling research or commercial scheduling tools. For example, it was observed that the schedulers studied planned prototype or new or changed tasks for Tuesday, Wednesday, and Thursday day shifts, avoiding afternoon and night shifts, and the weekend. If this information and heuristic is not included in the mathematical representation, it might be possible to claim that a schedule generated with heuristics is better than that of the human scheduler. It might be better mathematically, but not better operationally. There is *mathematical feasibility* where finite resource constraints are acknowledged and quantitative rules such as shift times are obeyed. This is different from *operational feasibility*. If the research is purely theoretical, then operational feasibility is not an issue. However, any claims about application and the real world should address operational feasibility as well. In the above example, operational feasibility dictates the choice of shift for work that is considered odd or risky. Mondays are not typically good days to do something new or risky in a factory. For example, on Monday, the factory is recovering from the end of the previous week, any issues that arose from the weekend, and getting the week sorted out. Fridays also present challenges as people and organizations deal with the weekly objectives and prepare for the weekend and upcoming week.

All these issues can create variance and uncertainty. The factory might not have the right people at the right time to address issues, or their concentration is divided. On Friday, people are trying to get the week sorted out and prepare for the weekend and the beginning of the next week. Hence, Friday is also not a good choice for starting any special work. Afternoon and night shifts are also not good choices for new or different work processes to be introduced. The engineering and support staff typically do not work in evenings, and the early morning hours around midnight.

Some factories want to introduce work only when the support staff is available – else it is not operationally feasible in their context. Some factories also do not want to assign new or critical work to a crew who did not pay attention during the training session. From these examples, it is clear that additional knowledge and data are used by the planners when making a sequence. There will be some work that is not sensitive to such issues, but in any dynamic or rapidly changing situation, there will be some work that is likely to be very sensitive. Some of this data can be teased out of the corporate databases (e.g., such as repair plans, upgrade histories, changes in vendor, and changes in material), but other data is culturally based (e.g., attitude during training). These types of issues can also be observed in business process studies as not all data or inputs going through the process have the same value or risk to the factory.

If part of the research goal is to understand or compare a schedule generated by a human with a schedule generated by an algorithm or heuristic, then the comparison has to use all the constraints driving the human's decision. In the example presented, if the algorithm did not acknowledge the possible risks associated with doing new or risky work on the second and third shifts, and on the Monday and Friday, the comparison would be similar to comparing apples and kumquats.

There are certain assumptions that can be made about empirical work involving schedulers and planners that help design the field methodology. Six assumptions to consider are:

- Planning and scheduling is a cognitive skill involving various degrees of competence and expertise. Once this assumption is made, it is then possible to view the situation as possibly having different levels of scheduler performance. The assumption also helps to view how the scheduler learns and improves. To become a grand master at a cognitive skill it can take approximately two person decades of experience and learning. Thus, it is important to recognize that truly expert schedulers will be scarce and that a journeyman level will be the most common level of skill for mature, established planners.
- The decision-making process is really about solving problems, avoiding problems, and minimizing problems. To a scheduler, a late job is a problem, excess inventory can be a problem, having the right amount of inventory is a problem, incurring shipping penalties is a problem, and assigning risky work to a less than desired worker is a problem. Constraints and objectives can all be viewed in the problem-solving perspective (McKay 1987, 1992).
- The decision making is ongoing and overlapping. The planners and schedulers do their task every day and come to work with expectations about what was planned, what was supposed to happen, and knowledge about what happened and what did not happen yesterday or last week. Unless they are returning from vacation or a sick day, they do not come to work without memory or expectations. There is also a great deal of plan revision and review as the time horizon constantly moves forward. Production control decision making is not the same as project management; structurally or in execution. It is not the same as planning a trip to the local grocery. Nor is it the same as planning single major tasks involving many decision makers or assistants such as docking an aircraft carrier (e.g., Hutchins 1995).

- The majority of production control is repetitive and has many instances – orders, processes, resources, vendors, materials, products, operations, personnel, and so forth. The decision makers have to deal with hundreds of part numbers, dozens or hundreds of resources, and many operations per part. They also possibly deal with hundreds of open work orders, partially completed jobs, and queues of work at each resource. They live the experience and may be directly involved with reactive rescheduling and problem solving when any of the operational details goes awry.
- The decision making may be iterative, partial, and heterogeneous. While it is possible to find situations where all the plannings are done first, then all the scheduling, and then all the dispatching, it is also possible to find situations where this is not the case. It is possible that one person will do one or more of these tasks and do a little bit of one, then switch to a different planning problem, and then switch back. The decision maker might also make partial decisions using part of the input and state information without waiting for all the information to be delivered. In these cases of mixed decision making, knowledge and decisions at the dispatch level can guide scheduling decisions which can then guide planning decisions. Not everything is top down in a nice hierarchy (McKay 2000).
- Decision makers do not exist in isolation. They exist within a culture and organizational structure. The culture will have terminology, explicit and implicit norms and expectations, and often various parts of the culture are dependent or codependent on each other. The environment will influence the information flows, decision processes, and the actual decisions.

These six assumptions are supported by initial empirical research in the field (McKay 1987, 1992) and have been used in subsequent theoretical and field-based research. They can also be used to critique the validity of simplified laboratory studies or other claims of equivalence. For example, what is missing from the scheduling process when a laboratory study is done in which schedulers are brought in and given a set of inputs and asked to create a sequence? It might be possible to discuss the laboratory results relative to another laboratory study, but it is not clear what a laboratory study can say about the real situation of scheduling in the manufacturing setting.

Sanderson (1989) reviewed 25 years of research on the human role in scheduling. In the review, two types of studies are discussed: laboratory studies and field studies. The laboratory studies summarized in Sanderson's review focused on three main themes: comparing unaided humans with scheduling techniques, studying interactive systems of humans and techniques, and studying the effect of display types on scheduling performance. The tasks studied in the research were quite varied, as were the study methods with few replications or general results. Moreover, the research questions mainly focused on comparisons of humans and mathematical techniques. As pointed out in Crawford and Wiers (2001), these laboratory studies lack the richness and context that define scheduling and that the majority of research conducted since 1990 on humans and scheduling has been more field based.

Before discussing the specifics of deriving, gathering, testing, applying, and training in the context of sociotechnological research, a discussion about background preparation is required.

10.4.1 Education of the Researcher

It is probably fair to state that most production control researchers have been well schooled in topics such as operations research, operations management, scientific programming, algorithmic design, complexity, statistics, computational experiments, simulation, database design, computer science, and artificial intelligence. On the other hand, how many have studied or have taken courses in survey design, field methods (e.g., Gummesson 1988; Easterby-Smith et al. 1991), interview methods, skill and expertise, field data analysis, and ethnographic techniques? Some like Crawford (2000) have used a background in psychology to then engage in a study of planners and schedulers, but their initial education was not in industrial engineering or operations management. Assuming that a researcher has a typical mathematics or engineering undergraduate student starting graduate studies, what can be recommended?

If the research is to involve speculation about scheduler skill and expertise, three topics should be addressed – field methods, social science statistics, and skill and expertise. For more general field work, the first two topics will possibly suffice. It is not a perfect situation though. Ideally, two full graduate degrees are almost needed if the human planner or scheduler is to be studied as the main topic. If the human is a smaller part of the research, compromises must and will be made.

As noted above, there are two courses or sets of subject material which are recommended for any field-based empirical work on production planning. The first is field study methods including ethnographic methods (e.g., Schensul et al. 1999; Grills 1998; Spradley 1979). This area of research provides a solid foundation for how to observe and how to understand a field situation. In addition to such basic knowledge, additional material on interviewing and survey methods is recommended (e.g., Lavrakas 1993; Stewart and Cash 2003). Courses and material can often be found in sociology departments and this type of knowledge is useful in looking at specific situations or larger processes. While undergraduate material or courses might be offered, a graduate level course is recommended because of the emphasis on research and rigor. The focus on ethnographic methods is also very strong in many sociology and anthropology departments. In the ethnographic approach, the concentration is on learning from the subjects and understanding their culture and situation; not on studying them with a previously generated theory per se. Observer bias is dealt within the methods, and this is important in studying schedulers in the field.

A researcher in an ethnographic setting will likely be a participant with a role to play. The role may be in the action science sense of introducing a causal effect or may be passive. For example, a researcher learning what independent spreadsheets are used and creating an integrated system to do the same functions is not really changing or altering the structure of a situation. By way of contrast, moving from an ad hoc planning activity performed by one person using magnets on a wall board to separate dispatching and scheduling tasks using specialized software which alters the order and timing of the daily tasks is in the spirit of action science (Argyris et al. 1985). In essence, an ethnographic researcher almost becomes a member of the community being studied. This is how terminology, the meaning behind terminology, implicit rules of conduct, factors that influence compromise, and similar are discovered. References such as Jorgensen (1989), Delaney (2004), and Yin (1989) provide valuable insights about conducting such research.

The second recommended area of study is that of statistical methods specifically designed for social science settings. This includes qualitative data analysis and techniques such as nonparametric methods. Simulation courses and engineering-oriented courses on statistics teach certain concepts and methods. The social sciences use a different subset. Sources such as Nachmias and Nachmias (1987) and Siegel and Castellan (1988) provide useful material for field-based empirical studies.

These two areas of study, field methods and social science statistics, are recommended for any field study that involves humans as subjects and will be using data from the study. If the research is going to probe or use information pertaining to what the human decision maker does (the what, the why, and the how), then additional learning is required from the researcher. This is the third area of study necessary for rigorous field research and might not be as important for business process studies, but this will depend on the focus of the study itself. While the following paragraphs will once again use the scheduler as an example, the points apply equally well to any knowledge- and skill-rich situation.

In this third case, the researcher should understand something about cognitive psychology, specifically skill and expertise (e.g., Chi et al. 1988; Ericsson and Smith 1991). It is possible to find graduate level courses on this topic or arrange for a directed readings course. There are differences among novices, personnel at the journeyman level, and true experts. The study of expertise is important to understand (e.g., see Camerer and Johnson 1991; Colley and Beech 1989; Galotti 2004). The differences relate to a variety of behaviors ranging from what information is used, what it is used for, how it is used, and how problems are reasoned through. It depends if the problem is ill-structured like planning and scheduling, or well-structured like a chess game (Voss and Post 1988). There are also differences in how a novice or expert explains their behavior. It is important to know or assess the level of expertise of a human decision maker. The number of years scheduling is not sufficient proof of expertise. This is a naïve assumption. The managers call him or her an expert. This is another naïve assumption. The scheduler knows all the parts and all the operations and impresses the researcher with their encyclopedic knowledge of the products and processes. Hence, the scheduler is considered an expert – yet another naïve assumption.

A good memory alone does not make someone an expert in planning and scheduling. There is the problem-solving aspect to consider. For example, how much anticipating and reading of the situation is done? How are the resources allocated, and does the scheduler have strategic and tactical reasoning that guides those decisions? How quickly (and how accurately) can the scheduler size up a situation and devise a solution? How many times does the scheduler overlook a key constraint in the process plan? How many unnecessary setups are performed? How many times resources are primed for work before work arrives? How often are the wrong resources allocated? General techniques for how to capture and analyze expertise in a situation such as production control are described in Olson and Biolsi (1991) and specifically explained in the production control context in McKay et al. (1995).

The following sections discuss the deriving, gathering, testing, applying, and training aspects related to the sociotechnological situations.

10.4.2 Deriving

It is possible that the most difficult research to undertake is that intended to uncover aspects of the problem and understand subtle and possibly hidden relationships. In this research, the researcher is not likely to be standing on the shoulders of others and it will be difficult to leverage existing results in the literature. The researcher will be working from first principles in the actual situation, which will require the greatest amount of preparation and care. The research will be difficult, and it will be equally or more difficult to get the work published in credible journals. The latter is the case because such research is rarely done and even more rarely done well. Furthermore, the research is likely to be unlike that of other papers published in the journal, and the journal will have difficulty in reviewing the paper. In one case, a submission to a typical operations management style journal was successful through the referee and issue editor levels but was rejected as being too risky for the journal by a senior editor – nothing like it had appeared before. When faced by such obstacles, the researcher can decide to publish the work based on a method perspective and publish in cognitive or social science outlets which will be used to the method and approach. Unfortunately, if the researcher wants others in production control to be aware of the research, the social science outlets might not be the best choice. Eventually, it is possible to publish field-based research on the scheduler and planner in suitable journals, but the process might not be quick and easy.

There are some steps that will help with this type of research – both in the doing and in the dissemination of results. Four steps are given as follows:

- First, the researcher must establish credibility in understanding the appropriate field methodology and human aspects being studied. This may require, for example, suitable courses in sociology for field methods and recognized instruction in cognitive skill acquisition and expertise, if the research is to tackle the human problem-solving process. It is possible to self-study this material, but this is risky and is not recommended for the key sciences. Whether it is admitted or recognized, derivative research is interdisciplinary in many cases and it is necessary to establish credentials in each area and to recognize the key research contributions in each field.

- Second, the field methods and research must have checks and balances to detect bias on the part of the research and method itself. A longitudinal design is suggested for this purpose with a third, independent party used to check any interpretations and encodings. A narrow case study of limited duration and focus is prone to sampling bias since the situation is likely to be context sensitive. A longitudinal study should pick up organizational linkages as well as the life cycle phenomena associated with various parts of the puzzle including products, processes, and resources.
- Third, an ethnographic approach to the initial phases of the research will avoid premature creation of theories and self-limiting assumptions. Researchers learn and discover in the initial phases from the people and situation. Going in with a bank of questions and a fully developed field instrument implies that the science is already developed and the researcher is actually testing it. The ethnographic methods also deal with observer–subject relationships and dynamics such as how to fit into the community and be accepted. For example, if the researcher is studying the scheduler, and the scheduler wears blue jeans and comes into work at 5:30 AM in the morning, the researcher should too. Coming in at 9:30 AM in nice office clothes is not the way to do it. If the scheduler has to come in Saturday or Sunday mornings at 7 AM occasionally, the researcher should also come in on the odd occasion to understand what is different about planning and scheduling on the weekend.
- Fourth, the derivation must be supported by appropriate data and reasoning; otherwise it is just a nice story. Quick or simple answers need to be carefully thought through and challenged. It is important to observe the "what" and understand the "so what?" This is the difference between superficial knowledge or partial data, and deeper knowledge. Consider one of the over 200 episodes captured in McKay (1992).

The scheduler was observed to schedule a particular order on a Monday. Based on the due date and other obvious data, a Tuesday date would have been predicted using previously observed heuristics. *Why was a special decision taken to schedule the work on a Monday instead of a Tuesday?* The work was considered a type of job that required special attention when it was run. *Fine, but why schedule it on Monday?* The scheduler said that the increasing emails about an issue in the factory were the reason. *Why was this the reason?* The scheduler said that once the emails start flying hot and heavy, they usually call a staff meeting. *So?* The staff meetings are usually held on Tuesday mornings. *So, why is this a problem?* All the key supervisors and technicians are usually called to the staff meeting to discuss such issues. *OK, so why is this a problem?* The supervisors and key people will not be on the manufacturing line and when they are not on the line, quality and other problems often arise. *So?* When this part is run, I want all the normal production people around and doing their jobs and I want to avoid any possible problems with this one part.

Such an example is rich with information. There are trigger events, and signals about the trigger events. This is the meeting and the clues about a meeting possibly being called. There is also the meaning or implication of the event that is predicted to take place. The implication is a lowered level of supervision and technical support which, in turn, implies a risk to quality and performance. For any special decision or heuristic observed, it is important to follow the causal chain and observe the initiation conditions, execution constraints, and performance implications. Note that it is possible for a scheduler to start at the key reason and to work back to the initial clue (e.g., how do you know that this will happen or be a problem?). It is also possible for a scheduler to start the explanation in the middle of the chain (e.g., there will be a meeting on Tuesday morning). The knowledge acquisition must be flexible enough to move up or down the reasoning chain.

Once captured, there are a variety of ways to encode and analyze such data. With such data, it is possible to start the construction of reasonable models of what and perhaps why. Without such data, any model or concept is inherently weak and will not stand up to rigorous challenge.

While the above four suggestions will not guarantee that a study is rigorous and done properly, the suggestions address the six assumptions noted earlier and increase the likelihood of good empirical science.

One goal of derivative research is to understand the necessary and sufficient factors that define the problem or capture a solution. To achieve this, it is necessary to understand the context or situation, and create sufficient science to defend any claims of bias, limitations, and generalization. For example, if the study site was a modern high-tech electronics factory at the start of the supply chain supplying one major customer with a few parts, will the results apply to a low-tech metal fabricator supplying many different companies' assembly plants with a large number of parts using old equipment? There are many factors to consider when thinking about the derivation of the problem characteristics and how the situation and solution can be generalized.

10.4.3 Gathering

The gathering aspect of field research involving human subjects is perhaps a little less strenuous compared with the derivation purpose, but is still very important. The derivation aspect deals with major theory or substantial conceptual elements of the science – e.g., problem structure and solution formulation. In the gathering phase, the researcher may be seeking information about objectives, constraints, copies of schedules, input data, statistical information on productivity or execution, or heuristics used for creating sequences. The researcher is seeking data to perhaps guide research or to be used in creating simulations, numerical experiments, or analyzes. In general, the problem has already been formulated, and parameters need to be tuned possibly, or actual data is needed to establish reasonable data distributions (for claims of applicability).

While the researcher does not need to go undercover and join the community, other skills are still required from the social sciences. In a number of cases, the gathering will be of hard numbers from reports and data files – the historical facts.

There is not much guesswork when it is clear what resource was assigned, how many hours were taken, and what the yield was. The field methods apply to the interpretation and understanding of the data gathered. If the data is from a single sample, it is important to know how representative the data is and how much can be generalized. That is, is there variance, what does the variance look like, and what causes the variance? Is variance possible because of a machine upgrade, material substitution, new work method, product change, poorly maintained equipment, *ad nauseam*? It is also important to know if there are any issues with the data and what the scheduler thinks about the data. For example, how accurate is the shop floor reporting mechanism? What are the possible errors in the data to be aware of? Being aware of human bias in interpreting data is critical when asking the scheduler to speculate. This is where courses or some training in interviewing and field methods can assist. It is important to tease out of the process possible bias attributed to recent experiences or other influences.

If the data being gathered is more speculative or deals with the future, additional care is needed to process the data. It is desirable to capture the expected value, possibly bounds for variance, and the normative value with its bounds. If there is a difference between the normative and expected, the data acquisition process should probe the reasons why. Similarly, if a heuristic is being gathered for later simulation or analysis, it is important to decompose the heuristic carefully to understand what logic controls the use or guides the heuristic, and how the heuristic deals with exceptions.

10.4.4 Testing

One form of field research is taking a theory or idea from the purity of the academic situation and testing it in a real situation. As in a context-free situation, the strongest research will establish a benchmark situation as the control point, and then perform the intervention controlling all other aspects constant. This is very hard to do in the best situation imaginable and possibly impossible to do where a human is part of the actual science. For example, is the researcher investigating what heuristic is used when, and under what conditions by the scheduler? The researcher might have a theory that predicts when a scheduler will relax a certain constraint. How is this going to be tested? Alternatively, the researcher might be doing a study to compare how the mathematical algorithm compares to a human scheduler. First, what schedule is going to be compared against? Second, what is the measure to determine what is better? Are tardiness or the average number of late jobs going to be used? Was this the objective used by the scheduler? Perhaps the scheduler's objective was to create a schedule that could actually be executed by the shop floor. How can the algorithm be tested? Is it possible to have exactly the same situation twice and try two different schedules and see how they play out? These were discussed briefly in the technical or context-free situation, but are also important considerations when the human is involved.

It is possible to test for existence of concepts and ideas, but it might be very hard to test for causal relationships or magnitude of impact when humans are part of the process. If a completed schedule is part of the experimental design, a classic problem occurs for which there is no current answer: what is a good schedule? If the researcher is going to rate and compare schedules where one has a human element in the construction, the objectives need to clearly stated. It is not appropriate to claim that a schedule generated by an algorithm is better because the average tardiness is superior to the schedule generated by the human when the human was creating a schedule for the end of the month that maximized shipping dollars. Similarly, it is not valid to use late jobs as a measure when the plant will not have late jobs planned – creative problem solving is used to avoid any late job. This is the case in many automotive supply chains where the suppliers do everything possible to avoid a late delivery to an assembly plant. For these plants, a schedule generated by a human will NOT show any late jobs and if compromises need to be made on all other factors, so be it.

If the research is to test the scheduler's ability and knowledge, then the researchers need to be skilled in the cognitive skill domain and understand the methods for isolating and analyzing skill and expertise (see Ericsson et al. 2006). There is tacit knowledge that cannot be conveyed, explicit knowledge that has been codified and documented, and explicit knowledge that is verbal or culturally maintained. There are varying degrees of expertise and it is not always appropriate to use the same methods when dealing with a novice versus a true expert. In general, this type of research will require the greatest effort to perform, and will also require the most effort in the analysis phase. For example, true experts and grand masters have great difficulty (often impossible) in describing how their decisions are arrived at. When pushed, they will often tell the researcher how they were instructed as a novice which need not bear any resemblance to the actual practice. Novices will also focus on superficial data or issues and not see the deeper relationships. In general, novices proceed from concrete to abstract reasoning while experts proceed from abstract to concrete.

10.4.5 Applying

The application phase of field research is not the simple testing of ideas in an actual setting, but the adoption and full use of the ideas. The challenge is similar to the testing type of research – what is the researcher or practitioner going to say about the research? Is it possible to claim a descriptive, normative, or predictive relationship? In a context-free situation with computerized manufacturing, it might be possible to give credit to the new idea if other factors are controlled for.

If the human is not part of the subject, it is also easier to do, as described in the technological section of this chapter (10.3). There have been many case studies and excellent stories written about the application of a certain method or concept. It is true that in some cases, a gain is very obvious after the application of a new method or concept. The situation might have been so bad before that any organized approach to manufacturing will result in a gain. It might also be true that the new application was used as an agent of change and allowed other changes to occur. Careful field work and data gathering can be of immense benefit in these situations. The goal is to create a clear and justified link between the science and the benefits. The benefits should also be studied long term to ensure that the science was sustainable. All the points noted in the technological section (10.3) should be considered as a starting point.

It does not come as a surprise that the greatest difficulties will be with the research that involves the human as subject. The appropriate field methods from cognitive science and the social sciences will assist with isolation and control issues. Establishing controls and measurement techniques will be problematic in any event if any claims are to be asserted. Consider the type of research that could be applied in production control that involves the human as part of the research material. One type would be to replace the human element with a technology; eliminating the human. Another type is a partial or hybrid solution in which the human works with an enhanced technology to improve production control. Yet another type of research could attempt to improve the human's decision making while leaving other production control technology alone. The researcher has to think about what will be measured before the application, what can be measured while the new science is being used, what can be measured at milestones, and what the measurements can say about the results and causal relationships.

Production control in a factory is a mission critical task and this creates a problem when a full application of new science is contemplated. If the new science does not work, what are the possible negative impacts? A field application in the real sense must have the appropriate procedures in place to detect issues and to control any negative impact. In the cases where the human is marginalized, the researcher is also faced by many of the "management of change" issues in acceptance and deployment. For example, will the scheduler feel threatened by the technology? Will the scheduler trust it? Will the technology create new work? In one case study (McKay and Wiers 2004), the initial scheduling technology changed the time horizon being considered, increased expectations about smoothing or capacity loading, lowered the granularity of analysis (down to the hour from a shift level), and increased expectations about schedule quality. The initial pilot test was a failure and further development was needed to address the changes imposed upon the scheduler and provide better support.

10.4.6 Training

Occasionally, it is possible that researchers use a field situation as a training situation for students to learn about production control and what happens in production control. If part of the intent of the learning is to understand the schedulers and planners, what they do and why they do it, then it is necessary to properly train the student in advance. The student will need the field study methods' background to observe properly and understand what they see. If this is not undertaken, there is a risk that the observations will not be valid, and weak or erroneous models and solutions created.

In the cases where the student is then used to further educate the main researcher (the student represents the extended eyes, ears, and feet of the instructor), it is possible that erroneous data or misguided observations can further distort science.

10.5 Supply Chains and Interorganizational Research

It is safe to say that the majority, if not all, of the above applies to supply chain studies, or studies that involve multiple organizations. However, this is not sufficient for business process studies or those which cross firm boundaries. In this section, several additional aspects will be discussed which should be considered.

It is important to understand all operational agreements and mechanics that control the flow of material from suppliers to the customers. This is somewhat obvious, but different organizations will supply firm and forecasted demand patterns in different ways and with different variances. It is necessary to know when information is available and what happened to the data before it arrived. For example, in the automotive supply chain, the assembly plant forecasts might go through a divisional office for review and adjustment before they arrive at the fabricating plant. This can result in delays and possible alteration (intended and unintended) of the data. In one case, the sales force accidentally added three zeros to a forecast. Imagine what happened to a related part that had a normal demand of 10,000 per month. This large error was eventually discovered when the MRP logic was executed and very large material requirements were noted. However, it is not clear if smaller errors are ever found in such processes. This example illustrated to the author the importance of knowing how the demand requirements really move from the customer to the plant. If the divisional level is monitoring and adjusting the forecasts, how is the plant expected to make sense of the forecasts? The data moving between organizations also needs to be analyzed since stated policies and practices may not be followed. For example, how often do firm forecasts change? To complicate matters, terminology might change between customers and between suppliers. Care must be taken to verify all definitions and objectives that will constrain or guide decision making.

Larger studies imply that the researchers must also be aware of strategic and tactical issues which might not be obvious in day-to-day operations. For example, if a factory is supplying multiple assembly plants owned by competing firms, what policies and promises have been made about factory resources and responsiveness? These policies might place constraints on schedules and plans which might be taken for granted or which might only appear when multiple customers increase demand simultaneously. There might also be an informal corporate policy that one plant will act as a loss leader to help other plants gain business with a major customer.

This might not be stated in writing, but it is understood that the one plant will do everything possible to make the customer happy. This will affect any penalty or benefit analysis as the numerical value of a decision is not the real business value.

If possible, it is useful to spend time and study both ends of any relationship. For example, it might be possible to visit and study a supplier. This will aid in understanding what a supplier needs to know in order to supply the plant in a better fashion. It can also provide knowledge about how a supplier will respond in different situations. It might also help in understanding how the supplier might be able to supply material or parts in a different state of processing, or in a different configuration or packaging to make the overall situation more cost effective for both parties. Understanding how a supplier fits within other supply chains involving customers or sister plants is also vital. For example, it has been observed that large final customers might negotiate directly with different parts of the supply chain to obtain favorable prices. This can happen when the supplier supplies materials to multiple fabricating or assembly plants controlled by the large, final manufacturer. This creates interesting dynamics as the supplier does not have the direct contract or negotiating position with a receiving plant. This can affect normal business relationships as the relationships become twisted – the fabricating plant having no power or influence. For example, the supply emphasis might be placed on a different assembly plant for the large manufacturer, but the supply chain metrics and performance objectives for each of the other plants (possibly in different corporations) do not reflect this bias.

In summary, it is not possible to give a definitive set of guidelines for the larger business process types of studies, but it is important to consider assumptions made about tactical and operational policies and mechanical processes. When possible, spend time and study both ends of any relationship and observe how each end of the relationship relates to other relationships in each organization. For example, study how one customer is dealt with versus another by a supplier. Obtaining and following a matched set of data following a demand order throughout a chain is also recommended. This will document the organizations involved and the processes used by each to fulfill the demand. Use the concepts outlined in the sociotechnological section (10.4) when dealing with the individual decision makers.

10.6 Conclusion

In this chapter, a number of topics have been discussed which relate to empirical research involving production control. The focus has been on the production control department, but many of the same issues relate to larger business process studies. A categorization schema was presented to help clarify the situation regarding research involving humans and research not involving humans as subjects. In addition, five forms of empirical activities were discussed in each case: deriving, gathering, testing, applying, and training. Each type of empirical activity, with and without humans as subjects, requires careful planning and careful inspection of assumptions.

It is easy to dismiss this preplanning and preactivity as being unnecessary. It is also easy to dismiss the education and training needed to perform field work.

One purpose of this chapter was to present suggestions and rationale for why care and training is needed if empirical work is to be rigorous. As noted in the introduction, a goal of scientific endeavor is to make good science, science that can explain, and science that can predict. Without rigor in the field methods, it is hard to justify claims of generality or of causality. It is possible to tell good stories and document good case studies, but that is the limit of informal empirical studies.

If a study is purported to be an empirical study, is it clear what science is actually being performed and how it was conducted? What biases are possible? How strong is the study? Were there proper safeguards instituted to ensure that claims could be made and causal relationships stated? Did the researchers know what to look for, know how to gather it, know how to analyze it, and know what it indicated? Hopefully this chapter has provided some insights into how empirical research can be conducted and viewed. It is possible to do good empirical research but it does not happen by accident or by dismissing bothersome field methods as unnecessary. Empirical research is messy and in many ways harder to conduct than pure theoretical research of a computational nature or simplified laboratory studies. However, the possible insights into the problem formulation and possible solutions are well worth the investment.

References

Argyris C, Putnam R, McLain Smith D (1985) Action science. Jossey-Bass Publishers, San Francisco

Browne MN, Keeley SM (2004) Asking the right questions. A guide to critical thinking, 7th edn. Pearson, New Jersey

Camerer CF, Johnson EJ (1991) The process-performance paradox in expert judgment – how can experts know so much and predict so badly. In: Ericsson KA, Smith J (eds) Toward a general theory of expertise – prospects and limits. Cambridge University Press, Cambridge, pp 195–217

Chi MTH, Glaser R, Farr MJ (eds) (1988) The nature of expertise. Lawrence Erlbaum, Hillsdale

Colley AM, Beech JR (eds) (1989) Acquisition and performance of cognitive skills. Wiley, New York

Crawford S (2000) A field study of schedulers in industry: understanding their work, practices and performance. PhD thesis, University of Nottingham, UK

Crawford S, Wiers VCS (2001) From anecdotes to theory: reviewing the knowledge of the human factors in planning and scheduling. In: MacCarthy BL, Wilson JR (eds) Human performance in planning and scheduling. Taylor & Francis, London, pp 15–44

Delaney C (2004) Investigating culture, an experiential introduction to anthropology. Blackwell, Malden, MA

Easterby-Smith M, Thorpe R, Lowe A (1991) Management research – an introduction. Sage Publications, London

Ericsson KA, Smith J (1991) Toward a general theory of expertise – prospects and limits. Cambridge University Press, Cambridge

Ericsson KA, Charness N, Feltovich PJ, Hoffman RR (2006) The Cambridge handbook of expertise and expert performance. Cambridge University Press, New York

Galotti KM (2004) Cognitive psychology in and out of the laboratory, 3rd edn. Thomson, Belmont, CA
Grills S (ed) (1998) Doing ethnographic research – fieldwork settings. Sage Publications, London
Gummesson E (1988) Qualitative methods in management research. Chartwell-Bratt, Sweden
Hutchins E (1995) Cognition in the wild. MIT Press, Cambridge
Jorgensen DL (1989) Participant observation: a methodology for human studies. Sage Publications, London
Kempf K, Uzsoy R, Smith S, Gary K (2000) Evaluation and comparison of production schedules. Comput Ind 42:203–220
Lavrakas PJ (1993) Telephone survey methods, sampling, selection, and supervision, 2nd edn. Sage Publications, Thousand Oaks, CA
McKay KN (1987) Conceptual framework for job shop scheduling. MASc Dissertation, University of Waterloo, Waterloo, Ontario, Canada
McKay KN (1992) Production planning and scheduling: a model for manufacturing decisions requiring judgement. PhD thesis, University of Waterloo, Ontario, Canada
McKay T (2000) Reasons, explanations, and decisions – guidelines for critical thinking. Wadsworth, Belmont, CA
McKay KN, Buzacott JA (2000) The application of computerized production control systems in job shop environments. Comput Ind 42:79–97
McKay KN, Buzacott JA, Charness N, Safayeni FR (1992) The scheduler's predictive expertise: an interdisciplinary perspective. In: Doukidis GI, Paul RJ (eds) Artificial intelligence in operational research. Macmillan Press, London, pp 139–150
McKay KN, Safayeni F, Buzacott JA (1995) Schedulers & planners: what and how can we learn from them. In: Brown DE, Scherer WT (eds) Intelligent scheduling systems. Kluwer Publishers, Boston, pp 41–62
McKay KN, Wiers VCS (2004) Practical production control: a survival guide for planners and schedulers. J.R. Ross & APICS Co-publishers, Boca Raton, Florida
Nachmias D, Nachmias C (1987) Research methods in the social sciences, 3rd edn. St. Martin's Press, New York
Olson JR, Biolsi KJ (1991) Techniques for representing expert knowledge. In: Ericsson KA, Smith J (eds) Toward a general theory of expertise – prospects and limits. Cambridge University Press, Cambridge
Sanderson PM (1989) The human planning and scheduling role in advanced manufacturing systems: an emerging human factors domain. Hum Factors 31(6):635–666
Schensul SL, Schensul JJ, LeCompte MD (1999) Essential ethnographic methods. Altamira Press, New York
Siegel S, Castellan NJ (1988) Nonparametric statistics for the behavioral sciences, 2nd edn. McGraw-Hill, New York
Spradley JP (1979) The ethnographic interview. Holt, Rinehart and Winston, New York
Stewart CJ and Cash WB (2003) Interviewing principles and practices, 10th edn. McGraw-Hill, New York
Voss JF, Post TA (1988) On the solving of ill-structured problems. In: Chi MTH, Glaser R, Farr MJ (eds)The nature of expertise. Lawrence Erlbaum, Hillsdale, pp 261–265
Yin RK (1989) Case study research: design and methods. Sage Publications, London

Chapter 11
Collaborative Supply Chain Management

Feryal Erhun and Pinar Keskinocak

11.1 Introduction

The management of supply chains has become progressively more complex and challenging due to higher customer expectations for better service, higher quality, lower prices, and shorter leadtimes; ongoing demand uncertainty; an increase in product variation; and shorter product life cycles. The increasing importance of supply chain efficiency as a key competitive advantage has changed the nature of many intra- and inter-firm relationships from adversarial to collaborative. An industry survey by Forrester Research reveals that 72% of firms say supplier collaboration is "critical to their product development success" (Radjou et al. 2001). An AMR Survey at Microsoft Engineering and Manufacturing Executive Summit, which was conducted with the participation of CEO, CFO, CIO, and Sr. VPs of Fortune 1,000 companies, shows that 58% of the participants consider collaboration as a strategic necessity, while another 32% consider it to be very important. 57% of the participants state that they are directly involved in leading collaboration efforts (Caruso 2002).

11.1.1 What is Collaboration?

According to Cohen and Roussel (2005), collaboration is "the means by which companies within the supply chain work together toward mutual objectives through the sharing of ideas, information, knowledge, risks, and rewards." Hence, collaboration requires that companies in a supply chain work actively together as one toward common objectives. Collaboration includes the sharing of information,

F. Erhun (✉)
Department of Management Science and Engineering, Stanford University,
Stanford, CA 94305-4026, USA
e-mail: ferhun@stanford.edu

K.G. Kempf et al. (eds.), *Planning Production and Inventories in the Extended Enterprise*, International Series in Operations Research & Management Science 151,
DOI 10.1007/978-1-4419-6485-4_11,

	Intra-enterprise	Inter-enterprise
Vertical	• Manufacturing & Sales/Marketing • Manufacturing & Distribution • Design & Marketing & Manufacturing	• Supplier & Manufacturer (e.g., Toyota and its suppliers) • Manufacturer & Retailer (e.g., P&G and Wal-Mart) • Distributor & Retailer
Horizontal	• Procurement functions across multiple divisions (e.g., Dial Corp.)	• Multiple manufacturers (e.g., Covisint) • Multiple logistics providers (e.g., Transplace)

Fig. 11.1 Different facets of collaboration

knowledge, risk, and profits/benefits in a consistent fashion for all participants, and entails understanding how other companies operate, how they make decisions, and what is important to them. Everyone involved must benefit or it is not true collaboration.

Collaboration can have many different facets (Fig. 11.1). *Intra-enterprise* collaboration takes place in numerous areas, including procurement, product design, and logistics. For example, by replacing their decentralized purchasing functions with centralized procurement, many companies were able to reduce the associated administrative costs, better utilize their procurement managers' time, and more importantly, leverage their buying power with their suppliers, resulting in significant savings. Using centralized procurement, Dial Corp. was able to eliminate $100 million in total costs in 5 years (1996–2001) (Reilly 2002), Siemens Medical Systems cut its costs by 25% over a 3-year period (1998–2001) (Carbone 2001), and Fujitsu has detailed plans to reduce its spending on components and materials by approximately $3.85 billion over 2 years.[1]

These examples illustrate *intra-enterprise horizontal collaboration* where a company coordinates or centralizes the activities of multiple entities which are responsible for the same function, e.g., purchasing. Alternatively, a company can achieve significant benefits by coordinating the decisions of different functional areas, i.e., *intra-enterprise vertical collaboration*. An example of intra-enterprise vertical collaboration is the net landed cost approach, where a company coordinates several functions, such as purchasing and logistics. Traditionally, purchasing's goal in many companies has been to procure goods at minimum cost, without necessarily considering the impact of their decisions on the overall profitability of the company. Such cost focus sometimes leads to high-volume less-frequent purchases driven by trade promotions, resulting in excess inventory and increased logistics costs due

[1] "Fujitsu cuts procurement costs and suppliers." *Purchasing Magazine Online*. February 21, 2002.

to limited storage space. An extensive survey by Bain & Company shows that "the cost of excess inventory in stores, driven by 'silo' planning and misaligned trade promotions, amounts to more than 25% of annual sales" (Cook and Tyndall 2001). The net landed cost approach eliminates such inefficiencies by coordinating the procurement decisions with other functions, such as inventory and logistics, throughout the enterprise (Erhun and Tayur 2003).

Inter-enterprise collaboration occurs when independent companies work together, synchronize and modify their business practices for mutual benefits, thereby shifting the nature of traditional relationships from adversarial to collaborative (Lapide 1998). As in the case of intra-enterprise collaboration, inter-enterprise collaboration can be *horizontal* where companies with similar characteristics, which are potentially competitors, collaborate on a particular business function, such as procurement. Examples include Covisint (founded by Daimler-Chrysler, Ford, General Motors, and Renault-Nissan) (http://www.covisint.com), and numerous group purchasing organizations in the health care industry (http://www.firstmark.com/fmkdirs/gpo_hsys.htm). Collaborative procurement helps the buyers to leverage more value-added pricing, service, and technology from their external suppliers than could be obtained individually (Hendrick 1997; Langley 2001). Alternatively, it can be *vertical*, where partners in a supply chain, e.g., a supplier and a manufacturer, collaborate to improve the overall efficiency of the chain.

11.1.2 Why Do We Need Collaboration?

Supply chain management covers a whole spectrum of activities from product and process design to manufacturing, procurement, planning and forecasting, order fulfillment, and distribution. Managing such complex systems requires complex tradeoffs. Many times the subsystems rely on local optimization. However, different entities in the chain may have different and often conflicting objectives. Moreover, the output of one system is the input of another. Hence even though they are decentralized, supply chain activities are interconnected. Therefore, it is essential to consider the *entire* system and coordinate decisions (Fig. 11.2).

Given that each participant in a supply chain acts on self-interest and does not necessarily have access to the same information (e.g., the retailer may have more information about the end consumer demand compared with the supplier), the individual choices of the participants collectively do not usually lead to an "optimal" outcome for the supply chain. That is, the total profits of a typical "decentralized" supply chain which is composed of multiple, independently managed companies, are less than those of the "centralized" version of the same chain, if such a chain could exist and be managed optimally by a single decision-maker to maximize its profits.

One possible strategy for reducing such inefficiencies is "vertical integration," where a company owns every part of its supply chain, including the raw materials,

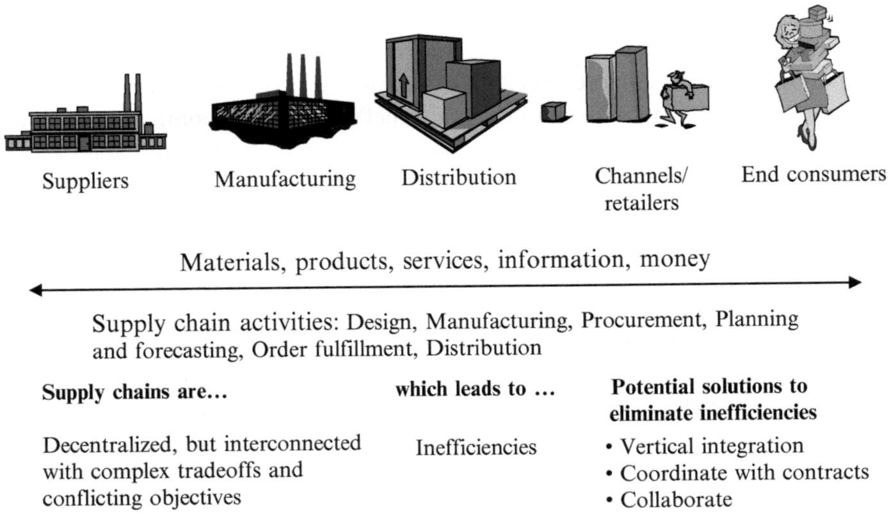

Fig. 11.2 A typical supply chain

factories, and stores. An excellent example of vertical integration was Ford Motor Co. early in the twentieth century (Swanson 2003). In addition to automobile factories, Henry Ford owned a steel mill, a glass factory, a rubber tree plantation, an iron mine, and railroads and ships used for transportation. Ford's focus was on "mass production," making the same car, Model T, cheaper and faster. This approach worked very well in the beginning. The price of Model T fell from $850 in 1908 to $290 in 1924. By 1914, Ford had a 48% share of the American market, and by 1920 it was producing half the cars made worldwide. Vertical integration allows a company to obtain raw materials at a low cost, and exert more control over the entire supply chain, both in terms of leadtimes and quality. However, we do not see many examples of vertically integrated companies today. Why? Mainly because in today's fast paced economy, where customers' needs and tastes change overnight, companies that focus on core competencies and are nimble are more likely to stay ahead of their competition and succeed. Hence, we see an increasing trend towards "virtual integration," where supply chains are composed of independently managed but tightly linked companies. Innovative practices, such as information sharing or vendor managed inventory (VMI), are successfully used by some companies such as Dell Corporation (Magretta 1998) and Cisco Systems to get closer to virtual integration. All of these steps companies take toward virtual integration fall under the umbrella of *collaboration*, which is the main focus of this chapter.

While companies increasingly believe in the potential benefits of collaboration, most remain reluctant to change their supply chain practices, and in such cases it is desirable to design contracts (defining the terms of trade) or change the terms of the existing contracts, to align incentives and reduce inefficiencies. This is known as "supply chain coordination" and is discussed in Sect. 11.2. Similar concepts apply to independently managed divisions within a company as well. One should

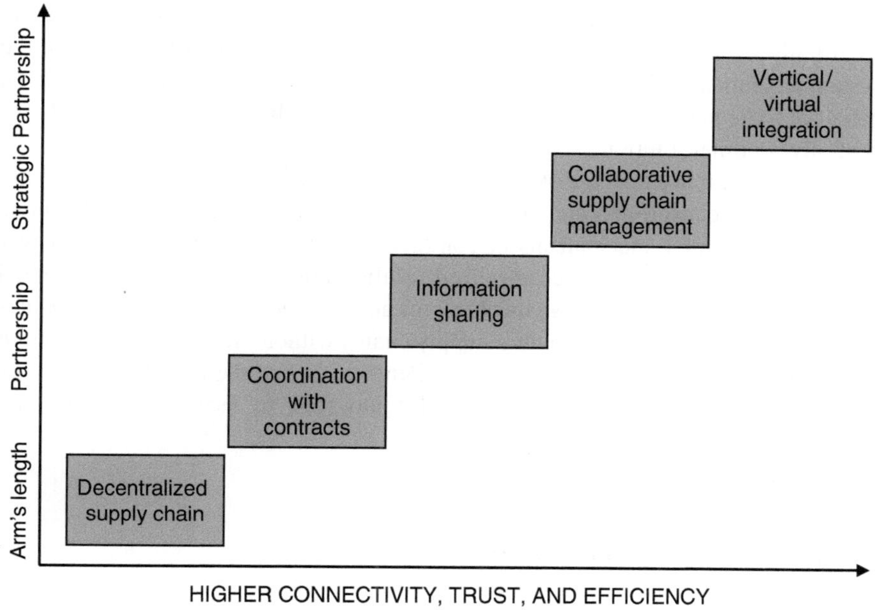

Fig. 11.3 A spectrum of collaboration in supply chains. This chapter provides an overview of concepts in "coordination with contracts" and "collaborative supply chain management"

keep in mind that even though supply chain coordination is a step toward supply chain collaboration, it does not entail all benefits as well as complications that collaborative environments create.

In summary, collaboration may be as simple as sharing information, or as involved as joint product design (Fig. 11.3). Supply chain activities with a high potential of benefiting from collaboration include new product introduction (planning, pricing, product design, and packaging), procurement, logistics, replenishment planning, and demand forecasting. We discuss these activities in some detail in the following sections.

11.2 Intra-firm and Supply Chain Coordination Through Incentives and Contracts

In a supply chain (SC), there are multiple firms owned and operated by different parties, and each of these firms take decisions which are in line with their own goals and objectives. Similarly, often times large firms consists of multiple "divisions," such as marketing, production, procurement, logistics, and finance, each having their own goals and incentives and focusing on different aspects of the firm's operations. As in all decentralized systems, the actions chosen by SC participants or different divisions of a firm might not always lead to the "optimal" outcome if one considers the entire system as one entity. That is, since each player acts out of self-interest, we usually see inefficiencies in the system, i.e., the results look different than if the system were managed "optimally" by a single decision-maker who could decide on behalf of these players and enforce the type of behavior dictated by this globally (or centrally) optimal solution.

In this section, we take a look at the nature of inefficiencies that may result from decentralized decision-making within a firm or a supply chain, and if and how one can design incentive mechanisms or contracts such that even though each player acts out of self-interest, the decentralized solution might approach the centralized optimal solution. Such incentive mechanisms and contracts are useful if companies want to reduce inefficiencies in their supply chains without necessarily engaging in elaborate collaborative activities. For excellent reviews of the literature on supply chain contracts and coordination, the reader may refer to Tsay et al. (1998) and Cachon (2003).

11.2.1 Intra-firm Coordination

Examples of "disconnect" between different divisions (or functional units) or a firm are abundant. "It's a familiar scenario. Orders spike and manufacturing can't keep up. The people on the manufacturing floor blame purchasing for the production line shutting down because purchasing didn't buy enough materials and logistics didn't get the materials to the plant in time. Purchasing and logistics look at the sales organization and say they did not get any forecasting information to predict a big spike in demand so inventories could be built to cover the demand" (Hannon 2003).

In addition to communication inefficiencies between different divisions of a firm, inefficiencies or suboptimal decisions can also arise due to misaligned incentives. "Rewards and recognition systems misaligned with corporate objectives can result in behavior that is not anticipated or desired by management. These unanticipated actions may be personally beneficial to front-line sales reps, manufacturing floor managers, or even senior executives, yet they move the company away from its overall goals or cause systemic harm."[2] In an executive roundtable at Tuck School of Business at Dartmouth, Steve Stone from Lowe's recalled an oversupply of lawn mowers pushed out to the stores in an effort to minimize distribution center inventory, only to have them linger unprofitably as Spring arrived unseasonably late.[3] Kirk Drummond of Sysco (a leading foodservice distributor) described situations where salespeople brought in huge last minute orders for next-day fulfillment

[2] "Aligning incentives with strategic and operational goals critical to performance management success" http://www.pharmalive.com/News/Index.cfm?articleid=342812.

[3] "Making the link between sales and operations planning." http://mba.tuck.dartmouth.edu/digital/Programs/CorporateRoundtables/ElusiveIntegration/Overview.pdf.

without any advance warning to operations. Staples' Kevin Holian described an overenthusiastic promotion where merchants dramatically underestimated potential demand at a deeply discounted price point.

Misaligned incentives also lead to conflicts in decision-making in an organization. For example, sales/marketing is often evaluated based on revenues or sales volume, whereas manufacturing is often evaluated based on cost or operational efficiency. When it comes to decisions such as quoting leadtimes to customers, marketing would prefer shorter leadtimes with the goal of attracting more customers whereas manufacturing would prefer longer leadtimes with the goal of completing the orders on time, without using overtime or other costly options, and to avoid delay penalties.

Pekgun et al. (2008) study the incentive alignment between marketing and manufacturing in a setting where marketing is responsible for price and manufacturing is responsible for leadtime decisions. They find that coordination among these departments can be achieved with a transfer price contract with bonus payments. Under coordination, both production and marketing are better off, i.e., costs are lower and revenues are higher, leading to higher overall profitability for the firm. de Groote (1994) studies product variety versus process flexibility in the marketing/operations interface. Balasubramanian and Bhardwaj (2004) model a duopoly in which firms with decentralized marketing and manufacturing functions with conflicting objectives compete on the basis of price and quality. Teck and Zheng (2004), Chatterjee et al. (2002), and Erkoc and Wu (2002) study the leadtime quotation problem within the marketing/operations interface.

It is crucial for a firm to adjust the incentives of different divisions so that they are better aligned with the firm's overall goals and strategy. While incentive alignment (or sharing information) by itself is not a cure for all inefficiencies that arise in a firm, it contributes significantly toward better processes and higher motivation for collaboration.

11.2.2 Supply Chain Coordination

To illustrate the inefficiencies that might result in a decentralized supply chain (DSC), we consider a simple stylized two-stage supply chain with one supplier and one retailer (Fig. 11.4), where the retailer buys goods from the supplier and sells them in the end market.[4]

For simplicity, assume that (a) the supplier is uncapacitated and has unit cost of production c, (b) the retailer faces a market where the price is inversely related to the quantity sold, (c) there is a linear demand curve $P = a - bq$ where P is the

[4] The material in this section is based on Erhun and Keskinocak (2003).

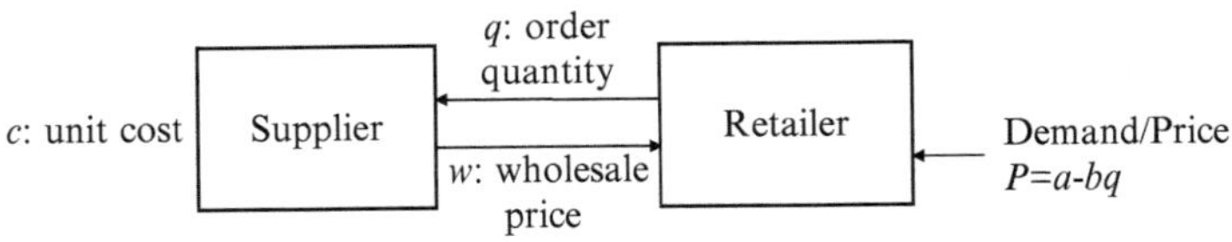

Fig. 11.4 A simple supply chain

Table 11.1 Wholesale price contract

	DSC	CSC
Supplier's wholesale price (w)	$w = (a + c)/2$	w
Retailer's quantity (q)	$q = (a - c)/(4b)$	$Q^* = (a - c)/(2b)$
Market price (P)	$P = (3a + c)/4$	$P^* = (a + c)/2$
Supplier's profit (Π_S)	$\Pi_S = (a - c)^2/(8b)$	$\Pi_S^* = (w - c)q^*$
Retailer's profit (Π_R)	$\Pi_R = (a - c)^2/(16b)$	$\Pi_R^* = (P^* - w)q^*$
Total SC profits (Π)	$\Pi = 3(a - c)^2/(16b)$	$\Pi^* = (a - c)^2/(4b)$

market price and q is the quantity sold by the retailer, and (d) all of this information is common knowledge.[5]

First, let us consider the simple *wholesale price contract* where the supplier charges the retailer w per unit. The supplier's and the retailer's profits are $\Pi_S = (w - c)q$ and $\Pi_R = (a - bq - w)q$, respectively. The supply chain's profits are $\Pi = \Pi_S + \Pi_R = (a - bq - c)q$. Note that the choice of w only indirectly affects the total SC profits, since the choice of w impacts the choice of q.

Decentralized supply chain: As in most real-world supply chains, suppose that the supplier and the retailer are two independently owned and managed firms, where each party is trying to maximize his/her own profits. The supplier chooses the unit wholesale price w and after observing w, the retailer chooses the order quantity q. The equilibrium solution for this decentralized supply chain (DSC) is given in the second column of Table 11.1. In this contractual setting, the supplier gets two-thirds of the SC profits, the retailer gets only one-third. This is partly due to the first-mover advantage of the supplier.

Now, let us consider a centralized (integrated) supply chain (CSC) where both the retailer and the supplier are part of the same organization and managed by the same entity.

Centralized supply chain: In this case, there is a single decision-maker who is concerned with maximizing the entire chain's profits $\Pi = (a - bq - c)q$. The solution for the CSC is given in the third column of Table 11.1.

From Table 11.1, we see that the quantity sold as well as the total SC profits are higher and the price is lower in the CSC than in the DCS. Hence, both the

[5] Note that in such a deterministic environment, the retailer will always purchase from the supplier exactly as much as he will sell in the market. The "common knowledge" assumption may imply some information sharing between the supply chain partners.

supply chain and the consumers are better off in the CSC. What about the retailer and the supplier? Are they both better off, or is one of them worse off in the CSC? What is the wholesale price? How does the choice of w affect the market price, quantity, and the supply chain profits? A closer look would reveal that w has no impact on these quantities. Any positive w would result in the same outcome for the CSC because the firm would be paying the wholesale price to itself! However, the choice of w in the CSC is still very important as it determines how the profits will be allocated between the supplier and the retailer. We can interpret w as a form of transfer payment from the retailer to the supplier. What is the minimum w that is reasonable? For positive supplier profits, we need $w \geq c$. If we set $w = c$, the supplier's profits are zero, whereas the retailer captures the entire supply chain's profits. What is the w that splits the SC profits equally between the retailer and the supplier? If we set $w = (a + 3c)/4$, $w - c = P - w = (a - c)/4$ and each party's profits are $(a - c)^2/(8b)$. Note that this is the same as the supplier's profits in the DSC. Hence, if the supplier and the retailer split the profits equally in the CSC, the supplier is at least as well off, and the retailer strictly better off, than in the DCS.

In the DSC, the outcomes are worse for all the parties involved (supplier, retailer, supply chain, and consumer) compared with the CSC because in the DSC both the retailer and the supplier independently try to maximize their own profits, i.e., they each try to get a margin, $P - w$ and $w - c$, respectively. This effect is called *double marginalization* (DM).

> In a serial supply chain with multiple firms there is coordination failure because each firm charges a margin and neither firm considers the entire supply chain's margin when making a decision.
>
> Spengler (1950)

In this stylized model, the profit loss in the DSC due to DM is 25% (also referred to as the DM loss). It is clearly in the firms' interest to eliminate or reduce double marginalization, especially if this can be done while allocating the additional profits to the firms such that both firms benefit. This simple model suggests that vertical integration could be one possible way of eliminating double marginalization. However, for reasons we discussed at the beginning of this chapter, vertical integration is usually not desirable, or not practical. Then the question is, can we change the terms of the trade so that independently managed companies act as if they are vertically integrated? This is the concept known as *supply chain coordination*. In this stylized model, the retailer should choose $q^* = (a - c)/(2b)$ in any coordinating contract.

One can easily think of some very simple alternative contracts to eliminate double marginalization:

Take-it-or-leave-it-contract: The supplier offers the following contract to the retailer: Buy q^* at the wholesale price $w = (a + c)/2$, or nothing. In this case, the supplier's profit is Π^*, i.e., the supplier captures 100% of the CSC profits.

Marginal pricing: The supplier sets $w = c$. In this case, the retailer's profit is Π^*, i.e., the retailer captures 100% of the CSC profits.

Note that the take-it-or-leave-it contract would require a very powerful supplier, whereas the marginal pricing contract would require a very powerful retailer. In practice, neither the supplier nor the retailer is so powerful in general to dictate such contract terms. Hence, we need to consider alternative contracts that coordinate the supply chain. The following aspects are important in (coordinating) contracts (Cachon 2003): (a) *Profitability*: achieve profits close to optimum. (b) *Fairness and flexibility*: allow for flexible division of profits. (c) *Implementability*: should be easy and low cost to administer.

Revenue sharing contract: In a revenue sharing contract, the retailer pays a unit wholesale price w to the supplier and shares a fraction $\alpha < 1$ of his revenues with the supplier. For the simple supply chain in Fig. 11.4, one can show that the revenue sharing contract can coordinate the supply chain with a w lower than what is charged under the wholesale price contract, and α can be chosen such that both the supplier and the retailer are better off.

A well-known example of the revenue sharing contract has been implemented between Blockbuster and movie studios.[6] Blockbuster is a retailer which purchases movies from the studios (suppliers) and rents them to customers. The supplier's wholesale price impacts how many videos Blockbuster orders and hence, how many units are eventually rented by customers. Before 1998, the price of purchasing a tape from the studio was very high, around $65. Given that rental fees are in the order of $3–$4, Blockbuster could purchase only a limited number of videos which resulted in lost demand; especially during the initial release period, where the demand was high (peak demand usually lasts less than 10 weeks), 20% of customers could not find what they were looking for on the shelf. Hence, the studio's high wholesale price impacted the quantity purchased by Blockbuster, and in turn, the revenues and the profitability of both firms. Seeing this problem, Blockbuster and the studios went into a revenue sharing agreement. According to this, Blockbuster pays only $8 per tape initially, but then gives a portion (somewhere around 30–45%) of the revenues generated from that tape back to the supplier. Since this agreement reduces Blockbuster's initial investment in movies, it orders more tapes from the studio, is able to meet more demand, generates more revenues, and gives back a portion of those revenues back to the supplier. Blockbuster increased its overall market share from 25 to 31% and its cash flow by 61% using this agreement. This is clearly a win-win situation. The supplier might be better off even if he sells each unit below its production cost. A similar agreement is used between studios and theaters as well.

In case of multiple retailers, it turns out that coordination is not guaranteed under revenue sharing unless the supplier has the flexibility to offer different contracts to different retailers (Cachon 2003). Unfortunately, such "discrimination" might not always be possible due to legal considerations. Another issue is the impact of such an agreement on the behavior of a retailer who sells competing goods and also sets prices. In such a case, the retailer may have an incentive to use the goods under revenue sharing agreement as loss leaders, to drive the traffic to the store, and increase

[6] "Revenue-sharing contracts boost supply chain performance." *CNet News.com*, October 18, 2000.

the overall sales. Finally, revenue sharing loses its appeal if the revenues depend on the retailer's sales effort. For a retailer who is taking only a fraction of the revenues he generates, the incentive to improve sales goes down. While revenue sharing helps to ensure that the retailers buy and sell the "right" quantity, it hurts their sales effort. This is especially true in retail industries such as automobile sales, where a retailer's sales effort makes a big difference in the overall sales rate.

Buyback contract: We observed that in the decentralized chain under wholesale price contract, the retailer orders less than the optimal quantity. This was mainly due to double marginalization. An additional reason for ordering less than the optimal quantity could be the risk of excess inventory. For example, consider a newsvendor type model, where there is a single selling opportunity and the seller needs to procure/produce before knowing the demand. This is the case for most fashion retailers, where products need to be ordered months in advance, due to long leadtimes, well before the demand is known and the actual selling season begins. Whatever is not sold at the end of the selling season is salvaged, and the retailer bears the cost of having excess inventory.

In a buyback contract, the retailer can return unsold goods to the supplier at the end of the selling season and get some money back. Specifically, the supplier purchases leftover units at the end of the selling season for a per unit price that is less than w. Buyback contracts allocate the inventory risk between the supplier and the retailer and motivate the retailer to purchase more than what he would in a typical wholesale price contract, which can benefit both the retailer and the supplier.[7] Buyback contracts are commonly used in industries with perishable products and short product lifecycles, such as high-tech industry.

Two-part tariff: In a two-part tariff, the supplier charges a fixed fee F and a wholesale price w per unit. The following is an example of two-part tariffs from fractional aircraft ownership:

> For travelers who value flexibility and the increased security of knowing everyone on the flight, there is a compelling incentive for opting for fractional ownership.... Under NetJets' scheme, a one-sixteenth share of a small Cessna Encore, which seats seven passengers, costs $487,500 plus a monthly management fee of $6,350 and an occupied hourly fee of $1,390 for each of the allotted 50 hours.
>
> Foster (2001)

For the simple supply chain in Fig. 11.4, one can show that the two-part tariff can coordinate the supply chain with $w = c$, and the choice of F determines how the profits are allocated between the supplier and the retailer.

---

[7] The ability to share the risk of excess inventory is not unique to buyback contracts. In fact, revenue sharing also allows inventory risk sharing since the retailer in this case commits to a smaller initial capital expense for inventory.

Quantity discount contract: In the examples we discussed above, we assumed that w is fixed per unit. However, in many applications, suppliers offer quantity discounts.

> "We offer a quantity discount for orders of 10 pieces and more of the same products." (http://www.decor24.com)

> "Quantity discounts on miscellaneous accessories: $0-4 = 0\%$; $5-9 = 5\%$; $10-24 = 10\%$; $25-49 = 15\%$; 50-up $= 20\%$." (http://www.frye.com)

In quantity discount contracts, if the retailer purchases more units, the unit price goes down. Therefore, the supplier charges $w(q)$ where w is a decreasing function of q. Again, for the simple supply chain in Fig. 11.4, one can show that the quantity discount contract can coordinate the supply chain with the appropriate choice of $w(q)$.

Our focus in this section was on incentive mechanisms or contracts which partially "align" the incentives of different players in a firm or a supply chain to achieve results which are better than in a purely decentralized setting. We restricted ourselves to environments where all players have access to the same information. However, this is rarely the situation in supply chains. Complex contract structures, which most probably will be renegotiated,[8] may be required in environments with asymmetric information. Hence, contracts and incentive mechanisms that do not require true collaboration may have limited effectiveness in complex supply chains. In the following section, we discuss collaborative practices which usually require more interaction among participants than setting contractual terms.

11.3 Supply Chain Collaboration

As we discussed earlier, supply contracts are one alternative to eliminate inefficiencies in supply chains. Yet another alternative is collaboration. One should note that collaboration is more than managing relations with written contracts. In the following sections, we discuss collaborative activities that have a high potential for improving the performance of supply chains.

11.3.1 Collaboration in Design and New Product Introduction

Reducing time-to-market (TTM) and shortening product life cycles are commonly cited as sources of competitive advantage (Bower and Hout 1988; Brown and Karagozoglu 1993). By launching new products faster than competitors, firms can rapidly respond to customer feedback, establish new standards, improve brand

---

[8] Gartner Research estimates that "80 percent of outsourcing relationships will be renegotiated during lifetime of contract." ("80 percent of outsourcing relationships will be renegotiated during lifetime of contract." *AEC Online*. April 27, 2005).

recognition, quickly introduce technical innovations, and capture a bigger share of the market, resulting in higher profits (Zirger and Hartley 1996). Following this trend, several companies in different industries have gained market share, increased profits, and built strong brands. For instance, Intel leapfrogged an entire technological generation in 2001 to maintain leadership in the semiconductor manufacturing industry; the company named its strategy One Generation Ahead (Erhun et al. 2005).

Aberdeen Group reports that companies can reduce costs by nearly 18% by involving suppliers and procurement in new product development processes during the design inception and development phase compared with companies delaying such collaboration until the product prototype phase. Early involvement speeds TTM 10–20%. For example, Microsoft was able to launch the first generation of XBox in 16 months (6 months shorter than Sony's PlayStation 2), winning 3.6% market share in 4 months by creating a collaborative environment with its manufacturing partner and nearly 200 component vendors (Shah and Serant 2002). General Motors (GM) saved $1 billion in information technology (IT) expenses by reducing its TTM from 48 months to 18 months (Richardson 2003).

Even small steps such as information sharing enable improvements in design and new product introduction by enabling standardization, eliminating repetition, and shortening the leadtime of the product without lowering the quality. Such improvements are especially valuable in industries such as apparel, which are characterized by very short product lifecycles but long development cycles. Boyd Rogers, VF's (the world's largest apparel maker) VP of supply chain and technology states that "It takes as long as 9 months to design a new pair of jeans and get them on the shelves.... If you look at the cycle times from design to retail shelf, about two-thirds is spent in product development" (Sullivan 2005). According to Johnson et al. 2004), the biggest challenges in the apparel industry are: (1) Ensuring that everyone in the supply chain has an accurate and up to date description of the product. (2) Visibility of the processes to the entire product and sourcing team with a documented history of product changes. Long development times, combined with the above problems, create an excellent environment for collaborative design. For example, VF believes that collaboration on design and logistics could result in savings of $100 million a year and significantly reduce the TTM for new designs. Intra-enterprise communication through a Web-enabled system improves the design process by reducing cycle times, reducing communication costs and errors, speeding TTM, improving pricing, and standardizing design process through templates of past designs (Johnson 2002; Sullivan 2005). Collaboration brings additional improvements by combining the knowledge bases of partners.

Without doubt, collaborative product design and new product introduction has its challenges. Trust is a big issue – adversarial relationships may emerge due to lack of trust and failure to generate win-win solutions. In that sense, partner selection is a key for achieving the greatest benefits (Mentzer et al. 2000). Partners should be comfortable with the relationship, as well as their roles and responsibilities within the relationship, to make collaboration work. Yet another dimension is the availability of tools and processes to enable collaboration. With vendors such as UGS

Corp., Dassault Systèmes, Agile Software, MatrixOne, and Parametric Technology, Product Lifecycle Management (PLM) suites provide such tools and processes and take their place among other important enterprise software (Malykhina 2005).

The academic literature on collaboration in new product design includes studies on collaborative prototyping (Terwiesch 2004), value of information exchange (Browning et al. 1996; Clark and Fujimoto 1991; Krishnan 1996; Krishnan et al. 1997; Loch and Terwiesch 2005), how to respond to uncertain market conditions (Krishnan and Bhattacharya 2002; Loch and Terwiesch 1998), how to integrate the development (i.e., new product design) and planning (i.e., supply chain management) processes (Swink 2006), and negotiation and contracting (Erat 2006; Plambeck and Taylor 2004). Erat (2006) studies the negotiation and contracting process in a codevelopment setting with a focus on the information acquisition and uncertainty resolution characteristics of processes. The author's goal is to explain the rationales for delayed contracts in the joint product development process. With a two-player, two-period model, Erat considers how individual efforts map onto the final joint value of the development project, features of information acquisition, and the negotiation process and the type of contracts that may be agreed on. The contracts increase the effort levels and the performance of the product. However, the author shows that under conditions of high market or development, uncertainty firms may still delay signing a contract till at least a part of the uncertainty is resolved. Plambeck and Taylor (2004) address the question of who should own the capability to produce in a two-stage, three-player [two OEMs and a single contract manufacturer (CM)] supply chain, where the OEMs invest in innovation. The authors analyze two models for capacity investment. In the first model, the OEMs delegate the capacity investment to the CM, who then allocates capacity between them. In the second model, the OEMs retain plant ownership and pool capacity through supply contracts. Plambeck and Taylor show that OEMs might do better to trade capacity among themselves rather than to outsource to a CM. They also show that with contract manufacturing an OEM will underinvest in innovation.

Erhun et al. (2007) study new product introductions and provide a framework and a process to promote the alignment of actions and decisions across different internal groups and across organizations. The process of managing product transitions begins by identifying specific market objectives, which might include meeting profit or market share goals or maintaining technology leadership. Once these have been selected, companies need to understand the product drivers and risks, and conduct a factor assessment, which involves monitoring and measuring the factors affecting both old and new products. The process also necessitates a detailed analysis of the risks arising from interactions between products and the development of a transition playbook, which amounts to a catalog of primary and contingency strategies for preventing and mitigating transition risks. As market conditions change, managers need to be prepared to initiate the process again. The framework helps level expectations and synchronize responses across the various teams involved in product transitions, thereby improving the company's ability to anticipate and react to environmental changes, which is a critical aspect of managing product transitions.

The notion of socially responsible and eco-friendly supply chains opens a new venue for future research in collaborative new product design. Product design forms the basis of sustainable supply chains and collaboration in design has tremendous benefits. For example, Toyota did not have the technology for a Nickel Metal Hydride (NiMH) battery, one of the key components of the Prius, and collaborated with Matsushita on battery development (Carlson and Rafinejad 2006). As regulations become more stringent, sustainable manufacturing practices will be inevitable. Without collaboration, however, companies face serious risks. The incident of illegal cadmium found in outsourced PlayStation cables in 2001 by Dutch authorities (Engardio 2007) caused Sony to miss the Christmas season and lose millions of dollars as a result ("Reman ENews," http://www.reman.org/news/2005-05.htm). The alignment of incentives and responsibilities in new product development under stringent regulations is one possible research area in this venue.

11.3.2 Collaboration in Planning, Forecasting, and Inventory Management

The bullwhip effect, which refers to the phenomenon where the order variability amplifies as one moves upstream in a supply chain from retailers to distributors to manufacturers to suppliers (Fig. 11.5), is a well-documented source of inefficiency in supply chains (Lee et al. 1997). There are many factors that contribute to the bullwhip effect: demand forecasting and inventory management, order batching, price fluctuation, and rationing and shortage gaming.

In order to eliminate inefficiencies due to demand forecasting and inventory management, many supply chain partners now rely on collaborative processes such as Quick Response (QR), Collaborative Planning, Forecasting, and Replenishment (CPFR), and VMI. QR, which can be viewed as the simplest of all collaborative planning activities, entails suppliers to receive Point-of-Sale (POS) data from retailers and use this information to improve forecasting, and to synchronize production and inventory activities. Well-publicized success stories of QR include Zara (Ghemawat and Nueno 2003) and Benetton (Heskett and Signorelli 1984). These

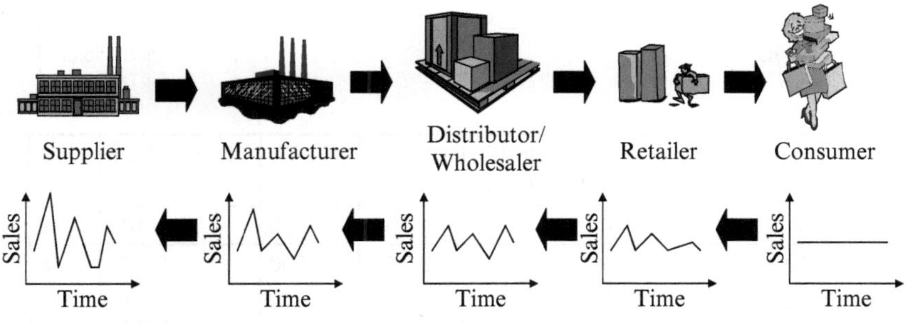

Fig. 11.5 The bullwhip effect

companies, among others, successfully use QR to operate their effective and responsive supply chains. However, companies can take additional collaborative efforts to further improve their operations. For example, they can choose to participate in collaborative forecasting in order to arrive at an agreed upon forecast between all partners, which is the idea behind CPFR. This requires sharing of additional information on future demand, such as pricing, promotions, and release of new products. Finally, retailers may leave inventory decision to suppliers within agreed-upon limits, as in VMI.

11.3.2.1 Collaboration in Planning and Forecasting

Collaborative forecasting is an iterative forecasting process in which all participants in the supply chain collaborate to arrive at an agreed upon forecast. It entails sharing not only forecasts but also information about other factors which affect future demand, such as, pricing, promotions, and release of new products. Such information sharing significantly reduces, but does not completely eliminate, the bullwhip effect (Simchi-Levi et al. 2000).

CPFR is a standard set of processes and a protocol, developed by Voluntary Inter-Industry Commerce Standards Committee (http://www.vics.org), for sharing a wide range of data over the Internet. It is a platform for negotiation before agreeing on a forecast (Fig. 11.6). In one of the first CPFR pilots, Wal-Mart and Warner-Lambert eliminated 2 weeks of inventory and cut cycle times in half for Listerine. On a successful pilot by Nabisco and Wegmans, the total snack nut category sales went up from 11%, as opposed to a 9% decline for other retailers. In particular, Planters sales went up 40%, as efficient replenishment enabled more promotions and discounting. The warehouse fill rate increased from 93 to 97% and inventory dropped

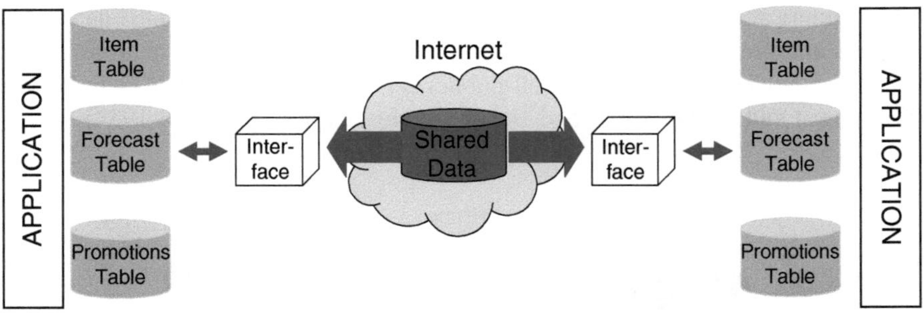

ITEM NUMBER	RETAILER'S FORECAST	MANUFACTURER'S FORECAST	DELTA	TOLERANCE	OK?
567890001	1,200	1,150	50	100	√
567890002	14,000	9,000	5,000	2,000	X
567890003	330	350	20	50	√

Fig. 11.6 CPFR: Shared process and data model (Source: CPFR Committee)

by 18%. Henkel KgaA, a German-based manufacturer of household cleaners and home care products, announced that from October 1999 to March 2000, the number of forecasts with average error of more than 50% declined from nearly half to 5% and the number of forecasts with error rate of less than 20% grew from 20 to 75% as a result of a CPFR implementation (Andraski and Haedicke 2003).

The benefits of CPFR are numerous. For retailers, the benefits include improved forecast accuracy, better store shelf stock rates and higher sales, lower inventories, better promotions planning, lower logistics costs, and lower process costs. For suppliers, the benefits include lower inventory costs, smoother demand patterns, faster replenishment cycles, better customer service, and lower production planning and deployment costs (CPFR Committee, http://www.cpfr.org). Given these benefits, the academic literature on collaborative planning and forecasting is flourishing (Aviv 2007, 2002, 2007; Kurtulus and Toktay 2004; Miyaoka 2003). Sheffi (2002) provides an overview of the development of collaboration in supply chains and discusses the possible benefits of collaboration based on case studies focusing on CPFR.

In a series of papers, Aviv (2007, 2002, 2007) studies the potential value of CPFR in a cooperative supply chain using stylized models with a single retailer and a single manufacturer. The author shows that the benefits of collaborative forecasting depend on (a) the relative explanatory power of the supply chain partners, (b) the supply side agility, and (c) the internal service rate. He identifies cases where a partnership does not appear to be valuable to the manufacturer. Miyaoka (2003) considers a decentralized two-stage supply chain with a single selling season and addresses the incentive issues associated with implementing collaborative forecasting. She introduces a concept she calls collaborative forecasting alignment (CFA) that aligns the parties' incentives so that demand information can be shared credibly with a simple transfer price agreement. Kurtulus and Toktay (2004) investigate the conditions that favor the use of collaborative forecasting between a supplier and a retailer in a newsvendor setting. Both the supplier and the retailer can exert independent, costly effort to improve the quality of their local demand forecasts. The authors characterize the existence and stability conditions of an equilibrium in which both parties invest in improving forecast quality.

An interesting research direction in CPFR is N-tier relationships, i.e., analyzing either multiple partners in one tier or multiple tiers of partners. Mechanisms that enable truthful information sharing for multiple partners in one tier would be especially valuable. To that extent, behavioral studies on CPFR would be useful in understanding the drivers of failure and success in CPFR applications. Current practices would also greatly benefit from empirical research on the value, adoption, and success factors of CPFR.

11.3.2.2 Collaboration in Inventory Management

VMI, also called the continuous replenishment program (CRP), is an agreement between a supplier and a buyer where the supplier is responsible for maintaining adequate inventory levels at the buyer's site (Hausman 2001). Both the academic literature (Achabal et al. 2000; Axsater 2001; Çetinkaya and Lee 2000; Cheung and Lee 2002; Disney et al. 2003: Disney and Towill 2003a, b; Fry et al. 2001; Kaipia et al. 2002) and industrial implementations of VMI (Wheelwright and Gill 1990; Peleg 2003) are growing at a steady pace.

VMI allows coordination of production and distribution between suppliers and retailers. There are several benefits of this type of coordination that are documented by successful implementations. One such success story is due to Campbell Soup Company. In early 1990s, Campbell Soup Company was enjoying a stable business environment: only 5% of their products were changing every year and the demand was predictable. More than 98% of demand was satisfied from stocks of finished goods. Replenishment leadtime for new products was 1 month and the minimum market life cycle was 6 months. However, the company suffered from low margins and in 1991 implemented CRP. They started monitoring demand and inventory levels daily [via electronic data interchange (EDI) links] and jointly decided on inventory policy and parameters with their retailers. As a result of these changes, inventory levels went down from 4 weeks to 2 weeks of supply, in-stock availability went up to 99.2% from 98.5%, and the company recognized the negative impacts of the overuse of price promotions (Wheelwright and Gill 1990; Hausman 2001). Wal-Mart and P&G, Kmart and First Brands, Inc., and Kmart and Whitehall Robbins, who have successfully incorporated VMI into their operations, have observed similar benefits (Simchi-Levi et al. 2000).

A critical question in any VMI implementation is: Who owns inventory? There are different ways of handling this issue of ownership. Traditionally in VMI implementations, the retailer owned inventory. This creates obvious incentive misalignments between retailers and suppliers. Recently, we see a movement toward a consignment relationship where the supplier owns inventory until the goods are sold. Based on the Institute of Management and Administration's 2004 Inventory Management Report Survey, Mullen reports that 37.3% of respondents have consignment agreements with their suppliers. In the list of top ten best practices, the respondents ranked inventory consignment in third place in 2004, up from fifth place in 2003 and tenth place in 2002 (Mullen 2006). Wal-mart, for example, has such consignment relationships with some of its suppliers, including most of its grocery purchases (Taylor 2004). As a part of the E-Chain Optimization Project (eChO), STMicroelectronics and one of its strategic partners implemented CPFR, based on a VMI model with consignment, which transformed the collaborative forecasting process from being superficial to having more depth (Peleg 2003).

The benefits of VMI discussed above have created interest in academia as well. Mishra and Raghunathan (2004) identify yet another benefit of VMI for the retailer by showing that VMI intensifies the competition between manufacturers of competing brands due to brand substitution, and the increased competition benefits a retailer who stocks these brands as manufacturers stock more inventories. Çetinkaya and Lee (2000) analyze a model that coordinates inventory and transportation decisions in a VMI setting. In their model, a vendor, who uses a certain kind of (s, S) policy, satisfies demands for several retailers in close proximity. In order to benefit from the scale economies in transportation, the vendor may batch orders. Hence, the vendor's inventory should take into account the replenishment frequency. The authors jointly compute the optimum replenishment quantity as well as dispatch frequency. Cheung and Lee (2002) study the relative benefits of two VMI-initiatives, i.e., joint/coordinated replenishment systems and joint/coordinated replenishment systems with stock rebalancing, in a setting close to Çetinkaya and Lee's. These authors conclude that shipment coordination and rebalancing reduce costs as the number of retailers who can participate, i.e., many customers in close proximity, increases. In a related work, Altintas et al. (2008) show that the increasing burden of transportation costs is forcing suppliers to eliminate transportation inefficiencies by motivating buyers to place full-truckload orders and suggest discount schemes that suppliers may use to moderate buyers ordering behaviors under different transportation costs.

Fry et al. (2001) model a type of VMI agreement called a (z, Z) VMI contract. In a (z, Z) VMI contract, inventory at the customer is reviewed periodically. If upon review, the inventory level is below z, then the supplier has to pay the retailer a penalty of b^-, and if upon review, the inventory level is above Z, then the supplier has to pay the retailer a penalty of b^+. The authors formulate a Markov decision process (MDP) in which the supplier makes three types of decisions: production quantity, outsourcing (expediting) quantity, and delivery quantity to the retailer. The retailer takes the supplier's optimal policy in the MDP into account, and chooses the values of z and Z that minimizes the retailer's expected cost. They show that there exist values of z, Z, b^-, and b^+ (that could be chosen by a central decision maker) that would optimize the overall supply chain performance. They also formulate a MDP model of a conventional retailer managed inventory (RMI) setting, in which the supplier has complete knowledge of the consumer demand distribution, the retailer's inventory level, and the retailer's ordering policy, thus modeling RMI with complete information sharing. They compare the overall performance of the supply chain under both settings numerically and show that the results are inconclusive. They find that depending on the contract parameters (such as b^-, b^+, and Q) VMI can lead to overall supply chain performance that can be significantly better or significantly worse than RMI. However, by choosing the contract parameters carefully, the performance of VMI can be improved.

VMI implementations are not always successful, especially when communication and trust are not built into the partnership. Spartan Stores, a grocery chain, shut down their VMI program after just 1 year. Buyers did not trust suppliers and spent no less time in ordering. Suppliers could not incorporate promotional information into forecasts, hence, delivery levels were unacceptably low during promotion times (Simchi-Levi et al. 2000). This example highlights a very common mistake in VMI implementations. Retailers should always keep in mind that because they no longer manage the inventory, inventory does not disappear from the supply chain. Hence, they should not expect suppliers to incorporate big swings in demand to their operations. Working with suppliers collaboratively can lead to a long-term partnership with mutual benefits, such as reduced inventory throughout the supply chain (Mullen 2006).

11.3.3 Collaboration in Production Management

Collaborative production management (CPM) – an application of collaborative concepts to the factory floor – includes intra- and inter-enterprise collaboration and real-time information sharing on "production planning, finite scheduling, material and recipe management, data collection, document control, lot and work order tracking, plant floor and enterprise system interfacing, messaging and alarming, performance analysis, genealogy, dispatching, and workflow management."[9] It thus "synchronizes, executes, tracks, reports, and optimizes manufacturing processes."[10]

One of the dimensions of CPM is collaborative scheduling. According to a manager of Military Technology and Operations at a multinational aerospace manufacturer, "In order to reduce production cost and customer lead time, it is very important to coordinate daily manufacturing schedules closely between our own plants and those of our suppliers. Since our suppliers typically have different objectives from ours, conflicts often arise and have to be resolved through scheduling coordination" (Hall 2005).

In a series of papers, Hall and his colleagues study the benefits and challenges of coordination within supply chain scheduling models. Hall and Potts (2003) and Dawande et al. (2006) analyze conflict and cooperation issues arising in an arborescent supply chain.[11] In Hall and Potts's model, a supplier makes deliveries to several manufacturers, who also make deliveries to customers. In Dawande et al.'s model, a manufacturer makes products which are shipped to customers by a distributor. Hall and Potts show that cooperation between the supplier and the manufacturer may reduce the total system cost by at least 20–25% and up to 100%, depending upon the scheduling objective. Dawande et al. argue that the ideal schedule of the manufacturer and the distributor (determined by cost and capacity considerations) are in general not well coordinated, which leads to poor overall performance. Hence, Dawande et al. consider the extent to which one decision maker's cost is larger than optimal when the other decision maker imposes its locally optimal schedule. The authors show that the cost of conflict can be eliminated and the parties benefit from cooperation when the dominant player agrees not to use its individually optimal schedule. The authors recommend a perfectly equitable split of the surplus, when the parties accurately and continuously share all cost data in a verifiable way. Alternately, they recommend negotiation for a transfer payment from the dominated player to the dominant player. Chen and Hall (2005) study the same issues in an assembly system where suppliers provide parts to a manufacturer who performs a nonbottleneck operation for each product. The authors computationally demonstrate that the cost saving realized by cooperation between the decision makers is significant in many cases in assembly systems as well.

---

[9] "Collaborative production management solutions on the rise." http://www.ferret.com.au/.

[10] "Discrete manufacturing CPM market to double; will top $1 billion by 2008." http://www.mhmonline.com/.

[11] In arborescent supply chains, each player has only one supplier but can act as a supplier to one or many players.

According to the ARC Advisory Group, the market for CPM systems for discrete manufacturing will hit $1.4 billion by the end of 2010 (The market was $860.5 million in 2006).[12] The benefits of CPM are significant: reduced errors, increased production rates, improved capacity utilization, increased equipment reliability, improved efficiency and productivity of staffing, improved responsiveness to demand changes, improved quality, and continuous improvement. To achieve these benefits, CPM should integrate with business, engineering, and maintenance systems.

11.3.4 Collaboration in Logistics

Recent studies show an increase in logistics costs; the estimated logistics costs in 2005 totaled $1.183 trillion, a $156 billion increase over 2004. The 17th Annual State of Logistics Report now places logistics expenditures at 9.5% of US's gross domestic product (GDP) (Cooke 2006). Since logistics costs constitute 5–50% of a product's total landed cost (Hart 2005), it is critical to control these costs in supply chains and collaborative logistics can provide the means to do that.

According to John Sobczak, the supply chain manager for Cogistics, "the most interesting and exciting topic in the industry right now is collaborative logistics. The collaborative approach, in contrast to traditional third-party logistics outsourcing, is becoming the preferred way of doing business" (Malone 2003). There are several aspects of collaborative logistics, such as collaborative transportation and collaborative warehousing. For example, Land O'Lakes started a collaborative warehousing initiative recently. For Land O'Lakes, warehousing constitutes about three-quarters of their $40 million logistics budget (Stepanek 2003). Hence, sharing warehouse space can potentially decrease inventory costs and increase their supply chain efficiency considerably. However, "transportation continues to be the biggest component of overall logistics cost and accounts, on average, for 6% of a company's annual expense budget .... Technology and shipper–carrier collaboration are opening new doors to cost/price reductions in all areas of the transportation process – procurement, planning, execution, and monitoring" (Murphy 2002). Hence, collaborative practices have seen an increased adoption in the transportation industry in recent years and we concentrate on collaborative transportation in the rest of this section.

11.3.4.1 Collaboration in Transportation

To transport the shipments of different shippers, a carrier often has to reposition its assets, e.g., trucks in case of a trucking company and containers in case of an ocean carrier. A recent industry report estimates that 17 and 22% of all truck movements in the USA are empty for large and small carriers, respectively,

[12] "Collaborative production management for the discrete market grows 15%." *IndustryWeek*. September 12, 2005.

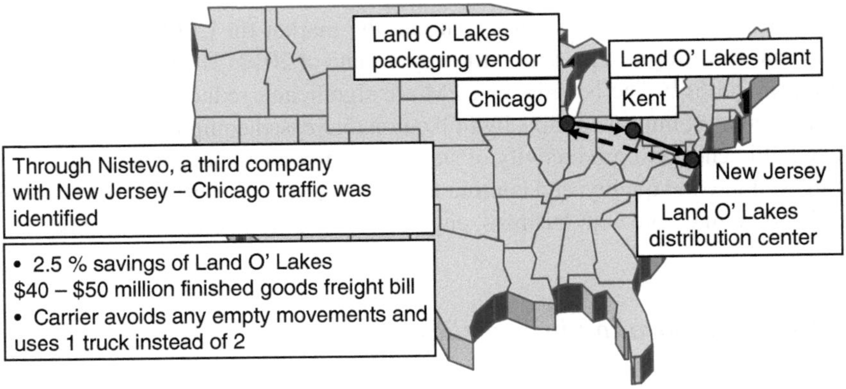

Fig. 11.7 Collaborative routing example from Nistevo Network

resulting in approximately 35 million empty miles monthly and a loss of billions of dollars (ATA 2005). These repositioning costs are reflected in the prices paid by the shipper, and eventually translate into higher prices for the goods sold in the market, impacting the entire economy. To reduce these costs, shippers and carriers can get together and collaborate on managing the timing and frequency of the shipments to better utilize the truck capacity of a carrier. Examples of such collaborative logistics networks include Nistevo, Transplace, and One Network Enterprises. By participating in collaborative activities through these networks, a number of companies such as General Mills, Georgia-Pacific, and Land O'Lakes have been able to identify cost efficient routes and realize considerable savings (Fig. 11.7). For example, Georgia-Pacific's percentage of empty movements decreased from 18 to 3% after forming collaborative partnerships with other companies in the Nistevo Network, where each 1% reduction in empty moves corresponds to savings of $750,000 annually (Strozniak 2003).

For similar reasons as in trucking, collaboration has also been embraced by the sea-cargo industry. Repositioning empty containers is very expensive: According to ROI Container Cargo Alliance (July 2002), a 10% reduction in equipment and repositioning costs can potentially increase profitability by 35–50%. Several ocean carriers have formed alliances (e.g., Sea-Land and Maersk share vessels in the Atlantic and Pacific oceans), which allow them to realize economies of scale, extend customer base, and increase asset utilization (reducing empty container moves) while providing customers with more frequent sailings and faster transit times (Agarwal and Ergun 2005).

There are several important issues that need to be addressed for achieving successful collaborative transportation: (1) How to form routes to reduce empty moves? (2) How to share costs among the different participants? (3) How to establish trust among the participants and overcome cultural differences? The third question is of utmost importance, since collaborative networks cannot exist without trust and agreement among participants. For example, Kellogg's and General

Mills, major competitors in food manufacturing, send most of their products to the same stores and could greatly benefit from sharing truck capacity or routes. However, they were not able to collaborate because of cultural roadblocks (Strozniak 2003). While we acknowledge the importance of trust, our focus in this section will be primarily on analytical approaches which can help answer the first two questions.

Generating collaborative tours: Identifying tours to minimize asset repositioning costs in a collaborative logistics network can be challenging, especially as the size of the network (i.e., the number of participants and the number of lanes) grows. Adding to the challenge are various timing constraints, including: (1) dispatch time windows, i.e., the time interval in which the load to be moved should be dispatched to arrive at its destination on time, and (2) Department of Transportation Hours of Service regulations, which limit the driving and duty hours of truck drivers.

Erhun et al. (2007) discuss optimization methodology for the identification of repeatable, dedicated truckload continuous move tours, which is relevant for companies that regularly send truckload shipments and are interested in collaborating with each other. Considering the constraints mentioned above, they focus on the time-constrained lane covering problem (TCLCP), which is defined as follows: For a given set of lanes, find a set of tours covering all lanes such that the total duration of the tours is minimized and the dispatch windows are respected.

Given the large size of practical instances, Erhun et al. focus on developing an effective and efficient heuristic for TCLCP. They implement a greedy heuristic that generates a large number of time-feasible cycles (potentially all) and greedily selects a subset of those cycles to cover the lanes based on some criterion measuring the desirability or attractiveness of a cycle. After all lanes are covered they perform a local improvement step to improve the solution. They test their methods on randomly generated test problems and also conduct a case study for a group purchasing organization to assess the potential value of collaborative transportation procurement for individual member companies. In the case study, a typical all-member instance for a single week involves about 750 locations and 5,500 lanes. The potential savings due to continuous moves are estimated to be in the order of 9–10%. They observe that the smaller the dispatch window width the smaller the savings, due to the increased chance of waiting between two consecutive moves.

Sharing costs/savings among collaborators: One of the most challenging aspects of collaboration is devising "fair" mechanisms to allocate the costs/savings among the participants such that the resulting collaborative arrangement is sustainable.[13] According to Kevin Lynch, President and CEO of Nistevo Corporation, "The key to understanding Collaborative Logistics lies in recognizing how costs are

[13] In a *stable* or *fair* cost allocation, no coalition of members can find a better way of collaborating on their own. Hence, the grand coalition is perceived as fair and is not threatened by its subcoalitions. Thus, stability is the key concept that holds a collaboration together.

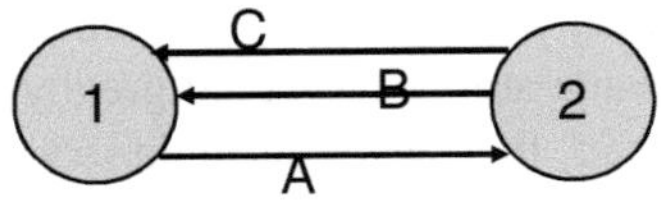

Fig. 11.8 Cost sharing in a simple collaborative network

distributed in a logistics network" (Lynch 2006). In a recent paper, Ozener and Ergun (2008) discuss desirable properties of cost allocation mechanisms in collaborative transportation.

In current practice, collaborative networks allocate benefits in proportion to the base cost (cost before collaboration) of the participating shipper's lanes. That is, the savings (the difference between the total base cost and the total cost of the collaborative tours) are distributed based on the percentage of the base cost each shipper has contributed to the collaborative transportation network. The example in Fig. 11.8 (based on Ozener and Ergun) shows that although such "proportional cost allocation" schemes are easy to implement, they are not fair from a game theoretic point of view. In this example, suppose that the cost of covering a lane is equal to 1. If only shippers A and B are in the network, the total cost of covering the lanes is 2, and the proportional cost allocation method allocates a cost of 1 to each shipper. With the addition of the new shipper C, the total cost of covering the lanes in the network becomes 4, and a cost of 4/3 is allocated to each shipper. However, with this allocation, it is easy to see that shippers A and B (or A and C) are better off collaborating on their own with a total cost of 2. Therefore, the proportional cost allocation in this case is not fair and the *grand coalition* that consists of all the shippers in the network could be replaced by a subgroup of its members.[14] Note that the only allocation where the grand coalition is not threatened by any subgroup of its members is the allocation of (0, 2, 2) to shippers A, B, and C, respectively. Since shipper A creates a positive value for the other two shippers, it is charged less than B and C. However, charging shipper A nothing makes A a *free-rider* which may not be desirable in a collaboration. Furthermore, in the only fair allocation, where the grand coalition is maintained together, both shippers B and C are allocated their stand-alone costs, so being in a collaboration brings no positive value for these two shippers. This simple example illustrates some of the challenges in finding a robust mechanism for allocating costs and savings in a collaborative framework.

As our simple example indicates that a cross monotonic cost allocation, i.e., an allocation where no member's benefit decreases with the addition of a newcomer, may not exist in the core of the game representing the shipper collaboration problem. Hence, increased synergies due to the addition of new members to a coalition do not necessarily create additional benefits for the participants. Therefore, Ozener and Ergun study cross monotonic and stable allocations that recover a good percentage of the total cost, even if not the entire cost.

---

[14] Ozener and Ergun (2008) note that due to the costs associated with managing collaborations, limited rationality of the players and membership fees, a subcoalition might not be formed even though it offers additional benefits to its members. Therefore, relaxing the stability restriction in a limited way might be acceptable for a cost allocation method.

A collaborative game where the players compensate each others' costs with side payments is called a *transferable payoffs* game. A collaborative game is called a *nontransferable payoffs* game if positive transfers between members of the collaboration are not allowed. In the shippers' collaboration problem, seeking a cost allocation method which distributes the total transportation costs corresponds to a transferable payoffs game, whereas seeking an allocation which only distributes the asset repositioning costs corresponds to a nontransferable payoffs game. To ensure that each shipper pays at least its original lane cost (i.e., to avoid the situation in the example in Fig. 11.8 where shipper A pays nothing), allocations with nontransferable payoffs are of interest. Furthermore, it is desirable that each shipper is guaranteed an allocation less than its stand-alone cost so that being a member of the collaboration offers a positive benefit. Ozener and Ergun show that when either of these two restrictions is imposed, it is not possible to have a budget balanced and stable cost allocation for the shippers' collaboration problem. Hence, they relax the efficiency and stability properties in a limited way and develop allocations with the above two restrictions.

Despite its challenges, collaboration can offer tremendous benefits to its participants. In addition to reducing transportation costs, collaborative logistics can also lower inventories and at the same time eliminate stockouts resulting in lower inventory holding costs and better customer service. Consider AIT, a leading distributor of industrial products in North America, which operates 450 service centers that sell maintenance, repair, and operational industrial products to large and small manufacturers. By sharing truck space with its partners, AIT has seen its dedicated freight charges drop by nearly 30%. In addition, since deliveries are now made daily (due to shared truck capacity) rather than weekly as before, service centers do not have to carry as much safety stock and can order products as late as 5 P.M. to be delivered the next day. As a result, customer service has improved and the need for the company's service centers to hold safety stock inventory has declined by 15–20% (Strozniak 2003). In this example, in addition to reducing the transportation costs, collaboration also helped the company to reduce leadtimes and increase delivery frequencies, resulting in better demand forecasts and allowing the company to better match demand and supply with less inventory in a timely manner.

Traditionally, supply chain partners have focused their attention on controlling and reducing their own costs to increase profitability, but now they realize that a system-wide collaborative focus offers opportunities that cannot be achieved by any one company alone. Collaborative transportation not only reduces the shipping costs for the participants, but can also result in major cost savings due to lower inventory levels, shorter leadtimes, lower stockouts, and better customer service.

11.3.5 Collaboration in Procurement

Collaborative procurement has emerged as one of the many initiatives for achieving improved intra- and inter-firm coordination and collaboration. With intra-enterprise vertical and horizontal collaboration, a company streamlines and coordinates its procurement functions which can lead to significant savings. For example, by aggregating its spending, Lucent reduced the number of invoices that amount to less than $1,000 by 23% in 2003, according to Joe Carson, chief procurement executive. In addition, Lucent has reduced its supplier base by about half, placing a larger part of its spend with fewer suppliers to leverage its buying power.[15] Similarly, by simplifying its supply base, working closely with its key suppliers, leveraging its buying power with those suppliers across different business units, and developing a system for implementing innovative cost-savings ideas, Dial Corp. was able to lower its purchasing costs by $100 million in 5 years (Reilly 2002).

Inter-enterprise vertical collaborative procurement, i.e., collaboration between buyers and suppliers, has been successfully adopted by companies with world-class procurement practices. According to the Purchasing Magazine, using supply base rationalization and partnering with key suppliers, world-class procurement organizations incur procurement costs that are 20% less than typical companies (0.68% of procurement spending vs. 0.85%) and operate with nearly half the staff (44.9 staff per $1 billion of spend vs. 89.2).[16] One of the best known examples of collaborative intra-enterprise procurement is between Japanese auto makers and their suppliers, which are based on long-term purchasing relationships, intense collaboration, cross-shareholding, and the frequent exchange of personnel and technology (Ahmadjian and Lincoln 2000).

Quoting a senior executive of a major supplier to Ford, GM, Chrysler, and Toyota: "Toyota helped us dramatically improve our production system. We started by making one component, and as we improved, [Toyota] rewarded us with orders for more components. Toyota is our best customer." According to Ahmadjian and Lincoln, Toyota and Honda have built great supplier relationships by consistently following six steps: they understand how their suppliers work, turn supplier rivalry into opportunity, monitor vendors closely, develop those vendors' capabilities, share information intensively but selectively, and help their vendors continually improve their processes.

Inter-enterprise horizontal collaboration in procurement has also seen much interest and success in recent years. This practice is also known as cooperative purchasing, group purchasing, and consortium purchasing, where two or more independent organizations (potential competitors) with similar products or services, join together, either formally or informally, or through an independent third party, for the purpose of combining their requirements for the purchase of materials, services, and capital goods.

Group purchasing is not a new concept; cooperatives and other nonprofit organizations (mainly educational and social) have practiced it for many years. There are numerous examples of cooperatives and for-profit firms making purchases such

---

[15] "Buyers use more than one way to cut component costs." *Purchasing*. March 17, 2005: http://www.purchasing.com/article/CA510893.html?industryid=2147&nid=2419.

[16] "Hackett report finds best procurement orgs see greater ROI." *Purchasing*. December 8, 2005: http://www.purchasing.com/article/CA6289412.html?text=procurement+spending.

as office equipment and supplies, tooling, software, engineering and consulting services, air freight, and other items through purchasing consortia. Group purchasing has been especially prevalent in the healthcare industry, perhaps because decreasing supply expenses plays an important role in increasing the already low profit margins in this industry and offers a unique competitive advantage. The Group Purchasing Organizations (GPO) Directory (http://www.firstmark.com/fmkdirs/gpo_hsys.htm) contains information on more than 700 GPOs and multihospital systems which own, manage, or provide purchasing services to hospitals. Some of the other industry sectors where big group purchasing organizations exist are telecommunications, transportation, and services.

One of the major benefits of collaborative procurement for buyers is reduced purchasing costs due to the quantity discounts offered by the suppliers. Through collaborative procurement, multiple customers can combine their orders, present a single face to supplier, and given that the supplier has to process one large order rather than several small orders, it can respond with lower prices (Melymuka 2001). Such collaboration is sometimes enabled through intermediaries or e-markets (Griffin et al. (2005)). For example, Transplace (http://www.transplace.com), an Internet-based global logistics alliance formed by six of the largest US-based transportation companies, offers shippers and carriers efficiencies and cost savings from combined purchasing power on items such as fuel, equipment, maintenance and repair parts, insurance, and other services. Under its Fuel Program, Transplace negotiates the price of fuel with suppliers of petroleum products, allowing member carriers to procure fuel at lower prices at designated fuel stops across the country. In a quantity discount scheme, the supplier sets price break(s) (Lam and Wong 1996) or uses continuous pricing (Ladany and Sternlieb 1974; Lal and Staelin 1984; Dave et al. 1996). Dolan (1987) provides a thorough analysis and categorization of the studies on quantity discounts.

The extent of actual cost savings due to collaborative procurement may lead potential members of group purchasing programs to question whether it is actually worth joining such a program, i.e., whether the savings will justify the efforts on reshaping the purchasing process. The supplier has parallel concerns: Does the potential increase in sales volumes justify offering lower prices to GPOs? In particular, under what market conditions (such as the demand structure and uncertainty, number of buyers and suppliers in the market, the size and market power of the participants, uncertainty in supply availability and quality, etc.) is collaboration beneficial to each participant? These questions become particularly important when the collaborating buyers are competitors in the end market (Keskinocak and Savasaneril 2008).

Despite its increasing adoption in practice, the effects of collaborative procurement on buyer and supplier profitability have not been studied systematically. Collaboration between *multiple buyers who purchase from multiple suppliers* is studied in Mathewson and Winter (1996) and Griffin et al. (2005). Mathewson and Winter study a problem where a group of buyers negotiates and makes a contract with a group of suppliers to get lower prices. In turn, the buyer group gets supplies only from the contracting group which implies a tradeoff between low price and low product availability. The authors conclude that as the number of suppliers increases, the buyer group is more likely to benefit from contracts and the formation of buyer groups might be welfare increasing or decreasing depending on the model's parameters. Griffin et al. study alternative buyer strategies in markets where procurement costs are affected by economies of scale in the suppliers' production costs and by economies of scope in transportation. They consider buyer strategies with different types of collaboration, namely, (a) no collaboration among buyers or buyer divisions, (b) intra-enterprise collaboration among the purchasing organizations of the same buyer enabled by an internal intermediary, and (c) inter-enterprise (full) collaboration among multiple buyers enabled by a third-party intermediary. They find that when the potential benefits from economies of scope are high, intra-enterprise collaboration performs very well. When the potential benefits from economies of scale are high, they observe that buyer strategies need to consider potential future trades in the market by other buyers while contracting with a supplier. Their computational analysis indicates that the potential benefits of collaboration are highest in capacitated markets with high fixed production and/or transportation costs.

Collaboration between *multiple buyers who purchase from a single supplier* is studied in Anand and Aron (2003) and Keskinocak and Savasaneril (2008). Anand and Aron study the optimal design of an online business-to-customer group-buying scheme under demand uncertainty. In their model, the buyers arrive and demand single units, and as the number of units demanded increases the price drops. The demand function is not known to the supplier before he decides on price-quantity tuples. Under this setting, the supplier's benefit from group-buying increases as demand heterogeneity (the difference in the slopes of the demand curves of the buyers) increases. Furthermore, group-buying outperforms single pricing when the goods are produced after total demand is realized under scale economies. Keskinocak and Savasaneril study collaboration in a business-to-business (B2B) setting, where buyer companies participate in group purchasing for procurement, but produce independently and remain competitors in the end market. When the buyers are uncapacitated and identical in terms of costs, the authors show that buyers and end consumers are better off under joint procurement as compared to independent procurement, and rather than selling a large quantity to a single buyer, the supplier is better off by selling smaller and equal quantities to both buyers. Next, they consider the case of different size buyers, i.e., buyers with different capacity availability. Intuitively, one might think that a "large" buyer would have less incentive to collaborate with a smaller buyer on procurement, since the large buyer already has enough volume to obtain a good price from the supplier. The "small" buyer, on the other hand, might prefer to collaborate with a large buyer, since it will obtain additional price breaks due to the volume of the large buyer. Given these conflicting incentives, one might expect that joint procurement would occur mainly among roughly equal size buyers (in terms of capacity and purchase volume). However, the authors find that depending on the market characteristics, collaboration may occur among different size buyers. Furthermore, depending on the capacity of the large buyer, the small buyer may not always be willing to collaborate. Similar to Keskinocak and

Savasaneril, Spiegel (1993) also focuses on competition, by studying production subcontracting between two rival firms operating at the same horizontal stage in the supply chain. He shows that this arrangement, if it occurs at all, always increases production efficiency.

There is still a lot of room for research in advancing our understanding of how collaborative procurement impacts supply chains and developing decision technologies to aid companies in effectively utilizing new opportunities provided by collaboration. Developing models and decision support tools for coordinated replenishment with quantity discounts, buyers' procurement strategies in the face of quantity discounts under demand uncertainty, coordinating intra-enterprise collaborative procurement with logistics to simultaneously optimize the net landed cost, and understanding the dynamics of intra- and inter-enterprise collaborative procurement under stochastic demand in a dynamic setting are just a few of the many potential research directions.

11.4 Role of Information Technology in Collaboration

In the last two decades, we have witnessed tremendous changes in how enterprises manage their intra- and inter-enterprise operations. Up until the 1990s, the main focus of companies was on improving cost and quality. During 1990s, with the development of ERP systems – which are central repositories for information about an enterprise that facilitate real-time information exchange and transactions within and between enterprises – companies made huge progress completing their internal integration and started going beyond internal integration and integrating with their external business partners. Over the past decade, there has been considerable advancement on software products, such as instant messaging and virtual network computing, and enabling technologies, such as Meta Markup Language (XML), to further support collaborative processes. However, true end-to-end visibility and collaboration in supply chains can only be possible with the development of common data standards. According to Vinay Asgekar, an AMR Analyst, "Common data standards that are readily embraced can let leading companies truly achieve a streamlined extended supply chain" (Schoonmaker 2004).

EDI is a "communication standard that enables the electronic transfer of routine documents between business partners" (Turban et al. 2005). EDI enables computers to talk to each other by sending standardized messages over (traditionally) a value-added network (VAN) or (more and more) the Internet. EDI facilitates and fosters collaborative relationships. In addition, it minimizes data entry errors, secures message/data transfer, reduces cycle time, improves inventory management and increases productivity, and increases customer service. Based on an Aberdeen Group study, while 90% of all invoices among the Fortune 1,000 companies are handled by EDI, this is less than 10% of all business invoices. Even though it is very mature (EDI has been around for more than three decades) and very secure (through the use of VAN), EDI is also very costly and difficult to afford for most medium-sized and smaller companies (Roberts 2001). Thus, the adoption of traditional EDI has been slower than expected. However, the XML/EDI framework and the Internet-enabled EDI are closing the gap.

RosettaNet (http://www.rosettanet.org) is a global consortium-based standards organization, which develops a common XML-based platform for communication to enable collaboration and automation of transactions in global supply chains. RosettaNet defines business processes, semantics, and a framework for how data gets passed over the Internet. According to Joseph Matysik, Materials Manager at Intel Corporation, Assembly/Test Materials Operations "Intel has significantly extended its supply chain visibility and agility through its RosettaNet implementations and process re-engineering efforts" (Schoonmaker 2004). Among the benefits of RosettaNet (and such standards in general), one can count error-free forecast-to-cash procurement processes, reduction in manual transactions, reduction in contract costs, reduction in inventory levels, decrease in change orders, reduction in administrative costs, reduction in logistics costs, and reduction in planning time (http://www.rosettanet.org). Other standards organizations include (Peleg and Lee (2006)): the AIAG (Automotive Industry Action Group) for the automotive industry (http://www.aiag.org), the WWRE (WorldWide Retail Exchange) for retailers and suppliers in the food, general merchandize,[17] textile/home, and drugstore sectors (http://www.wwre.org), the AIA (Aerospace Industries Association of America) for the defense and aerospace supply chains (http://www.aia-aerospace.org), and the GS1 for global supply chains across industries (http://www.gs1.org).

In spite of all these advances in IT, the biggest mistake companies can make today is to view the new IT products and tools alone as a silver bullet. IT by itself is not enough to lead to successful collaboration. Firms must know how to use IT to reap the benefits of collaboration. Human contribution, through data analysis and information utilization, is where the true benefits of IT lie. Therefore, the enabling and supporting role of IT to collaborative processes can only be realized if the technology is employed effectively. When used effectively, IT enables collaboration by providing the necessary tools to make it feasible, such as real-time data transfer and automated communication. IT supports collaborative inter-organizational relationships by reducing the transaction costs and risks with automated process and provides the opportunity for outsourcing of processes between partners.

11.5 Concluding Remarks

Any improvement in the design of integrated and collaborative supply chains by better coordination between involved parties can be expected to have significant economic and social impacts. The advantages of a successful collaboration are numerous, including reduced inventory, increased sales, lower costs, increased

[17] WWRE is now a part of Agentrics LLC, due to a merger with GNX.

revenue, better forecast accuracy, improved customer service, and more efficient use of resources (CPFR Committee, CTM Sub-Committee of CPFR, Cohen and Roussel 2005). In spite of these well-documented advantages, companies are still reluctant to open up their supply chains to collaboration. This fact is most obvious from the AMR Survey at Microsoft Engineering and Manufacturing Executive Summit: even though only 20% of the participants were concerned about trading partner acceptance, 44% had concerns about trading partner readiness (Caruso 2002).

There are excellent examples of companies such as Dell, Microsoft, Cisco, Wal-Mart, who have successfully implemented supply chain collaboration. Even though there is not a single recipe for such a success, there are several common factors that we observe:

- Collaboration is about sharing of ideas, information, knowledge, risks, and associated costs and rewards. In a collaborative environment, unless all parties benefit, it is not a true relationship.
- Collaboration should not be the new flavor of the month; companies should know why they want to collaborate. Essentially, the relationship should fit with the partners' strategies, processes, and technologies (Matchette and Seikel 2005).
- Many times companies do not effectively collaborate internally (Cohen and Roussel 2005). However, collaboration should often start with intra-enterprise collaboration. Based on an AMR study, Sabath and Fontanella (2002) argue that "Enterprises that have learned to collaborate internally are the most successful in creating collaborative relationships."
- The effort required to collaborate with partners means that companies will only be able to do this with only a handful of strategic partners, hence, it is of key importance to identify the partners correctly. Partners should trust each other and should be ready and willing to share practices, processes, and information, even when this means sharing proprietary information (CTM Sub-Committee of CPFR, Matchette and Seikel 2005).
- The partners should have a clear understanding of their and others' roles and responsibilities in the relationship (Mentzer et al. 2000). To enable this, the expectations should be set up-front clearly, which requires "a formal, documented front-end agreement that defines the scope (i.e., the steps, measures, terms, and protocols that define the nature of the collaboration) and goals (i.e., the specific benefits that the collaboration is expected to deliver) of the relationship" (CTM Sub-Committee of CPFR). Sabath and Fontanella (2002) identify misaligned expectations and using different definitions as a cause of disappointment in collaboration. Unless all parties are open, committed, truthful, and have a stake in outcome, collaborations are doomed to failure.
- The scope and the goals of collaboration should be open to an evolving collaborative relationship. According to Langley, collaboration "must allow members to dynamically create, measure, and evolve collaborative partnerships" (Langley 2001). This will enable companies to learn and adapt, and create new opportunities for future collaborations.
- As with any implementation, targeting both short-term and long-term objectives and developing appropriate short-term and long-term performance measures is

important. This will help people understand the dynamics and value of the relationship, and potentially increase commitment at every level.

- Making good use of information infrastructure and technology enables a smoother relationship. By itself, IT will not result in a successful collaboration; however, it is an *enabler* of collaboration (Mentzer et al. 2000). When successfully combined with strategies and processes, IT creates an environment that fosters timely reporting, interaction, and visibility.

While collaboration promises great value, most companies lack the vision on how it would change their business processes and impact key performance metrics such as inventory turns, sales, and margins. Currently available academic research on collaborative supply chain management is still at its infancy, and does not provide the most needed foundational insights on when and how collaboration would benefit the participants. Hence, there is a rapidly increasing need for a better understanding on how to transform businesses into collaborative partners in supply chains, and for professionals who can work with companies as they sooner or later go through such transformation.

Acknowledgments This research is partially funded by NSF Grants DMI-0400345, DMI-0427446, DMI-0400301, and ITR 0427446. We would like to acknowledge the research assistance of Özlem Bilginer, Department of Management Science and Engineering, Stanford University.

References

Achabal DD, McIntyre SH, Smith SA, Kalyanam K (2000) A decision support system for Vendor Managed Inventory. J Retailing 76(4):430–455

Agarwal R, Ergun O (2005) Collaborative logistics in the sea cargo industry. OR/MS Tomorrow. http://ormstomorrow.informs.org/archive/summerfall05/sea cargo.pdf

Ahmadjian CL, Lincoln JR (2000) Keiretsu, governance, and learning: case studies in change from the Japanese automotive industry. Working paper Institute of Industrial Relations. University of California, Berkeley. http://repositories.cdlib.org/cgi/viewcontent.cgi?article=1013&context=iir

Altintas N, Erhun F, Tayur S (2008) Quantity discounts under demand uncertainty. Manage Sci 54(4):777–792

Anand KS, Aron R (2003) Group buying on the web: a comparison of price discovery mechanisms. Manage Sci 49(11):1547–1564

Andraski JC, Haedicke J (2003) CPFR: time for the breakthrough? Supply Chain Manage Rev 7(3):54–60

ATA (2005) Trucking activity report. American Trucking Associations, www.truckline.com

Aviv Y (2007) The effect of collaborative forecasting on supply chain performance. Manage Sci 47(10):1326–1344

Aviv Y (2002) Gaining benefits from joint forecasting and replenishment processes: the case of auto-correlated demand. Manuf Serv Oper Manage 4(1):55–75

Aviv Y (2007) On the benefits of collaborative forecasting partnerships between retailers and manufacturers. Manage Sci 53(5):777–794

Axsater S (2001) A note on stock replenishment and shipment scheduling for Vendor-Managed Inventory systems. Manage Sci 47(9):1306–1311

Balasubramanian S, Bhardwaj P (2004) When not all conflict is bad: manufacturing marketing conflict and strategic incentive design. Manage Sci 50(4):489–450

Bower JL, Hout TM (1988) Fast-cycle capability for competitive power. Har Bus Rev 66:110–118

Brown WB, Karagozoglu N (1993) Leading the way to faster new product development. Acad Manage Exec 7(1):36–47

Browning TR, Deyst JJ, Eppinger SD, Whitney DE (1996) Adding value in product development by creating information and reducing risk. IEEE Trans Eng Manage 49(4):443–458

Cachon GP (2003) Supply chain coordination with contracts. In de Kok AG, Graves S (eds) Supply chain management-handbook in OR/MS. Chapter 6, Pages: 229–340, North-Holland, Amsterdam, The Netherlands

Carbone J (2001) Strategic purchasing cuts costs 25% at Siemens. Purchasing (Magazine Online) 130(18):29

Carlson R, Rafinejad D (2006) Business and environmental sustainability at Toyota Motor Corporation: development of Prius hybrid vehicle. Stanford MS&E Case, Stanford

Caruso D (2002) Collaboration moves up the org chart. AMR Research Report (May 31)

Çetinkaya S, Lee C-Y (2000) Stock replenishment and shipment scheduling for vendor-managed inventory systems. Manage Sci 46(2):217–232

Chatterjee S, Slotnick A, Sobel MJ (2002) Delivery guarantees and the interdependence of marketing operations. Prod Oper Manage 11(3):393–409

Chen Z-L, Hall NG (2005) Supply chain scheduling: conflict and cooperation in assembly systems. Oper Res 55(6):1072–1089

Cheung KL, Lee H (2002) The inventory benefit of shipment coordination and stock rebalancing in a supply chain. Manage Sci 48(2):300–207

Clark KB, Fujimoto T (1991) Product development performance. Harvard Business School Press, Boston, MA

Cohen S, Roussel J (2005) Strategic supply chain management: the five disciplines for top performance. McGraw-Hill, Boston, MA, pp 139–167

Cook M, Tyndall R (2001) Lessons from the leaders. Supply Chain Manage Rev 5(6):22–32

Cooke JA (2006) Logistics costs under pressure. Logist Manage (July 1). http://www.logisticsmgmt.com/article/CA6352889.html

Dave DS, Fitzpatrick KE, Baker JR (1996) An advertising-inclusive production lot size model under continuous discount pricing. Comput Ind Eng 30(1):147–159

Dawande M, Geismar HN, Hall NG, Sriskandarajah C (2006) Supply chain scheduling: distribution systems. Prod. Oper. Manage. 15(2):243–261

de Groote X (1994) Flexibility and marketing/manufacturing coordination. Int J Prod Econ 36(2):153–167

Disney SM., Potter AT, Gardner BM (2003) The impact of vendor managed inventory on transport operations. Transp Res: Part E 39(5)

Disney SM, Towill DR (2003a) The effect of vendor managed inventory (VMI) dynamics on the bullwhip effect in supply chains. Int J Prod Econ 85(2):199–216

Disney SM, Towill DR (2003b) Vendor-managed inventory and bullwhip reduction in a two-level supply chain. Int J Oper Prod Manage 23(6):625–652

Dolan R (1987) Quantity discounts: managerial issues and research opportunities. Mark Sci 6(1):1–22

Engardio P (2007) Beyond the green corporation. BusinessWeek (January 29)

Erat S (2006) Joint product development and inter-firm innovation. Unpublished Ph.D. Dissertation. College of Management, Georgia Institute of Technology, Atlanta, GA

Ergun O, Kuyzu G, Savelsbergh M (2006) The time-constrained lane covering problem. Transportation Science 41, 206–221, 2007

Erhun F, Gonçalves P, Hopman J (2007) Managing new product transitions. MIT Sloan Manage Rev 48(3):73–80

Erhun F, Hopman J, Lee HL, Murphy-Hoye M, Rajwat P (2005) Intel Corporation: product transitions and demand generation. Stanford Graduate School of Business Case No: GS43

Erhun F, Keskinocak P (2003) Game theory in business applications. Working paper. Department of Management Science and Engineering. Stanford University, Stanford and School of Industrial and Systems Engineering, Georgia Institute of Technology, Atlanta

Erhun F, Tayur S (2003) Enterprise-wide optimization of total landed cost at a grocery retailer. Oper Res 51(3):343–353

Erkoc M, Wu D (2002) Due date coordination in an internal market via risk sharing. Working paper. Department of Industrial and Systems Engineering, Lehigh University, Bethlehem, PA

Foster L (2001) Reaping the benefits while sharing costs. Financial Times (December 12)

Fry M, Kapuscinski R, Olsen T (2001) Coordinating production and delivery under a (z,Z)-type vendor-managed inventory contract. Manuf Serv Oper Manage 3(2):151–174

Ghemawat P, Nueno JL (2003) ZARA: fast fashion. Harvard Business School Case 9-703-497

Griffin P, Keskinocak P, Savasaneril S (2005) The role of market intermediaries for buyer collaboration in supply chains. In: J. Geunes, E. Akcali, P.M. Pardalos, H.E. Romeijn, and Z.J. Shen (editors), Kluwer Academic Publishers, Chapter 3, 87–118

Hall NG (2005) Supply chain scheduling. Presentation. The Ohio State, Bethlehem, PA

Hall NG, Potts CN (2003) Supply chain scheduling: batching and delivery. Oper Res 51:566–584

Hannon D (2003) Demand forecasting rises on logisticians' radar screens. Purchasing (Magazine Online) 132(12):55

Hart B (2005) 10 tips for reducing supply chain logistics costs. Logist Today (August 9). http://www.logisticstoday.com/sNO/7355/LT/displayStory.asp

Hausman WH (2001) Vendor-managed inventory. Supply Chain Online. http://www.supplychainonline.com

Hendrick TE (1997) Purchasing consortiums: horizontal alliances among firms buying common goods and services – what? who? why? how? Center for Advanced Purchasing Studies (CAPS) Survey Report, Tempe, AZ

Heskett JL, Signorelli S (1984) Benetton. Harvard Business School, Boston, Case 9–685–014

Johnson ME (2002) Product design collaboration: capturing lost supply chain value in the apparel industry. Ascet 4(May 16)

Johnson ME (2004) Harnessing the power of partnerships. FT.com (October 7)

Kaipia R, Holmström J, Tanskanen K (2002) VMI: what are you losing if you let your customer place orders? Prod Plann Control 13(1):17–26

Keskinocak P, Savasaneril S (2008) Collaborative procurement among competing buyers. Naval Research Logistics 55(6):516–540

Krishnan V (1996) Managing the simultaneous execution of coupled phases in concurrent product development. IEEE Trans Eng Manage 43(2):210–217

Krishnan V, Bhattacharya S (2002) Technology selection and commitment in new product development. Manage Sci 48(3):313–328

Krishnan V, Eppinger SD, Whitney DE (1997) A model-based framework to overlap product development activities. Manage Sci 43(4):437–451

Kurtulus M, Toktay LB (2004) Investing in forecast collaboration. Working paper 2004/48/tm. INSEAD

Ladany S, Sternlieb A (1974) The interaction of economic ordering quantities and marketing policies. AIIE Transa 6:35–40

Lal R, Staelin R (1984) An approach for developing an optimal discount pricing policy. Manage Sci 30(12):1524–1539

Lam M, Wong DS (1996) A fuzzy mathematical model for the joint economic lot size problem with multiple price breaks. Eur J Oper Res 95(3):611–622

Langley CJ Jr (2001) Internet logistics market analysis: clarification and definition for an emerging market. Nistevo white paper

Lapide L (1998) Are we moving from buyers and sellers to collaborators? AMR Res Rep Supply Chain Manage (July 1)

Lee HL, Padmanabhan V, Whang S (1997) Information distortion in a supply chain: the bullwhip effect. Manage Sci 43(4):546–558

Loch CH, Terwiesch C (1998) Communication and uncertainty in concurrent engineering. Manage Sci 44(8):1032–1048
Loch CH, Terwiesch C (2005) Rush and be wrong or wait and be late? A model of information in collaborative processes. Prod Oper Manage 14(3):331–343
Lynch K (2006) Collaborative logistics networks – breaking traditional performance barriers for shippers and carriers. Nistevo white paper. http://www.nistevo.com/v1/pdfs/lynch.pdf
Magretta J (1998) The power of virtual integration: an interview with Dell Computer's Michael Dell. Harv Bus Rev 76(2):73–84
Malone R (2003) Harnessing the power of collaboration. Inbound Logistics (July). http://www.inboundlogistics.com/articles/supplychain/sct0703.shtml
Malykhina E (2005) PLM is a multibillion-dollar business. InformationWeek (February 15)
Matchette JB, Seikel MA (2005) Inquiries and insights on supply chain collaboration. Ascet 7(September 13)
Mathewson F, Winter RA (1996) Buyer groups. Int J Ind Organ 15:137–164
Melymuka K (2001) Efficient? Become superefficient. Computerworld (September 10)
Mentzer J, Foggin J, Golicic S (2000) Collaboration: the enablers, impediments, and benefits. Supply Chain Manage Rev 4(4):52–58
Mishra BK and Raghunathan S (2004) Retailer- vs. vendor-managed inventory and brand competition. Manage Sci 50(4):445–457
Miyaoka J (2003) Implementing collaborative forecasting through buyback contracts. Working paper. Department of Management Science and Engineering, Stanford University, Stanford, CA
Mullen J (2006) Fact or fiction: the reality of consignment inventory. APICS Mag 16(1)
Murphy JV (2002) Want to know how to save on transportation? Ask your carrier. Supplychainbrain.com: http://www.supplychainbrain.com/archives/02.02.win.htm?adcode=90
Ozener OO, Ergun O (2008) Allocating costs in a collaborative transportation procurement network. Transportation Science 42(2):146–165
Pekgun P, Griffin P, Keskinocak P (2008) Coordination of marketing and production for price and leadtime decisions. IIE Transactions 40(1):12–30
Peleg B (2003) STMicroelectronics E-chain optimization project: achieving streamlined operations through collaborative forecasting and inventory management. Stanford Global Supply Chain Management Forum Case SGSCMF-001-2003
Peleg B, Lee HL (2006) Impacts of standardization on business-to-business collaboration. In: Shaw MJ (ed) E-commerce and the digital economy. M.E. Sharpe, Inc. Advances in Management Information Systems edited by V. Zwass
Plambeck EL, Taylor TA (2004) Sell the plant? The impact of contract manufacturing on innovation, capacity, and profitability. Manage Sci 51:133–150
Radjou N, Orlov LM, Child M (2001) Apps for dynamic collaboration. Forrester research report. (May)
Reilly C (2002) Central sourcing strategy saves Dial $100 million. Purchasing (Magazine Online) (January 15)
Richardson B (2003) PLM evolution: from product to lifecycle. Presentation in The Stanford Global Supply Chain Management Forum Conference on Product Life Cycle Management in Increasingly Dynamic and Complex Supply Chains, Stanford, CA (April 3)
Roberts B (2001) Banking on B2B. Line56 Magazine (April). http://www.line56.com/articles/default.asp?ArticleID=2440
Sabath RE, Fontanella J (2002) The unfulfilled promise of supply chain collaboration. Supply Chain Manage Rev (July/August) 6(4):24–29
Schoonmaker M (2004) Rosettanet: driving E-business processes on a global scale. Presentation. Department of Management Science and Engineering, Stanford University, Stanford, CA
Shah JB, Serant C (2002) Microsoft's Xbox sets supply chain standard. ESM (March 11). http://www.my-esm.com/showArticle?articleID=2915125
Sheffi Y (2002) The value of CPFR. RIRL Conference Proceedings (October 13–16). Lisbon, Portugual

Simchi-Levi D, Kaminsky P, Simchi-Levi E (2000) Designing and Managing the Supply Chain. Irwin McGraw-Hill, New York

Spengler J (1950) Vertical integrations and anti-trust policy. J Polit Econ 58:347–352

Spiegel Y (1993) Horizontal subcontracting. RAND J Econ 24(4):570–590

Stepanek M (2003) Land O'Lakes exec talks warehousing. CIO Insight (May)

Strozniak P (2003) Collaborative logistics: overcoming its challenges can lower transportation and inventory costs and reduce stockouts. Frontline Solutions. http://www.frontlinetoday.com

Sullivan L (2005) Designed to cut time. InformationWeek (February 28)

Swanson A (2003) Ford marks 100th anniversary. http://www.hawaiireporter.com/story.aspx?431cf5aa-eb6c-4433-a358–297b8889781e

Swink M (2006) Building collaborative innovation capability. Res Technol Manage 49(2):37–47

Taylor DA (2004) The problem with programs. Logist Today (March). http://www.logisticstoday.com/sNO/6361/iID/20876/LT/displayStory.asp

Teck H, Zheng Y (2004) Setting customer expectation in service delivery: an integrated marketing-operations perspective. Manage Sci 50(4):479–488

Terwiesch C, Loch CH (2004) Collaborative prototyping and the pricing of custom designed products. Manage Sci 50(2):145–158

Tsay AA, Nahmias S, Agrawal N (1998) Modeling supply chain contracts: a review. In: Tayur, S, Ganeshan, R Magazin M (eds) Quantitative models of supply chain management. Kluwer Academic Publishers, Norwell, MA, Chapter 10 Pages: 229–336

Turban E., Leidner D, McLean E, Wetherbe J (2005) Information technology for management: transforming organizations in the digital economy, 5th edn. Wiley, New York

Wheelwright SC,Gill GK (1990) Campbell Soup Co. HBS Case 9–690–051

Zirger J., Hartley JL (1996) The effect of acceleration techniques on product development time. IEEE Trans Eng Manage 46(2):143–152

Chapter 12
Sequencing Strategies and Coordination Issues in Outsourcing and Subcontracting Operations

Tolga Aydinliyim and George L. Vairaktarakis

12.1 Introduction

Managing the supply chain has recently been the most significant task for manufacturing companies toward cost efficiency and customer satisfaction. Globalization not only increased the size of supply chains, but also created an environment where competing suppliers, manufacturers, and distributors need to cooperate. Even when firms own a significant portion of their supply chain, multiple parties with different performance measures and goals are involved in the decision-making processes. Examples from the automotive industry include General Motors, Ford Motor Company and others. Each division in such organizations tries to reach optimal plans for the portion of the overall system under its authority. However, the extent to which the individual goals are achieved depends on the decisions made by other parties involved in supply chains. Therefore, the quality of the strategic decisions made by such decision makers depends on the amount of information they have regarding other members of the supply chain. The more information is shared among the decision makers, the greater the resulting coordination benefits.

Realizing the importance of coordination in supply chains, researchers have worked on problems involving the decisions made by multiple parties within the supply chain and the effects of coordination among members of the supply chain. Numerous quantitative models addressing these issues have emerged in recent years, which include excellent reviews by Lariviere (1998), Cachon (2003), Chen (2003), and Swaminathan and Tayur (2003). The research methods applied by operations management researchers addressing supply chain coordination issues include mathematical modeling techniques and tools from economics such as game theory. Game theory is most useful when multiple tiers of the supply chain are involved in the decision-making process, and the outcome of their decisions depends on the decisions made by other chain members. In this case, the solution to the capacity

T. Aydinliyim (✉)
Decision Sciences Department, Charles H. Lundquist College of Business,
1208 University of Oregon, Eugene, OR 97403-1208, USA
e-mail: tolga@uoregon.edu

K.G. Kempf et al. (eds.), *Planning Production and Inventories in the Extended Enterprise*, International Series in Operations Research & Management Science 151,
DOI 10.1007/978-1-4419-6485-4_12,

planning problems requires not only determining the strategies of the players, but also setting incentives for coordination among supply chain members. For a successful review of game theoretical analysis in supply chain issues, see Cachon and Netessine (2004).

Concepts of coordination have received significant attention since the 1950s, originally in the context of economics in industrial organization and subsequently in operations management. Since the 1980s, these concepts are responsible for a large body of literature on supply chain management where the primary focus is on *coordination of inventories*. However, research at the shop floor level is conspicuously scarce even though supply chain (and inventory) coordination necessitates the coordination of production activities. Besides, the broad use of the internet and recent advances in information technologies have created new coordination opportunities at the production planning level by means of centralized production scheduling and capacity management. Therefore, a significant research opportunity exists, which should address supply chain coordination issues at an *operational* level (Thomas and Griffin 1996).

Most of the models presented in this review focus on coordination issues across multiple tiers of supply chains, in outsourcing and/or subcontracting contexts. Unlike large manufacturers of the past century, the most powerful firms of today are vertically disintegrated, sometimes acting externally as mere coordinators in the production process of their partners. As a result, subcontracting and outsourcing have recently become prominent business practices across many industries. Subcontracting of industrial production is generally based on the short-term need for additional production capacity. When the available production capacity of a manufacturing firm is not sufficient to cope with the total volume of production necessary to execute an order on time, or when further creation of in-house capacity is neither feasible nor desirable, the main contractor has to depend on a subcontractor (third-party) to produce the balance of an order. In such situations, firms make outsourcing decisions after the demand is realized and known with certainty and hence further research that addresses coordination issues in deterministic outsourcing settings is needed.

In an attempt to address the aforementioned issues, this chapter focuses on recent coordination research that focuses on *deterministic and operational* issues rather than the *stochastic and strategic* ones. Specifically, we present models where multiple agents share a common third resource for part of their operations. Conflicting interests among the parties result in a serious capacity allocation problem. Hence, three important questions arise:

1. Can significant benefits be achieved as a result of centralized decision making? In other words, are coordination benefits worth the effort to achieve centralization?
2. Do all parties involved improve their individual performances under centralized control? If not, how should the savings due to coordination be allocated so that individual agents accept the centralized solution?
3. Can centralization be achieved without centralized control, i.e., does there exist an instrument, e.g., contract, mechanism, priority rule, etc., invoking strategic decision makers to act the way they would under centralized control?

Various models presented in this chapter are described in outsourcing and/or subcontracting contexts, where a number of manufacturers process part or all of their operations using the resources of the same third-party. This mode of operation has become prominent since the 1970s when focused factories created the opportunity for excellence by capitalizing on a small number of operations done with high quality and quick response and outsourcing all others that are not core competencies of the firm. Also, as subcontracting to third-parties is becoming common across many industries, many powerful third-parties that excel in *contract manufacturing* are emerging.

According to Day (1956), subcontracting refers to "the procurement of an item or service that a firm is normally capable of economic production in its own facilities." Outsourcing, on the other hand, is a special case where the firm has no means to produce on its own. Firms strategically outsource their non-core operations and focus on their core competencies to enhance their effectiveness in the long-term (Greaver 1999). In contrast, firms use subcontracting as a short-term solution to increase their flexibility, reduce their exposure to risk, improve their response to unexpected increases in demand, and reduce costs. However, it is not easy to make decisions on when to outsource/subcontract and by how much, because the contract manufacturers (referred to as third-parties throughout this chapter) have scarce resources and their limited capacity is to be shared with all of their customers demanding timely completion of their outsourced/subcontracted operations. In an effort to deal with this problem, to leverage cooperation opportunities, and to provide better service to their customers, many companies make broad use of information technologies – in particular the internet – by creating secure information sharing portals and online capacity booking systems. In what follows, we demonstrate relevant examples from various industries.

12.1.1 *Applications from Industry*

Today, many industry sectors have created portals which are used to share key information among networks of suppliers and manufacturing partners. Such networks may consist of the divisions of a parent company, the smaller companies of a large corporation, the customers of a contract manufacturer, or different tiers of the extended supply chain. These information sharing portals help network members to cooperate so as to reduce the costs of the entire chain. Using these secure online platforms, firms achieve end-to-end visibility across the supply chain and are able to provide their customers with sensitive capacity information. In what follows, we discuss three such applications.

12.1.1.1 Cisco's *eHub*

Cisco Systems is among the first to create an information sharing portal known as *eHub*. It is a private trading e-marketplace providing a central point for planning and execution of tasks across the company's extended manufacturing supply chain

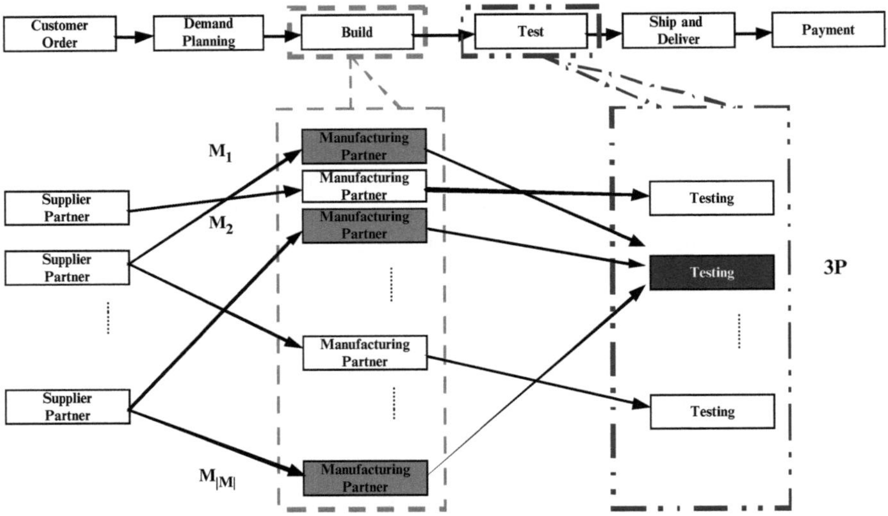

Fig. 12.1 Order fulfillment process of the Cisco's networked supply chain

(Grosvenor and Austin 2001). It also creates opportunities to perform coordinated planning and production among the members of Cisco's extended supply chain and has led to dramatic cost and inventory reductions. By 2001, successful implementation of eHub had resulted in inventory reductions of 45%, order cycle time reductions of 70%, and subsequently, productivity across the entire supply chain increased.

In Fig. 12.1, we explain the flow of orders in Cisco's network as they relate to the outsourcing models presented in this section. Customer orders are first stored in Cisco's enterprise resource planning (ERP) database and then sent to the related manufacturing partners over the virtual private network (VPN). Within Cisco's supply chain there are supply partners and other contractors who can see information on the network because their own production systems are also connected to Cisco's ERP system. In this framework, one would expect situations where multiple manufacturing partners use the testing services of a common third-party in the same supply chain. This creates a time-sensitive capacity allocation problem depending on the outsourcing decisions made by all manufacturing partners as well as the availability at the testing facility. In Sect. 12.3, we consider such models and discuss opportunities to coordinate production schedules at the third-party.

12.1.1.2 MyUMC by UMC

MyUMC,[1] UMC's total online supply chain portal, is another example that demonstrates how a large electronics contract manufacturer (ECM) provides information

[1] http://my.umc.com

sharing and real-time capacity booking to its customers (i.e., the manufacturers who outsource). UMC is a world-leading semiconductor foundry, specializing in the contract manufacturing of customer designed integrated circuits for high performance semiconductor applications. Founded in 1980 as Taiwan's first semiconductor company, UMC currently employs over 12,000 people worldwide at its manufacturing factories and offices in Taiwan, Japan, Singapore, Europe, and the USA. To facilitate close collaboration with its customers as well as partners throughout the entire supply chain, UMC developed MyUMC in 1998. This is a full-service information portal, offering customers 24-h our access to detailed account information such as manufacturing, engineering, design, and financial data. In particular, MyUMC's capacity booking engine ATP (Available-to-Promise) allows customers to receive instant capacity availability information and to book production capacity in UMC's fabs online. Evidently coordinated capacity and production planning opportunities exist within this framework.

12.1.1.3 SPADE by HKUST

The operational protocol of the Semiconductor Product Analysis and Design Enhancement (SPADE) Center[2] of the Hong Kong University of Science and Technology provides another concrete example of online capacity booking systems that are used in the semiconductor industry. SPADE provides various services to local and nearby semiconductor companies such as analysis and optimization of designs and products when their silicon prototypes are available in the form of silicon wafers or silicon dies. SPADE has a group of specialized facilities, including Focused Ion Beam, Emission Microscope, ESD Tester, Backside Preparation System (Chip UnZip), Laser Cutting System, etc., which can be booked at different prices. Rules of the charging scheme are available to customers,[3] and include:

(a) *Jobs performed on Sundays and public holidays will be charged at 2 × basic rate.*
(b) *Jobs performed during nonoffice hours will be charged at 1.5 × basic rate.*

The varying rates correspond to the arbitrary booking prices in some of the outsourcing models discussed in Sect. 12.3. SPADE has also built a mechanism to allow a user to preempt another one, provided that a higher cost is paid according to the following scheme:

(c) *Jobs preempting other jobs on queue will be charged at 1.5 × basic rate.*

This rule reflects a reservation adjustment which can be made by the customers of the third-party (SPADE in this example). As we will see in the sequel, instead of price-based preemption, one can devise allocation schemes based on cooperative

---
[2] http://www.ust.hk/spade

[3] http://www.ust.hk/spade/pricelist.html

game theory to achieve coordination among the parties involved and introduce a notion of fairness (see the discussion on core of cooperative games in Sect. 12.3 and the dynamic cost allocation model in Sect. 12.4).

12.1.1.4 Subcontracting and Partnership Exchanges

Coordination among multiple manufacturers subcontracting to the same third-party is more easily achieved when the third-party such as UMC has the internet-based technologies to provide capacity and production schedule information to his customers, or when the manufacturers are members of the extended supply chain of the same parent company, such as Cisco (Grosvenor and Austin 2001). However, such opportunities are not always available. Having recognized this fact and the increasing trend of subcontracting and outsourcing, the United Nations Industrial Development Organization (UNIDO) has formed Subcontracting and Partnership Exchanges (SPX).[4] SPX's are technical information, promotion, and match making centers for industrial subcontracting and partnership among main contractors, suppliers, and subcontractors, aiming at the optimal utilization (the most complete, rational, and productive) of the manufacturing capacities of the affiliated industries. Today more than 44 SPX's in over 30 countries facilitate production linkages among small, medium, and large manufacturing firms and connect such firms with global markets and supply chain networks.

SPX's members frequently subcontract to the same third-party if the third-party is a proven leader because of its expertise, size, or technological capabilities such as UMC in silicon-on-chip (SoC) designs. In such situations, the manufacturers served by the third-party are sometimes of equal importance to the third-party, and the third-party acts as a rational profit maximizer who does not prioritize over customers. In case the third-party has limited capacity, the customers compete for a share of this capacity (see Sect. 12.2 for a review of literature in capacity allocation in supply chains). Even when the demand for third-party capacity is less than the available capacity, there is competition for earlier production capacity at the third-party resulting in a time-sensitive capacity allocation problems which we consider in Sect. 12.5.

12.1.1.5 Boeing 787 Dreamliner

Another example of many primary contractors subcontracting to the same third-party can be found within the diverse supply chain of Boeing used in the production of their newest airplane, 787 Dreamliner. Boeing authorized a team of parts suppliers to design and build major sections of its 787 Dreamliner[5] (see Fig. 12.2). The plan calls for suppliers to ship mostly completed fuselage sections, already stuffed

[4] *http://www.unido.org/doc/4547*

[5] Source: "Boeing Scrambles to Repair Problems With New Plane," *Wall Street Journal,* December. 7, 2007.

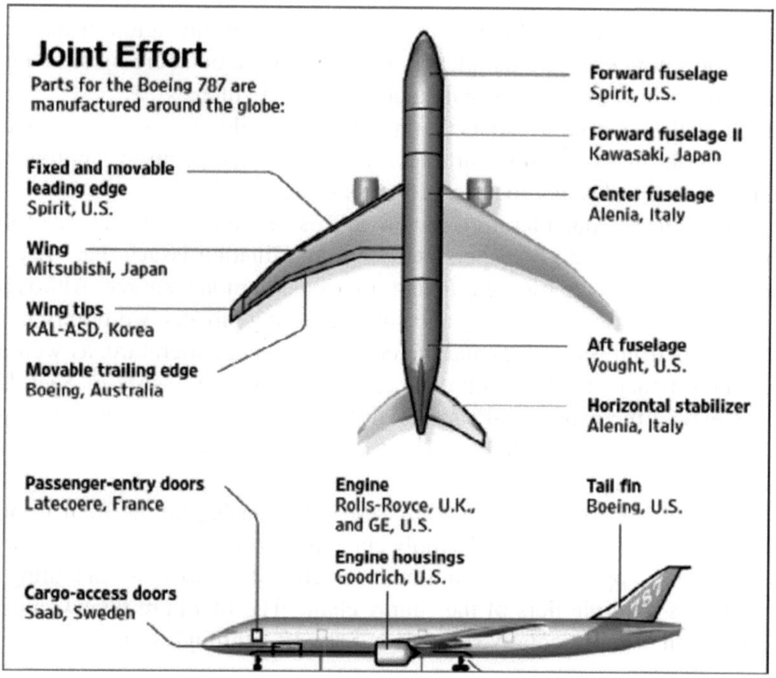

Fig. 12.2 Manufacturing Sites for Boeing 787 Parts

with wiring and other systems, to Boeing facilities around Seattle so they could be put together in as few as 3 days. This is a tall order given that the existing production methods can keep a plane, the size of the Dreamliner, in the final-assembly area for a month. Hence, the timely delivery from suppliers is extremely important. However, many of these suppliers, instead of using their own engineers to do the design work, farmed out these key tasks to even-smaller companies. Some of those subcontractors ended up overloading themselves with work from multiple 787 suppliers, and Boeing had to announce delays twice during the second half of 2007. Therefore, a careful assessment of the time-sensitive subcontracting activities and the coordination opportunities in this framework is crucial for all parties involved. The competitive models in Sect. 12.5 highlight the decentralization costs resulting from such subcontracting activities; and analyze this setting from (a) the common third-party's, (b) the primary contractors', and (c) Boeing's (the overall performance of all of Boeing's primary contractors) point of view.

12.2 Literature Review

In this section, we review research streams related to the models discussed in the remainder of this chapter. We start our review by discussing cooperative sequencing games. Specifically, we review literature on games in outsourcing contexts and observe that the essence of this research is to find a fair allocation of the centralized solution savings among the players involved in the outsourcing situation. We conclude surveying cooperative game theory models in queues.

Subsequently, we review papers that highlight the value of outsourcing and subcontracting, starting with those articles that follow a scheduling approach, and continuing with articles that focus on coordination issues.

More specifically, our literature review in this section is organized as follows. Section 12.2.3 is devoted to papers studying coordination issues that arise when multiple players compete for the efficient use of a common resource. Although our focus is essentially on outsourcing and subcontracting models, we briefly mention similar models from economics and the computer science literature, as well as the supply chain management literature on capacity allocation issues in Sect. 12.2.4. Then, we extend the capacity allocation discussions to the time-sensitive capacity allocation problems such as the ones that are observed by the SPX users and the contractors of Boeing.

In Sect. 12.2.5, we survey scheduling research focusing on mechanism or contract design to achieve centralized supply chain performance. The last category that we survey is the supply chain scheduling research that focuses on scheduling coordination across multiple tiers of the supply chain. The literature review concludes with a selection of representative models in Sect. 12.3 through 12.6.

12.2.1 Cooperative Games

12.2.1.1 Sequencing Games

At the interface of cooperative game theory and scheduling, Curiel et al. (1989) was the first to introduce *sequencing games*. They considered the simplest case; i.e., a single machine scheduling situation with no restrictions on the jobs with weighted flow-time cost criterion. They showed that the corresponding sequencing game is convex. Thus, the core of the game is guaranteed to be nonempty (Shapley 1971). For arbitrary regular cost criterion and for a special class of games (referred to as σ_0 *Component Additive Games*), Curiel et al. (1994) proposed a core allocation known as the *β-rule*.

Hamers et al. (1995) extended the class of single machine sequencing situations considered by Curiel et al. (1989) by imposing ready times on the jobs. In this case, the corresponding sequencing game is called an *r-sequencing game,* and is not convex except for the special subclass with unit processing times or with unit weights. Instead of ready times, Borm et al. (2002) considered the case with due dates, i.e. *d-sequencing games*. For three different due date-related cost criteria, they showed that the corresponding games have a nonempty core. However, convexity was proven for only a special subclass. Hamers et al. (2002) imposed precedence relations on the machines to come up with precedence sequencing situations and proved that the corresponding game is convex if the precedence relations form a network of parallel chains.

Calleja et al. (2004) studied another extension of one-machine sequencing games where each player might have more than one job to be processed and each job might be of interest to more than one player. The authors showed that a core allocation is guaranteed to exist only if the underlying cost functions are additive with respect to the initial order of jobs. However, convexity does not necessarily hold.

Hamers et al. (1999) considered games with m parallel identical machines and no restrictions on the jobs. These games are referred to as *m-machine sequencing games*. The cost criterion is weighted flow time. The authors prove the existence of a core allocation for $m = 2$ and for some special cases when $m > 2$. Another extension on multiple machine situations is by Calleja et al. (2002). In their model, every player has precisely two operations, each processed on a specific machine. The cost criterion is the maximal weighted completion time of the two operations. It is shown that the game is not convex but has a nonempty core.

Curiel et al. (2002) present an excellent survey of sequencing games and consider core allocation and convexity issues up to 2002. Most of the research after this date was focused on relaxing the constraints imposed by Curiel et al. (1989). van Velzen and Hamers (2003) studied one such extension where they introduced the concept of *weak-relaxed sequencing games*. In their formulation, a weak-relaxed sequencing game arises when a specific player can switch with any player in a coalition provided that the players outside this coalition do not suffer from this switch, a move that was not allowed in earlier sequencing games. The authors showed that a nonempty core exists even though the cooperative game is not convex. In this direction, Slikker (2006) made the most significant contribution by relaxing the common assumption of restricting cooperation to players that are connected according to the initial order and defined the *relaxed sequencing games*. He proved that relaxed sequencing games have stable profit divisions, i.e., a nonempty core exists, and proposed using the concept of *permutation games* (Tijs et al. 1984) to find a core allocation.

12.2.1.2 Sequencing Games Approach to Outsourcing

Literature on sequencing games is of value for its theoretical contributions, but application of such concepts to practical real-life applications had not been the primary focus. However, in most outsourcing situations, the initial capacity reservations at the third-party can be revisited, and hence a resequencing of the outsourced workloads leading to a savings game is possible (e.g., see the booking rules and the pricing scheme at SPADE in Sect. 12.1.1.3). The overall benefits to all supply chain members are apparent, but individual member performances could improve or worsen. Therefore, a reallocation of the supply chain savings is essential and the cooperative sequencing game concept is most applicable.

Applying cooperative sequencing game concepts to a practical outsourcing context motivated by the coordination of manufacturing operations at Cisco's supply chain, Aydinliyim and Vairaktarakis (2010) studied a setting where a group of manufacturers outsource operations to the same third-party whose limited capacity is available in terms of manufacturing windows. Cai and Vairaktarakis (2007) considered a similar model where tardiness penalties are considered, and optional overtime capacity is available for windows booked by the manufacturers. In all these studies, allocating savings or costs is essential in achieving coordination by using the concept of the *core* which is formally defined by Gillies (1959). Further details of these two models are presented in Sect. 12.3.

12.2.1.3 Cooperation in Queues

Queueing problems are closely related to the sequencing problems. Hence, we briefly review some of the work in this area that adopts the cooperative game approach. Maniquet (2003) studied a problem of sequencing agents who require service at a common resource. This problem is mathematically equivalent to minimizing the total completion time in nonpreemptive single machine scheduling. The author considered two cooperative game formulations where the value of a coalition is (1) the total flow time cost incurred if the members of the coalition collectively arrived first or (2) the total flow time cost incurred if the members of the coalition collectively arrived last. A cost allocation rule which is shown to be the Shapley value (Shapley 1953) is proposed as well as an axiomatic characterization of the Shapley value. Motivated by Maniquet's (2003) results, Katta and Sethuraman (2006) studied the same cooperative game setting, but instead of an axiomatic approach, they focused on the notion of *fairness* by proposing two solution concepts, namely random priority (RP) core and constrained random priority (CRP) core.

Schulz and Uhan (2007) studied a special class of cooperative games with supermodular costs, which also included a queueing game with total weighted completion time as a special case. The notion of supermodularity implies that, as a coalition grows, the cost of adding a particular agent increases, hence it is not possible to motivate the formation of a grand coalition. In such situations, it is necessary to penalize smaller coalitions for defecting from the grand coalition. The minimum such penalty which ensures the existence of a core solution is called *least core value*. The authors developed a fully polynomial time approximation scheme for computing the least core value of queueing games with supermodular costs.

We believe that the results in the aforementioned articles, can be used to develop solutions for many practical settings that have been outlined in Sect. 12.1. Similar adoptions of cooperative game theory can be found in the supply chain management literature (e.g., capacity pooling, group buying, and centralized warehousing). For example, see Yu et al. (2007) for a cooperative game theory model in a capacity pooling setting, where the authors provided an core allocation rule which distributes the costs of pooled production among the parties involved. For an extensive review of game theoretic analyses of cooperation among supply chain agents, the reader may refer to an excellent survey by Nagarajan and Sošič (2008). We expect similar adoptions in future work that focuses on issues at the interface of outsourcing/subcontracting and scheduling.

12.2.2 *Value of Outsourcing and Subcontracting*

For the most part, literature on outsourcing in supply chains focuses on inventory management issues. Cachon and Lariviere (2001) provide an overview of various contracts that allow supply chain members to share demand forecasts credibly under a compliance regime. Cachon and Harker (2002) presented a model of competition between two firms that face scale economies and outsource to a supplier by means of a queueing game and an economic order quantity game.

More closely related to the models in this chapter are articles in which it is attempted to coordinate the production and subcontracting decisions to assess the value of outsourcing and subcontracting. In Van Mieghem (1999), the option of subcontracting to improve financial performance is considered. The author analyzes a sequential stochastic investment game. The manufacturer and the subcontractor first decide independently on their capacity investment levels and then the manufacturer has the option to subcontract part of his production to the subcontractor. Kamien and Li (1990) presented a multiperiod competitive model with capacity constraints on the aggregate production level.

Similarly, Atamturk and Hochbaum (2001) provided a multiperiod treatment of subcontracting that focuses on the production aspect. They consider the trade-off among acquiring capacity, subcontracting, production, and holding inventory, by providing analytical models, structural results on the optimal solutions, and algorithms that simultaneously optimize these interrelated decisions. Following a Markov Decision Process approach, Tan (2004) considered a model with a subcontractor who guarantees long-term availability. Similarly, using a stochastic optimal control problem formulation, Tan and Gershwin (2004) analyzed the production and subcontracting strategies for a manufacturer with limited capacity and volatile demand where there are a number of subcontractors available.

The focus of this chapter differs from the issues emphasized in the aforementioned studies where the products are assumed to be identical, customers are assumed to be equally important, and product demand is aggregated into units. The goal is usually to minimize certain cost that is a function of the production, outsourcing, inventory, and backlog orders. In contrast, the models presented in this chapter are more closely related to the scheduling models where customers' orders are differentiated based on the characteristics of the jobs that make up such orders. Examples of such characteristics include processing times, due dates, weights that represent relative importance, etc. In what follows, we review such scheduling models that study decentralized and/or coordinated scheduling models, and attempt to quantify the value of scheduling coordination in certain outsourcing and subcontracting situations.

12.2.2.1 Scheduling Models that Deal with Outsourcing and Subcontracting

At the interface of scheduling and subcontracting, Bertrand and Sridharan (2001) considered a make-to-order manufacturing environment with randomly arriving orders, which can be processed in-house on a single machine, or subcontracted. The objective is to maximize the utilization of in-house capacity while minimizing tardiness in fulfilling orders. The authors propose heuristics the validity of which is tested by a computational experiment. Lee et al. (2002) considered a multistage scheduling model where each order requires multiple operations. Each operation can be processed on a number of alternative machines in-house, or it is subcontracted to a third-party provider. They proposed genetic algorithms to find schedules that minimize the makespan of all the orders. Similar to Lee et al. (2002), Qi (2007) studied a multistage flow-shop scheduling problem with the option of subcontracting and considered the makespan objective. Chung et al. (2005) considered the subcontracting costs in a job-shop scheduling problem, where due-dates are imposed on the orders.

Qi (2008) studied a problem where there is a single in-house machine and a single subcontractor with a single machine. Subcontracted orders need to be shipped back in batches. The objective is to minimize the weighted sum of a delivery lead time performance measure of orders plus the total subcontracting and transportation cost. He proposed dynamic programming algorithms for four problems where the time-based performance measure is total completion time, makespan, maximum lateness, and number of tardy orders, respectively. Similarly, Chen and Li (2008) considered a model where a manufacturer receives orders from many customers and has the option to process these orders in-house or to subcontract them at a different cost. Their model is more complex than Qi (2008) in the sense that there are parallel resources both in-house and at the subcontractor and they considered subcontracting costs subject to a target makespan for all the orders. The authors presented complexity results, developed a heuristic, and compared the performance of the manufacturer with and without the option of subcontracting.

12.2.2.2 Coordination Issues

The aforementioned models are at the interface of scheduling and outsourcing/subcontracting, and focus on the optimization problems that arise when there is an opportunity of outside processing in addition to the in-house resources. Additional issues that arise when we seek to coordinate production across manufacturers competing for third-party capacity, include allocation of savings when switching to a centralized solution, bargaining, and the incentives for truth-telling. Below, we survey articles that address such issues in a scheduling context. One of the first such papers that deal with detailed scheduling and outsourcing is by Biskup and Simons (1999). The authors examined a dynamic total tardiness problem - motivated by a common repair facility used by multiple computer dealers. In such a case, an incentive problem arises as the updated schedules in case of new arrivals may result in some jobs being late. Hence, in addition to providing optimal algorithms, the authors focused on cost allocation issues. Details of their analysis are presented in Sect. 12.4.

Another treatment of decentralized scheduling is by Chen and Cai (2006) who considered the problem of two manufacturers who negotiate to partition a set of jobs that they have jointly been awarded by a customer. Each manufacturer has to take into account his own processing capacity and the job requirements, while considering whether a partition is beneficial to him. In their models, the overall objective is to achieve a partition of the jobs into two subsets, which are considered fair and acceptable to both players. They applied the Nash bargaining solution (NBS) (Nash 1950) to this problem and provided algorithms that combine the calculations of NBS and optimal processing schedules. Gan et al. (2007) also applied the NBS concept to a common due-window (CDW) scheduling problem of jobs on a single machine to minimize the sum of common weighted earliness and weighted number of tardy jobs when only one manufacturer processes these jobs. They proposed a novel dynamic programming algorithm to obtain a reasonable set of processing utility distributions on the bipartition of these jobs.

12.2.3 Capacity Allocation Problems

The issue of coordination where multiple parties compete for common resources has also been studied in various fields such as economics, computer science, and supply chain management. For example, *congestion games* were introduced by Rosenthal (1973), where it is shown that congestion games admit a potential function, therefore a Nash equilibrium (Nash 1951) exists. Congestion games can be used to model coordination and pricing issues in highways and communication networks (e.g., Internet pricing). For example, Sandholm (2002) considered a problem on the efficient use of a highway network and suggested pricing schemes. Ganesh et al. (2007) developed a congestion pricing mechanism for allocating bandwidth in communication networks. On the other hand, literature on load balancing in communication networks focuses on finding bounds on the ratio between the total cost of the uncoordinated equilibrium and the cost of the centralized solution. Papadimitriou (2001) used the term *price of anarchy* to denote the worst-case ratio between a Nash outcome and the social optimum. A similar term *cost of decentralization* is commonly used in supply chain coordination literature (Cachon 2003).

12.2.3.1 Aggregate Capacity Allocation Problems in Supply Chains

Capacity allocation issues in supply chains have been studied extensively. When many retailers demand limited supplier capacity, competition leads to a serious capacity allocation problem. In such a situation, the supplier must allocate his capacity in some manner. For example, Cachon and Lariviere (1999) considered a two-tier stage supply chain with one supplier and two retailers and analyzed three allocation schemes: *proportional, linear, and uniform.* They found that with either proportional or linear allocation, a retailer receives less than his order whenever capacity binds. They presented methods to find Nash equilibria in the capacity allocation game with either proportional or linear allocation. Despite the discouraging result that a Nash equilibrium may not exist with proportional or linear allocation scheme, with uniform allocation there always exists a unique Nash equilibrium. In a related paper, Cachon and Lariviere (1999) considered a supply chain with one supplier and several downstream retailers. In their model, the retailers are privately informed about their optimal stocking levels. The most striking finding is that, depending on the capacity allocation mechanism announced by the supplier, (a) the retailers may manipulate the system by ordering more than they actually need in order to receive a favorable allocation and (b) the mechanism may be truth-inducing. However, they also showed that a coordinating mechanism, i.e., that maximizes the retailers' profit, which is also truth-inducing, does not exist.

Motivated by the problem of allocating capacity to product lines at a semiconductor manufacturer, Mallik and Harker (2004) considered a similar incentive problem. They observed that some product managers inflated their demand forecasts on purpose, in an attempt to gain greater capacity allocation. The authors developed (a) a capacity allocation and (b) an incentive payment scheme that induces truth-telling. The proposed allocation rule is a modified lexicographic allocation rule according to which each product line receives the minimum of the so far unallocated capacity and the product managers' optimal newsvendor quantity. In addition, a payment scheme is required to induce truth-telling.

Similar to Mallik and Harker (2004), Karabuk and Wu (2005) studied incentive issues in capacity allocation for semiconductor manufacturing. They also observed that product managers have an incentive to inflate their demand estimates to increase their capacity allocations, hence a specific bonus payment to the product managers is proposed as a way to resolve the order inflation. The authors modeled the capacity allocation stage as a noncooperative game. Their main finding is that keeping a fraction of the available capacity on-hand is both necessary and sufficient for making the appropriate bonus payments to the product managers.

The aforementioned papers successfully identified the need for an incentive payment scheme that should accompany capacity allocation decisions. They also showed that the capacity allocation decisions at the strategic level may be modified at the tactical level in an attempt to improve the overall performance of all the parties involved, and transfer payments can be used to resolve the incentive issues that may arise. There has been a lot of research done at this interface, under the broader domain of *hierarchical production planning*. The interested reader may refer to a survey by Bitran and Tirupati (1989) and the book by Miller (2002).

12.2.3.2 Time-Sensitive Capacity Allocation Problems

Among the papers that study capacity allocation issues, of particular interest to us are the ones that address the timing of the production activities as well. The papers that we reviewed in the previous subsections analyzed different allocation schemes, but none of them considered the detailed execution of the allocated amount of capacity. Specifically, suppose that the available capacity of a resource is allocated to parties A, B, and C according to a specific allocation mechanism, say 50% to party A, 30% to party B, and the remaining 20% to party C. When delivery times to the customers of A, B, and C matter, the different ways of allocating the available capacity are not equivalent to each other. Hence, a *time-sensitive capacity allocation* problem arises in such situations.

One of the first studies that consider aggregate capacity allocation mechanisms with scheduling decisions is Hall and Liu (2007). They considered a supply chain with one manufacturer who receives orders from several distributors. In case the available capacity at the manufacturer is not enough for all orders, the distributors are given the opportunity to form coalitions, i.e., a subset of distributors, by rescheduling all the orders of the coalition within the total capacity allocated to the coalition. Within this setting they considered three coordination issues by calculating (a) the benefit to the manufacturer from considering scheduling costs, (b) the additional profit that the distributors achieve when they share their allocated capacity, and (c) the value of coordination between the manufacturer and the distributors. In this study, the distributors' capacity sharing problem is modeled as a cooperative game.

Vairaktarakis (2008) considered a time-sensitive capacity allocation problem in a competitive subcontracting context. In his model, a number of manufacturers who are capable of in-house production also have the opportunity to subcontract part of their workload to common third-party. The objective for each manufacturer is to complete his entire workload, on both resources, as soon as possible. The allocation mechanism by the third-party is not static as in aggregate capacity allocation mechanisms and depends on the amount of workload subcontracted by each manufacturer. Hence, manufacturers compete for earlier positions at the third-party schedule by adjusting their subcontracted workloads. The author derived Nash equilibrium schedules for three production protocols-overlapping, preemption, and nonpreemption (see Pinedo 2002)–and four different information protocols.

Vairaktarakis and Aydinliyim (2007) considered the same subcontracting setting and focused on three additional issues. First, they recognized the conflicts between the manufcturers' and the third-party's objectives, i.e., minimizing the overall completion time and maximizing the total profit, respectively. Hence, (a) they provided algorithms to find schedules that maximize the third-party's workload and (b) estimated the loss in profits by the third-party when the manufacturers compete according to the incentive rules suggested in Vairaktarakis (2008). Second, they allowed the manufacturers to form coalitions and revise their subcontracting strategies, and showed that the corresponding cooperative game is balanced, i.e., a core allocation of savings exists. They also provided closed-form expressions that would result in a core allocation and performed a computational study to gain insights on the value of information in avoiding the costs of the decentralization.

Bukchin and Hanany (2007) also analyzed a competitive subcontracting setting and considered scheduling costs; namely, flow time costs. They formulated a noncooperative game where a job that belongs to an agent can either be processed on a common resource, which they referred to as the in-house resource, or be subcontracted at a per-unit subcontracting cost. They estimated the *decentralization cost*, i.e., the ratio between the Nash equilibrium cost and the cost attained at the centralized optimum. They derived some of the properties of Nash equilibrium schedules and subsequently used them to develop bounds on the decentralization costs. Finally, they proposed a scheduling-based coordinating mechanism that provides incentives invoking the agents to follow centralized schedules.

The models of Vairaktarakis (2008), Vairaktarakis and Aydinliyim (2007), and Bukchin and Hanany (2007) as well as their main findings are discussed in Sect. 12.5.

12.2.4 Coordination by Contracts and Other Mechanisms

12.2.4.1 Coordination by Contracts

Contractual agreements between suppliers and buyers regarding inventory management issues are heavily studied with emphasis on the cost of the (decentralized) equilibrium solution as compared to the centralized optimal; see Lariviere and Porteus (2001) for wholesale price contracts, Pasternack (1985) for buy-back contracts, Tsay (1999) for quantity-flexibility contracts, Taylor (2002) for sales-rebate contracts, Bernstein and Federgruen (2005) for price-discount contracts, Cachon and Lariviere (2005) for revenue sharing contracts, etc. Lariviere (1998) and Cachon (2003) surveyed related results. All these studies are related to inventory management decisions and represent production capacity in aggregate units, whereas our survey differs from this huge body of literature in that it addresses the *timing of production activities*. According to a 2003 survey in *Wall Street Journal*, the original equipment manufacturers rated the ability to meet delivery schedules as the most significant factor in choosing contractors, whereas price was ranked only fifth (Ansberry and Aeppel 2003). As noted in Anupindi and Bassok (1999), delivery commitments are of crucial importance. Li (1992) noted that, in semiconductor manufacturing, the delivery delays for chips provided by subcontractors cause costs to increase significantly as the production lines need to be shut down when chips are not available. Therefore, it is of crucial importance to study the contractual agreements that include delivery time-related terms and penalty clauses. Below is an example contract with such clauses.

> *...The weekly delivery deadlines agreed on in the binding production and delivery plan shall be binding. If SWS (Schoeller Wavin Systems AG) shall default in delivery, then SWS, without exclusion of other rights and claims by Ifco, shall pay the following default penalties to Ifco:*
>
> - *for default of 2 weeks (8 to 14 days), 0.05 euros per crate,*
> - *for default of 3 weeks (15 to 21 days), 0.10 euros per crate,*
> - *after the third week (more than 22 days), 0.15 euros per crate.*
>
> *Ifco (Ifco Systems GmbH) shall only be authorized to invoice these contractual penalties at the end of the year with the invoices for deliveries by SWS ...*

In Sect. 12.6, we present a model by Aydinliyim and Vairaktarakis (2010) who investigate the coordination properties of contracts between a manufacturer and a third-party contractor. The main finding of this paper is that, by setting the unit processing charge and the tardiness penalties appropriately, this two-tier supply chain can coordinate its actions and achieve the performance of a centralized system.

12.2.4.2 Coordination by Other Mechanisms

Achieving centralized supply chain performance is ideal, but is also costly. Although contracts are good instruments to manipulate strategic behavior, agreeing upon contract terms or checking to see if the players obey the contract terms may not be easy. Alternatives such as auction and other mechanisms that approximate centralized control and/or induce truth-telling by agents are of interest. In this review, we constrain ourselves to articles that focus on mechanism design in scheduling applications which is closely related to the time-sensitive capacity allocation problems on a common resource, as motivated in previous sections.

Hain and Mitra (2004) considered a model where each manufacturer outsources a single job to a third-party who is committed to process jobs in nondecreasing order of processing times (or shortest processing time – SPT order). In an effort to gain processing priority, each manufacturer has the incentive to quote smaller than actual processing time for his job, the validity of which cannot be verified by the third-party. To resolve this problem, the authors develop a money transfer mechanism based on the job durations announced by the manufacturers. The mechanism is such that every chain member is better off announcing his true processing requirement thus ensuring the third-party of the SPT order on her facility.

A recent application of decentralized scheduling in a factory environment is by Wellman et al. (2001). They investigated the existence and the quality of equilibrium prices, showed that the price equilibrium and the system optimum coincide in a class of scheduling problems, and presented auction mechanisms. Their model is closely related to the capacity booking applications in a subcontracting context, which is among the focal points of the analysis in this chapter. Details of their study are presented in Sect. 12.6. Hall and Liu (2008) considered a similar problem where a number of competing agents each with a job that requires processing for an arbitrary amount of time and has to be processed on a common resource. Agents bid for capacity according to an ascending auction. The authors provided dynamic programming algorithms to find the winning combination of bids from the owner of the common resource's perspective.

Prior to the two papers mentioned above, Kutanoglu and Wu (1999) designed a combinatorial auction mechanism for allocating capacity in a job shop environment. In a similar production planning setting, Reeves et al. (2005) studied a simple ascending auction and showed that straightforward bidding policies and their variants cannot approximate the optimal solution well. For a survey of the literature of agent-based allocation mechanisms for intelligent manufacturing systems, for traditional scheduling, and for decentralized online machine scheduling problems, the reader may refer to Shen et al. (2006), Heydenreich et al. (2007), and Heydenreich et al. (2007), respectively.

12.2.5 Supply Chain Scheduling

A growing number of papers have appeared in the area of *supply chain scheduling*. The focus of these papers is to quantify the benefits that can be attained when the scheduling issues in multiple tiers of the supply chain are handled in a centralized manner. Although the models differ in terms of the problem setting, the objectives, and the constraints considered, the approach followed in all supply chain scheduling papers is similar and requires (a) solving the problem of a single firm in the supply chain – usually when one party is more dominant than the other, (b) solving the problem by considering the overall supply chain objective, and (c) comparing the two settings by a computational experiment.

The first study of supply chain scheduling by Hall and Potts (2003) evaluates the benefits of cooperative decision making in a supply chain where a supplier makes deliveries to several manufacturers, who may in turn supply to several customers. They develop models which minimize the sum of scheduling and batch delivery costs. Agnetis et al. (2006) study models for resequencing jobs, using limited storage buffer capacity, between a supplier and several manufacturers with different ideal sequences. They describe efficient solution procedures for a variety of objectives including total interchange cost and the total interchange cost plus buffer costs. Dawande et al. (2006) study conflict and cooperation issues in a supply chain where a manufacturer makes products which are shipped to customers by a distributor. They evaluate conflict and describe efficient solution procedures.

This line of research successfully addressed the inefficiencies that arise in the supply chain due to the lack of centralization and the conflict of interest among different members of the supply chain, but does not consider incentive issues. In particular, the fact that some of the supply chain members may be individually worse-off in the centralized solution is mostly ignored. In that regard, Chen and Hall (2007) considered a similar problem in assembly systems and evaluated the cost of conflict and the benefit of cooperation. Their study is different than the previous work in this area in the sense that, they not only consider the optimization problem that arises, but also a compensation scheme that uses the NBS concept (Nash 1950).

12.2.6 This Chapter

In this chapter, we mainly focus on the papers that investigate the scheduling coordination issues in an outsourcing/subcontracting context. In particular, we want to survey the time-sensitive capacity allocation problems that arise when multiple agents seek services at a common third-party resource. Sections 12.3 through 12.6

Table 12.1 Summary

	Sect. 12.3	Sect. 12.4	Sect. 12.5	Sect. 12.6
Scheduling coordination	√	√	√	√
Outsourcing	√	√		√
Subcontracting		√	√	√
Time-sensitive capacity allocation	√	√	√	√
Allocation of costs/savings	√	√		
Cooperative game theory	√		√	
Competition (non-cooperative game theory)			√	
Coordination by incentives			√	√

are devoted to recent advances in this area that follow an interdisciplinary approach by utilizing tools such as mathematical modeling, algorithm design, game theory, and mechanism design. The chapters differ from one another by the specific criterion used. Some follow cooperative game theory approaches, while others use noncooperative game theory. Some compare centralized and decentralized solutions and find ways to allocate savings due to centralization, whereas others address incentives and contracts to induce strategic behavior that mimics centralized control. In Table 12.1, we summarize the similarities and differences among the papers discussed in the following four sections.

It is clear from the table that, every section reflect at least five of the eight issues indicated. In contrast, the papers that we surveyed in Sect. 12.2.1 through 12.2.6 reflect at most three issues.[6]

Next, we describe the organization of the rest of the chapter. In Sect. 12.3, we discuss the basics of sequencing games and present variants of a cooperative outsourcing model. Section 12.4 is devoted to a dynamic outsourcing model for which an alternative cost allocation scheme is presented. In Sect. 12.5, we consider models that focus on competition and coordination issues in a subcontracting context. The cost of decentralized planning and incentive schemes to leverage centralization are also discussed. In Sect. 12.6, instruments such as price mechanisms, auctions, and contracts are introduced, which are used to achieve centralization with minimal centralized control. We conclude the chapter in Sect. 12.7 with possible future research directions.

12.3 Cooperative Outsourcing Models

In this section, we consider models where a group M of manufacturers book third-party ($3P$) capacity for their operations, say $N = \{J_1, \ldots, J_j, \ldots, J_n\}$, with deterministic processing time requirements p_j and job specific parameters such as

[6] For example, the "Supply Chain Scheduling Research" focuses on scheduling coordination and is sometimes about outsourcing, but except for Hall and Chen (2007) who consider cost allocation issues, the other six criteria are ignored.

weights w_j (the relative importance of the job), and/or due dates d_j (a job incurs a penalty if it is completed later than its due date). These jobs may correspond to certain industrial finishing processes, testing of mechanical parts, quality control operations, and some assembly operations.

Consider a supplier in Cisco's supply chain network that provides subassemblies to manufacturing partners. We consider a group of manufacturing partners who sole source from a single supplier. This supplier plays the role of the third-party $3P$ in our model. Similarly, one could think of a contractor in the same network as a firm which performs the testing operations for the assemblies or components produced by the manufacturing partners. This situation is equivalent to a model where manufacturing partners M outsource their testing operations to a single contractor in Cisco's network. Recall that the production schedules of all parties involved are transparent to everyone because they are all connected to Cisco's information sharing portal. Therefore, coordinated capacity and production planning opportunities exist within this framework. We consider such coordination possibilities and provide incentive schemes to make everyone benefit from coordination.

First, we describe the sequence of events. The $3P$ announces its capacity availability as well as the booking cost for each day of production, referred to as a *manufacturing window* of L hours (a measure of capacity or the length of a shift). The finite set of manufacturing windows is $\Gamma = \{W_1, \ldots, W_T\}$, where T is the number of windows in the next planning period. The booking price h_k of each window W_k for $1 \leq k \leq T$ reflects peak demand periods or timeliness with respect to the beginning of the planning horizon. Knowing the availability and the costs of the manufacturing windows, each manufacturer books $3P$ capacity for its jobs in N_m, $m \in M$, independently in a first-come-first-served (FCFS) basis, say in order $1, 2, 3, \ldots, |M|$, to construct his initial schedule ${\sigma_0}^m$ with the objective of jointly minimizing his individual costs expressed as the total internal costs incurred according to a performance measure, i.e., $\sum_{J_j \in N_m} c_j({\sigma_0}^m)$, plus booking costs which are paid to the $3P$, i.e., $\sum_{W_k \in \mathcal{W}_{{\sigma_0}^m}} h_k$. Therefore, for each manufacturer, there is a trade-off between booking expensive early windows and cheap late windows. Let ${\sigma_0}^m$ be the optimal schedule and $\mathcal{W}_{{\sigma_0}^m}$ the collection of windows for manufacturer $m \in M$. Note that the FCFS ordering of the manufacturers implies that:

$$\mathcal{W}_{{\sigma_0}^m} \subseteq \{W_1, \ldots, W_T\} \setminus (\mathcal{W}_{{\sigma_0}^1} \cup \mathcal{W}_{{\sigma_0}^2} \cup \mathcal{W}_{{\sigma_0}^3} \cup \cdots \cup \mathcal{W}_{{\sigma_0}^{m-1}}). \quad (12.1)$$

Following the determination of $\mathcal{W}_{{\sigma_0}^m}$ for each $m \in M$, the $3P$ reschedules all jobs in N as if they belong to one party so as to obtain the best schedule σ^* and collection $\mathcal{W}_{\sigma^*}$ of windows. Schedule σ^* minimizes the total cost over all manufacturers and hence is at least as well off as the schedule σ_0 obtained by concatenating ${\sigma_0}^1, {\sigma_0}^2, {\sigma_0}^3, \ldots, {\sigma_0}^{|M|}$. Manufacturer $m \in M$ would agree to go along with σ^* only if his cost minus a side payment is not greater than that of ${\sigma_0}^m$. Therefore, $3P$ must find an allocation of the savings (an incentive scheme) produced by σ^* such that every manufacturer is at least as well off. Normally, σ^* is expected to utilize fewer windows than σ_0 because of its better utilization of idle times across all windows initially booked by manufacturers. On the other hand, the $3P$ also creates a mechanism that generates benefits for himself via *booking refunds* and *rebookings*. Specifically, some of the manufacturing windows are emptied due to the coordinated schedule. The $3P$ refunds a fraction ρ $(0 \leq \rho \leq 1)$ of the booking costs of these windows and keeps the rest for himself in addition to generating additional booking revenues by reselling released windows.

12.3.1 Variants of the Outsourcing Model

In what follows, we describe possible variants of the aforementioned general outsourcing model where the internal cost measure $c_j({\sigma_0}^m)$ (the completion time-related cost of job J_j of manufacturer m in his initially booked schedule ${\sigma_0}^m$) for each manufacturer depends on operational factors such as the cost criterion, the shipment protocol, set-up times between different jobs, and production capacity at the third-party:

- Cost Criterion
  - *Total weighted flow time*: a widely used measure to evaluate the WIP (work-in-process) costs of a production schedule.
  - *Total (weighted) tardiness*: a penalty is applied for each job that is completed after its due date. In case of weighted tardiness, a job-specific penalty is applied proportional to the amount of time by which a job is past its due date.
  - *Total delivery time*: captures the order-to-delivery time and is a measure of customer service.
  - *Total (weighted) earliness and (weighted) tardiness*: an earliness or a tardiness penalty is applied for each job that is not completed exactly at its due date. Tardiness captures late delivery penalties, whereas earliness captures holding costs for finished goods due to earlier than expected completions. It is an important performance measure related to the just-in-time production principles.
- Shipment protocols
  - *Batch shipment*: all jobs in a batch are shipped as a group at the end of the manufacturing window in which they are finished.
  - *Immediate shipment*: Every job is shipped individually upon completion.
- Set-up times
  - *Negligible set-up times*: set-up times between consecutive jobs from different manufacturers are negligible or can be done during off-hours without affecting the schedule. Examples include painting, finishing, testing, and certain assembly operations.
  - *Significant set-up times*: set-up times between jobs that belong to different manufacturers are large and hence they have to be processed in different manufacturing windows.

- Production capacity
  - *Regular production*: the regular production costs is included in the booking cost and does not account for overtime.
  - *Hourly overtime production*: overtime is available up to a daily limit.
  - *All-or-nothing overtime production*: Overtime is booked at the daily limit (or not used at all) at a fixed cost charged regardless of the amount of overtime actually utilized.

In what follows we will demonstrate a number of models with some of these aforementioned operational characteristics. Our analysis will include complexity results and optimal or heuristic algorithms. Also, we will present structural properties of the game among the manufacturers as well as incentive payment schemes according to which the 3P should allocate the savings to leverage coordination.

12.3.2 Cooperative Sequencing Games

Sequencing games have widely been used to study coordination issues in outsourcing. Hence, we first discuss the basic principles related to the sequencing games. Seminal papers in sequencing games include Curiel et al. (1989) and Curiel et al. (1994). Given a set of jobs $N = \{J_1, \ldots, J_j, \ldots, J_n\}$, an initial permutation of the jobs σ_0 and the initial total cost $f(\sigma_0)$, the first problem is to find the best schedule σ^* with the minimum total cost of $f(\sigma^*)$. Let coalition $S \subseteq N$ be a set of jobs and $v(S)$ be the maximum amount of savings that can be achieved by coalition S when $J_j \in S$ are rearranged. The second part of a sequencing game requires finding an allocation vector $\mathbf{x} = \{x_1, x_2, x_3, \ldots, x_n\}$ of the total savings $f(\sigma_0) - f(\sigma^*)$ such that the following core (in)equalities are satisfied:

$$\sum_{J_j \in N} x_j = v(N), \quad \sum_{J_j \in S} x_j \geq v(S) \ \forall\, S \subseteq N, \quad x_j \geq 0 \ \forall\, J_j \in N. \quad (12.2)$$

The second set of inequalities imply that the savings allocated to each player are nonnegative, whereas this set of constraints are needed to ensure that every coalition S is allocated to savings that are at least as much as they can save by themselves.

The first constraint guarantees that all savings generated by coordination are allocated to the players.

The set of allocations where savings are distributed to players in such a way that no subset of players can be better off by seceding from the rest of the players or by acting solely on their own behalf is called the *core* of a game. Each cooperative game does not necessarily have a core. Mathematically, it is possible that no such vector $\mathbf{x} = \{x_1, \ldots, x_{|M|}\}$ satisfies the (in)equalities in (12.2). Therefore, one might consider looking for a structural property that guarantees the existence of the core of a cooperative game. It is known that *convex* games have nonempty core (Shapley 1971).

Definition 12.1. Cooperative game (M, v) is said to be convex when

$$\forall\, S, T \subseteq M, \quad v\{S \cup T\} + v\{S \cap T\} \geq v\{S\} + v\{T\}.$$

However, nonconvexity does not necessarily imply that the core is empty, see for example the *permutation games* introduced by Tijs et al. 1984. A necessary but not sufficient property for the existence of the core is *superadditivity*.

Definition 12.2. Cooperative game (M, v) is said to be superadditive when

$$\forall\, S, T \subseteq M \text{ with } S \cap T = \emptyset, v\{S \cup T\} \geq v\{S\} + v\{T\}.$$

Intuitively this means that two disjoint coalitions can do no worse by forming a larger coalition which is just the union of the two.

Coordinating the operations of various manufacturers provides significant benefits to the manufacturing chain. This coordination can be modeled as a cooperative game among manufacturers who are willing to trade changes in their schedule for a larger portion of the resulting savings. In what follows, we describe how we apply the cooperative sequencing games concept to the outsourcing models we described eariler. The optimal schedule for all manufacturers, say σ^* (or $\sigma^*(M)$), can only be utilized if all manufacturers agree to schedule their jobs as suggested by $3P$. For this to happen, $v(M)$ must be allocated so that all manufacturers are better off following σ^* rather than using their previously reserved windows or forming smaller coalitions $S \subset M$.

Unlike in (12.2) where each job belongs to a different player, in the outsourcing models presented in this section, each manufacturer m owns a subset N_m of jobs, i.e., $N = \{J_1, \ldots, J_n\} = \cup_{m \in M} N_m$. Therefore, a coalition $S \subseteq M$ is specified as a subset of manufacturers in M. Next, we introduce further notation and make some assumptions:

- $\mathcal{W}_{\sigma_0(S)} \subseteq \Gamma$; the set of windows utilized in the initial schedule by the members of S. Mathematically, $\mathcal{W}_{\sigma_0}(S) = \cup_{m \in S} \mathcal{W}_{\sigma_0^m}$.
- $\mathcal{W}_{\sigma^*(S)} \subseteq \mathcal{W}_{\sigma_0(S)}$; the optimal set of windows utilized by all manufacturers in S. Coalition S cannot utilize the windows that do not initially belong to its members in schedule $\sigma_0(S)$.
- $c_j(\sigma_0(S))$ and $c_j(\sigma^*(S))$; the completion time-related costs associated with job $J_j \in N$ in $\sigma_0(S)$ (the initial schedule for manufacturers in S) and $\sigma^*(S)$ (the optimal schedule for manufacturers in S), respectively.
- A cooperative game over the set of manufacturers (M, v) where $v : 2^{|M|} \rightarrow \mathbf{R}$, and

$$v(S) = \sum_{m \in S} \sum_{j \in N_m} \left[c_j(\sigma_0(S)) - c_j(\sigma^*(S))\right] + \rho \cdot \left[\sum_{W_k \in \mathcal{W}_{\sigma_0(S)}} h_k - \sum_{W_k \in \mathcal{W}_{\sigma^*(S)}} h_k\right], \quad \forall\, S \subseteq M \tag{12.3}$$

where $0 \leq \rho \leq 1$ is the refund percentage.

The value function $v(S)$ for a coalition $S \subseteq M$ is the difference between the initial total production cost of coalition S due to $\sigma_0(S)$ and the optimal total production costs of coalition S due to $\sigma^*(S)$ plus a fraction of the difference between the initial and optimal booking costs for coalition S. Intuitively, $v(S)$ represents the maximum amount that the members of coalition S can collectively save by rescheduling the jobs in $N_S = \cup_{m \in S} N_m$ over manufacturing windows $\mathcal{W}_{\sigma_0}(S)$.

12.3.3 *The Weighted Flow Time Criterion*

The analysis of this section is drawn from Aydinliyim (2007) and Aydinliyim and Vairaktarakis (2010). The internal costs of each manufacturer are expressed as a measure of the WIP costs. Manufacturing windows have arbitrary booking prices, i.e., h_k for each window W_k, each representing a regular production capacity of L hours. Note that, for each manufacturer, there is a trade-off between booking expensive early windows and cheap late windows. By using the aforementioned composite objective function, we consider both *outsourcing costs* incurred by each manufacturer, i.e., booking costs actually paid to the third-party, and *internal costs* that relate to the timeliness of the delivery of the outsourced workload back to each manufacturer, i.e., the monetary equivalence of holding WIP inventory of nonoutsourced workload represented as a linear function of the delivery time. Booking costs make practical sense as it is the current practice for using outside production capacity (e.g., SPADE, Cisco, and UMC). For the motivation of the WIP costs, consider an assembler sourcing a critical component from a third-party. Until the third-party delivers the critical component to the assembler, the assembler incurs the WIP costs associated with the production of other components which are not outsourced, and this cost is a function of the delivery time – the completion (shipment) time – of the outsourced jobs at the third–party facility. Throughout this section, we consider models to schedule jobs that are of similar nature and hence set-up times during changeovers from one job to another (even they belong to different manufacturers) are assumed insignificant. Furthermore, preemptive resume schedules are considered to allow it for a job to start in one manufacturing window and resume in a future window without additional processing. Aydinliyim (2007) considers two distinct shipment protocols for completed jobs, namely *batch shipments* and *immediate shipments*.

12.3.3.1 The Batch Shipment Protocol

In this case, if $J_j \in N$ is completed during manufacturing window W_k, it incurs a cost $w_j D_k$ where D_k is the completion time of W_k. Minimizing the total weighted flow time subject to batch shipments is strongly NP-complete. The proof uses a reduction from 3-Partition (Aydinliyim 2007). Near-optimal schedules can be attained by first producing a single machine sequence π of jobs in N and then finding the optimal collection of windows to process the jobs according to π, resulting in σ^*. Suppose ${F_j}^{\pi}$ is the completion time of the job J_j in π. Define $t_i = i \cdot L$ as the total processing capacity of the first i windows booked by the manufacturers and let B_i be the set of jobs that finish at time ${F_j}^{\pi}$, $t_{i-1} < {F_j}^{\pi} \leq t_i$. We want to partition π in $\omega = \lceil (\sum_{J_j \in N} p_j)/L \rceil$ parts, where ω is the minimum number of windows needed to process the jobs in N, and find the total weight $w(i)$ of the completed jobs in each part B_i. Define

$$w(i) = \sum_{\substack{J_j \in N \\ t_{i-1} \leq {F_j}^{\pi} \leq t_i}} w_j. \tag{12.4}$$

Then, given π we obtain batches $B_1, B_2, B_3, \ldots, B_\omega$. The WIP cost associated with processing B_i in window W_k is $\sum_{J_j \in B_i} w_j \cdot D_k = w(i) \cdot D_k$. Let ${y_j}^i$ be an indicator variable ${y_j}^i$ that shows whether J_j belongs to B_i; and ${z_i}^k$ is 1 if jobs in B_i utilize W_k; 0 otherwise. Therefore, the objective function for the problem of minimizing total WIP plus booking costs subject to batch shipments can be expressed as:

$$\min_{\sigma} \sum_{W_k \in \Gamma} \sum_{i=1}^{\omega} \sum_{J_j \in N} w_j \cdot D_k \cdot {y_j}^i \cdot {z_i}^k + \sum_{W_k \in \Gamma} h_k \cdot \sum_{i=1}^{\omega} {z_i}^k \tag{12.5}$$

Aydinliyim (2007) proposed three heuristics to produce sequence π:

- **Heuristic 1:** *WSPT rule*; simply process jobs in decreasing w_j/p_j order.
- **Heuristic 2:** ω-partition; solve a series of $\omega - 1$ knapsack problems by taking processing times p_j and weights w_j for $J_j \in N = \{J_1, \ldots, J_n\}$ as input and producing the single machine sequence π, resulting in batches $B_1, B_2, B_3, \ldots, B_\omega$.

ω-Partition:

(0) Let $k = \omega - 1$, $B_1 = B_2 = B_3 = \ldots = B_\omega = \emptyset$ and $NU = N$
(1) Solve knapsack problem: $M(k) = \max \sum_{J_j \in NU} w_j \cdot x_j$ s.t. $\sum_{J_j \in NU} p_j \cdot x_j \leq k \cdot L$, $x_j \in \{0, 1\}$ where $x_j = 1$ if J_j is in the knapsack and 0 otherwise.
(2) Set $B_{k+1} = \{J_j : x_j = 0\}$, $NU = NU - B_{k+1}$
If $k = 0$ then STOP, else set $k = k - 1$ and go to (1).

- **Heuristic 3:** ω-Ratio-Partition; identical to *ω-Partition* except that the objective function of the knapsack problem in line (1) now becomes $\max \sum_{J_j \in NU} (w_j/p_j) \cdot x_j$.

Let *WSPT*, *Knap*, and *RKnap* denote the heuristic solutions produced by *DP* (described next) when the initial batching is produced by the WSPT rule, *ω-Partition*, and *ω-Ratio-Partition*, respectively. The application of *WSPT* takes only $O(n \log n)$ time, whereas *ω-Partition* and *ω-Ratio-Partition* both take $O(nT \sum_{J_j \in N} p_j)$ time.

Given $B_1, B_2, B_3, \ldots, B_\omega$, the following dynamic programming formulation, referred to as *DP*, assigns batches to manufacturing windows. Let $f_i(k)$ be the minimum total booking plus WIP costs for $B_1, \ldots, B_i$ when B_i is processed in window W_k, and $B_1, \ldots, B_{i-1}$ are processed prior to W_k.
Recursive relation:

$$f_i(k) = \min_{i-1 \le r < k} \{f_{i-1}(r) + h_k + w(i) \cdot D_k\} \quad \text{for} \quad i \le k \le T - \omega + i. \quad (12.6)$$

Boundary condition: $f_0(k) = 0$ for $k = 0$ and ∞ otherwise.
Optimal value: $f_{\text{opt}} = \min_{\omega \le k \le T}\{f_\omega(k)\}$.
The worst-case asymptotic complexity of *DP* is $O(T^3)$.

To test the quality of the heuristics, Aydinliyim (2007) proposes two different lower bound schemes. The first one is a variation of *ω-Partition* called *Max-ω-Partition*. This scheme is called *MaxKnap* and is identical to *Knap* except that $w(i)$ in (12.6) is replaced by $M(i) - M(i-1)$ which is obtained by solving the knapsack problems in line (1) of *ω-Partition* and NU is set to N for all $\omega - 1$ knapsack iterations. Namely, if $\hat{f}_i(k)$ denote the state variables associated with *Max-ω-Partition*, then

$$\hat{f}_i(k) = \min_{i-1 \le r < k} \{\hat{f}_{i-1}(r) + h_k + [M(i) - M(i-1)] \cdot D_k\} \quad \text{for} \quad i \le k \le T - \omega + i \quad (12.7)$$

where $M(0) = 0$ and $M(\omega) = \sum_{J_j \in N} w_j$.

An alternative lower bound can be attained by assuming that each delivery includes not only the batch completed within a window, but also the completed portion of a preempted job. Let p_{jk} be the portion of p_j completed in W_k. Then to each job $J_j \in N$ we assign weight $f_{jk} = (p_{jk}/p_j w_j)$. Likewise, we assign p_{jk} fraction (p_{jk}/Lh_k) of the booking cost of W_k. Equivalently we define c_{jk} = $(w_j/p_j) \cdot D_k + (h_k/L)$ as the cost of processing one unit of J_j in W_k. For any collection of windows, the WSPT rule with respect to weights f_{jk} provides an optimal batching to process the jobs in N. Aydinliyim (2007) presented results of an extended computational experiment which evaluates these heuristics against the two lower bounds. Statistics for 400 instances over 40 parameter combinations show that the best heuristic result differ from the tightest lower bound by 0.6% on average. Hence, the heuristics perform well, and the lower bounds are tight.

12.3.3.2 The Immediate Shipment Protocol

In this case, if $J_j \in N$ is completed during window W_k at time C_j due to single machine sequence π, it incurs a production cost $w_j \cdot C_j$, where $C_j = D_k - L + F_j{}^\pi - t_{i-1}$. Adapting the notation from the previous case, the objective function becomes:

$$\min_\sigma \sum_{W_k \in \Gamma} \sum_{i=1}^{\omega} \sum_{J_j \in N} w_j \cdot [D_k - L + F_j{}^\pi - t_{i-1}] \cdot y_j{}^i \cdot z_i{}^k + \sum_{W_k \in \Gamma} h_k \cdot \sum_{i=1}^{\omega} z_i{}^k. \quad (12.8)$$

Minimizing the total weighted flow time subject to immediate shipments is also strongly NP-complete, which can be proved by a reduction from 3-Partition (Aydinliyim and Vairaktarakis 2010). Three heuristics have been proposed for the problem of minimizing weighted flow time plus booking costs subject to immediate shipments. All three heuristics consist of two steps: finding a batching of the jobs and then finding an optimal collection of windows to allocate the batches. They all start by arranging the jobs in WSPT order. Without the loss of generality, suppose $J_1, J_2, J_3, \ldots J_n$ is the WSPT order. Finding sequence π can be done by three procedures adopted from those mentioned in the previous section: WSPT rule, *ω-Partition*, and *ω-Ratio-Partition*.

Subsequently, a modified version of the dynamic program DP, say DP', is applied where (12.6) is replaced by

$$\tilde{f_i}(k) = \min_{i-1 \leq r < k} \{\tilde{f}_{i-1}(r) + h_k + \sum_{J_j \in N} w_j \cdot [D_k - L + F_j{}^{\pi} - t_{i-1}] \cdot y_j{}^{i}\}$$

$$\text{for} \quad i \leq k \leq T - \omega + i.. \tag{12.9}$$

The resulting heuristics are called *WSPT′*, *Knap′*, and *RKnap′*, corresponding to the batching procedures used. Aydinliyim and Vairaktarakis (2007) noted that, following the statistics obtained by running the heuristics for 400 instances over 40 parameter combinations, the best heuristic result differ from the tightest lower bound by 3.1% on average. Hence, the heuristics for this shipment protocol also perform well, and the lower bounds are tight.

12.3.3.3 Coordination Results

This subsection discusses the coordination results related to the models described in Sect. 12.3.3.1. Before we begin to deal with the specifics, we decompose the value function (total savings) $v(S)$, $S \subseteq M$ into the booking savings $v_h(S)$ and the completion time-related internal cost measure savings $v_c(S)$ (referred to as WIP savings). Then,

$$v_c(S) = \sum_{m \in S} \sum_{J_j \in N_m} [c_j(\sigma_0(S)) - c_j(\sigma^*(S))], \quad v_h(S) = \sum_{k \in \mathcal{W}_{\sigma_0}(S)} h_k - \sum_{k \in \mathcal{W}_{\sigma^*}(S)} h_k. \tag{12.10}$$

Therefore,

$$v(S) = v_c(S) + \rho \cdot v_h(S). \tag{12.11}$$

Note that the windows initially owned by a coalition S of manufacturers $\mathcal{W}_{\sigma_0}(S)$ constitute blocks of windows. Let $[a,b]$ denote the set of windows $\{W_a, W_{a+1}, W_{a+2}, \ldots, W_b\}$. One can express $\mathcal{W}_{\sigma_0}(S)$ as $[a_1, b_1] \cup [a_2, b_3] \cup \cdots \cup [a_r, b_r]$ where

$1 \leq a_1 < b_1 < a_2 < \cdots < a_r < b_r \leq T$. Then, $[a_1, b_1], [a_2, b_2]$, etc. are called *maximally connected* components of the coalition $\mathcal{W}_{\sigma_0}(S)$ of windows. WIP savings produced by the windows in $[a, b]$ is defined similar to $v_c(S)$. Let $N_{[a,b]}$ be the set of jobs that complete in windows $W_k \in [a, b]$. Then we can express $\omega_c([a, b])$ as follows:

$$\omega_c([a,b]) = \sum_{J_j \in N_{[a,b]}} [c_j(\sigma_0([a,b])) - c_j(\sigma^*([a,b]))]$$

$$v_c(S) = \sum_{k=1}^{r} \omega_c([a_k, b_k]) \tag{12.12}$$

Aydinliyim (2007) considers the scenario where $3P$ does not refund booking savings to any coalition other than the grand coalition M, i.e., refund percentage $\rho = 0$ unless $S = M$. The $3P$ poses such stipulation to leverage his refund policy so as to affect coordination among all manufacturers and propose to manufacturers the following allocation of the coordination savings. For $0 \leq \lambda \leq 1$, let $x_m{}^{\text{wip}}$ and $x_m{}^{\text{book}}$ be the WIP savings allocation and the booking savings allocation offered by $3P$ to manufacturer m, respectively, and define

$$\begin{aligned} x_m{}^{\text{wip}} = \sum_{W_k \in \mathcal{W}_{\sigma_0 m}} & [\lambda(\omega_c([1,k]) - \omega_c([1,k-1])) \\ & + (1-\lambda)(\omega_c([k,T]) - \omega_c([k+1,T]))] \end{aligned} \tag{12.13}$$

and

$$x_m = x_m{}^{\text{wip}} + \rho \frac{1}{|M|}(v(M) - v_c(M)). \tag{12.14}$$

The following coordination results hold true for the outsourcing model with weighted completion time cost criterion subject to batch shipments:

(1) *Games* (M, v_c), (M, v_h), *and* (M, v) *are superadditive but not convex.*
(2) *Expressions* (12.13) *and* (12.14) *provide a core allocations for the games* (M, v_c) *and* (M, v), *respectively.*

Proofs, similar structural results, and an alternative allocation rule for immediate shipments can be found in Aydinliyim and Vairaktarakis (2010).

12.3.3.4 Value of Coordination: A Numerical Example

The transfer payments in our model are explained in detail via the following example. Consider the following instance with three manufacturers each having six jobs (see Table 12.2). There are 20 manufacturing windows each of length 8 time units having 16 time units between them (W_1 from 0 to 8, W_2 from 24 to 32, and so on). Booking prices take on two values, i.e., $h_k \in \{2{,}000, 2{,}600\}$, $1 \leq k \leq 20$ where W_1, W_3, W_4, W_7, W_{13}, W_{15} and W_{19} are expensive windows representing peak demand periods. The manufacturers book windows on a FCFS basis and optimize their

Table 12.2 Parameters for the illustrative example

M	1						2						3					
Jobs	J_1	J_2	J_3	J_4	J_5	J_6	J_7	J_8	J_9	J_{10}	J_{11}	J_{12}	J_{13}	J_{14}	J_{15}	J_{16}	J_{17}	J_{18}
w_j	9	20	12	12	14	4	19	1	16	9	16	5	9	3	12	7	7	5
p_j	6	6	3	4	6	6	6	4	6	2	5	2	6	2	5	4	2	6

schedules to minimize the combined WIP costs plus booking costs. The following initial schedules and costs are obtained:

- $\mathcal{W}_{\sigma_0{}^1} = \{W_1, W_2, W_3, W_5\}$; ${\sigma_0}^1$: $J_3 \to J_4 \to J_2 \to J_5 \to J_1 \to J_6$, WIP cost = 2,968, booking cost = 9,200, Total cost = 12,168.
- $\mathcal{W}_{\sigma_0{}^2} = \{W_4, W_6, W_8, W_9\}$; ${\sigma_0}^2$: $J_{10} \to J_{11} \to J_7 \to J_{12} \to J_9 \to J_8$, WIP cost = 8,088, booking cost = 8,600, Total cost = 16,688.
- $\mathcal{W}_{\sigma_0{}^3} = \{W_7, W_{10}, W_{11}, W_{12}\}$; ${\sigma_0}^3$: $J_{17} \to J_{18} \to J_{15} \to J_{16} \to J_{13} \to J_{14}$, WIP cost = 10,816, booking cost = 8,600, total cost = 19,416.

Therefore, the initial overall schedule σ_0 is the concatenation of ${\sigma_0}^1$, ${\sigma_0}^2$, and ${\sigma_0}^3$ with WIP cost = 21,872, booking cost = 26,400, and total cost = 48,272. On the other hand, if all manufacturers accept to cooperate and follow the coordinated 3P schedule σ^*, then W_{12} will be released decreasing the booking cost to 24,400 which is a reduction of 7.58%.

Also, better sequencing of the jobs and better utilization of the windows will reduce WIP costs to 17,648 which is an improvement of 19.31%. At the individual player level, M_1's WIP cost increases by 3,240 (to 6,208 from 2,968), whereas the WIP costs of M_2 and M_3 decreases by 3,792 (to 4,296 from 8,088) and 3,672 (to 7,144 from 10,816), respectively. Hence, 3P collects 7,464 (=3,792 + 3,672) from players M_2, and M_3, and uses 3,240 to compensate for M_1's losses. The net savings of 4,224 (=3,792 + 3,672 – 3,240) is allocated to the players according to the core allocation proposed. Specifically, M_1 gets 2,653, M_2 gets 1,020, and M_3 gets 546. The rewards are even greater when a fraction of the booking savings are allocated to manufacturers (2,000/3 more to each manufacturer). After coordination, the overall chain costs are reduced by 6,224 (a 12.89% improvement) as the total cost decreases to 42,048. As a result, M_1 is awarded a bonus of 152% (= 5,877/3,840) for accepting a WIP loss of 3,840; M_2 receives superior performance plus a bonus of 3552 (=3840 – 288); and M_3 receives superior performance at 13% discount (= 2,250/15,509).

12.3.4 The Number of Tardy Jobs Criterion

In this section, a variant of the outsourcing model is presented where the internal costs of the manufacturers are tardiness penalties. If a job $J_j \in N$ is completed after its due date d_j, then a penalty β is incurred. Let ${C_j}^{\sigma}$ denote the completion time of job J_j in schedule σ; ${y_j}^k$ be the binary variable that takes a value of 1, if

J_j is completed in $W_k = [a_k, b_k]$, and 0 otherwise; $U(z)$ be a binary variable which is 1 if $z > 0$ and 0 otherwise. Then the objective function for the problem of jointly minimizing tardiness penalties plus booking costs is:

$$\min_{\sigma} \left\{ \sum_{k=1}^{|\Gamma|} \left[h_k \cdot U\left(\sum_{J_j \in N} y_j^k \right) \right] + \beta \cdot \left[\sum_{J_j \in N} U(C_j{}^{\sigma} - d_j) \right] \right\}. \quad (12.15)$$

A special case of the problem with two booking prices is studied by Cai and Vairaktarakis (2007). For every window W_k, $k = 1, \ldots, |\Gamma|$ the booking cost h_k may only take on the values $\{h, h_{\mathrm{P}}\}$ where h is the booking cost on a regular demand day while h_{P} is the cost on a peak demand day, and $h_{\mathrm{P}} > h$. For every job $J_j \in N_m$, $m \in M$, its due-date d_j coincides with the end of a manufacturing window, i.e., $d_j = b_k$ for some $k = 1, \ldots, |\Gamma|$. This is consistent with practice where jobs are delivered in batches at the end of the day. The following are some optimal properties of the optimal schedule σ^* and the optimal collection of ω windows utilized by σ^*.

(1) There exists a critical window W_{c} with booking cost $h_{\mathrm{c}} = h_{\mathrm{P}}$ such that windows $W_1, \ldots, W_{\mathrm{c}-1}$ are also booked.
(2) All windows in $\mathcal{W}_{\sigma^*}$ following W_{c} have booking cost h.
(3) All windows in $\mathcal{W}_{\sigma^*}$ except possibly the last are fully utilized.
(4) Nontardy jobs are scheduled in earliest-due-date (EDD) order and precede all tardy jobs.

Properties (1)–(3) can be proved trivially and Property (4) is true because on any collection of ω windows, Moore-Hodgson algorithm (MH) (see Moore 1968) determines the optimal sequence of processing the jobs. The optimal properties suggest that an optimal collection of windows can be found by comparing at most ω schedules. One can start by utilizing the first ω windows, and then iteratively replace the latest expensive utilized window with the earliest empty cheap window. At each iteration, an optimal sequence can be found by MH in $O(n \log n)$ time. By comparing the total costs of the schedules at each iteration, the optimal schedule can be found. Cai and Vairaktarakis (2007) suggest the following optimal algorithm *Regular* to solve the problem of minimizing tardiness penalties plus booking costs.

Algorithm Regular

(0) Book the earliest ω windows with booking price h. If the number of h-windows is less than ω, fill in with early h_p-windows. Sequence the jobs according to MH, let Z^* be the total tardiness plus booking costs.
(1) Replace the latest utilized h-window with the earliest free h_p-window and sequence the jobs using MH. Calculate the total cost. If it is less than Z^*, update Z^* and repeat (1); otherwise, **STOP**.

12.3.4.1 Optional Overtime Capacity

In this extension, the availability of overtime provides the opportunity to finish more jobs earlier thus reducing the number of tardy jobs. Each unit of overtime costs α dollars which is assumed to be more expensive than the regular hourly cost in a peak window, i.e., $\alpha > h_P/L$. This assumption is consistent with practice where overtime is significantly more expensive than regular production. It is assumed that, the amount of overtime O_k booked in window W_k is not more than a fixed upper limit of O hours, i.e., $0 \leq O_k \leq O$. Furthermore, batch shipment for overtime is assumed such that the departure time for orders completed during W_k is $b_k + O_k$. All other operational characteristics remain the same as before. In addition to Properties (1)–(4), the following property holds true for an optimal schedule σ^*:

(5) In an optimal schedule, there exists a critical window $W_f \in \mathcal{W}^*$ which utilizes overtime, and all preceding windows in $\mathcal{W}^*$ have $O_k = O$.

Finding an optimal schedule for the model with optional overtime capacity, one needs to make use of *Regular*. Let $\mathcal{W}_{\sigma^*(\mathrm{OT})}$ be the optimal collection of windows for the model where overtime is available and $\sigma^*(\mathrm{OT})$ be the corresponding schedule. Assume that $\mathcal{W}_{\sigma^*(\mathrm{OT})}$ is known. The following two schedules help obtaining the optimal schedule $\sigma^*(\mathrm{OT})$.

σ_R: Optimal schedule when no overtime is used and MH-Algorithm is applied assuming that the last window in $\mathcal{W}_{\sigma^*(\mathrm{OT})}$ has unlimited capacity.

σ_{OT}: Optimal schedule when the overtime of each window in $\mathcal{W}_{\sigma^*(\mathrm{OT})}$ is fully utilized before allocating work to the following window.

Let T_R and T_{OT} be the set of tardy jobs for σ_R and σ_{OT}, respectively. The optimal set of tardy jobs $T_{\sigma^*(\mathrm{OT})}$ satisfy $T_{\mathrm{OT}} \subseteq T_{\sigma^*(\mathrm{OT})} \subseteq T_R$. Moreover, if $J_j \in T_R \backslash T_{\mathrm{OT}}$ and J_j is nontardy in the optimal schedule $\sigma^*(\mathrm{OT})$, then all jobs with smaller p_j will also be nontardy in the optimal schedule $\sigma^*(\mathrm{OT})$. Using those properties, Cai and Vairaktarakis (2007) propose the following optimal algorithm that takes $O(n^2\omega^2)$ time.

Algorithm Overtime

(0) Apply *Regular* to find $\mathcal{W}_{\sigma^*}$ and T_R. Apply *Regular* to find $\mathcal{W}_{\sigma^*(\mathrm{OT})}$ and T_{OT} when each window has length $L + O$. Order jobs in $T_R \backslash T_{\mathrm{OT}}$ in non-decreasing order of processing times. Set $Z^* := \infty$. Also set $|\mathrm{OT}| := 0$ and $|P| := 0$; these are counters for the number of OT-windows and h_p-windows used, respectively.

(1) If $|\mathrm{OT}| <= \lceil \left(\sum_{J_j \in N} p_j\right)/(L + O) \rceil$ then go to (2); otherwise, **STOP**.

(2) If $|P| <= \lceil \left(\sum_{J_j \in N} p_j\right)/(L) \rceil$ then go to (3). Otherwise, set $|\mathrm{OT}| := |\mathrm{OT}| + 1$ and go to [1].

(3) Find the critical window $W_f \in \mathcal{W}_{\sigma^*(\mathrm{OT})}$. Apply MH to determine the sequence. Delay tardy jobs as much as possible without using additional windows. If the total cost of the schedule is less than Z^*, update Z^* and the tardy set $T_{\sigma^*(\mathrm{OT})}$. Set $|P| := |P| + 1$ and go to (2).

12.3.4.2 Coordination Results

Recall the definition of production cost savings $v_c(S)$ and booking savings $v_h(S)$ from Sect. 12.3.3.3. Production costs for this variant are expressed as tardiness penalties, i.e., $v_c(S) = v_T(S)$. Let $\tau(\sigma_0(S))$ and $\tau(\sigma^*(S))$ be the number of tardy jobs in the initial schedule and the optimal schedule for coalition $S \subseteq M$, respectively. Then,

$$v_T(S) = \beta \cdot [\tau(\sigma_0(S)) - \tau(\sigma^*(S))] \tag{12.16}$$

and $v(S) = v_T(S) + \rho \cdot v_h(S)$.

The following example demonstrates that for the variant of the outsourcing general model with tardiness costs, the cooperative game (M, v) defined in this section is not convex. Consider the instance where $M = \{1, 2, 3\}$, $N_1 = \{J_1\}$, $N_2 = \{J_2\}$, $N_3 = \{J_3\}$, $p_j = 2$, and $d_j = 24$ for $j = 1, 2, 3$, $|\Gamma| = 3$, $L = 8$, $h_1 = h_2 = 0$, $h_3 = 50$, $\rho = 1$, and $\beta = 100$. There is no idle time between windows. Before coordination, manufacturers 1 and 2 incur no cost while manufacturer 3 incurs cost of 50. No tardiness penalties are incurred as all jobs finish before their due dates. If all manufacturers cooperate, then all three jobs can be scheduled in window W_1. Consider the coalitions $S = \{1, 3\}$ and $T = \{2, 3\}$. Then, $v(T) = v(S) = 50$ because both release W_3 thus earning an refund of 50. The same is true for coalition $S \cup T = \{1, 2, 3\}$. However, $v(S \cap T) = v(\{3\}) = 0$. Therefore, $v(S \cup T) + v(S \cap T) = 50 < 100 = v(S) + v(T)$, and (M, v) is not convex. However, Cai and Vairaktarakis (2007) show that cooperative games (M, v_h), (M, v_T) and (M, v) are superadditive. They define $[a, b] = \{a, a + 1, \ldots, b\}$, $1 \le a \le b \le |M|$ as a coalition of manufacturers (numbered according to the FCFS order). With some constraints on the admissible coalitions, they prove that

$$x_m^T = v_T([m, |M|]) - v_T([m + 1, |M|]) \quad \text{for} \quad m \in M \tag{12.17}$$

is a core allocation for the tardiness savings game (M, v_T) and

$$x_m^h = v_h([1, m]) - v_h([1, m - 1]) \quad \text{for} \quad m \in M \tag{12.18}$$

is a core allocation for the booking savings game (M, v_h). Combination

$$x_m = x_m^T + \rho \cdot x_m^h \quad \text{for} \quad m \in M \tag{12.19}$$

is proved to be a core allocation for cooperative game (M, v). These results are valid for both the base model and the case where optional hourly overtime capacity is available at the third-party (see Cai and Vairaktarakis 2007 for further details).

12.4 A Dynamic Cost Allocation Model with Total Weighted Tardiness Cost

In this section, we present a dynamic total tardiness problem (D-TTP) following the analysis by Biskup and Simons (1999). Their study is motivated by the operations at a central repair office serving multiple computer dealers. The customers of the computer dealers contact them and specify the requirements of their repair needs as well as the due date. Then an agent transmits this request to the central headquarters. The central repair office then schedules the jobs according to their due dates, and the importance of the customers. This operational setting is very similar to the outsourcing model presented in earlier sections. The computer dealers play the role of the manufacturers, and the central repair facility plays the role of the third-party.

Common business practice calls for the FCFS processing of the repair requests as they are received at discrete times, say t, over time. However, significant overall savings are possible if the new jobs and the jobs waiting in the queue at time t are rescheduled in an attempt to minimize the weighted tardiness costs of all jobs. Overall savings are obvious, but it is possible that some jobs are completed unusually late because it may be frequently delayed by rescheduling. Hence, it is necessary that the central planner devise a fair cost allocation scheme.

Suppose at any discrete time point, t, a new job arrives when there are already $n_t - 1$ jobs waiting for service at the central repair facility. Each job J_j has a known processing time p_j and a promised due date d_j for its completion, as well as an associated weight w_j which reflects the importance of the customer who owns it. Let $c(\sigma^*(t))$ be the minimum weighted total tardiness cost incurred by the processing of these n_t jobs on or after time t according the optimal schedule $\sigma^*(t)$, i.e.,

$$c(\sigma^*(t)) = \min_{\sigma(t)} \left\{ \sum_{j=1}^{n_t} w_j [\max\{0\,,\, C_j(\sigma(t)) - d_j\}] \right\} \tag{12.20}$$

where $C_j(\sigma(t))$ denotes the completion time of job J_j according to schedule $\sigma(t)$. As noted before, $\sigma^*(t)$ may be optimal for the entire system but it may result to inferior performance by a collection of jobs. Let $M_a(t)$ denote the jobs (among the n_t jobs) transmitted by agent a. Also, let $\sigma_0(t)$ denote the schedule obtained if the job that arrived at time t was placed at the end of the queue of $n_t - 1$ jobs. If arriving jobs were just scheduled according to the FCFS rule, then no incentive problem would have occurred. However, due to the lack of optimal rescheduling, the overall service of the entire repair system would have deteriorated which might result in excessive financial tardiness penalties and loss of customer goodwill in the long run. Therefore, compensation amount $x_a(t)$ is allocated to every agent $a \in A$.

The central repair facility is faced with (a) a serious capacity allocation problem and (b) an incentive problem, which can be jointly formulated as:

$$\begin{aligned}
\min \; & \sum_{j=1}^{n_t} w_j [\max\{0\,,\, C_j(\sigma(t)) - d_j\}] \\
s.t. \; & \textstyle\sum_{J_j \in M_a(t)} w_j [\max\{0\,,\, C_j(\sigma(t)) - d_j\}] - x_a(t) \\
& \quad \leq \sum_{J_j \in M_a(t)} w_j [\max\{0\,,\, C_j(\sigma_0(t)) - d_j\}]\,,\; \forall a \in A \\
& \sum_{a \in A} x_a(t) = c(\sigma_0(t)) - c(\sigma^*(t))
\end{aligned} \tag{12.21}$$

The first set of constraints ensure that all agents are better off after rescheduling either by means of improved tardiness costs or as a result of the compensation that they receive for giving up earlier positions in the sequence, and hence are called *participation constraints*. The second constraint ensures that the system is *budget balanced*, i.e., the compensations that are allocated are generated by the cost savings as a result of rescheduling.

12.4.1 Solving the Scheduling Problem

The D-TTP problem and its static version are well-known scheduling problems. The static version was shown to be NP-hard by Du and Leung (1990). For optimal and/or approximate algorithms for the static version, see Emmons (1969), Schrage and Baker (1978), Potts and van Wassenhove (1982), and others. References for D-TTP include Baker and Bertrand (1982) and Raghu and Rajendran (1993).

Biskup and Simons (1999) suggested a simple heuristic for the online version. If the facility is occupied at time t, a waiting queue further builds up until the resource becomes free. The first arriving job, say J_1, is positioned first in the queue. The second, J_2, is either positioned first or second, whichever costs less. Suppose $J_2 \rightarrow J_1$ is the least cost sequence. Then J_3 can be positioned first, second, or third without changing the already fixed precedence relation between J_1 and J_2, i.e., the possible sequences are $J_2 \rightarrow J_1 \rightarrow J_3$, $J_2 \rightarrow J_3 \rightarrow J_1$, and $J_3 \rightarrow J_2 \rightarrow J_1$. In general, the ith arriving job is positioned according to the minimum cost sequence, without affecting the already established precedence relation among the jobs (first arriving through $(i-1)$th arriving) which joined the queue earlier. As soon as the resource becomes free, the job in first position at the current queue starts processing and is no longer considered in later rescheduling attempts.

12.4.2 Cost Allocation

The cost allocation suggested by Biskup and Simons (1999) iteratively allocates the savings or costs, every time a job is inserted in the waiting queue according to the algorithm described before. In other words, when job J_k is inserted before job J_j, a transfer payment of

$$x_{kj} = w_j[\max\{0, C_j(\sigma^*(t)) - d_j\} - \max\{0, C_j(\sigma_0(t)) - d_j\}] + f_j[c(\sigma_0(t)) - c(\sigma^*(t))] \tag{12.22}$$

is allocated from job J_k to job J_j. The first term in x_{kj} ensures that the participation constraints are satisfied, whereas the second term ensures that a fraction f_j of the overall savings are allocated to the agent that owns the delayed job J_j. Note that $0 \leq f_j \leq 1$ and $\sum_{J_j \in B_k} f_j = 1$ where B_k is the collection of jobs that are delayed after the insertion of job J_j, and hence all the generated savings $(c(\sigma_0(t)) - c(\sigma^*(t)))$ are allocated to the resequenced jobs. Also note that job J_j may later be allocated similar savings, y_{lk}, from job J_l that is inserted to a position before job J_k. Hence every job J_k holds a cost balance of

$$L_k(\sigma^*(t)) = w_k \max\{0, C_k(\sigma^*(t)) - d_k\} + \sum_{J_j \in B_k} x_{kj} - \sum_{J_k \in \{m \mid m \in B_l\}} y_{lk}. \tag{12.23}$$

Note that

$$\sum_{k=1}^{n_t} L_k(\sigma^*(t)) = c(\sigma^*(t))$$

and hence the allocation scheme is budget balanced.

12.5 Noncooperative Subcontracting Models

In this section, we present models related to another popular business model where multiple manufacturers, each capable of processing his entire workload in his own facility, have the option to subcontract some of their operations to a common third-party who has the flexibility of processing the subcontracted jobs from all manufacturers. Each customer order is delivered when the entire batch of jobs is completed. Therefore, each manufacturer divides his workload between his own resource and the third-party resource with the objective of completing his workload the soonest possible, i.e., minimizing his makespan. This creates an important time-sensitive capacity problem at the third-party as the sequence in which the subcontracted workloads are processed would affect the completion of the work subcontracted by each manufacturer. Currently, in practice, manufacturers book third-party capacity on a first-come-first-served (FCFS) basis using online booking systems which serve as information sharing portals.[7] Alternatively, the manufacturers can compete for earlier processing at the third-party while making their initial bookings. In this case, the third-party should announce the rules of engagement regarding the use of its capacity. Then the manufacturers, given the amount of information available from competitors (i.e., other manufacturers competing for the third-party capacity), act strategically and decide on the amount of workload to subcontract.

Part of the analysis of this section is drawn from Vairaktarakis (2008), who studies competition in this setting and reports Nash equilibrium outcomes. The strategies where each manufacturer optimizes locally are not optimal for the entire system, i.e., do not minimize the sum of the makespans of all manufacturers. Moreover, subcontracting requires contactual agreements with a third-party who may have conflicting interests (Kamien and Li 1990). In particular, the aforementioned subcontracting strategies do not optimize the utilization of the third-party capacity. In most practical settings, the relationship between manufacturers and providers is usually managed by *price-only contracts* due to their simplicity even though they are proven inefficient in many supply chain settings (see Lariviere and Porteus 2001; Cachon 2003). Being a rational profit maximizer, the third-party seeks to maximize, his utilization, i.e., maximize the sum of subcontracted workloads over all manufacturers. This centralized problem is studied in Vairaktarakis and Aydinliyim (2007) who suggest a savings allocation rule which is in the core of the corresponding cooperative game. A similar setting is also studied by Bukchin and Hanany (2007) whose main results will be discussed shortly.

Vairaktarakis (2008) and Vairaktarakis and Aydinliyim (2007) both study three production protocols – overlapping, preemption, and nonpreemption (see Pinedo 2002). Consider set M of manufacturers (also referred to as *players*) each having a known total workload of $P_i : i \in M$. Player i has to determine the amount of work to be subcontracted to third-party $3P$. The remaining amount will be processed on resource M_i owned by player $i \in M$. The subcontracted amount is referred to as the *strategy* of player i. Overlapping allows processing parts of a job of player i simultaneously on M_i and at the $3P$. Preemption allows processing part of a job of player i on M_i and the rest at the $3P$, however, not simultaneously. Nonpreemption stipulates that preemption is not possible for jobs. Four different information protocols are considered.

(IP1) Value $|M|$ is disclosed to all manufacturers.
(IP2) Values $P_i : i \in M$ are disclosed to all manufacturers.
(IP3) Values P_i and $p^i_{\max} : i \in M$ are disclosed to all manufacturers.
(IP4) Job processing profiles $\{p_{ij} : j \in N_i\}, i \in M$ are disclosed to all manufacturers,

where N_i denotes the job set of player $i \in M$, $p^i_{\max}$ is the processing time of a longest job in N_i, and p_{ij} is the processing time of job $J_j \in N_i$.

[7] See Sect. 12.1 for a discussion of Cisco's eHub, MyUMC, and SPADE of HKUST.

In all variants of the subcontracting models described, player i's objective is to minimize his maximum completion time of the in-house and subcontracted portions of his workload, i.e., the *makespan* denoted by C_i. All players compete for earlier processing at $3P$ who announces the rules of engagement. Possible objectives for the third-party include optimizing the overall service to manufacturers expressed as $\sum_{i \in M} C^i$. Alternatively, the third-party may look out for her own interest and seek to maximize the total workload subcontracted by manufacturers in M, i.e., $\sum_{i \in M} x_i$.

12.5.1 Incentive Rules to Manage Competition

12.5.1.1 Incentive Rules of Bukchin and Hanany (2007)

Bukchin and Hanany (2007) are the first to study a decentralized, nonpreemptive scheduling problem and propose incentives that will manipulate strategic behavior so that all players follow centralized strategies. In their model, the authors considered the flow time objective, i.e., sum of the completion times of all jobs of a player. Similar to the aforementioned subcontracting models, they model competition for the common resource which they refer to as the in-house resource. However, their alternative resource has unlimited capacity at a constant cost, k per unit processing time subcontracted. For simplicity, the authors choose k such that $1 \leq k \leq |N_i|$ for all $i \in M$. They measure the difference between the centralized system cost and the overall cost resulting from competition, a ratio they refer to as the *decentralization cost*.

In Bukchin and Hanany (2007), all jobs that are processed on the common resource (those that are not subcontracted) are processed in the shortest processing time (SPT) order because this sequence maximizes overall service, (Smith 1956). Alternatively, each subcontracted job incurs a cost that is a constant multiple of its processing time, i.e., $k \cdot p_{ij}$ regardless of the player who actually owns it. Had all jobs belonged to a single party, the centralized solution would be to process the k longest jobs among all jobs in N. Instead, each player assigns a set M_i of jobs to the common in-house resource and subcontracts the rest implying a different outcome which may be different than the centralized schedule. This results in an optimality gap, i.e., cost of decentralization, between the optimal centralized cost and the sum of the individual players' cost with competition.

The authors provide heuristics that generate upper bounds on the decentralization cost and demonstrate that it could be as high as 20%. On the other hand, correlated equilibrium bounds are calculated following the framework of Aumann (1974), which results in 5% gap between the lower bounds and the upper bounds of the overall cost. This finding suggests that it is possible for players to approximate centralized system performance without centralized control. However, whether this outcome occurs in practice is questionable as (a) it requires information sharing at a very detailed level, (b) calculating correlated equilibrium is computationally intractable except for a few small instances with limited number of jobs, and (c) the correlated equilibrium might be achieved as a result of mixed strategies which does not reflect practical subcontracting strategies Bukchin and Hanany (2007) proposed the following scheduling-based coordinating mechanism which they refer to as the *incentive compatible SPT rule* (IC-SPT) to incentivize players to act according to the centralized schedule.

Definition 12.3. *IC-SPT* rule: Let J_S be the set of shortest jobs of each player i, $i \in M$. Schedule all jobs in $(N \backslash J_S) \cap \cup_i M_i$ in SPT order. If $| \cup_i M_i| > k$ or if there exists a player who subcontracts his longest job which is longer than the shortest job that has not processed on the common resource, then this job j^* will be processed last.

IC-SPT prevents the players from occupying the common resource for their smaller jobs. Otherwise, they bear the risk that a considerably cheap-to-subcontract job may be delayed on the common resource. As a result, *IC-SPT* makes the Nash strategies of players perfectly coincide with the centralized solution, i.e., a player subcontracts all his jobs unless it is among the k longest jobs.

IC-SPT has one important drawback. In order to make it work, there has to be a centralized controller of the common resource, who keeps track of the jobs that are subcontracted by each player. One needs to know if there exists a subcontracted job that is longer than j^*. This is very difficult to achieve in practice. A managerially feasible incentive rule preferably achieves what is intended with the least overhead possible.

12.5.1.2 Incentive Rules of Vairaktarakis (2008)

In this section, we focus on the models by Vairaktarakis (2008) and Vairaktarakis and Aydinliyim (2007). Consider an arbitrary order [1], [2], ..., [$|M|$] of manufacturers. If the $3P$ announces that player [1] will be processed first, then [1] will subcontract as much workload of his as possible to the $3P$ resource, say F, without regard to other manufacturers. Similar will be the strategy of the next few manufacturers who will occupy all early processing capacity at the $3P$. Consequently, subsequent manufacturers cannot benefit from the services of the $3P$ and will not subcontract at all. Therefore, the $3P$ would like to announce priority rules which incentivizes manufacturers to compete for its capacity more productively. In practice, the third-party has the power to impose such policies.[8] Player strategies depend on the information and production protocols which in turn affect the rule-making process of the $3P$. Keeping these rules in mind, the manufacturers determine their subcontracted workloads $\{y_i^{\text{prod}} : i \in M\}$, where prod $\in \{O, P, N\}$ refers to the three production protocols considered.

[8] Outsourcing is the act of transferring some of a company's recurring internal activities and "decision rights" to outside providers (Greaver 1999).

To facilitate competition for the processing order at his resource, the $3P$ announces the following *incentive rules* for the protocols O, P, N, respectively:

IRO: If $y_i^O \le y_k^O$ then player i precedes k on F (i.e., in the shortest processing time order of outsourced workloads); break ties with smaller P_i.

IRP: Manufacturer workloads will be processed in quasi-SPT order, i.e., player i precedes k on F if $y_i^P \le y_k^P$ (break ties with smaller P_i), unless $y_i^P = P_i - p_{\max}^i \le y_k^P$ and $P_i > P_k$.

IRN: Manufacturer workloads y_i^N will be processed in nondecreasing order of P_i subject to the workload constraints $y_{[i]}^N \ge f_{[i]}(\max_{k<i} y_{[k]}^N),\ i \in M$,

where $f_i(w)$ is the maximum workload of any subset $A_i \subseteq N_i$ that does not exceed w.

In all production/IP combinations, every player chooses his strategy under the following assumptions:

(1) every player has complete information of his own job profile,
(2) the workload of all other players is infinitely divisible, and
(3) the total workload of other players is the same as his own.

These three assumptions yield equilibrium strategies for each player that would not hurt its *worst-case* makespan performance where "worst" is interpreted in terms of the (unrevealed portion of the) processing time profile of other players. Evidently, IP4 corresponds to complete information. However, when overlapping is allowed, IP2 is equivalent to complete information. When preemption is allowed but overlapping is not, then IP3 is equivalent to complete information because detailed job processing information does not provide additional preemption opportunities. On the other hand, detailed job information is useful in nonpreemptive schedules. Let $y_i^{\text{prod}}(IP)$ denote the equilibrium subcontracted workload amount for player i under production protocol prod and information protocol IP. In particular, let us denote $y_i^O(\text{IP2}) = y_i^O$, $y_i^P(\text{IP3}) = y_i^P$, and $y_i^N(\text{IP4}) = y_i^N$. Then, rules IRO, IRP, and IRN yield the Nash equilibria summarized in Table 12.3:

Note that the algorithms for obtaining the competitive strategies under perfect information, i.e., $y_i^O(\text{IP2}) = y_i^O$, $y_i^P(\text{IP3}) = y_i^P$, and $y_i^N(\text{IP4}) = y_i^N$ are quite technical and are beyond the scope of this chapter. See Vairaktarakis (2008) for details.

Table 12.3 Equilibrium strategies

$y_i^{\text{prod}}(\text{IP})$	IP1	IP2	IP3	IP4
O	$\frac{P_i}{\lvert M\rvert+1}$	$y_{[i]}^O$	–	–
P	$\min\left\{P_i - p_{\max}^i, \frac{P_i}{\lvert M\rvert+1}\right\}$	$\min\left\{P_i - p_{\max}^i, y_{[i]}^O\right\}$	$y_{[i]}^P$	–
N	$f_{[i]}\left(\frac{P_{[i]}}{\lvert M\rvert+1}\right)$	$f_{[i]}\left(y_{[i]}^O\right) : i < \lvert M\rvert$	$f_{[i]}\left(y_{[i]}^P\right) : i < \lvert M\rvert$	$y_{[i]}^N$

12.5.1.3 A Computational Study by Vairaktarakis and Aydinliyim (2007)

Ideally, the incentive priority rules lead to subcontracting strategies that mimic the centralized schedule. In Vairaktarakis and Aydinliyim (2007), the merits of the aforementioned incentive rules are measured by the decentralization cost which results from a comparison of the centralized schedules and the competitive Nash schedules. It is found that Nash equilibrium schedules under-utilize the third-party capacity compared to the centralized schedules. The average loss in utilization is 7.5% for complete information. Table 12.4 summarizes the findings from a computational experiment where averages are calculated over ten instances for each combination of parameters $|M|, |N_i|, p_j$. Observe from Table 12.4 that, compared to the centralized schedules, third-party capacity under-utilization is severe for IP1 (on average 22.76, 22.76, and 34.70% for O, P, N respectively.), improves dramatically for IP2 and IP3 (on average 7.05, 7.05, and 14.81% for O, P, N respectively.), and further improves by another 6% for N/IP4 (from 14.81 down to 8.54%). Evidently, advances in information sharing dramatically increase the benefits of competition; however, these benefits dissipate rather quickly. This finding suggests that a high level of overall service for all players and the third-party utilization under centralization can be achieved by appropriate incentive rules imposed by the third-party provided that a high level of horizontal information sharing is achieved among the manufacturers.

Table 12.4 The cumulative percentage workload deviation from the centralized schedule

			O/P		N		
$\|M\|$	$\|N_i\|$	p_j	IP1	IP2/IP3	IP1	IP2/IP3	IP4
4	5	(1,5)	23.76	6.72	34.55	9.88	6.73
		(1,10)	23.20	7.08	35.33	13.49	5.59
	10	(1,5)	20.55	8.91	24.14	11.48	8.37
		(1,10)	21.33	8.36	23.43	10.10	6.57
	15	(1,5)	20.21	9.25	22.81	13.57	15.32
		(1,10)	20.48	9.39	21.93	10.40	11.33
6	5	(1,5)	25.62	6.14	54.40	24.55	8.31
		(1,10)	25.52	6.59	47.82	20.38	2.14
	10	(1,5)	21.69	7.41	28.82	13.78	10.25
		(1,10)	23.01	6.53	27.42	9.57	5.46
	15	(1,5)	21.00	8.25	26.07	13.26	15.56
		(1,10)	x	x	x	x	x
8	5	(1,5)	26.36	4.96	59.56	19.98	6.44
		(1,10)	26.06	5.34	62.26	29.43	2.05
	10	(1,5)	22.10	5.15	31.63	13.63	8.34
		(1,10)	22.89	5.91	27.82	10.78	5.29
	15	(1,5)	20.31	6.83	27.13	12.75	12.41
		(1,10)	x	x	x	x	x
Averages			**22.76**	**7.05**	**34.70**	**14.81**	**8.14**

12.6 Other Mechanisms and Instruments for Centralized Control

As demonstrated in earlier sections, competition for common resources that are of interest to multiple parties with conflicting interests create serious capacity allocation and incentive problems. The increasing outsourcing/subcontracting trend has brought up these two problems to the attention of numerous manufacturers which need extra capacity or special expertise for their various operations. This inclination has increased the demand for contract manufacturing which has led to the growth of firms with core competence in providing third-party services for a number of customers in the same industry. Therefore, coordination mechanisms that yield centralized optimal (or near-optimal) outcomes without centralized control are of crucial importance. In what follows, we discuss two such instruments.

12.6.1 Price Mechanisms and Auctions

Wellman et al. (2001) consider a decentralized scheduling problem of allocating resources to autonomous agents. They investigate the existence of equilibrium prices for some general classes of scheduling problems, the quality of equilibrium solutions, and the behavior of an ascending auction mechanism and bidding protocol. The following analysis follows from their findings.

Consider a factory with an unscheduled day shift, which might be represented by a number of 1-hour time slots, say $\Gamma = \{1, \ldots, k, \ldots, T\}$. These production slots can be allocated to the production of customer orders that are associated with different agents. Following the common business practice, assume that each slot has a booking price referred to as the *reserve price* q_k, representing the minimum price that the factory is willing to accept in exchange for that time slot. Assume each agent j, $j \in M$, has one job he wants completed. A job is defined by its duration, p_j, its deadline d_j, and the value v_j (expressed in dollars) that the agent places on the job. An agent is willing to spend up to this value to complete the job. To do so, the agent must acquire a number of slots not less than the duration of his job (not necessarily contiguous), which are not later than the jobs's deadline. The agent gets no value if his job cannot be completed before its deadline. The total value of a solution is the sum of the values of all agents, i.e., the sum of the reserve prices for time slots not sold plus the value associated with agents who meet their job deadlines.

Given prices $\mathbf{p} = \{p_1, \ldots, p_k, \ldots, p_T\}$ offered for the time slots in Γ, each agent tries to achieve the maximum value by processing his jobs on a collection $X \subseteq \Gamma$ of time slots, i.e.,

$$H_j(\mathbf{p}) = \max_{X \subseteq \Gamma} \left[v_j(X) - \sum_{k \in X} p_k \right]. \tag{12.24}$$

Let F_j be the collection of the spots allocated to agent j, and let F_0 be the collection of slots that are not booked by any agent, i.e., $\Gamma = \cup_{j \in M} f_j \cup F_0$. The owner of the factory would like to maximize his own benefit, i.e.,

$$v(\sigma) = \sum_{k \in F_0} q_k + \sum_{j=1}^{|M|} v(j), \tag{12.25}$$

which is referred to as the *global value* of schedule σ, and is determined by the choices F_j of agents $\{j \mid j \in M\}$. With this in mind, the owner of the factory tries to find a mechanism which satisfies the following conditions:

- Self-interested agents can make effective decisions based on local private information, without knowing the private information and strategies of other agents.
- The method requires minimal communication overhead.
- The method reaches closure in reasonable time at reasonable computational expense.
- Solutions do not waste resources. If there is some way to make some agents better off without harming others, it should be done. A solution that cannot be improved in this way is called *Pareto* optimal.

To demonstrate the importance of optimal prices on achieving the globally optimal solution (the factory owner's optimal solution), we provide the following definition.

Definition 12.4 (Price Equilibrium). A solution σ is an equilibrium at prices $\mathbf{p}$ if and only if

(1) For all agents $j \in M$, $v_j(F_j) - \sum_{k \in F_j} p_k = H_j(p)$;
(2) For all production slots $k \in \Gamma$, $p_k \geq q_k$;
(3) For all unbooked production slots $k \in F_0$, $p_k = q_k$.

Then we state the following theorem:

Theorem 12.1. *For the decentralized scheduling problem, if there exists a* $\mathbf{p}$ *such that* σ *is in equilibrium, then* σ *is a globally optimum solution.*

However, it is possible that no equilibrium prices exist. In fact Wellman et al. (2001) show that, one can easily construct such an instance if any job duration requires more than one production slot. On the other hand, they show that, when each processing time requirement equals one production slot – a problem referred to as the *single-unit scheduling problem*, then a price equilibrium always exists. Hence, one needs to look for ways to make the agents propose prices $\mathbf{p}$, that will generate a price equilibrium schedule for the single-unit scheduling problem.

The authors propose *ascending auctions* to find equilibrium prices for the single-unit scheduling problem. According to McAfee and McMillan (1987), an auction is a market institution with an explicit set of rules determining resource allocation and prices on the basis of bids from market participants. The rules for the ascending auctions are fairly simple.

At any point in time, the *bid price* in the auction for production slot k, denoted β_k, is the highest bid in the auction thus far. If auction has received no bids, β_k is undefined. Production slot k's *ask price*, denoted α_k, is $\beta_k + \epsilon$, for some fixed ϵ (the bidding increment), if β_k is defined. Otherwise, the ask price is q_k. The ascending auction rejects any bid less than its ask price. Agents are not allowed to withdraw bids. An agent may replace its bid with another, but the new bid must be at least the current ask price. These rules guarantee that prices do not decrease and that the bidding process terminates.

Even though the ascending auction may fail to generate a global optimum for the factory owner and equilibrium prices, these values can be approximated with tight worst-case errors. In particular, according to Wellman et al. (2001), we have the following two theorems for the single-unit scheduling problem.

Theorem 12.2. *The final price of any production slot determined by the ascending auction protocol will differ from the unique minimum equilibrium prices by at most* $\kappa\epsilon$, *where* $\kappa = \min\{|\Gamma|, |M|\}$.

Theorem 12.3. *The ascending auction protocol with a given* ϵ *produces a solution to the single-unit scheduling problem that is suboptimal by at most* $\kappa\epsilon(1 + \kappa)$.

The authors also suggest *combinatorial auctions* and *generalized vickery auctions* for the multiple-unit scheduling problem, for which ascending auctions may produce solutions that are arbitrarily far from the optimal.

12.6.2 Contracts

In this section, we discuss the merits of another instrument, contracts, in achieving system-wide optimal performance in certain subcontracting situations. Cachon (2003) notes that optimal chain performance is achievable if the firms coordinate by contracting on a set of transfer payments such that each firm's objective becomes aligned with the supply chain objective. In an excellent survey, he reviews and extends the supply chain literature on the management of incentive conflicts with contracts. The focus of his work and many others in the supply chain contracting literature is the coordination of inventories at an aggregate level, with not much emphasis on the timeliness of the production activities and production chain coordination possibilities.

In what follows, we present the analysis by Aydinliyim and Vairaktarakis (2007) who study a subcontracting model where manufacturer m cannot process all his workload P before the customer due date, d. Therefore, he subcontracts part of his workload x to a third-party with prior customer commitments y. These prior commitments follow a general distribution, i.e., $y \sim F(\cdot)$, and hence the available capacity at the third-party is uncertain. The manufacturer wants to maximize his expected profits by partitioning his workload among his in-house production capacity and the third-party (see Fig. 12.3).

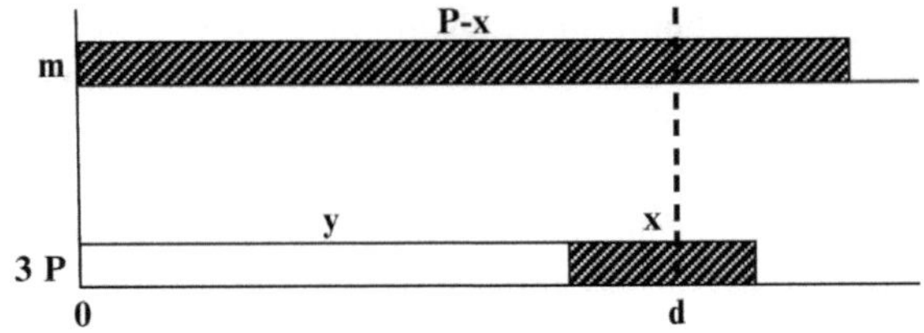

Fig. 12.3 A typical schedule for m and $3P$

This is a commonly observed business model, where a manufacturer who is in need of short-term extra capacity is matched with a third-party (committed to long-term customers) which is in need of additional customers willing to pay for the potential extra capacity at $3P$. This third-party is sometimes a committed provider of a significantly important customer (e.g., a small bumper manufacturer who has a long-term contract with Toyota), and hence the orders coming from this important customer is always prioritized.

In this section, we present two different contracts that illustrate the relationship between the manufacturer who subcontracts for extra capacity and the third-party who is already committed to other customers:

(1) *Unit processing charge contracts*: The third-party charges u for each unit of m's subcontracted workload, i.e.,

$$T(x, u) = u x. \tag{12.26}$$

(2) *Tardiness penalty sharing contracts*: In addition to the terms of the first contract, the third-party agrees to share the tardiness penalties incurred due to the delays on his schedule, i.e.,

$$T(u, x, \lambda) = u x - \lambda g_m \left(E_y[L_{3P}(x)]\right),\ \lambda \in (0, 1) \tag{12.27}$$

where g_m is the unit tardiness cost that manufacturer m incurs, λ is the penalty sharing fraction, and $E_y[L_{3P}(x)]$ is the expected tardiness incurred at $3P$, when manufacturer m subcontracts x units. The third-party acts as the Stackelberg leader, decides on the contract type and the contract parameters, and then offers the contract to the manufacturer as a take-it-or-leave-it deal. If the manufacturer takes the offer, he responds by subcontracting part of his workload to the manufacturer.

Aydinliyim and Vairaktarakis (2007) show that this production chain cannot be coordinated under the first contract unless the third-party accepts to merely break even, i.e., the manufacturer subcontracts x^* units which is the amount that maximizes the total chain profit only when $u = c_{3P}$ where c_{3P} is the unit production cost at the third-party. On the other hand, under the second contract, coordination is possible if the unit processing fee and the tardiness sharing fraction are determined jointly, i.e., the manufacturer subcontracts x^* units when

$$u = c_{3P} + \lambda (g_m - c_m),$$

where $g_m - c_m$ is the manufacturer's opportunity cost of not subcontracting a unit when it is certain that it will incur tardiness. In this case, the coordinating contract

Pareto dominates noncoordinating contracts, as it allocates the additional chain profits arbitrarily between the manufacturer and the third-party. Moreover, under the coordinating contract, the tardiness sharing fraction coincides with the third-party's share of the additional profits generated by the coordinated chain. So, the coordinating contract can easily be administered for managerial interpretation. However, the issue of trutfully revealing the cost parameters, and the question of whether the parties can exactly negotiate the coordinating (u, λ) pair still remain. The authors show that even when there is coordination failure, the Nash equilibrium strategies under the second contract achieve a smaller optimality gap (difference between the optimal chain profit and the total profit at the decentralized equilibrium) than what the parties would incur under the unit processing charge contract. In supply chain contracting literature this is equivalent to stating that the tardiness penalty sharing contracts achieve higher *efficiency* than the unit processing charge contracts.

12.7 Future Research Directions

To the best of our knowledge, this is a new line of research which combines scheduling theory with game theory to model outsourcing and subcontracting operations and investigate coordination opportunities at the production scheduling level. Various extensions of the models described in this chapter are attracting research opportunities including multiple-resource models with many other practical objective functions and industry-specific restrictions. Also note that, with the recent advances in information technologies, information sharing capabilities of supply chain members have improved substantially. Hence, many coordination problems that had been studied in supply chain management literature can be revisited to investigate coordination benefits at the detailed scheduling level. In what follows, we provide some possible directions extending the cooperative and noncooperative game theory models, and the mechanism and contract design problems that are surveyed in this chapter.

12.7.1 Cooperative Games and Cost Allocation Models

Sequencing games and cooperative outsourcing models assume that the initial booking of the common resource is often times done in a FCFS manner. However, strategic behavior that takes the resequencing stage into account may yield interesting insights. Another important extension is the consideration of nontransferable utility. Although side payments are common in practice, current models assume that one dollar lost in booking costs can be compensated by one dollar worth of improvements in the delivery time, which is an important drawback. With more emphasis on multicriteria optimization, where sensitivity of agents to monetary costs are different than their sensitivity to delivery time performance, one can find Pareto optimal schedules which explains the trade-offs in a more meaningful way.

Another important assumption that should be tackled is full information sharing among players. Bayesian models with asymmetric information, investigating the benefits of sharing information, and the mechanisms that induce truth-telling by players are all fruitful directions.

Most manufacturing systems are more complex than the single common resource models which were almost always common in all cooperative outsourcing models we surveyed in this chapter. Parallel resource environments, flow-shop, and job-shop settings should receive more attention in future studies.

Finally, cost sharing is of crucial importance in large and complex projects where many different contractors are responsible for different group of tasks. In such models, those that finish their tasks earlier can move their resources to other tasks in return for some side payments, which can create win-win situations in improving project completions times and in reducing overall project costs by minimizing penalties.

12.7.2 Competitive Models of Outsourcing and Subcontracting

Although aggregate capacity allocation issues in supply chains have been studied extensively, existing research for time-sensitive common-resource capacity allocation problems is limited. The growing trend of contract manufacturing and the recent problems of Boeing with its contractors for its 787 Dreamliner project proved that there is still a lot to be investigated.

Possible directions include but are not limited to the study of dynamic models of competition for third-party capacity, which is a promising research direction considering the growing number of online capacity booking systems. This level of end-to-end visibility requires revising the production and subcontracting decisions frequently.

Coordinated production, subcontracting, and pricing decisions in case of time-sensitive competition is a fruitful topic as well. With the recent increases in fuel prices and high storage costs, coordinating the production and distribution decisions to reduce finished goods inventory and not to fall behind the delivery schedules at the least possible cost are more important than ever. Moreover, there are many other issues such as the subcontracting costs or the technological advantages at the third-party that should be incorporated in the subcontracting models that we discussed in Sect. 12.6.2.

Finally, two-staged models that include decisions involving the creation of capacity and related investments, followed by the production and subcontracting decisions after the realization of demand have recently begun to gauge interest from researchers. Although, a limited number of studies exist that investigate single-manufacturer, single-subcontractor models, the issue of competition for the subcontractor capacity in a time-sensitive manner is a research opportunity that is yet to be exploited.

12.7.3 Contract and Mechanism Design

The subcontracting model with uncertain third-party capacity can be extended in many directions, which include the analysis of the multiple manufacturers' case where one can expect competition for earlier capacity at the third-party. Conflicting interests of the manufacturer and the third-party will create a capacity allocation subproblem. We believe that the issue of competition with contract design will reveal interesting and nonintuitive insights which might help primary contractors to make more efficient time-sensitive subcontracting decisions. Similarly, the case with multiple third-parties is also worth considering as it leads to more subcontracting choices for the manufacturer. One can revisit the supplier selection problem in the context of a third-party selection problem with more emphasis on the contractors' timely delivery capabilities. From the third-party's point of view, one can study the due-date quotation problem which is a promising direction at the interface of marketing and operations. To capture the effect of more complex penalty structures, one can study multiple due date contracts with incremental penalties. Finally, one can investigate the change in the third-party's scheduling decisions when *performance-based bonus schemes* are included in contract terms.

Regarding the issue of designing auctions and pricing mechanisms that rely on equilibrium concepts, there are significant research opportunities as well. In case the manufacturers bid for multiple units of production capacity at the third-party, the combinatorial auctions have to be considered, for which the winner determination or the optimal bidding problems are computationally intractable. Therefore, developing heuristic algorithms which utilize simple and practical auction mechanisms to approximate price equilibrium schedules provide decent research opportunities.

Acknowledgments We would like to thank Professor Reha Uzsoy and two anonymous reviewers for their insightful comments on the earlier versions of this chapter.

References

Agnetis A, Hall NG, Pacciarelli D (2006) Supply chain scheduling: sequence coordination. Discrete Appl Math, 154:2044–2063

Ansberry C, Aeppel T (2003) Battling imports: surviving the onslaught; U.S. companies customize, rethink strategies to compete with products from abroad. The Wall Street Journal, New York, NY, October 6

Anupindi R, Bassok Y (1999) Centralization of stocks: retailer vs. manufacturer. Manage Sci, 45(2):178–191

Atamturk A, Hochbaum DS (2001) Capacity acquisition, subcontracting and lot sizing. Manage Sci, 47:1081–1100

Aumann RJ (1974) Subjectivity and correlation in randomized strategies. J Mathe Econ 1:67–96

Aydinliyim T (2007) Coordination and competition in outsourcing operations. Doctoral Dissertation. Case Western Reserve University, Cleveland, OH

Aydinliyim T, Vairaktarakis GL (2010) Coordination of Outsourced Operations to Minimize Weighted Flow Time and Capacity Booking Costs. Manuf Serv Oper Manag 12(2):236–255

Aydinliyim T, Vairaktarakis GL (2007) Subcontracting strategies with stochastic third-party capacity and tardiness penalty contracts. Working Paper, TM-817, Weatherhead School of Management, Case Western Reserve University
Baker KR, Bertrand JWM (1982) A dynamic priority rule for scheduling against due dates. J Oper Manag 3:37–42
Bernstein F, Federgruen A (2005) Decentralized supply chains with competing retailers under demand uncertainty. Manage Sci 51(1):18–29
Bertrand JWM, Sridharan V (2001) A study of simple rules for subcontracting in make-to-order Manufacturing. Eur J Oper Res 128(3):509–531
Biskup D, Simons D (1999) Game theoretic approaches to cost allocation in the dynamic total tardiness problem. IIE Trans 31(9):899–908
Bitran GR, Tirupati D (1989) Hierarchical production planning. Working Paper, Sloan School of Management, Massachusetts Institute of Technology, Cambridge, MA
Borm P, Fiestras-Janeiro G, Hamers H, Sánchez E, Voorneveld M (2002) On the convexity of games corresponding to sequencing situations with due dates. Eur J Oper Res 136:616–634
Bukchin Y, Hanany E (2007) Decentralization cost in scheduling: a game-theoretic approach. Manuf Serv Oper Manag 9(3):263–275
Cachon GP (2003) Supply chain coordination with contracts. In: Graves S, de Kok T, (eds) Handbooks in operations research and management science: supply chain management. North-Holland, Amsterdam, The Netherlands
Cachon GP, Harker PT (2002) Competition and outsourcing with scale economies. Manag Sci 48(10):1314–1333
Cachon GP, Lariviere M (1999) An equilibrium analysis of linear, proportional and uniform allocation of scarce capacity. IIE Trans 31:835–849
Cachon GP, Lariviere M (2001) Contracting to assure supply: how to share demand forecasts in a supply chain. Manag Sci 47(5):629–646
Cachon GP, Lariviere M (2005) Supply chain coordination with revenue sharing contracts: strengths and limitations. Manag Sci 51(1):30–44
Cachon GP, Netessine S (2004) Game theoretic applications in supply chain analysis. In: Simchi-Levi D, Wu SD, Shen ZJ (eds) Supply chain analysis in e-business era, Kluwer, New York, NY
Cai XQ, Vairaktarakis GL (2007) Cooperative strategies for manufacturing planning with negotiable third-party capacity. Working Paper, TM-820, Weatherhead School of Management, Case Western Reserve University, Submitted to *Manag Sci*
Calleja P, Borm P, Hamers H, Klijn F, Slikker M (2002) On a new class of parallel sequencing situations and related games. Ann Oper Res, 109:263–276
Calleja P, Estévez-Fernández MA, Borm P, Hamers H (2004) Job scheduling, cooperation and control. CentER Discussion Papers, 2004–65, Tilburg University, Tilburg, The Netherlands
Chen F (2003) Information sharing and supply chain coordination. In: Graves S, de Kok T, (eds) Handbooks in operations research and management science: supply chain management, North-Holland, Amsterdam, The Netherlands
Chen Q, Cai X (2006) Application of nash bargaining solution to a scheduling problem involving negotiation between two manufacturers. Working Paper, Systems Engineering and Engineering Department, Chinese University of Hong Kong, Shatin, Hong Kong SAR
Chen ZL, Hall NG (2007) Supply chain scheduling: conflict and cooperation in assembly systems. Oper Res 55:1072–1089
Chen ZL, Li CL (2008) Scheduling with subcontracting options. IIE Trans 40(12):1171–1184
Chung D, Lee K, Shin K, Park J (2005) A new approach to job shop scheduling problems with due date constraints considering operation subcontracts. Int J Prod Econ 98:238–250
Curiel I, Pederzoli G, Tijs S (1989) Sequencing games. Eur J oper Res 40:344–351
Curiel I, Potters J, Rajendra Prasad V, Tijs S, Veltman B (1994) Sequencing and cooperation. Oper Res 42:566–568
Curiel I, Hamers H, Klijn F (2002) Sequencing games: a survey. In: Borm P, Peters H (eds) Chapters in game theory: in honor of Stef Tijs. Kluwer Academic Publishers, Boston, pp. 27–50

Dawande M, Agnetis A, Hall NG, Pacciarelli D (2006) Supply chain scheduling: distribution systems. Prod Oper Manag 15:243–261

Day JS (1956) Subcontracting policy in the airframe industry. Graduate School of Business Administration, Harvard University, Boston, MA

Du J, Leung J (1990) Minimizing total tardiness on one machine is NP-hard. Math Oper Res 15:483–495

Emmons H (1969) One machine sequencing to minimize certain functions of job tardniess. Oper Res 17:701–715

Gan X, Gu Y, Vairaktarakis GL, Cai X, Chen Q (2007) A scheduling problem with one producer and the bargaining counterpart with two producers. In: First International Symposium, ESCAPE 2007, Hangzhou, China, 7–9 April, Revised Selected Papers

Ganesh A, Laevens K, Steinberg R (2007) Congestion pricing and noncooperative games in communication networks. Oper Res 55:430–438

Gillies DB (1959) Solutions to general non-zero-sum games. In: Tucker AW, Luce RD (eds) Contributions to the theory of games IV (Ann Math Stud 40), Princeton University Press, Princeton, pp. 47–85

Greaver MF (1999) Strategic outsourcing. AMA Publications, New York, NY

Grosvenor F, Austin TA (2001) Cisco's eHub initiative. Supply Chain Manag Rev, July/August, 28–35

Hain R, Mitra M (2004) Simple sequencing problems with interdependent costs. Game Econ Behav 48:271–291

Hall NG, Liu Z (2007) Capacity allocation and scheduling in supply chains. Working Paper, Fisher College of Business, The Ohio State University, Submitted for publication

Hall NG, Liu Z (2008) Auctions for competitive capacity allocation and scheduling. Working Paper, Fisher College of Business, The Ohio State University, Submitted for publication

Hall NG, Potts CN (2003) Supply chain scheduling: batching and delivery. Oper Res 51:566–584

Hamers H, Borm P, Tijs S (1995) On games corresponding to sequencing situations with ready times. Math Program 70:1–13

Hamers H, Klijn F, Suijs J (1999) On the balancedness of multimachine sequencing games, Euro J Oper Res 119:678–691

Hamers H, Klijn F, van Velzen B (2002) On the convexity precedence sequencing games. CentER Discussion Papers 2002–112, Tilburg University, Tilburg, The Netherlands.

Heydenreich B, Muller R, Uetz M (2007) Games and mechanism design in machine scheduling – an introduction. Forthcoming in Prod Oper Manag

Heydenreich B, Muller R, Uetz M (2007) Mechanism design for decentralized online machine scheduling. Working Paper. Maastricht University, Maastricht, The Netherlands

Kamien MI, Li L (1990) Subcontracting, coordination, flexibility, and production smoothing in aggregate planning. Manag Sci 36(11):1352–1363

Karabuk S, Wu DS (2005) Incentive schemes for semiconductor capacity allocation: a game theoretic analysis. Prod Oper Manag 14:175–188

Katta AK, Sethuraman J (2006) Cooperation in queues. Working Paper, Department of Industrial Engineering and Operations Research, Columbia University, New York, NY

Kutanoglu E, Wu DS (1999) On combinatorial auction and lagrangean relaxation for distributed resource scheduling. IIE Trans 31:813–826

Lariviere M (1998) Supply chain contracting and co-ordination with stochastic demand. In: Tayur S, Ganeshan R, Magazine M (eds) 'Quantitative models for supply chain management'. Kluwer, Dordrecht, The Netherlands

Lariviere M, Porteus E (2001) Selling to the newsvendor: an analysis of price-only contracts. Manuf Serv Oper Manag 3(4):293–305

Li L (1992) The role of inventory in delivery-time competition. Manag Sci 38(2):182–197

Mallik S, Harker PT (2004) Coordinating supply chains with competition: capacity allocation in semiconductor manufacturing. Eur J Oper Res 159:330–347

Maniquet F (2003) A characterization of the shapley value in queueing problems. J Econ Theor 109:90–103

McAfee RP, McMillan J (1987) Auctions and bidding. J Econ Lit 25:699–738
Miller T (2002) Hierarchical operations and supply chain planning. Springer, New York, NY
Moore JM (1968) An n job one machine sequencing algorithm for minimizing the number of late jobs, Management Science 18(1):102–109
Nagarajan M, Sošič (2008) Game-theoretic analysis of cooperation among supply chain agents Review and Extension. Eur J Oper Res 187(3):719–745
Nash JF (1950) The bargaining problem. Econometrica 18:155–162
Nash JF (1951) Non-cooperative games. Ann Math 54:286–295
Papadimitriou C (2001) Algorithms, games, and the Internet Proc 33rd Annual ACM Sympos Theor Comput. Association for Computing Machinery, Hersonissos, Greece, 749–753
Pasternack B (1985) Optimal pricing and returns policies for perishable commodities. Mark Sci 4(2):166–176
Pinedo M (2002) Scheduling: theory, algorithms and systems. Prentice Hall, Englewood Cliffs, NJ
Potts CN, van Wassenhove LN (1982) A decomposition algorithm for the single machine total tardiness problem. Oper Res Lett 1:177–181
Qi X (2007) Outsourcing and production scheduling for a two-stage flow shop. Working Paper, Hong Kong University of Science and Technology, Kowloon, Hong Kong SAR, PROC.
Qi X (2008) Coordinated logistics scheduling for in-house production and outsourcing. IEEE Trans Autom Sci Eng 5(1):188–192
Raghu TS, Rajendran C (1993) An efficient dynamic dispatching rule for scheduling in a job shop. Int J Prod Econ 32:301–313
Reeves DM, Wellman MP, MacKie-Maosn JK, Osepayshvili A (2005) Exploring Bidding Strategies for Market Based Scheduling. Decis Support Syst 39:67–85
Rosenthal RW (1973) A class of games possessing pure-strategy nash equilibria. Int J Game Theory 2:65–67
Sandholm WH (2002) Evolutionary implementation and congestion pricing. Rev Econ Stud 69:667–689
Schrage L, Baker KR (1978) Dynamic programming solution of sequencing problems with precedence constraints. Oper Res 26:444–449
Schulz AS, Uhan NA (2007) Encouraging cooperation in sharing supermodular costs. Working Paper, Sloan School of Management, Massachusetts Institute of Technology, Cambridge, MA
Shapley LS (1953) A value for n-Person games. In Kuhn HW, Tucker TW (eds) Contributions to the theory of games II. Annals of Mathematics Study. Princeton, New Jersey: Princeton University Press, 28:307–317
Shapley L (1971) Cores of convex games. Int J Game Theor 1:11–26
Shen W, Hao Q, Yoon HJ, Norrie DH (2006) Applications of agent-based systems in intelligent manufacturing: an updated review. Adv Eng Informat 20:415–431
Slikker M (2006) Relaxed sequencing games have a nonempty core. Nav Res Logist 53:235–242
Smith WE (1956) Various optimizers for single stage production. Nav Res Logist 3(1):59–66
Swaminathan JM, Tayur SR (2003) Tactical planning models for supply chain management. In: Graves S, de Kok T (eds) Handbooks in operations research and management science: supply chain management. North-Holland, Amsterdam, The Netherlands
Tan B (2004) Subcontracting with availability guarantees: production control and capacity decisions. IIE Trans 36:711–724
Tan B, Gershwin SB (2004) Production and subcontracting strategies for manufacturers with limited capacity and volatile demand. Ann Oper Res 125:205–232
Taylor T (2002) Coordination under channel rebates with sales effort. Manag Sci 48(8):992–1007
Thomas DJ, Griffin PM (1996) Coordinated supply chain managament, Eur J Oper Res 94(1):1–15
Tijs S, Parthasarathy T, Potters J, Prasad VR (1984) Permutation games: another class of totally balanced games. OR Spektrum, 6:119–123
Tsay A (1999) Quantity-flexibility contract and supplier-customer incentives. Manag Sci 45(10): 1339–1358

Vairaktarakis GL (2008) Non-cooperative outsourcing games. Working Paper, TM-822, Weatherhead School of Managament, Case Western Reserve University. Submitted to Manuf Serv Oper Manag

Vairaktarakis GL, Aydinliyim T (2007) Competition vs. centralization in subcontracting operations. Working Paper, TM-819, Weatherhead School of Managament, Case Western Reserve University. Submitted to Oper Res

Van Mieghem JA (1999) Coordinating investment, production and subcontracting. Manag Sci 45(7):954–971

van Velzen B, Hamers H (2003) On the Balancedness of Relaxed Sequencing Games. Math Meth Oper Res 57:287–297

Yu Y, Benjaafar S, Gerchak Y (2007) Capacity pooling and cost sharing among independent firms in the presence of congestion. Working Paper, Department of Mechanical Engineering, University of Minnesota, Minneapolis, MN, Submitted for publication.

Wellman MP, Walsh WE, Wurman PR, MacKie-Mason JK (2001) Auction protocols for decentralized scheduling. Game Econ Behav 35:271–303

Chapter 13
Inventory Management: Information, Coordination, and Rationality

Özalp Özer

13.1 Introduction

Inventory control problems have attracted researchers for many years[1]. Fundamentally, the problem is one of matching supply and demand by efficiently coordinating the production and the distribution of goods. Recent developments in information technology have equipped managers with the means to obtain better and timely information regarding, for example, demand, lead times, available assets, and capacity. Technology has also enabled customers to obtain vast amounts of information about a product, such as its physical attributes and availability. In today's increasingly competitive marketplace, consumers are constantly pressuring suppliers to simultaneously reduce costs and lead times and increase the quality of their products. Good inventory management is no longer a competitive advantage. It is an essential capability to survive in a global market.

An important aspect of good inventory management is effective use of information. Knowing how to use information effectively also enables a manager to decide what data to collect, buy, and store, and what information technology to invest in. Note that information has no value, if it is not used effectively. For example, an inventory manager can obtain order progress information through the use of a tracking technology. If this information is not used to improve replenishment decisions, then neither the information nor the technology used to obtain it has any value. In this chapter, we provide some examples of how information is incorporated into classical inventory management problems.

The second important aspect of good inventory management is to quantify the value of information. A manager may need to invest in a technology that collects and stores information relevant for effective inventory management. The cost of obtaining information is often not difficult to analyze. Quantifying the benefits, however,

[1] Throughout the chapter, we use the terms inventory/production,control,replenishment/production, and order/produce interchangeably.

Ö. Özer (✉)
Columbia University, New York, NY 10027, USA
e-mail: oozer@columbia.edu

K.G. Kempf et al. (eds.), *Planning Production and Inventories in the Extended Enterprise*, International Series in Operations Research & Management Science 151,
DOI 10.1007/978-1-4419-6485-4_13,

requires thorough analysis and modeling. Consider, for example, the recent tracking technology known as radio frequency identification (RFID). Quantifying the cost of RFID implementation is relatively straightforward. But the benefit of this technology for the management of inventory is not clear. Comparing inventory models with and without the information obtained through RFID enables an inventory manager to quantify the value of RFID. In this chapter, we provide modeling examples that illustrate how an inventory manager can quantify the value of information.

The third important aspect of good inventory management is to coordinate decentralized operations. The coordination of information and inventory management has become increasingly more difficult with recent increases in supply chain complexity. Such complexities are the result of dramatic changes in manufacturing and distribution, including globalization and outsourcing. As a result, independent firms manage inventory allocated across different parts of the global supply chains. Each firm in the supply chain individually and myopically sets strategic and operational goals to minimize inventory- and production-related costs. Firms also maximize profits by using local information such as local cost structures, profit margins, and forecasts. As a result, the supply chain is suboptimized and not synchronized.

We have observed in the past that inability to optimize and synchronize these very complex inventory management issues can lead to catastrophic supply chain failures that make top business news. In 2001, Solectron, a major electronics manufacturer, had $4.7 billion in excess component capacity due to inflated forecasts provided by its customers. For exactly the same reason, Cisco, a major telecommunication equipment manufacturer, held $2.1 billion in excess inventory during the same year. Anticipating such inflation, manufacturers may discount the forecast information. Unfortunately, this caution, e.g., second guessing the forecast, may also lead to huge losses. In 1997, Boeing's suppliers were unable to fulfill Boeing's large orders because they did not believe in Boeing's forecasts. In this chapter, we provide examples of research that show such catastrophic outcomes are due to misaligned incentives and lack of coordination. These research works consider the interaction among multiple inventory managers and illustrate how these managers can align incentives through structured agreements and avoid (or mitigate) the adverse effects of lack of coordination.

Finally, good inventory management requires decision tools that can be embraced by their users. The formulations and the methodologies developed in multiechelon production and distribution systems are often very difficult to explain to nonmathematically oriented students and practitioners. In addition, data fed to these tools are not always accurate. Systems and people are bounded by limited information. In this chapter, we provide a discussion of some efforts to efficiently control multiperiod, multiproduct supply chains by developing easy-to-describe, near-optimal, and robust heuristics that can be implemented on a spreadsheet by solving, for example, newsvendor type problems.

To summarize, the chapter aims to provide a discussion of various topics and concepts from the centralized and decentralized inventory management literature. The emphasis will be on the use of information, and the role of new information technologies in inventory management. We provide examples of some ongoing research work. Our focus is on the modeling aspect rather than the detailed analysis. We do not state all the assumptions, the results nor the proofs. We deliberately trivialize and simplify the models so as to make the discussions easier to follow. We aim to bring together separate but inherently related research in inventory literature. By doing so, we hope to highlight potential research opportunities that lie on the boundaries. We focus primarily on the author's previous work. The chapter does not aim to provide a review of the rich volume of publications. For that purpose, where possible, we refer the reader to comprehensive reviews.

The rest of the chapter is organized as follows. In Sect. 13.2, we provide some examples of how managers can use information to better control inventory. In Sect. 13.3, we consider the interaction between multiple inventory control managers and the economics of contracting. In Sect. 13.4, we provide a discussion on large-scale inventory systems and rationality. In Sect. 13.5, we provide some concluding thoughts and possible future research directions.

13.2 Information in Centralized Inventory Management

We first discuss the use of information in centralized inventory management systems. An inventory management system is centralized when the system has access to credible information collected in a central location and managed by a single decision maker. Such a system is ideal; it does not have to coordinate disparate decisions and information. The manager needs to incorporate available information into the inventory control problem, identify the best replenishment policy and manage the system accordingly.

There are at least four reasons for studying centralized inventory systems. First, the results provide a benchmark against which decentralized inventory systems are measured. Second, the results enable us to quantify and understand the role and value of information in inventory management. Third, small-scale inventory systems are often centralized and are common in practice. Hence, it is necessary to know how to manage these systems. Industry has also learned the importance of centralized decision making such as the vendor managed inventory (VMI) initiatives. Fourth, the results also provide building blocks for large-scale systems with decentralized operations.

To effectively manage inventory, a manager must have access to three fundamental sets of information (1) information about demand such as forecasts; (2) information about assets such as the inventory available for sales, on order and where they are located; and (3) information about replenishment lead times. In Sect.13.2.1–13.2.4, we discuss single-location inventory control problems, which are the minimal building blocks for multilocation centralized inventory systems. We illustrate how the three fundamental sets of information are incorporated to develop effective production and inventory policies. We also show how managers can quantify the value of information by means of numerical computations. In Sect.13.2.5, we provide a discussion on how these single-location inventory control models are used to study multilocation inventory systems.

13.2.1 Current Demand Information

We refer to demand information as *current* when the information is based on current data such as point of sales information and when it does not provide future information such as a promotion scheduled for next period, or advance order information. Here, we briefly review the classical single-location inventory literature as a bridge to more recent work that incorporates the dynamic nature of demand information, such as forecast updates.

Early inventory models addressed the problem of minimizing ordering, holding, and backlogging costs for a single product at a single location over either a finite or an infinite horizon. Demand uncertainty is modeled as independent and identically distributed over time, i.e., demand D_t at each period t is an iid random variable. This modeling assumption uses current demand information. Historical data, such as, forecast errors can be used to estimate demand distribution for each period. Here, we will not provide a discussion on such estimation procedures.

In particular, the sequence of events for such a system is as follows. At the beginning of each period t, the manager reviews on-hand inventory I_t, any backorders B_t, and the pipeline inventory. The manager decides whether to produce $z_t \geq 0$. She incurs a nonstationary production cost of $K_t\delta(z_t) + c_t(z_t)$, where $\delta(z) = 1$ if $z > 0$, K_t is the fixed production cost, and c_t is the variable production cost. The production initiated at period $t - L$ is added to the inventory, i.e., L periods are required to complete the production. Demand D_t is observed. The demand for period t is satisfied through on-hand inventory; otherwise it is backordered. The manager incurs holding and penalty costs based on end-of-period net inventory.

Completing production takes L periods; hence, the manager needs to protect the system against uncertain demand during the production lead time, i.e., $D_t^L = \sum_{s=t}^{t+L} D_s$. We let

$$\begin{aligned} x_t :&\ \text{inventory position } \textit{before} \text{ the production decision is made} \\ &= I_t + \sum_{s=t-L}^{t-1} z_s - B_t, \\ y_t :&\ \text{inventory position } \textit{after} \text{ the production decision is made} \\ &= x_t + z_t. \end{aligned}$$

The expected holding and penalty costs charged to period t are given by $\tilde{G}_t(y_t) = \alpha^L E g_{t+L}(y_t - D_t^L)$, where α is the discount factor and $g_t(x)$ is the single period holding and penalty cost based on inventory on hand at the end of period t. The expectation is with respect to the lead time demand D_t^L. It is assumed that g_t is convex and coercive for all t.[2] These properties are satisfied, for example, when a positive holding cost is charged per unit of inventory on hand and a positive penalty cost is

[2] A function $g: \mathcal{R} \rightarrow \mathcal{R}$ is coercive if $\lim_{|x|\rightarrow\infty} g(x) = \infty$.

charged per unit of backlog. The solution to the following dynamic programming recursion minimizes the cost of managing this single item, single-location system for a finite horizon problem with $T - t$ periods remaining until termination.

$$J_t(x_t) = \min_{y_t \geq x_t} \{K_t \delta(y_t - x_t) + G_t(y_t) + \alpha E J_{t+1}(y_t - D_t)\},$$

where $J_{T+1}(\cdot) \equiv 0$ and $G_t(y_t) = (c_t - \alpha c_{t+1}) y_t + \tilde{G}_t(y_t)$.[3]

Scarf (1959) characterizes the optimality of an (s, S) policy. Under this policy, the manager orders up to S_t whenever the inventory position x_t falls below a critical level s_t. Veinott (1966) proves the optimality of (s, S) policies under different conditions. Infinite horizon results are due to Iglehart (1963). When the fixed cost of ordering is negligible, i.e., $K = 0$, an optimal policy is the base-stock policy with base-stock level S_t. Karlin (1960) and Veinott (1965) generalize the problem to account for seasonal variations in demand and nonstationary data and prove the optimality of period-dependent base-stock policy. We refer the reader to Porteus (1990a) for a review of classical inventory models[4].

Such policy parameters can often be obtained by a backward induction algorithm. A remarkable result that significantly reduces the computational burden is the optimality of a myopic policy that minimizes the current period inventory cost. Karlin (1960) and Veinott (1965) show that a myopic policy is optimal when the problem is stationary[5]; demand is stochastically increasing over time; or the myopic base-stock levels are increasing[6]. Morton and Pentico (1995) provide numerical evidence of how a myopic policy performs under various nonstationary environments. They also propose close-to-optimal, near-myopic policies. Iida (2001) also shows that myopic policies are effective when data change "slowly".

Noticing that historical demand information might be used to understand uncertain customer demand, several authors incorporated demand history into inventory control problems. Three groups of work capture this idea. The first group uses Bayesian models. Under these models, Bayes' rule defines a procedure to update the distribution of demand as new information becomes available. To the best of our knowledge, Dvoretzky et al. (1952) were the first to use this approach. Scarf (1960),

---

[3] It is often assumed that leftover inventory at the end of the planning horizon T is salvaged for c_{T+1} per item. Veinott (1965) shows that the inventory control problem with linear salvage value can be converted into an equivalent problem with zero salvage. Here we report the result of this conversion.

[4] Notice that the manager controls replenishment decisions to minimize inventory-related costs given an exogenous demand process. In certain cases, the manager may be able to adjust prices to shape demand at each period. This requires an inventory manager to have control over both replenishment and pricing decisions. For more discussion on this topic, we refer the reader to the reviews by Bitran and Caldentey (2003), Elmaghraby and Keskinocak (2003) and to a more recent paper by Huh et al. (2008).

[5] An inventory problem is said to be stationary if the cost and demand distributions are time invariant.

[6] We use the terms increasing and decreasing in the weak sense. Increasing means nondecreasing.

Azoury and Miller (1984), and Azoury 1985 extended this approach. The second group, Johnson and Thompson (1975), Miller (1989), and Lovejoy (1990), realized that the demand over consecutive periods might be correlated and used time series models to subsume demand dynamics. The third group incorporates Markov-modulated demand to the above inventory control problem (see, for example, Song and Zipkin 1993, Beyer and Sethi 1997, Abhyankar and Graves 2001 and Atali and Özer 2005).

13.2.2 Advance Demand Information

Most businesses rely heavily on demand forecasts for production and inventory planning. Demand over time can be highly correlated. Forecasting methods can help identify such patterns. A group of scholars have incorporated the dynamic nature of forecast revisions into inventory control problems. Papers in this group include those of Hausman (1969), Graves et al. (1986), Heath and Jackson (1994), Güllü (1996), and Toktay and Wein (2001). All of these works show that incorporating demand updates to control problems reduces the cost of managing inventories by proposing control methods that are responsive to forecast information.

Recent advances in information technology have enabled managers to be more proactive and obtain *advance* demand information in addition to improving demand forecast. Different customers have varying willingness to wait for the orders they placed. A good example of this concept is Dell's online Intelligent Fulfillment initiative, which allows four different levels of response time to customer orders: (1) standard (conventional or 5 day promised order lead time); (2) value delivery (slower but lower shipping cost); (3) premium delivery (same day delivery); and (4) precision delivery (specific date). A portfolio of online customers with differing response time preferences gives rise to advance demand information (ADI). Comparing inventory models with and without ADI, a manager can quantify the value of demand information contained in ADI (Özer 2003).

Several plausible strategies can be used to obtain advance demand information. When people order a customized product, they expect to wait for the product to be customized to their request. This can be called a built-in ADI. Alternatively, a discount could be offered for early orders to segment the customer based on their willingness to wait. If pricing is not an option, special service incentives could be offered for early orders. For example, a major truck manufacturer in North America provides free maintenance (up to 10 years) for third party logistic providers (such as UPS) who purchase trucks a few years in advance of using them. Essentially, we are seeking those customers who have a high sensitivity to customization, price or service, and who also have a lower sensitivity to lead time or waiting time. These are denoted by "A" in Fig. 13.1. They are the possible source of ADI.

Designing effective strategies to collect this information requires one to quantify the benefit of ADI. To do so, Gallego and Özer (2001a), Özer (2003), and Özer

Fig. 13.1 Source of ADI

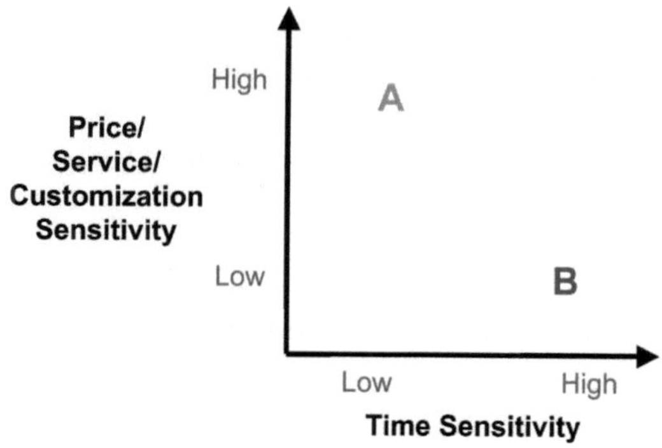

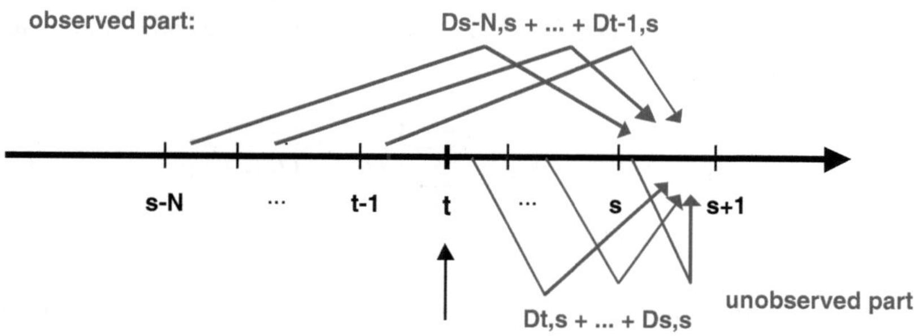

Fig. 13.2 Observed and unobserved part of the demand

and Wei (2004) show how to use this information optimally. In particular, they incorporate advance demand information into periodic-review inventory control problems.

ADI is obtained when a customer places an order in any period t for delivery in a future period $s \in \{t+1,\ldots,t+N\}$. From the perspective of the production manager, the demand stream during period t is a vector:

$$D_t = (D_{t,t},\ldots,D_{t,t+N}),$$

where $D_{t,s}$ represents the nonnegative demand for period s placed during period t and N is the length of the *information horizon*. Note that when $N=0$, the problem reduces to the inventory problem with current demand information. This is a random vector and its uncertainty is resolved at the end of period t. Under this demand model, at the beginning of each period t, demand for a future period $s>t$ can be decomposed into two parts as illustrated in Fig. 13.2: the observed part $O_{t,s} \equiv \sum_{r=s-N}^{t-1} D_{r,s}$ and the unobserved part $U_{t,s} \equiv \sum_{r=t}^{s} D_{r,s}$.

The sequence of events is similar to the one described in the previous section. Completing production takes L periods; hence, the manager should protect the system against the lead time demand. Because of advance demand information, the manager knows part of the lead time demand, that is, $\sum_{s=t}^{t+L} O_{t,s}$. The expected cost charged to period t is based on the net inventory at the end of period $t + L$. Let

$$x_t^a : \textit{modified} \text{ inventory position } \textit{before} \text{ the production decision is made}$$
$$= x_t - \sum_{s=t}^{t+L} O_{t,s},$$
$$y_t^a : \textit{modified} \text{ inventory position } \textit{after} \text{ the production decision is made}$$
$$= x_t^a + z_t.$$

Notice that these variables subtract the observed part of the lead time demand, hence the name *modified*. In addition to x_t^a, the manager also keeps track of observations beyond the lead time, $O_t = (O_{t,t+L+1}, \ldots, O_{t,t+N-1})$. At the end of the period t, we update the state space by

$$x_{t+1}^a = y_t^a - D_{t,t} - \sum_{s=t+1}^{t+L+1} D_{t,s} - O_{t,t+L+1}, \tag{13.1}$$
$$O_{t+1,s} = O_{t,s} + D_{t,s}. \tag{13.2}$$

The expected holding and penalty cost charged to period t is given by $\tilde{G}_t(y_t) = \alpha^L E g_{t+L}(y_t - \sum_{s=t}^{t+L} U_{t,s})$. The solution to the following dynamic programming recursion minimizes the cost of managing this system for a finite horizon problem with $T - t$ periods remaining to the termination.

$$J_t(x_t, O_t) = \min_{y_t \geq x_t} \{K_t \delta(y_t - x_t) + G_t(y_t) + \alpha E J_{t+1}(x_{t+1}, O_{t+1})\}, \tag{13.3}$$

where $J_{T+1}(\cdot,\cdot) \equiv 0$ and $G_j(y_j) = (c_j - \alpha c_{j+1})y_j + \tilde{G}_j(y_j)$.

Gallego and Özer (2001b) characterize the optimality of (1) a state-dependent (s, S) policy for an inventory system with positive fixed (set-up) costs and (2) a state-dependent base stock policy for an inventory system without set-up costs both for finite and infinite horizon problems. The policy parameters depend on customer commitments made beyond the production leadtime. For example, if the production lead time is four periods, optimal policy parameters depend on the total customer commitments made today for delivery after four periods. Under this policy, the manager produces up to S whenever the *modified* inventory position x_t^a drops to or below s. Gallego and Özer provide monotonicity results and characterize conditions when myopic policies are optimal. They also determine conditions under which ADI has no operational value. Through numerical studies and by comparing models with and without ADI, the authors quantify the benefit of inducing and obtaining ADI.

We note that incorporating advance demand information not only yields better practices through reducing inventories, but also enables companies to have control policies that are more responsive to changes in demand patterns. This information allows a shift from make-to-stock to make-to-order production. There is a growing body of research that shows how ADI can be used to improve costs in a capacity constrained system, continous review problems, or multiechelon structures (Hariharan and Zipkin 1995; Schwarz et al. 1998; Gallego and Özer 2002; Karaesmen et al. 2002; Özer 2003; Zhu and Thonemann 2004; Özer and Wei 2004; Hu et al. 2004; Benjaafar et al. 2005; Marklund 2006; Wang and Toktay 2008; Gayon et al. 2009). These models can be used to quantify the value of advance demand information in various settings. Being able to quantify its value, an inventory manager can decide how to optimally acquire advance demand information through pricing and advance sales and how to use this information in, for example, capacity decisions (Boyaci and Özer 2004). Such research also bridges the revenue management literature with the capacity management literature.

ADI and capacity management: We discuss how the results from the ADI literature were used to quantify the value of capacity and advance demand information for a global telecommunications equipment manufacturer. During the last quarter of 2002, this equipment manufacturer explored the strategy of advance selling to improve long-range forecasting for planning the capacity of a new factory. Accordingly, before securing the capacity the firm considered preselling wireless base stations to its regional cellular phone operators.

The traditional view of capacity planning is that capacity is fixed, and lead times will vary to compensate for surges and gaps in orders. A different viewpoint is that one can fix and guarantee a lead time; this requires the ability to flex capacity as needed. Fig. 13.3 summarizes these two approaches. Suppose we had several types of customers as in Fig. 13.4, where each class had different lead-time requirements. Then we could guarantee lead times by customer segment, and implement this through careful scheduling of the facility. This strategy enables the firm to obtain advance demand information which can be used for better inventory and capacity planning.

The next issue is the management of the production system given the available capacity Q and the advance demand information. In order to minimize the cost of

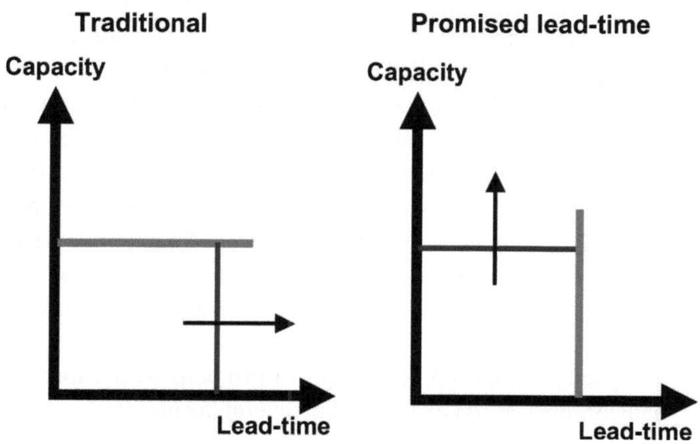

Fig. 13.3 Resource planning: always having capacity

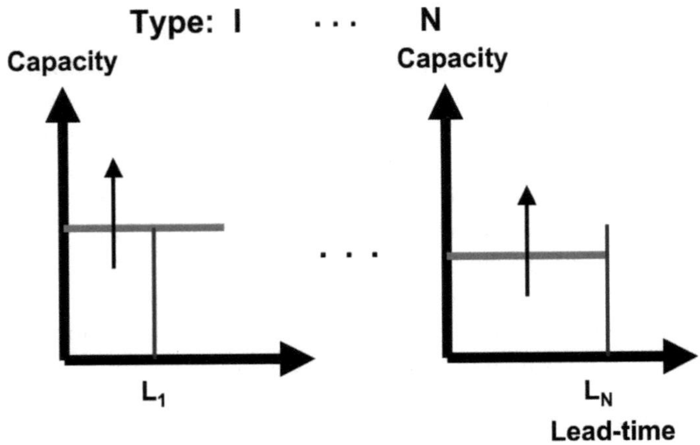

Fig. 13.4 Resource planning: fixed capacity with advance demand information

managing this production system, the manager maintains a safety stock. Recall that the manager would also like to satisfy some customers who have short lead times (even shorter than the production lead time) in addition to those who book well in advance. Hence one needs to maintain a safety stock. But what is the optimal level of inventory? More inventory means more money tied up, while less inventory may result in loss of customers and loss of goodwill. The solution to a dynamic program similar to the one in (13.3) (but significantly more difficult to analyze) provides the best level of safety stock that minimizes the cost of underage and overage for a given planning horizon (see Özer and Wei 2004 for details).

Through an extensive numerical studies one can quantify the benefits gained through ADI for a capacity constrained system. In Fig. 13.5 we provide one such example. The x-axis shows the available capacity for production. The y-axis shows the total inventory management cost (which is rescaled so that the cost of managing a system with infinite capacity and 100% ADI is zero). The three curves illustrate different levels of ADI. Curve A is the base case where all customers demand the product as soon as they place an order, whereas curve C has some customers who place their orders well in advance, and curve B is in between.

We observe from each of these curves that additional capacity has diminishing returns. This suggests that there is an optimal level of capacity beyond which increasing capacity has limited operational value in managing this system. We also observe the reduction in the inventory management cost as a function of ADI and capacity. The vertical difference between these curves (for example between curves A and C) depicts the reduction in inventory costs due to employing advance demand information. Note that this reduction is more valuable when the firm is working under tight capacity.

Consider a capacity expansion (contraction) problem with capacity increment ΔQ. The expansion cost $C(\Delta Q)$ may take several forms, such as linear, power or step cost function (Luss 1982). If capacity expansion is a one time decision,

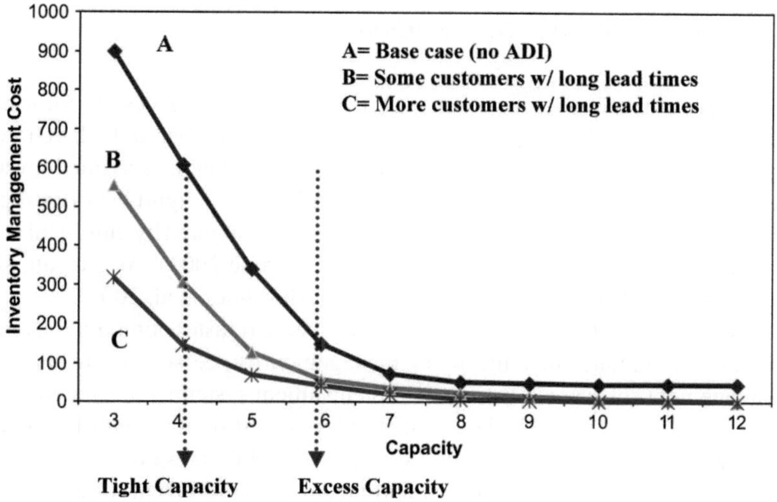

Fig. 13.5 ADI versus capacity

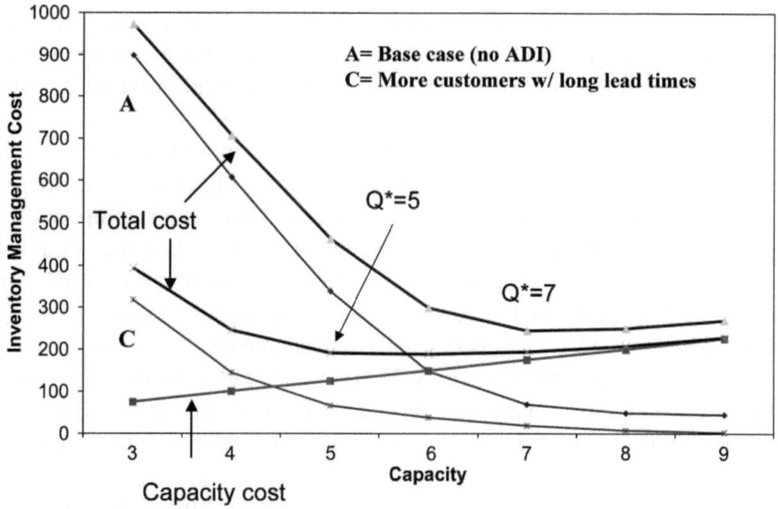

Fig. 13.6 Optimal capacity size

then the manufacturer's problem is to solve $\min_{\Delta Q}\{C(\Delta Q) + J_t(x, O|Q + \Delta Q)\}$. Figure 13.6 provides an example of this problem, when $C(\Delta Q) = 100 * \Delta Q$ under two advance demand information scenarios. For this particular example, convincing customers to place orders in advance reduces the optimal capacity expansion decision from $Q^* = 7$ to 5 units. This is another example illustrating how advance demand information can be a substitute for capacity (Özer and Wei 2004).

13.2.3 Imperfect Asset Information

To effectively manage inventory, a manager must also have access to information about assets, such as, the inventory available for sales, on order and where they are located. Recent surveys and empirical work have shown that unaccounted inventory due to, for example, theft or misplacement, can lead to a significant discrepancy between inventory records and actual inventory (as documented by empirical studies such as Rinehart 1960; Raman et al. 2001; ECR Europe 2003). As a result, stock-outs are widespread at retailers and distributors (Alexander et al. 2002).

Since the early 1980s, the availability of cheaper and faster computation enabled companies to automate their inventory management processes and to use inventory management softwares. Automatic replenishment systems track the number of products in stock and place replenishment orders based on the control policies set by the underlying software. A crucial assumption used by these inventory management systems is that inventory record and actual on-hand inventory are identical. Similarly, the standard inventory control literature has never differentiated between inventory *record* and *actual* inventory. The two have always been considered to be the same. In the previous sections, we assumed that the manager knows the exact value of, for example, the inventory position. The implicit assumption was that all demand sources are visible. Next we provide modeling examples through which we aim to illustrate the impact of inaccuracies and asset information on effective inventory management.

Early modeling approach for inventory control under imperfect asset information is due to Iglehart and Morey (1972). They study the impact of transaction errors only and do not consider misplacement or shrinkage. They also decompose the error management problem from inventory management. In particular, they establish the approximate buffer stock required to hedge against transaction errors independent of the buffer stock necessary to hedge against paying customer demand. Kang and Gershwin (2005) consider discrepancy due to shrinkage and its impact on inventory management through a simulation study. We refer the reader to Lee and Özer (2007) for a detailed discussion of inaccuracy problems. Through model-based analysis, the authors quantify the benefit of a tracking technology known as RFID in supply chain management.

In a recent paper, Atali et al. (2004, 2006) characterize three different kinds of demand streams that result in inventory discrepancy. Some demand streams result in permanent inventory shrinkage (such as theft and damage). They refer to this stream as *shrinkage*. Some demand streams are temporary and can be recovered by physical inventory audit and returned to inventory (such as misplacement). They refer to this demand stream as *misplacement*. The final type of demand stream (such as scanning error) affects *only* the inventory record and leaves actual inventory unchanged. They refer to this stream as *transaction errors*. It is necessary to characterize these sources separately because each of these sources affects the system in a unique way. For example, misplaced items can be returned back to inventory after an inventory audit, whereas stolen items cannot.

Control \ Visibility	*Without*	*With*
Passive	*Base Case* POS-driven replenishment	*RFID I Case* Replenishment based on accurate inventory
Active	*Smart Case* Adjustment based on statistical parameters	*RFID II Case* Actively correct misplaced items

Fig. 13.7 Inventory management cases under imperfect asset information

There are four ways to manage an inventory system that faces an inventory discrepancy problem, summarized in Fig. 13.7 (Atali et al. 2004, 2006). The first way is to ignore the discrepancy problem and use only the point of sales data information to drive the replenishment process. The second way is to use the statistics about unobservable demand sources in driving the replenishment process, for example, by carrying additional buffer stock to hedge against transaction errors. The third way is to invest in a technology such as RFID that enables complete visibility of inventory movement and use the actual information (instead of the statistics) to drive the replenishment process. The fourth way is to go one step further and use the visibility to prevent or reduce unobservable demand sources, for example, by locating and reshelving misplaced items as soon as a customer misplaces the item.

Here we discuss the formulation of the third case (RFID I case in Fig. 13.7) and refer the reader to Atali et al. (2004, 2006) for a detailed treatment of the other cases. The sequence of events for this case is as follows. At the beginning of period t, the inventory manager reviews the state of the system and decides to order $z_t \geq 0$ units from an outside supplier with ample supply. The replenishment lead time is assumed to be zero. The cost of ordering is c_t per unit. There is no fixed cost for placing an order. Purchasing customer D_t^p, misplacement D_t^m, shrinkage D_t^s, transaction errors D_t^τ arrive in any sequence. Note that the realizations of these error sources are observed because we are considering the third way to manage the inventory system. At the end of the period, the manager incurs a linear holding cost h_t and a linear lost-sales cost p_t based on the end of period physical on-hand inventory. Holding cost is incurred for the misplaced items even though they are not available for sales. No lost-sales cost is incurred for unmet demand from nonpaying customers. If the period is a counting (audit) period, an inventory audit is conducted at the end of that period. The inventory record is reconciled: error is corrected, and all misplaced items are returned to inventory. Otherwise, errors continue to accumulate. The planning horizon is a multiple of counting cycle length, that is, $T \in \{N, 2N, 3N, \ldots\}$. At the end of the planning horizon T, the inventory left over is sold for a linear salvage value of c_{T+1}.

At the beginning of period t, the manager knows the inventory *record* x_t^r, the accumulated error terms e_t^s, e_t^m, e_t^τ and the number of periods elapsed since the last inventory count, i_t. The state space of such as system can be summarized by (x_t, e_t^m, i_t), where

$$x_t = x_t^r - e_t^m - e_t^s - e_t^\tau$$

is the *sales-available* on-hand inventory, and $i_t \in \{0, 1, \ldots, N-1\}$. The state of the system evolves according to the following equations.

$$x_{t+1} = \begin{cases} [y_t - D_t]^+, & \text{if } i_t \neq N-1 \\ [y_t - D_t]^+ + e_t^m + m_t, & \text{if } i_t = N-1 \end{cases} \quad (13.4)$$

$$e_{t+1}^m = \begin{cases} e_t^m + m_t, & \text{if } i_t \neq N-1 \\ 0, & \text{if } i_t = N-1 \end{cases} \quad (13.5)$$

$$i_{t+1} = (i_t + 1) \bmod N, \quad (13.6)$$

where $y_t = x_t + z_t$ and $D_t = D_t^p + D_t^s + D_t^m$ and m_t is the realized misplacement. The single period expected holding and penalty cost charged to period t is based on *sales-available* on-hand inventory and the accumulated misplacement.

$$\tilde{G}_t(y_t, e_t^m) = h_t E_{D_t, m_t}([y_t - D_t]^+ + e_t^m + m_t) + p_t E_{D_t^p, a_t}(D_t^p - a_t). \quad (13.7)$$

Transaction errors are random observation disturbances and they have no direct impact on the *sales-available* on-hand inventory x_t.

With perfect visibility, the manager optimizes the stock levels in full awareness of the inventory errors that take place during period t. Let J_t^v be the cost of managing this system for a finite horizon with $T - t$ periods remaining to the end of the planning horizon. The optimal replenishment policy would be to select the value of y_t that minimizes the following dynamic programming algorithm.

$$J_t^v(x_t, e_t^m, i_t) = \min_{y_t \geq x_t} \{G_t(y_t, e_t^m) + \alpha E J_{t+1}^v(x_{t+1}, e_{t+1}^m, i_{t+1})\}, \quad (13.8)$$

where $J_{T+1}^v(x_{T+1}, \ldots, .) = 0$ and $G_t(y_t, e_t^m) = c_t y_t - \alpha c_{t+1} E x_{t+1} + \tilde{G}_t(y_t, e_t^m)$. The leftovers at the end of the planning horizon $T+1$ are salvaged for a linear price.

To calculate the aforementioned expectations in the dynamic programming algorithm, one needs to obtain the distribution of sales a_t and misplacement m_t during any period t. However, the realization of these variables and their distribution depend on the sales-available on-hand inventory x_t and the order in which misplacement, shrinkage, and paying customer demands arrive.

Consider a modified model in which the paying customer demand always arrives first, demand for shrinkage arrives next and demand for misplacement arrives last. With this sequence, sales during any period are maximized while misplacement is minimized. The transaction error can arrive in any where in the sequence because it does not affect the physical inventory. Given this sequence, the sales and the misplacement during period t are

$$a_t = \min\{D_t^p, y_t\}, \tag{13.9}$$

$$m_t = \min\{D_t^m, [y_t - D_t^p - D_t^s]^+\}. \tag{13.10}$$

The state of the system evolves according to (13.4–13.6), but with m_t replaced by its new definition above. Similarly, the single-period cost function is the same as in (13.7) but with a_t and m_t replaced by their respective definitions. Using similar demand prioritization ideas, one can construct bounds and effective solutions for the above dynamic programming problem. They enable effective inventory control methods when the manager uses RFID or a similar technology that provides complete visibility of inventory movement in the store.

Atali et al. (2004) characterize efficient replenishment policies for all four cases in Fig. 13.7. Using these models and comparing the resulting cost of each scenario, they quantify the *true* value of visibility provided by RFID. Consider, for example, the value of visibility. When the system does not use a technology such as RFID, the manager can use; either the *informed* policy that corresponds to the smart case, or an *ignorant* policy that corresponds to the base case in Fig. 13.7. Recall that in the base case the replenishment policy is obtained without consideration of the discrepancy problem, whereas informed policy uses some statistics about discrepancy. The *true* value of visibility is given by the cost difference between the informed policy and the policy that uses visibility, i.e., RFID I or II.

Figure 13.8 compares the resulting cost for a problem instance as a function of total error with respect to paying customer demand, i.e., total average error divided by average paying customer demand. The lowest curve is the cost of following an effective replenishment policy when the manager has complete visibility of inventory and follows an active control strategy. This figure illustrates that by

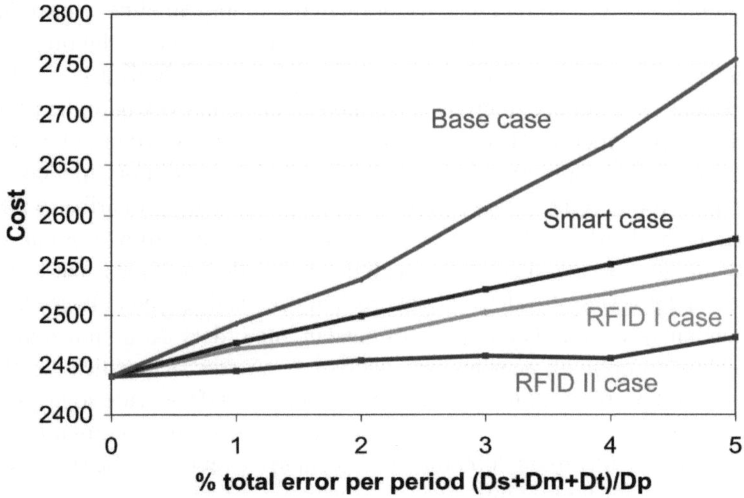

Fig. 13.8 Value of visibility and active control as a function of total error source

using an informed policy to compensate for the discrepancy problem, the manager can reduce costs significantly. The value of visibility also increases with the total percentage errors. For a particular example, when compared with the ignorant policy (base case), the visibility enabled system reduces cost by 9.1% and increases sales by 1.8%. However, when compared with the smart policy, the cost is reduced by 3.1% and the sales is increased by 0.1%. For this system, assuming that visibility also enables one to reduce the shrinkage rate by 50%, the manager can save (the difference between the visibility enabled systems with different shrinkage rates) an additional 2.6%, and increase sales by 0.1%, both of which can be interpreted as the value of prevention due to visibility brought by a technology such as RFID.

Recently, Atali et al. (2006) model demand streams using a random disaggregation model. In particular, let D_t denote the random customer demand during period t. An arriving customer buys the product with probability θ_p; misplaces the item with probability θ_m; or damages/steals the item with probability θ_s such that $\theta_p + \theta_m + \theta_s = 1$ for all t. Both demand modeling approaches have their own appeal. Random disaggregation approach simplifies the previous analysis. In particular, one does not need to construct bounds through demand prioritization. Calibrating the model and fitting data are relatively simpler as well. However, the previous approach allows for independent demand streams for paying and nonpaying customers.

13.2.4 Lead-Time Information

Effective management of inventory also requires one to have information about replenishment lead time and products' location in the supply pipeline. Note that when replenishment lead time is certain, i.e., when the manager knows the time required to replenish her inventory with certainty, she can follow methods discussed in previous sections. In this case, there is really no reason to know where the product is within the supply pipeline. However, when lead times are uncertain, information on the location of the supply plays a critical role. Classical inventory models assume that lead time for an order is independently drawn from a given distribution. Kaplan (1970) and Ehrdhardt (1984) discuss two assumptions that allow optimal control policies analogous to classical results with deterministic lead times. These assumptions are (1) deliveries of orders cannot cross in time and (2) the delivery lead time is independent of the number and size of outstanding orders. Anupindi et al. (1996) provide close-to-optimal, near-myopic heuristics to solve stochastic lead time inventory control problems. Janakiraman and Roundy (2004) provide some convexity results that enables the use of search procedures to determine optimal base-stock levels (see Chap. 7 of Zipkin 2000 for a comprehensive review of the stochastic lead time inventory control problems).

Song and Zipkin (1996) Song and Zipkin (1996) model the supply process as a Markov chain. The lead time L_t for an order in period t is a Markov chain with a finite state space. The transition matrix is such that orders are received in the order they were shipped (i.e., order cross over is not allowed). They assume that the inventory manager has visibility of the supply process (i.e., the Markov state) at the beginning of each period. She uses this information to revise the inventory ordering decision. The authors show that a state-dependent replenishment policy is optimal. Chen and Yu (2005) consider the problem in which the manager does not know the status of the system, but knows that the lead time is generated by a Markov process. They show that the value of lead time information is small for slow-moving items. However, it can be as high as 40% for fast-moving items. To demonstrate this, they numerically compare the model in which lead time is observable to that of Song and Zipkin.

The conventional modeling approaches for stochastic lead times generally assume that the statistical information essentially boils down to the mean and standard deviation of the lead time, and the safety stock takes into consideration such statistics. Recent information technologies, however, enable a manager to collect some advanced knowledge about the lead time as the product progresses over intermediate points, known in logistics as "choke points". Through tracking technologies and well-connected computer networks, a manager can follow the progress of a supply before it reaches the store. Gaukler et al. (2008) quantify the benefit of this supply progress information. They propose and evaluate a replenishment policy that uses order progress information for emergency ordering together with the (Q, R) policy. In particular, the manager places a *regular* replenishment order of size Q when the inventory position drops to the reorder level R. They model the sojourn time for a regular order to move from one choke point to the next with a general nonidentical distribution and provide additional results for the exponential distribution case. In addition, the manager also has the option to place an *emergency* order at a cost premium $K(l)$ of size αQ, which arrives after a deterministic lead time l. They characterize the optimality of a state-dependent threshold policy for releasing an emergency order. In particular, the retailer monitors the outstanding regular orders location in the supply system, that is, the last choke point where the regular order was registered (of course, if a regular order is outstanding). If the inventory position is less than a state-dependent threshold $\bar{y}$, the retailer places the *emergency* order. The threshold depends on where the regular order was registered last. Through a numerical study, the authors report overall cost savings ranging from 2.8 to 5.5% due to supply progress information. They show that the emergency ordering option eliminate up to 99% of the cost due to backlogging a customer. See also Moinzadeh and Schmidt (1991) and Moinzadeh and Aggarwal (1997) for the use of emergency ordering in single- and multiechelon inventory systems.

13.2.5 Multi-location Inventory Systems

So far, we discussed the role of three fundamental sets of information (demand, asset, and lead time information) in effective inventory management for single-location systems. Next, we provide a brief discussion on how these single-location inventory control models form the foundations for studying multi-location inventory systems. We still focus on systems that are managed by a single decision maker. There are three fundamental multi-location systems: locations in series, assembly, and distribution (arborescence or divergent) systems. In a series system, each location replenishes its inventory form its immediate predecessor. In an assembly system, each location replenishes from its immediate predecessors, i.e., from multiple locations but each location ships its product to only a single-downstream location (successor). In both series and assembly systems, customer demand is satisfied only by the last location. In distribution systems, each location can order inventory only from its immediate predecessor but can ship its inventory to multiple-downstream locations. Below we provide some references that study these inventory systems without providing a detailed discussion of the model and problem setup.

Clark and Scarf (1960) initiated the study of multiechelon inventory systems. They show that a serial system can be optimally decomposed into *single*-location problems and characterize the optimality of echelon base-stock policies. Hence, one needs to know how to manage a single-location problem to effectively manage a serial system. Under this policy, a central inventory manager observes the echelon inventory position of each location and places an order from the outside supplier if the first echelon's inventory position is below its base-stock level. The manager also pushes inventory (as much as possible) to the downstream location j from its immediate predecessor if location j's echelon inventory position is less than its echelon base-stock level. Federgruen and Zipkin (1984c), extend the results for stationary infinite horizon problems. Chen and Zheng (1994) establish lower bounds on the average cost and construct feasible policies that achieve these bounds. Unlike the single-location inventory control literature, the multiechelon in series literature lacks models that incorporate historical demand information. Chen and Song (2001) write the first paper to study the serial system with a nonstationary demand process, which is modulated by a finite state, exogenous Markov chain. Graves et al. 1998 provide heuristic allocation of inventories across a serial system that obtains a forecast over a finite horizon. Gallego and Özer (2003) incorporate advance demand information into multiechelon, inventory systems in series and prove the optimality of state-dependent, echelon base-stock policies for finite and infinite horizon problems. The authors show that under certain conditions a *myopic* policy is optimal for a finite horizon multiechelon inventory problem in series with and without advance demand information. This result significantly reduces the computational burden required to solve such serial systems. Muharremoglu and Tsitsiklis (2008) prove the optimality of state-dependent base-stock policies for serial systems with Markov-modulated demand and Markov-modulated stochastic lead times without order crossing. These systems are also fundamental to the study of assembly and distribution systems. For example, Rosling (1989) shows how to convert an *assembly* system to an equivalent serial system. See also Muckstadt (1973), Van Houtum et al. (1996).

One of the most common multiechelon structures in practice are the distribution systems, which are also known as one-warehouse-multiretailer systems. Products enter the system from an outside supplier to the warehouse, which in turn replenishes various retailers. Stochastic demand is satisfied as much as possible through on hand inventory at the retailers. Clark and Scarf (1960) show that optimal control policies, if they exist, would be very complex for distribution systems. Since then the research on distribution systems has shifted towards identification of close-to-optimal heuristics and evaluation of a plausible class of policies. There are two approaches to solve this problem: approximation by relaxation as by Federgruen and Zipkin (1984a), Aviv and Federgruen (2001a), Özer (2003) and approximation by restriction as by Eppen and Schrage (1981), Federgruen and Zipkin (1984b) and Özer (2003). The first approach considers relaxing a constraint set to obtain a simpler problem with lower-dimensional state space. It develops a heuristic based on this lower bound problem to solve the original problem. The second approach restricts the policy space to a class of policies and optimizes over this class under additional assumptions. The restriction approach, unlike the relaxation approach, does not guarantee any bound on the optimal solution. Other researchers that use approaches that do not guarantee any bounds include Diks and de Kök (1998). Comprehensive earlier reviews can be found by Axsäter (1993) for continuous review (also known as pull) and Federgruen (1993) for periodic review (also known as push) inventory systems.

In distribution systems, the "warehouse" may serve as the coordination center. It may also help negotiate lower procurement prices. Eppen and Schrage (1981) illustrate that the warehouse also serves an important enabler for *statistical economies of scale*, commonly known as risk pooling, that is, the portfolio effect of coordinating inventory decisions and holding inventory at the distribution center rather than at the retailers. Aviv and Federgruen (2001a) incorporate a Bayesian framework into the demand process and introduce the concept of *learning effect* to the benefit of having a central distribution center. The ability to obtain information about the demand during the first periods enables updating the demand process, resulting in improved allocation to retailers. Özer (2003) incorporates advance demand information structure obtained from customers through each retailer. The author establishes a close-to-optimal state-dependent replenishment and allocation policy that responds to the changes in customer demand. The author also provides a closed-form solution to approximate the system-wide inventory level. Using such explicit solutions, the model and the heuristic, he quantifies, for example, the benefit of advance demand information and its impact on allocation decisions and the joint role of risk pooling and advance demand information. For a review of these approaches, we also refer the reader to Özer (2003).

The distribution system described here can also be interpreted as a multi-item production system with a common intermediate product. In this interpretation, the warehouse represents the differentiation point. During the first phase of the production a common batch is produced. At the end of this phase, the manager must decide on how much of each differentiated item to produce from the batch of the common intermediate product. This interpretation forms the basis of postponement strategies; see the papers by Lee et al. (1993) and Lee and Tang (1997). We conclude this section by noting that at the heart of all complex inventory systems lies the single-location (stage, product, item) model that we addressed in the previous subsections.

13.3 Information in Decentralized Inventory Management

So far, we discussed the first two important aspects of good inventory management: effective use of available information and how to quantify the value of this information. Here, we focus on the third important aspect of good inventory management, that is, to coordinate decentralized operations in an extended supply chain. Global operations involve several locations managed by several inventory managers. The decisions and information are often decentralized. Many experts have heralded advances in information technology and Internet infrastructure, both of which enable better visibility and information sharing, as the key to effective management of inventory. Suppliers and manufacturers can share private information regarding, for example, costs or forecasts, but will they want to? Firms may be reluctant to collect, process, and share information because of conflicting incentives. Aligning incentives improves firms profits and sustains the use of information technology.

Inventory managers can use formal contracts to align incentives and induce information sharing. There are two forms of information asymmetry. The informed party may withhold information to gain strategic advantage. In such cases, the uninformed partner can propose a menu of contracts to extract this information; this interaction is known as *adverse selection* or *screening*. Alternatively, the informed party may signal his information to gain cooperation. However, he needs to signal private information in a *credible* way; this interaction is known as *signaling* game. Another form of information asymmetry arises as moral hazard where one partner influences system profit through an action or choice not observable to the other. The nonacting partner designs a contract to maximize his own profit (Fudenberg and Tirole 1991 and Salanie 1997). This section provides examples of such interactions in inventory management.

We attribute incentive problems in supply chains to lack of credible information sharing and three major risk imbalances: capacity risk, inventory risk, and quality risk (Özer 2004). Because of lack of credible information sharing, the adverse effects of inventory and quality risks are more severe for a decentralized supply chain than for a vertically integrated supply chain. Here, we discuss some recent and ongoing research in designing contracts to eliminate or mitigate these adverse effects. We typify a two level supply chain by referring to an upstream member as the supplier and the downstream as the manufacturer.

13.3.1 Capacity Risk

Forecasting demand is inherently difficult due to short product life cycles and long production lead time. Hence, supply chains face the risk of either excess capacity due to low demand realization (downside risk) or lack of product availability due to high demand realization (upside risk). Consider a manufacturer who builds to order and requires the supplier to deliver just in time. To deliver on time, the supplier secures component capacity or inventory in advance of a manufacturer order. If consumer demand turns out to be high, both the supplier and the manufacturer face upside capacity risk. However, if consumer demand turns out to be low, only the supplier faces downside capacity risk. Lack of proper risk sharing exacerbates the cost of capacity risk. Next we provide a model to quantify and illustrate how structured agreements can be used to align incentives and avoid (or mitigate) the adverse effects of capacity risk. For a more detailed discussion see Özer and Wei (2006).

Double Marginalization

The severity of capacity risk for each party depends on the contractual agreements. Under a wholesale price contract, for example, the manufacturer pays a wholesale price w to the supplier for each unit ordered. The supplier decides on the component capacity K to maximize his profit prior to observing demand. Let c_k be the unit cost of capacity. This cost could also represent an equivalent annual cost of capacity. Demand D is realized and the manufacturer places an order. The supplier fills the order as much as possible, at a unit cost c; that is, he delivers $\min(D, K)$. The manufacturer receives the order and sells at a fixed price $r > 0$.[7] Suppose unmet demand is lost without additional stock out penalty, and unsold inventory has zero salvage value without loss of generality.

Note that demand D is uncertain at the time when the supplier builds capacity. Suppose the demand forecast is such that $D = \mu + \epsilon$, where μ is the mean, which is a positive constant, and ϵ is a zero mean random variable with a cdf $G(\cdot)$, which represents the market or forecast uncertainty. Such information can be constructed by using information obtained, for example, through a third-party market research firm (such as Dataquest services of Gartner group). For a given capacity K, the manufacturer's and the supplier's expected profit before demand is realized are given by

$$\Pi^m(K) = (r - w)E \min(D, K), \tag{13.11}$$

$$\Pi^s(K) = (w - c)E \min(D, K) - c_k K. \tag{13.12}$$

The supplier maximizes his profit in (13.12) by setting capacity to

$$K^w = \mu + G^{-1}\left(\frac{w - c - c_k}{w - c}\right).$$

[7] The manufacturer may carry out some value added operations that cost, say m per unit. She sells at a fixed unit price $r' > 0$. So her effective sales price is $r = r' - m$. Hence, without loss of generality, we assume $m = 0$. Of course, the story would be different if the manufacturer was building to stock as we will discuss in Sect.13.3.2.

Next consider the centrally integrated supply chain in which a single firm owns the manufacturer and the supplier. This centralized firm's expected profit and its optimal capacity would be

$$\Pi^{cs}(K) = (r-c)E\min(D,K) - c_k K, \tag{13.13}$$

$$K^{cs} = \mu + G^{-1}\left(\frac{r-c-c_k}{w-c}\right). \tag{13.14}$$

Note from (13.12) and (13.13) that the supplier's marginal profit is less than the vertically integrated supply chain's marginal profit. This difference is due to double marginalization. The supplier, therefore, secures less capacity than what would be optimal for a vertically integrated supply chain, that is $K^w \leq K^{cs}$. Note that $\Pi^{cs}(K^{cs}) \geq \Pi^m(K^w) + \Pi^s(K^w)$. Hence, both the manufacturer and the supplier are leaving money on the table due to decentralized operations. The magnitude of this inefficiency depends on the parameters.

The manufacturer may encourage the supplier to build more capacity by providing some protection against the downside risk, the risk of having excess capacity. Observe that the manufacturer's payoff (the realized profit) is always nonnegative, while the supplier faces the risk of a negative payoff. The manufacturer can share this risk by providing a payment in case of excess capacity after demand is realized. One such contract is the *payback* contract (w, τ), under which the manufacturer pays the supplier w per unit for its order and τ per unit for unused capacity $(K-D)^+$.

The manufacturer's and supplier's expected profit functions for this case are

$$\Pi^m(K) = (r-w)E\min(D,K) - \tau E(K-D)^+,$$
$$\Pi^s(K) = (w-c)E\min(D,K) + \tau E(K-D)^+ - c_k K.$$

The supplier solves $\max_{K\geq 0}\Pi^s(K)$. The optimal capacity is $K^\tau \equiv \mu + G^{-1}((w-c-c_k)/(w-c-\tau))$. To achieve channel coordination, we equate K^τ with K^{cs} and solve for τ, resulting in $\tau = (r-w)c_k/r-c-c_k$. Hence, the payback contract (w,τ) can coordinate the channel, that is the sum of the manufacturer's and supplier's profit under this contract is equal to the profit of the centralized firm. The supplier captures $(w-c-c_k)/(r-c-c_k)\times 100\%$ of the total profit; and the manufacturer captures $(r-w)/(r-c-c_k)\times 100\%$ of the total profit. Notice that arbitrary profit division among the parties is also achievable by changing the wholesale price w. Therefore, with the appropriate choice of (w,τ) the payback contract results in mutually beneficial terms, that is, the manufacturer's and supplier's expected profits are at least as large as their profits under any wholesale price contract.[8]

[8] The payback contract provides a reward mechanism that induces the supplier to secure more capacity. Another mechanism is to penalize the supplier for every unit of order that he is unable to satisfy due to capacity shortage.

Asymmetric Forecast Information

Another issue in the aforementioned supply chain is the forecast sharing problem. The wholesale price is often set during the product design stage, which takes place long before the manufacturer ramps up the production. Component capacity or inventory commitments, however, are often made closer to production. Hence, the forecast sharing problem is often decoupled from product design and wholesale price negotiations. The manufacturer often has better forecast information than the supplier due to her proximity to consumers. Lee et al. (1997) provide four reasons, such as order batching, for why the downstream member distorts the demand forecast when sharing it with the upstream member. Özer and Wei (2006) show that another key reason for the bullwhip is the form of the contract.

Suppose that the aforementioned manufacturer has new forecast information before the supplier sets the capacity. Let ξ denotes the manufacturer's *private* information about demand forecast. Suppose ξ is a deterministically known quantity to the manufacturer. The manufacturer's new demand forecast information is $D = \mu + \epsilon + \xi$. If the supplier has access to the manufacturer's private forecast information ξ, he maximizes (13.12) by setting the capacity to

$$K^w = \mu + \xi + G^{-1}\left(\frac{w - c - c_k}{w - c}\right). \tag{13.15}$$

However, ξ is known *only* to the manufacturer. Can the manufacturer share this forecast information *credibly*? The answer is no because the manufacturer has an incentive to inflate her report of ξ. This incentive arises because the manufacturer's profit in (13.11) is increasing in the supplier's capacity choice K and the suppliers optimal capacity K^w is increasing in the manufacturer's forecast information ξ. Hence, by sharing an inflated forecast the manufacturer may increase her expected profit if the supplier believes this information. Anticipating the manufacturer's incentive to exaggerate her forecast, the supplier would not consider the forecast information to be *credible* regardless of whether the manufacturer reports accurate information. Instead, the supplier would resort to his *prior* belief about the manufacturer's private forecast information. For example, the supplier may perceive ξ to be a zero mean random variable that takes values in $[\underline{\xi}, \bar{\xi}]$ with cdf $F(\cdot)$.

This concept leads to what is known as *asymmetric* forecast information. The supplier and the manufacturer have asymmetric information about ξ and hence the overall demand forecast. The manufacturer knows ξ deterministically, whereas the supplier has a prior *belief* about its possible value. Hence, the supplier's expected profit is

$$E_\xi \Pi^s(K, \xi) = (w - c)E \min(\mu + \xi + \epsilon, K) - c_k K, \tag{13.16}$$

where the uncertainty is due to *both* ξ and ϵ. The supplier maximizes (13.16) by setting capacity level

$$K^{wa} \equiv \mu + (F \circ G)^{-1}\left(\frac{w - c - c_k}{w - c}\right), \tag{13.17}$$

where $F \circ G$ is the distribution function of $\xi + \epsilon$.

Comparing the supplier's capacity decision when she has and does not have access to the manufacturer's forecast information reveals the source of inefficiency. The supplier's capacity choice without having access to ξ under asymmetric information K^{wa} is not a function of ξ. Without credible forecast information sharing, the supplier cannot adjust the capacity to account for the manufacturer's private forecast. The consequences of this inefficiency could be severe for *both* parties. When the manufacturer's private forecast is very high, both parties may lose sales, resulting in lower profits (as the Boeing case in Cole 1997). When the manufacturer's private forecast information is low, the supplier may suffer from excess capacity (as the Solectron case in Hibbard 2003). The remedy for *this* inefficiency is to induce *credible* information sharing.

Özer and Wei (2006) show that the supplier can hold the manufacturer accountable for her private forecast information by requiring a *monetary* commitment before securing component capacity. This accountability can be achieved by designing a menu of prices for reserving capacity. The menu should be designed in a way that the supplier can *screen* the manufacturer's forecast information. To do so, the supplier offers this menu any time before setting the capacity.

The sequence of events is as follows. The supplier provides a *menu of contracts* $\{K(\xi), P(\xi)\}$ for all $\xi \in [\underline{\xi}, \bar{\xi}]$. Both capacity and corresponding payment are functions of private forecast information ξ. Here, the supplier's objective is to find the optimal menu that maximizes his profit. Given this menu, the manufacturer chooses a particular contract $(K(\hat{\xi}), P(\hat{\xi}))$ that maximizes her profit. By doing so, she announces her forecast information to be $\hat{\xi}$, which could differ from her true forecast information ξ. The supplier receives the payment $P(\hat{\xi})$ and builds capacity $K(\hat{\xi})$ at unit cost c_k. The manufacturer observes demand D and places an order. The supplier produces as much of the order as possible given the capacity constraint; that is, he delivers $\min(D, K(\hat{\xi}))$. The manufacturer receives the order and sells at unit price $r > 0$. Two decisions are the supplier's choice for the optimal *menu of contracts* that maximizes his profit; the manufacturers' choice from this menu is the optimal *contract* that maximizes her profit.

By choosing a contract, the manufacturer defines her profit, the supplier's profit and the total supply chain profit as

$$\Pi^m(K(\hat{\xi}), P(\hat{\xi}), \xi) = (r - w)E \min(\mu + \xi + \epsilon, K(\hat{\xi})) - P(\hat{\xi}), \tag{13.18}$$

$$\Pi^s(K(\hat{\xi}), P(\hat{\xi}), \xi) = (w - c)E \min(\mu + \xi + \epsilon, K(\hat{\xi})) + P(\hat{\xi}) - c_k K(\hat{\xi}). \tag{13.19}$$

The supplier's challenge is to elicit truthful information and to maximize his profit by choosing a menu of contracts while ensuring the manufacturer's participation. To identify an optimal menu of contracts, the supplier solves

$$\begin{aligned}
\max_{K(\cdot),P(\cdot)} \quad & E\Pi^s(K(\xi), P(\xi), \xi) \\
s.t. \quad & \text{IC:} \;\; \Pi^m(K(\xi), P(\xi), \xi) \geq \Pi^m(K(\hat{\xi}), P(\hat{\xi}), \xi), \quad \text{for all } \hat{\xi} \neq \xi \\
& \text{PC:} \;\; \Pi^m(K(\xi), P(\xi), \xi) \geq \pi^m_{min}, \quad \text{for all } \xi \in [\underline{\xi}, \bar{\xi}].
\end{aligned} \tag{13.20}$$

The expectation in the supplier's objective is with respect to ξ. The first set of constraints is the incentive compatibility (IC) constraints. These constraints ensure that the manufacturer maximizes her profit only by truthfully revealing her forecast information. The second set of constraints is the participation constraints (PC). They ensure a minimum profit π_{min}^{m} to the manufacturer regardless of her forecast information. This minimum profit could be the manufacturer's profit from her outside option, or her profit under other contracts. Note that this problem is a difficult one to solve because it involves optimization over functions. Through obtaining structural results, this problem can be converted to equivalent but simpler formulations that are easier to solve (see Özer and Wei 2006)).

The authors provide closed form solutions/formulas as a solution to the the optimization problem in (13.20). They also show that the optimal $P^{cr}(\xi)$ and $K^{cr}(\xi)$ are monotone in ξ. Hence, one can construct a function $P(K)$ by setting $P(K) = P^{cr}(\xi)$, if $K = K^{cr}(\xi)$. This function can be interpreted as a *capacity reservation* contract; i.e., pay $P(K)$ to reserve K units of capacity. Note that the optimal contract is independent of the manufacturer's forecast information. The supplier simply gives this contract as a menu of fees for the corresponding capacity level that the manufacturer may reserve. Essentially, the supplier delegates the capacity decision right to the manufacturer, who has superior forecast information.

The supplier can also hold the manufacturer accountable for her private forecast information by requiring a *quantity* commitment before the supplier secures component capacity. Özer and Wei structure the *advance purchase* contract under which the manufacturer pays the supplier w_a for each unit she orders before the supplier secures capacity; hence the name, advance purchase. This agreement provides an option to the manufacturer to place firm orders at an advance purchase price before the supplier secures capacity. The advance purchase could be costly to the manufacturer if the realized demand turns out to be smaller than the advance purchase quantity. Intuitively, this commitment prevents a manufacturer with a low forecast from communicating a high forecast. Özer and Wei (2006) show that the manufacturer can credibly *signal* her forecast through placing an advance purchase before the supplier decides the capacity. The authors also show that channel coordination is possible *even* under asymmetric forecast information by combining the advance purchase contract with an appropriate payback agreement. The formulation and analysis of the advance purchase contract leads to a *signaling* game, whereas the capacity reservation contract is a *screening* game. By comparing these models and analysis, the authors also show analytically when to use these contracts.

Which Contract Form to Adopt?

Özer and Wei (2006) identify two key drivers of the (supplier's, manufacturer's, and supply chain's) expected profits under different contracts: the *risk adjusted profit margin* and the *degree of forecast information asymmetry*.

Recall that the supplier's profit margin is less than the integrated supply chain's profit margin per unit of capacity investment. Hence, the supplier builds less than the supply-chain-optimal capacity. Two factors determine the impact of this inefficiency on the supply chain: the market uncertainty modeled by ϵ, and the supplier's profit margin per unit *sold* or per unit of capacity built, that is $w - c - c_k$. Hence, this inefficiency can be measured by the *risk-adjusted profit margin* $(w - c - c_k)/\sigma_\epsilon$, that is, the supplier's profit margin per unit sold per unit of market uncertainty.

The severity of supply chain inefficiency also depends on how much the supplier knows about demand as compared to the manufacturer. This knowledge disparity is measured by the *degree of forecast information asymmetry.* Let σ_ξ and σ_ϵ be the standard deviations of $F(\cdot)$ and $G(\cdot)$, respectively. Consider a supply chain with $\sigma_\xi >> \sigma_\epsilon$. For this supply chain, the inefficiency due to the lack of credible forecast information sharing would be large because the supplier's knowledge of market demand is much less certain than that of the manufacturer's. One possible measure of *degree of forecast information asymmetry* is the ratio of the standard deviations $\sigma_\xi/\sigma_\epsilon$.

Özer and Wei (2006) show that the supplier and the manufacturer can choose among structured agreements that enable a mutually beneficial partnership depending on the risk-adjusted profit margin and the degree of forecast information asymmetry. The results are summarized in Fig. 13.9. For example, when forecast information between the parties is highly imbalanced, and the risk-adjusted profit margin is high, then their analysis shows that the advanced purchase contract generates higher profits for both parties. Based on anecdotal evidence and private conversations with executives from several industries, we conjecture that the industry can also be mapped along the same dimensions. Figure 13.10 maps the level of these drivers for industries. For example, in the semiconductor industry, compared to the manufacturer, the supplier knows very little about the manufacturer's private forecast. Further empirical and field research is needed to verify Fig. 13.10.

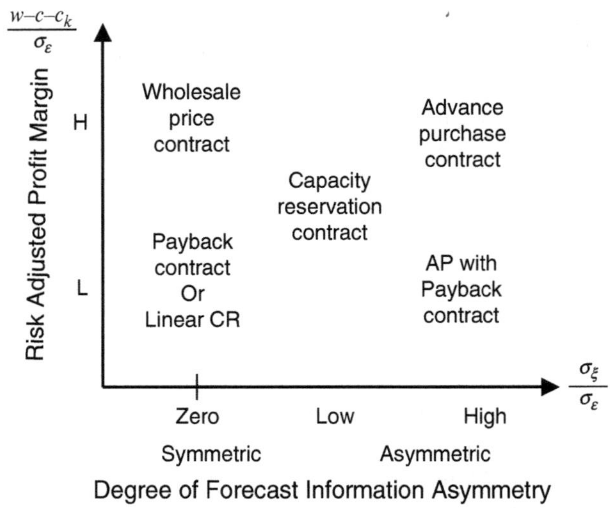

Fig. 13.9 Mutually beneficial contracts

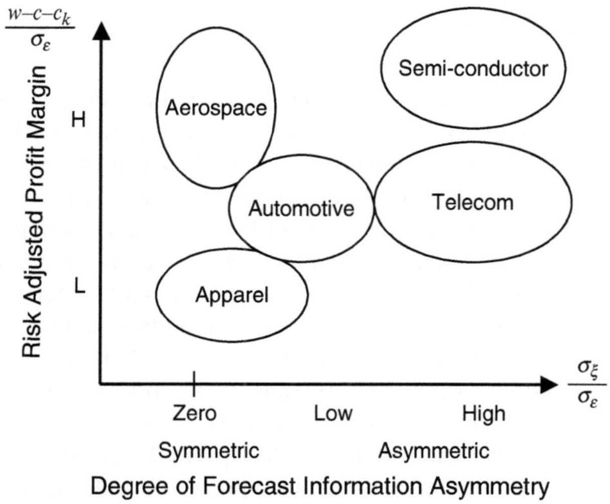

Fig. 13.10 Capacity risk drivers across different industries

A Brief Review

Research exploring contracts that coordinate the supply channel under *full (or symmetric)* forecast information falls into two groups. In the first group, contracts align incentives by inducing the supplier and manufacturer to share the risk of low demand, resulting in excess capacity or inventory. Buyback contracts (Pasternack 1985), quantity flexibility contracts (Tsay 1999), and capacity reservation contracts (Erkoc and WU 2005) are a few examples. The second category of contracts aligns incentives by sharing the risk of high demand, resulting in capacity or inventory shortage. Revenue sharing contracts (Cachon and Lariviere 2000) and quantity premium contracts (Tomlin 2003) are two examples from this category. Cachon (2003) provides a comprehensive review of supply chain contracting and coordination[9]. The supply chain literature that explicitly models *asymmetric information* can be classified into two groups. A group of researchers (Corbett et al. 2001) focus on information asymmetry in production cost and another group (Porteus and Whang 1991, Cachon and Lariviere 2001, Özer and Wei 2006) focuses on information asymmetry in market demand and forecasts. Chen (2003) provides an excellent review of the use of these models in supply chains.

[9] In this literature, mainly the downstream firm is assumed to face demand uncertainty, while the upstream firm "builds to order", unlike the interaction discussed in this section. Nevertheless, the results are analogous.

13.3.2 Inventory Risk

Here we provide an example model of how structured agreements can be used to align incentives to mitigate the adverse affect of inventory risk. To do so, we summarize some results from Lütze and Özer (2008). These authors study the incentive problems in a multi period, two-echelon supply chain with a manufacturer and a retailer both of whom build or procure to stock. Note that the manufacturer in this case faces inventory risk, unlike the previous section's build-to-order manufacturer. Both the manufacturer and the retailer hold inventory to satisfy their respective customers. They review inventory periodically, i.e., at the beginning of each period t. The manufacturer produces at a per unit cost $c_m > 0$ and the retailer places an order at a per unit ordering cost $c_r > 0$. Suppose all cost and demand parameters are stationary, i.e., independent of period t. There is no fixed cost for production or placing an order. The manufacturer has ample capacity for production, which takes L periods to complete. The retailer orders are processed and shipped in l periods. Customer demand D_t is realized. The retailer satisfies customer demand through on-hand inventory. Unsatisfied demand is backlogged. Backorders of end customer demand incur a unit penalty cost p_r per period only at the retailer. The manufacturer incurs a shortage cost for unsatisfied retailer order based on the contractual agreement we specify later. The manufacturer and the retailer incur unit holding cost $h_m > 0$ and $h_r > 0$, respectively, where $h_m \leq h_r$, for any inventory remaining at the end of each period. Both the retailer and the manufacturer choose an optimal inventory replenishment policy to minimize their respective total expected inventory costs over T periods. At the end of period T, leftover inventory (resp., backlog) is salvaged (resp., purchased) at a linear per unit value of c_m and c_r, at each stage, respectively.

The manufacturer needs to protect himself against the retailer's demand over the production lead time L, and the retailer needs to protect herself against the consumer demand during the processing lead time l and supply shortage at the manufacturer. Hence, to reduce inventory exposure, the manufacturer prefers the retailer to commit to purchase in advance and wait for delivery (commit and wait). However, the retailer prefers to delay her order and have immediate product availability and delivery (now or never). To address these opposing interests, Lütze and Özer consider a *promised lead time* contract with two parameters: promised lead time τ and corresponding per period lump-sum payment K.

Under a promised lead time contract, when the retailer places an order, the manufacturer promises to ship this order, in full, after τ periods. To guarantee this delivery, the manufacturer arranges an alternate sourcing strategy to fill retailer demand that exceeds the manufacturer's on-hand inventory. That is, the manufacturer *borrows* emergency units from an alternative source and incurs penalty p_m per unit per period until the alternative source is replenished[10]. The effect of promised lead

[10] Similar alternative sourcing strategies are also discussed in Lee et al. (2000) and Graves and Willems (2000).

time is to shift the responsibility for demand uncertainty from the manufacturer to the retailer. Note that if the retailer agrees to a promised lead time $L+1$, exceeding the manufacturer's production lead time, the manufacturer builds to order for the retailer and does not carry any inventory.

Under this agreement, each firm independently solves a periodic-review inventory control problem discussed in Sect. 13.2.1. Let x_t^j and y_t^j be firm $j \in \{m, r\}$ inventory position before and after ordering, respectively, in period t, where m and r stand for the manufacturer and the retailer. The following dynamic program recursion minimizes the cost of managing the inventory system for a finite horizon problem with $T-t$ periods remaining until termination.

$$J_t^j(x_t^j|\tau) = \min_{y_t^j \geq x_t^j} \{G^j(y_t^j|\tau) + \alpha E_D J_{t+1}^j(x_{t+1}^j|\tau)\}$$

where $J_{T+1}^j(x_{T+1}^j|\tau) \equiv 0 \quad$ for $j \in \{m, r\}$, and

$$G^m(y_t^m|\tau) = (1-\alpha)c_m y_t^m + E[h_m(y_t^m - D^{L+1-\tau})^+ + p_m(D^{L+1-\tau} - y_t^m)^+] \quad \text{and}$$
$$G^r(y_t^r|\tau) = (1-\alpha)c_r y_t^r + E[h_r(y_t^r - D^{l+1+\tau})^+ + p_r(D^{l+1+\tau} - y_t^r)^+].$$

For the stationary finite horizon inventory control problems, a myopic base stock policy is known to be optimal (Veinott 1965). These myopic base stock levels for the manufacturer and retailer are the minimizers of their respective single-period cost functions and are defined as

$$Y^m(\tau) = F_{L+1-\tau}^{-1}\left(\frac{p_m - (1-\alpha)c_m}{h_m + p_m}\right) \quad \text{and}$$
$$Y^r(p_r, \tau) = F_{l+1+\tau}^{-1}\left(\frac{p_r - (1-\alpha)c_r}{h_r + p_r}\right).$$

Hence, with promised lead time τ, firm j orders up to an optimal base stock level $Y^j(\tau)$ if its inventory position x_t^j is below this level at the beginning of period t. The expected discounted inventory cost over T periods equals the sum of the discounted single-period costs, that is,

$$J^m(x_1^m|Y^m(\tau), \tau) = \sum_{t=1}^{T} \alpha^{t-1} G^m(\tau),$$

where $G^m(\tau) \equiv c_m\mu + E\left\{h_m[Y^m t(\tau) - D^{L+1-\tau}]^+ + p_m[D^{L+1-\tau} - Y_t^m(\tau)]^+\right\}$, and

$$J^r(x_1^r|Y^r(p_r, \tau), \tau) = \sum_{t=1}^{T} \alpha^{t-1} G^r(p_r, \tau),$$

where $G^r(p_r, \tau) \equiv c_r\mu + E\{h_r[Y_t^r(p_r, \tau) - D^{l+1+\tau}]^+ + p_r[D^{l+1+\tau} - Y_t^r (p_r, \tau)]^+\}$.

When the manufacturer has full information about the retailer's inventory-related costs, she can determine the optimal promised lead time contract (τ, K) by solving the following problem.

$$\begin{aligned} \text{minimize}_{\tau,K} \quad & \sum_{t=1}^{T} \alpha^{t-1}\{G^m(\tau) - K\} \\ \text{subject to} \quad & K + G_r^*(p_r, \tau) \le \pi_{\max}^r \\ & \tau \in \{0, \ldots, L+1\}. \end{aligned} \tag{13.21}$$

The constraint ensures that the retailer is not charged a cost larger than her *maximum reservation cost.* Note also that the summation over T periods does not affect the solution of this problem, hence it can be dropped from the objective function for optimization purposes. The constraint must be binding at optimality. Otherwise, we can increase K and reduce the objective function. Substituting $K = \pi_{\max}^r - G_r^*(p_r, \tau)$ one can solve for the optimal contract parameters.

To solve the above problem, the manufacturer needs to know the retailer's cost information. He can perhaps estimate h_r fairly accurately because he knows the value of the product. The same may not necessarily be true for p_r. Companies often state penalty cost as a strategic cost parameter never to be revealed. Lütze and Özer show that the retailer has every incentive to conceal her service level to end consumers (or equivalently the penalty cost structure). Intuitively, the retailer has an incentive to exaggerate the service level, thereby shortening the promised lead time for the same agreed upon price and reducing his expected inventory cost per period. Hence, it is often not possible for the manufacturer to know the retailer's penalty cost.

Suppose that there are two types of retailers in the market: one with a low-penalty cost p_r^L and the other one with a high p_r^H. Suppose also that the belief is such that with probability q, she is a high-penalty cost retailer and with probability $(1-q)$, she is a low-penalty cost retailer. To determine the optimal contract mechanism $\{(\tau^L, K^L), (\tau^H, K^H)\}$, the manufacturer solves the following problem.

$$\begin{aligned} &\text{minimize}_{(\tau_i, K_i)_{i=L,H}} \quad q[G^m(\tau^H) - K^H] + (1-q)[G^m(\tau^L) - K^L] \\ &\text{subject to} \\ &\qquad IC_1: \; K^H + G^r(p_r^H, \tau^H) \le K^L + G^r(p^H, \tau^L) \\ &\qquad IC_2: \; K^L + G^r(p_r^L, \tau^L) \le K^H + G^r(p_r^L, \tau^H) \\ &\qquad IR_1: \; K^H + G^r(p_r^H, \tau^H) \le \pi_{\max}^r \\ &\qquad IR_2: \; K^L + G^r(p_r^L, \tau^L) \le \pi_{\max}^r \\ &\qquad\qquad \tau_i \in \{0, \ldots, L+1\} \text{ for } i = L, H. \end{aligned}$$

The first two incentive compatibility constraints ensure that a retailer with a high-penalty cost voluntarily chooses the contract (τ^H, K^H) and the low-penalty cost retailer chooses (τ^L, K^L). The next two individual rationality constraints guarantee the retailer finds a satisfactory contract regardless of his service level. This problem can be solved once we show certain properties of the cost function G^r and the result is in closed form solution. For example, IC_2 and IR_1 imply that IR_2 is redundant when we show G^r is increasing in p_r. Note also that IC_2 must be binding at optimality otherwise the manager can increase K^L and reduce the objective function until IC_2 binds. By showing that $G^r(p_r, \tau)$ has *single crossing property*,[11] we can also show that $\tau^H \leq \tau^L$. Intuitively, it is optimal to offer a shorter promised lead time to a retailer that has higher- penalty cost. Together with this observation, the binding IC_2 implies that IC_1 is redundant. At optimality

$$
\begin{aligned}
\tau^H &= \text{minimizer of } q[G^m(\tau^H) + G^r(p_r^H, \tau^H)] \\
&\quad +(1-q)[G^r(p_r^H, \tau^H) - G^r(p_r^L, \tau^H)] \\
K^H &= \pi_{\max}^r - G^r(p_r^H, \tau^H) \\
\tau^L &= \text{minimizer of } (1-q)[G^m(\tau^L) + G^r(p_r^L, \tau^L)] \\
K^L &= [\pi_{\max}^r - G^r(p_r^L, \tau^L)] - [G^r(p_r^H, \tau^H) - G^r(p_r^L, \tau^H)]
\end{aligned}
$$

Lütze and Özer (2008) discuss properties of optimal promised lead time contracts and the resulting inventory levels under both full and asymmetric service information with multiple discrete types. We caution that the results for mechanism design problems with multiple discrete types do not simply follow from two-type case. The study of the more general case requires intricate analysis and may lead to different solutions. Lovejoy's (2006) paper is an excellent reference that clarifies related issues. Lütze and Özer (2008) also show how the ensuing inventory risk sharing strategy changes under asymmetric service information. They also compare the performance of a supply chain operating under a central decision maker to one with independent firms operating under a promised lead time contract. By comparing different control mechanisms and information scenarios, they provide insight into stock positioning and how the promised lead time affects the system performance. They quantify, for example, how much and when the manufacturer and the retailer over- or under-invest in inventory as compared to centralized supply chain, which operates as a serial system, discussed in Sect. 13.2.5 and later in Sect. 13.4.

13.3.3 Quality Risk

So far, we discussed the two risk imbalances in supply chains, leading to incentive problems. They were namely capacity and inventory risk. Next we discuss the third one: quality risk. The quality literature in operations management focuses mainly on centralized inventory management problems with random yields. Yano and Lee (1995) provide a review of more than 70 academic papers, such as the works of Porteus (1990b) and Pentico (1994). The focus of these papers is on establishing production and stocking policies when production or procurement yields are

[11] When $f(x, y) - f(x, y-1)$ is increasing in x, then function f is said to satisfy single crossing property.

random. They address, for example, optimal time and size for inspection. Only a handful of researchers study the effect of product quality in decentralized supply chains. In Reyniers and Tapiero (1995), the supplier determines the effort invested in quality, where high effort causes a lower probability of defect. The manufacturer decides whether to conduct costly inspection. Lim (2001) uses a similar setting, with asymmetric information on the supplier's quality type. Baiman, Fischer and Rajan (2000) analyze the effects of different assumptions regarding the contractibility of quality and inspection efforts. All of these papers define quality as the percentage of the products that are not defective.

Today, manufacturers are outsourcing advance functions such as procurement, design, and even research and development. The manufacturer can use inspection techniques to measure yield and, hence, can enforce a certain yield in the contract. However, when the supplier undertakes more advanced tasks, measuring either the supplier's quality effort or his cost to achieve the desired quality level is difficult. This difficulty precludes the manufacturer from enforcing the desired quality level with a legal contract.

Not being able to foresee all possible contingencies and time to market pressures are two other reasons that make quality difficult to measure. Quality requirements may be better understood after the supplier builds a prototype, but this step typically occurs after an outsourcing agreement is signed. According to a Toshiba manager, if Toshiba waited until they were absolutely sure of every final detail and then wrote a complete contract, they would be 6–12 months late to the marketplace. Therefore, in addition to structured and legally binding agreements, establishing strategic relationship management systems between the manufacturer and the supplier is probably a good idea. This strategic relationship may encourage, for example, the implementation of quality programs such as TQM or Six Sigma.

Kaya and Özer (2004) refer to the adverse effect of inefficiencies caused by the immeasurability of both quality effort level and the quality cost as the quality risk. Consider an original equipment manufacturer that outsources the design and production of a custom component to a supplier and sells the final product at a price p. The market demand is a function of the manufacturer's sales price p, the supplier's quality effort e and the market uncertainty ϵ, i.e.,

$$q = a - bp + e + \epsilon,$$

where $a > 0$ and $b > 0$ are the intercept and slope of the downward sloping demand curve. The manufacturer offers a procurement contract to the supplier. If the supplier accepts the contract, the parties establish a supply chain. Next, the supplier determines the product's quality level e by exerting costly quality effort. The quality cost is the supplier's private information. The manufacturer determines the sales price p to maximize her expected profit. The market shock ϵ realizes and the firms observe the quantity demanded. Finally, the supplier produces to satisfy the manufacturer's order, which is equal to the consumer demand. Note that the manufacturer cannot *verify* the quality level set by the supplier due to the market uncertainty ϵ. Hence, the manufacturer cannot directly link the CM's compensation to the quality level the CM sets. Instead, the manufacturer needs to offer a contract and indirectly influence the CM's quality decision. The authors model this interaction as *screening* and *moral hazard* problems embedded into a three stage game.

Kaya and Özer (2004) design procurement contracts that improve the supplier's and the manufacturer's profits by inducing the supplier to exert effort to produce better quality products when parties cannot explicitly contract on quality. The authors answer and quantify broad questions of managerial interest. They quantify the value of being able to contract on quality. They study the effects of the manufacturer not knowing the supplier's cost of quality. They investigate the value of an enterprise-wide quality management system, a recent information technology tool that enables accounting of quality-related activities across the supply chain. The authors also study the effect of the manufacturer's product-pricing policy on the resulting quality of the product. They report the outcome of two opposing product-pricing strategies: setting market price for the final product in the contract terms with suppliers versus pricing the product after receiving components from the supplier.

13.4 Large–Scale Systems and Rationality

Global supply chains (or perhaps networks) have multiple locations to carry inventory; multiple products to manage; several decisions to coordinate; various sources and flows of information; and uncertain demand and processes. The management of inventory and information in such systems is difficult, and reviewing the related literature is even more so! We refer the reader to the books by Zipkin (2000) and Muckstadt (2005) for a systematic treatment of fundamental inventory control methods. First similar models and analysis can be applied to study multilocation systems and multiproduct systems. Second, by allocating decoupling inventories,[12] complex supply chain structures can be decomposed into fundamental structures, such as serial systems, of which we have a very good understanding. What is the best (if not the efficient) way to decompose a complex structure into smaller problems is an open research question. There is also an extant literature on the supply chain configuration problem (see, for example, Graves and Willems 2003). Here we will provide some discussion on how to allocate inventory effectively across two fundamental structures, serial and distributions systems, to minimize inventory-related costs while keeping an eye on *rationality*.

Despite considerable progress over the years, existing optimization and policy evaluation algorithms for multiechelon systems remains fragmented and opaque to nonexperts. The computational methods involved are intricate and require voluminous data. Data fed to these tools are not always accurate, as discussed in Sect. 13.2.3. Systems and people have limitations. Users are more likely to embrace decision tools when they understand what is in the black box. Therefore, it is

[12] Decoupling stock is used to permit separation of inventory decisions at different locations in the supply chain. Having a large inventory between two locations would make it possible for the downstream location to make an inventory decision independent of any supply problem at the upstream location.

necessary to develop *easy-to-describe*, *close-to-optimal*, and *robust* heuristics that can be implemented on a spreadsheet by solving, for example, newsvendor type problems[13]. Unlike multiechelon results, the newsvendor problem is widely known, commonly used in practice and a standard component of any production and operations curriculum.

The above discussion suggest that heuristics and approximations can collectively enable better inventory management if they pass all or some of the following tests: (1) Is it close to optimal? (2) Is it simple to describe and use? (3) Can it be used to test system design issues accurately? (4) Is it robust? (5) Is it computationally easy? Note, however, that focusing narrowly on the one criterion overlooks other important aspect and leads to a gap between theory and practice (see Özer and Xiong 2008 for more discussions). For example, the computational methods used for exact solutions can be intricate and may require voluminous data. They may require advance knowledge. They may not provide explicit information regarding the key factors that drive performance. Recently, researchers have realized this gap and started to focus on developing easy-to-use, robust heuristics, and approximations that are insightful (see, for example, Lee et al. 1993; Hopp et al. 1997, 1999; Gallego et al. 2007; Shang and Song (2003); Gallego and Özer 2003; Caglar et al. 2004; Watson and Zhang 2005; Özer and Xiong 2008 and references therein). In the following two subsections, we provide some examples from Gallego and Özer (2003) and Gallego et al. (2007).

13.4.1 Serial Systems

Consider a serial system consisting of J stages. Stage $j < J$ procures from Stage $j+1$ and Stage J replenishes from an outside supplier with ample stock. Customer demand occurs only at Stage 1 and follows a (compound) Poisson process, $\{D(t), t \geq 0\}$ with arrival rate λ. It takes L_j units of time for a unit to arrive at Stage j once it is released by its predecessor. Unsatisfied demand is backordered at each stage, but only Stage 1 incurs a linear backorder penalty cost p, per unit, per unit of time. We assume, without loss of generality, that each stage adds value as the item moves through the supply chain, so echelon holding costs h_j^e are positive. The local holding cost for stage j is $h_j \equiv \sum_{i=j}^{J} h_i^e$. The system is operated under continuous review. The following random variables describe the state of Stage j in equilibrium: D_j is the lead-time demand, I_j the on-hand inventory, and B_j the backorders. The total long-run average cost for any policy can be expressed as

$$E\left[\sum_{k=1}^{J} h_k I_k + pB_1 + \sum_{k=2}^{J} h_k D_{k-1}\right].$$

[13] This problem is a simple single period, single-location inventory control problem faced by a newsvendor. The vendor has to decide how much to order from the publisher so as to satisfy uncertain demand. The model is used to teach the risk of overstocking and understocking.

Optimality of an echelon base-stock policy $(s_J, \ldots, s_1)$ for this serial system is well known (see the original work by Clark and Scarf 1960). Gallego and Özer (2003) provide the following *new* recursive algorithm to obtain optimal base-stock levels. Let $c_1(s) = E[h_1(s - D_1)^+ + p(D_1 - s)^+]$ and for $j = 2, \ldots, J$ define

$$c_j(s) = \min_{x \in \{0,\ldots,s\}} c_j(x;s)$$
$$c_j(x;s) = E[h_j(x - D_j)^+ + c_{j-1}(\min(s - x, s - D_j)) + h_j D_{j-1}]. \quad (13.22)$$

Let $\mathcal{N} = \{0, 1, \ldots, \}$ be the set of nonnegative integers and let

$$s_j^* \equiv \min\{s \in \mathcal{N} : c_j(s+1) - c_j(s) > h_{j+1}\} \quad \text{for } j = 1, \ldots, J.$$

Function $c_j(s)$ is the long-run average cost of optimally managing the subsystem $\{j, \ldots, 1\}$ given echelon base-stock level s and s_j^* is the optimal base-stock levels. The recursion is somewhat intuitive. Suppose $c_j(\cdot)$ has been computed and consider the subsystem $\{j+1, \ldots, 1\}$. The goal is to compute $c_{j+1}(\cdot)$ from the knowledge of $c_j(\cdot)$. Note the link between the two subsystems. We allocate x units to Stage $j+1$ and the remaining $s - x$ units of echelon base-stock to subsystem $\{j, \ldots, 1\}$. Given this allocation, the net inventory at Stage $j+1$ will be $(x - D_{j+1})^+$ which accrues at cost rate h_{j+1}. Since Stage $j+1$ will face a shortage when $D_{j+1} - x > 0$, the effective echelon inventory for subsystem $\{j, \ldots, 1\}$ is $s - x - (D_{j+1} - x)^+ = \min(s - x, s - D_{j+1})$. Thus, a finite local base-stock level at Stage $j+1$ imposes an externality to the subsystem $\{j, \ldots, 1\}$ whose expected cost is now $Ec_j(\min(s - x, s - D_{j+1}))$. As a result, when we allocate $x \leq s$ units of local base-stock level to Stage $j+1$, the cost of managing a serial system with $j+1$ stages is given by (13.22).

Note that the classical recursive formulation presented in Chen and Zheng (1994) or Gallego and Zipkin (1999) has no intuitive interpretation. Although the above new algorithm has an interpretation and is intuitive, it is still difficult to explain to nonexperts. It also does not provide any transparent relationship. Using this formulation, Gallego and Özer provide a fast exact algorithm based on gradient updates and a close-to-optimal heuristic that requires solving one newsvendor problem per stage! The heuristic is based on the approximate holding cost rate

$$h_j^{GO} \equiv \sum_{k=1}^{j} \frac{L_k}{L_1 + \cdots + L_j} h_k.$$

The idea is based on adding the holding cost as the product goes through the stages without delay and then dividing by the total lead time that it spends before reaching the end customer. This approximate holding cost minus the cost associated to upstream operations that is h_{j+1}, is charged to any excess inventory in echelon j that faces demand uncertainty over the leadtime $L_0 + \cdots + L_j$. This cost is charged per excess inventory because it is the approximate value that the echelon j is responsible for. Similarly, the penalty cost $p + h_{j+1}$ is charged to echelon j because it is the approximate opportunity cost. The resulting problem then has a newsvendor type cost structure of

$$\tilde{c}_j(s) = E\left[(h_j^{\mathrm{GO}} - h_{j+1})\left(s - \sum_{k=1}^{j} D_k\right)^+ + (p + h_{j+1})\left(\sum_{k=1}^{j} D_k - s\right)^+\right], \quad (13.23)$$

$$s_j^{GO} \equiv \min\left\{s \in \mathcal{N} : Pr\left(\sum_{k=1}^{j} D_k \le s\right) > \frac{p + h_j^{\mathrm{GO}}}{p + h_{j+1}}\right\}. \quad (13.24)$$

Gallego and Özer (2003) show that over 1,000 experiments, the optimality gap[14] is less than 0.25%. Note that the newsvendor problem is known to be somewhat robust in that small changes in data would not change the optimal solution significantly. This heuristic can easily be implemented with a simple spreadsheet. So, the heuristic is close-to-optimal, easy-to-describe, and robust.

The authors also consider an approximation by approximating the lead-time demand distribution using Normal with mean μ and σ. This approximation would be good in particular when the mean of lead-time demand is large. The resulting cost $c_j(s_j^*) \approx (p + h_j^{\mathrm{GO}})\sigma\phi(z)$, where $z = \Phi^{-1}((p + h_{j+1})/(p + h_j^{\mathrm{GO}}))$. The base stock level is $s_j^* \approx \mu + z\sigma$. They are both in closed form. The authors also provide a distribution free cost upper bound in the approximate sense, that is $c_J(s_J^*) \le \sqrt{p(h_1 L_1 + \cdots + h_J L_J)\lambda}$. The bound does not depend on any distribution. So it is quite robust with respect to demand parameter estimation. Such bounds, heuristics, and approximations can also be used to quantify the value of system design issues. Using these results, it is easy to show, for example, that management should focus on reducing the lead time at the upstream stages, while reducing the holding cost at the downstream stages. If process resequencing is an option, the lowest value added processes with the longest processing times should be carried out sooner rather than later. More importantly, a manager can easily quantify the impact of such changes using these simple heuristics and bounds.

13.4.2 Distribution Systems

Consider a two-level distribution system. All items enter the system from an external supplier and proceed first to location $j = 0$, called the warehouse. The warehouse in turn supplies J retailers, where the customer demands occur, indexed by $j = 1, \ldots, J$. Shipments from the external supplier arrive at the warehouse after time L_0. Shipments arrive at retailer j after time L_j. The retailers satisfy the customer demand from on-hand inventory, if possible. Unsatisfied demand at retailer j is backordered at a linear penalty cost rate b_j. All locations are allowed to carry

[14] The gap is defined as percentage difference between the optimal cost and the cost of the heuristic.

inventory. The local holding cost is h_j per unit at retailer j. Holding inventory at the retailer is more expensive than holding it at the warehouse $h_j \geq h_0$ for $j > 0$. On the other hand, inventory located closer to the customer enables a quick response, hence reduces the possibility of a backorder at each retailer. Demand at each retailer j follows a Poisson process $\{D_j(t), t > 0\}$ with rate λ_j, and these are independent across retailers. The problem is, where to locate the inventory and how to control the system, so as to minimize the long-run average holding and penalty costs. No one knows the optimal policy for distribution systems, yet.

Under any policy, the total average cost can be expressed as

$$h_0 E[I_0] + h_0 \sum_{j>0} E[IT_j] + \sum_{j>0} (h_j E[I_j] + b_j E[B_j]),$$

where I_j is the on hand inventory, B_j is backorders, and IT_j is the inventory in transit at location j at equilibrium.

Gallego et al. (2007) distinguish two modes of control: central and local. Under central control, all information flows to one point, where all decisions are made. Local control means that each location observes local information and makes decisions accordingly. However, even under local control, a single decision maker provides operating rules to all locations, which the locations then implement in real time. The locations do not have their own distinct objectives, as they do in contracting models of Sect. 13.3.

Here we focus on the local control case and on a class of simple replenishment policies, base-stock or one-for-one policies (see Gallego et al. 2007 for central control policies). Under local policy, whenever the inventory position at location j falls below the local base-stock level s_j, the retailer orders from the upstream location to raise the inventory position up to s_j. The sum of the retailers' orders constitutes the warehouse's demand process. The warehouse satisfies the retailers' requests on a first-come-first-served basis. Notice that information and control are decentralized or localized, in that each location sees its own demand and monitors its own inventory-order position. The exact analysis of this system is due to Simon (1971) and Graves (1985). Graves (1985) derives the steady-state distributions of inventory levels and backorders by disaggregating the backorders at the warehouse. Axsäter (1990) provides a recursive method to calculate the average holding and penalty cost associated with every supply unit that is matched with the demand that triggers it. The following random variables describe the system at equilibrium.

$$B_0 = [D_0 - s_0]^+, \tag{13.25}$$

$$I_0 = [s_0 - D_0]^+, \tag{13.26}$$

$$B_j = [B_{0j} + D_j - s_j]^+ \quad \text{for } j > 0, \tag{13.27}$$

$$I_j = [s_j - B_{0j} - D_j]^+ \quad \text{for } j > 0. \tag{13.28}$$

Here, B_{0j} and D_j are independent, and $(B_{0j}|B_0)$ is binomial with parameters B_0 and $\theta_j = \lambda_j/\lambda_0$. Given the s_j, one can compute the $E[I_j]$ and $E[B_j]$ and thus

$$c(s_0, s_1, \ldots, s_J) = h_0 E[I_0] + \sum_{j>0} c_j(s_0, s_j), \tag{13.29}$$

$$c_j(s_0, s_j) = h_j E[I_j] + b_j E[B_j]. \tag{13.30}$$

Let $s^* = (s_j^*)_{j=0}^J$ denote the policy that achieves the minimum average cost c^*.

For fixed s_0, the total average cost in (13.29) separates into a constant, plus functions c_j of one variable each (s_j), each convex in its variable. This separation is quite useful computationally. On the other hand, the remaining problem is still not trivial. To compute $E[B_j]$ and $E[I_j]$ requires numerical convolution of B_{0j} and D_j. Also, the cost $c(s_0, s_1^*(s_0), \ldots, s_J^*(s_0))$ is not convex in s_0. Finding the optimal s_0, therefore, requires an exhaustive search.

Gallego et al. (2007) provide various heuristics based on restriction and decomposition ideas. Note, for example, that restricting the warehouse not to carry inventory decomposes the system to J retailers facing longer replenishment lead times, i.e., L_0+L_j. This heuristic is referred to as cross-docking. To obtain the base-stock level at each retailer, they solve newsvendor type problems as in (13.23). The other extreme is to assume that the warehouse always has ample stock. Doing so, decomposes the system into individual retailers with lead time L_j. The authors solve for the warehouse's base-stock level by assuming that the retailers base-stock levels are fixed to zero. In this case the warehouse's problem is a newsvendor type. The solution provides the maximum possible stock needed at the warehouse. Hence, they refer to this heuristic as stock-pooling. Another heuristic allocates zero safety stock to the warehouse, hence named the zero-safety heuristic. The authors show that a combination of these heuristics yields asymptotically optimal results, i.e., the combined heuristics yields optimal results as the number of retailer increases. Through an extensive numerical study involving all plausible distribution system parameters, the authors show that the optimality gap for the restriction and decomposition-based heuristic is less than 2%. The authors also provide several other heuristics, bounds, and approximations both for central and local control systems.

13.5 Ending Thoughts and Future Directions

In this chapter, we provided a discussion around four fundamentals of effective inventory management. First we discussed how to effectively use information in centralized inventory systems. Such inventory systems are managed by a single decision maker who possesses all relevant information. As we discussed in Sect. 13.2, this line of research will always be necessary even though global inventory systems are decentralized in practice. Advances in technology, cheaper computational, and storage devices will continue to enable managers to obtain more information. Inventory managers would need to quantify the value of information and the technology even more so than before. These systems also serve as a benchmark for decentralized systems. They are the building blocks for large-scale systems as discussed in Sect. 13.4. How to use and quantify new information in inventory management will continue to be an important area for future research.

Note also that there are still open questions. For example, we do not know the impact of imperfect inventory information on multiechelon inventory systems. Intuitively, the adverse affect of inventory record inaccuracy will amplify as we go up in the echelon. Perhaps RFID technology has more value in such systems. But we simply do not know. Another example is the centralized distribution system, for which we still do not have an optimal inventory policy. As discussed in Sect. 13.2.1, researchers have realized that an optimal policy would be very complex, if one exists. Hence, they have developed close-to-optimal heuristics, but none of these heuristics have worst case performance bounds. Developing such bounds is an interesting research direction. We have started to see recent research in this direction (Levi et al. 2006, 2007).

Designing contracts to align incentives and coordinate inventory decisions will continue to be an important research area given global supply chains. Decentralized inventory management systems consist of managers with asymmetric information and separate objectives. We discussed several inefficiencies due to decentralized operations. Inventory managers need to keep an eye on inefficiencies introduced due to decentralization. Most of the work in this area consists of single-period interactions between two inventory managers. Future work is needed to consider the effect of repeated interactions and reputation. This line of work also assumes that inventory managers are responsible for single-location systems. In reality, however, an inventory manager could be responsible for a serial system or a distribution system (as discussed in sections Sect. 13.2.1 and Sect. 13.4). Hence, it is also important to study the interaction between two such managers. For example, the manufacturer in Lütze and Özer (2004) might offer a shorter promised lead time when he is managing a multiproduct inventory system due to perhaps the risk pooling effect. Studying the impact of supply chain design strategies such as postponement on contract terms would contribute to our understanding of these systems and bring us one step closer to real systems. The approximations, bounds, and closed form solutions developed for centralized systems discussed in Sect. 13.4 may also help us to study complex decentralized inventory systems that are controlled by several managers.

Many years of research also suggest that large-scale, centralized stochastic inventory systems are even more difficult to deal with and are not amenable to a simple optimal policy. As a research community, we need to develop close-to-optimal, easy-to-describe, robust heuristics for solving large scale systems. To make a heuristic universally acceptable, we need to test its performance against a lower bound or an optimal solution. For large-scale systems, however, we lack optimal solutions. Developing sensible lower bounds could be difficult as well. As an alternative, such heuristics can be tested on real systems. However, real systems differ from each other, making it difficult to compare plausible heuristics proposed by researchers. Perhaps one potential research area is to design test problems that are universally acceptable to qualify as difficult, real, and large scale.

Acknowledgment The authors research has been partially supported by National Science Foundation Grant No. 0556322. This chapter was written in 2006 when the author was a faculty member at the department of Management Science and Engineering at Stanford University.

References

Abhyankar HS, Graves S (2001). Creating an inventory hedge for Markov-modulated Poisson demand: Application and model. Manuf Service Oper Manag 3(4):306–320.

Alexander KT, Gilliam K, Gramling C, Grubelic H, Kleinberger S, Leng D, Moogimane C (2002). Applying auto-ID to reduce losses associated with shrink. IBM Business Consulting Services, MIT Auto-ID Center White Paper, November.

Anupindi R, Morton T, Pentico D (1996). The non-stationary stochastic lead time inventory problem: Near-myopic bounds, heuristics, testing. Manag Sci 42(1):124–129.

Arrow KJ (1985). The economics of agency in principals and agents. In: Pratt JP, Zeckhauser RJ (eds) Harvard Business School Press, Cambridge, MA.

Atali A, Özer Ö (2005). Multi-Item inventory systems with Markov-modulated demands and production quantity requirements. Working paper, Stanford University.

Atali A, Lee H, Özer Ö (2004). Inventory control under imperfect information: Bounds, heuristics and approximations with demand prioritization. Working paper, Stanford University, Stanford, California.

Atali A, Lee H, Özer Ö (2006). If the inventory manager knew: Value of visibility and RFID under imperfect inventory information. Working paper, Stanford University.

Aviv Y (2001). The effect of collaborative forecasting on supply chain performance. Manag Sci 47(10):1326–1343.

Aviv Y, Federgruen A (2001a). Design for postponement: a comprehensive characterization of its benefits under unknown demand distributions. Oper Res 49(4):578–598.

Aviv Y, Federgruen A (2001b). Capacitated Multi-item inventory systems with random and seasonally fluctuating demands: Implications for postponement strategies. Manag Sci 47: 512–531.

Axsäter S (1990). Simple solution procedures for a class of two-echelon inventory problems. Oper Res 38:64–69.

Axsäter S (1993). Continuous review policies for multi-level inventory systems with stochastic demand. Chapter 4 in Handbooks in Operations Research and Management Science, vol 4, North-Holland.

Azoury KS (1985). Bayes solution to dynamic inventory models under unknown demand distribution. Manag Sci 31:1150–1160.

Azoury KS, Miller BS (1984). A comparison of the optimal levels of Bayesian and non-Bayesian inventory models. Manag Sci 30:993–1003.

Baiman S, Fischer PE, Rajan MV (2000). Information, contracting and quality costs. Manag Sci 46:776–789.

Beyer D, Sethi SP (1997). Average cost optimality in inventory models with markovian demands. J Optimiz Theory App 92:497–526.

Benjaafar S, Cooper WL, Mardan S (2005). Production-inventory systems with imperfect advance demand information and due-date updates. Working paper.

Bitran GR, Caldentey R (2003). An overview of pricing models for revenue management. Manuf Service Oper Manag 5(3):203–229.

Boyacı T, Özer Ö (2004). Information acquisition via pricing and advance sales for Capacity Planning: When to stop and act? To appear in Operations Research.

Buzacott JA, Shantikumar JG (1994). Safety stock versus safety time in MRP controlled production systems. Manag Sci 40:1678–1689.

Cachon G (2003). Supply chain coordination with contracts, Ch. 6 in Handbook of Operations Research and Manag Sci vol. 11, Elsevier, Amsterdam.

Cachon G, Lariviere M (2001). Contracting to assure supply: How to share demand forecasts in a supply chain. Manag Sci 47:629–646.

Cachon G, Lariviere M (2000). Supply chain coordination with revenue-sharing contracts: Strengths and limitations. Manag Sci 51(1):30–34.

Caglar D, Li C, Simchi-Levi D (2004). Two-echelon spare parts inventory system subject to a service constraint. IIE Trans 36:655–666.

Chen F (2003). Information sharing and supply chain coordination. In: de Kok AG, Graves S (eds) Chapter 7 in Handbooks in Operations Research and Management Science, vol 11, Elsevier, Amsterdam.

Chen F (2001). Market segmentation, advanced demand information, and supply chain performance. Manuf Service Oper Manag 3:53–67.

Chen F, Song J (2001). Optimal policies for multiechelon inventory problems with Markov modulated demand. Oper Res 49:226–234.

Chen F, Yu B (2005). Quantifying the value of leadtime information in a single-location inventory system. Manuf Service Oper Manag 7:144–151.

Chen F, Zheng Y-S (1994). Lower bounds for multiechelon stochastic inventory systems. Manag Sci 40:1426–1443.

Clark A, Scarf H (1960). Optimal policies for a multiechelon inventory problem. Manag Sci 6:475–490.

Cohen M, Ho T, Ren J, Terwiesch C (2003). Measuring imputed costs in the semiconductor equipment supply chain. Manag Sci 49:1653–1670.

Cole J, (1997). Boeing, pushing for record production, finds part shortages, delivery delays. Wall Street J June 26.

Corbett CJ, DeCroix GA, Ha AY (2001). Optimal shared savings contracts in supply chains: linear contracts and double moral hazard. Working paper.

Diks EB, de Kök AG (1998) Optimal control of a divergent multiechelon inventory system. EJOR, 111:75–97.

Dvoretzky A, Keifer J, Wolfowitz J (1952). The inventory problem: II. Case of unknown distributions of demand. Econometrica 20:450–466.

Elmaghraby W, Keskinocak P (2003). Dynamic pricing in the presence of inventory considerations: Research overview, current practices, and future directions. Manag Sci 49:1287–1309.

ECR Europe (2003). Shrinkage: A collaborative approach to reducing stock loss in the supply chain.

Eppen G, Schrage L (1981). Centralized ordering policies in a multi-warehouse system with lead times and random demand. TIMS Stud Manag Sci 16:51–67.

Erdhart R (1984). (s, S) policies for a dynamic inventory model with stochastic lead times. Oper Res 32:121–132.

Erkip N, Hausman WH, Nahmias S (1990). Optimal centralized ordering policies in multiechelon inventory systems with correlated demands. Manag Sci 36:381–392.

Erkoc M, Wu D (2005). Managing high-tech capacity expansion via reservation contracts. Prod Oper Manag 14(2):232–251.

Federgruen A (1993). Centralized planning models for multiechelon inventory systems under uncertainty. Chapter 3 in Handbooks in Operations Research and Management Science vol 4, North-Holland.

Federgruen A, Zipkin P (1984a). Approximations of dynamic multilocation production and inventory problems. Manag Sci 30:69–84.

Federgruen A, Zipkin P (1984b). Allocation policies and cost approximations for multi-location inventory systems. Nav Res Logist Q 31:97–129.

Federgruen A, Zipkin P (1984c). Computational issues in an infinite horizon, multiechelon inventory model. Oper Res 32:818–836.

Files J (2001). Economic downturn leaves Cisco with stacks of excess inventory. Knight-Ridder/Tribune Business News, April 27.

Fudenberg D, Tirole J (1991). Game theory. MIT Press, Cambridge, MA.

Gallego G, Özer Ö (2001a). Integrating replenishment decisions with advance demand information. Manag Sci 47:1344–1360.
Gallego G, Özer Ö (2001b). Optimal use of demand information in supply chain management. In: Song J, Yao D (eds) Ch. 5 in Supply Chain Structures 119–160.
Gallego G, Özer Ö (2002). A new algorithm and a new heuristic for serial supply systems. Oper Res Lett 33(4):349–362.
Gallego G, Özer Ö (2003). Optimal replenishment policies for multiechelon inventory problems under advance demand information. Manuf Service Oper Manag 5(2):157–175.
Gallego G, Özer Ö, Zipkin P (2007). Bounds, Heuristics and Approximations for Distribution Systems. Oper Res 55(3):503–517.
Gallego G, Zipkin P (1999). Stock positioning and performance estimation for multi-stage production-transportation systems. Manuf Service Oper Manag 1:77–88.
Gaukler G, Özer Ö, Hausman WH (2008). Order progress information: Improved dynamic emergency ordering policies. Prod Oper Manag 17(6):599–613.
Gayon JP, Benjaafar S, de Véricourt F (2009). Using imperfect demand information in production-inventory systems with multiple demand classes. Manuf Service Oper Manag 11(1):128–143.
Graves S (1985). A multiechelon inventory model for a repairable item with one-for-one replenishment. Manag Sci 31:1247–1256.
Graves S, Meal Hc, Dasu S, Qiu Y (1986). Two-stage production planning in a dynamic environment. In: Axsater S, Schneeweiss C, Silver E (eds) Multi-Stage Production Planning and Control. Springer-Verlag, Berlin, Germany, 9–43.
Graves S, Kletter DB, Hetzel WB (1998). A dynamic model for requirement planning with application to supply chain optimization. Oper Res 46:S35–49.
Graves G, Willems S (2003). Supply chain design: Safety stock placement and supply chain configuration. In: de Kok AG, Graves S (eds) Handbooks in Operations Research and Management Science vol 11, Ch. 3 Elsevier, Amsterdam.
Güllü R (1996). On the value of information in dynamic production/ inventory problems under forecast evolution. Naval Res Logist 43:289–303.
Ha A (2001). Supplier-buyer contracting: Asymmetric cost information and cutoff level policy for buyer participation. Naval Res Logist 48:41–64.
Hariharan R, Zipkin P (1995). Customer-order information, leadtimes and inventories. Manag Sci 41:1599–1607.
Hausman WH (1969). Sequential decision problems: A model to exploit existing forecasts. Manag Sci 16:B93–B111
Heath D, Jackson P (1994). Modeling the evolution of demand forecasts with application to safety stock analysis in production/distribution systems. IIE Transact 26:17–30.
Hibbard J, (2003). The case of the obsolete inventory. Red Herring 123:34–38.
Hopp W, Spearman M, Zhang RQ (1997). Easily implementable inventory control policies. Oper Res 45:327–340.
Hopp W, Zhang RQ, Spearman ML (1999). Easily implementable hierarchical heuristic for a two-echelon spare parts distribution system. IIE Trans 31:977–988.
Hu X, Duenyas I, Kapuscinski R (2004). Advance demand information and safety capacity as a hedge against demand and capacity uncertainty. Working paper. University of Michigan.
Huh W, Janakiraman G (2008). Inventory management with auctions and other sales channels: Optimality of (s,S) Policies. Manag Sci 54:139–150.
Iglehart (1963). Optimality of (s,S) policies in the infinite horizon dynamic inventory problem. Manag Sci 9:259–267.
Iglehart D, Morey R (1972). Inventory systems with imperfect asset information. Manag Sci 18:B-388–394.
Iida T (2001). The infinite horizon non-stationary stochastic multiechelon inventory problem and near myopic policies. Eur J of Oper Res 134(3):525–539.
Janakiraman G, Roundy RO (2004). Lost-sales problems with stochastic lead times: Convexity results for base-stock policies. Oper Res 52:795–803.

Johnson O, Thompson H (1975). Optimality of myopic inventory policies for certain depended processes. Manag Sci 21:1303–1307.
Kaminsky P, Swaminathan J (2001). Utilizing forecast band refinement for capacitated production planning. Manuf Service Oper Manag 3:68–81.
Kang Y, Gershwin S (2005). Information inaccuracy in inventory systems-stock loss and stockout. IIE Trans 37:843–859.
Kaplan R (1970). A dynamic inventory model with stochastic lead times. Manag Sci 16:491–507.
Karaesmen F, Buzacott JA, Dallery Y (2002). Integrating advance order information in make-to-stock production systems. IIE Trans 34(8):649–662.
Karlin S (1960). Dynamic inventory policy with varying stochastic demands. Manag Sci 6:231–258.
Kaya M, Özer Ö (2004). Quality risk in outsourcing: Noncontractible product quality and private quality cost information. Working paper, Stanford University.
Lee H, Billington C, Carter B (1993). Hewlett-Packard gains control of inventory and service through design for localization. Interfaces 23:1–11.
Lee H, Özer Ö (2007). Unlocking the value of RFID. Prod and Oper Manag 16(1):40–64.
Lee H, Padmanadbhan P, Whang S (1997). Information distortion in a supply chain: The bullwhip effect. Manag Sci 43:546–558.
Lee H, Tang C (1997). Modeling the costs and benefits of delayed product differentiation. Manag Sci 43:40–53.
Levi R, Roundy R, Truong VA (2006). Provably near-optimal balancing policies for multiechelon stochastic inventory control models. Working paper.
Levi R, Pal M, Roundy R, Shmoys D. (2007). Approximation algorithms for stochastic inventory control models. Math of Oper Res 32(2):284–302.
Luss H (1982). Operations research and capacity expansion problems: A survey. Oper Res 5:907–947.
Lütze H, Özer Ö (2004). Multi-product manufacturing and market segmentation with advance demand information. Working paper, Stanford University.
Lütze H, Özer Ö (2008). Promised lead time contracts under asymmetric information. Oper Res 56:898–915.
Lim WS (2001). Producer-supplier contracts with incomplete information. Manag Sci 47:709–715.
Lovejoy W (1990). Myopic policies for some inventory models with uncertain demand distributions. Manag Sci 6:724–738.
Lovejoy W (2006). Optimal mechanisms with finite agent types. Manag Sci 47:264–279.
Marklund J (2006). Controlling inventories in divergent supply chains with advance-order information. Oper Res 54:998–1010.
Miller LM (1989). Scarf's state reduction method, flexibility, and a dependent demand inventory model. Oper Res 34:83–90.
Moinzadeh K, Schmidt C (1991). An (S-1,S) inventory system with emergency orders. Oper Res 39:308–321.
Moinzadeh K, Aggarwal P (1997). An information based multiechelon inventory system with emergency orders. Oper Res 45:694–701.
Morton T, Pentico D (1995). The finite horizon non-stationary stochastic inventory problem: Near-myopic bounds, heuristics, testing. Manag Sci 41(2):334–343.
Muckstadt J (1973). A model for a multi-item, multiechelon, multi-indenture inventory system. Manag Sci 20:472–481.
Muckstadt J (2005). Analysis and algorithms for service parts supply chain. Springer, New York, NY.
Muharremoglu A, Tsitsiklis J (2008). A single-unit decomposition approach to multiechelon inventory systems. Oper Res 56:1089–1103.
Özer Ö (2003). Replenishment strategies for distribution systems under advance demand information. Manag Sci 49:255–272.
Özer Ö (2004). Capacity and quality risk in decentralized supply chains. Supply Chain Connect, vol 10, Issue 4.

Özer Ö, Wei W (2006). Strategic commitment for optimal capacity decision under asymmetric forecast information. Manag Sci 52(8):1239–1258.

Özer Ö, Wei W (2004). Inventory control with limited capacity and advance demand information. Oper Res 52:988–1000.

Özer Ö, Xiong H (2008). Stock positioning for distribution systems with service constraints. IIE Trans 40:1141–1157.

Pasternack B (1985). Optimal pricing and return policies for perishable commodities. Market Sci 4:166–176.

Pentico DW (1994). Multistage production systems with random yield and recourse. Int J Prod Res 32:2455–2462.

Porteus E (1990a). Stochastic inventory theory. Chapter 12 in Handbooks in Operations Research and Manag Sci 2:605–652.

Porteus E (1990b). The impact of inspection delay on process and inspection lot sizing. Manag Sci 36:999–1007.

Porteus E, Whang S (1991). On manufacturing/marketing interface. Manag Sci 37:1166–1181.

Raman A, DeHoratious N, Ton Z (2001). Execution: The missing link in retail operations. California Manag Rev 43:136–152.

Reyniers DJ, Tapiero CS (1995). Contract design and the control of quality in a conflictual environment. Eur J Oper Res 82:373–382.

Rinehart RF (1960). Effects and causes of discrepancies in supply operations. Oper Res 8(4):543–564.

Rosling K (1989). Optimal inventory policies for assembly systems under random demands. Oper Res 37:565–579.

Salanie B (1997). The economics of contracts: A primer MIT Press, Cambridge, MA.

Scarf H (1959). The optimality of (s, S) policies in the dynamic inventory problem. Chapter 13 In: Arrow K, Karlin S, Suppes P (eds) Mathematical Methods in the Social Sciences. Stanford University Press, CA.

Scarf H (1960). Bayes solutions of the statistical inventory problem. Naval Logist Res Q 7:591–596.

Schwarz LB (1989). A model for assessing the value of warehouse risk-pooling: Risk-pooling over outside-supplier leadtimes. Manag Sci 35:828–842.

Schwarz LB, Petruzzi NC, Wee K (1998). The value of advance order information and the implication for managing the supply chain: An information/control/buffer portfolio perspective. Working paper Purdue University.

Shang K, Song J (2003). Newsvendor bounds and heuristics for optimal policies in serial supply chains. Manag Sci 49:618–638.

Simon RM (1971). Stationary properties of a two-echelon inventory model for low demand items. Oper Res 19:761–773.

Simchi-Levi D, Kaminsky P, Simchi-Levi E (2000). Designing and managing the supply chain: concepts, strategies, and case studies. Irwin/McGraw-Hill series, Operations and decision sciences.

Song J, Zipkin P (1993). Inventory control in a fluctuating demand environment. Oper Res 41:351–370.

Song J, Zipkin P (1996). Inventory control with information about supply condition. Manag Sci 42:1409–1419.

Tan T, Güllü R, Erkip N (2006). Modeling imperfect advance demand information and analysis of optimal inventory policies. To appear in EurJ Oper Res

Toktay LB, Wein LW (2001). Analysis of a forecasting-production-inventory system with stationary demand. Manag Sci 47:1268–1281.

Tomlin B (2003). Capacity investment in supply chains: Sharing-the-gain rather than sharing-the-pain. Manuf Service Oper Manag 5(4)317–333.

Tsay A (1999). The quantity flexibility contract and supply-customer incentive. Manag Sci 45:1339–1358.

Van Houtum GJ, Inderfurth K, Zijm WHM (1996). Materials coordination in stochastic multiechelon systems. EJOR 95:1–23.
Veinott A (1965). Optimal policy for a multi-product, dynamic, non-stationary inventory problem. Manag Sci 12:206–222.
Veinott A (1966). On the optimality of (s, S) inventory policies: New conditions and a new bound. J. SIAM Appl Math 14:1067–1083.
Wang T, Toktay LB (2008). Inventory management with advance demand information and flexible delivery. Manag Sci 5(4):716–732.
Watson N, Zhang Y (2005). Decentralized serial supply chains subject to order delays and information distortion: Exploiting real time sales data. Manuf Service Oper Manag 7:152–168.
Yano C, Lee H (1995). Lot sizing with random yields: A review. Oper Res 43:311–334.
Zhu K, Thonemann UW (2004). Modeling the benefits of sharing future demand information. Oper Res 52:136–147.
Zipkin PH (2000) Foundations of inventory management, McGraw Hill, Boston.

Chapter 14
Pricing, Variety, and Inventory Decisions for Product Lines of Substitutable Items

Bacel Maddah, Ebru K. Bish, and Brenda Munroe

14.1 Introduction and Motivation

Integrating operations and marketing decisions greatly benefits a firm. Marketing actions drive consumer demand, which significantly influences operations management (OM) decisions in areas such as capacity planning and inventory control. On the other hand, the marketing department of a firm relies on OM cost estimates in making decisions concerning pricing, variety, promotions, etc. In this chapter, we review recent research on pricing, assortment (or variety), and inventory decisions in retail operations management, which contribute to the growing literature on joint marketing/OM models (e.g., Eliashberg and Steinberg 1993; Griffin and Hauser 1992; Karmarkar 1996; Pekgün et al. 2006, 2008; Porteus and Whang 1991). Other important contributions of the reviewed works account for inventory costs in pricing and variety models and utilize realistic demand models based on consumer choice theory. These contributions are discussed below.

Before detailing the contribution of the research surveyed in this chapter, we define the scope. We focus on retailer settings because of a large number of recent works, including ours, in retail operations management. In addition, the research in retail settings is strongly connected to product line design and production planning problems in manufacturing (see Sect. 14.3). Within the retail setting, we consider decisions involving a "product line" or a "category," which refers to a family of substitutable items that serve the same need for the consumer but differ in secondary aspects. Thus, a product line may consist of different brands with the same usage (e.g., different brands of coffee or yogurt) as well as different variants of the same brand (such as different sizes, colors, or flavors). When faced with a purchasing decision from a product line, consumers select their most preferred item, given the trade-off between price and quality. Pricing has a major impact on consumers'

B. Maddah (✉)
Engineering Management Program, American University of Beirut, Beirut 1107 2020, Lebanon
e-mail: bacel.maddah@aub.edu.lb

K.G. Kempf et al. (eds.), *Planning Production and Inventories in the Extended Enterprise*, International Series in Operations Research & Management Science 151,
DOI 10.1007/978-1-4419-6485-4_14,

choice among the available alternatives. However, the assortment or variety level, in terms of the items offered in the product line, and the shelf inventory levels of these items are equally important.

Under an integrated marketing/OM framework, the retailer sets two or all of the above decisions simultaneously. This seems to be a successful business practice for many retailers. For example, JCPenney received the "Fusion Award" in supply chain management for "its innovation in integrating upstream to merchandising and allocation systems and then downstream to suppliers and sourcing." A JCPenney vice president attributes this success to the fact that, at JCPenney, "assortments, allocations, markdown pricing are all linked and optimized together" (Frantz 2004). Canadian retailer Northern Group managed to get out of an unprofitable situation by implementing a merchandise optimization tool. Northern Group's chief financial officer credits this turnaround to "assortment planning" and the attempt to "sell out of every product in every quantity for full price" (Okun 2004). Moreover, our experience with Hannaford, one of the largest chains of grocery stores in New England, on various aspects of pricing, variety, and shelf inventory decisions attests to the strong interdependence among these decisions.

An important contribution of the reviewed research is to include inventory costs within pricing and assortment optimization models. Most of the classical literature along this avenue assumes that demand is known with certainty and therefore excludes inventory considerations (Dobson and Kalish 1993; Green and Krieger 1989; Kaul and Rao 1995; and the references therein). We believe this is due, in part, to the complexities introduced by modeling inventory. For example, the review paper by Petruzzi and Dada (1999) indicates a high level of difficulty associated with joint pricing and inventory optimization even for the single item case. These difficulties do not, however, justify ignoring inventory effects in modeling. For example, in 2003, the average End-of-Month capital invested in inventory of food retailers (grocery and liquor stores) in the USA was approximately 34.5 billion $s, with an inventory/sales ratio of approximately 82% (US Census Bureau). On the other hand, the net 2003 profit margin in food retailing is estimated to be 0.95% Food Marketing Institute. These numbers indicate that food retailers can significantly increase their profitability by reducing their inventory costs.

In addition to utilizing an integrated approach which accounts for inventory costs, the reviewed works adopt demand models from the marketing and economics literature that reflect the actual manner consumers make their buying decisions. These "consumer choice" models are based on the classical principle of utility maximization (e.g., Anderson et al. 1992; Ben-Akiva and Lerman 1985; Manski and McFadden 1981; McFadden 1974).

Given the complexity of the product line problem, most research focuses on two of the three essential decisions involved (pricing, variety, and inventory), with the exception of one recent work (Maddah and Bish 2007) that considers an integrated model involving all three decisions. This chapter is structured to offer a representative cross section of this line of research, reviewed in Sects. 14.5–14.7. The reviewed works are streamlined according to their decision variables (see Table 14.1 for an overview). Note that this review is not intended to be exhaustive. Section 14.2

Table 14.1 Reviewed papers classified according to their decision variables

Decisions	Papers	Reviewed in
Inventory and variety	Bish and Maddah (2008)	Sect. 14.5
	Cachon et al. (2005)	
	Gaur and Honhon (2006)	
	van Ryzin and Mahajan (1999)	
Inventory and pricing	Aydin and Porteus (2008)	Sect. 14.6
	Bish and Maddah (2008)	
	Cattani et al. (2010)	
Inventory, pricing, and variety	Maddah and Bish (2007)	Sect. 14.7

briefly reviews additional (broadly related) works. Other supporting sections in this chapter are as follows: In Sect. 14.3, we discuss how the research reviewed here is related to manufacturing. In Sect. 14.4, we present the key ideas of a set of consumer choice models that are commonly used in developing product line demand functions. In Sect. 14.8, we present a critique and a comparison of the insights and methodologies across the different works reviewed in Sects. 14.5–14.7, with a focus on their connection to practical applications. In Sect. 14.9, we summarize our observations on the current practice of retail pricing, inventory, and variety management. Finally, in Sect. 14.10, we conclude and provide suggestions for future research.

We note that Mahajan and van Ryzin (1999) wrote an excellent book chapter on a similar topic, which we complement by reviewing the research that mostly appeared after the publication of Mahajan and van Ryzin (1999).

14.2 Background Literature

The decisions considered in this chapter lie at the interface of economics, marketing, and OM disciplines. As a result, there are many other works in these disciplines, which, although not reviewed here in detail, are nevertheless broadly related. The economics literature approaches this topic from the point of view of product differentiation (see Lancaster [1990] for a review) and focuses on developing consumer choice models that reflect the way consumers make their purchasing decisions from a set of differentiated products (e.g., Hotelling 1929; Lancaster 1966; McFadden 1974).

The marketing literature emphasizes the process of fitting appropriate choice models to data collected on actual consumer behavior (e.g., Besanko et al. 1998; Guadagni and Little 1983; Jain et al. 1994). Such data are typically compiled from scanner data (i.e., log of all sales transactions in a store) or panel data (obtained by tracking the buying habits of a selected group of customers). These are then utilized to address product line design (i.e., variety) and pricing decisions (see Green and Krieger (1989) and Kaul and Rao (1995) for reviews). A typical approach utilizes the data collected on consumer behavior to obtain deterministic estimates of utilities for each consumer segment. The deterministic utilities can then be used to formulate an integer program that gives the optimal assortment and its pricing (e.g., Dobson and Kalish 1993, 1988; Green and Krieger 1985). See also the two recent papers by Hall et al. (2010) Yücel et al. (2009) for the utilization of mathematical programming approaches to address the pricing, assortment, and inventory decisions of a product line but under endogenous demand (rather than demand aggregated from consumer preferences).

Another important research problem not reviewed here is concerned with shelf space allocation among substitutable items (e.g., Corstjens and Doyle 1981); some of this work also integrates assortment planning with shelf space allocation within a mathematical programming framework. (See Martín-Herrán et al. 2006 for a review of research in this area and interesting competitive analysis in a manufacturing duopoly setting as well as Irion et al. 2006, who propose an efficient linearization technique).

Finally, the research on single item inventory models with price-dependent demand is also relevant to the research reviewed in this chapter (see Petruzzi and Dada 1999 and Elmaghraby and Keskinocak 2003 for comprehensive reviews).

14.3 Relevance to Manufacturing

While most of the research reviewed in this chapter is presented within the retailing context, this research is relevant to the manufacturer's product line design and production planning problems in two important aspects.

First, manufacturers need to make decisions regarding the composition (assortment), pricing, and production planning of their product lines, and these decisions revolve around similar trade-offs to those discussed in this chapter. As a result, manufacturers can benefit from the models and insights presented here in answering these questions. In fact, many recent papers address the manufacturer's problems in this context (e.g., Alptekinoglu and Corbett 2008; Cattani et al. 2010; Hopp and Xu 2003, 2005). In addition, consumer choice models (such as the ones considered in this chapter) are gaining popularity among manufacturers, in an attempt to more realistically model their demand. We are aware of such efforts at several leading automotive manufacturers such as General Motors and Honda. On the cost side, implementing the models presented here in a manufacturing setting may necessitate certain adjustments. This is because the cost structure for a manufacturing firm may involve additional terms not considered here (e.g., costs related to product development, launch, and marketing, and fixed setup costs). A manufacturer also has capacity constraints for its production resources (e.g., plant, labor), which are of a different nature than the retailer's shelf space capacity constraints.

Second, in supply chain settings, manufacturers' and retailers' product line design, pricing, and inventory decisions are intimately related in that they impact one another. These dependencies are also impacted by cooperation and contractual agreements on profit sharing between retailers and manufacturers. For example,

Aydin and Porteus (2009) investigate the dependence between the manufacturer's rebates and the retailer's pricing and inventory decisions in a setting where the retailer's demand is generated from the multinomial logit model (MNL) reviewed in Sect. 14.4. Aydin and Hausman (2009) discuss supply chain coordination in assortment planning between a manufacturer and a retailer with end customers making their purchase decisions based on the MNL choice model.

Another supply chain related issue is the attempt of retailers to benefit from the manufacturers' (suppliers') specialized expertise through adopting "category captainship" schemes where one lead manufacturer is responsible for managing the *whole* category (product line). Kurtuluş and Toktay (2009) discuss in detail the theory and practice of category captainship. Among the works reviewed in Kurtuluş and Toktay (2009), Kurtuluş and Toktay (2005), and Wang et al. (2003) consider delegation of pricing decisions to a category captain while assuming exogenous assortments and ignoring inventory costs (under a deterministic demand assumption), whereas Kurtuluş and Toktay (2006) assume exogenous prices and consider delegating the assortment decision under random MNL-based demand and limited inventory. One drawback of category captainship is that the category captain may make decisions in a way that place the competing manufacturers at a disadvantage, which raises concerns of violating antitrust legislation. Drake and Swann (2006) suggest and study "vendor-specific category management" as an alternative which avoids antitrust issues, where every manufacturer is responsible for managing the product(s) she supplies.

Finally, the channel selection problem widely studied in the marketing literature is also related, since it refers to the manufacturer's problem of what type of retailers to select for her product line (e.g., Choi 1991).

14.4 Overview of Consumer Choice Models

In this section, we briefly review some of the discrete choice models that represent the consumer preference as a stochastic utility function. We refer the readers interested in an in-depth treatment of the choice theory to Anderson et al. (1992) and Ben-Akiva and Lerman (1985).

Let $\Omega = \{1, 2, \ldots, n\}$ be the set of possible variants (substitutable items) from which the retailer can compose her product line. Let $S \subseteq \Omega$ denote the set of items stocked by the store. Demand for items in S is generated from customers arriving to the retailer's store. A customer chooses to purchase *at most* one item from set S so as to maximize her utility. The consumer utility, U_i, for each item $i \in S$, is assumed to be a random variable with a known distribution. The randomness can be either due to the fact that the seller does not have access to this information or due to the differences in tastes among consumers as well as inconsistencies in individual consumer behavior on different shopping occasions, see Anderson et al. (1992) for further discussion. A similar assumption is made concerning U_0, the consumer utility for the "no-purchase" option. Then, the probability that a consumer buys item $i \in S$ is $q_i(S) = Pr\{U_i = \max_{j \in S \cup \{0\}} U_j\}$. Several consumer choice models are derived based on the distribution of $U_i, i \in S \cup \{0\}$. We discuss two of these models in Sects. 14.4.1 and 14.4.2.

14.4.1 The Multinomial Logit Model

The MNL model is among the most popular consumer choice models (e.g., Anderson et al. 1992 and Ben-Akiva and Lerman 1985). Interestingly, the MNL has its roots in mathematical psychology (e.g., Luce 1959 and Luce and Suppes 1965).

Under the MNL, the utility for $i \in S \subseteq \Omega$ is $U_i = u_i + \epsilon_i$, and the utility for the no-purchase option is $U_0 = u_0 + \epsilon_0$, where u_i (u_0) is the expected utility for item i (the no-purchase option), and $\epsilon_i, i \in S \cup \{0\}$ are independent and identically distributed (i.i.d.) Gumbel (double exponential) random variables with mean 0 and shape factor μ. The cumulative distribution function for a Gumbel random variable is $F(x) = e^{-e^{-(x/\mu+\gamma)}}$, where $\gamma \approx 0.5772$. The Gumbel distribution is utilized mainly because it is closed under maximization (i.e., the maximum of several independent Gumbel random variables is also a Gumbel random variable). This property leads to closed-form expressions for purchase probabilities, given below, which facilitate their use in analytical models:

$$q_i(S) = \frac{v_i}{\sum_{j \in S \cup \{0\}} v_j}, \quad i \in S, \tag{14.1}$$

where $v_j \equiv e^{u_j/\mu}$, $j \in S \cup \{0\}$.

Another reason for the widespread use of MNL is that it has been shown to be a good fit to actual store transaction data, and its parameters can be estimated somewhat easily, especially with the wide use of information systems that track such transactions (e.g., Guadagni and Little 1983, Hauser 1978, McFadden 1974, McFadden et al. 1978).

One drawback of the MNL is that it suffers from the *independence from irrelevant alternatives* (IIA) property, i.e., the ratio of purchase probabilities of two items is the same regardless of the choice set they are in. This implies that MNL is suited for modeling situations where all items in the product line are "broadly similar" or "close substitutes," which is not always the case in practice. For example, in an ice-cream product line, a chocolate flavor is a closer substitute for other chocolate flavors than for vanilla flavors. The nested multinomial logit model (NMNL), a variation of the MNL, has been proposed by Ben-Akiva (1973) as a remedy for the IIA property.

14.4.2 Locational Choice Model

This model is attributed to (e.g. Lancaster 1990).[1] Items in Ω are assumed to be located on the interval $[0, 1]$ representing the attribute space, with the location of product i denoted as b_i. Consumers have an ideal product in mind, with location X. (This location may vary among consumers and is therefore considered a random variable.) Then, the consumer's utility for item i is $U_i = U - g(|X - b_i|)$, where U represents the utility of a product at the ideal location and $g(\cdot)$ is a strictly increasing function representing the disutility associated with deviation from the ideal location ($|X - b_i|$ is the distance between the location of item i and the ideal location). The purchase probabilities of items in Ω are then derived for a given probability distribution of X, based on the principle of utility maximization.

In the remainder of this chapter, we present several product line models that utilize the MNL and locational choice models. We then provide a discussion in Sect. 14.10 on the attractiveness of utilizing the NMNL model for future research.

14.5 Variety and Inventory Under Exogenous Prices

In this section, we review models that assume that prices of items in the choice set Ω are *exogenously* set. The retailer's problem is to decide on the subset of items, $S \subseteq \Omega$, to offer in her product line, together with the inventory levels for items in S. The papers we review make the following modeling assumptions:

(A1): Demand is generated from consumers arriving to the retailer's store in a single selling period, and behaving according to one of the choice models discussed in Sect. 14.4.

(A2): Inventory costs are derived based on the classical single-period newsvendor model.

(A3): Demand and cost functions are derived under "static choice" assumptions. That is, consumers make their purchasing decisions independently of the inventory status at the moment of their arrival, and they will leave the store empty-handed if their preferred item (in S) is out of stock.

Assumptions (A2) and (A3) simplify the analysis. Although they are somewhat restrictive, they can be justified in certain settings. For example, (A2) holds in the case of retailers utilizing electronic data exchange and computer-assisted ordering, whereas (A3) holds for retailers who sell based on catalogs or floor models. We refer the reader interested in a more in-depth discussion of (A2) and (A3) to Smith and Agrawal (2000) and van Ryzin and Mahajan (1999), respectively. We note here that

[1] This model is a generalization of Hotelling (1929). It is based on perceiving a product as a "bundle of characteristics" rather than only utilizing location and transportation costs as in Hotelling (1929), (see Lancaster 1990 for details).

recent works that relax (A3) present numerical evidence suggesting that the static choice model provides a reasonable approximation. (see Gaur and Honhon 2006; Mahajan and van Ryzin 2001).

The first works on joint variety and inventory decisions under assumptions (A1)–(A3) are those of van Ryzin and Mahajan (1999), with exogenous prices, and Smith and Agrawal (2000). Many recent papers build on the work of van Ryzin and Mahajan (1999). Bish and Maddah (2008) study a simplified version of the van Ryzin and Mahajan model by considering a product line of similar items having the same cost and demand structure but endogenize the pricing decision. Cachon et al. (2005) focus on the effect of consumer search on the assortment decision in the van Ryzin and Mahajan framework, and Gaur and Honhon (2006) utilize the Lancaster consumer choice model (instead of the MNL). In the remainder of this section, we review van Ryzin and Mahajan's work, i.e., the basic van Ryzin and Mahajan model and these subsequent works in detail.

Consider a product line under the MNL consumer choice process within a newsvendor inventory setting. The probability that a customer buys item $i \in S \subseteq \Omega$ is given by $q_i(S)$ in (14.1). The mean number of customers visiting the store in the selling period is λ. The demand for item $i \in S$ is assumed to be a Normal random variable, X_i, with mean $\lambda q_i(S)$ and standard deviation $\sigma(\lambda q_i(S)^{\beta})$, where $\sigma > 0$ and $0 \leq \beta < 1$. The reason behind this choice of parameters is to have a coefficient of variation of X_i that is decreasing in the mean store volume λ (this seems to be the case in practice). The special case with $\sigma = 1$ and $\beta = 1/2$ represents a Normal approximation to demand generated from customers arriving according to a Poisson process with rate λ.

van Ryzin and Mahajan assume that all items in Ω have the same unit cost, c, and are sold at the same price, p (or have the same c/p ratio). This assumption holds, for example, in the case of a product line having different flavors or colors of the same variant. On the cost side, items of the product line do not have a salvage value and no additional holding or shortage costs apply. This cost structure captures the essence of inventory costs in terms of overage and underage costs. By utilizing the well-known results for the newsvendor model under Normal demand, the optimal inventory level for item $i \in S$, $y_i^*(S)$ and the expected profit from S at optimal inventory levels, $\Pi(S)$, can be written as

$$y_i^*(S) = \lambda q_i(S) + \Phi^{-1}(1 - \frac{c}{p})\sigma(\lambda q_i(S))^{\beta}, \quad i \in S\,, \tag{14.2}$$

$$\Pi(S) = \sum_{i \in S}\Big[\lambda q_i(S)(p-c) - p\theta\sigma(\lambda q_i(S))^{\beta}\Big]\,, \tag{14.3}$$

where $\theta \equiv \phi(\Phi^{-1}(1 - c/p))$, $\phi(\cdot)$ and $\Phi(\cdot)$ are the probability density function and the cumulative distribution function of the standard Normal distribution, respectively.

The retailer's objective is to find the optimal assortment S^* yielding the maximum profit Π^*:

$$\Pi^* = \Pi(S^*) = \max_{S \subseteq \Omega}\{\Pi(S)\}\,. \tag{14.4}$$

The main factor involved in determining the optimal assortment is the trade-off between the sales revenue and the inventory cost. High variety leads to a high sales revenue as well as a high inventory cost (due to "thinning" of item demand). As a result, the optimal assortment should not be too small (to generate enough sales revenue) or too large (to avoid the excessive inventory cost). The following result from van Ryzin and Mahajan (1999) is a consequence of the trade-off between the sales revenue and the inventory cost.

Lemma 14.1. *Consider an assortment $S \subseteq \Omega$. Then, the expected profit from S, $\Pi(S, v_i)$, is quasi-convex in v_i, the preference of item $i \in S$.*

Lemma 14.1 allows deriving the structure of the optimal assortment, the main result in van Ryzin and Mahajan (1999).

Theorem 14.1. *Assume that the items in Ω are ordered such that $v_1 \geq v_2 \geq \cdots \geq v_n$. Then, an optimal assortment is $S^* = \{1, 2, \ldots, k\}$, for some $k \leq n$.*

Theorem 14.1 states that an optimal assortment contains the k most popular items for some $k \leq n$. Thus, the structure of the optimal assortment is quite simple. van Ryzin and Mahajan then study the factors that affect the variety level. Assuming $v_1 \geq v_2 \geq \cdots \geq v_n$, they consider assortment of the optimal form $S_k = \{1, 2, \ldots, k\}$ and use k as a measure of variety. They derive asymptotic results stating that (1) $\Pi(S_{k+1}) > \Pi(S_k)$ for sufficiently high selling price, p; (2) $\Pi(S_{k+1}) < \Pi(S_k)$ for sufficiently low no-purchase preference, v_0; and (3) $\Pi(S_{k+1}) > \Pi(S_k)$ for sufficiently high store volume, λ. For example, (3) implies that stores with high volume such as "super stores" should offer a high variety.

14.5.1 Inventory and Assortment Decisions with Similar Items

This section is based on Bish and Maddah (2008). Consider the model in van Ryzin and Mahajan (1999) under the additional assumption that $v_1 = v_2 = \cdots = v_n = v$. However, now the consumer utility depends on the price by assuming that $u_1 = u_2 = \cdots = u_n = \alpha - p$ (equivalently $v = e^{(\alpha - p)/\mu}$), where α is the mean reservation price (quality index) of an item. Such a structure, which allows a better understanding of the effect of pricing, is common in the literature (e.g., Guadagni and Little 1983).

This is a stylized model with "similar" items. It may apply in cases such as a product line with different colors or flavors of the same variant, where consumer preferences for the items in the product line are quite similar. The main research question here is to characterize the optimal assortment size (i.e., the number of similar items to carry in the store). In addition, this simple setting allows a comprehensive study of the factors that affect the variety level through a comparative statics analysis.

While all the results in Bish and Maddah hold under the general demand model in van Ryzin and Mahajan, to simplify the exposition, the following results are given in terms of $\sigma = 1$ and $\beta = 1/2$ (which corresponds to demand generated from a Poisson process). Then, for an assortment of k items, the optimal inventory level of each item in (14.2) and the expected profit at the optimal inventory levels in, (14.3) respectively, reduce to

$$y^*(p,k) = \lambda q(p,k) + \Phi^{-1}\left(1 - \frac{c}{p}\right)\sqrt{\lambda q(p,k)}\,, \tag{14.5}$$

$$\Pi(p,k) = k\Big[\lambda q(p,k)(p-c) - p\theta(p)\sqrt{\lambda q(p,k)}\Big]\,, \tag{14.6}$$

where $q(p,k) = (e^{(\alpha-p)/\mu})/(v_0 + ke^{(\alpha-p)/\mu})$ and $\theta(p) \equiv \phi(\Phi^{-1}(1-c/p))$.[2]

Despite the simplified form of the expected profit function in (14.6), it is still difficult to analyze it because of the complicating term $\theta(p)$. Bish and Maddah develop the following approximation to simplify the analysis:

$$\theta(x) \approx ax(1-x)\,, \tag{14.7}$$

where $a = 1.66$. (See Maddah (2005) for more details on this approximation.) With this approximation, $\Pi(p,k)$ in (14.6) simplifies to the following:

$$\Pi(p,k) = k(p-c)\Big[\lambda q(p,k) - a\frac{c}{p}\sqrt{\lambda q(p,k)}\Big]\,. \tag{14.8}$$

Under a regularity assumption (see Bish and Maddah for details), aimed at eliminating trivial cases where demand is too low and the retailer is better off selling nothing, the expected profit in (14.8) is well behaved in the assortment size k, as the following theorem indicates.

Theorem 14.2. *The expected profit $\Pi(p,k)$ is strictly pseudo-concave and unimodal in k.*

Let $k_p^* \equiv \arg\max_k \Pi(p,k)$. Theorem 14.2 states that the expected profit increases with variety (k) up to $k = k_p^*$, and then decreases. Thus, Theorem 14.2 implies that there exists an upper limit on the variety level. This is not the case in the riskless case (with deterministic demand $\lambda q(p,k)$ and profit $k(p-c)\lambda q(p,k)$) where there is no upper bound on variety in the product line. That is, inventory cost limits the variety level of the product line. Theorem 14.2 formally proves this last statement. The intuition behind this result is linked to the trade-off between the sales revenue and the inventory cost and their implications on variety level discussed above.

Based on Theorem 14.2, one can perform a comparative static analysis on the optimal assortment size, k_p^*, as presented in Theorem 14.3.

---

[2] We write $y^*(p,k)$ and $\Pi(p,k)$ as functions of both p and k because we will refer to this model later, in Sect. 14.6.2, to present the pricing analysis.

Theorem 14.3. *The optimal assortment size k_p^* is:*

(i) Decreasing in the unit cost per item, c;
(ii) Increasing in the expected store volume (arrival rate), λ.

Theorem 14.3 states that the higher the unit cost per item, the lower the optimal variety level. That is, retailers selling costly items should not offer a wide variety. On the other hand, retailers with low-cost items should diversify their assortments. This result should not be seen as conflicting with the asymptotic result of van Ryzin and Mahajan (1999) indicating that the expected profit is increasing in the variety level at a price that is large enough since high prices are not necessarily based on high costs. Theorem 14.3 also indicates that a higher store volume allows the retailer to offer a wide variety. This extends the asymptotic result of van Ryzin and Mahajan.

14.5.2 Inventory and Assortment Decisions Under Consumer Search

This section is based on Cachon et al. (2005), whose model extends van Ryzin and Mahajan (1999) by accounting for "consumer search" and by considering a general concave inventory cost function. Consumer search refers to the phenomenon that consumers may not purchase their most preferred item in the retailer's product line if it is possible for them to search other retailers for, perhaps, "better" items.

The expected profit function considered in Cachon et al. (2005) is a generalization of (14.3), with an inventory cost function that is concave and increasing. Cachon et al. show that the structure of the optimal assortment of van Ryzin and Mahajan (1999) in Theorem 14.1 continues to hold with this more general cost function.

Cachon et al. present two consumer search models that differ from the "no-search" model (van Ryzin and Mahajan 1999) in the expressions for the purchase probabilities. In the "independent assortment" search model, it is assumed that no other retailer in the market carries any of the items offered in the product line of the retailer under consideration. This applies, for example, to product lines of jewelry or antiques. In the "overlapping assortment" model, a limited number of variants are available in the market, and the same variant can be offered by many retailers. This applies, for example, to product lines of digital cameras. In this case, the consumer's expected value from search decreases with the assortment size. Offering more items in an assortment reduces the search value for the consumer.[3]

The purchase probabilities in the independent assortment model are derived based on Gumbel utilities as in Sect. 14.4.1, with an additional utility from the search, $U_r = u_r + \epsilon_r$. In addition, the search cost is b. Then, the

[3] We are using the term "assortment size" loosely here to refer to variety level in terms of number of items in an assortment. Cachon et al. use a more precise measure.

purchase probabilities, $q_i^{si}(S)$, are derived as a function of the no-search purchase probabilities, $q_i(S)$ in (14.1), as

$$q_i^{si}(S) = q_i(S)(1 - H(\overline{U}, S)) \ , \ \ i \in S, \tag{14.9}$$

where $\overline{U} \equiv u_r - b$ and $H(\overline{U}, S) = e^{-(v_0 + \sum_{j\in S} v_j)e^{-(\overline{U}/\mu+\gamma)}}$.

In the overlapping assortment model, the purchase probabilities, $q_i^{so}(S)$, are derived as

$$q_i^{so}(S) = q_i(S)(1 - H(\overline{U}(S), S)) \ , \ \ i \in S, \tag{14.10}$$

where $\overline{U}(S)$ is the unique solution to $\int_{\overline{U}(S)}^{\infty}(x - \overline{U}(S))w(x, S)\mathrm{d}x = b$, $w(x, S)$ is the density function of $\overline{U}_{\max} = \max_{j\notin S} U_j$, and $H(\cdot)$ is as defined in (14.9).

Cachon et al. show that the independent consumer search model does not change the structure of the optimal assortment in the no-search (van Ryzin and Mahajan) model in Theorem 14.1. However, this result does not hold under the overlapping assortment search model. Nevertheless, Cachon et al. point out that restricting the search to "popular assortments" having the structure given by Theorem 14.1 provides reasonable results in most of their numerical test cases.

14.5.3 Inventory and Assortment Decisions Under Locational Choice

This section is based on Gaur and Honhon (2006). Rather than choosing from a finite set of variants Ω in composing the product line, Gaur and Honhon determine the locations and the number of items to be offered in the [0, 1] interval. Specifically, the problem in Gaur and Honhon (2006) is to determine the number, n, and the locations of items in an assortment given by $\mathbf{b} = (b_1, \ldots, b_n)$, where $b_j \in [0, 1]$, $b_j < b_{j+1}$, is the location of item j. All items have the same unit cost, c, and are sold at the same price, p (as in Bish and Maddah 2008 and van Ryzin and Mahajan 1999).

The demand function under the locational choice model is derived as discussed in Sect. 14.4.2 with $U = Z - p$. The utility of the no-purchase option is assumed to be zero. The *coverage distance* of item j is defined as $L = \max_x\{|x - b_j| : Z - p - g(|x - b_j|) > 0\}$. The *first choice* interval containing the ideal item locations for customers who purchase item j (i.e., customers who obtain maximum positive utility from j) is then given by $[b_j^-, b_j^+]$, where $b_j^- = \max\{b_j - L, (b_j + b_{j-1})/2\}$ and $b_j^+ = \min\{b_j + L, (b_j + b_{j+1})/2\}$. Finally, the purchase probability of item j is given by

$$q_j(\mathbf{b}) = Pr\{b_j^- \le X \le b_j^+\} = F_X(b_j^+) - F_X(b_j^-) \ ,$$

where $F_X(\cdot)$ is the cumulative distribution function of X. Gaur and Honhon consider that the distribution of X is such that its density function is either unimodal or uniform. Then, similar to Bish and Maddah (2008), demand for item j is considered to be a Normal random variable with mean $\lambda q_j(\mathbf{b})$ and standard deviation $\sqrt{\lambda q_j(\mathbf{b})}$.

Gaur and Honhon derive a newsvendor-type expected profit function similar to van Ryzin and Mahajan (1999) but they include a fixed cost, f, for adding an item to an assortment. Gaur and Honhon then determine the product locations $\mathbf{b} = (b_1, \ldots, b_n)$, $b_j \in [0, 1]$, together with the number of items to offer, n, so as to maximize the expected profit. Their main result is that items should be equally spaced on a subinterval of [0, 1], with the distance between any two adjacent items equal to $2L$. They also derive the optimal number of items to offer, n^*, as a function of b_1^* and propose a line search method to find b_1^*. The main insight here is that the optimal assortment contains products that are equally spaced, with no substitution between the products, regardless of the distribution of consumer preference on the attribute space (given by $F(\cdot)$). Gaur and Honhon obtain further insights through a numerical study by showing that, unlike the MNL case (in van Ryzin and Mahajan [1999]), the optimal assortment may not contain the most popular items. They also find, contrary to the MNL case (e.g., Theorem 14.2 of Bish and Maddah 2008), that it is always optimal to cover the entire market (i.e., the [0,1] interval) in the absence of fixed costs.

14.6 Inventory and Pricing for a Given Assortment

In this section, we review models that assume that the retailer's assortment is *exogenously* determined, and study the retailer's problem of determining the prices and inventory levels for the items in the assortment. These models are based on the papers by Aydin and Porteus (2008), Bish and Maddah (2008), and Cattani et al. (2010), which, to our knowledge, are the only papers that consider joint pricing and inventory decisions for substitutable items under consumer choice. All these models continue to consider a single-period setting and static choice assumptions, as discussed in Sect. 14.5, and they all consider demand functions generated by the MNL choice model. In that sense, these papers can be seen as extentions to van Ryzin and Mahajan (1999), but without considering the assortment decision, as well as variants to the price-setter newsvendor model (e.g., Petruzzi and Dada 1999).

Bish and Maddah (2008) and Cattani et al. (2010) essentially consider the same model as van Ryzin and Mahajan, with endogenous prices and price-dependent demand functions derived from the MNL choice model. However, Bish and Maddah (2008) consider a stylized model with similar items having the same cost and demand structure, and obtain stronger analytical results that lead to useful insights on the effect of accounting for inventory costs on the pricing of product lines. Aydin and Porteus (2008) consider a model similar to Cattani et al. (2010), but simplify the demand function by utilizing a "multiplicative" structure, which assumes that the coefficient of variation of the demand is independent of pricing. This allows establishing important structural properties of the optimal prices of items in the assortment (the main result is that these prices are the unique solution to the first-order optimality conditions). Such regularity results could not be derived with the "mixed multiplicative-additive" demand function in Bish and Maddah (2008) and Cattani et al. (2010) (where both the standard deviation and coefficient variation of the demand are functions of prices)[4]. The multiplicative model of Aydin and Porteus (2008) is commonly used in the academic literature (e.g., Petruzzi and Dada 1999) and may be applicable in certain practical situations. However, it does not apply directly to many practical situations, such as demand being generated from Poisson arrivals. This is the motivation for the mixed multiplicative–additive model of Bish and Maddah (2008) and Cattani et al. (2010).[5]

14.6.1 Inventory and Pricing Decisions with a Multiplicative Demand Function

Aydin and Porteus (2008) assume that the demand for item $i \in S$ is given by $X_i = Mq_i(S,\mathbf{p})\xi_i$, where M is a positive constant representing the expected market size, and ξ_i, $i \in S$, are i.i.d. random variables with positive support and with a cumulative distribution function which is IFR[6], and $q_i(S,\mathbf{p})$ is the MNL purchase probability obtained from (14.1) by assuming that $u_j = \alpha_j - p_j$, (similar to Bish and Maddah model [2008] of Sect. 14.5.1), i.e.,

$$q_i(S,\mathbf{p}) = \frac{\mathrm{e}^{(\alpha_i - p_i)/\mu}}{v_0 + \sum_{j \in S} \mathrm{e}^{(\alpha_j - p_j)/\mu}}, \quad i \in S, \tag{14.11}$$

with $\mathbf{p} = (p_1, \ldots, p_{|S|})$ denoting the prices of items in S, with $|S|$ denoting the cardinality of set S.

Aydin and Porteus write the expected profit at the optimal inventory levels as

$$\Pi(S,\mathbf{p}) = \sum_{i \in S} \Pi_i(S,\mathbf{p}) = \sum_{i \in S} p_i \int_0^{y_i^*(S,\mathbf{p})} x f_{X_i}(S,\mathbf{p},x)\mathrm{d}x\,, \tag{14.12}$$

---

[4] In a "multiplicative" demand model, demand is of the form $D(p) = f(p)\epsilon$, where ϵ is a random variable and $f(p)$ is a function of the price, p. In an "additive" demand model, demand is of the form $D(p) = f(p) + \epsilon$. In a "mixed multiplicative–additive" demand model, $D(p) = g(p) + f(p)\epsilon$, where $g(p)$ is also a function of the price.

[5] Another practical setting where the mixed multiplicative-additive demand model is a good approximation for the actual demand is when the total number of customers visiting the store follows a Negative Binomial distribution (see Maddah and Bish 2007 for details).

[6] $F(.)$ is IFR if its failure rate, $f(x)/(1-F(x))$, is increasing in x, where $f(x)$ is the corresponding density function.

where $\Pi_i(S, \mathbf{p})$ is the expected profit from item $i \in S$, $y_i^*(S, \mathbf{p})$ is the optimal inventory level for item $i \in S$, i.e., $y_i^*(S, \mathbf{p}) = F_{X_i}^{-1}(S, \mathbf{p}, 1 - c_i/p_i)$, with $F_{X_i}(\cdot)$ and $f_{X_i}(\cdot)$ respectively, denoting the cumulative distribution function and density functions of X_i. The objective is then to find the optimal prices,

$$\Pi^* = \Pi(S, \mathbf{p}^*) = \max_{\mathbf{p} \in \Gamma_S} \Pi(S, \mathbf{p}) , \tag{14.13}$$

where $\Gamma_S = \{(p_1, \ldots, p_{|S|}) \mid p_1 > c_1, \ldots, p_{|S|} > c_{|S|}\}$.

Under a regularity assumption ensuring that the optimal price vector is an internal point solution (see Aydin and Porteus 2008 for details), Aydin and Porteus prove the following result.

Theorem 14.4. *There exists a unique price vector* $\mathbf{p}^*$ *that satisfies* $\partial \Pi(S, \mathbf{p})/\partial p_i = 0$, $i \in S$. *Furthermore,* $\mathbf{p}^*$ *maximizes* $\Pi(S, \mathbf{p})$.

Theorem 14.4 states that the expected profit is well behaved in the sense that the optimal prices are the unique solution to the first-order optimality conditions. Aydin and Porteus also develop comparative statics results on the behavior of the optimal prices as a function of demand and cost parameters.

14.6.2 Inventory and Pricing Decisions with Similar Items

Bish and Maddah (2008) study the retailer's pricing decision of a product line within the setting of their similar items model presented in Sect. 14.5.1, with the expected profit at optimal inventory levels given by (14.6). In this section, we discuss the analytical properties of the optimal price, $p_k^* = \arg\max_{p>c} \Pi(p, k)$, assuming that the assortment size, k, is fixed.

Under a regularity assumption, which ensures that the retailer will not be better off by not selling anything, Bish and Maddah observe numerically that the expected profit, $\Pi(p, k)$, is well behaved (pseudo-concave) in the price, p, for reasonably low prices where $\Pi(p, k) > 0$. Bish and Maddah then focus on developing comparative statics results on the behavior of the optimal price as a function of the problem parameters. The following result studies the effect of store volume on pricing.

Lemma 14.2. *If* $p_k^* > 2c$ ($p_k^* < 2c$) *at some* $\lambda = \lambda_0$, *then* p_k^* *is decreasing (increasing) in* λ *for all* λ.

Lemma 14.2 asserts that the optimal price as a function of the expected store volume moves in one direction only, all else held constant. This might be the case of an expensive store with a low volume where the price decreases as a result of an increase in volume, or the case of a low-price high-volume store where the price increases with volume. The condition $p_k^* > 2c$ may be seen as an indicator of the nature of the marketplace and the store.

Bish and Maddah also compare the optimal "riskless" price when assuming deterministic demand p_k^0, to the optimal "risky" price, p_k^*. This is an important question in the literature on single item joint pricing and inventory problem. For an additive demand function, Mills (1959) finds that $p^* \leq p^0$, where p^* and p^0, respectively, denote the risky and riskless price. On the other hand, for the multiplicative demand case, Karlin and Carr (1962) prove that $p^* \geq p^0$. Recall that in the case of Bish and Maddah model, the demand is mixed multiplicative–additive.

Lemma 14.3. *If* $p_k^0 < 2c$, *then* $p_k^* \leq p_k^0$. *Otherwise,* $p_k^* \geq p_k^0$.

Lemma 14.3 is based on the insight that the risky price is adjusted from the riskless price in a way as to reduce demand variability, measured by *both* demand variance $(\lambda q(p,k))$ and coefficient of variation, CV $(1/\sqrt{\lambda q(p,k)})$ (see Petruzzi and Dada 1999.) If $p_k^0 < 2c$, then variability due to demand variance is critical, and, accordingly, p_k^* is adjusted up from p_k^0 to reduce variance. Otherwise, variability due to demand CV is critical, and p_k^* is adjusted down from p_k^0 to reduce CV.

14.6.3 Inventory and Pricing Decisions with a Mixed Multiplicative–Additive Demand Function

Cattani et al. (2010) consider the pricing and inventory (capacity) decisions for two substitutable products, which can be produced either by a single flexible resource or by two dedicated resources. Their expected profit function is similar to that of van Ryzin and Mahajan (1999) in (14.3), but the purchase probabilities are replaced by those in (14.11). The main finding of Cattani et al. relevant to our discussion here is a heuristic for setting the prices and inventory levels of a given assortment, referred to as "cooperative tattonement" (CT). The idea of the CT heuristic is to develop a near-optimal solution by iterating between a marketing model, which sets prices, and a production model, which determines the inventory cost. At each iteration of CT, the prices (having equal profit margins at optimality, as Cattani et al. formally prove) of the marketing model are used to develop the demand for the production model, and the inventory cost from the production model is utilized to develop an equivalent unit cost for the marketing model (see Cattani et al. 2010 for details).

Cattani et al. report a good performance of the CT procedure in obtaining near-optimal solutions. The expected profit from CT is found to be within 0.1% of the optimal expected profit in many cases. One possible reason behind the good performance of the CT heuristic is that the equal profit margin property adopted by this heuristic does in fact hold, approximately (see Sect. 14.7).

14.7 Joint Variety, Pricing, and Inventory Decisions

Finally, we consider a retailer who jointly sets the three key decisions for her product line: variety, pricing, and inventory. This is a realistic integrated setting, which is sought to enhance retailers' profitability. To the best of our knowledge, the only work that addresses this joint setting is the recent paper by Maddah and Bish (2007), which considers the joint problem under assumptions **(A1)–(A3)** of Sect. 14.5. We devote the remainder of this section to the model Maddah and Bish (2007).

The Maddah and Bish model can be seen as an extension to the van Ryzin and Mahajan (1999) model by *endogenizing* the prices similar to Cattani et al. (2010). Maddah and Bish adopt a Normal demand model, which depends on pricing in a mixed multiplicative–additive fashion, similar to Cattani et al. (2010). With this demand model, the optimal inventory level for item $i \in S$, $y_i^*(S, \mathbf{p})$ and the expected profit from S at optimal inventory levels, $\Pi(S, \mathbf{p})$, in (14.2) and (14.3) are given by

$$y_i^*(S,\mathbf{p}) = \lambda q_i(S,\mathbf{p}) + \Phi^{-1}\left(1 - \frac{c_i}{p_i}\right)\sqrt{\lambda q_i(S,\mathbf{p})}, \quad i \in S\,, \tag{14.14}$$

$$\Pi(S,\mathbf{p}) = \sum_{i \in S}\left[\lambda q_i(S,\mathbf{p})(p_i - c_i) - p_i \theta_i(p_i)\sqrt{\lambda q_i(S,\mathbf{p})}\right], \tag{14.15}$$

where $q_i(S, \mathbf{p})$ is given by (14.11). The retailer's objective of maximizing the expected profit is then given by

$$\Pi^* = \Pi(S^*, \mathbf{p}^*) = \max_{S \subseteq \Omega} \quad \max_{\mathbf{p} \in \Gamma_S}\{\Pi(S,\mathbf{p})\}\,. \tag{14.16}$$

Under a regularity assumption, which guarantees that the retailer will not be better off not selling anything, Maddah and Bish develop the following result on the behavior of the expected profit as a function of the mean reservation price of an item.

Lemma 14.4. *Consider an assortment $S \subseteq \Omega$. Assume that prices of items in S are fixed at some price vector* $\mathbf{p}$. *Then, the expected profit from S, $\Pi(S, \mathbf{p}, \alpha_i)$, is strictly pseudo-convex in α_i, the mean reservation price of item i.*

Lemma 14.4 extends the result of Van Ryzin and Mahajan in Lemma 14.1. The intuition behind this lemma is related to the trade-off between the sales revenue and the inventory cost discussed in Sect. 14.5. Lemma 14.4, together with intuitive results on monotonicity in the unit costs, leads to Maddah and Bish's main "dominance result" presented next.

Lemma 14.5. *Consider two items $i, k \in \Omega$ such that $\alpha_i \le \alpha_k$ and $c_i \ge c_k$, with at least one of the two inequalities being strict, that is, item k "dominates" item i. Then, an optimal assortment containing i must also contain k.*

The number of assortments to be considered in the search for an optimal assortment can be significantly reduced if there are a few dominance relations like the one described in Lemma 14.5 (see Maddah and Bish 2007 for details). In addition, Lemma 14.5 allows the development of the structure of an optimal assortment in a special case, as stated in Theorem 14.5.

Theorem 14.5. *Assume that the items in Ω are such that $\alpha_1 \ge \alpha_2 \ge \cdots \ge \alpha_n$,* and *$c_1 \le c_2 \le \cdots \le c_n$. Then, an optimal assortment is $S^* = \{1, 2, \ldots, k\}$, for some $k \le n$.*

The most important situation where Theorem 14.5 applies is the case in which all items in Ω have the same unit cost. Theorem 14.5 extends the result of van Ryzin and Mahajan in Theorem 14.1 to a product line with items having distinct *endogenous* prices.

Theorem 14.5 greatly simplifies the search for an optimal assortment in the special case where it applies, as it suffices to consider only n assortments out of $(2^n - 1)$ possible assortments. For cases where Theorem 14.5 does not apply, one may expect an optimal assortment to have the $k, k \leq n$, items with the largest "average margin," $\alpha_i - c_i$. However, Maddah and Bish present several counter-examples of optimal assortments not having items with the largest average margins, indicating that a result similar to Theorem 14.5 does not hold in general. Nevertheless, Maddah and Bish (2007) also report that popular assortments, consisting of items with the largest average margins, return expected profits that are very close to the optimal expected profit.

Maddah and Bish also analyze the structure of optimal prices by exploiting the first-order optimality conditions. They find (analytically) that the optimal prices are characterized by *approximately* equal profit margins (i.e., $(p_i^* - c_i) \approx (p_j^* - c_j)$, $i, j \in S^*$). It is to be noted that in the absence of inventory costs it has been shown that the profit margins are *exactly* equal (e.g., Anderson et al. 1992 and Aydin and Ryan 2002). That is, accounting for inventory costs complicates pricing.

The analytical and numerical results that popular assortments and prices with equal profit margins are near-optimal motivate an efficient "equal-margin heuristic" (EMH), which builds on these insights. In particular, EMH exploits a limited number of assortments utilizing single-variable line searches to determine the optimal prices. Extensive numerical results presented by Maddah and Bish suggest an excellent performance of EMH, with an average optimality gap of 0.5% in terms of the expected profit.

14.8 A Critique and a Comparison of Insights from the Reviewed Models

Most of the research reviewed in Sects. 14.5–14.7 can be seen as extensions of the seminal work by van Ryzin and Mahajan (1999). The use of consumer choice theory to generate the demand function in a realistic manner paved the way for a meaningful integration of marketing and operations decisions in operations research models. We believe that the works reviewed in Sects. 14.5–14.7 are just the beginning of this promising stream of research, which is strongly connected to practice (see Sects. 14.1 and 14.9). The applicability and importance of this research particularly holds in today's uncertain and unstable business environments.

Assumptions (**A2**)–(**A3**), adopted by the papers reviewed here, may seem restrictive at first. However, the insights gained on the effect of consumers' preference and substitution attitude on the retailer's pricing, inventory, and variety decisions are valuable as they enhance our understanding of the substitution and cross-price effects in isolation of other factors in the complex problem of product line management. Relaxing these assumptions does not seem to be too challenging in practice, especially if one is interested in developing practical tools that generate "good" decisions that are not necessarily optimal. The CT and EMH heuristics developed, respectively, by Cattani et al. (2010) and Maddah and Bish (2007), clearly demonstrate how heuristics grounded on a sound theoretical foundation yield "good" product line management solutions.

All the reviewed models adopt a single-period newsvendor-type framework on the supply side. The main motivation behind this is analytical tractability (as well as applicability in certain situations). An exact analytical treatment of other realistic stochastic inventory policies, such as the well-known (s, S) policy, does not appear to be possible within the integrated framework considered here. (For these situations, one can develop heuristics that build on the single-period newsvendor insights.) On the demand side, there is more leeway, however. Indeed, the models in Sects. 14.5–14.7 are distinguishable mainly by their demand functions. This raises the important question of which demand model to use. The straightforward answer, as in any other mathematical modeling situation, is to use the model that best fits the situation at hand. Ad hoc goodness of fit tests, particularly from the economic and marketing literature (e.g., Guadagni and Little 1983; Hauser 1978; McFadden 1974; and McFadden et al. 1978), can help with this process. Nonetheless, the demand functions based on the MNL logit choice are particularly suited for many situations, as discussed above. In contrast, the locational choice model utilized by Gaur and Honhon (2006) appears to be too stylized (e.g., by assuming the availability of a continuum of products) to be of direct practical use.

An interesting and useful feature of the logit demand functions is that they can be (somewhat) easily refined to capture additional factors. The adaptation to incorporate cross-price effects by Aydin and Porteus (2008), Bish and Maddah (2007, 2008), and Cattani et al. (2010) is borrowed from a vast literature in marketing and economics (e.g., Anderson et al. 1992 and Guadagni and Little 1983).[7] Other innovative adaptations, such as incorporating consumer search by Cachon et al. (2005), seem to be also possible. Finally, the "classical" objection to logit demands due to the IIA property can be addressed by another refinement leading to the nested logit model, as discussed in Sect. 14.4.

Another useful aspect of logit demand models is that the results obtained in simplified frameworks seem to continue to hold with minor "perturbations" in more complex settings. That is, logit models are robust. For example, the fundamental result of van van Ryzin and Mahajan (1999), that an optimal assortment contains the most popular items, continues to hold, "approximately," when Cachon et al. (2005)

[7] One issue that may wrongly appear as a mere technicality is how to incorporate price sensitivity in logit demand functions. We prefer natural adaptations based, for example, on assuming a Poisson or a negative binomial market size, which lead to mixed multiplicative-additive demand models, as in Bish and Maddah (2007, 2008) and Cattani et al. (2010). Simplified pure multiplicative or pure additive adaptations, as in Aydin and Porteus (2008), while popular in the academic literature, seem to be difficult to justify in practice.

and Maddah and Bish (2007) endogenize consumer search and pricing, respectively. In fact, these authors find that assortments with the most popular items provide very good solutions that are close to optimal. A similar observation can be made for the equal-margin property of the optimal prices that holds exactly in the riskless case (which ignores inventory); this result also continues to hold "approximately" when inventory is accounted for, as shown in Cattani et al. (2010) and Maddah and Bish (2007). Our recent work with nested logit choice also indicates that the most-popular-item assortments and equal-margin pricing lead to solutions of high quality (see Kalakesh 2006)[8]. This robustness property facilitates the development of effective heuristics for practical purposes.

Finally, some of the reviewed models take the pricing decision as exogenous and focus on the other two (i.e., variety and inventory), see Sect. 14.5. In practice, whether or not the pricing decision should be assumed exogenous depends on the setting. Exogenous pricing is somewhat justified for popular, fast moving, competitive items, for which pricing is a complex matter, see Sect. 14.9. On the other hand, endogenizing the pricing decision proves to be quite useful for slow movers for which consumers are less price sensitive. Hence, price optimization is a valuable tool, especially for retailers with high variety levels that extend beyond the basic fast movers.

14.9 Current Practice

While a few retailers have started integrating assortment, pricing, and inventory decisions for their product lines (see Sect. 14.1), most retailers still make these decisions separately. However, there has been an increase in interest in this type of integrated decision-making in both the academic community and the retailing industry. This interest is due to (a) advances in information technology, which make it possible to collect detailed data on store performance (e.g., sales, inventory levels) and (b) progress in academic research that is at the interface of marketing, economics, and OM. In addition to a sequential approach to these decisions, most retailers still use heuristic, rule-of-thumb approaches, rather than optimization-based methodology to make these decisions. This is because of the complexities involved (e.g., assessing competition effects, developing demand and cost functions, managing a large number of items, difficulties in obtaining reliable data, etc.). In this section, we briefly summarize our observations on pricing, inventory, and variety management based on our experience with Hannaford, a large chain of grocery stores in the North East of the USA.

In Hannaford, variety, pricing, and inventory management are currently separated in most stores. However, some stores, with sophisticated inventory systems, are in

[8] The robustness of the MNL models seems to be due to the fact that the essential demand splitting and cross-price effects dominate other factors in most cases.

the process of incorporating their inventory data into the pricing decision. We first discuss the pricing practice.

Pricing strategy in a supermarket can be very challenging, mainly due to the size of the problem. There are thousands of items to price, and the number of new items continues to increase. Product lines also have a large number of items, which may be available in different forms, such as frozen biscuits, refrigerated biscuits, biscuit mix, and bakery fresh biscuits, in the supermarket. The large size of the problem makes the decision set confusing to both consumers and retailers. On the one hand, the pricing strategy needs to be tailored to each individual product. On the other hand, it should be scalable such that it is a manageable task. Furthermore, there are additional complexities not considered in any of the stylized models discussed above. For example, the store brand needs to represent a better value than the national brand, but not so discounted that quality is questionable. Another major consideration is competition coming from the multiple channels that carry similar products, i.e., supercenters, dollar stores, other supermarkets, and C-stores.

As a result, pricing strategy in most retailers is done in a rule-of-thumb manner, and different approaches are used to reduce the size of the problem. For example, it is typical for retailers to utilize more sophisticated pricing strategies only for a small portion (100–300) of their items, which includes the best selling and most price sensitive items. These items are divided into groups (e.g., meat, bread, milk), and analyst(s) assigned to each group decide on the prices by carefully compiling available data on competitors' pricing, historical records, and other aspects.

Items not considered best sellers are typically priced according to a simple rule (e.g., by applying a constant percentage markup or by using a simple constant price elasticity model). Line pricing, which refers to the process of grouping hundreds of SKUs into one master item and charging a common price for the group, is a means of managing like items. For example, all Pepsi 2-L products may be priced the same regardless of their individual costs. Private label retails can be managed simply by using the national brand as a benchmark. For example, a retailer may decide to price its coffee at 10% below the national brand. Fresh foods (e.g., produce, deli, and bakery) have the additional consideration of "shrink" (i.e., loss of inventory mainly due to deterioration and spoilage). Some supermarkets use shrink estimates in pricing. However, our experience is that it is quite difficult to predict shrink values, especially at the individual-item level.

As mentioned above, pricing and inventory decisions are typically managed separately. However, some stores with sophisticated inventory systems that track inventory levels continuously and present useful statistics are attempting to leverage the inventory data in pricing. In addition, these inventory management systems can help the retailer reduce supply chain costs, which can then be reflected in prices. Inventory decisions for most items are managed in a heuristic manner, depending on whether they are fast or slow movers and whether they have short or long leadtimes.

Regarding variety management, this decision is separated from pricing and inventory management. Store assortments are typically determined based on the store size and layout (there are few formats that are adopted by stores in the chain).

The store format, in turn, is determined based on the demographical composition of the area where the store is located as well as the type of competitor stores available.

14.10 Conclusions and Directions for Further Research

In this chapter, we review recent works on pricing, variety, and inventory decisions for a retailer's product line composed of substitutable items. This stream of literature contributes to both the theory and the practice of operations management and marketing in two aspects. First, demand models are developed in a realistic way from consumer choice models, based on the classical principal of utility maximization. Second, decisions pertaining to traditionally separated departments in a firm (e.g., inventory and variety, inventory and pricing) are integrated and optimized jointly, an approach that can eliminate inefficiencies due to lack of coordination between marketing and operations. We must note, however, that the current retail practice, for the most part, has not yet caught up with this integrated paradigm that utilizes sophisticated demand models. Nevertheless, the widespread use of information systems and the recent progress made in the academic literature are fueling an interest in this type of approach.

The area of research considered in this chapter is relatively new and there remain many open questions. One area for further research is to use more refined (i.e., more applicable) consumer choice models. Most of the reviewed papers utilize the popular MNL model. However, as discussed in Sect. 14.4, due to the IIA property, the MNL model implicitly assumes that items in a product line are broadly similar. Needless to say, this may not apply in many practical situations. One remedy is to adopt a nested logit choice (NMNL) model, which, as discussed in Sect. 14.4, overcomes this shortcoming, while maintaining a reasonable level of analytical tractability. Hopp and Xu (2003) utilize NMNL in a competitive setting to understand the effect of modular design on variety and pricing competition within a stylized duopoly environment. They do not address the issue of utilizing the NMNL as a refined choice model for joint pricing, assortment, and inventory optimization, which we believe is an important direction for future research. Cachon and Kök (2007) utilize the NMNL for modeling consumer choice between retailers offering multiple categories in a competitive duopoly setting, where some customers are "basket shoppers" who buy items from multiple "complementary" categories. (Agrawal and Smith [2003] consider a similar basket shopping behavior within a monopoly setting and a general consumer choice model.) While Cachon and Kök (2007) offer an interesting application of the NMNL, we believe that there is theoretical and practical value of studying the NMNL as a refined choice model in monopoly settings.

Another important area for further research (that would make this line of research more applicable) is to develop effective decision tools under fairly general assumptions. For example, most of the reviewed works assume a "static choice" setting, where consumers do not substitute another item for their preferred item in the event of a stock out. While the limited results discussed in Sect. 14.5 indicate that solutions based on static choice assumptions do not perform too badly in handling the dynamic situation, more research is needed in this direction. Specifically, numerically efficient heuristics that can generate good solutions for the dynamic situation and that can be used in practice are needed.

A promising area for future research involves studying the coordination of retailers' and manufacturers' pricing, variety, and inventory decisions in supply chains. The problem of coordinating order decisions has been widely studied. As discussed in Sect. 14.3, some research has already started in this direction but more research is needed.

Finally, further investigation of the analytical properties of the existing models is also worthwhile. For example, the structural properties of the optimal assortment and prices for the joint problem studied by Maddah and Bish (2007) deserve more study.

Acknowledgment We thank Dr. Walid Nasr from the American University of Beirut for his prompt help in proofreading and editing the final version of this chapter.

References

Agrawal N, Smith SA (2003) Optimal retail assortment for substitutable items purchased in sets. Naval Res Logist 50:793–822.

Alptekinoglu A, Corbett CJ (2008) Mass customization versus mass production: variety and price competition. Manufact Serv Oper Manage 10:204–217.

Anderson SP, de Palma A (1992) Multiproduct Firms: a nested logit approach. J Ind Econ 40: 264–276.

Anderson SP, de Palma A, Thisse J (1992) Discrete choice theory of product differentiation. MIT Press, Cambridge, MA.

Aydin G, Ryan J (2002) Product line selection and pricing under the multinomial logit choice model. Working paper, Purdue University.

Aydin G, Porteus EL (2008) Joint inventory and pricing decisions for an assortment. Oper Res 56:1247–1255.

Aydin G, Porteus EL (2009) Manufacturer-to-retailer versus manufacturer-to-consumer rebates in a supply chain. In: Agrawal N, Smith S (eds) Retail supply chain management, Springer, Boston, MA.

Aydin G, Hausman WH (2009) The role of slotting fees in the coordination of assortment decisions. Prod Oper Manag 18:635–652.

Ben-Akiva M (1973) Structure of passenger travel demand models. Ph.D. Dissertation, MIT.

Ben-Akiva M, Lerman SR (1985) Discrete choice analysis. MIT Press, Cambridge, MA.

Besanko D, Gupta S, Jain D (1998) Logit demand estimation under competitive pricing behavior: an equilibrium framework. Manage Sci 44:1533–1547.

Bish EK, Maddah B (2008) On the interaction between retail pricing, assortment, and inventory decisions. Working paper, Virginia tech.

Cachon PG, Terwiesh C, Xu Y (2005) Retail assortment planning in the presence of consumer search. Manufact Serv Oper Manage 7:330–346.

Cachon PG, Kök AG (2007) Category management and coordination in retail assortment planning in the presence of basket shopping consumers. Manage Sci, 53:934–951.

Cattani K, Dahan E, Schmidt G (2010) Lowest cost may not lower total cost: Using "spackling" to smooth mass-customized production.

Choi SC (1991) Price Competition in a Channel Structure with a Common Retailer. Market Sci 10:271–296.

Corstjens M, Doyle P (1981) A model for optimizing retail space allocation. Manage Sci 27: 822–833.

Dobson G, Kalish S (1988) Positioning and pricing of a product-line: formulation and heuristics. Market Sci 7: 107–125.

Dobson G, Kalish S (1993) Heuristics for pricing and positioning a product-line using conjoint and cost data. Manage Sci 39:160–175.

Drake MJ, Swann JL (2006) Mechanisms to improve decentralized category management with vendor-specific product portfolios. Working paper, Georgia Institute of Technology,

Eliashberg J, Steinberg J (1993) Marketing production joint decision making. In: Eliashberg J, Lilien GL (eds) Handbooks in operations research and management science, vol. 5. Elsevier, NY.

Elmaghraby W, Keskinocak P (2003) Dynamic pricing in the presence of inventory considerations: research overview, current practices and future directions. Manage Sci 49:1287–1309.

Frantz M RIS news recognizes JCPenney in 1st annual fusion awards. RIS News, October, 2004.

Food Marketing Institute. Food retailing in the 21st century: riding a consumer revolution. http://www.fmi.org/media/bg/FoodRetailing2003.pdf.

Gaur V, Honhon D (2006) Assortment planning and inventory decisions under a locational choice model. Manage Sci 52:1528–1543.

Green PE, Krieger AM (1985) Models and heuristics for product-line selection. Market Sci 4:1–19.

Green PE, Krieger AM (1989) Recent contributions to optimal product positioning and buyer segmentation. Eur J Oper Res 41:127–141.

Griffin A, Hauser JR (1992) Patterns of communication among marketing, engineering and manufacturing - a comparison between two new product teams. Manage Sci 38:360–373.

Guadagni PM, Little JDC (1983) A logit model of brand choice calibrated on scanner data. Market Sci 2:203–238.

Irion J, Al-Khayyal F, Liu JC (2006) A piecewise linearization framework for retail shelf space management models. Working paper, Georgia Institute of Technology

Jain D, Vilcassim N, Chintagunta PA (1994) A random-coefficients logit brand-choice model applied to panel data. J Bus Econ Statistics, 12:317–328.

Hall JM, Kopalle PK, Krishna A (2010) Retailer dynamic pricing and ordering decisions: category management versus brand-by-brand approaches. J Retailing 86:172–183.

Hauser JR (1978) Testing the accuracy, usefulness, and significance of probabilistic choice models: an information theoretic approach. Oper Res, 26:406–421.

Hopp WJ, Xu, X (2003) A competitive marketing/operations model of differentiated product decisions. Working paper, Northwestern University.

Hopp WJ, Xu X (2005) Product line selection and pricing with modularity in design. Manufact Serv Oper Manage, 7:172–187.

Hotelling H (1929) Stability in competition. Econ J, 39:41–57.

Kalakesh SR (2006) Pricing, variety and inventory decisions under nested logit consumer choice. Master Thesis, American University of Beirut.

Karlin S, Carr CR (1962) Prices and optimal inventory policy. In: Arrow KJ, Karlin S, Scarf H (eds) Studies in Applied Probability and Management Science. Stanford University Press, Stanford, CA.

Karmarkar US (1996) Integrative research in marketing and operations management. J Market Res 33:125–133.

Kaul A, Rao VR (1995) Research for product positioning and design decisions: an integrative review. Int J Res Market 12:293–320.

Kurtuluş M, Toktay LB (2005) Category captain: outsourcing retail category management. Working paper, Georgia Institute of Technology.

Kurtuluş M, Toktay LB (2006) Retail assortment planning under category captainship. Working paper, Georgia Institute of Technology.

Kurtuluş M, Toktay LB (2009) Category captainship practices in the retail industry. In: Agrawal N, Smith S (eds) retail supply chain management, Springer, Boston, MA.
Lancaster K (1966) A new approach to consumer theory. J Pol Econ, 74:132–157.
Lancaster K (1990) The economics of product variety: a survey. Market Sci, 9:189–206.
Luce R (1959) Individual choice behavior: a theoretical analysis. Wiley, New York.
Luce R, Suppes P (1965) Preference, utility, and subjective probability. In: Luce R, Bush R, Galanter E (eds) Handbook of mathematical psychology, Wiley, New York.
Maddah B (2005) Pricing, variety, and inventory decisions in retail operations management. Ph.D. Dissertation, Virginia Tech.
Maddah B, Bish EK (2007) Joint pricing, variety, and inventory decisions for a retailer's product line. Naval Res Logist 54:315–330.
Mahajan S, van Ryzin G (1999) Retail inventories and consumer choice. In: Tayur S, Ganeshan R, Magazine MJ (eds) Quantitative Models for Supply Chain Management, Kluwer Academic Publishing, Boston.
Mahajan S, van Ryzin G (2001) Inventory competition under dynamic consumer choice. Oper Res 49:646–657.
Manski CF, McFadden D (eds). (1981) Structural analysis of discrete data and econometric applications MIT Press, Cambridge, MA.
Martín-Herrán G, Taboubib S, Zaccour G (2006) The impact of manufacturers wholesale prices on retailers shelf-space and pricing decisions. Decis Sci, 37:71–90.
McFadden D (1974) Conditional logit analysis of qualitative choice behavior. In: Zarembka P (eds). Frontiers in econometrics, Academic Press, New York. pp. 105–142.
McFadden D, Train K, Tye WB (1978) An Application of diagnostic tests for the independence from irrelevant alternatives property of the multinomial logit model Transportation Research Record: forecasting passenger and fright travel. Transport Res Board 637:39–46.
Mills ES (1959) Uncertainty and price theory. Quart. J. Econom., 73:116–130.
Okun A (2004) Assortment agility. RIS News, October.
Petruzzi NC, Dada M (1999) Pricing and the newsvendor problem: a review with extensions. Oper Res, 47:183–194.
Pekgün, P, Griffin P, Keskinocak P (2006) Centralized vs. decentralized competition for price and lead-time sensitive demand. Working paper, Georgia Tech.
Pekgün P, Griffin P, Keskinocak P (2008) Coordination of marketing and production for price and leadtime decisions. IIE Trans, 40:12–30
Porteus EL, Whang SJ (1991) On manufacturing/marketing incentives. Manage Sci, 37: 1166–1181.
Van Ryzin G, Mahajan S (1999) On the relationship between inventory costs and variety benefits in retail assortments. Manage Sci, 45:1496–1509.
Smith SA, Agrawal N (2000) Management of multi-item retail inventory system with demand substitution. Oper Res, 48:50–64.
US Census Bureau. Annual benchmark report for retail trade and food services: January 1992 through February 2004. http://www.census.gov/prod/2003pubs/br02-a.pdf.
Wang Y, Raju JS, Dhar SK (2003) The choice and consequences of using a category captain for category management. Working paper, University of Pennsylvania.
Yücel E, Karaesmen F, Salman FS, Türkay M (2009) Optimizing product assortment under customer-driven demand substitution. Eur J Oper Res 199:759–768.

Chapter 15
Managing Perishable and Aging Inventories: Review and Future Research Directions

Itir Z. Karaesmen, Alan Scheller-Wolf, and Borga Deniz

15.1 Introduction

Over the years, several companies have emerged as exemplary of "best practices" in supply chain management; for example, Wal-Mart is frequently cited as using unique strategies to lead its market. One significant challenge for Wal-Mart is managing inventories of products that frequently *outdate*: A significant portion of Wal-Mart's product portfolio consists of *perishable* products such as food items (varying from fresh produce to dairy to bakery products), pharmaceuticals (e.g., drugs, vitamins, cosmetics), chemicals (e.g., household cleaning products), and cut flowers. Wal-Mart's supply chain is not alone in its exposure to outdating risks – to better appreciate the impact of perishability and outdating in society at large, consider these figures: In a 2003 survey, overall unsalable costs at distributors to supermarkets and drug stores in consumer packaged goods *alone* were estimated at $2.57 billion, and 22% of these costs, over 500 million dollars, were due to expiration in *only* the branded segment (Grocery Manufacturers of America 2004). In the produce sector, the $1.7 billion US apple industry is estimated to lose $300 million annually to spoilage (Webb 2006). Note also that perishability and outdating are a concern not only for these consumer goods, but for industrial products (for instance, Chen (2006), mentions that adhesive materials used for plywood lose strength within 7 days of production), military ordnance, and blood – one of the most critical resources in health care supply chains. According to a nationwide survey on blood collection and utilization, 5.8% of all components of blood processed for transfusion were outdated in 2004 in the USA (AABB 2005).

In this chapter, we provide an overview of research in supply chain management of products that are perishable or that outdate, i.e., products that *age over time*. Thus, we largely exclude single-period models which are commonly used to represent perishable items (with an explicit cost attached to expected future outdates).

I.Z. Karaesmen (✉)
Kogod School of Business, American University, 4400 Massachusetts Avenue NW, Washington DC 20016
e-mail: karaesme@american.edu

K.G. Kempf et al. (eds.), *Planning Production and Inventories in the Extended Enterprise*, International Series in Operations Research & Management Science 151,
DOI 10.1007/978-1-4419-6485-4_15,

While these newsvendor-type models can convey important insights, they are often too simple to provide answers to some of the more complex questions that arise as inventory levels, product characteristics, markets and customer behaviors change over time. Furthermore, we exclude research that models decay or deterioration of inventories assuming certain functional forms (e.g., exponential decay). The reader can refer to Goyal and Giri (2001) and Raafat (1991), which is supplemented by Dave (1991), for a bibliography and classification of research on that topic. Finally, we also exclude the research on management, planning or allocation of *capacity* which is commonly referred as perishable inventory, for instance, in the airline revenue management literature (see Talluri and van Ryzin (2004), for information on revenue management). Our emphasis is on models where the inventory level of a product must be controlled over a horizon taking into account demand, supply, and a finite shelf-life (which may be fixed or random).

Modeling in such an environment implies that at least one or both of the following holds: First, demand for the product may change over time as the product ages; this could be due to a decrease in the utility of the product because of the reduced lifetime, lessened quality, or changing market conditions. Second, operational decisions can be made more than once (e.g., inventory can be replenished by ordering *fresher* products, or prices can be marked down) during the lifetime of the product. Either of these factors makes the analysis of such systems a challenge, owing to the expanded state space.

Our goal in this chapter is not to replicate surveys of past work (such as Nahmias (1982), Prastacos (1984), and Pierskalla (2004); in fact we refer to them as needed in the remainder of this chapter) but to discuss in more detail *directions for future research.* We provide a selective review of the existing research and focus on those papers which, in our view, constitute crucial stepping stones or point to promising directions for future work. We also refer to several papers in the supply chain management literature that do not specifically study perishable inventories in order to highlight analogous potential research areas in the supply chain management of perishable goods.

The chapter is organized as follows: We first provide a discussion of common challenges in production planning of perishables in Sect. 15.2. We review the research on single product, single location models in inventory management of perishables in Sect. 15.3. We next focus on multi-echelon and multi-location models in Sect. 15.4. Research with novel features such as multiple products, multiple-types of customers and different demand models are reviewed in Sect. 15.5. We conclude by detailing a number of open research problems in Sect. 15.6.

15.2 Challenges in Production Planning

One of the most comprehensive studies on production planning and perishable goods is the doctoral dissertation of Lütke Entrup (2005), which focuses on the use of leading advance planning and scheduling (APS) systems (such as

PeopleSoft's EnterpriseOne and SAP's APO) to manage products with short shelf-lives. According to Lütke Entrup (2005), the use of APS in perishable supply chains remains low in contrast to the supply chains of non-perishable goods. He describes how shelf-life is integrated into the current APS, and identifies particular weaknesses of APS systems for perishables by carefully studying the characteristics and requirements of the supply chains of three different products. Based on these case studies, he proposes customized solutions which would enable the APS systems to better match the needs of the fresh produce industry. We refer the reader to this resource for more information on practical issues in production planning for perishable goods. In the remainder of this section, we provide an overview of the *analytical* research in production planning, focusing mainly on the general body of knowledge in operations management and management science disciplines.

Capacity planning is one of the key decisions in the production of perishables. Research on this topic is primarily focused on agricultural products: Kazaz (2004) describes the challenges in production planning by focusing on long-term capacity investments, yield uncertainty (which is a common problem in the industries that involve perishables), and demand uncertainty. He develops a two-stage problem where the first stage involves determining the capacity investment and the second-stage involves the production quantity decision. Jones et al. (2001) analyze a production planning problem where after the initial production there is a second chance to produce, still facing yield uncertainty; Jones et al. (2003) describe the real-life capacity management problem for a grower in more detail. Allen and Schuster (2004) present a model for agricultural harvest risk. Although these papers do not model the aging of the agricultural product once it is produced, they highlight the long term investment and planning challenges. We refer the reader to Kazaz (2004) for earlier references on managing yield uncertainty and to Lowe and Preckel (2004) for research directions within the general domain of agribusiness.

In addition to capacity planning and managing yield uncertainty, product-mix decisions are integral to the management of perishables. Different companies in the supply chain (producer, processor, distributor, and retailer) face this problem in slightly different ways. For example, in agribusiness, a producer decides on the use of farm land for different produce, hence deciding on the capacity-mix. A supermarket can offer a fresh fruit or vegetable as is or as an ingredient in one of their ready-to-eat products (e.g., pre-packed fruit salad). Similarly, fresh produce can be used in ready-to-eat, cooked products (e.g., pre-cooked, frozen dishes) or it can be sold as an uncooked, frozen product. Note that, in these examples, one perishable component can be used to produce products that have different shelf lives. Consider another example from blood inventories: whole blood outdates in 42 days, whereas a critical blood component, platelets, outdate in 5–7 days. Given their prevalence in practice, the strategic management of interrelated products with explicitly different shelf-lives stands as an important open problem. We discuss the limited research in this area in Sect. 15.5.

For perishables, both capacity planning and product-mix decisions are typically driven by target levels of supply for a final product. In that respect, production plans are tightly linked to inventory control models. We therefore discuss research in inventory control for perishables in the rest of this chapter, treating single location models in Sect. 15.3, the multi-location and multi-echelon models in Sect. 15.4, and modeling novelties in Sect. 15.5. Once again, we attempt to emphasize those areas most in need for further research.

15.3 Managing Inventories at a Single Location

Single-location inventory models form the basic building blocks for more complex models with multiple locations and/or echelons, information flows, knowledge of market or customer behavior, and/or additional logistics options. The work is pioneered by Veinott (1960), Bulinskaya (1964), and Van Zyl (1964), who consider discrete review problems without fixed cost under deterministic demand, stochastic demand for items with a one-period lifetime, and stochastic demand for items with a two-period lifetime, respectively. We first discuss the discrete review setting, before expanding our scope to continuous review systems. There is a *single product* in all the models reviewed in this section, and inventory is depleted starting from the oldest units in stock, i.e., *first-in-first-out (FIFO) inventory issuance* is used.

15.3.1 Discrete Review Models

We provide an overview of research by classifying the work based on modeling assumptions: Research assuming no fixed ordering costs or lead times is reviewed in Sect. 15.3.1.1, followed by research on models with fixed ordering costs but no lead times in Sect. 15.3.1.2. Research on models with positive lead times, which complicate the problem appreciably, is reviewed in Sect. 15.3.1.3.

Table 15.1 provides a high-level overview of some of the key papers within the discrete review arena. Categorization of the papers and the notation used in Table 15.1 are described below:

- Replenishment policy: Papers have considered the optimal control policy (Opt), base stock policies that keep a constant order-up-to-level for total items in system (TIS), summed over all ages, or only new items in system (NIS) (see Sect. 15.3.1.1), other heuristics (H), and, when fixed ordering costs are present, the (s,S) policy. When (s,S) policy is annotated with a ‡, it is implied to be the optimal replenishment policy based on earlier research (however a formal proof of optimality of (s,S) has not appeared in print for the model under discussion).
- Excess demand: Excess demand is either backlogged (B) or lost (L).
- Problem horizon: Planning horizon is either finite (F) or infinite (I).
- Replenishment lead time: All papers, save for one, assume zero lead time. The exception is a paper that allows a deterministic (det) replenishment lead time.
- Product lifetime: Most papers assume deterministic lifetimes of general length (D) although two papers assume the lifetime is exactly two time periods (2).

Table 15.1 A summary of discrete time single item inventory models

Article	Replenishment policy	Excess demand	Planning horizon	Lead time	Life-time	Demand distribution	Costs
Nahmias and Pierskalla (1973)	Opt	*B/L*	*F/I*	0	2	cont	*p,m*
Fries (1975)	Opt	*L*	*F/I*	0	*D*	cont	*c,h,b,m*
Nahmias (1975a)	Opt	*B*	*F*	0	*D*	cont	*c,h,p,m*
Cohen (1976)	TIS	*B*	*I*	0	*D*	cont	*c,h,p,m*
Brodheim et al. (1975)	NIS	*L*	*I*	0	*D*	disc	
Chazan and Gal (1977)	TIS	*L*	*I*	0	*D*	disc	
Nahmias (1975b,1975c)	*H*	*B*	*F*	0	*D*	E_k	*c,h,p,m*
Nahmias (1976)	TIS	*B*	*F*	0	*D*	cont	*c,h,p,m*[1]
Nahmias (1977a)	*H*	*B*	*F*	0	*D*	cont	*c,h,p,m*[1]
Nahmias (1977c)	TIS	*B*	*F*	0	disc[7]	cont	*c,h,p,m*[1]
Nandakumar and Morton (1993)	*H*	*L*	*I*	0	*D*	cont	*c,h,b,m*
Nahmias (1978)	(*s,S*)	*B*	*F*	0	*D*	cont	*c,h,p,m*,K[1]
Lian and Liu (1999)	(*s,S*)‡	*B*	*I*	0	*D*	batch geo	*h,p,b,m*,K
Lian et al. (2005)	(*s,S*)‡	*B*	*I*	0	PH[2]	batch PH	*h,p,b,m*,K
Williams and Patuwo (1999,2004)	*H*	*L*	*F*	det	2	cont	*c,h,b,m*

[1]Denotes papers in which holding and shortage costs can be general convex functions.
[2]Denotes use of a *disaster* model in which all units perish at once.
[7]Denotes the assumption that items, even though they have random lifetimes, will perish in the same sequence as they were ordered.
‡Denotes the case when (*s,S*) policy is implied to be the optimal replenishment policy based on earlier research (however, a formal proof of optimality of (*s,S*) has not appeared in print for the model under discussion).

One paper allows for general discrete phase-type lifetimes, (PH), and assumes that all items perish at the same time, i.e., a *disaster* model. This latter is denoted by [2] in Table 15.1. Another permits general discrete lifetimes, but assumes items perish in the same sequence as they were ordered. This is denoted by [7] in the table.

- Demand distribution: In most cases, demand in each period has a continuous density (cont) function. In others, it is discrete (disc), Erlang – which may include exponential – (E_k) or batch demand with either geometric (geo) or general discrete phase-type renewal arrivals (PH). Oftentimes the continuous demand is assumed for convenience – results appear to generalize to general demand distributions.
- Costs: Costs include unit costs for ordering (c), holding per unit time (h), perishing (m), shortage per unit time (p), or one-time cost for shortage (b) and fixed cost for ordering (K). The annotation [1] in Table 15.1 indicates papers that allow the unit holding and shortage costs to be generalized to convex functions. Any paper that does not list cost parameters is concerned with the general properties of the model, such as expected outdates, without attaching specific costs to these.

15.3.1.1 Discrete Review Models Without Fixed Ordering Cost or Lead Time

Early research in discrete time models without fixed ordering costs focus on characterizing optimal policies. Fundamental characterization of the (state dependent) optimal ordering policy is provided by Nahmias and Pierskalla (1973) for the two period lifetime problem, when penalty costs per unit short per unit time, and unit outdating costs are present. Fries (1975) and Nahmias (1975a), independently, characterize the optimal policy for the general lifetime problem. Both of these papers have per unit outdating costs, per unit ordering costs, and per unit per period holding costs. In addition, Fries (1975) has per unit shortage costs, and Nahmias (1975a) per unit per period shortage costs; this difference arises because Fries (1975) assumes lost sales and Nahmias (1975a) backlogging of unsatisfied demand. The cost structure in Nahmias (1975a) is the "standard" for discrete time models without fixed ordering cost, and we refer to it as such within the rest of this section, although in later papers the per unit per period penalty and holding costs are extended to general convex functions. The incorporation of separate ordering and outdating costs allows modeling flexibility to account for salvage value; use of only one or the other of these costs is essentially interchangeable.

Treatment of the costs due to outdating is the crucial differentiating element within the perishable setting. In fact, while Fries (1975) and Nahmias (1975a) take alternate approaches for modeling the costs of expiration – the former paper charges a cost in the period items expire, while the latter charges an expected outdating cost in the period items are ordered – Nahmias (1977b) shows that these two approaches were essentially identical, modulo end of horizon effects. The policy structures outlined in Fries (1975) and Nahmias (1975a) are quite complex; perishability destroys the simple base-stock structure of optimal policies for discrete review models without fixed ordering costs in the absence of perishability.

This complexity of the optimal policy is reinforced by Cohen (1976), who characterizes, the stationary distribution of inventory for the two-period problem with the *standard* costs, showing that the optimal policy for even this simple case was quite complex, requiring state-dependent ordering. To all practical extents, this ended academic study of optimal policies for the discrete review problem under FIFO issuance; to find an optimal policy dynamic programming would be required, and implementation of such a policy would be difficult owing to its complex, state dependent nature (with a state vector tracking the amount of inventory in system *of each age*). Therefore many researchers turned to the more practical question of seeking effective heuristic policies that would be (a) easy to define, (b) easy to implement, and (c) close to optimal.

Very soon after the publication of Nahmias (1975a), a series of approximations appeared for the backorder and lost sales versions of the discrete review zero-fixed ordering cost problem. Initial works (Brodheim et al. 1975; Nahmias 1975b) examine the use of different heuristic control policies; Brodheim et al. (1975) propose a fascinating simplification of the problem – making order decisions based only on the amount of new items in the system; what we call the NIS heuristic. They show that if a new order size is constant, Markov chain techniques can be used to derive exact expressions, or alternately very simple bounds, on key system statistics (which is the focus of their paper). Nahmias (1975b) used simulation to compare multiple heuristics for the problem with *standard* costs, including the "optimal" TIS policy, a piecewise linear function of the optimal policy for the non-perishable problem, and a hybrid of this with NIS ordering. He found that the first two policies outperform the third, effectively disrupting further study of NIS and its variants for a few decades.

Nahmias (1975c, 1976, 1977a), explicitly treats the question of deriving good approximation policies (and parameters) for the problem with the *standard* costs; Nahmias (1977a) can be considered as the culmination of these initial heuristic efforts. This latter computationally compares the heuristic that keeps only two states in the inventory vector, new inventory, and inventory over one day old, with both the globally optimal policy (keeping the entire inventory vector) and the optimal TIS policy, which is easily approximated using the techniques in Nahmias (1976). For Erlang or exponential demand and a lifetime of three periods, the performance of both heuristics is exceptional – always within 1% of optimal, with the reduced state-space heuristic uniformly outperforming the optimal TIS policy. Note also that using the two-state approximation eliminates any need to track the age of inventory other than new versus old, significantly reducing both the computational load and complexity of the policy.

All of these heuristics consider backorder models. The first heuristic policy for discrete time lost sales models is provided by Nandakumar and Morton (1993). They incorporate the properties and bounds on the expected outdates under TIS in lost sales systems, originally provided by Chazan and Gal (1977), into heuristics following the framework of Nahmias (1976) for the backorder problem. Nandakumar and Morton (1993) compare these heuristics with alternate "near-myopic" heuristics which use a newsvendor-type logic, for the problem with *standard* costs. They found that all of the heuristics performed within half a percentage of the optimal, with the near myopic heuristics showing the best performance, typically within 0.1% of the optimal. (These heuristics could likely be improved even further, with the use of even tighter bounds on expected outdates derived by Cooper (2001).)

Note that both the backorder and lost sales heuristics should be quite robust with respect to items having even longer lifetimes than those considered in the papers: Both Nahmias (1975a) and Fries (1975) observed that the impact of newer items on ordering decisions is greater than the impact of older items. (This, in fact, is one of the factors motivating the heuristic strategies in Nahmias (1977a).) Thus, the *standard* single-item, single-location, discrete-time, fixed-lifetime perishable inventory model, with backorders or lost sales has for all practical purposes been solved – highly effective heuristics exist that are well within our computational power to calculate. One potential extension would be to allow items to have random lifetimes; if items perish in the same sequence as they were ordered, many of the fixed lifetime results continue to hold (Nahmias 1977c), but a discrete time model with random lifetimes that explicitly permits items to perish in a different sequence than the one they were ordered remains an open, and likely challenging question. The challenge arises from the enlarged state space required to capture the problem characteristics – in this case the entire random lifetime vector – although techniques from continuous time models discussed in Sect. 15.3.2.1 could prove useful in this endeavor.

15.3.1.2 Discrete Review Models Without Lead Times Having Positive Fixed Ordering Cost

Nahmias (1978) was the first to analyze the perishable inventory problem with no lead time but positive fixed ordering cost in addition to the *standard* costs described in Sect. 15.3.1.1. In his paper, *for the one-period problem*, Nahmias (1978) established that the structure of the optimal policy is (s,S) only when the lifetime of the object is two; lifetimes of more than two periods have a more complex, non-linear structure. Extensions of the two-period (s,S) result to the multi-period problem appeared quite difficult, given the analytical techniques of the time. The fact that the costs *excluding* the fixed ordering cost are not convex appears to render hopes of extending the K-convexity of Scarf (1960) to the perishable domain to be in vain. Nevertheless, Nahmias (1978) reports extensive computational experiments supporting the conjecture that the general ((s,S) or non-linear) optimal policy structure holds for the multi-period problem. One possible avenue to prove such structure could be through the application of recent proof techniques involving decomposition ideas, such as those in Muharremoglu and Tsitsiklis (2003).

Lian and Liu (1999) consider the discrete review model with fixed ordering cost, per unit per period holding cost, per unit *and* per unit per unit time shortage costs, and per unit outdating costs. Crucially, their model is comprised of discrete time epochs where demand is realized or units in inventory expire (hence discrete time refers to distinct points in time where change in the inventory levels occur) as opposed to distinct time periods in Nahmias (1978). Lian and Liu (1999) analyze an (s,S) policy which was shown to be optimal under continuous review by Weiss (1980) for Poisson demand. The instantaneous replenishment assumption combined with the discrete time model of Lian and Liu (1999) ensures that the optimal reorder level s will be no greater than -1 because any value larger than -1 will add holding costs without incurring any shortage cost in their model. The zero lead time assumption and cost structure of Lian and Liu (1999) are common in the continuous review framework for perishable problems (see Sect. 15.3.2.3); it was a desire to develop a discrete time model to approximate the continuous review that motivated Lian and Liu (1999). In the paper they use matrix analytic methods to analyze the discrete time Markov chain and establish numerically that the discrete review model is indeed a good approximation for the continuous review model, especially as the length of the time intervals gets smaller.

Lian et al. (2005) follow Lian and Liu (1999), sharing similar costs and replenishment assumptions. This later paper allows demands to be batch with discrete phase type interdemand times and lifetimes to be general discrete time phase type, but requires that all items of the same batch perish at the same time. They again use matrix analytic methods to derive cost expressions, and numerically show that the variability in the lifetime distribution can have a significant effect on system performance. Theirs is a quite general framework, and provides a powerful method for the evaluation of discrete time problems in the future, provided the assumption of the instantaneous replenishments is reasonable in the problem setting.

Of great practical importance again is the question of effective heuristic policies. Nahmias (1978) establishes, computationally, that while not strictly optimal, (s,S) type policies perform very close to the optimal. Moreover, he also presents two methods of approximating the optimal (s,S) parameters. Both of these are effective: On average their costs are within 1% of the globally optimal cost for the cases considered, and in all cases within 3%. Thus, while there certainly is room for further refinements of these heuristic policies (along the lines of the zero fixed cost case, for example), the results in Nahmias (1978) are already compelling.

15.3.1.3 Discrete Review Models with Positive Lead Time

As optimal solutions to the zero lead time case require the use of dynamic programming, problems with lead times are in some sense no more difficult, likewise requiring dynamic programming, albeit of a higher dimension. Thus problems considering optimal policies for discrete review problems with lead times are also often considered in the context of other generalizations to the model, for example, different selling prices depending on age (Adachi et al. 1999). Interestingly, there is very little work on extending discrete review heuristics for the zero lead time model to the case of positive lead times, although many of the methods for the zero lead time case should in principle be applicable, by expanding the vector of ages of inventory kept to include those items on order, but which have not yet arrived.

One possible direction is provided in the work of Williams and Patuwo (1999, 2004) who derive expressions for optimal ordering quantities based on system recursions for a one-period problem with fixed lead time, lost sales, and the following costs: per unit shortage costs and per unit outdating costs, per unit ordering costs and per unit per period holding costs. Williams and Patuwo (2004) in fact state that their methods can be extended to finite horizons, but such an extension has not yet appeared. Development of such positive lead time heuristics, no matter what their genesis, would prove valuable both practically and theoretically, making this a potentially attractive avenue for future research. The trick, essentially, is to keep enough information so as to make the policy effective, without keeping so much as to make the policy overly cumbersome. More complex heuristics have been proposed for significantly more complex models with lead times, such as in Haijema et al. (2005, 2007), but these likely are more involved than necessary for standard problem settings.

15.3.2 Continuous Review Models

We now turn our attention to continuous review models. These are becoming increasingly important with the advent of improved communication technology (such as radio frequency identification, RFID) and automated inventory management and ordering systems (as a part of common enterprise resource planning, ERP, systems). These two technologies may eventually enable management of perishables in real time, potentially reducing outdating costs significantly. We first consider continuous review models without fixed ordering costs or lead times in Sect. 15.3.2.1, models without fixed ordering costs but positive lead times in Sect. 15.3.2.2, and finally the models which incorporate fixed ordering costs in Sect. 15.3.2.3.

Tables 15.2 and 15.3 provide a high-level overview of some of the key papers within the continuous review framework, segmented by whether a fixed ordering cost is present in the model (Table 15.3) or not (Table 15.2). Categorization of the papers and the notation used in these tables are described below:

- Replenishment policy: In Table 15.2, base stock policy is denoted by a z. Papers which focus on characterizing system performance have a blank in the replenishment policy column. In Table 15.3, all papers either use the (s, S) policy, annotated with a $*$ when proved to be optimal, with a $\ddagger$ when only implied to be the optimal policy based on earlier research (however a formal proof of optimality of (s, S) has not appeared in print for the model under concern) or the (Q, r) or (Q, r, T) policies when batch sizes are fixed. The (Q, r, T) policy orders when inventory is depleted below r or when items exceed T units of age.
- Excess demand: Papers may assume simple backlogging (B), some sort of generalized backlogging in which not all backlogged customers wait indefinitely (B^g), or lost sales (L). Two papers assume all demand must be satisfied; this field is blank in that case.
- Problem horizon: Planning horizon is either finite (F) or infinite(I).

Table 15.2 A summary of continuous time single item inventory models without fixed inventory cost

Article	Replenishment policy	Excess demand	Planning horizon	Lead time	Life-time	Demand distribution	Costs
Graves (1982)		*B/L*	*I*	*M*	*D*	batch *M*	
Kaspi and Perry (1983)		*L*	*I*	*M*	*D*	*M*	
Kaspi and Perry (1984)		B^g/L	*I*	renewal	*D*	*M*	
Perry and Posner (1990)		*L*	*I*	M^3	*D*	M^3	c,h,b,c_s
Perry (1997)		*L*	*I*	M^3	*D*	M^3	c,h,b,c_s^8
Perry and Stadje (1999)		B^g/L	*I*	M^4	exp/*D*	M^4	c,h,p,m,r,h',p'
Perry and Stadje (2000a)		*L*	*I*	*M*	D^0	*M*	
Perry and Stadje (2000b)		*L*	*I*	*M*	gen	*M*	
Perry and Stadje (2001)		*L*	*I*	*M*	D/M^2	*M*	
Nahmias et al. (2004a)		*L*	*I*	M^4	*D*	M^4	
Nahmias et al. (2004b)		*L*	*I*	batch *M*	*D*	renewal	c,b,m,r^9
Pal (1989)	*z*	*B*	*I*	exp	exp	*M*	h,p,b,m
Liu and Cheung (1997)	*z*	B^g/L	*I*	exp/ample	exp	*M*	constraints
Kalpakam and Sapna (1996)	*z*	*L*	*I*	exp	exp	renewal	c,h,b,m
Kalpakam and Shanthi (2000)	*z*	B^g/L	*I*	M^4	exp	*M*	c,h,p,b,m
Kalpakam and Shanthi (2001)	*z*	*L*	*I*	cont	exp	*M*	c,h,b,m
Schmidt and Nahmias (1985)	*z*	*L*	*I*	det	*D*	*M*	c,h,b,m
Perry and Posner (1998)	*z*	B^g/L	*I*	det	*D*	*M*	

[g]Denotes a generalized backorder model.
[0]Denotes the characteristic that after items perish from a first system they have use in a second system.
[2]Denotes use of a *disaster* model in which all units perish at once.
[3]Denotes the ability to control the arrival and/or demand rates.
[4]Denotes state-dependent arrival and/or demand.
[8]Denotes slightly modified cost definitions to fit within the context of a Brownian control framework.
[9]Denotes actuarial valuations based on expected costs and age-dependent revenues.

Table 15.3 Continuous time single item inventory models fixed ordering cost

Article	Replenishment policy	Excess demand	Planning horizon	Lead time	Life-time	Demand distribution	Costs
Weiss (1980)	$(s,S)^*$	*B/L*	*I*	0	*D*	*M*	c,h,p,m,r,K[1]
Kalpakam and Arivarignan (1988)	$(s,S)^\ddagger$	*L*	*I*	0	exp	*M*	c,h,m,K
Liu (1990)	$(s,S)^\ddagger$	*B*	*I*	0	exp	*M*	h,p,b,m,K
Moorthy et al. (1992)	$(s,S)^\ddagger$	*L*	*F/I*	0	E_k	batch iid	
Liu and Shi (1999)	$(s,S)^\ddagger$	*B*	*I*	0	exp	renewal	h,p,b,m,K
Liu and Lian (1999)	$(s,S)^\ddagger$	*B*	*I*	0	*D*	renewal	h,p,b,m,K
Lian and Liu (2001), Gürler and Özkaya (2003)	$(s,S)^\ddagger$ / (s,S)	*B*	*I*	0 / det	*D*	batch renewal	h,p,b,m,K
Gürler and Özkaya (2006)	(s,S)	*B*	*I*	0 / det	gen[2]	batch renewal	h,p,b,m,K
Kalpakam and Sapna (1994)	(s,S)	*L*	*I*	exp	exp	*M*	c,h,b,m,K
Ravichandran (1995)	(s,S)	*L*	*I*	cont[5]	*D/D*[6]	*M*	c,h,b,m,K
Liu and Yang (1999)	(s,S)	*B*	*I*	exp	exp	*M*	h,p,b,m,K
Nahmias and Wang (1979)	(Q,r)	*B*	*I*	det	decay	cont	h,b,m,K
Chiu (1995)	(Q,r)	*B*	*I*	det	*D*	general	c,h,b,m,K
Tekin et al. (2001)	(Q,r,T)	*L*	*I*	det	*D*[6]	*M*	h,m,K,constraint
Berk and Gürler (2006)	(Q,r)	*L*	*I*	det	*D*[2]	*M*	h,b,m,K

[1]Denotes papers in which holding and shortage costs can be general convex functions.
[2]Denotes use of a *disaster* model in which all units from the same order perish at once.
[5]That only one outstanding order is allowed at a time.
[6]The assumption that product aging starts only once all inventory from the previous lot has been exhausted or expired.
*Denotes the case when (s,S) policy is proved to be optimal.
‡Denotes the case when (s,S) policy is implied to be the optimal replenishment policy based on earlier research (however, a formal proof of optimality of (s,S) has not appeared in print for the model under concern).

- Replenishment lead time: In Table 15.2, many papers assume that items arrive according to a Poisson process, *outside of the control of the system manager*; these are denoted by (M). This may be generalized to the case of batch replenishments (batch), the case when the replenishment rates can be controlled (M^5), or are state dependent (M^6). Other papers in the no fixed ordering cost regime assume either exponential (exp) with a single or ample (ample) servers, renewal (renewal), continuous (cont) or fixed (det) lead times. Most papers in Table 15.3 assume either zero lead time or a deterministic (det) lead time, although some allow exponential (exp), or general continuously distributed (cont) lead times.
- Product lifetime: Most lifetimes are deterministic (D), possibly with the assumption that the item can be used for a secondary product after it perishes (D^0), that all perish at once, a *disaster* model, indicated by (D^2), or that items within a lot only begin to age after all items from the previous lot have left inventory (D^6). Item lifetimes may also be exponentially distributed (exp), generally distributed (gen), Erlang (E_k) and in one case they decay (decay).
- Demand distribution: As we are within a continuous model, this column describes the assumptions on the demand *interarrival* distribution. Most are Poisson (M), although in some cases they may have rates that can be controlled by the system manager (M^3) or which are state-dependent (M^4). Selected papers allow independent and identically distributed arrivals (iid), continuously distributed interarrivals (cont), some allow arrivals in batches (batch), and others have renewal (renewal), or general (general) demand interarrival distributions.
- Costs: Typically, the costs introduced in Sect. 15.3.1, Table 15.1 are used. When demand and interarrival rates can be controlled, there is a cost (c_s) for adjusting these rates. Some models calculate profits, via using a unit revenue (r). One paper not only charges the standard per unit time holding and penalty costs (h) and (p), but also per unit time costs based on the average age of the items being held (h') or backlogged (p'). Two papers optimize subject to constraints (possibly with other costs), and several (but not all) that concern themselves only with characterizing system performance have this column left blank. In addition, one paper utilizes a Brownian control model with two barriers; it defines shortage and outdate costs slightly differently to account for the infinite number of hits of the barriers. This is denoted by ^{8}. One paper calculates *actuarial* valuations of costs and expected future revenues, where the latter depend on the age of an item when it is sold. This is denoted with a ^{9}. Finally, when annotated with a ^{1}, a paper allows the unit holding and shortage costs to be generalized to convex functions.

15.3.2.1 Continuous Review Without Fixed Ordering Cost or Lead Time

The continuous review problem without fixed ordering costs is unique in that under certain modeling assumptions, direct parallels can be drawn between the perishable inventory problem and stochastic storage processes in general, and queueing theory in particular. These parallels provide structural results as well as powerful analytical tools.

One of the earliest papers to make these connections is written by Graves (1982), who focuses on characterization of the system behavior, as in the earlier papers. Graves (1982), without an explicit cost structure or replenishment policy, shows that the virtual waiting time in an M/M/1 queue with impatient customers and a M/D/1 with finite buffer can be used to model the inventory in a perishable inventory system with Poisson demands of exponential, or unit size, respectively, with either lost sales or backlogging. A crucial assumption is that items are replenished according to what Graves (1982) calls "continuous production"; the arrival of items into inventory is modeled as a Poisson process, essentially out of the control of the inventory manager. Under this convention, Graves (1982) notes that the key piece of information is the age of the oldest item currently in stock, an observation that is used by many subsequent researchers. For example, Kaspi and Perry (1983, 1984) model systems with Poisson demand and Poisson or renewal supply, as might be the case, for example in blood banks that rely on donations for stock. For their analysis Kaspi and Perry (1983 1984) track what they call the virtual death process, which is the time until the next death (outdate) if there would be no more demand. This, of course, is just a reformulation of the age of the oldest item in stock, as used by Graves (1982).

These papers mark the start of a significant body of work by Perry (sometimes in conjunction with others) on continuous review perishable inventory systems with no fixed ordering cost, some sort of Poisson replenishment, and nearly all using the virtual death process. Only Perry and Posner (1990) and Perry and Stadje (1999) contain explicit cost functions; the rest of the papers concern themselves solely with performance characteristics, as in Graves (1982). Perry and Posner (1990) develop level crossing arguments for storage processes to capture the effects of being able to control supply or demand rates within the model of Kaspi and Perry (1983); the limiting behavior of this system was analyzed by Perry (1997) using a diffusion model and martingale techniques. Perry and Stadje (1999) depart from the virtual death process, using instead partial differential equations to capture the stationary law of a system which now may have state-dependent arrival and departure rates with deterministic or exponential lifetimes and/or maximum waiting times, as well as finite storage space. This work is generalized in Nahmias et al. (2004a) using the virtual death process. Likewise using the virtual death process, Perry and Stadje (2000a, b, 2001) evaluate systems where after perishing the item can be used for a secondary product (such as juice for expired apples); items may randomly perish before the expiration date; or, in addition to their fixed lifetime, items may all perish before their expiration date (due to disasters, or obsolescence). Still using the virtual death process, Nahmias et al. (2004b) provide actuarial valuations of the items in system and future sales, when item values are dependent on age. Recently Perry and Stadje (2006) solved a modified M/G/1 queue and showed how it related, again, to the virtual death process in a perishable system, this time with lost sales. Thus work on extensions to what can reasonably be called the *Perry model* continues.

Note that the work in these papers is concerned with performance analysis of systems with random input and output. This is in contrast to the models in Sect. 15.3.1

which typically assume input is completely controllable. Thus most of the papers in this section do not focus on optimization. However, a few do include the ability to control the input rate (typically at a cost); see Table 15.2.

15.3.2.2 Continuous Review Without Fixed Ordering Cost Having Positive Lead Time

A defining characteristic of the *Perry model*, mentioned in Sect 15.3.2.1, is that the supply of perishable items arrives according to a (possibly state dependent) Poisson process. When replenishment decisions and lead times must be included in a more explicit manner, researchers need to develop other analytical methods.

When lifetimes and lead times are exponentially distributed the problem is simplified somewhat, as this allows the application of renewal theory, transform methods, and Markov or semi-Markov techniques, often on more complex versions of the problem. Pal (1989) looks at the problem with exponential lead times and lifetimes, Kalpakam and Sapna (1996) allow renewal demands with lost sales, Kalpakam and Shanthi (2000, 2001) consider state dependent Poisson lead times and then general continuous lead times, and Liu and Cheung (1997) are unique in that they consider fill rate and waiting time constraints. All of these papers consider base-stock, or $(S-1, S)$ inventory control; Kalpakam and Sapna (1996) and Kalpakam and Shanthi (2001) consider per unit shortage costs and per unit outdating costs, per unit ordering costs and per unit per period holding costs. To these Kalpakam and Shanthi (2000) add a per unit per period shortage cost, while Pal (1989) also adds the per unit per unit time shortage cost, but disregards the per unit ordering cost. Liu and Cheung (1997) take a different approach, seeking to minimize the inventory subject to a service level constraint. In all of these cases, the cost function appears to be unimodal in S, but no formal proofs have been provided, owing to the difficulty in proving unimodality. Furthermore, nowhere has the performance of base-stock policies been formally benchmarked against the optimal, possibly due to the difficulty in dealing with a continuous state space – time – within the dynamic programming framework. This remains an open question.

The case of fixed lead times and lifetimes is arguably both more realistic and analytically more difficult. Schmidt and Nahmias (1985) consider a system operating under a base-stock policy with parameter S, lost sales, Poisson demand, per unit shortage, outdating and ordering costs, and per unit per period holding costs. They define and solve partial differential equations for the S-dimensional stochastic process tracking the time since the last S replenishment orders. Numerical work shows that again cost appears to be monotonic in S (in fact convex), although surprisingly, the optimal value of S is *not* monotonic in item lifetime.

Perry and Posner (1998) generalize this work to allow for general types of customer impatience behavior, using level crossing arguments to derive the stationary distribution of the vector of times until each of the S items in the system outdate (reminiscent of their *virtual death process*). They also show that the distribution of the differences between the elements of this vector follows that of uniform order statistics, which enables them to derive expressions for general customer behavior. These expressions may, as a rule, require numerical evaluation.

Perry and Posner (1998) are concerned with general system characteristics; they do not include explicit costs in their paper. While Perry and Posner (1998) provide rich material for future research – for example exploring how different customer behavior patterns affect different echelons of a supply chain for perishable products, – there is still a need for research following the work of Schmidt and Nahmias (1985), with the aim of minimizing costs in the continuous review setting under fixed lead times and fixed lifetimes.

15.3.2.3 Continuous Review with Fixed Ordering Cost

Within the continuous review fixed ordering cost model we make a distinction between those models that assume fixed batch size ordering, leading to (Q,r) type models, and those that assume batch sizes can vary, leading to (s,S) type models. We consider the (s,S) models first.

Initial work in this setting assumed zero lead time, which simplifies the problem considerably, as there is no need to order until all the items are depleted. In this case when considering fixed ordering costs, unit revenues, unit ordering and outdate costs, and convex holding and penalty costs per unit time, the optimal policy structure under Poisson demand was found by Weiss (1980) who showed that for the fixed lifetime problem over an infinite horizon, with lost sales or back ordering, an (s,S) policy is optimal. Weiss (1980) also established that the optimal s value is zero in the *lost sales* case, and in the *backorder* case no larger than -1 (you never order if you have items in stock). Thus the optimal policy structure was established, but the question of efficiently finding the optimal parameters is open.

The publication of Weiss (1980) initiated a series of related papers, all having in common the assumption of immediate supply. Kalpakam and Arivarignan (1988) treat the lost sales model as Weiss (1980), but assume exponential lifetimes, consider only costs (not revenues), and of these costs disregard the shortage costs as they are irrelevant, as Weiss shows that the optimal is $s = 0$. They also show that in this case the cost, assuming an optimal s value, is convex in S. Liu (1990) takes the setting of Weiss (1980), assumes exponential lifetimes, disregards the unit ordering costs and revenues, but considers penalty costs per unit and per unit time. His focus is on providing closed-form expressions for system performance, based on transform analysis. Moorthy et al. (1992) perform a similar analysis using Markov chains theory under Erlang lifetimes, assuming no shortage is permitted – or equivalently lost sales, as in this case Weiss (1980) shows that it is optimal not to allow any shortage. Liu and Shi (1999) follow Liu (1990), but now allow for general renewal demand. They focus their analysis on the reorder cycle length, using it as a vehicle to prove various structural properties of the costs with respect to the parameters. The assumption of exponential lifetimes is crucial here – the results would be very unlikely to hold under fixed lifetimes. Liu and Lian (1999) consider the same problem as Liu and Shi (1999) and Lian but with fixed, rather than exponential lifetimes. They permit renewal demands, and derive closed-form cost expressions, prove unimodality of costs with respect to parameters, and show that the distribution of inventory level is uniform over (s,S) (as in the non-perishable case).

All of the previous papers make the zero lead time assumption of Weiss (1980), which simplifies the problem. A few papers include positive *fixed* lead times, Lian and Liu (2001) treat the model of Lian and Liu (1999) incorporating batch demands to provide a heuristic for the fixed lead time case, but a proof in that paper contained a flaw, which was fixed by Gürler and Özkaya (2003). These papers show how to efficiently find good (s,S) parameters for the fixed lead time problem, but do not provide any benchmark against the optimal. Thus while efficient heuristics exist for the fixed lead time case, they have not as yet been benchmarked against more complex control policies.

Random lead times have appeared in several papers within the continuous review, fixed ordering cost framework. Kalpakam and Sapna (1994) allow exponential lead times, while also assuming exponential lifetimes. As in the fixed lead time case, this implies that the (s,S) policy is no longer necessarily optimal. They nevertheless assume this structure. In addition to fixed ordering costs, they account for unit purchasing, outdate and shortage costs, as well as holding costs per unit per unit time. Under the assumption of only one outstanding order at a time, they derive properties of the inventory process and costs. Ravichandran (1995) permits general continuous lead times, deriving closed-form expressions for costs under the assumption that only one order is outstanding at a time, and items in an order only begin to perish after all items from the previous order have left the system. Liu and Yang (1999) generalize Kalpakam and Sapna (1994) to allow for backlogs and multiple outstanding orders, using matrix analytical methods to generate numerical insights under the assumption of an (s,S) policy. For random lifetimes, still within the (s,S) structure, the most comprehensive work is by Gürler and Özkaya (2006), who allow a general lifetime distribution, batch renewal demand, and zero lead time with a heuristic for positive lead time. Their cost structure follows Lian and Liu (2001); Gürler and Özkaya (2006) can be thought of as generalizing Lian and Liu (2001) to the random lifetime case. Gürler and Özkaya (2006) argue that the random lifetime model is important for modeling lifetimes at lower echelons of a supply chain, as items arriving there will have already begun to perish. To this end they demonstrate the importance of modeling the variability in the lead time distribution on costs, including a comparison to the fixed lifetime heuristic of Lian and Liu (2001).

In general, within the fixed ordering cost model with variable lot sizes, for the zero lead time case research is quite mature, but for positive lead times and/or batch demands there are still opportunities for research into both the optimal policy structure and effective heuristics. These models are both complex and practically applicable, making this yet another problem that is both challenging and important.

If lot sizes must be fixed, initial work for "decaying" goods by Nahmias and Wang (1979), using a (Q,r) policy, was followed two decades later by Chiu 1995), who explicitly considers perishable (rather than decaying) items by approximating the outdating and inventory costs to get heuristic (Q,r) values. Nahmias and Wang (1979) consider unit ordering and shortage costs, and holding costs per unit per unit time, as well as unit outdate costs. To these Chiu (1995) adds unit ordering costs.

Tekin et al. (2001) take a slightly different approach; they disregard unit ordering costs and consider a service level constraint, rather than a shortage cost. They also simplify the problem by assuming that a lot only starts aging after it is put into use, for example moved from a deep freezer. To combat the effects of perishability, they advocate placing an order for Q units whenever the inventory level reaches r or when T time units have elapsed since the last time a new lot was unpacked, whichever comes first, giving rise to a (Q,r,T) policy. Not surprisingly, they find that inclusion of the T parameter is most important when service levels are required to be high or lifetimes are short. Berk and Gürler (2006) define the "effective shelf life" of a lot: The distribution of the remaining life at epochs when the inventory level hits Q. They show that this constitutes an embedded Markov process (as they assume Poisson demand), and thus via analysis of this process they are able to derive *optimal* (Q,r) parameters, when facing unit outdate and penalty costs, holding costs per unit per unit time, and fixed ordering costs. They compare the performance of their policy with that of Chiu (1995) and also with the modified policy of Tekin et al. (2001). Not surprisingly, the optimal (Q,r) policy outperforms the heuristic of Chiu (1995), sometimes significantly, and is outperformed by the more general modified policy of Tekin et al. (2001). Nevertheless, this latter gap is typically small, implying that use of the more simple (Q,r) policy is often sufficient in this setting.

Note that throughout these papers the (Q,r) structure has only been assumed, and once again there does not appear to be any benchmarking of the performance of the (Q,r) policy against the optimal, as optimal policies are difficult to establish given the increased problem complexity the continuous time setting with perishability causes. If such benchmarking were done, it would help identify those problem settings for which the (Q,r) or (Q,r,T) policy is adequate, and those which would most benefit from further research into more complex ordering schemes. Essentially, the underlying question of how valuable lifetime information is, in what degree of specificity, and when it is most valuable, remains.

15.4 Managing Multi-Echelon and Multi-Location Systems

Analysis of multi-echelon inventory systems dates back to the seminal work of Clark and Scarf (1960); the reader can refer to Axsäter (2000) for a unified treatment of the research in that area, and Axsäter (2003) for a survey of research on serial and distribution systems. In serial and distribution systems, each inventory location has one supplier. In contrast, multi-location models consider flow of products from various sources to a particular location (possibly including transshipments). See, for instance, Karmarkar (1981) for a description of the latter problem. Managing multi-echelon and/or multi-location systems[1] with aging products is a challenge because of the added complexity in:

- *replenishment and allocation decisions,* where the *age* of goods replenished at each location affects the age-composition of inventory and the system performance, and the *age* of goods supplied/allocated downstream may be as important as the *amount* supplied,
- *logistics-related decisions such as transshipment, distribution, collection*, which are complicated by the fact that products at different locations may have different remaining lifetimes,
- *centralized vs. decentralized planning,* where different echelons/locations may be managed by different decision-makers with conflicting objectives, operating rules may have different consequences for different decision-makers (e.g., retailers may want to receive LIFO shipments but suppliers may prefer to issue their inventory according to FIFO), and system-wide optimal solutions need not necessarily improve the performance at each location.

Given the complexity in obtaining or characterizing optimal decision structures, analytical research in multi-echelon and multi-location systems has mainly focused on particular applications (modeling novelties) and heuristic methods. We review analysis of replenishment and allocation decisions in Sect. 15.4.1, logistics and distribution related decision in Sect. 15.4.2, and centralized vs. decentralized planning in Sect. 15.4.3.

15.4.1 Research on Replenishment and Allocation Decisions

Research on multi-echelon systems has been confined to two-echelons[2] except in the simulation-based work (e.g., van der Vorst et al. (2000)). Motivated by food supply chains and blood banking, the upstream location(s) in the two-echelon models typically involve the supplier(s), the distribution center(s) (DC), or the blood banks, and the downstream location(s) involve the retailer(s), the warehouse(s), or the hospital(s). In this section, we use the terms *supplier* and *retailer* to denote the upstream and downstream parties, respectively. We call the inventory at the retail locations the *field inventory*.

We first classify the research by focusing on the nature of the decisions and the modeling assumptions. Table 15.4 provides a summary of the *analytical* work

---

[1] Note that multi-location models we review here are different from the two-warehouse problem described in Section 10 of Goyal and Giri (2001); that model considers a decision-maker who has the option of renting a second storage facility if he/she uses up the capacity of his/her own storage.

[2] The model of Lin and Chen (2003) has three echelons in its design: A cross-docking facility (central decision maker) orders from multiple suppliers according to the demand at the retailers and the system constraints, and allocates the perishable goods to retailers. However, the replenishment decisions are made for a single echelon: The authors propose a genetic algorithm to solve for the single-period optimal decisions that minimize the total system cost.

Table 15.4 A summary of the analytical models on replenishment and/or allocation decisions in multi-echelon and multi-location systems

Article	Replenishment policy		Inventory issuance	Excess demand	Allocation rule	Trans-shipments	No. of retailer(s)	Fixed costs	Centralized planning
	Supplier	Retailer							
Yen (1975)	TIS	TIS	FIFO	Backlog	Proportional, fixed	–	2	–	Yes
Cohen et al. (1981a)	TIS	TIS	FIFO	Backlog	Proportional, fixed	–	2	–	Yes
Lystad et al. (2006)	TIS	TIS	FIFO	Backlog	Myopic	–	>2	–	Yes
Fujiwara et al. (1997)	TIS (per cycle)	TIS (per period)	FIFO = LIFO	Lost at supplier, expediting for retailer	–	–	2	–	Yes
Kanchanasuntorn and Techanitisawad (2006)	(s,S)	TIS	FIFO	Backlog at supplier, lost at retailer	FIFO	–	2	Supplier	Yes
Prastacos (1978)[†]	–	–	FIFO	Lost at retailer	Myopic	Rotation	>2	–	Yes
Prastacos (1981)	–	–	FIFO	Lost at retailer	Myopic	Rotation	> 2	–	Yes
Prastacos (1979)[†]	–	–	LIFO	Lost at retailer	Segregation	Rotation	>2	–	Yes
Nose et al. (1983)[†]	–	–	LIFO	Lost at retailer	Based on convex programming	Rotation	>2	–	Yes
Federgruen et al. (1986)[†]	–	–	FIFO	Lost at retailer	Based on convex programming	Rotation	>2	–	Yes
Abdel-Malek and Ziegler (1988)	EOQ	EOQ	–	–	–	–	1	Retailer, supplier	Yes
Ketzenberg and Ferguson (2006)	Order 0 or Q	Order 0 or Q	FIFO	Expediting for Supplier, lost at retailer	–	–	1	–	Yes, no

[†]Indicates single-period decision making.

that determines heuristic replenishment policies for a supplier and retailer(s), and/or effective allocation rules to ship the goods from the supplier to the retailers; analytical research has been rather limited, modeling assumptions have varied, and some problems (such as replenishment with fixed costs or decentralized planning) have received very little attention.

Notice that the centralized, multi-echelon models with perishables are no different from the serial or distribution systems studied in classical inventory theory; the stochastic models involve the one-supplier, multi-retailer structure reminiscent of Eppen and Schrage (1981) or the serial system of Clark and Scarf (1960). For the one-supplier, multi-retailer system with non-perishables, it is known that order-up-to policies are optimal under the "balance assumption" (when the warehouse has insufficient stock to satisfy the demand of the retailers in a period, available stocks are allocated such that retailers achieve uniform shortage levels across the system; this might involve "negative shipments" from the warehouse, i.e., transshipments between retailers at the end of that period to achieve system balance), and optimal policy parameters can be determined by decomposing the system and solving a series of one-dimensional problems (see e.g., Diks and de Kok (1998)). There is no equivalent of this analysis with perishable inventories, mainly due to the complexity of the optimal ordering policy at a single location. Similarly, there is no work that investigates continuous replenishment policies with perishables for multi-echelons or multi-locations, although this is a well-studied problem for single-location models with perishables (see Sect. 15.3.2) and for non-perishables in multi-echelon supply chains.

15.4.1.1 Analytical Research on Allocation Decisions

All the models listed in Table 15.4 involve no capacity restrictions, and have a single supplier who receives the freshest goods upon replenishment (although the goods may not have the maximal lifetime at the time of arrival when lead time is positive). However, the shipments from the supplier to the retailers can involve stock of any age depending on the supplier's inventory. Allocation and/or transshipment decisions are simplified, for instance, when there is a single retailer (see Table 15.4) or when the supplier's inventory consists of goods of the same age. The latter happens in the following two cases: (a) The goods start perishing at the retailer but not at the supplier, i.e., all retailers are guaranteed to receive fresh goods from the supplier (see Fujiwara et al. (1997) and Abdel-Malek and Ziegler (1988)). (b) The lifetime of the product is equal to the length of the periodic review interval. In that case, all the goods at any location are of the same age and the goods perish at the end of one cycle, as is the case for the supplier in the model of Fujiwara et al. (1997).

The focus on multi-echelon problems is on allocation decisions when the supplier is assumed to receive a random amount of supply at the beginning of every period (Prastacos 1978, 1979, 1981) – we refer to this as the *Prastacos model*. The random supply assumption is motivated by the application area – blood products which rely on donations for supply. Two special distribution systems are considered in these papers: (a) A *rotation* (or *recycling*) system where all unsold units that have not expired at the retailers are returned to the supplier at the end of each period – these units are distributed among the retailers along with the new supply of freshest goods at the beginning of the next period; (b) A *retention* system where each retailer keeps all the inventory allocated. In the *Prastacos model*, the supplier does not stock any goods, i.e., all the inventory is allocated and shipped to the retailers at the beginning of a period, there are no inventory holding costs at any location, and the goal is to minimize shortage and outdating costs that are uniform across all the retailers.

In a rotation system, the total number of units to outdate in a period depends only on how the units with only one period of lifetime remaining are allocated in the previous period, and the total amount of shortage depends only on how the inventory is allocated, regardless of the age. Based on these observations, Prastacos (1978) proposes the following *myopic allocation policy that minimizes one-period system-wide outdate and shortage costs*: Starting with the oldest, the stocks of a given age are allocated across the retailers so that the probability that the demand at each location exceeds the total amount allocated to that retailer up to that point in the algorithm are equalized, and this is repeated iteratively for items of all ages. Prastacos (1978) also analyzes a retention system where the supplier only ships the fresh supply to the retailers at the beginning of each period. In the one-period analysis of the retention system, the amount to outdate at the end of a period depends on the amount of oldest goods in stock and demand at each retailer, but does not depend on the supply allocated in that period. Based on this observation, Prastacos (1978) suggests a myopic allocation rule that equalizes the one-period shortage probability at each retailer to minimize the one-period system-wide shortage and outdate costs. Prastacos (1981) extends the analysis of the *Prastacos model* to the multi-period setting and shows that the myopic allocation policy preserves some of the properties of the optimal allocation that minimizes expected long-run average shortage and outdating costs, and is, in fact, optimal in numerical examples with two retailers and product lifetime of two periods. Since the cost parameters are the same for all the retailers, the allocation resulting from the myopic rule is *independent* of the unit costs of shortage and outdating.

Prastacos (1979) analyzes essentially the same single-period model assuming LIFO issuance, as opposed to FIFO in Prastacos (1978). In case of LIFO, the optimal myopic allocation policies in both rotation and retention systems depend on the unit costs of shortage and outdating. In addition, the optimal allocation policies *segregate* the field inventory by age as opposed to a *fair* allocation where each retailer receives goods of each age category. Under segregation, some retailers have only newer goods and some only older goods so that system-wide expected outdates are minimized. The optimal myopic allocation policy under LIFO for both rotation and retention systems can be determined by solving a dynamic program with stages corresponding to retail locations. Prastacos (1979) obtains the optimal allocation rule for specific demand distributions. For rotation, he proposes a heuristic: First, allocate the stock in order to equalize the probability of shortage at each retailer and then swap the inventory among retailers to obtain field inventories that are segregated by age.

There are several practical extensions of the *Prastacos model*: The assumption on uniform outdating and shortage costs can be relaxed, shipments/transportation costs can be added, penalties can be incurred on leftover inventory (e.g., end of period holding costs), transshipments among retailers can be enabled, and the supplier may keep inventory as opposed to shipping all units downstream. The first three of these issues have been addressed in the literature for only single-period decision-making: Nose et al. (1983) and Federgruen et al. (1986) generalize the FIFO model of Prastacos (1978) by assuming that there is a unit transportation cost for each item shipped from the supplier to the retailers, and the unit outdating, shortage, and transportation costs are retailer-specific. They both develop convex programming formulations for the single-period inventory allocation problem and propose algorithms based on the Lagrangean relaxation to determine the optimal allocation. Their models rely on the observation that the costs are only a function of the amount of old vs. fresh (i.e., stock that will outdate in one period vs. the inventory that has more than one period of lifetime remaining) goods allocated to each retailer. In the rotation system considered by Nose et al. (1983), the retailers are also charged per unit inventory returned to the supplier at the end of each period (i.e., there is an end-of-period penalty on leftover inventory at each location).

In the model of Yen (1975) and Cohen et al. (1981a), the supplier uses the FIFO rule to determine which goods are to be shipped downstream and then uses one of the following two allocation rules to determine the age-composition of the shipments to the retailers in each period: (a) *proportional allocation* and (b) *fixed allocation*. In proportional-allocation, each retailer receives a proportion of goods of each age category based on their share of the total demand. In *fixed allocation*, each retailer receives a pre-determined fraction of goods in each period. Both of these allocation rules are *fair* in that the shipments to retailers involve goods of each age category. Yen (1975) and Cohen et al. (1981a) explore the optimality conditions for the parameters associated with these allocation rules. They show that, under certain conditions, there exists a fixed allocation rule that yields the same expected shortage, outdating, and holdings costs as a system operating under the optimal proportional allocation rule. Their analysis can be extended to include multiple (>2) retailers. Prastacos (1981) shows that his myopic allocation rule is the same as proportional allocation for certain probability distributions of demand.

15.4.1.2 Analytical Research on Replenishment and Allocation Decisions

Analysis of optimal replenishment policies for a serial, two-echelon system is presented in Abdel-Malek and Ziegler (1988) assuming *deterministic* demand, zero lead times, and price that linearly decreases with the age of the product. They determine the economic order quantities (EOQs) for the retailer and the supplier by restricting the order cycle lengths to be no more than the product lifetime.

Under demand uncertainty, several heuristic, discrete review replenishment policies have been considered and the focus has been on determining the optimal parameters for these restricted policies. These heuristic replenishment policies include the

TIS policies (Yen (1975), Cohen et al. (1981a), Lystad et al. (2006), Fujiwara et al. (1997) and Kanchanasuntorn and Techanitisawad (2006)), a "zero-or-fixed quantity" ordering policy in Ketzenberg and Ferguson (2006) – denoted as "0 or Q" in Table 15.4, – and (*s,S*) policy for the retailers in Kanchanasuntorn and Techanitisawad (2006), analysis which is restricted to the case where the retailers' demand is Normal. The replenishment lead times are no longer than one period (i.e., goods are received no later than the beginning of the next period) with exceptions being Lystad et al. (2006) and Kanchanasuntorn and Techanitisawad (2006); the latter assumes that the lifetime of the product is a multiple of the replenishment cycle lengths of the retailers and the supplier, and that the supplier responds to retailers' orders in a FIFO fashion (hence allocation decisions are trivial).

Research that analyzes replenishment and allocation decisions jointly is confined to the work of Yen (1975), its extension in Cohen et al. (1981a), and Lystad et al. (2006). Yen (1975) and Cohen et al. (1981a) explore the structural properties of the expected total cost function that includes expected holding, outdating, and shortage costs, when both the supplier and the retailer use TIS policies to replenish inventory. They investigate fixed and proportional allocation rules and identify conditions on the existence of unique target inventory levels. Yen (1975) also identifies conditions for the optimality of the proportional allocation rule for this system. These conditions are satisfied, for instance, when the lifetime of the product is restricted to two or three periods, or when the demand at each location and each period is i.i.d and the target inventory levels of the retailers are the same. However, the analysis relies on one simplification: the finite lifetime of the perishable product is not taken into account explicitly, and goods that remain in inventory beyond their lifetime can be used to satisfy excess demand, but are charged a unit outdating cost. Analysis of replenishment *and* allocation policies that explicitly take these factors into account remains an open problem of theoretical interest.

Lystad et al. (2006) use the myopic allocation rule of Prastacos (1981), and propose heuristic echelon-based TIS policies. For a given system, they first determine the best TIS policy via simulation. Then, they do a regression analysis to establish the relationship between the order-up-to levels of this best policy and two heuristic order-up-to levels: One heuristic is based on the newsvendor-based, approximate echelon-stock policies for non-perishables and the other is the single-location heuristic of Nahmias (1976) for perishables. The resulting regression model is then used in computational experiments to study the effect of the lifetime of a product on system costs, and to compare the performance of the approximation against policies that are derived assuming the product is non-perishable. Thus Lystad et al. (2006) take a first step in developing approximate echelon-based policies for perishables, and this topic deserves more attention.

Notice that the effectiveness of the proposed allocation rules combined with good replenishment policies have not been benchmarked in any of these studies, and various simplifying assumptions have been made to derive the policies. There is a need for further research on the analysis of replenishment and allocation decisions in multi-echelon systems. Interesting research directions include investigation of the "balancing" of echelon inventories for perishables (as in Eppen and Schrage (1981)), analysis of systems without making simplifying assumptions on inventory recursions (as in Yen (1975)), analysis of different system designs (e.g. rotation and retention systems have been studied to some extent), or incorporation of different cost parameters to the models (e.g. the *Prastacos model* excludes holding costs).

15.4.1.3 Simulation Models of Multi-Echelon Inventory Systems

In addition to the analytical research, simulation models have also been used in analyzing multi-echelon, multi-location systems with perishable goods. For this complex problem, simulation models present more opportunities in terms of model richness, which we highlight in this section.

The earlier research (e.g., Yen (1975), Cohen and Pierskalla (1979), Cohen et al. (1981b)) is motivated mainly by managing regional blood centers; see also Prastacos (1984) and Pierskalla (2004).

More recently, van der Vorst et al. (2000) describe a discrete-event simulation model to analyze a fresh produce supply chain with three echelons. Among other factors, van der Vorst et al. (2000), test the system performance – measured in terms of inventory levels at the retailers and distribution centers, and product freshness – using several scenarios. Katsaliaki and Brailsford (2007) present results of a project to improve procedures and outcomes by modeling the entire supply chain for blood products in the UK. Their simulation model includes a serial supply chain with the product flow that includes collection of supply, processing/testing and storage at a service center, and shipment to a hospital where blood is crossmatched/transfused[3] for patients use. The model includes multiple products with different shelf-lives. Six different policies varying in (a) the type of products that are stocked at the hospital, (b) the target inventory levels, (c) the time between crossmatching and release which can influence the amount of unused and still usable inventory that is returned, (d) the order trigger points for expedited deliveries, (e) the inventory issuance rules for releases and returns, (f) the order and delivery lead times, and (g) the number of daily deliveries to the hospital. Performance is measured in terms of number of expired units, mismatched units, amount of shortage, and number of routine and expedited deliveries. Note that allocation decisions are not included in Katsaliaki and Brailsford (2007) because they model a serial supply chain. Mustafee et al. (2006) provide the technical details of the simulation model and the distributed simulation environment used in this latter project.

In contrast to Katsaliaki and Brailsford (2007), the simulation model of the single supplier, multiple-retailer system in Yen (1975) includes returns of unused units

[3] One common practice in managing blood inventories is cross-matching, which is assigning units of blood from inventory to particular patients. Jagannathan and Sen (1991) report that more than 50% of blood products held for patients are not eventually transfused (i.e., used by the patient). The release of products that are cross-matched enable re-distribution of inventories in a blood supply chain. See Prastacos (1984), Pierskalla (2004), and Jagannathan and Sen (1991) for more information on cross-matching.

from retailers to the supplier, variations of the fixed and proportional allocation rules, expedited shipments to retailers, transshipments between retailers, and limited supply at the supplier. In addition to analyzing the impact of inventory levels, allocation and transshipment rules on system costs, Yen (1975) also looks at the impact of magnitude of demand at the retailers, and observes that the system cost in his centralized model is more sensitive to the sizes of the retailers rather than the number of retailers. Cohen et al. (1981b) also study different allocation rules in a centralized system: In the first one, the supplier chooses a retailer and fills its demand and goes on to fill the demand of the next retailer until all stock is depleted or all demand is satisfied. In the second one, the supplier uses the proportional allocation rule, and in the third one, the myopic allocation rule. Cohen et al. (1981b) suggest using the second method in a practical setting because it has less information needs (i.e., does not need the probability distribution of demand at each retailer like the third method) and advise against using the first method in a centralized system because it will lead to an imbalanced distribution of aging inventory. In addition, the outdate probabilities of the retailers will vary significantly with the first method; this increases the possibility of costly transshipments which are discussed in the next section.

15.4.2 Logistics: Transshipments, Distribution, and Routing

Other than replenishment and allocation decisions, three of the critical logistics activities in managing perishables in multi-location systems are transshipments, distribution and collection (particularly for blood). Research that focuses on these three decisions has been limited. Within the analytical work cited in Table 15.4, the rotation system is the only form in which excess inventory is exchanged among the retailers, and the exchange happens with a one period delay. Rotation systems are also the basis for the goal programming model developed by Kendall and Lee (1980). Prastacos and Brodheim (1980) develop a mathematical programming model for a hybrid rotation-retention system to efficiently distribute perishables in a centralized system. Both of these papers are motivated by operations of regional blood centers; see Prastacos (1984) for a review of these and other earlier work on the distribution and transshipment problems.

Note that rotation encompasses only indirect transshipments among the retailers. The simulation model of Yen (1975) includes transshipments between retailers after each location satisfies its own demand and serves as a guideline for the practical inventory control and distribution system described in Cohen et al. (1981b). Cohen et al. (1981b) suggest using transshipments in this practical setting if (a) the supplier is out-of-stock, one retailer has an emergency need, and transshipping units from other retailers do not significantly increase the probabilities of shortage at those retailers, (b) the difference between the shortage probabilities of two retailers when a unit is transshipped from one to the other is greater than the ratio of the unit transportation cost to the shortage cost, or (c) the difference between outdate probabilities of retailers when a unit is shipped from one to the other is greater than the ratio of transportation cost to the outdate cost. They use the terms *emergency*, *shortage-anticipating* and *outdate-anticipating* transshipments, respectively, to denote these three cases. Cohen et al. (1981b) point out that when all the retailers use optimal TIS policies to manage their inventories, the amount of transshipments is insignificant based on simulation results. This emphasizes the need for effective replenishment policies in multi-echelon and multi-location systems.

Federgruen et al. (1986), in addition to their analysis of the allocation decision, consider the distribution of goods from the supplier to the retailers by formulating a combined routing and inventory allocation problem. The decisions involve assigning each location to a vehicle in the fleet and allocating fresh vs. old products among the locations. The allocations do not affect the transportation costs and the routes of vehicles do not affect shortage and outdating costs. They propose exact and heuristic solution methods. They also compare their combined routing and allocation approach to a more hierarchical one where the allocation problem is solved first, and its solution is used as an input to the distribution problem. Based on computational experiments, the combined approach provides significant savings in terms of total transportation costs, although these savings may not lead to a significant decrease in total costs depending on the magnitude of the inventory related costs. However, the combined approach has significant benefits when the number of vehicles used is few (where the hierarchical approach may yield an infeasible solution). In addition to inventory levels and allocation, the simulation-based research conducted by Gregor et al. (1982) also examines the impact of the number of vehicles used in distribution on system-wide costs.

Or and Pierskalla (1979) consider daily vehicle routing decisions as a part of a regional location-allocation problem where they also determine the optimal number and location of blood centers, and the assignment of hospitals to the blood centers that supply the hospitals on a periodic basis. They develop integer programming models and propose heuristic solution methods. However, their model is designed at an aggregate level and age of inventory is not considered. A similar problem is studied by Hemmelmayr et al. (2006) where periodic delivery schedules and vehicle routes are determined to distribute blood across a region.

Recent research on supply chain scheduling has addressed the need to effectively distribute time-sensitive goods; Chen (2006) provides a survey of research in this growing area. However, perishability is not modeled explicitly in this literature, rather production orders are assumed to come from customers along with information on delivery time windows and delivery due dates. Similarly, there are several articles that model and solve real-life distribution problems of perishable products such as dairy products, or food (e.g., Adenso-Diaz et al. (1998), Golden et al. (2001)) where aging or perishability is not explicitly modeled but is implicit in the time-windows. More recently, Yi (2003) developed a model for daily vehicle routing decisions to bring back supply (blood) from collection sites to a central location in order to meet the daily target level of platelets (that can only be extracted within 8 hours of blood donation); this is a *vehicle routing problem with time windows and time-dependent rewards.*

Notice that the research in this area has been confined to a single product, or multiple products without age considerations. Interestingly, the distribution problem posed by Prastacos (1984) still remains open: *How can a distribution plan for a centralized system be created to include shipments for multiple products, each with a different lifetime and supply-demand pattern?* Katsaliaki and Brailsford's (2007) simulation model provides only a partial answer to this question; their model involves only one supplier and one retailer. Although challenging, analysis of the centralized problem with multiple retailers and multiple products definitely deserves attention.

15.4.3 Information Sharing and Centralized/Decentralized Planning

Information technology paved the way for various industry-wide initiatives including Efficient Consumer Response in the grocery industry; these initiatives aim to decrease total system costs and inventories while improving availability of products and customer satisfaction. A critical component of these initiatives is the sharing of demand and product flow information among the suppliers, distributors, and retailers. Fransoo and Wouters (2000) discuss the benefit of sharing electronic point of sale (EPOS) information for supply chains of two perishable products (salads and ready-made pasteurized meals). Their empirical analysis suggests that the benefit of EPOS would be higher for the supply chain of salad because of the magnitude of the bullwhip effect observed. The reason for the higher bullwhip effect appears to be the larger fluctuations in the demand for salad (e.g., when there is a sudden increase in temperature, there is a spike in demand), associated shortage-gaming by the retail franchisees, and the additional order amplification by the DC.

Information sharing and the value of information has been widely studied in the general supply chain literature. Chen (2002) provides a survey of research on this topic by focusing on where the information is coming from (such as the point-of-sale data from *downstream*, or capacity information from *upstream* in the supply chain), quantity, accuracy and speed of information, and centralized vs. decentralized planning in the supply chain (specifically he considers incentives for sharing information and whether the environment is competitive or not). However, this rich literature studies non-perishable goods or single-period models; the unique characteristics of perishables are ignored, except by Ferguson and Ketzenberg (2006) and Ketzenberg and Ferguson (2006).

Ferguson and Ketzenberg (2006), motivated by the grocery industry, investigate the value of information for a retailer managing inventory. They focus on the retailer's replenishment problem and consider an infinite-horizon periodic review inventory model for a single product with finite lifetime, lost sales, one-period lead time and no outdating cost (see Sect. 15.3.1.1). The age of all units in a replenishment are the same. The age of stock at the supplier is a random variable, and its distribution is known to the retailer. In case of information sharing, the retailer knows exactly the age of stock prior to giving an order. Ferguson and Ketzenberg

(2006) quantify the value of information on the age of stock under heuristic replenishment policies with FIFO, LIFO or random issuing of inventory. Numerical experiments reveal profits increase and outdates decrease, on the average, when information is shared. An interesting finding is that investments that extend the product lifetime provides a greater benefit than information sharing.

There are complications associated with different parties operating by different rules in managing supply chains of perishables: For instance, the supplier can presumably induce the retailers to order more frequently by adapting an issuance and/or replenishment policy that leads to more frequent outdates (e.g., using FIFO issuance and/or having older stock in its inventory will enable the supplier to sell goods that have a smaller shelf-life to the retailer). This is only mentioned in Ketzenberg and Ferguson (2006) – but not analyzed – and is ignored by other researchers. In that paper, Ketzenberg and Ferguson (2006) study the value of information in a two-echelon setting with one retailer and one supplier. Both parties replenish inventory heuristically, the retailer's order quantity in each period is either 0 or Q (which is an exogenous fixed batch size) and issues inventory in a FIFO fashion, whereas the supplier uses the same quantity Q in giving orders but need not give an order every period. In fact, the supplier determines the timing of his replenishments by considering a safety lead time. The retailer knows the supplier's inventory state – including the age of items in stock – and the supplier knows the retailer's replenishment policy. Ketzenberg and Ferguson (2006) quantify (a) the value of information regarding the inventory state and replenishment policies in a decentralized system via numerical examples, and (b) the value of centralized planning. The value of information for perishables can be significant, and increase in supply chain profit, due to centralization, is not always Pareto improving for both parties.

Note that all the papers we introduced so far focus on a single decision maker. Likewise, Hahn et al. (2004) derive the optimal parameters of a TIS policy for a retailer under two different contracts offered by the supplier; however, the supplier's optimal decisions are disregarded. Among the few studies that mention decentralized decision-making, Popp and Vollert (1981) provide a numerical comparison of centralized vs. decentralized planning for regional blood banking. Problems that involve multiple decision-makers, decentralized planning (vs. centralized) and coordination of supply chains have been widely studied for non-perishables and/or using single-period models (see, Chen (2002) and Cachon (2003)). In practice, perishable products share the same supply chain structure as many non-perishables, and decentralized planning and/or coordination issues are just as critical. Furthermore, there are more challenges for perishables due to cost of outdating, and possibly declining revenues due to aging. However, research in this area has been scarce, and this issue remains as one of the important future research directions.[4]

[4] There is research on coordination issues in supply chains with deteriorating goods: A *permissible delay in payment* agreement between a retailer and a supplier is proposed in the deterministic model of Yang and Wee (2006) to coordinate the supply chain. Chen and Chen (2005) study centralized and decentralized planning for the joint replenishment problem with multiple deteriorating goods.

15.5 Modeling Novelties: Demand and Product Characteristics, Substitution, Pricing

The research we have reviewed so far includes models where the inventory of a single product is depleted either in a LIFO or FIFO manner. Analysis of single location models in Sect. 15.3 is confined to FIFO. Earlier research on single location models has shown the difficulty in characterizing stationary distribution of stock levels under LIFO even when the lifetime of the product is only two periods (see the references and comments in Nahmias (1982)). Nahmias (1982) mentions that when the lifetime is two periods, the order up to level in a TIS policy is insensitive to the choice of FIFO vs. LIFO despite the difference in total system costs. Therefore, the replenishment policies/heuristics developed under FIFO can also be used effectively for LIFO. However, replenishment and issuance decisions may be interconnected – hence a more a careful analysis is needed – under more general demand models.

Typically, excess demand is treated via backlogging or lost sales, with some papers incorporating expedited delivery in their models (e.g., Fujiwara et al. (1997), Ketzenberg and Ferguson (2006), Yen (1975), Bar-Lev et al. (2005) and Zhou and Pierskalla (2006)). In practice, there is another way to fulfill the excess demand for a product: *Substitution*. In the case of perishables, products of different ages often coexist in the market place, and inventory can be issued using rules more complicated than FIFO or LIFO, allowing items of different ages or shelf-lives to be used as substitutes for each other. This idea first appeared in the perishable inventory literature in Pierskalla and Roach (1972) who assume there is demand for any category (age) and that the demand of a particular category can be satisfied from the stocks of that category or using items that are fresher. Pierskalla and Roach (1972) show that FIFO is optimal in this model with respect to two objectives: FIFO minimizes total backlog/lost sales and minimizes outdates. The model has an important simplification: The demand and supply (replenishment) are assumed to be independent of the issuing policy. Since issuing can potentially affect demand – fresher goods could lead to more loyal customers – the study of models in which this assumption of independence is relaxed will be important.

The motivation for many of the papers that involve substitution and age-dependent demand streams come from health care. Cohen et al. (1981a) mention hospitals doing special surgeries get higher priority for fresh blood. Haijema et al. (2005, 2007) mention that platelets have 4–6 days of effective shelf-life, and 70% of the patients requiring platelets suffer from platelet function disorder and need a fresh supply of platelets (no older than 3 days) on a regular basis whereas the remaining 30% of the patients who may lack platelets temporarily due to major trauma or surgery do not have a strong preference with respect to the age of the platelet up to the maximal shelf-life. For supply chains involving perishable goods other than blood, substitution usually depends on customers' choice and/or a retailer/supplier's ability to influence customers' purchasing decisions. In their empirical research, Tsiros and Heilman (2005) study the effect of expiration dates on the purchasing behavior of grocery store customers. They conducted surveys to investigate consumer behavior across different perishable product categories. They find that consumers check the expiration dates more frequently if their perceived risk (of spoilage or health issues) is greater. They also determine that consumers' willingness to pay decreases as the expiration date nears for all the products in this study; again finding that the decrease varies across categories in accordance with customer perceptions. Tsiros and Heilman's (2005) findings support the common practice of discounting grocery goods that are aging in order to induce a purchase. However, they find that promotions should differ across categories and across customer groups in order to exploit the differences in customers' tendencies to check the expiration dates and the differences in their perceived risks across categories. In light of these motivating examples and empirical findings, we provide below an overview of analytical research that considers substitution, multiple products and pricing decisions.

15.5.1 Single Product and Age-Based Substitution

Research like that of Pierskalla and Roach (1972), where products of different shelf-lives are explicitly modeled is limited. For a single product with a limited shelf-life, substitution has been considered to sell goods of different ages: Parlar (1985) analyzes the single-period problem for a perishable product that has two periods of lifetime, where a fixed proportion of unmet demand for new items is fulfilled by unsold old items and vice-versa, but his results do not extend to longer horizons. Goh et al. (1993) consider a two-stage perishable inventory problem. Their model has random supply and separate, Poisson-distributed demand streams for new and old items. Their analysis relies on an approximation and they *computationally* compare a restricted policy (where no substitution takes place) and an unrestricted policy (where stocks of new items are used to fulfill excess demand for old). Considering only shortage and outdating costs they conclude that the unrestricted policy is less costly, unless the shortage cost for fresh units is very high. Ferguson and Koenigsberg (2007) study a problem in a *two-period setting* with pricing and internal competition/substitution. In their model, the demand for each product in the second period is given by a linear price-response curve which is a function of the price of both products as well as the quality deterioration factor of the old product, and their decisions include the prices of both products as well as the number of leftover units of old product to keep in the market. They investigate whether a company is better off by carrying both or only the new product in the second period.

Ishii (1993) models two types of customers (high and low priority) that demand only the *freshest* products or products of *any age,* respectively, and obtains the optimal target inventory level that maximizes the expected profits in a single period for a product with finite lifetime. The demand of high priority customers is satisfied from the freshest stock first, and then inventory is issued using FIFO in this model. Ishii and Nose (1996) analyze the same model under a warehouse capacity constraint. More recently, Haijema et al. (2005, 2007) study a finite horizon problem for blood platelet production with a demand model of two types of customers similar to Ishii (1993) and Ishii and Nose (1996). Haijema et al. (2005, 2007) formulate a Markov Decision Process (MDP) model to minimize costs associated with holding, shortage, outdating and substitution (incurred when the demand for a "fresh" item is fulfilled by older stock) costs. They assume inventory for the any-age demand is issued in a FIFO manner from the oldest stock and fresh-demand is issued using LIFO from the freshest stock. Haijema et al. (2005, 2007) propose a TIS and a combined TIS-NIS heuristic, i.e., there is a daily target inventory level for total inventory in stock and also the new items in stock. Computational experiments show that the hybrid TIS-NIS policy is an improvement over TIS and that these heuristics provide near-optimal inventory (production) policies.

Deniz et al. (2008) provide a detailed analysis of the interplay between replenishment policies and inventory issuance. Their infinite horizon, periodic review formulation for a product with two periods of lifetime includes lost sales, holding, outdating as well as substitution costs (both new-to-old and old-to-new). They assume two separate demand streams for new and old items; demand can be correlated across time or across products of different ages. Deniz et al. (2008) study different substitution options: The excess demand for a new item is satisfied from the excess stock of old, and/or the excess demand for an old item is satisfied from the excess stock of new, or not, including the no-substitution case. Both LIFO and FIFO inventory issuance, as is common in the literature, can be represented using this substitution model. The inventory is replenished using either TIS or NIS in Deniz et al. (2008), and they identify conditions for the cost parameters under which the supplier would indeed benefit from restricted (only old-to-new, only new-to-old, or no substitution) or unrestricted forms of substitution while using a practical replenishment policy. They show that even when substitution costs are zero, substitution can be economically inferior to no-substitution for a supplier using a TIS policy. Alternately, even when substitution costs are very high, no-substitution is not guaranteed to be superior for a supplier using TIS. These counter-intuitive properties are the side-effects of the TIS policy which constrains reordering behavior. In contrast, more intuitive results under the NIS policy exist and conditions on cost parameters establish the economic benefit of substitution for this replenishment policy.

Deniz et al. (2008) and Deniz (2007) do extensive computational experiments to quantify the benefits of substitution and to compare TIS and NIS. Deniz et al. (2008) and Deniz (2007) find that NIS – the policy that uses *no information* on the level and age of inventory – proves more effective than TIS and provides lower long-run average costs except when the demand for new items is negligible. The effectiveness of NIS in their model is in contrast with the observations in earlier research papers. This is because inventory is depleted in a FIFO manner – there is no demand for new items as long as old items are available – in the classical literature. Similar to the observation of Cohen et al. (1981b) on the limited need for transshipments (see our discussion in Sect. 15.4.2), Deniz et al. (2008) show that the amount of substitution is small when inventory is replenished using effective policies. This latter paper really only *begins* the consideration of managing age-dependent demand and effect of different inventory issuance rules (via substitution) for perishable items

– items with longer lifetimes, other issuance rules, or substitution between different perishable products within the same category (e.g., different types of fruit) remain to be investigated. Note that substitution is only modeled as a *recourse* in these papers, and dynamic substitution where one strategically sells a product of an age diggerent from that requested before the stocks of the requested item are depleted, has not been studied.

15.5.2 Multiple Products

Nahmias and Pierskalla (1976) study the optimal ordering policies for an inventory system with two products, one with a fixed, finite shelf-life and the other with an infinite lifetime. The problem is motivated by the operation of a blood bank storing frozen packed red cells.[5] Demand is satisfied from the inventory of perishable product first in a FIFO manner, any remaining demand is fulfilled from the inventory of non-perishable products. Nahmias and Pierskalla (1976) analyze the structural properties of the expected cost function in a finite-horizon, dynamic, discrete review system and show that the optimal ordering policy in each period is characterized by three choices: Do not order, order only product with the finite lifetime, or order both products. Their results include monotonicity of order-up-to level of the perishable product, e.g., the decrease in order up-to-level is higher with the increase in the stock levels of newer items as opposed to old ones. They also show that if it is optimal to order both products in a given period, then it is optimal to bring the total system-wide inventory up to a level that does not vary with the on-hand inventory levels, but with the time remaining until the end of the planning horizon.

Multiple perishable products are also considered by Deuermeyer (1979, 1980). Deuermeyer (1979) determines the one-period optimal order-up-to-levels for two products. In his model, the products are produced by two processes, one of which yields both products and the other yields only one product. Deuermeyer (1980) determines the single-period, optimal order-up-to levels for multiple perishable products, each with a different lifetime. A critical assumption in the latter is the *economic substitution* assumption where the marginal total cost of a product is assumed to be nondecreasing in the inventory levels of other products. Using the resulting properties of the single-period expected total cost function, Deuermeyer (1980) is able to obtain the monotonicity results on order-up-to-levels for the single-period, multi-product problem. His results mimic that of Fries (1975) and Nahmias (1975a) for the single product problem. Specifically, these results show that the optimal order-up-to level of a product is more sensitive to changes in stock levels of newer items (as discussed above for Nahmias (1976)), and that the optimal order quantity decreases with an increase in the on-hand stock levels, while the optimal target inventory level remains nondecreasing.

[5] We refer the reader to Prastacos (1984) for earlier, simulation-based research on the effect of freezing blood products on inventory management.

15.5.3 Pricing of Perishables

Pricing, in general, has become one of the most widely studied topics in the operations management literature in the last decade. There is a significant body of research on dynamic pricing and markdown optimization for "perishables." One well-studied research problem in that domain involves determining the optimal price path for a product that is sold over a finite horizon given an initial replenishment opportunity and various assumptions about the nature of the demand (arrival processes, price-demand relationship, whether customers expect discounts, whether customers' utility functions decrease over time etc.). We refer the reader to the book by Talluri and van Ryzin (2004) and survey papers by Elmaghraby and Keskinocak (2003), and Bitran and Caldentey (2003), for more information on pricing of perishable products. In that stream of research, all the items in stock at any point in time are of the same age because there is only one replenishment opportunity. In contrast, Konda et al. (2003), Chandrashekar et al. (2003), and Chande et al. (2004, 2005) combine pricing decisions with periodic replenishment of a perishable commodity that has a fixed lifetime, and their models include items of different ages in stock in any period. They provide MDP formulations where the state vector includes the inventory level of goods of each age. Their pricing decisions are simplified, i.e., they only decide whether to promote all the goods in stock in a period or not. Chande et al. (2004) suggest reducing the size of the state of space by aggregating information of fresher goods (as opposed to aggregation of information on older goods as in Nahmias (1977a)). Performance of this approximation is discussed via numerical examples in Chande et al. (2004, 2005), and sample look-up tables for optimal promotion decisions are presented for given inventory vectors.[6]

Based on Tsiros and Heilman (2005) observations on customers' preferences and close-substitutability of products in fresh-produce supply chains, it is important to analyze periodic promotion/pricing decisions across age-groups of products. There are several research opportunities in this area in terms of demand management via pricing to minimize outdates and shortages across age-groups of products and product categories.

15.6 Summary and Future Research Directions

We presented a review of research on inventory management of perishable and aging products, covering single-location inventory control, multi-echelon and multi-location models, logistics decisions and modeling novelties regarding demand and

[6] Another paper that considers prices of perishable products is by Adachi et al. (1999). Items of each age generate a different revenue in this model, demand is independent of the price, and the inventory is issued in a FIFO manner. The work entails obtaining a replenishment policy via computation of a profit function given a price vector.

product characteristics. We identified or re-emphasized some of the important research directions in Sect. 15.2 to 15.5, ranging from practical issues such as product-mix decisions and managing inventories of multiple, perishable products, to technical ones such as the structure of optimal replenishment policies with fixed costs in the single-product, single-location problem. We provide some final comments on possible research topics below.

Multiple products (joint replenishment and product-mix): The research within the perishable domain has largely been confined to inventory management of a *single product* as the survey in this chapter shows. However, grocery or blood supply chains involve multiple perishable products with possibly differing lifetimes. Joint replenishment is a typical practice in these industries, and analytical research that studies the interaction between multiple items in ordering decisions – focusing on economies of scale, or substitution/complementarity effects of products with different lifetimes and in different categories – has not been studied. These interactions provide opportunities for more complex control policies, which make such problems both more challenging analytically, and potentially more rewarding practically.

Considering multiple products, another problem that has not attracted much attention from the research community is determining the optimal product-mix when one type of perishable product can be used as a raw material for a second type of product, possibly with a different lifetime, as we mentioned in Sect. 15.2. Decisions regarding when and how much of a base product to sell/stock as is, vs: how much to process in order to obtain a final product with a different lifetime or different potential value/revenue are quite common in blood and fresh produce supply chains. Consideration of multiple products will lead to more realistic decision problems, for which practical and effective solutions are needed.

Capacity, freshness, disposal, and outdating: A significant majority of the research on inventory management or distribution of perishable goods disregards capacity constraints. Models with limited capacity are better representative of the challenges in practice and require innovative heuristic policies.

The practical decision of when (if at all) to *dispose of the aging inventory* has not received much attention even in single location models, possibly because capacity is assumed to be unlimited and/or demand is assumed to be satisfied with FIFO inventory issuance. However, disposal decisions are especially critical when *capacity is constrained* (e.g., a retailer has limited shelf-space to display the products), and customers choose the products based on their (perceived) *freshness/quality*. Veinott (1960), in his deterministic model, included disposal decisions for a retailer of perishable products with fixed lifetime. Martin (1986) studied optimal disposal policies for a perishable product where demand is stochastic. His queueing model considers the trade-off between retaining a unit in inventory for potential sales vs. salvaging the unit at a constant value. Vaughan (1994) models an environment where a retailer decides on the optimal parameter of a TIS policy and also a "sell-by" date that establishes an effective lifetime for the product with a random shelf life; this may be considered a joint ordering and outdating policy. Vaughan (1994) discusses that his model would be useful for retailers if they were to select suppliers based on their potential for ordering and outdating, but no analysis is provided.

When customers prefer fresher goods, disposal and outdating are key decisions that affect the age-composition (freshness) of inventory, and can influence the demand. Analysis of simple and effective disposal and outdating policies, coordination of disposal with replenishment policies, and analysis of inventory models where customers (retailers) choose among suppliers and/or consider risk of supply/freshness remain among the understudied research problems.

Inventory issuance and demand models: The majority of research on perishables assumes demand for a product is either independent of its age, or that the freshest items are preferred. These typical assumptions motivate the primary use of FIFO and LIFO issuance in inventory control models. However, one can question how realistic these issuance policies are especially in a business-to-business (B2B) setting. A service level agreement between a supplier (blood center) and its retailer (hospital) may not be as strict as "freshest items must be supplied" (motivating LIFO) or as loose as "items of any age can be supplied" (motivating FIFO), but rather "items that will not expire within a specified time-window must be supplied".[7] Faced with such a demand model, and possibly with multiple retailers, a supplier can choose his/her optimal issuance policy which need not necessarily be LIFO or FIFO. Similarly, Pierskalla (2004) notes that for a regional blood supply chain, FIFO issuance may not be the most appropriate for a supplier who distributes blood to multiple locations; if certain locations receive shipments infrequently, then it is better to use LIFO for those locations to extend the lifetime of the product. To the best of our knowledge, few researchers have shown an awareness of the heterogeneity among the customers/locations (see Sect. 15.5.1). For future research on perishables to be of more practical use, we need demand models and inventory issuance rules that are representative of the more general business rules and policies today.

Competition: While problems that involve competition (among retailers, suppliers, or supply chains) have received a lot of attention in the last decade (see, for e.g., Cachon (1998)), models that include competition involving perishable and aging products have not appeared in the literature. One distinct feature of competition in a perishable commodity supply chain is that suppliers (retailers) may compete not only on availability and/or price but also on freshness.

Contracting: In the produce industry, a close look at the relationship between suppliers and buyers reveals several practical challenges. Perosio et al. (2001) present survey results that indicate that about 9% of the produce in the USA. is sold through spot markets, and about 87.5% of product purchases are made under contracts with suppliers. Perosio et al. (2001) make the following observation which is essentially a call for further research: "*Despite a number of considerable disadvantages, in general, todays buyers and sellers alike appear to be won over by the greater price certainty that contracting makes possible...However, high*

[7] We thank Feryal Erhun from Stanford University for bringing this practical issue, which she has witnessed in blood supply chains, to our attention.

degrees of product perishability, weather uncertainty and resulting price volatility, and structural differences between and among produce buyers and sellers create significant challenges to the design of the produce contract."

Recently, Burer et al. (2006) introduced different types of contracts used in the agricultural seed industry and investigated – via single-period models – whether the supply chain can be coordinated using these contracts. In their ongoing work, Boyabatli and Kleindorfer (2006) study the implications of a proportional product model (where one unit of input is processed to produce proportional amounts of multiple agricultural outputs) on the optimal mix of long-term and short-term (spot) contracting decisions. We believe further analysis of supplier-retailer relations, and the design of contracts to improve the performance of a supply chain that involves perishable products remain fruitful research topics.

Pricing and blood supply chains: Pricing was mentioned as one of the important research directions by Prastacos (1984) to encourage collaboration between hospitals and blood banks/centers; this is also echoed in Pierskalla (2004). There seems to be almost no research in this direction to date. According to a recent survey in the U.S., the mean cost of 250 ml of fresh frozen plasma to a hospital varied from \$20 to \$259.77, average costs of blood components were higher in Northeastern states compared to the national average, and hospitals with higher surgical volume typically paid less than the national average for blood components in 2004 (AABB, 2005). Given the importance of health care both for the general welfare and the economy, there is a pressing need for further research to understand what causes such variability in this environment, and whether pricing can be combined with inventory management to better match demand for perishable blood components with the supply. The potential relevance of such work reaches well beyond the health care industry.

Technology: Advances in technology have increased the efficiency of conventional supply chains significantly; for perishable goods, technology can potentially have an even greater impact. Not only is there the potential to enable information flow among different parties in a supply chain, as has proved to be valuable in conventional chains, but there is also the possibility of detecting and recording the age of the products in stock (e.g., when RFID is implemented). This information can be used to affect pricing decisions, especially of products nearing their usable lifetimes. Moreover, advances in technology can potentially increase the freshness and extend the lifetime of products (e.g., when better storage facilities or packaging equipment is used). The relative magnitudes of these benefits calibrated to different product and market characteristics remains an important open problem.

The majority of the work on the analysis of inventory management policies assume that the *state* of the system is known completely; i.e., inventory levels of each age of product at each location are known. However, this may not be the case in practice. While effective heuristic policies, such as TIS and NIS, for inventory management at a single location reduce the information need and do not require a complete characterization of the state of the inventory, availability of information can be pivotal for applications with features that are not represented in typical single-location models (e.g., models that emphasize freshness of inventory and/or consider disposal). Cohen et al. (1981a) discuss the need for detailed demand and inventory information to apply shortage or outdate anticipating transshipment rules in a centralized system, and argue that the system would be better without transshipments between the retailers if accurate information is not available. Chande et al. (2005) presents an RFID architecture for managing inventories of perishable goods in a supply chain. They describe how the profile of current on-hand inventory, including the age, can be captured on a real-time basis, and conclude by stating that *"there is a need for measures and indicators ... to determine ... (a) whether such development would be beneficial, and (b) when implemented, how the performance of the system compares to the performance without auto ID enhancements."*

Final Note: With the acceleration of product life cycles, the line between "perishable" and "durable" products continues to be blurred. Strictly speaking, goods such as computers and cell phones obsolesce rather than perish, but many of the same questions we raised about "perishable" inventory currently are, and will continue to become increasingly relevant in this category as well.

References

AABB (2005) The 2005 nationwide blood collection and utilization survey report. The American Association of Blood Banks (AABB), Bethesda, Maryland, MD.

Abdel-Malek LL, Ziegler H (1988) Age dependent perishability in two-echelon serial inventory systems. Comput Oper Res 15:227–238.

Adachi Y, Nose T, Kuriyama S (1999) Optimal inventory control policy subject to selling prices of perishable commodities. Int J Prod Econ 60-61:389–394.

Adenso-Diaz B, Gonzalez M, Garcia E (1998) A hierarchical approach to managing dairy routing source. Interfaces 28:21–31.

Allen SJ, Schuster EW (2004) Controlling the risk for agricultural harvest. Manufact Serv Oper Manage 6:225–236.

Axsäter S (2000) Inventory Control. Kluwer international series in operations research and management science, Kluwer Academic Publishers, Boston, MA.

Axsäter S (2003) Supply chain operations: Serial and distribution inventory systems. In: de Kok AG, Graves S (eds) Handbooks in OR & MS, Vol. 11, Supply Chain Management: Design, Coordination and Operations, Elsevier, Amsterdam, The Netherlands.

Bar-Lev SK, Perry D, Stadje W (2005) Control policies for inventory systems with perishable items: Outsourcing and urgency classes. Prob Eng Inform Sci 19:309–326.

Berk E, Gürler Ü (2006) Analysis of the (Q,r) inventory model for perishables with positive lead times and lost sales. Working paper. Faculty of Business Administration, Bilkent University, Ankara,Turkey.

Bitran G, Caldentey R (2003) An overview of pricing models for revenue management. Manufact Serv Ope Manage 5:203–229.

Boyabatli O, Kleindorfer P (2006) Integrating long-term and short-term contracting in fed-cattle supply chains: Proportional product model. Presented at the INFORMS Annual Meeting, Pittsburgh, November, 2006.

Brodheim E, Derman C, Prastacos GP (1975) On the evaluation of a class of inventory policies for perishable products such as blood. Manage Sci 22:1320–1325.

Bulinskaya EV (1964) Some results concerning optimum inventory policies. Theory Prob Appl 9:389–402.

Burer S, Jones PC, Lowe TJ (2006) Coordinating the supply chain in the agricultural seed industry. Working paper. Department of Management Sciences, University of Iowa, Iowa City, IA.

Cachon GP (1998) Competitive supply chain inventory management. In: Tayur S, Ganeshan R, Magazine M (eds) Quantitative Models for Supply Chain Management, Kluwer, Boston.

Cachon GP (2003) Supply chain coordination with contracts,In: Graves S, de Kok T (eds) Handbooks in Operations Research and Management Science: Supply Chain Management North Holland.

Chande A, Hemachandra N, Rangaraj N, Dhekane S (2004) Fixed life perishable inventory problem and approximation under price promotion. Technical report. Industrial Engineering and Operations Research, IIT Bombay, Mumbai, India.

Chande A, Dhekane S, Hemachandra N, Rangaraj N (2005) Perishable inventory management and dynamic pricing using RFID technology. Sādhanā, 30:445–462.

Chandrashekar K, Dave N,Hemachandra N, Rangaraj N (2003) Timing of discount offers for perishable inventories. In: Rao MR, Puri MC (eds) Proceedings of sixth Asia Pacific operations research society. Allied Publishers, New Delhi.

Chazan D, Gal S (1977) A markovian model for a perishable product inventory. Manag Sci 23: 512–521.

Chen F (2002) Information sharing and supply chain coordination. In: de Kok T, Graves S (eds) Handbook of Operations Research and Management Science: Supply Chain Management North-Holland.

Chen J-M, Chen T-H (2005) Effects of joint replenishment and channel coordination for managing multiple deteriorating products in a supply chain. J Oper Res Soc 56:1224–1234.

Chen Z-L (2006) Integrated production and outbound distribution scheduling in a supply chain: Review and extensions. Working paper, Robert H. Smith School of Business, University of Maryland, College Park. MD.

Chiu HN (1995) An approximation to the Continuous review inventory model with perishable items and lead times. Eur J Oper Res 87:93–108.

Clark AJ, Scarf H (1960) Optimal policies for a multi-echelon inventory problem. Manag Sci 45:475–490.

Cohen M (1976) Analysis of single critical number ordering policies for perishable inventories. Oper Res 24:726–741.

Cohen M, Pierskalla WP (1979) Simulation of blood bank systems. ACM SIGSIM Simulation Digest 10:14–18.

Cohen M, Pierskalla WP, Yen H (1981a) An analysis of ordering and allocation policies for multi-echelon, age-differentiated inventory systems. TIMS Studies Manage Sci 16:353–378.

Cohen M, Pierskalla WP, Sassetti RJ (1981b) Regional blood inventory control and distribution. Proceedings of the 1980 conference on the management and logistics of blood banking, National Heart, Lung and Blood Institute, vol.5 (October 1981), pp 21–88.

Cooper W (2001) Pathwise properties and performance bounds for a perishable inventory system. Oper Res 49:455–466.

Dave U (1991) Survey of literature on continuously deteriorating inventory models – A rejoinder. J Oper Res Soc 42:725.

Deniz B, Karaesmen I, Scheller-Wolf A (2008) Managing perishables with substitution: Inventory issuance and replenishment. Working paper, Tepper School of Business, Carnegie Mellon University, Pittsburgh PA.

Deniz B (2007) Essays on perishable inventory management. PhD dissertation, Tepper School of Business, Carnegie Mellon University, Pittsburgh, PA

Deuermeyer BL (1979) A multi-type production system for perishable inventories. Oper Res 27:935–943.

Deuermeyer BL (1980) A single period model for a multi-product perishable inventory system with economic substitution. Naval Res Logis 27:177–185.

Diks EB, de Kok AG (1998) Optimal control of a divergent multi-echelon inventory system. Eur J Oper Res 111:75–97.

Elmaghraby W, Keskinocak P (2003) Dynamic pricing in the presence of inventory considerations: Research overview, current practices, and future directions. Manage Sci 49:1287–1309.

Eppen G, Schrage L (1981) Centralized ordering policies in A multi-warehouse system with lead times and random demand. In: Schwarz LB, (ed) Multi-level Production/Inventory Control Systems: Theory and Practice, North-Holland, Amsterdam.

Federgruen A, Prastacos GP, Zipkin P (1986) An allocation and distribution model for perishable products. Oper Res 34:75–82.

Ferguson M, Koenigsberg O (2007) How should a firm manage deteriorating inventory? Prod Oper Manag 16:306–321.

Ferguson M, Ketzenberg ME (2006) Sharing information to improve retail product freshness of perishables. Prod Oper Manag 15:57–73.

Fransoo JC, Wouters MJF (2000) Measuring the Bullwhip effect in the supply chain. Supply Chain Manag: An Int J 5:78–89.

Fries B (1975) Optimal ordering policy for A perishable commodity with fixed lifetime. Oper Res 23:46–61.

Fujiwara O, Soewandi H, Sedarage D (1997) An optimal ordering and issuing policy for a two-stage inventory system for perishable products. Eur J Oper Res 99: 412–424.

Goh C, Greenberg BS, Matsuo H (1993) Two-stage perishable inventory models. Manag Sci, 39:633–649.

Golden BL, Assad AA, Wasil EA (2001) Routing vehicles in the real world: Applications in the solid waste, beverage, food, dairy, and newspaper industries. In: Toth P, Vigo D (eds) The vehicle routing problem society for industrial and applied Mathematics, Philadelphia, PA, pp 245–286.

Goyal SK, Giri BC (2001) Recent trends in modeling of deteriorating inventory. Eur J Oper Res 134:1–16.

Graves SC (1982) The application of queueing theory to continuous perishable inventory systems. Manag Sci 28:400–406.

Gregor PJ, Forthofer RN, Kapadia AS (1982) Evaluation of inventory and transportation policies of a regional blood distribution system. Eur J Oper Res 10:106–113.

Grocery Manufacturers of America (2004) 2004 unsalables benchmark report. http://www.gmabrands.com/industryaffairs/docs/benchgma2004.pdf.

Gürler Ü, Özkaya BY (2003) A Note on “Continuous review perishable inventory systems: models and heuristics.” IIE Trans 35:321 – 323.

Gürler Ü, Özkaya BY (2006) (s, S) policy for perishables with random shelf life. Working Paper, Bilkent University, Ankara, Turkey.

Hahn KY, Hwang H, Shinn SW (2004) A returns policy for distribution channel coordination of perishable items,. Eur J Oper Res 152:770–780.

Haijema R, van der Wal J, van Dijk N (2005) Blood platelet production: A multi-type perishable inventory problem. Research Memorandum, FEE, University of Amsterdam.

Haijema R, van der Wal J, van Dijk N (2007) Blood platelet production: Optimization by dynamic programming and simulation, Comput Oper Res Special Issue OR Health Care 34:760–779.

Hemmelmayr V, Doerner KF, Hartl RF, Savelsbergh MWP (2006) Delivery strategies for blood products supplies. Working paper, School of Industrial and Systems Engineering, Georgia Institute of Technology, Atlanta, GA.

Ishii H (1993) Perishable inventory problem with two types of customers and different selling prices. J Oper Res Soc Jpn, 36:199–205.

Ishii H, Nose T (1996) Perishable inventory control with two types of customers and different selling prices under the warehouse capacity constraint. Int J Prod Econ 44:167–176.

Jagannathan R, Sen T (1991) Storing crossmatched blood: A perishable inventory model with prior allocation Manag Sci 37:251–266.

Jones PC, Lowe T, Traub RD, Keller G (2001) Matching supply and demand: The value of a second chance in producing hybrid seed corn. Manufact Serv Oper Manage 3:122–137.

Jones PC, Keller G, Lowe T, Traub RD (2003) Managing the seed-corn supply chain at syngenta, INTERFACES 33:80–90.

Kalpakam S, Arivarignan G (1988) A continuous review perishable inventory model. Statistics 19:389–398.

Kalpakam S, Sapna KP (1994) Continuous review (*s,S*) inventory system with random lifetimes and positive lead times. Oper Res Lett 16:115–119.

Kalpakam S, Sapna KP (1996) A lost sales (S-1,S) perishable inventory system with renewal demand. Naval Res Logist 43:129–142.

Kalpakam S, Shanthi S (2000) A perishable system with modified base stock policy and random supply quantity. *Comput Math with Appl* 39:79–89.

Kalpakam S, Shanthi S (2001) A perishable system with modified $(S-1, S)$ policy and arbitrary processing times. Comput Oper Res 28:453–471.

Kanchanasuntorn K, Techanitisawad A (2006) An approximate periodic model for fixed-life perishable products in A two-echelon inventory-distribution system. Int J Prod Econ 100:101–115.

Karmarkar U (1981) The multi-period multi-location inventory problem. Oper Res 29:215–228.

Kaspi H, Perry D (1983) Inventory systems of perishable commodities. Adv Appl Prob 15: 674–685.

Kaspi H, Perry D (1984) Inventory systems of perishable commodities with renewal input and poisson output. Adv Appl Prob 16:402–421.

Katsaliaki K, Brailsford SC (2007) Using simulation to improve the blood supply Chain. J Oper Res Soc 58:219–227.

Kazaz B (2004) Production planning under yield and demand uncertainty with yield-dependent cost and price. Manufact Serv Oper Manage 6:209–224.

Kendall KE, Lee SM (1980) Formulating blood rotation policies with multiple objectives. Manage Sci 26:1145–1157.

Ketzenberg ME, Ferguson M (2006) Managing slow moving perishables in the grocery industry. Working paper, College of Business, Colorado State University, Fort Collins, CO.

Konda C, Dave N, Hemachandra N, Rangaraj N (2003) Timing of discount offers for perishable inventories. In: Rao MR, Puri MC (eds) Proceedings of sixth Asia Pacific Operations Research Society. Allied Publishers, New Delhi.

Lian Z, Liu L (1999) A discrete-time model for perishable inventory systems. Ann Oper Res 87:103–116.

Lian Z, Liu L (2001) Continuous review perishable inventory systems: Models and heuristics. IIE Trans 33:809–822.

Lian Z, Liu L, Neuts M (2005) A discrete-time model for common lifetime inventory systems. Math Oper Res 30:718–732.

Lin C-WR, Chen H-YS (2003) Dynamic allocation of uncertain supply for the perishable commodity supply chain. Int J Prod Res 41:3119–3138.

Liu L (1990) (s, S) continuous review inventory models for inventory with random lifetimes. Oper Res Lett 9:161–169.

Liu L, Cheung KL (1997) Service constrained inventory models with random lifetimes and lead times. J Oper Res Soc 48:1022–1028.

Liu L, Lian Z (1999) (s, S) continuous review models for products with fixed lifetimes. Oper Res 47:150–158.

Liu L, Shi D (1999) An (s, S) Model for inventory with exponential lifetimes and renewal demands. Naval Res Logist 46:39–56.

Liu L, Yang T (1999) An (s, S) random lifetime inventory model with a positive lead time. Eur J Oper Res 113:52–63.

Lowe T, Preckel PV (2004) Decision technologies for agribusiness problems: A brief review of selected literature and a call for research. Manuf Serv Oper Manage 6:201–208.

Lütke Entrup M (2005) Advanced planning in fresh food industries: Integrating shelf life into production planning. Contributions to Management Science Series, Physica-Verlag, Heidelberg.

Lystad E, Ferguson M, Alexopoulos C (2006) Single stage heuristics for perishable inventory control in two-echelon supply chains. Working paper, The College of Management, Georgia Institute of Technology, Atlanta, GA.

Martin GE, (1986) An optimal decision model for disposal of perishable inventories. Int J Prod Res 24:73–80.

Moorthy AK, Narasimhulu YC, Basha IR (1992) On perishable inventory with Markov chain demand quantities. Int J Infor Manage Sci 3:29–37.

Muharremoglu A, Tsitsiklis J (2003) A single-unit decomposition approach to multi-echelon inventory systems. Working paper, Graduate School of Business, Columbia University, New York, NY.

Mustafee N, Katsaliaki K, Taylor SJE, Brailsford S (2006) Distributed simulation with cots simulation packages: A case study in health care supply chain simulation.In: Perrone LF, Wieland FP, Liu J, Lawson BG, Nicol DM, Fujimoto RM (eds) Proceedings of the 2006 Winter Simulation Conference, pp 1136–1142.

Nahmias S (1975a) Optimal ordering policies for perishable inventory-II. Oper Res 23:735–749.

Nahmias S (1975b) A comparison of alternative approximations for ordering perishable inventory. INFOR 13:175–184.

Nahmias S (1975c) On ordering perishable inventory under Erlang demand. Naval Res Logist Quar 22:415–425.

Nahmias S (1976) Myopic approximations for the perishable inventory problem. Manag Sci 22:1002–1008.

Nahmias S (1977a) Higher-order approximations for the perishable inventory problem. Oper Res 25:630–640.

Nahmias S (1977b) Comparison between two dynamic perishable inventory models. Ope Res 25:168–172.

Nahmias S (1977c) On ordering perishable inventory when both demand and lifetime are random. Manag Sci 24:82–90.

Nahmias S (1978) The fixed charge perishable inventory problem. Oper Res 26:464–481.

Nahmias S (1982) Perishable inventory theory: a review. Oper Res 30:680–708.

Nahmias S, Pierskalla W (1973) Optimal ordering policies for a product that perishes in two periods subject to stochastic demand. Naval Res Logist Quar 20:207–229.

Nahmias S, Pierskalla W (1976) A two product perishable/non-perishable inventory problem. SIAM J Appl Math 30:483–500.

Nahmias S, Wang SS (1979) A heuristic lot size reorder point model for decaying inventories. Manag Sci 25:90–97.

Nahmias S, Perry D, Stadje W (2004a) A perishable inventory systems with variable input and demand rates. Math Meth Oper Res 60:155–162.

Nahmias S, Perry D, Stadje W (2004b) Actuarial valuation of perishable inventory systems. Prob Eng Inform Sci 18:219–232.

Nandakumar P, Morton TE (1993) Near myopic heuristics for the fixed-life perishability problem. Manage Sci 39:1490–1498.

Nose T, Ishii T, Nishida T (1983) LIFO allocation problem for perishable commodities. J Oper Res Soc Jpn 26:135–146.

Or I, Pierskalla W (1979) A transportation location-allocation model for regional blood banking. AIIE Trans 11:86–95.

Pal M (1989) The $(s-1, S)$ inventory model for deteriorating items with exponential lead time. Calcutta Statist Assoc Bull 38:83–91.

Parlar M (1985) Optimal ordering policies for a perishable and substitutable product - A markov decision-model. INFOR, 23:182–195.

Perosio DJ, McLaughlin EW, Cuellar S, Park K (2001) FreshTrack 2001 supply chain management in the produce industry. Department of Agricultural, Resource, and Managerial Economics College of Agriculture and Life Sciences Cornell University, Ithaca, NY 14853.

Perry D (1997) A double band control policy of a Brownian perishable inventory system. Proba Eng Inform Sci 11:361–373.

Perry D, Posner M (1990) Control of input and demand rates in inventory systems of perishable commodities. Naval Res Logis 37:85–97.

Perry D, Posner M (1998) An $(S-1,S)$ inventory system with fixed shelf life and constant lead times. Oper Res 46:S65–S71.

Perry D, Stadje W (1999) Perishable inventory systems with impatient demands. Math Meth Oper Res 50:77–90.

Perry D, Stadje W (2000a) An inventory system for perishable items with by-products. Math Meth Oper Res 51:287–300.

Perry D, Stadje W (2000b) Inventory systems for goods with censored random lifetimes. Oper Res Lett 27:21–27.

Perry D, Stadje W (2001) Disasters in markovian inventory systems for perishable items. Adv Appl Prob 33:61–75.

Perry D, Stadje W (2006) A controlled M/G/1 workload process with an application to perishable inventory systems. Math Meth Oper Res 64:415–428.

Pierskalla WP (2004) Supply chain management of blood banks. In: Brandeau M, Sainfort F, Pierskalla WP (eds) Operations research and health care, a handbook of methods and applications Kluwer Academic Publishers, New York, pp 104–145.

Pierskalla WP, Roach CD (1972) Optimal issuing policies for perishable inventory. Manag Sci 18:603–614.

Prastacos GP (1978) Optimal myopic allocation of a product with fixed lifetime. J Oper Res Soc 29:905–913.

Prastacos GP (1979) LIFO distribution systems, J Oper Res Soc 30:539–546.

Prastacos GP (1981) Allocation of a perishable product inventory. Oper Res 29:95–107.

Prastacos GP (1984) Blood inventory management: An overview of theory and practice. Manag Sci 30:777–800.

Prastacos GP, Brodheim E (1980) PBDS: Decision support system for regional blood management.Manag Sci 26:451–463.

Popp W, Vollert H (1981) A comparison of decentralized and centralized disposition systems for red cell concentrates. In: Mohr JR, Kluge A (eds) The Computer and Blood Banking. Springer-Verlag, Lecture Notes on Medical Informatics, vol. 13, pp 141–154.

Raafat F (1991) Survey of literature on continuously deteriorating inventory models, J Oper Res Soc 42:27–37.

Ravichandran N (1995) Stochastic analysis of a continuous review perishable inventory system with positive lead time and poisson demand. Eur J Oper Res 84:444–457.

Scarf H (1960) The optimality of (s,S) policies in dynamic inventory problems. In: Arrow K, Karlin S, Suppes P (eds) Mathematical Models in the Social Sciences, Stanford University Press, Stanford, CA.

Schmidt CP, Nahmias S (1985) (S -1,S) policies for perishable inventory. Manage Sci 31:719–728.

Talluri K, van Ryzin G (2004) The Theory and Practice of Revenue Management. Kluwer Academic Publishers, Boston.

Tekin E, Gürler Ü, Berk E (2001) Age-based vs. stock-level control policies for a perishable inventory system. Eur J Oper Res, 134:309–329.

Tsiros M, Heilman CM (2005) The effect of expiration dates and perceived risk on purchasing behavior in grocery store perishable categories. J Marketing 69:114–129.

van der Vorst JGAJ, Beulens AJM, van Beek P (2000) Modelling and simulating multi-echelon food systems. Eur J Oper Res 122:354–366.

Van Zyl GJJ (1964) Inventory control for perishable commodities. PhD Thesis, University of North Carolina, Chapel Hill, NC.

Vaughan T (1994) A model of the perishable inventory system with reference to consumer- realized product expiration. J Oper Res Soc 45:519–528.

Veinott AF (1960) Optimal ordering, issuing and disposal of inventory with known demand. PhD Thesis, Columbia University, New York, NY.

Webb A (2006) Ripeness sticker takes the guesswork out of picking and eating. Albuquerque J Thursday, May 25, 2006, URL: http://www.abqjournal.com/biz/462732business05-25-06.htm.

Weiss H (1980) Optimal ordering policies for continuous review perishable inventory models Oper Res 28:365–374.

Williams CL, Patuwo BE (1999) A perishable inventory model with positive order lead times. Eur J Oper Res 116:352–373.

Williams CL, Patuwo BE (2004) Analysis of the effect of various unit costs on the optimal incoming quantity in a perishable inventory model. Eur J Oper Res 156:140–147.

Yang PC, Wee HM (2006) A collaborative inventory system with permissable delay in payment for deteriorating items. Math Comput Model 43:209–221.

Yen H (1975) Inventory management for a perishable Product: Multi-echelon system. PhD Thesis, Northwestern University, Evanston, IL.

Yi J (2003) Vehicle routing with time windows and time-dependent rewards: A problem from the American Red Cross. Manufact Serv Oper Manag 5:74–77.

Zhou D, Pierskalla W (2006) Perishable product inventory policies with emergency replenishments. Working paper, Chinese University of Hong Kong.

Chapter 16
Optimization Models of Production Planning Problems

Hubert Missbauer and Reha Uzsoy

16.1 Introduction

Mathematical programming formulations have been proposed for a wide range of production-related problems since the 1950s, addressing problems of long-term aggregate production planning, medium-term allocation of capacity to different products, lot sizing and product cycling, and detailed short-term production scheduling. The range of problems addressed by these methods spans a variety of managerial levels, problem environments, and time scales, often confusing even experienced practitioners and researchers as to what exactly is meant by the term "production planning." Hence, it is necessary to begin a chapter of this nature by specifying exactly what type of problems we plan to address, especially since an overview of all related areas is clearly beyond the scope of any single chapter. Indeed, we note in passing that there appears to have been no effort to collect all the production-related operations research literature in a single volume since the book by Johnson and Montgomery (1974), which remains an excellent reference for the basic concepts in the field.

In this chapter, we shall focus on manufacturing planning and control (MPC) systems for discrete parts manufacturing, where products are assembled from a variety of components, each of which in turn is produced by a multistage process. Manufacturing structures of this type occur in many industries, such as mechanical products, electrical appliances, electronics, and automotive manufacturing. Most facilities are not dedicated to specific products, and thus may require time-consuming changeovers when switching from one product type to another. Many insights in this chapter are also relevant for industries that only partly show these characteristics (e.g., semiconductor manufacturing). Process industries (e.g., steelmaking, continuous chemical processing or paper industry) often have substantially different

R. Uzsoy (✉)
Edwards P. Fitts Department of Industrial and Systems Engineering,
North Carolina State University, Raleigh, NC 27695-7906, USA
e-mail: ruzsoy@ncsu.edu

K.G. Kempf et al. (eds.), *Planning Production and Inventories in the Extended Enterprise*, 437
International Series in Operations Research & Management Science 151,
DOI 10.1007/978-1-4419-6485-4_16,

characteristics, so we will not address these cases. For a discussion of MPC systems in process industries, see Gunther and Van Beek (2003); for steel plants as a special case, see Tang et al. (2001) and Missbauer et al. (forthcoming).

For the types of manufacturing systems under consideration, a planning logic has evolved over the last 45 years starting from bill-of-material explosion and leading to the Material Requirements Planning (MRP) (Orlicky 1975) and Manufacturing Resource Planning (MRPII) systems (Wight 1983), which form the basis for most of the MPC and supply chain planning systems in industrial use today (Vollmann et al. 2005). Today's Advanced Planning and Scheduling (APS) Systems (Stadtler and Kilger 2008) aim at complementing MPC systems by concentrating on planning and coordinating the material flow between companies or manufacturing plants, leveraging the data collection and organization capabilities of the Enterprise Resource Planning (ERP) and Manufacturing Execution Systems (MES) used by many companies today. The chapter by Fordyce et al. in this volume gives an excellent view of such a system.

The basic problem of production planning in these environments is essentially one of matching supply to demand. This involves viewing the production system as a network of resource groups, which we shall refer to as work centers, and allocating the capacity of production resources at these work centers among different products over time, coordinating the associated inventories and raw material inputs so that known or predicted customer demand is met in the best possible manner. The "best possible manner," of course, requires more precise definition in order to form the basis of an optimization model, and can vary widely based on the specific production environment being considered. However, the objective functions are generally aimed at minimizing the total expected costs of production and inventories over the time horizon considered. There are also a number of models where demand is influenced by management decisions such as pricing, in which case the objective takes the form of profit maximization.

The decision variables emerging from the production planning activity depend on the decision structure of the MPC system, mainly on the extent to which detailed (mainly scheduling) decisions are made at the planning level. In the following, we assume that detailed scheduling decisions are made on the shop-floor level. This is the usual MPC structure and is described below in more detail. Given this structure of the MPC system, the basic decision variables emerging from the production planning activity are the amount of material (work orders) for each type of product that is to be released to each production resource over time, together with the required due dates. We shall focus on these decisions throughout this chapter since ultimately these decisions are the only actionable ones resulting from the production planning process. Estimates of a number of other quantities, such as production quantities at each production work center over time and inventory levels over time, are also obtained from production planning models, but we will take the position that all these ancillary quantities result from applying the work release decisions to the constraints defining the operation of the production resources (capacity constraints) and the material flows through the production system. We will assume that capacity determination over the long term, the domain of the classical literature on capacity expansion (represented by, e.g., Luss 1982) is not part of this problem domain.

Due to the complexity of the supply chain, the manufacturing process and the organization responsible for managing and coordinating them, the MPC task is usually structured hierarchically (Anthony 1966; Bitran et al. 1981, 1982; Bitran and Tirupati 1993; Hax and Candea 1984). Essentially two planning levels can be distinguished as follows:

- *Upper level (central MPC):* The complexity of the supply chain or manufacturing process is generally addressed by aggregating segments of the production process into departments or production units (Bertrand et al. 1990). This level then involves planning of the material flow over the entire logistic chain at an appropriate level of aggregation, without determining detailed schedules (production sequences) within the production units.
- *Lower level:* Detailed scheduling of the work orders within the production units, which involves determining the start and finish dates of the operations and their production sequence at the facilities.

Our specification of these levels follows the use of the terms "goods flow control" and "production unit control" in (Bertrand et al. 1990). Note that the upper level frequently is hierarchical in itself; see Vollmann, Berry et al. (2005), for its structure in today's MPC systems, and Bertrand et al. (1990) and de Kok and Fransoo (2003), for advanced planning architectures. Many of these concepts date back at least to the book by Anthony (1966), and an extensive literature has been developed around these issues (for conceptual issues of hierarchical production planning, see Schneeweiß 2003). When a hierarchical MPC system is designed, the data and the decision variables are aggregated with respect to products (aggregating products into product groups or families), capacity (aggregating resources to resource groups), and time (Stadtler 1996). Decisions taking place at a similar level of detail – hence often over similar time intervals or frequencies – are considered together in the same level of the hierarchy, and a mechanism is devised to propagate the implications of these decisions up and down the hierarchy. However, the two levels outlined above will be sufficient to motivate the work we wish to accomplish in this chapter.

The essential interface between the two planning levels is *order release*. The central MPC system coordinating the production units issues *work orders* that specify the particular product or component type to be produced, the amount to produce, and the due dates of the orders. Control over the work order then passes to the lower, detailed scheduling level within the production unit concerned. Clearly this approach requires *coordination norms* between the planning levels, which is most commonly implemented using *planned lead times*. The upper level creates its plans based on some assumptions as to the lead time, the time between the order being released into production and its being completed. The production unit must behave in a manner consistent with this assumption. The most common approach is for the upper level to assume a constant lead time for the released orders, which the production unit commits to meeting, at least as long as capacity utilization remains within reasonable limits. Thus the management of lead times by the production unit, and of the norms defining the upper level's view of the defined lead times, are of utmost importance.

In any hierarchical planning system, the upper level can evaluate its decision alternatives accurately only if it has access to a model that predicts the behavior of the system controlled by the lower level. Thus this *anticipation function* of the upper level (Schneeweiß 2003, p. 33 ff.) is a crucial element, allowing the upper level to anticipate the consequences of its decisions for the lower level(s). In our case, the central MPC system at the upper level needs a model that predicts the performance indicators of the manufacturing system (work in process inventory, flow time, due-date performance, etc.) resulting from an order release plan. *The formulation of mathematical programming models for the upper level of the MPC system described above that use a variety of different anticipation functions is the central topic of this chapter.*

Perhaps the most obvious consequence of shortcomings in the design of the upper level of the MPC system with respect to the anticipation of the dynamic behavior of the manufacturing system is insufficient management of lead times and work-in-process (WIP). In both Europe and the USA, a considerable body of work has approached these issues using the terminology of workload control. This type of approach usually implements the hierarchy discussed above, with the upper level (central MPC) performing order acceptance and deciding what demand will be served and which demand will be delayed in order to balance load and capacity in the medium term. The tradeoff between low flow times and high throughput is managed by the release of work into the production units over the short term based on some statistics reflecting the state of the production system.

In the following, we describe the workload control concept that provides the basis for Order Review and Release (ORR) systems that usually do not incorporate optimization models. In the subsequent sections we concentrate on optimization models for planning the aggregate material flow and order release in MPC systems.

16.2 Nonoptimization Workload Control

Since most of the MPC systems in industrial use today have largely evolved in practice (see McKay chapter in this volume) and have not been designed from an underlying theoretical basis, it is not surprising that their approaches to lead time management are often problematic. A well-known example is the *lead time syndrome* (Wight 1983, p. 108 ff.) that in extreme cases can inflate lead times beyond any arguable level. This can occur when lead time is considered to be a *forecast* variable, which is the usual practice in MRP systems: Lead times are estimated from realized values from the past, rather than being determined by the state of the production system. It is a crucial insight that lead times are *workload-dependent* and thus should be regarded as *control* variables (for this comparison, see Tatsiopoulos and Kingsman 1983): Lead times are determined mainly by the waiting times at the work centers which are determined by the utilization level of the resources, which are determined by the amount of released work. This insight, supported by a very extensive literature on queuing models of manufacturing systems (Hopp and Spearman 2001; Buzacott and Shanthikumar 1993), motivates the workload control (WLC) concept as a way to improve lead time management.

The workload control (WLC) concept considers flow times as output variables that can be controlled by the manner in which work is released into the shop over time. If the amount of work released (measured, e.g., in standard hours) is large, this leads to longer queues at the work centers and hence to longer waiting times and flow times. Reliable flow times can only be maintained if the amount of work released into the system and its output (determined by the available capacity) are balanced such that the queues of work in process inventory (WIP) at the work centers, usually measured in hours of work, are kept at a predefined level. This makes *order release* an essential decision function and a core component of WLC. Implementing any form of WLC thus requires the solution of two subproblems: (1) determining the target WIP level and (2) determining the release dates of orders, i.e., how work will be released into the shop over time.

The target WIP level must be a compromise between the goals of maintaining low WIP level and short flow times on the one hand and high output on the other. A high output requires an average level of WIP (queues at the work centers) that prevents idleness of the work centers, buffering against *variability* in the material flow between stations (see Hopp and Spearman 2001, p. 287 ff., for the "corrupting influence of variability"). This can be formalized as a functional relationship: Once the average WIP level is determined, the average flow time and the output (capacity utilization) are also determined, given the order and shop characteristics. The functional relationships between average WIP and other important performance indicators are usually expressed as *characteristic curves* that are often determined by simulation. Figure 16.1 shows an example. For this figure the manufacturing system has been simulated several times with different average levels of WIP in the shop. The average WIP levels were obtained by simulating the load-oriented order release approach of Wiendahl (1995) with different target WIP levels. The realized mean WIP values are shown on the x-axis, and the mean flow time and output on the y-axis.

It is important to note that the characteristic curves in Figure 16.1 are not completely determined by the technical properties of products and manufacturing system. They also depend on sequencing rules, lot sizes, capacity flexibility, order release frequency, etc., that is, they are a result of the long-term characteristics of the planning system. Thus WLC can be considered as a means of *complexity reduction*: the specification of long-term decision rules in the MPC system leads to stable operational characteristics of the manufacturing system it controls (expressed as characteristic curves), which, in turn, provide the basis for order release decisions that maintain the desired flow times. Consequently, WLC is an architecture for the entire MPC system (for conceptual issues, see Bertrand et al. 1990; Zäpfel and Missbauer 1993b; Missbauer 1998).

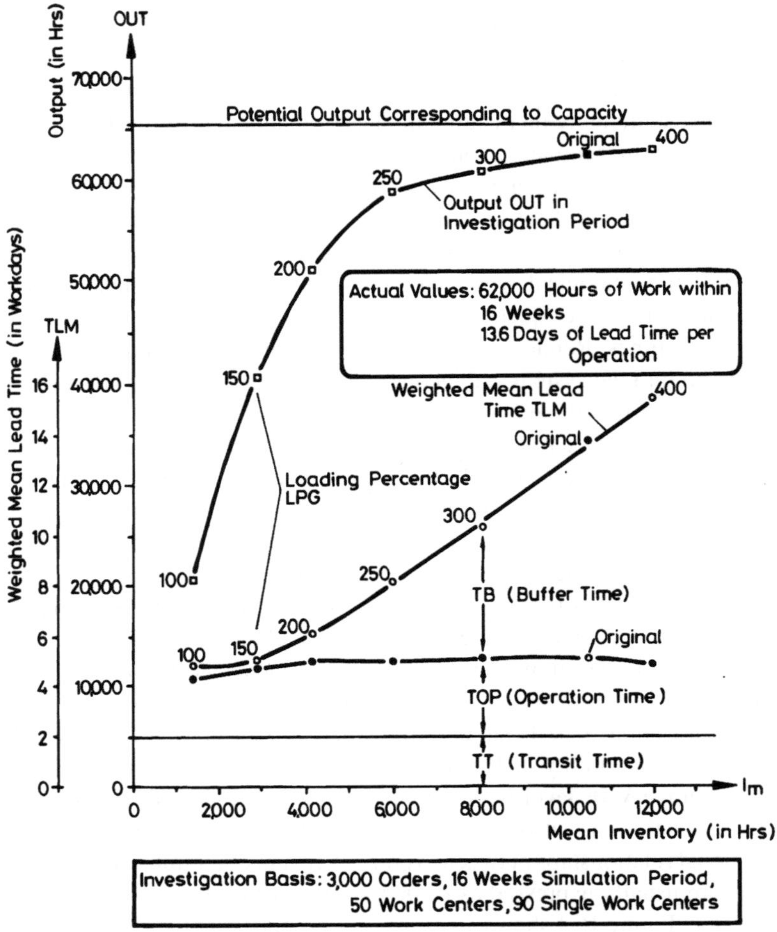

Fig. 16.1 Curves for output and mean flow time per operation with change in the work-in-process Wiendahl (1995, p. 246)

The decision problem of determining how to release work over time to maintain a predetermined target WIP level has been a research topic since the 1970s. Two main approaches can be distinguished as follows:

(a) WLC order release mechanisms that determine the work orders to release for a short planning horizon. We use the term "traditional WLC order release mechanisms" since this has been an essential part of WLC research in the last 20 years (Land 2004; Stevenson and Hendry 2006). We describe this general approach and its limitations in the following subsection.
(b) Order release planning procedures based on an explicit model of the material flow and the time-dependent WIP level over a longer planning horizon, usually divided into periods. This approach, its present state, and research topics are

described in Sects. 16.3 and 16.4. The focus of this chapter is very much on order release methods that are explicitly based on the WLC concept and that are applicable for complex product/process characteristics.

16.2.1 Traditional Order Release Methods Based on Workload Control

Based on the insights described above, a number of short-term order release mechanisms have been designed, mainly in the 1980s and 1990s. Examples are CONWIP (Spearman et al. 1990), load-oriented order release (Wiendahl 1995), and the order release stage of the LUMS approach (Hendry and Kingsman 1991; Stevenson and Hendry 2006) and the method of Bertrand and Wortmann (1981). In the following, we describe the general concept of these methods. Overviews can be found in Bergamaschi et al. (1997), Land (2004) and Fredendall et al. (2010). Kanban, which can also be considered as a Workload Control technique, is based on more restrictive assumptions and is not included here. Drum-Buffer-Rope is based on a more detailed schedule of the bottlenecks ("drum beat") and thus is different from the hierarchical MPC concept that provides the basis for this chapter (for a discussion of MPC concepts, see Zäpfel and Missbauer 1993a; for Drum-Buffer-Rope, see Cohen 1988; Gupta 2005).

The order release mechanisms described in this section usually are based on the following situation: The work orders in the MPC-system are generated from customer orders or from the MRP system from net requirements, lot sizing, and infinite capacity loading. Since in WLC order release is a decision function, the work orders are initially held in an *order pool* of unreleased orders. These orders are specified by product or component type, lot size, and required due date. A planned start date, derived from the required due date and a planned flow time that is consistent with the WIP norm, is usually available. Orders from the pool are released according to their planned start dates and the load situation in the shop. All orders should be finished on time, and the WIP norms should be maintained. When an order is released, control over the order is transferred to the scheduling level of the production unit, which has to meet the required due date.

We define a *traditional WLC order release mechanism* as follows (see also the basic release procedure described in Land (2004, p. 36 ff.)); we roughly follow the classification framework in (Bergamaschi et al. 1997). Order release is load limited, that is, a load limit is determined for each work center or for the shop as a whole, based on the characteristic curves of Fig. 16.1, and the order release mechanism prevents the load limits from being exceeded (in some cases with minor tolerance). The planning horizon is one period (in practice, usually 1 day to 1 week), and there is usually no order release plan for multiple periods. The capacities of the work centers are usually fixed, although some mechanisms consider capacity adjustments. The timing of order release may be event-driven in continuous time, or periodic at the beginning of each planning period; both options can be used simultaneously. An order is released if its release does not violate these workload norms, defined as upper and/or lower bounds. The subset of orders to be released within these constraints is usually selected heuristically (e.g., according to a priority following the urgency of the order), but integer programming can also be used, as in, for example, Irastorza and Deane (1974). When order release is performed, the feasibility of releasing the selected orders is checked against the load limits. This feasibility check requires the definition of how WIP is to be measured (e.g., in number of orders or in hours of work) and what WIP is to be considered in the decision (WIP for the entire shop, for each work center separately or just for the bottleneck work centers). Many of these issues of how to aggregate and measure WIP will also arise in our discussion of clearing functions in Sect. 16.5.

If a target WIP level is defined for individual work centers, then controlling the load at the work centers (*direct load*) is the most detailed technique. The future value of this direct load has to be estimated because the work input to the WIP level at a work center is controlled by sequencing decisions at the upstream work centers and is not known at the time of order release. Alternatively, only the *aggregate load* of the work centers, defined as the released work in the shop that has to be processed by a work center irrespective of its current position, is controlled, which avoids estimation of the order arrival patterns at the work centers.

When an order release mechanism is designed, the design options mentioned above must be specified. Figure 16.2 summarizes a well-known classification of these design options.

Once the order release mechanism has been designed by specifying the design options discussed above, its realized performance is determined by the parameter settings. The most important parameters are as follows:

- Target WIP (WIP norms) at the work centers. This parameter, which is closely related to the target value of the average flow time, can be expressed in different ways depending on the order release mechanism (e.g., load limit or load percentage in load-oriented order release) and specifies the compromise between output and WIP-flow time.
- A time limit that prevents the premature release of orders whose planned start date is far in the future. This parameter plays an important role in determining the extent of production smoothing (see below).

Other parameters are the length of the planning period and the order release frequency if order release is performed periodically (Perona and Portioli 1998).

Most of the order release mechanisms developed since 1980 follow this logic and can be obtained by specifying the design options above. We call these methods traditional because they share a common structure and have formed the mainstream of the research on WLC-based order release mechanisms in the last 25 years. They have also been tested extensively and are partly available in standard software. A stream of such methods and their interactions with shop-floor dispatching has been explored by a body of work focusing on the semiconductor industry,

Dimensions	Options
Order release mechanism	Load limited Time phased
Timing convention	Continuous Discrete
Workload measure	Number of Jobs Work quantity
Aggregation of workload measure	Total shop load Bottleneck load Load by each workcentre
Workload accounting over time	Atemporal Probabilistic Time bucketing
Workload Control	Upper bound only Lower bound only Upper and lower bounds Workload balancing
Capacity planning	Active Passive
Schedule visibility	Limited Extended

Fig. 16.2 Design options in traditional order release mechanisms (Bergamaschi et al. 1997)

as discussed in Uzsoy et al. (1994). For a comprehensive description of order review/release methods, which is beyond the scope of this chapter, see Land (2004), Bergamaschi et al. (1997) and Philipoom and Fry (1992).

Order release mechanisms of this type have a short planning horizon, but the release decisions determine the order arrivals at the work centers for a period that equals the flow times of the orders and also influence the options available to future release. Thus it is important to gain insight in how order release mechanisms work and how the goal of maintaining a stable WIP level can be achieved over a longer time horizon. Three topics are relevant here as follows:

- Order release determines the start dates of the orders and thus the extent of workload smoothing if (e.g., seasonal) demand variations occur.
- The workload norms must be consistent with the desired output, which in turn can depend on the extent of workload smoothing.
- Especially in the case of multiple products with different routings and resource requirements, the order release mechanisms should be able to perform load balancing among work centers. That is, the sequence of order releases should avoid temporary over- or underloading of certain routes or work centers. This is

essential for achieving high throughput and low WIP at the same time without just shifting the waiting time to the job pool.[1]

These tasks require the capability to look ahead beyond the short planning horizon of the release mechanisms. There are two ways to accomplish this as follows:

- Short-term release mechanisms can be complemented by a medium-term planning level that balances load and capacity and thus determines the required output over time. The release mechanisms perform short-term control that keeps WIP and shop flow times under control. This shifts much of the planning task to the medium-term level, and especially if flow times are long (e.g., semiconductor manufacturing with hundreds of operations per order), integration of medium-term planning and short-term order release can be difficult.
- The dynamic behavior of order release mechanisms and the material flow that results from release decisions can be controlled by the design of the release mechanisms and by the parameter setting (especially WIP norms and time limit). For instance, if load leveling over a longer horizon is desired, the time limit should be long, which allows early release of orders in periods with low demand. If utilization is low, the time limit should be short, because this prevents early release and completion of orders, etc. (see Zäpfel et al. 1992). Research, mainly by simulation, has accumulated a body of knowledge on decision rules that provide support for design and parameter setting of release mechanisms for specified material flow structure, demand pattern, etc. (for reviews, see Land 2004; Stevenson and Hendry 2006). Hence the traditional order release mechanisms can be regarded as a rule-based approach to optimize the material flow through a production unit as discussed in Missbauer (2009).

In the following sections we focus on optimization models that determine the optimal aggregate material flow through a production unit, assuming that demand forecasts are available at least at an aggregate level (groups of products). This optimized aggregate material flow forms the basis for order release that leads to this planned material flow. It can be expected that the potential of optimization models that determine order release is higher than the potential of release mechanisms that are controlled by appropriate parameter setting. If no reliable demand forecasts are available (especially in the case of customer order-driven production), order release planning might be difficult. In this case, the essential problem is to coordinate customer enquiry/order acceptance and order release/WIP control (for this topic, see Hendry and Kingsman 1991; Stevenson and Hendry 2006). The planning models described in the remainder of the chapter perform many of the functions of

---

[1] Simulation results indicate that traditional order release mechanisms can in fact reduce average *total* throughput time of orders (pool waiting time plus shop flow time) and do not just shift the waiting time from the shop to the pool, possibly increasing total flow time due to reduced shop capacity (for this critique, see Kanet 1988). A possible reason is the load balancing effect. A good order release mechanism limits the shop load, but it also aims at keeping the level of WIP at the target level for each work center, thus balancing the load among the work centers by releasing orders with different routing. (For simulation results on total throughput time reduction, see Land 2004.)

order release mechanisms (WIP control, load leveling, load balancing) and thus can potentially replace the release mechanisms to a large extent. However, depending on the level of detail of the models, implementation of the order release plans by means of a traditional release mechanism may still be necessary or useful. In this case, the traditional order review and release mechanisms discussed in this section will operate in the very short term as part of the lower, detailed scheduling level of the MPC hierarchy described in Sect. 16.1. When applied in this way, the release mechanisms take as input the quantity of work that is required to be released over some planning period, say a week or a month, and control releases to the shop floor based on the state of the system in close to real time. Applying multiperiod optimization models to determine optimal time-varying parameter settings for release mechanisms (Zäpfel and Missbauer 1993a) might be a serious alternative but remains largely a topic for future research.

In the following section, we give a brief historical review of the development of optimization models for aggregate material flow planning in MPC systems.

16.3 Historical Review

While production planning has obviously been executed in some form since the beginning of even craft production, the development of quantitative methods for these problems has surprisingly of recent origins. The chapter by McKay in this volume gives an overview of the historical development of production planning since the beginning of the Industrial Revolution. While the well-known work of Harris (1915) launched the area of inventory modeling, it was not until the 1950s that this became a major research area. Similarly, the use of optimization models for planning production over time has its origins in the work of Modigliani and Hohn (1955), although the roots of this work reach back to the activity-based economic models developed by economists such as Leontief and Koopmans (e.g., Koopmans et al. 1951). The area of sequencing and scheduling also had its pioneering papers in this period, notably those of Jackson (1955) on single machine scheduling problems and Manne (1960) on job shop scheduling. However, we will focus the discussion here on production planning, leaving detailed discussion of the extensive field of sequencing and scheduling to specialized volumes in that area (Pinedo 1995; Pinedo and Chao 2005; Parker 1995).

It is interesting to note that the basic formulations of most classical production planning-related problems were essentially in place by 1960, when the book by Holt et al. (1960) collecting their previous work appeared. Hence, it is worthwhile examining these early papers in some detail to gain insight into why these formulations developed the way they did, and what the implications of these are relative to the current situation.

The work of Modigliani and Hohn (1955) views the problem of production planning over time as that of trading off production costs against inventory holding costs. Production costs are assumed to be convex, with increasing marginal production costs, while inventory holding costs are approximated by the time average of the ending period inventories, leading to a linear cost function. The problem is formulated on discrete time periods, the cost function is assumed to be stationary over time, demand in each period is known with certainty, and no backlogging is allowed. The monotone increasing marginal production cost makes it more economical to meet periods of high demand by producing in prior periods of low demand and holding inventory, giving the basic tradeoff in the problem. The authors develop an optimal solution based on calculus that essentially identifies planning horizons, allowing the problem to be decomposed along the time horizon into subproblems consisting of a certain number of consecutive periods that can be solved independently. This approach forms the basis for the later work of Holt et al. (1955, 1956), which subsequently led to the HMMS book (Holt et al. 1960). It is also interesting that in Chap. 6 of their book they explicitly address the extension of their decision rules to an environment with uncertain demand, and show that under a quadratic objective function of the type they assume, the deterministic equivalent of the stochastic problem is achieved by using expected demand values in their deterministic rule, which is equivalent to assuming an unbiased demand forecasting procedure. This insight appears to have motivated the heavy focus on deterministic models, although the proof they provide is valid for the specific case of a quadratic objective function. An interesting discussion of this body of work is given by Singhal and Singhal (2007).

A number of interesting points emerge from this work. It is interesting that capacity constraints are not modeled; the implicit assumption appears to be that capacity can be varied in the short run, and the costs of doing this contribute to the increasing marginal cost of production. This discussion is made more explicit in the context of labor costs by Charnes et al. (1955). It is also interesting that while the cost function is explicitly built up from holding production and fixed costs independent of production volume there is no discussion of how one might actually estimate these costs from existing business records. Finally, the basic paradigm is that of modeling the physical flows in the problem – production and inventories – and assigning costs to these, rather than modeling the cash flows explicitly as a means of capturing the financial impact. This paper also seems to have motivated the idea that in problems over time, perfect information of the entire planning horizon is not necessary, but rather just the first few periods on a rolling horizon basis is quite close to optimality. This has led to a long stream of papers using these and related ideas, including the well-known dynamic lot sizing model of Wagner and Whitin (1958).

In the mid-1950s, it began to be realized that the emerging linear programming technology could be applied to production planning problems quite directly. In particular, researchers realized that models of the type addressed by Modigliani and Hohn (1955) and Holt et al. (1956) could be formulated as linear programs. The two principal papers that appear to have accomplished this independently of each other are Hanssmann and Hess (1960), whose title very much resonates with the HMMS work, and Manne (1957). Another notable early paper is that of Bowman (1956), which appears to be one of the earliest to identify the extensive presence of network structure in production planning models.

At this point, the principal characteristics of the mathematical programming models used for production planning problems were now in place. The decisions cover a planning horizon that is divided into a number of discrete time periods, each of which has an associated set of decision variables reflecting the decisions made in that period. The decision variables represent the physical flows of material through the different production resources; the objective function is generally that of minimizing the variable costs of production, inventories and backlogs over the planning horizon, and capacity constraints on the production resources in each period are satisfied at an aggregate level. In the following section we shall investigate the basic assumptions of these models in more detail, focusing on the manner in which they model the dynamics of capacitated production resources.

16.4 Optimization, Planning, and Work Release

It is interesting to note that although the decisions made in well-known optimization models are referred to almost uniformly as "production" decisions, in reality these models usually are work release models; in hierarchical manufacturing planning and control systems as described in Sect. 16.1, the only way they can be implemented is through the release of work orders with specified due dates into the production facility being modeled. As such, these models are closely related to the extensive stream of work on work release, order review/release (ORR), and workload control discussed in the previous section. They also address the basic problem of planning time-phased work releases addressed by the well-known and extensively implemented Material Requirements Planning (MRP) procedure (Vollmann et al. 1988; Baker 1993) and related techniques. Finally, it is worth noting that these problems have also been addressed in several research streams out of the artificial intelligence community (e.g., Smith 1993; Zweben and Fox 1994).

The vast majority of the mathematical programming models of interest to this chapter approach the problem in the same manner. The time horizon being considered is divided into discrete time periods, usually but not necessarily of the same length. Decision variables are associated with each period, and the objective is to minimize the total cost, which may be defined in different ways depending on the specific environment being considered, over the planning horizon. Following Hackman and Leachman (1989), we can view these models as containing three basic sets of constraints:

1. *Inventory or material balance equations*, which capture the flows of material through both space and time. These will also enforce the satisfaction of demand, which is viewed as a flow of material from the production system to an external demand source.
2. *Capacity constraints*, which model how the production activities capture and consume production resources.
3. *Domain-specific constraints* reflecting the special structure and requirements of the particular production environment being modeled.

The first two sets of constraints are critical to the accurate reflection of the actual behavior of the production system, which in turn is essential to the optimality and feasibility of the production plans obtained from the model. Following the discussion in Sect. 16.1, their aggregate nature requires these models to use some type of anticipation function that predicts the effect of their decisions on the detailed scheduling level of shop operation. As discussed in Sect. 16.1, the manner in which production lead times (also referred to as *cycle times* or *flow times*)[2] are treated is a crucial aspect of this anticipation function, affecting both the capacity and flow balance constraints above. We now examine several different models of lead times used in production planning models, starting from the simplest, and discuss their advantages and disadvantages.

16.4.1 Fixed Delays Independent of Workload

The simplest model of lead times that is encountered quite commonly in inventory models is one of instantaneous replenishment where the quantity ordered at a given point in time becomes available immediately upon ordering. The closest equivalent to this in the domain of mathematical programming models is to assume that lead times are approximately equal to one period, i.e., that the quantity of material R_t that is released into the system at the start of period t is available to meet demand at the end of that period. In order to ensure continuity of the solution between periods, we need to model the flows of material in and out of the finished goods inventory, which yields the system dynamics equations

$$I_t = I_{t-1} + R_t - D_t \tag{16.1}$$

where I_t denotes the finished goods available on hand at the end of period t and D_t the demand at the end of that period.

However, this is clearly often not realistic, and a more commonly encountered model in both inventory and the mathematical programming literature is a fixed, deterministic replenishment lead time that is independent of the quantity ordered or released. The stochastic equivalent of this model is a random lead time with a time-stationary probability distribution that is independent of the order quantities, such as the case treated by Eppen and Martin (1988), or the class of models discussed in Chap. 7 of Zipkin (1997). In this case, the amount Q_t ordered at the start of period t becomes available at the beginning of period $t + \tau$, where the integer τ denotes the fixed lead time. In a production system, the amount R_t released into the system at the start of period t becomes available for use at the start of period $t + \tau$. If we denote

[2] In general, we use the term *flow time* when we consider this time span from a manufacturing perspective, and the term *lead time* when it is considered from a planning perspective; see Hopp and Spearman (2001, p. 321). The terms are not always clearly distinguished in the literature.

the output of the production system in period t by X_t, we have the relationship $X_t = R_{t-\tau}$. The system dynamics are now described by the relationship

$$I_t = I_{t-1} + X_t - D_t = I_{t-1} + R_{t-\tau} - D_t. \quad (16.2)$$

We note in passing that this is exactly the model of lead times used in MRP in its backward scheduling phase. It is common both in the literature and in practice to assume that the fixed lead time τ corresponds to an integer number of planning periods; Hackman and Leachman (1989) present a straightforward method for managing lead times that correspond to a fractional number of planning periods, which we will discuss later.

The difficulty with this model in the context of production systems is that it assumes that there is no limit on the amount of material the system can produce in the given lead time. Hence, most optimization models of capacitated production systems will limit the total output of the system in a given period by imposing a constraint of the form $X_t \leq C_t$, where C_t denotes the maximum possible output of the production system in a given period. For exposition, let us assume that each unit produced requires one time unit of the resource, and the resource capacity is expressed in terms of time units available per planning period. There is now the question of reconciling the capacity constraint with the system dynamics constraint, which involves determining at what point in time releases into the system in period t occupy the capacity of the resource. Three logical constructions can be distinguished here: (1) "lag before" models that assume that the resource capacity is occupied at the end of the lead time of this operation, (2) "lag after" models where the resource capacity is occupied at the beginning of the lead time, and (3) models that allocate the production date (and the resource requirements) within the lead time and hence allow production smoothing within the lead time. For "lag before" and "lag after" models we refer to Hackman and Leachman (1989); models of type (3) are formulated in de Kok and Fransoo (2003). For the discussion of the treatment of WIP, in the following we describe the "lag before" models that assume that the releases R_t in period t occupy the resource capacity in the period that the output is produced, implying $X_t = min\{R_{t-\tau}, C_t\}$. This corresponds to the "lag before" models in Hackman and Leachman (1989).

We can thus describe the behavior of this system in a given period t with the set of constraints

$$\begin{aligned} X_t &= R_{t-\tau}, \\ I_t &= I_{t-1} + X_t - D_t, \\ X_t &\leq C_t. \end{aligned} \quad (16.3)$$

The first constraint is explicitly included for exposition; clearly in practice we would make the substitution to eliminate one of the two variables from the formulation.

Most common LP models for production planning will refer to "production" variables X_t, but these generally correspond to releases into the production system, with resource capacity being occupied τ periods after the release has taken place. Note that the amount of production that can take place in a given period is limited by both the capacity C_t and the amount of work available for processing, given by past releases $R_{t-\tau}$ per the first constraint. Hence the amount of WIP W_t available for the resource to process in period t is simply $R_{t-\tau}$. However, if we define WIP as the inventory literature defines on-order inventory, as orders that have been released but not yet completed, at the end of period t the production system in this model will have a WIP level of

$$W_t = \sum_{k=t-\tau+1}^{t} R_k - \sum_{k=t+1}^{t+\tau} X_k \tag{16.4}$$

units of product. It is interesting to note that this quantity does not generally appear in the constraints or objective functions of most common LP models of production planning, although as seen above it is not difficult to model; an exception is the model of Riaño et al. (2003). It is also interesting to note that only a portion of this WIP, given by $R_{t-\tau}$, is actually available to the resource for processing in period t.

The deficiency of this model is rooted in its treatment of WIP. Essentially it assumes that WIP will not accumulate in the system over time; the releases in period $t-\tau$ constitute the entire WIP available to the resource for processing in period t. The releases are implicitly constrained not to exceed the capacity, so the system is always able to process all its available WIP in a single period. The remainder of the WIP, given by

$$\sum_{k=t-\tau+2}^{t} R_k, \tag{16.5}$$

has no effect on the cycle time of the resource, which is always equal to the prespecified parameter τ, and, as far as this model of production capacity goes, is completely unrelated to the capacity C_t of the resource in a given period. All the lead time τ accomplishes is to delay the arrival of work at the resource after its release into the system; it does not describe the behavior of the resource itself, which is assumed in the capacity constraint to be able to process any amount of material up to the capacity limit C_t in a given period. This also explains an interesting anomaly with this type of model that positive dual prices for capacity constraints result only when the capacity is fully utilized. However, queuing models repeatedly show that system performance, especially as related to WIP levels and cycle times, often begins to degrade at utilizations substantially below 1, implying the existence of situations where even though a resource is not fully utilized, additional capacity at the resource might be beneficial to system performance, although adding that capacity would not necessarily be economically desirable.

To summarize this discussion, the conventional view of production capacity used in MRP and most mathematical programming models results in an LP of the following form:

$$\text{minimize} \sum_{t=1}^{t} (h_t I_t + c_t R_t) \tag{16.6}$$

subject to

$$I_t = I_{t-1} + R_{t-\tau} - D_t, \quad \text{for } t = 1, \ldots, T, \tag{16.7}$$

$$R_{t-\tau} \leq C_t, \quad \text{for } t = 1, \ldots, T, \tag{16.8}$$

$$R_t, I_t \geq 0, \quad \text{for all } t = 1, \ldots, T. \tag{16.9}$$

As pointed out by Hackman and Leachman (1989), most LP models encountered in practice will involve additional constraints specific to the application domain under study, but the model above represents the essentials of inventory balance between periods and aggregate capacity within periods. Note that because all rates are uniformly distributed over a planning period, ensuring nonnegative inventory levels at the boundaries between periods is sufficient to ensure inventory is nonnegative throughout the period.

Models that involve lot-sizing considerations may involve integer variables, but the basic view of system capacity and lead times is usually not very different from this. We have chosen a simple objective function, that of minimizing the sum of production and inventory holding costs over the planning horizon. Clearly far more elaborate objective functions are possible, but our emphasis in this chapter is on the representation of production capacity and system dynamics. Finally, we note that backlogging of any form is not allowed, again in service of our focus on the ability of different sets of constraints to accurately represent the actual capabilities of the production system.

We now examine some other approaches that have extended this basic model without fundamentally altering its treatment of lead times.

16.4.2 Formulations Based on Lead Time Distributions

The basic assumption of a fixed lead time equal to an integer number of planning periods has been relaxed by a number of authors. Leachman and his coworkers (Hackman and Leachman 1989; Hung and Leachman 1996) have proposed models where the lead time associated with each planning period can be a fractional number of planning periods. Their approach is essentially equivalent to assuming a deterministic lead time distribution for the input in period t, which specifies the fraction $w_{it\tau}$ of the amount of product i released in period t that will emerge as finished product in period τ i.e., will have a lead time of $\tau - t$. An alternative approach is to consider the $w_{it\tau}$ values as random variables whose probability distribution depends on the state of the production system. We shall begin our discussion with the workload-independent approach of Hackman and Leachman (1989), which has formed the basis for several industrial implementations in the semiconductor industry (Leachman 1993; Leachman et al. 1996) and a number of other refinements (Kim and Leachman 1994; Kim et al. 1996; Dessouky and Leachman 1997). A number of authors have taken related approaches based on the use of planned lead times (Spitter et al. 2005a, b). We shall then discuss the workload-dependent model of (Riaño 2003; Riaño et al. 2003), and some related approaches (Voss and Woodruff 2003; Lautenschläger and Stadtler 1998). The recent review paper by Pahl et al. (2005) is also a good source for some of this material, in addition to its discussion of the clearing function related methods presented in the next section.

16.4.3 Workload-Independent Lead Time Distributions

In order to present the basic approach to modeling workload-independent lead time distributions, we shall use the work of Hung and Leachman (1996) as the focus of the discussion, although the approach was originally proposed by Hackman and Leachman (1989). The basic formulation used is essentially the Step-Separated formulation of Leachman and Carmon (1992), which requires estimated lead times L_{ij} required for a lot of product i to reach operation j after being released into the plant. However, instead of fixed lead times that remain constant over the entire planning horizon, the authors associate values of the lead time parameters with the start of each planning period. In the following $t = 0$ is the start of period 1, $t = 1$ is the start of period 2, etc., that is, a time unit is the period length. The lead time parameters are defined as follows:

L_{ijt} Lead time required for a lot of product i to reach operation j if the lot reaches operation j at the end of period $t (=$ at time $t)$.

These lead times are allowed to take on fractional values. It is interesting to think how one might estimate these time-dependent lead time parameters in the absence of a production plan that defines the workload of the production resources over time. The iterative technique of Hung and Leachman (1996) discussed below addresses this issue directly.

Given these lead times, the loading of the production resource in period t is defined by releases occurring in the time interval $Q = [(t-1) - L_{i,j,t-1}, t - L_{ijt}]$, assuming planning period t starts at time $(t-1)$. There are two cases to consider here. In the first, simpler case, the time interval Q lies within a single planning period $\lceil (t-1) - L_{i,j,t-1} \rceil = \lceil t - L_{ijt} \rceil$ where the $\lceil x \rceil$ notation denotes the smallest integer greater than or equal to x. In this case, the proportional share of the amount released in period $\lceil (t-1) - L_{i,j,t-1} \rceil$ arrives at operation j in period t. This is consistent with the basic assumption of linear programming models that activity intensities are uniformly distributed over the planning period, as discussed by (Hackman and Leachman 1989). Hence the amount Y_{ijt} of product i loading resources at operation j in period t is given by

$$Y_{ijt} = \left(\frac{(t - L_{ijt}) - ((t-1) - L_{i,j,t-1})}{\Delta} \right) e_{ijt} R_{i,\lceil (t-1) - L_{i,j,t-1} \rceil} \tag{16.10}$$

where Δ denotes the period length (which we set to 1 by definition), e_{ijt} denotes the overall yield of product i from the start of the process to step j in period t. If, on the other hand, the time interval Q spans multiple planning periods, we allocate the load due to releases in that period in proportion to the fraction of that period's total duration included in the interval Q (again assuming uniform release rates within the planning periods). This yields

$$\begin{aligned} Y_{ijt} = &\left(\frac{(\lceil (t-1) - L_{i,j,t-1}) \rceil - ((t-1) - L_{i,j,t} - 1)}{\Delta} \right) e_{ijt} R_{i,\lceil (t-1) - L_{i,j,t-1} \rceil} \\ &+ \sum_{\tau = \lceil (t-1) - L_{i,j,t} - 1 \rceil + 1}^{\lceil t - L_{ijt-1} \rceil} e_{ij\tau} R_{i\tau} \\ &+ \left(\frac{(t - L_{ijt})}{\Delta} \lceil t - L_{ijt-1} \rceil \right) e_{ijt} R_i, \lceil t - L_{ijt} \rceil \end{aligned} \tag{16.11}$$

The operation of this approach is illustrated in Fig. 16.3, from Hung and Leachman (1996). The upper part of the figure shows the uniform release rates in each planning period, while the lower graph shows the resource loading that results from these releases arriving at the resource after the specified fixed lead times. Releases in periods 2 and 3 contribute to the work input in period 3 at the work center performing operation j corresponding to the first and the third term

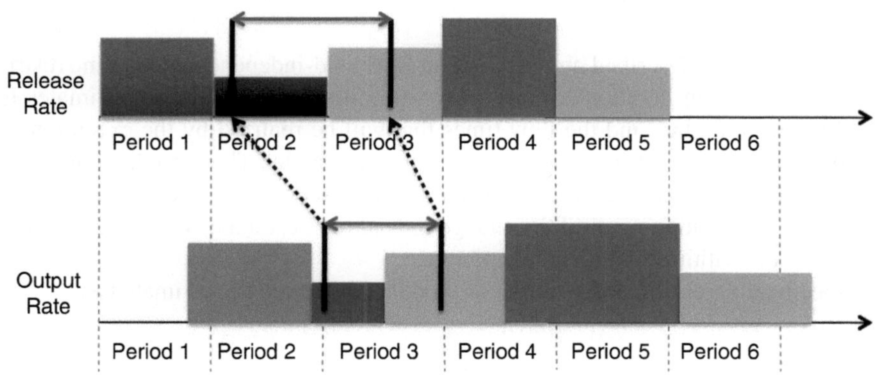

Fig. 16.3 Relationship between releases and loading with time-dependent lead times

in (16.11); the second term is not relevant here because the release interval Q only spans the two periods 2 and 3. Note that the lead times are associated with the boundary points between periods at the work center (not the boundary points between the release periods), and hence the lead time at the start of a period may not be the same as that at the end. The coloring indicates the correspondence between the releases and the arrival of the material at the resource.

Note that with this approach, we can write the output of product i in period t, denoted Y_{it}, as

$$Y_{it} = \sum_{\tau=1}^{T} R_{i\tau} w_{i\tau t} \tag{16.12}$$

where the $w_{i\tau t}$ values denote the fraction of releases in period τ that contribute to output in period t. This results in a linear constraint. Note that if we were able to obtain the $w_{i\tau t}$ values correctly, we would no longer need an explicit capacity constraint of the form

$$\sum_{i=1}^{N} a_{ij} Y_{it} \leq C_t \tag{16.13}$$

since the weights $w_{it\tau}$ would reflect the ability of the resource to produce output over time. However, queuing theory tells us that these weights depend on the resource utilizations, which are determined by both the WIP profile in the system and the releases that are determined by the planning model in use.

An obvious solution to this dilemma is to embed the LP models using these lead times in an iterative scheme where the releases obtained from the solution to the production planning models are fed into a simulation model of the production facility to evaluate the realized lead times they would impose on the production system. Such approaches have been suggested in the literature, which we shall discuss below.

16.4.4 Iterative Approaches

The formulations described above based on workload-independent lead time distributions suggest an iteration scheme where an initial set of lead time estimates is used to create a plan, and the flow times that will be realized by the execution of the plan are predicted using a simulation or queuing model. These predictions of the realized flow times are then used to generate a new set of lead time estimates, with the procedure continuing until the change in lead time estimates from one iteration to the next is within some specified tolerance.

Specifically, recall that the output of product i in period t is estimated as

$$Y_{it} = \sum_{\tau=1}^{T} R_{i\tau} w_{i\tau t} \tag{16.14}$$

where the $w_{i\tau t}$ values denote the fraction of releases in period τ that contribute to output in period t and are computed based on the lead time estimates L_{it}. At the k'th iteration of the procedure, lead time estimates L_{it}^k are used to compute weights $w_{i\tau t}^k$. The execution of the resulting plan is then simulated to obtain a new set of lead time estimates, and the procedure continues until convergence. It is worth noting in passing that similar iterative techniques have been used in the job shop scheduling literature to develop dispatching rules that consider the remaining time until completion of a job at an intermediate stage of processing (Vepsalainen and Morton 1987, 1988; Lu et al. 1994).

Hung and Leachman (1996) tested this iterative scheme using the workload-independent lead time distribution described above and a simulation model of a wafer fabrication facility. They examined the rate of convergence of the flow time estimates to the actual flow time values in the simulation, and found that convergence to the correct expected flow time values can be quite rapid, but that the procedure can fail to converge in some cases which are not fully understood. Irdem et al. (2008) present an experimental study of the convergence of the method, which suggests that for production systems at high utilization levels it can be difficult to confirm convergence. This approach has the advantage of combining two off-the-shelf modeling techniques, linear programming and simulation, that practitioners are likely to be familiar with, in an iterative scheme that addresses the complex interdependency of releases and lead times. However, the need for a simulation model of the facility being planned requires both large amounts of data to construct and validate, and also increases run time significantly. The authors discuss several ways to reduce the computational burden of the simulation by focusing on highly utilized work centers. Hung and Hou (2001) substitute an analytical flow time prediction model for the simulation in the iterative scheme, and report results that compare favorably in convergence performance with those of the scheme using the simulation model. Byrne and Bakir (1999) use an iterative technique (iterations between the production planning model and a simulation model) to determine realistic estimates for the available capacities, Byrne and Hossain (2005) present an extended production planning model within this framework. Kim and Kim (2001) simultaneously update flow times and available capacities using this iteration scheme.

A rather different iterative technique has been proposed by Riaño(Riaño 2003; Riaño et al. 2003), where the $w_{i\tau t}$ values are estimated using a model of the transient behavior of a queuing network. In order to present the basic idea of the approach, we shall consider its application to a work center consisting of a single server; the extension of this model to multiple stages and servers is discussed in Riaño et al. (2006).

Riaño's approach is to consider the system from a queuing perspective. A job released into the production system at time s will see $Q(s)$ jobs ahead of it in the queue or in process. Hence the flow time of that job will be given by

$$W(s) = \sum_{k=2}^{Q(s)} S_k + S_1, \tag{16.15}$$

where $S_k, k = 2, \ldots, Q(s)$ denote the processing times of jobs ahead of this job in the queue, and S_1 the residual (remaining) processing time of the job currently in process. The distribution function of the flow time of the job introduced into the system at time s is then given by

$$G(s,t) = \sum_{n=0}^{\infty} F_1 * F^{n*}(t) P\{Q(s) = n\}, \tag{16.16}$$

where F_1 denotes the distribution function of the residual processing time of the job currently in process, "$*$" denotes convolution, and F^{n*} the n-fold convolution of the processing time distribution F at the server. Note that $G(s,t)$ describes a state-dependent flow time distribution that depends on the number of jobs $Q(s)$ in the system at the time s the job was released. We wish to develop an approximation to this function that will allow us to calculate approximate values of the weights w_{st} that relate the input in the s-th interval with the cumulative output by time t. These weights can be used to estimate the output of the resource under a particular release pattern. To develop such an approximation, the author assumes that this time-dependent lead time distribution will have the same form as the steady-state distribution function of the waiting time for an $M/G/1$ queue, which is given by

$$(1-\rho) \sum_{n=0}^{\infty} \rho^n F_e^{n*}(t), \tag{16.17}$$

where F_e is the equilibrium residual processing time distribution, derived assuming that the time a new job enters the system is uniformly distributed over the service time. This suggests an approximation of the form

$$G(s,t) = F_1 * (1 - \beta(s)) \sum_{n=0}^{\infty} \beta(s)^n F_e^{n*}(t), \tag{16.18}$$

where $\beta(s)$ denotes a time-dependent traffic intensity. Noting that if we assume the service time distribution to be phase-type, then $G(s,t)$ will also be phase type, the author suggests heuristic estimates of $\beta(s)$, obtaining an approximation for $G(s,t)$ that depends only on the expected WIP level at time s, denoted by $\phi(s)$ and its time derivative $\phi'(s)$. Hence, to make the approximation to $G(s,t)$ we now need a viable technique for estimating $\phi(s)$ and $\phi'(s)$. This clearly depends on the pattern of releases into the production system, and so a recursive technique is used. Given a release pattern, we can compute estimates of $\phi(t)$ for every planning period t in a recursive manner, starting from period $t = 1$ and moving forward in time. If the processing time distribution at the server is phase-type (see, e.g., Neuts 1981), the author shows that these computations can be performed in an efficient manner. The resulting approximation to $G(s,t)$ yields approximate values of the w_{st}, which now correspond to the probability that a job released in period s will complete in period t. The author suggests a successive approximation method to compute the weights, where for a given release pattern the weights are first estimated and then a planning problem is solved to estimate WIP levels. These new WIP levels are used to estimate new weights until the estimates of weights converge.

The larger pattern of the iteration is now clear: we begin with an initial release pattern and calculate initial estimates of the w_{st}. We then calculate a new release pattern using these weights, and repeat until, hopefully, convergence is achieved. As with the approach of Hung and Leachman (1996), the convergence behavior of this procedure is not well understood; when it converges, it converges quite rapidly but in other cases it can cycle through a limited number of solutions. Further experimental and theoretical work is necessary to understand this convergence issue, but the overall approach stands as a very interesting and novel approach to modeling workload-dependent lead times in production planning, with a strong theoretical underpinning. Interesting discussions in this direction are given by Hackman (2008).

16.4.5 Workload-Dependent Lead Time Models

A major disadvantage of the approaches with fixed lead times or lead time distributions is their failure to consider the relationship between resource utilization and lead times. The iterative approaches described above attempt to restore this relationship by using a simulation or queuing model of the plant to estimate the effects of the planned releases on the lead times, but the underlying optimization formulation retains this basic flaw. A number of authors have developed models that allow lead times to vary according to the resource utilization, where the models include some mechanism that selects an appropriate lead time for each planning period based on the resource loading in that period.

An interesting area that is closely related to production planning but has not been widely explored in this context is that of dynamic traffic assignment models (Peeta and Ziliaskopoulos 2001). The objective of these models is to manage the routing of vehicles through a road network in order to optimize some measure of performance. Clearly, individual traffic links (road segments) are subject to congestion, and hence considerable effort has been devoted to developing formulations that capture this characteristic, as well as mathematical models of the relationship between the volume of flow on a traffic link and the velocity of that flow.

One way to model congestion in traffic links is *time–space links* (Carey and Subrahmanian 2000). If two nodes i and j of a traffic network are connected by a spatial link (e.g., a road segment), this two-node-network can be expanded over time, which yields a network of time–space nodes (see Fig. 16.4). A time–space link is a connection between time–space nodes. The flow through a time–space link represents the numbers of vehicles that pass the nodes at the respective times and hence require the respective traversal time.

If congestion is modeled as a link traversal time that increases with flow through the link, this can be represented as capacities of the time–space links leaving node i at time t (maximum flow on time–space link (i,t)) that depend on the flow through node i at time t, i.e., the inflow to the time–space links leaving node (i,t).

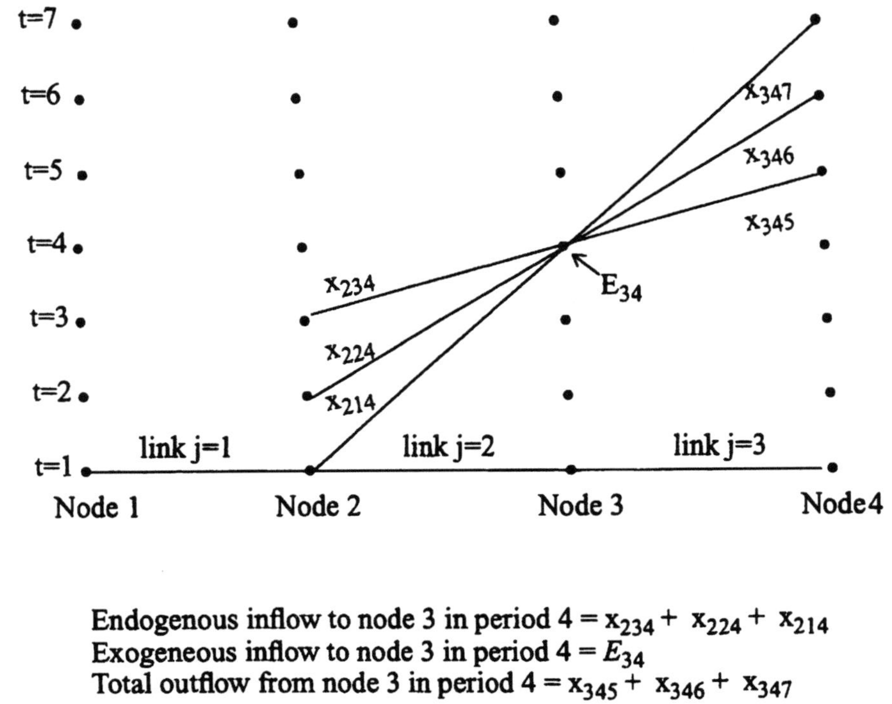

Fig. 16.4 Conservation of flows $x_{jt\tau}$ on a time expanded network (Carey and Subrahmanian 2000)

In the model of Carey and Subrahmanian (2000) the capacities of two time–space links leaving node (i, t) are usually greater than zero and the other time–space links are closed for the given inflow. As the inflow increases, the time–space links with positive capacities move to higher traversal times, implying a flow-dependent traversal time distribution that is stationary over time for a given inflow. The traversal time through a spatial link can be considered analogous to the flow time at a work center, and models with similar structure have been developed for order release planning in manufacturing.

A similar approach is described in Lautenschläger (1999). In order to consider load-dependent lead times for master production scheduling, the model determines the fraction of the planned production in a period t that has to be started one period ahead, in period $(t - 1)$. This fraction is a function of the planned utilization. Thus production on a resource can be performed in two modes, one with lead time of zero periods and the other with lead time of one period. The maximum production volumes that can take place in each mode are limited, which leads to a utilization-dependent lead time distribution.

Another related model has been proposed by Voss and Woodruff (2003). These authors assume a steady-state relationship between the utilization of a resource and the expected lead time at that resource. The basic idea is to discretize this curve and use integer variables to construct constraints that ensure that only one segment of the discretized curve is active in a given time period.

In order to implement this formulation, the relationship between utilization and expected lead time is evaluated at discrete utilization levels BP_r, $r = 1, \ldots, R$. Let L_r denote the expected lead time value associated with the r'th utilization level BP_r, i.e., the expected lead time of the resource is assumed to be L_r while the utilization level is between BP_r and BP_{r-1}. The authors suggest setting the breakpoints BP_r such that each lead time L_r corresponds to an integer number of periods. If we now define the fraction of the available capacity of the resource required for one unit of product j, $j = 1, \ldots, P$, as a_j, and R_{jt} to be the amount of product j released in period t, the workload (utilization) of the resource in period t is given by

$$\rho_t = \sum_{j=1}^{P} a_j R_{jt}. \tag{16.19}$$

We now define binary variables y_{tr} that select a particular lead time value L_r to be active in a given period t as follows:

$$L_t = \sum_{r=1}^{R} y_{tr} L_r, \quad \text{for all } t, \tag{16.20}$$

$$\sum_{r=1}^{R} y_{tr} = 1, \quad \text{for all } t. \tag{16.21}$$

Additional constraints of the form

$$\sum_{r=1}^{R} \mathrm{BP}_r y_{tr} \geq \sum_{j=1}^{P} a_j R_{jt}, \quad \text{for all } t \tag{16.22}$$

are required to ensure that the lead time selected is consistent with workload. In addition, for any given period t, we require $L_t - L_{t+1} \leq 1$, giving

$$\sum_{r=1}^{R} y_{t,r} L_r - \sum_{r=1}^{R} y_{t+1,r} L_r \leq 1, \quad \text{for all } t. \tag{16.23}$$

This latter constraint is interesting in that it restricts the decrease in lead time from one period to the next to at most one period to avoid overtaking, i.e., material released into the system at a later time emerging before material released earlier. Similar difficulties arise in dynamic traffic assignment problems (e.g., Carey and Subrahmanian 2000).

To complete the formulation, the authors present an objective function that includes an explicit holding cost for WIP, based on Little's Law (see, e.g., Hopp and Spearman 2001), leading to the term

$$\sum_{t=1}^{T} \sum_{j=1}^{P} h_{jt} \sum_{r=1}^{R} y_{tr} L_r R_{jt}. \tag{16.24}$$

This objective function is now nonlinear due to the product of the y_{tr} and R_{jt}, leading to a formulation that is computationally hard to solve.

16.4.6 Discussion of Models Based on Lead Time Distributions

While these models address the load-dependent nature of lead times directly, there are several shortcomings of these models:

- All the models described above assume the existence of a well-defined relationship between the workload or utilization of a resource and the expected lead time of that resource in that period. However, given that the planning models assume discrete planning periods of a fixed length and that the releases of work into the resource are varying over time, it is quite possible that the lead time incurred by work released in a given period may deviate quite substantially from that suggested by a long-run steady-state average relationship. The work of Riaño (2003) is a significant exception, explicitly addressing the transient nature of the queues involved, and thus merits further study.
- If the amount of work released decreases sharply from period t to period $t + 1$, the estimated lead time for the orders can decrease by more than one period from t to $t + 1$, implying overtaking (Voss and Woodruff 2003, p. 165; Carey and Subrahmanian 2000). This is unlikely to occur in practice and indicates that the models can lead to unrealistic results. Voss and Woodruff (2003) add a constraint that keeps the lead time from decreasing by more than one time bucket from t to $t + 1$, but this excludes decision alternatives and is not satisfactory from a theoretical point of view.

A number of researchers have proposed alternative approaches to these problems by developing formulations that do not consider the relationship between lead times and resource utilization explicitly, but instead use a relationship between the expected WIP level of a resource and its expected output in a given planning period. These clearing function based models will be discussed in the next section.

16.5 WIP-Based Models

The models discussed in the previous section all approach the problem of modeling the behavior of the production resource by computing a distribution of the lead times, a relationship between the time the work is released into the facility and the time it becomes available for consumption by the next stage. The distinguishing feature of this approach is the presence of a set of constraints implementing a lead time distribution used across different time periods, specifically in the material balance constraints. We now turn our attention to models where the lead time behavior of the production resources is not represented in the balance constraints, but by introducing nonlinear terms in the constraints or the objective function. The former class of models introduce nonlinear constraints (that may be linearized for computational purposes) representing the relationship between some measure of the expected WIP level (including jobs in queue and in process) in front of the production resource in question over a planning period and the expected output of the resource over the planning period. These formulations, which we shall term *Clearing Function* models, are discussed extensively in the next section, and generally result in models with linear objective functions and convex nonlinear constraint sets. The latter, which we shall term *WIP* Cost formulations, use queuing analysis to develop an expression for the expected WIP holding cost which is then added to the objective function of the model. These formulations, by contrast, tend to yield models with convex nonlinear objective functions.

A fundamental difference between these models and those discussed up to this point is their explicit representation of WIP. In the LP formulations discussed until this point, although it is possible to recapture planned WIP levels as the difference between cumulative releases and cumulative output, the WIP level has no effect on the behavior of the production resource. The exception to this is the transient queuing-based approach of Riaño (Riaño 2003; Riaño et al. 2003) and the respective simulation-based approaches, but even in this case the WIP enters the formulation only through its role in determining the lead time distribution. The WIP-based models, on the other hand, explicitly represent the WIP level in front of the resource in a planning period with distinct material balance equations, in addition to the balance equations for finished goods inventory (FGI) included in the conventional LP formulations.

We believe that this explicit distinction between WIP and FGI leads to substantially richer models, since the two types of inventory serve different purposes and are controlled in different ways. If we anticipate a seasonal surge in demand in the future, our production planning model must ensure that we have sufficient FGI in place in time to meet the demand. Thus, FGI levels can be planned, and are an output of the planning process. On the other hand, the WIP levels are a consequence of our release decisions, and determine to a great extent the flow time and throughput performance of the production resource. WIP levels must be managed to ensure timely production at minimum cost, while FGI must be planned to ensure effective satisfaction of demand at minimum cost. Thus the explicit separation and modeling of these two different types of inventory offers the potential for significantly richer production planning models. The separation of WIP and FGI also has implications for developing production planning models that consider uncertainties, since in practice some of the functions of safety stocks can be assumed by WIP; we shall discuss this aspect in more detail in Sect. 16.8.

The models described in this section explicitly represent the flow of WIP through a production unit. They differ from the previous models in that they relate the output from a resource in a given planning period to some measure of the WIP level at the resource during the period. Thus the nonlinear relationships between decision variables arising from the presence of congestion are represented in the constraints rather than in the objective function. We first describe the generic structure common to most models of this type and then concentrate on the technique used to incorporate the relationship between WIP, flow time, and output.

In models of this type, work centers are represented explicitly, and the material flow from order release through the required work centers and to the inventory of the final products or SKU's is represented by inventory balance equations. As in all models considered until now, material is modeled as a continuous medium, comparable to a fluid. Thus the approach is similar to fluid approximation in queuing theory (Kleinrock 1976, p. 60; Chen and Mandelbaum 1991) and to fluid relaxation in job-shop scheduling (Bertsimas and Sethuraman 2002; note that these fluid relaxation models are in continuous time and are usually deterministic). The planning horizon is again divided into discrete planning periods.

A variety of authors have discussed the relationship between system throughput and WIP levels. This is usually in the context of queuing analysis, where the quantities being studied are the long-run steady-state expected throughput rate and WIP levels. An example of this work is that of Agnew (1976), who studies this type of behavior in the context of optimal control policies. Spearman (1991) presents an analytic congestion model that describes a clearing function for closed production systems with processing time distributions with increasing failure rates. Standard texts on queuing models of manufacturing systems, such as Buzacott and Shanthikumar (1993), can be used to derive the clearing function – if not analytically, then at least numerically. Hopp and Spearman (2001) provide a number of illustrations of clearing functions for a variety of systems; for example, the relationship between WIP and throughput given in their practical worst case analysis represents a particular type of clearing function. Srinivasan et al. (1988) derive the clearing function for a closed queuing network with a product form solution. It is interesting to note that the concept of the clearing function has much in common with concepts developed in the dynamic traffic assignment literature. One such concept is that of the exit function, which defines the output of a traffic link in a time period as a function of the amount of traffic on the link at the start of the period (Merchant and Nemhauser 1978a, b; Carey 1987).

While the basic concept of a clearing function is quite intuitive, defining and implementing this concept in a rigorous and theoretically consistent manner is subject to some quite subtle difficulties that are not straightforward to resolve. Hence we will first discuss how optimization models of production planning problems may be formulated using clearing functions, assuming a valid clearing function can be generated. We shall then discuss the issues involved in estimating the clearing functions themselves.

16.5.1 Clearing Function Formulations

In its usual form the clearing function yields the expected aggregate output of a work center (e.g., hours of work, aggregated over the products) as a function of a suitable measure of WIP, aggregated over the products. This WIP measure can be the average WIP of the period, the average WIP over a longer time (e.g., two periods), or the total available work of the period (termed *load*, defined as initial WIP plus input during the period). It is commonly assumed, based on both theoretical and empirical arguments discussed in the next section, that the clearing function is a concave nondecreasing function of the WIP measure used to define it. Our discussion in this section will closely follow the development by Asmundsson et al. (2009).

To illustrate this approach, we begin by extending the single-product single-stage formulation of Karmarkar (1989) to multiple products, which can be stated using the following notation:

X_{it} = number of units of item i produced in period t,
R_{it} = number of units of item i released into the stage at the beginning of period t,
W_{it} = number of units of item i in WIP inventory at the end of period t,
$\hat{W}_{it}$ = WIP measure used for the clearing function and defined in separate constraint (e.g., $\hat{W}_{it} = W_{i,t-1} + R_{it}$),
I_{it} = number of units of item i in finished goods inventory (FGI) at the end of period t,
ξ_{it} = time required to produce one unit of item i at the resource.

Let $f_t(\hat{W})$ denote the clearing function that represents the resource in period t, with W denoting the WIP measured in units of time (i.e., $W_t = \sum_i \xi_{it}\hat{W}_{it}$), and D_{it} the demand for item i (in units) in period t. Then a naive extension of Karmarkar (1989)'s single-product formulation to multiple products is

$$\min \sum_t (\phi_{it}X_{it} + \omega_{it}W_{it} + \pi_{it}I_{it} + \rho_{it}R_{it}), \tag{16.25}$$

subject to

$$W_{it} = W_{i,t-1} - X_{it} + R_{it}, \quad \text{for all } i,t, \tag{16.26}$$

$$I_{it} = I_{i,t-1} + X_{it} - D_{it}, \quad \text{for all } i,t, \tag{16.27}$$

$$\sum_i \xi_{it}X_{it} \le f_t\left(\sum_i \xi_{it}\hat{W}_{it}\right), \quad \text{for all } t, \tag{16.28}$$

$$X_{it}, W_{it}, I_{it}, R_{it} \ge 0, \quad \text{for all } i,t, \tag{16.29}$$

where $\phi_{it}, \omega_{it}, \pi_{it}$, and ρ_{it} denote the unit cost coefficients of production, WIP holding, finished goods inventory holding, and releases (raw materials) respectively, and ξ_{it} the amount of the resource (machine time) required to produce one unit of product i in period t. Note that the argument of the clearing function in constraint (16.28) is the total WIP level over all products i expressed in units of time (or, equivalently, workload). The first two sets of constraints enforce flow conservation for WIP and FGI separately. Since the formulation distinguishes between WIP and FGI, flow conservation constraints are required for both. Constraints (16.28) represent the capacity constraint. Another interesting characteristic is that lead times do not appear in the formulation; they are represented implicitly by the nonlinear capacity constraints (16.28).

While this formulation appears intuitive, it can create significant modeling problems when applied in multiple product environments. Consider a situation where the system produces two products A and B, which consume capacity at the production resource in different amounts. The capacity constraint can be expressed as $X_A + X_B \le f(\hat{W}_A + \hat{W}_B)$.

A solution with

$$X_A > 0, X_B = 0, \hat{W}_A = 0, \hat{W}_B > 0$$

may exist, despite the fact there is no WIP in the system that can be converted into finished product A. Hence the optimal solution to this formulation can maintain high WIP levels of the product for which it is cheapest to do so, using the capacity generated by this device (i.e., the high value of the clearing function attained by holding high WIP of the cheap product) to hold very low or no WIP of all other products. An alternative way of expressing this difficulty is that there is no link between the mix of WIP available in the period and the output mix during the period.

Asmundsson and his coauthors (Asmundsson et al. 2006, 2009) propose an approach in which the no-passing requirement is enforced on average rather than at the level of individual jobs. To do this, they assume that the mix of output will reflect the mix of WIP. This is equivalent to assuming a service discipline at the queue representing the production resource where no product is given priority over another. After some analysis described in detail in Asmundsson et al. (2009) this yields the following formulation:

$$\min \sum_t (\phi_{it} X_{it} + \omega_{it} W_{it} + \pi_{it} I_{it} + \rho_{it} R_{it}), \tag{16.30}$$

subject to

$$W_{it} = W_{i,t-1} - X_{it} + R_{it}, \quad \text{for all } i, t, \tag{16.31}$$

$$I_{it} = I_{i,t-1} + X_{it} - D_{it}, \quad \text{for all } i, t, \tag{16.32}$$

$$\xi_{it} X_{it} \le Z_{it} f_t \left(\frac{\xi_{it} \hat{W}_{it}}{Z_{it}} \right), \quad \text{for all } i \text{ and } t, \tag{16.33}$$

$$\sum_i Z_{it} = 1, \quad \text{for all } t, \tag{16.34}$$

$$X_{it}, \mathrm{W}_{it}, I_{it}, R_{it}, Z_{it} \ge 0, \quad \text{for all } i, t. \tag{16.35}$$

This formulation will be referred to as the *Allocated Clearing Function* (ACF) model. The Z_{it} variables denote the fraction of the maximum possible output defined by the clearing function allocated to product i in period t. The intuition is that we wish to obtain a constraint that links the production of a given product to the

WIP level of that product alone, but the clearing function is defined in terms of the total WIP at the resource. This is, of course, what causes the difficulty described above: a high total WIP may result in a high maximum output level for the resource in a period, and the model may allocate this output in a manner that violates continuity of material flow. Assuming that all products at a resource will see the same expected lead time allows us to estimate the total WIP at the resource by the expression given as the argument of the clearing function in (16.33). The Z_{it} variables thus serve the dual purposes of scaling up the WIP for product i inside the parentheses of (16.33) to obtain a surrogate for total WIP on which the clearing function can operate, and then computing a fractional capacity for product i by multiplying the results. Asmundsson et al. (2009) prove that the total production of individual products will not exceed that suggested by the aggregate clearing function as suggested by (16.28).

The above formulation is a convex nonlinear program, due to the concave nature of the clearing function on the right hand side of constraints (16.33). However, an interesting and useful consequence of the partitioned formulation above arises when the concave clearing function is approximated using outer linearization. Since we assume the clearing functions are concave, they can be approximated by the convex hull of a set of affine functions of the form

$$\alpha^c \sum_i \xi_{it} \hat{W}_{it} + \beta^c \tag{16.36}$$

as

$$f(\hat{W}_t) = \min_c \{\alpha^c \hat{W}_t + \beta^c\}. \tag{16.37}$$

The $c = 1, \ldots, C$ index represents the individual line segments used in the approximation. In order to represent the concave clearing functions appropriately, we shall assume that the slopes of the line segments are monotonically decreasing, i.e.,

$$\alpha^1 > \alpha^2 > \ldots > \alpha^c = 0. \tag{16.38}$$

The slope of the last segment is set to zero to indicate that the maximum throughput capacity of the node has been reached, and adding WIP cannot increase throughput any further. To ensure that production cannot take place without some WIP being present, we impose the condition $\beta_1 = 0$ to ensure that the first line segment will pass through the origin. The capacity constraint in the CF formulation can now be replaced by the set of linear inequalities

$$\sum_i \xi_{it} X_{it} \leq \alpha^c \hat{W}_t + \beta^c, \quad \text{for all } c \text{ and } t. \tag{16.39}$$

Using this outer linearization to approximate the PCF model yields the following LP:

$$\min \sum_t \sum_i (\phi_{it} X_{it} + \omega_{it} W_{it} + h_{it} I_{it} + \rho_{it} R_{it}), \tag{16.40}$$

subject to

$$W_{it} = W_{i,t-1} - X_{it} + R_{it}, \quad \text{for all } i \text{ and } t, \tag{16.41}$$

$$I_{it} = I_{i,t-1} + X_{it} - D_{it}, \quad \text{for all } i \text{ and } t, \tag{16.42}$$

$$\xi_{it} X_{it} \leq \alpha^c \xi_{it} \hat{W}_{it} + Z_{it} \beta^c, \quad \text{for all } i, t, \text{ and }, c, \tag{16.43}$$

$$\sum_i Z_{it} = 1, \quad \text{for all } t, \tag{16.44}$$

$$Z_{it}, X_{it}, W_{it}, I_{it} \geq 0, \quad \text{for all } i \text{ and } t. \tag{16.45}$$

Notice that summing the set of constraints (16.43) over all i gives constraint (16.39), guaranteeing that the original constraint is satisfied. A consequence of the partitioning of the clearing function is that the clearing function constraint becomes linear, even with the Z-variables that were originally in the denominator since

$$Z_{it} f\left(\frac{\xi_{it}\hat{W}_{it}}{Z_{it}}\right) = Z_{it} \min_c \left\{\alpha^c \frac{\xi_{it}\hat{W}_{it}}{Z_{it}} + \beta^c\right\} = \min_c \{\alpha^c \xi_{it} \hat{W}_{it} + \beta^c Z_{it}\}. \tag{16.46}$$

In their experimental implementation, Asmundsson et al. (2006) assume the clearing function depends on the average WIP level over the time period, which they approximate as

$$\hat{W}_{it} = \frac{1}{2}(W_{i,t-1} + W_{it}).$$

16.5.2 Extensions to Multistage Systems

So far the clearing function formulations have been presented for a single-stage production system. In order to extend this to a multistage system, a number of extensions must be included. Since the model requires explicit modeling of WIP at each stage in order to compute the clearing function at each stage, we must represent the movement of material between stages; the output of one stage flows into the WIP of another. Another worthwhile enhancement is to distinguish between work centers that are potential bottlenecks and those that are unlikely to encounter significant congestion phenomena. The former can be represented explicitly using clearing functions, while the latter can be modeled using some form of workload independent delay, such as a fixed lead time, or a higher order delay as suggested by (Missbauer 2002a). While we assume in this model that groups of similar products are aggregated into product groups of families, the basic structure of the formulation remains the same if products are modeled individually, although, of course, a much larger formulation may result. We use the following notation:

Variables:

W_{jmt} – Work of product group j waiting at work center m at the end of period t.
I_{jt} – Finished goods inventory of product group j at the end of period t.
R_{jt} – Released work of product group j in period t.
X_{jmt} – Output of product group j from work center m in period t.

Note that for exposition we avoid the constants ξ of the above formulation and measure the variables in units of time (e.g., hours of work). Hence I_{jt} and R_{jt} are measured in hours of work, and W_{jmt} and X_{jmt} are measured in hours of work at work center m.
Parameters:

D_{jt} – Demand for product group j in period t (measured in hours of work).
C_{mt} – Capacity of work center m in period t.
$\tilde{p}_{jim}$– Average amount of work arriving at work center m when one unit of product group j is finished at work center i.
$z_{jim\tau}$ – Proportion of the output of product group j from work center i to work center m that arrives at m in τ periods after completion at i. These are not included to capture congestion-related lead times, but rather to represent delays such as shipping or flow through non-bottleneck work centers that are not subject to significant congestion effects, similar to those modeled by Hackman and Leachman (1989).

As for the single stage models, this formulation requires separate flow balance equations for WIP and FGI at each stage

$$W_{jmt} = W_{j,m,t-1} + \sum_{i=1}^{M}\sum_{\tau=0}^{\infty} X_{j,i,t-\tau}\tilde{p}_{jim}z_{jim\tau} + \sum_{\tau=0}^{\infty} R_{j,t-\tau}\tilde{p}_{j0m}z_{j0m\tau} - X_{jmt}, \quad \text{for all } j,m,t. \tag{16.47}$$

These constraints model the flow of WIP at the bottleneck work centers. The sources of input are the output of the other work centers and the release of orders. The work input can be delayed by transportation, non-bottlenecks (see below), etc. Work center $i = 0$ denotes the beginning of the line where order release takes place. The finished goods inventory is then represented as

$$I_{jt} = I_{j,t-1} + \sum_{m=1}^{M}\sum_{\tau=0}^{\infty} X_{j,m,t-\tau}\tilde{p}_{jm0}z_{jm0\tau} - D_{jt}, \quad \text{for all } j \text{ and } t. \tag{16.48}$$

Note that inflows into FGI can also be delayed after completing their last production operation. Work center index 0 in p and z denotes the completion of the product.

Constraints (16.47) and (16.48), together with an appropriate objective function, the clearing function-based capacity constraints (16.43) for the bottleneck work centers and the partitioning constraints (16.44), constitute a complete clearing function-based formulation for a multiproduct, multistage production system. Asmundsson et al. (2006, 2009) give a slightly different formulation that is based on Hackman and Leachman (1989), where an additional set of decision variables is added to represent material transfer between stages; note the formulation above assumes that material transfer to the next stage begins directly upon completion of processing at the current stage.

The computational results obtained using the clearing function formulations are quite promising, although more experimentation is clearly needed to be able to draw strong conclusions. Probably the most complete studies of these formulations at present are those of Asmundsson et al. (2006, 2009). The later study examines the performance of clearing function formulations in simple serial production systems, and concludes that when the clearing function is correctly estimated the clearing function formulations produce production plans that are much more aligned with the ability of the plant to execute them compared to models with fixed lead times. This study describes a systematic procedure for estimating the clearing functions from simulation data and fitting to the functional form (16.58) below using a nonlinear optimization algorithm. Asmundsson et al. (2006) compare the performance of the clearing function formulations to that of a conventional LP model in a reentrant line derived from a semiconductor wafer fabrication facility, and find that even when a simple visual technique is used to fit the clearing functions to the data, the CF models yield significantly better on time delivery than the LP models with fixed lead times.

16.5.3 Estimation of Clearing Functions

The reader will have noted that until now we have described formulations that make use of the clearing function concept but have not discussed how the clearing functions are estimated. In this section we provide a more formal definition of clearing functions and discuss this very important issue.

A *clearing function*, introduced by Graves (1986), can be defined as either the expected or maximum output of a work center (a relatively homogeneous group of production resources scheduled as a unit) in period t as a function of some measure of WIP (e.g., average WIP or planned available work) in period t and the maximum capacity of the work center C_{it}. In the following discussion we shall use as the WIP measure the *load* of work center i in period t, denoted by A_{it}, defined as the WIP at the beginning of period t plus the planned work release R_{it} in period t[3]. Thus, the load (total available work) at work center i in period t is given by

[3] This assumes a production unit consisting of a single work center, which we assume for exposition. For a multistage system as in Sect. 16.5.2, the work release R_{it} is replaced by the work input from release and from the other work centers as in (16.47).

$$\Lambda_{it} = W_{i,t-1} + R_{it}. \tag{16.49}$$

The clearing function for work center i is then a functional relationship of the form

$$X_{it} = f_i(W_{i,t-1} + R_{it}; C_{it}) \tag{16.50}$$

Note that in conventional LP models, only the maximum capacity C_{it} would be considered in a capacity constraint.

This approach has been followed by the majority of researchers proposing models of this type (e.g., Karmarkar 1989; Missbauer 2002a; Selçuk et al. 2007). The clearing function models the impact of the fact that the *actual* load (available work) is a random variable at the time of planning and thus can be lower than the planned value, and work arriving during the period can arrive later than expected and thus cannot be processed during the period. Uzsoy and his coworkers (Asmundsson et al. 2006; Hwang and Uzsoy 2005) have adopted a slightly different approach where the clearing function is defined as a function of the expected, time–average WIP level during the planning period (see Sect. 16.5.1). While this difference in approaches requires some modifications to the details of the formulations and the procedures used to estimate the clearing functions, the basic structure of the approach remains the same.

Figure 16.5, derived from Karmarkar (1989), depicts several examples of clearing functions considered in the literature to date, where X denotes the expected throughput in a planning period. The "constant proportion" linear clearing function of Graves (1986) and Parrish (1987) allows unlimited output in a planning period, but ensures fixed lead time. This type of clearing function is not generally applicable to order release models, because it can yield capacity-infeasible output levels at high levels of WIP. Alternatively, it requires an assumption that the production rate of resources can be managed such that the fixed lead time is always maintained.

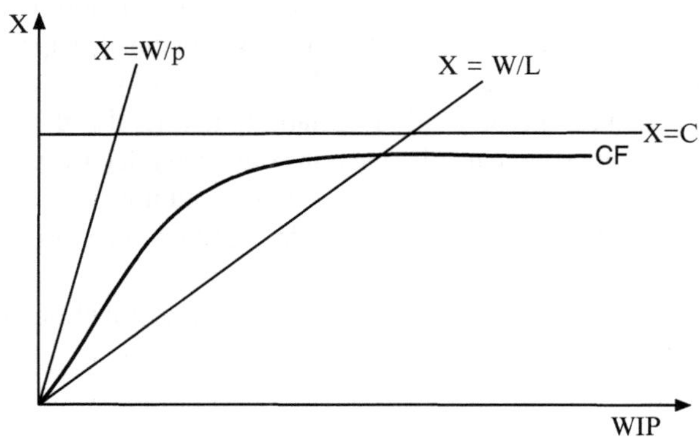

Fig. 16.5 Examples of clearing functions (from Karmarkar 1989)

However, by linking production rate to WIP level, it differs from the fixed delays used in most LP models, where the output of a production process is simply the input shifted forward in time by the fixed lead time. Orcun et al. (2006) illustrate the differences in the transient behavior of the production system under this and several other clearing function models.

The horizontal line $X = C$ corresponds to a fixed upper bound on output over the period, but without a lead-time constraint it implies that production can occur without any WIP in the system if work input and production are synchronized. This is reflected in the independence of output from the WIP level, which may constrain throughput to a level below the upper bound by starving the resource. This approach is implemented in, for example, the MRP-C approach of Tardif and Spearman (1997) and most LP approaches such as that of Hackman and Leachman (1989), but is supplemented with a fixed lead time as described in Sect. 16.3. The linear clearing function of Graves (1986) is represented by the $X = W/L$ line, which implies a lead time of L periods that is maintained independently of the WIP level. Note that if WIP and output are measured in the same time units (e.g., hours of work), the slope of the proportional part of the function is $1/L$, where L is the average lead time. However, as seen in the figure, this model may suggest infeasible output levels when WIP levels are high. If a fixed lead time is maintained up to a certain maximum output, we have the relationship $X = \min\{W/L, C\}$. When the parameters of the Graves clearing function are set such that the lead time is equal to the average processing time, with no queuing delays at all, we obtain the line $X = W/p$, where p denotes the average processing time. Assuming that lead time is equal to the average processing time up to a maximum output level gives the "Best Case" model $X = \min\{W/p, C\}$ described in Chap. 7 of Hopp and Spearman (2001). It is important to note that the workload-independent fixed lead time discussed in Sect. 16.4 differs from the linear model of Graves in that the former does not link output to WIP, while the latter does. Orcun et al. (2006) illustrate the differences between these models using system dynamics simulations.

It is apparent from the figure that the clearing function always lies below the $X = W/p$ and $X = C$ lines. For most capacitated production resources subject to congestion, limited capacity leads to a saturating (concave) shape of the clearing function. It is important to note that the nonlinear shape of the clearing function is not purely due to the presence of random variability in arrival and service processes at the production resource but can arise even in completely deterministic capacitated production systems, as shown in Karmarkar (1993). A number of recent papers (e.g., Asmundsson et al. 2009; Selçuk 2007) provide analytical support for the concave shape of the clearing function (see Sect. 16.5.3.1).

We now discuss techniques for estimating the clearing function associated with a set of production resources.

16.5.3.1 Analytical Approaches

A common approach to estimating clearing functions is to derive them using steady-state queuing analysis. It can be shown that for the $M/G/1$ model in steady state, the average throughput $E(X)$ is related to the expected WIP level $E(W)$ as follows:

$$E(X) = C \cdot \frac{E(W)}{E(W) + k} \tag{16.51}$$

where

$$k = \frac{\mu\sigma^2}{2} + \frac{1}{2\mu} \tag{16.52}$$

Here $1/\mu, \sigma^2$ denote the mean and variance of the processing time distribution and C the maximum capacity per period of the resource in hours of work. This is the same functional form as in Karmarkar (1989) (16.57 below), but (16.57) relates the output of the resource in period t to its load in period t, and the functional form of (16.57) is not supported by queuing models. (Missbauer 2002a) shows that for the $M/G/1$ model in equilibrium, the expected output $E[X_t]$ and expected load $E[\Lambda_t]$ of a work center are related as follows:

$$\begin{aligned} E[X_t] = {} & \frac{1}{2}(C + k + E[\Lambda_t] \\ & -\sqrt{C^2 + 2Ck + k^2 - 2C\,E[\Lambda_t] + 2k\,E[\Lambda_t] + E[\Lambda_t]^2}. \end{aligned} \tag{16.53}$$

The parameter k can be calculated analytically using (16.52) for the $M/G/1$ model, but can also be determined from empirical data or from simulation results.

For single-server work centers it seems reasonable to use (16.53) as a regression function with k and possibly also C as parameters. This need not be the case for multiple-server work centers, because in this case the first derivative of the average flow time with respect to the average WIP level is very low for low levels of WIP, where servers can be expected to be idle and the average waiting time is close to zero; this shape can be seen in Fig. 16.1). In this region the clearing function is nearly linear; the average flow time being insensitive to WIP implies a linear clearing function according to Little's Law (see Graves 1986 for the proof for the case of a discrete-time clearing function), which is not covered by (16.53). If the clearing function is approximated by piecewise linearization as discussed in the previous section, this is not a big problem.

The clearing function formulation (16.53) is derived from a steady-state model, which is a severe limitation. Selçuk (2007) derive a clearing function for a production resource with exponentially distributed service times, assuming the work available for the period is available at the time it is required for processing, without requiring steady state. They prove that this clearing function, which they refer to as a short-term nonlinear clearing function, is concave in the resource load defined above. Asmundsson et al. (2009) generalize this result. The general problem of determining a theoretically consistent clearing function for a single period without steady-state assumption for stochastic arrival and departure processes is largely unsolved and an important research topic, which we discuss further in Sect. 16.5.4.

Based on this discussion, we will be interested in clearing functions of the form $X_{it} \leq f_i(\Lambda_{it})$ with the following properties:

$$f_i(\Lambda_{it}) \leq \Lambda_{it} \quad \text{for } \Lambda_{it} \geq 0 \tag{16.54}$$

$$\frac{\mathrm{d} f_i(\Lambda_{it})}{\mathrm{d}\Lambda_{it}} \geq 0, \quad \text{for} \Lambda_{it} \geq 0, \tag{16.55}$$

$$\lim_{\Lambda_{it} \to \infty} f_i(\Lambda_{it}) = C_{it}. \tag{16.56}$$

The use of nonlinear clearing functions requires consideration of the optimization technique used to obtain solutions to the resulting formulations. Piecewise linearization is frequently used (Missbauer 1998, 2002; Asmundsson et al. 2006). If the clearing function is nonlinear and concave, models of a reasonable size can be solved by nonlinear programming since they generally result in formulations with convex constraints and objective functions. Hwang and Uzsoy (2005) discuss a clearing function model of this type including lot sizing, while Srinivasan et al. (1988) discuss possible solution techniques for their model that involves nonlinear clearing functions. Continuous-time models that assume a deterministic flow at the work centers (fluid relaxation) can be solved either by exact methods (e.g., for the makespan objective, see Bertsimas and Sethuraman (2002) or heuristically (e.g., for the holding cost objective; see Bertsimas et al. (2003). Continuous-time models are not considered because the integration of clearing functions into these models has not yet been studied and remains a topic for future research.

Empirical work to date (e.g., Asmundsson et al. 2009) suggests that optimization models of aggregate material flow can be quite sensitive to the properties of the clearing function, especially when at high WIP levels the maximum output decreases. This is often encountered in modeling traffic flows, where it takes the form of a flow-density relation as seen in Fig. 16.6 that relates the density of the traffic on the link to the flow velocity of traffic through the link. In our production planning context, density corresponds to WIP and flow velocity to output rate. The chapter in this handbook by Armbruster and Lefeber presents a number of models where this type of approach is used to develop continuum models of flow through a manufacturing system in a manner analogous to that used to study traffic flows.

In manufacturing systems a decrease in output at high WIP levels can occur in two cases: (1) when workers work less efficiently under high pressure or (2) when long queues, and hence long average flow times, threaten due-date performance and require preemptive sequencing rules in order to pull ahead urgent orders reducing capacity due to additional setups. This can lead to solutions that must be considered infeasible if an appropriate arrival rate control policy is not applied. Van Ooijen (1996, p. 139 ff., especially pp. 144–146) describe the effect on system behavior, while Van Ooijen and Bertrand (2003) present an arrival rate control policy for this case. Haxholdt et al. (2003) demonstrate the possibility of oscillating behavior and chaos under more sophisticated assumptions on the arrival and departure

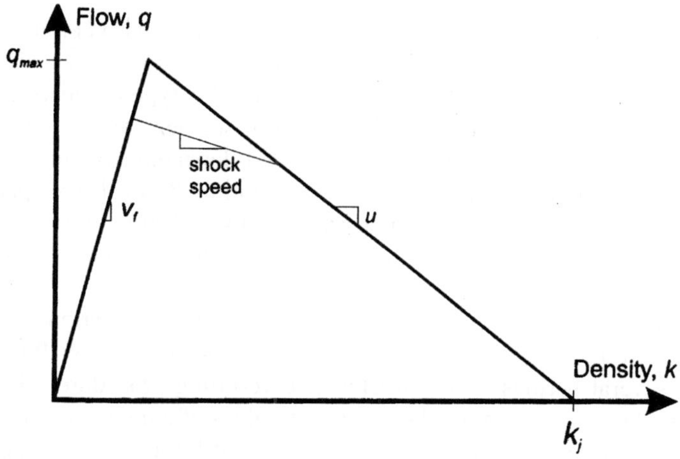

Fig. 16.6 Flow-density-relation (Cassidy 2003, p. 183)

process of queuing systems. The chapter by Elmaghraby in this volume also discusses possible shapes of clearing functions that may apply in this type of situation. Therefore, the shape of the clearing function has to be considered carefully in every case. In particular, a service rate that decreases if the level of WIP exceeds a certain threshold value can have extreme consequences on the system behavior, so this possibility has to be excluded or considered carefully when the clearing function is estimated. A literature review on the relationship between worker behavior and inventory level, especially in flow lines, can be found in Powell and Schultz (2004).

16.5.3.2 Empirical Estimation Techniques

A number of authors have suggested an empirical approach to estimating clearing functions, where a functional form with the desired properties is postulated, and then fit to data obtained either from an industrial facility or a simulation model using some form of regression analysis. Karmarkar (1989) uses the following functional form for the clearing function:

$$X_t = \min\left[C_t \frac{\Lambda_t}{\Lambda_t + k}; \Lambda_t\right], \tag{16.57}$$

where X_t denotes the output in period t, Λ_t the load of the resource at the start of period t, and C_t the maximum capacity of the resource available in period t. The shape parameter k is estimated by the user. The functional form of (16.53) above is an adaptation to models in discrete time. Srinivasan et al. (1988) suggest an alternative functional form

$$f(\Lambda_t) = C_t(1 - e^{-k\Lambda_t}) \tag{16.58}$$

where k is again a user-estimated shape parameter. Asmundsson et al. (2009) use this latter functional form, and give an extensive discussion of various issues in collecting simulation data for the purpose of fitting this type of clearing function. Asmundsson et al. (2006) use a visual fit of linear segments to simulation data to develop a clearing function formulation for a scaled-down semiconductor wafer fabrication facility with unreliable equipment and reentrant flows. There appears to be very little published literature using industrial data to fit clearing functions: the only paper we are aware of is Fine and Graves (1989), which motivated Graves' work on linear clearing functions.

If a saturating clearing function can be assumed, the estimation of the clearing function from empirical or simulated data (combinations of WIP or load and output for several periods) is essentially a curve-fitting procedure. The problem can be formulated as estimating the parameter values of a nonlinear function such as (16.58). If the clearing function is approximated by a set of N tangents, the parameters of the tangents can be derived from a nonlinear regression function (Missbauer 1998, p. 410 ff.), obtained directly from the observed data by numerical methods (Missbauer 1998, p. 407 f.) or by visual methods of curve fitting (Asmundsson et al. 2006). Estimating the clearing function can be difficult if the data include effects of machine downtimes. In this case, for some periods the average WIP is high and the output is low (because the work center has been down and the WIP could not be processed), and a simple curve fitting procedure would be misleading. Some sample data generated by simulation is shown in Fig. 16.7, which plots total throughput in a period against the average WIP in the period.

Asmundsson et al. (2006) describe a way to correctly estimate the clearing function in this case using multiple replications of simulation experiments. However, even in this case intuitive approaches can give poor results.

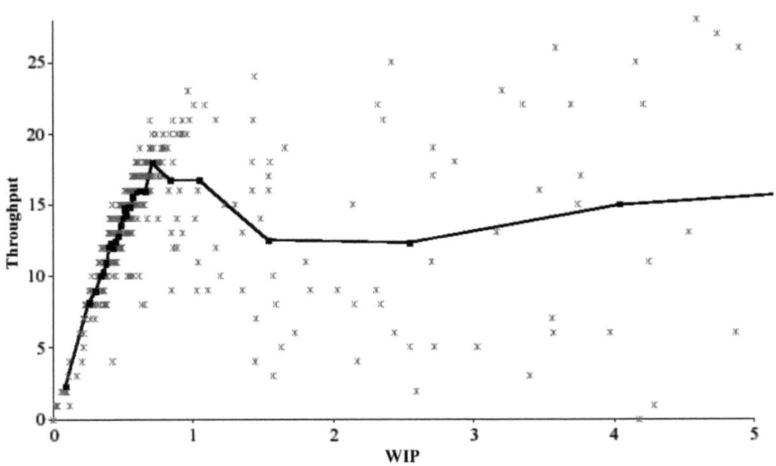

Fig. 16.7 Curve fitting in the case of machine downtimes – throughout vs. average WIP (Asmundsson et al. 2006)

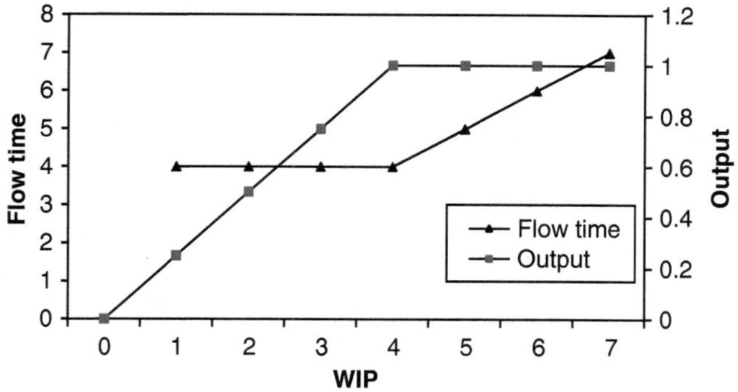

Fig. 16.8 Ideal clearing function for a paced flow line with four work centers (following Hopp and Spearman 2001, p. 221ff.)

An alternative way to determine the clearing function is the following: It is reasonable to assume that for very low WIP levels the clearing function is linear, and beyond a certain WIP level (that acts as buffer against variability) the work center can operate at full capacity. In an idealized situation, such as a deterministic paced flow line with equal processing times at each work center, the clearing function consists of these two parts (see Fig. 16.8).

In most practical cases the output will be lower for a certain range of WIP levels, and the problem of estimating the clearing function can be viewed as that of estimating the deviation of the clearing function from the ideal shape. Methods for this estimation are described in Nyhuis and Wiendahl (2003, p. 61 ff). Selçuk (2007) follows a similar approach and controls the shape of the clearing function by a parameter that reflects whether a more optimistic (overestimating throughput for a given WIP level) or conservative approach is applied (p. 115 f.).

The clearing function models described so far can become quite large if the number of work centers and product groups are large. If piecewise linearization of the clearing function is used, the number of linear constraints can be very high. If a reduction in model size is necessary, the non-bottleneck work centers can be eliminated from the model if a linear clearing function can be assumed for these work centers. In this case the non-bottlenecks can be represented as load-independent delay distributions (for the mathematical proof, see Missbauer 1998, p. 267 ff). This technique is similar to the delay functions in system dynamics models (Forrester 1962, p. 86 ff.).

16.5.4 Limitations of Clearing Function Models

It is evident from the last section that if the model of a clearing function as defined above is accepted, there are a number of open questions for future research. In this section, we examine the limitations of a clearing function that relates the expected or maximum output of period t to the planned WIP or load in period t. As discussed above two different approaches have been used to estimate these functions. Empirical methods postulate a particular functional form having the "right" properties and fitting a curve of this form to data obtained from historical observations of the production system or from simulation. Other researchers have used steady-state queuing models to derive closed-form expressions for clearing functions. Both these approaches implicitly assume that the form of the clearing function can be treated as invariant over at least a range of system operating conditions; in empirical methods, over the range of environmental conditions represented in the data set used to fit the clearing function, while in steady-state queuing methods, over the entire life of the system. Another way of phrasing this problem is that both these approaches to estimating clearing functions produce a clearing function that maps the expected value of one random variable describing the WIP level (at the start of the period, or the expected WIP level over the period) to the expected value of another, the total output of the resource in a given period, and this relationship is assumed to be time-invariant.

The two different approaches suffer from different difficulties under this paradigm. When empirical methods are used, it is assumed that the clearing function that is obtained from empirical or simulated data as the estimated functional relationship between the *actual* WIP and output is a valid estimation of the functional relationship between the *planned* WIP and expected or maximum output in the context of the planning model (Missbauer, forthcoming). The possible problems resulting from this are still not well understood today.

Next, there is a classical sampling issue – we assume that the fitted clearing function will be able to represent the behavior of the system under conditions not encountered in the data sets from which the function was generated. In addition, experimental work to date has shown that fitting clearing functions to empirical data is by no means a straightforward exercise. Asmundsson et al. (2009) present a detailed procedure for estimating clearing functions from simulation output, and apply their procedure to a production system with significant machine failures. Their approach consists of three basic stages: collect the simulation data, fit a functional form to this data using least-squares regression, and then piecewise linearize the resulting concave function using a nonlinear optimization model to minimize the deviation of the linearized model from the original concave function. An example of the data obtained in this experiment is shown in Fig. 16.9 below.

The profusion of points on the WIP axis, denoting periods in which WIP was present but the machine was unable to produce output due to being down, results in a least-squares approach giving a poor fit. (Note that the least-squares approach yields the mean value of the output conditional on the WIP level; see Davidson and MacKinnon 1993, p. 41.) Inspection of this figure, specifically the point at which the fitted line reaches its maximum, suggests that the fitted function significantly underestimates the amount of WIP required for the resource to achieve its maximum output as approximately 50 units as opposed to a reality of about 400 units.

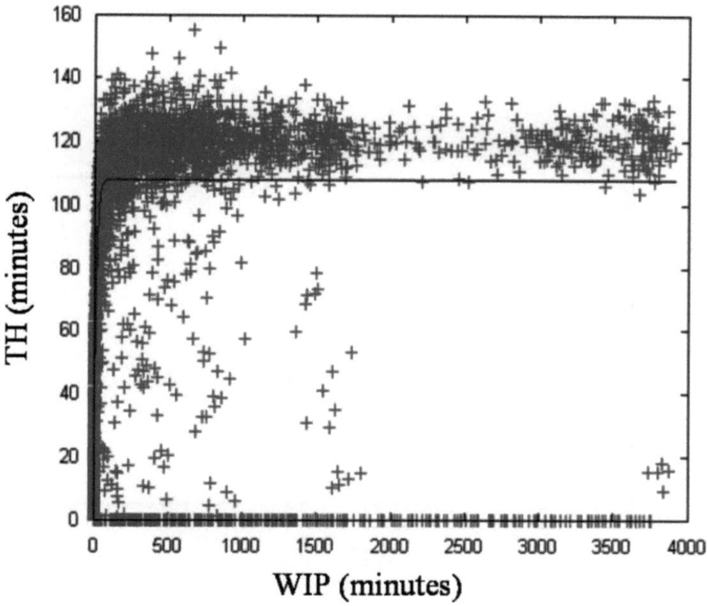

Fig. 16.9 CF data (WIP vs. TH) for a Work center from a Simulation Study

The effect of this error on the planning model is disastrous: it consistently releases too little material into the line too late, resulting in significant backlogs and missed demand.

In order to remedy this situation, the authors adopted an alternative fitting approach where they sought a curve such that a specified percentage of the data points fell above the curve, i.e., the fitted curve represents a percentile of the data, which corresponds to a quantile regression. The results of this approach for different percentile values are shown in Fig. 16.10. As the percentage of data points required to lie above the fitted curve increases, the fitted curve shifts to the right, providing a more realistic representation of how much WIP is required at this work center to achieve maximum output, based on the simulation results, and resulting in a planning model that yields better backlog results than an LP model. This heuristic approach proposed clearly needs a better theoretical justification, even though it works quite well in these experiments. These results highlight how a poorly fitted clearing function can result in poor performance of the planning models derived from it.

It is important to remember that the clearing function as defined in this paper is defined with relation to a planning period of a specific duration. When steady-state queuing models are used to derive an expression for the clearing function, we are assuming that the planning period is long enough that the behavior of the production resource being modeled is represented by a steady-state model to an acceptable degree of accuracy. However, in the production planning environment we are changing the releases, and therefore the workload, of the system in each planning period,

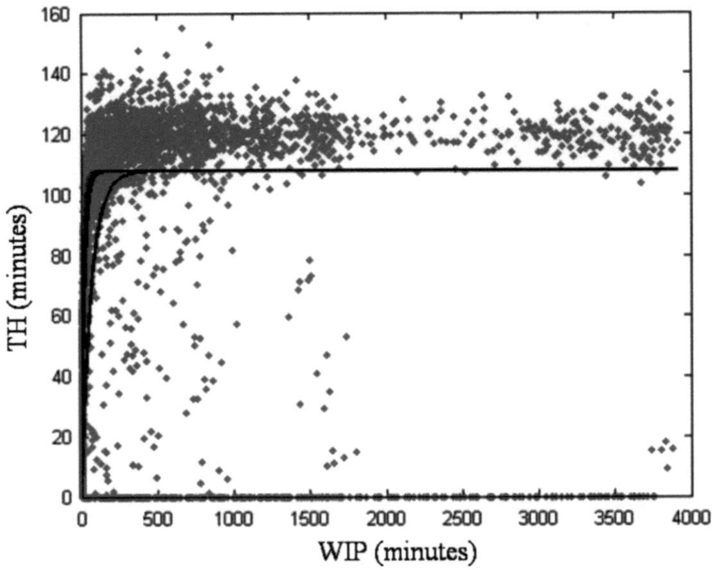

Fig. 16.10 Alternative percentile-based CF curves

calling into question whether the production resources ever attain the steady state required by conventional queuing analysis. We examine this question by comparing clearing functions derived for an $M/M/1$ queuing system in transient and steady state.

We consider an $M/M/1$ queuing system and define as a time unit the average operation time of an order, hence the service rate $\mu = 1$ without loss of generality. Hence the arrival rate λ is equal to the traffic intensity ρ. The length of a planning period is $\omega = 5$ time units. The clearing function for an arbitrary period t can be derived by recalling that the utilization u_t of the server in period t is the fraction of the period in which there is at least one order in the system and is given by

$$u_t = 1 - \frac{1}{\omega} \int_{\omega(t-1)}^{\omega \cdot t} p_0(\tau) d\tau \tag{16.59}$$

where $p_n(\tau)$ denotes the probability of n customers in the system at time τ, while $\omega(t-1)$ and $\omega \cdot t$ denote the beginning and the end of period t. The expected output $E[X_t]$ in period t is then

$$E\left[X_t\right] = u_t \cdot \omega \tag{16.60}$$

and the expected number of orders in the system at time τ, denoted $E\,[Ls(\tau)]$, is[4,5]

$$E\,[Ls(\tau)] = \sum_{n=0}^{\infty} n\; p_n(\tau). \tag{16.61}$$

This is also the average WIP, measured in units of time, at time τ, because the average service time of the orders in the queue is 1, and the expected remaining service time of the order at the server is also 1 due to the exponentially distributed service time. The expected load in period t, denoted by $E[\Lambda_t]$, is the average WIP at the beginning of period t plus the average input during period t:

$$E(\Lambda_t) = E\,[Ls(\omega(t-1))] + \rho\omega \tag{16.62}$$

The time-dependent state probabilities $p_n(\tau)$ for an $M/M/1$ system starting at the origin can then be calculated as (Stange 1964):

$$p_o(\tau) = 1 - \int_o^{\tau} \frac{e^{-(\rho+1)y}}{y} \cdot \sqrt{\rho}\,\text{Bessel}\; I_1(2\sqrt{\rho}y)\; dy \tag{16.63}$$

$$p_n(\tau) = \int_0^{\tau} \frac{e^{-(\rho+1)y}}{y} \left[n\rho^{\frac{n}{2}}\,\text{Bessel}\; I_n - (n+1)\rho^{\frac{n+1}{2}}\,\text{Bessel}\; I_{n+1} \right] dy, \quad \text{for } n > 0. \tag{16.64}$$

The expressions for the conditional probability of j customers in the system at time t given i customers in the system at time 0 can be found in Cohen (1969, p. 82 ff., p. 178).

The clearing function in period t can be calculated as a parametric curve with the average load in period t, $E[\Lambda_t]$ (16.61–16.64) on the x-axis and the expected output in period $tE[X_t]$ (16.59, 16.60, 16.63) on the y-axis. The arrival rate $\lambda = \rho$ is the control variable that yields the values for $E[\Lambda_t]$ and $E[X_t]$. The arrival rate is assumed to be the same from time $\tau = 0$ to the end of the period under consideration. Figure 16.11 shows the clearing functions for periods 1 and 2 and for a period after steady state has been reached. Note that in Fig. 16.11 the arrival process is stochastic for all work available. If work definitely will be available (e.g., at the beginning of the first planning period, where the actual WIP is known), the shape of the clearing function will be more extreme because the expected output cannot be lower than the minimum of the available work at the start of the period W_{t-1} (which we assume to be known) and available capacity. It can be shown (Missbauer, forthcoming) that

---

[4] Due to the computational complexity, the summation in (16.61) has been performed for $n = 0, \ldots, 80$ in the numerical examples below. This ignores at most 1.5% of the cases (for $\rho = 0.95$ in steady state), in most cases the error is close to zero.

[5] The index for the periods (discrete time) is denoted as subscript, the continuous time is denoted in parenthesis.

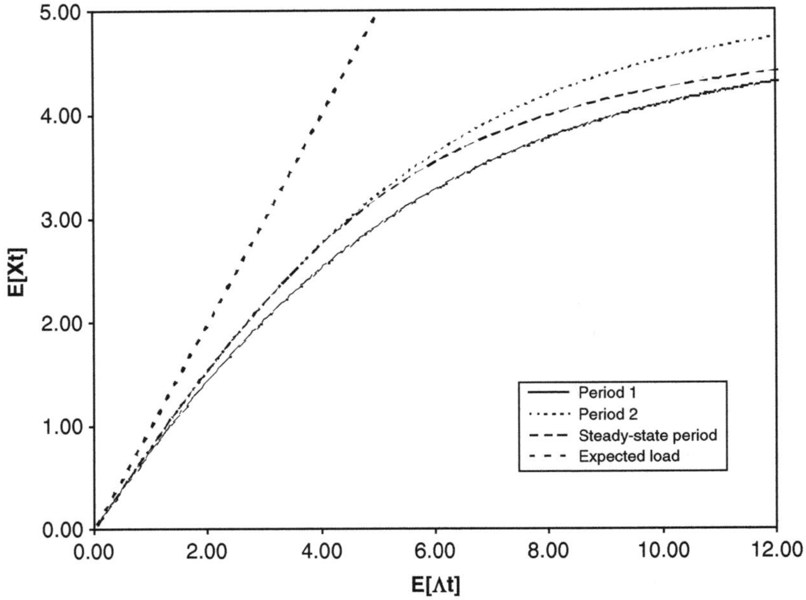

Fig. 16.11 Clearing functions for periods 1 and 2 and for a steady-state period of an *M/M/1* system starting at the origin. $\mu = 1$, length of a period $\omega = 5$. Note that identical values for $E[\Lambda_t]$ correspond to different values for the arrival rate λ for different periods

the shape of the clearing function (i.e., the functional relationship between expected load and expected output, with the expectation defined at the time of planning) depends on the variance of the initial WIP and of the planned input.

Figure 16.11 shows that if the system is not in steady state or in a specified transient phase there is no fixed functional relationship between expected load and expected output. The relationship changes with the phase of the transient state. This leads to a planning circularity that must be considered as a substantial problem of clearing function models. The estimated clearing function is based on assumptions about the dynamic behavior of the system. Hence, any order release plan derived using the clearing function can affect the dynamic behavior of the system, and hence the shape of the clearing function it is based on. This implies that *order release determines the validity of the assumption it is based on.* There is no evidence that the *assumed* shape of the clearing function is consistent with the *observed* shape(s) of the clearing function since the shape can change over time as seen in Fig. 16.11. We do not even know whether a consistent solution is possible, but it can be expected that this is not the case, because clearing function models assume that the clearing function is the same for each period, which need not be the case if transient/stationary phases occur during the planning horizon. Therefore the clearing function model can lead to systematic errors, and it is an empirical question whether or not the level of accuracy provided by the models is acceptable in practice. The experimental results of (Asmundsson et al. 2009) suggest that under some conditions

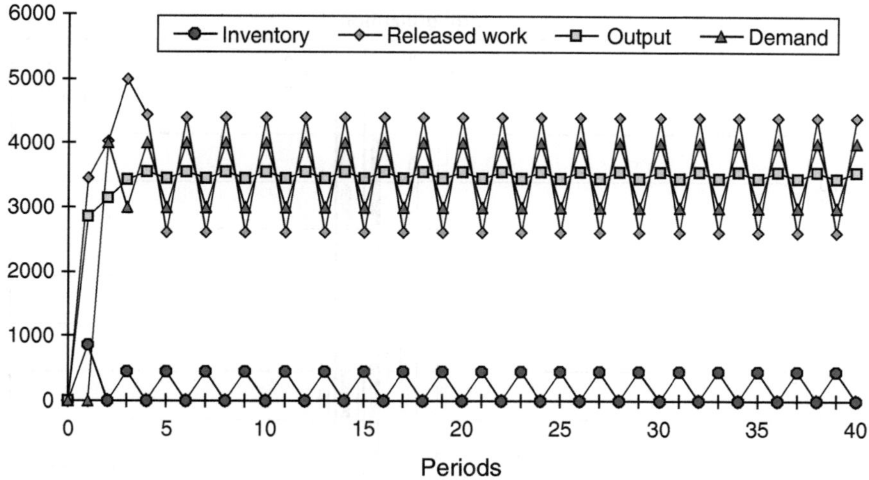

Fig. 16.12 Results of an optimization model for one work center and idealized underutilization

the performance of a planning model based on clearing functions can be very sensitive to how well the clearing function represents the actual system, which suggests caution in using these models until these issues are better understood.

Another limitation of clearing function models is that due to their use of discrete planning periods, they ignore any transient effects that arise at the boundaries between periods. While this limitation is shared with all planning models based on discrete time periods, it is worthy of note, and as far as we are aware has not been the subject of extensive research, as will be discussed further below.

Some characteristics of clearing function models can be seen from Fig. 16.12 which is an optimization result for one work center. The demand oscillates between 3,000 and 4,000 min of capacity per period, which is below the capacity (4,500 units/period). The variations in the planned output are much lower because of the nonlinear increase in the amount of WIP that is required. But the amount of released work exaggerates the demand variation. Figure 16.12 exhibits nervous behavior that has been reported as a property of optimization models if steady-state properties are assumed to hold for short periods (Lautenschläger 1999). Karmarkar (1993, p. 317), also states that "what happens in the transition between periods is not clear." In Fig. 16.12 it is difficult to decide whether this behavior is truly optimal, because this would require a model that incorporates the actual characteristics of the transient state.

Figure 16.13 exhibits the due-date deviations achieved by the clearing function model of Missbauer (2002a) for a highly utilized production unit with five bottleneck work centers. The average due-date deviation is quite low – lower than for load-oriented order release (Wiendahl 1995), which is used as reference, but the earliness/lateness of a small number of orders is high. A number of factors may contribute to this – only the aggregate clearing function is used (no

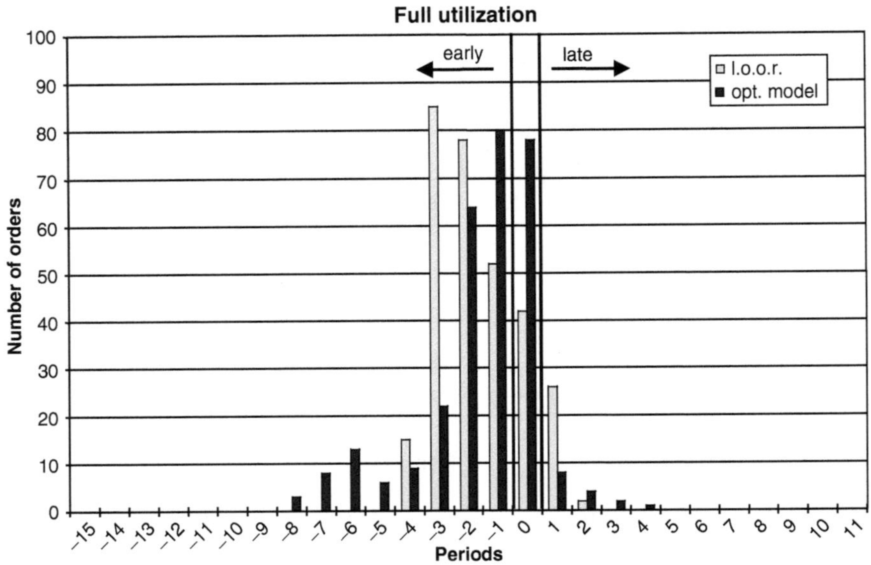

Fig. 16.13 Distribution of the due-date deviations (l.o.o.r.: load oriented order release)

partitioning; see Sect. 16.5.1), and non-bottlenecks are represented as delay functions (Missbauer 1998, p. 267 ff). However, since the clearing function only partly reflects the dynamic characteristics of the system and leads to counter-intuitive optimization results, it can be assumed that the due-date deviations in Fig. 16.13 are at least partly due to the shortcomings of the clearing function and the resulting nervousness. Future research must clarify to what extent this conclusion holds.

It is reasonable to assume that the performance of clearing function models can be improved if the history of the arrival and departure of orders at the work centers is analyzed in more detail – the clearing function aggregates the history to one dimension. Andersson et al. (1981) decompose the expected load into two parts: the expected WIP at the beginning of period t and the expected input in period t. A two-dimensional clearing function is formulated as:

$$X_t = \gamma W_{t-1} + \beta R_t, \tag{16.65}$$

if production does not exceed available capacity. Production, WIP and released work are measured in units of value. It can be argued that the expected WIP at the beginning of the period is actually available in period t with higher probability than the expected input during the period, which seems to be the reasoning behind this formulation. Numerical experiments based on analytical expressions for the transient $M/M/1$ queue confirm this (Missbauer, forthcoming) but the linear function in (16.65) is not derived explicitly from theory.

Conceptually, a clearing function expresses the expected WIP level that is required to obtain a certain output rate given the system variability and the production control policies that are applied. (Anli et al. in press) present a model for order release planning that takes into account load-dependent lead times that result from the stochastic material flow, and also considers lower bounds on the finished goods inventory that are required to maintain a desired service level in the face of stochastic demand process. The lower bound on the finished goods inventory (FGI) for each SKU is a nonlinear function of the planned production volumes (production targets) of the facilities that produce and require this SKU, and of parameters representing system variability and control policies. Likewise, the expected WIP level for each unfinished SKU, facility, and period is a nonlinear function of the planned production volumes (of all SKU's) of the facility and the variables representing system variability and control policies. Both nonlinear functions can be estimated either by simulation or by queuing models. Anli et al. (in press) present their paper as a proof-of-concept study and use queuing models, namely mean value analysis. The Queuing Network Analyzer is used in Caramanis et al. (2001). Optimization is performed iteratively. In each iteration the linear constraint set of the planning model is augmented using hyperplanes tangent to the nonlinear functions. These tangents are obtained from the tentative production targets (from the previous iteration) and from the required WIP and FGI levels and their sensitivities with respect to the production targets. The authors state that this iterative refinement of the local approximations leads to convergence under mild convexity or quasi-convexity conditions.

The approach can be classified as a WIP-oriented model since it traces the flow of WIP through the facilities and does not assign flow times to the orders. The functional relationship between WIP and the production targets (volumes) can be interpreted as a sophisticated clearing function formulation that addresses the product mix problem for which we have already presented the partitioning approach (see Sect. 16.5.1) However, the model is limited to steady-state relationships and does not consider transient effects. The paper provides optimization results, but no simulation experiments. It remains to be seen whether the approach can be applied efficiently in an environment where analytical models of the manufacturing system cannot be applied.

Our discussion of the limitations of clearing function models has been limited so far to single-stage systems. Another set of complications emerges when multiple stage systems are considered. Let us assume that we wish to derive a clearing function for a work center that is part of a multistage production system – i.e., the pattern of arrivals at the workstation over time depends on production and release decisions at other work centers. Jackson (1957) in his seminal paper showed that in an open Jackson queuing network in steady state each work center can be treated as an independent $M/M/s$ queuing system, but real-world manufacturing systems often do not meet these assumptions. Therefore, in a multistage production system there are likely to be correlations between the decisions at upstream stages and the pattern of arrivals at a downstream center, which will influence the shape of the clearing function. To illustrate the point, consider a single resource that can be modeled as a $G/G/1$ queuing system in steady state. The expression (16.66) describes the average number in system $E[Ls]$ as a measure of the expected WIP for a single server where the coefficient of variation for interarrival time and service time are denoted by c_a and c_s, respectively (see Medhi (1991) for the derivation), and ρ denotes the utilization of the server.

$$E[L_s] = \frac{c_a^2 + c_s^2}{2}\frac{\rho^2}{1-\rho} + \rho = \frac{c^2\rho^2}{1-\rho} + \rho \tag{16.66}$$

Solving for ρ and assuming $c = c_a^2 + c_s^2 > 1$, we obtain utilization as a function of WIP as:

$$\rho = \frac{\sqrt{(E[L_s]+1)^2 + 4E[L_s](c^2-1)} - (E[L_s]+1)}{2(c^2-1)} \tag{16.67}$$

If we consider the utilization as a surrogate measure of output, Fig. 16.14 illustrates the relationship for different c values, where c combines the coefficients of variation of the arrival and service (production) processes as seen in (16.67).[6] For a fixed c value, utilization, and hence throughput, increases with WIP but at a declining rate. This is because as WIP increases, the server becomes less likely to starve. Utilization decreases as c increases, due to variability in service time and interarrival time, which causes queues to build up and throughput to slow as customers are trapped behind a customer with an unduly long service time, or the number of customers arriving in a small time interval is unexpectedly high. Note that in a multistage system, the coefficient of variation of the arrival stream c_a will be determined at least in part by the production and lot sizing decisions made at the upstream stages (e.g., giving priority to orders with low WIP at the next work center). Thus the decisions made by

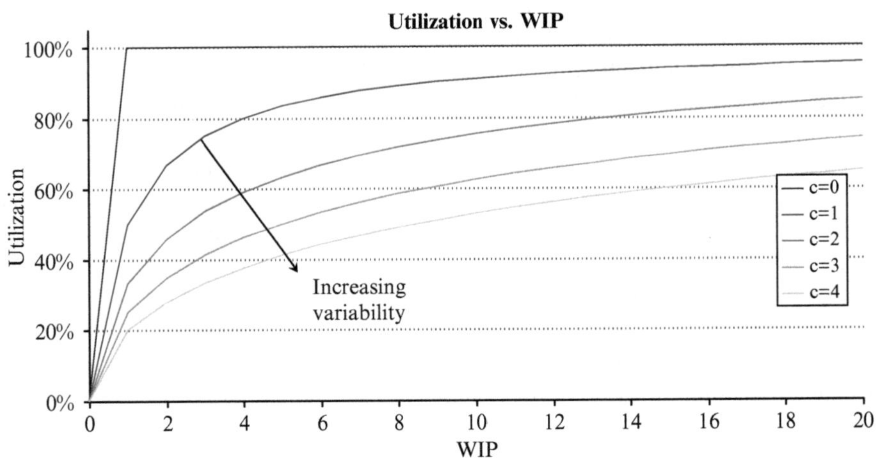

Fig. 16.14 Utilization (ρ) as a function of average WIP for different c values

[6] We are aware of the similar $GI/G/1$ approximation by Krämer and Langenbach-Belz (1976) that distinguishes between $c_a^2 \leq 1$ and $c_a^2 > 1$ (Tijms 1994), where c_a and c_s are not additive.

the model at one stage of the system affect the shape of the clearing function faced by the model at another stage. Clearing functions determined by empirical means are less susceptible to this criticism, as the correlations are captured at least in part in the data to which the functions are fitted.

16.6 Modeling Capacity with Multiple Products: Extensions to the Basic Model

In many production environments there are a number of alternative processes by which a product can be produced. These generally arise from the presence of a number of alternative machines that are capable of performing a given operation required by a product. In general, the costs of production may depend on the specific choice of equipment made for each stage. In addition, not all alternative machines for a given operation are equally efficient; a typical scenario is that there is newer equipment that can perform a given operation faster than the older equipment. Another typical scenario in high-technology industries such as semiconductor manufacturing is that the newer equipment can perform a wider range of operations than the older equipment, since the older equipment is incapable of meeting the finer tolerances that the newer equipment can handle.

In this situation, determining the optimal allocation of products to equipment over time becomes a complex decision. The dependence of costs on the particular sequence of operations followed means that a model must keep track of how much work is allocated to each possible sequence of operations, i.e., each possible path that a product can follow through the plant, as illustrated in Fig. 16.15. This figure represents a production system with four stages and a number of alternative machines, represented by the boxes. The decision variable R_{ij} denotes the amount of product i released for processing on operation sequence j. These can be

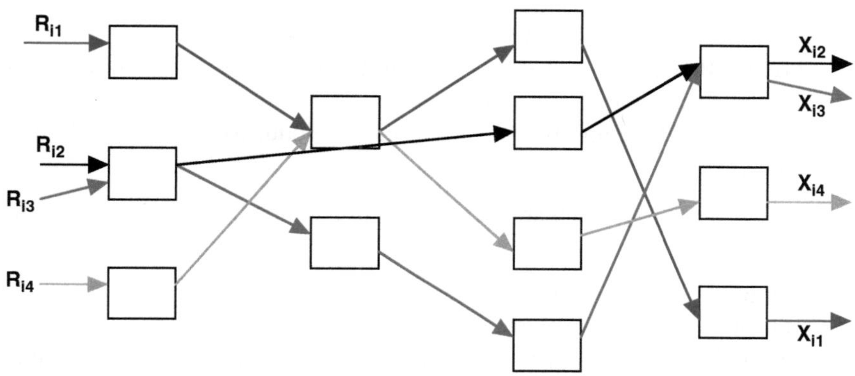

Fig. 16.15 Path-based formulation of alternative resources

considered path-based formulations, since we are explicitly specifying the amount of each product that will be launched on each possible path through the production system in each period.

These types of models, referred to as process selection models by Johnson and Montgomery (1974), have been known for quite some time, but have the obvious drawback that in a production system of any complexity, the number of possible paths through the system that a product can follow, and hence the number of decision variables in a time period, grows exponentially in the number of alternatives at each stage and the number of stages. As pointed out by Leachman and Carmon (1992), they also give us a great deal of redundant information. In order to implement the results of this model, all we need to know is the total releases of each product into the line in each period; we do not need to know the specific allocation of work to individual machines, since this is likely to change as the shop floor reacts to local circumstances and reallocates work among alternative machines. Hence a number of authors have developed models that are more compact in terms of the number of decision variables. It is not surprising that much of this work has been motivated by applications in the semiconductor industry, where a product may require several hundred unit processes, all of which may have a significant number of alternative machines.

Much of the work in this area takes as its starting point the LP models with fixed time lags treated by Hackman and Leachman (1989). Leachman and Carmon (1992) present a series of models that address this issue. We shall focus on this paper in some detail, and then outline the extensions proposed by other authors. We define the following notation:
Parameters:

a_{ijk} = time to process one unit of product i at step j on machine type k.
C_{kt} = capacity in time unit of machine type k in period t.
D_{it} = maximum cumulative demand for product i in period t, made up of the forecast and confirmed orders.
d_{it} = minimum cumulative demand for product i by time t (firm orders).
p_{it} = estimated net discounted cash flow from selling one unit of product i in period t.
h_{it} = estimated unit holding cost for product i in period t.
L_i = average flow time for product i from beginning to end of its entire process, i.e., the cycle time for the entire process.
L_{ij} = average lead time for product i from start of production process until start of step j.

The decision variables, which are common to all the three formulations presented, are as follows:

R_{it} = number of units of product i to be released in period t.
I_{it} = units of product i in inventory in period t.
B_{it} = shortfall of cumulative production vs. cumulative max demand for product i in period t.

The first model, referred to by the authors as the Step-Separated Formulation, introduces additional decision variables W_{ijkt} that denote the amount of workload of each product type i in each process step j that is assigned to machine type k in period t. The model can now be stated as follows:

$$\max \sum_{i=1}^{n} \sum_{t=1}^{T} [p_{it} R_{i,t-L_i} - h_{it} I_{it}], \tag{16.68}$$

subject to

$$R_{i,t-L_{ij}} = \sum_{k \in P(i,j)} \frac{W_{ijkt}}{a_{ijk}}, \quad \text{for all } i, j, t, \tag{16.69}$$

$$\sum_{\{(i,j)|k \in P(i,j)\}} W_{ijkt} \leq C_{kt}, \quad \text{for all } k, t, \tag{16.70}$$

$$\sum_{\tau=1}^{t} R_{i,\tau-L_i} - I_{it} + B_{it} = D_{it}, \quad \text{for all } i; \text{for all } t \leq T-1, \tag{16.71}$$

$$\sum_{\tau=1}^{T} R_{i,\tau-L_i} + B_{iT} = D_{iT}, \quad \text{for all } i, \tag{16.72}$$

$$B_{it} \leq D_{it} - d_{it}, \quad \text{for all } i, t, \tag{16.73}$$

$$R_{it}, B_{it}, I_{it}, W_{ijkt} \geq 0. \tag{16.74}$$

The critical constraints for the purposes of modeling capacity and workload are the first two. The first set of constraints converts the releases in period $t - L_{ij}$ into the current workload on the resources k in period t. The second set of constraints then ensures that no resource is loaded in excess of its capacity. Note that accumulation of WIP within the production process is not modeled at all; only finished goods inventory is represented in the decision variables. The objective function is to maximize the difference between the discounted revenue from sales and inventory holding costs. Note that a product is credited as producing revenue as soon as it is produced, even if it may not be sold immediately. Also note that in the last two constraint sets, output aimed at meeting forecasts may be delivered late, but firm orders must be met on time. This effectively creates two hierarchical demand classes, with the demand from firm orders having absolute priority over forecast demand in terms of allocating limited output to meet demand. Production costs are not considered, as they are independent of production routing and the revenue is assumed to be much greater than unit production cost, as is the case in much of the semiconductor industry. Hence the emphasis of the model is on allocating production capacity to meet demand, and thus maximize revenue.

This formulation is inefficient in that it gives a detailed allocation of workload to individual stations (the W_{ijkt} variables), even though all we really need is the total releases R_{it} of each product in each period. In order to arrive at a more efficient formulation, the authors make the additional assumption of uniform processing times across machines in a work center, i.e., $a_{ijk} = a'_{ij} S_k$, where a'_{ij} denotes the processing time of product i at step j on the standard machine whose speed is used as a baseline for the others. The available capacity of each set of alternative machines can then be rescaled in a similar manner, $C'_{kt} = C_{kt}/s_k$. Now since all workloads are expressed in terms of the baseline machine, we only need to track the total workload at step j that is assigned to machine type k. In order to do this, we define unique sets S_m of alternative machine types, and additional variables Z^m_{kt} that denote the workload on machine set S_m assigned to machine type k in period t. Defining the set $P(i, j)$ as the set of machines capable of processing step j of product i, this yields the following formulation:

$$\max \sum_{i=l}^{n} \sum_{i=l}^{n} [\rho_{it} R_{it-l_i} - h_{it} I_{it}], \tag{16.75}$$

subject to

$$\sum_{\{(i,j)|k \in P(i,j)\}} a'_{ij} R_{i,t-L_{ij}} = \sum_{k \in S_m} Z^m_{kt}, \quad \text{for all } m, t, \tag{16.76}$$

$$\sum_{m=1}^{M} Z^m_{kt} \leq C'_{kt}, \quad \text{for all } k, t \tag{16.77}$$

$$\sum_{\tau=1}^{t} R_{i,t-L_i} - I_{it} + B_{it} = D_{it}, \quad \text{for all } i; \text{for all } t \leq T-1, \tag{16.78}$$

$$\sum_{\tau=1}^{T} R_{i,t-L_i} + B_{iT} = D_{iT}, \quad \text{for all } i, \tag{16.79}$$

$$B_{it} \leq D_{it} - d_{it}, \quad \text{for all } i, t, \tag{16.80}$$

$$R_{it}, B_{it}, I_{it}, Z^m_{kt} \geq 0. \tag{16.81}$$

Using these assumptions, the authors construct a formulation where all allocation variables are eliminated, and capacity constraints are written for sets of alternative machines whose capacity is likely to be binding on the optimal solution. The structure of these sets depends on the problem data, and hence this formulation is not always the most compact in terms of the number of variables and constraints, but in many industrial situations yields substantially smaller formulations. The sets S of alternative machines for whom capacity constraints will be written are determined based on cut sets in the bipartite graph representing operation-machine requirements. The authors provide a procedure to identify the dominant cut sets whose time complexity is linear in the number of operations ij and the cardinality of the alternative machine sets, but exponential in the number of machines that occur in the connected component of the bipartite graph. Thus, if there are a very large number of machines that can process a large number of steps as alternatives, the time complexity of generating this formulation may be quite high, but such instances seldom arise in industrial data sets. Once the dominant cut sets S have been obtained, we can write capacity constraints for each set and each period of the form

$$\sum_{ij \in S} a_{ij} R_{i,t-L_{ij}} \leq \sum_{k \in S} C_{kt}. \tag{16.82}$$

The complete formulation is now as follows:

$$\max \sum_{i=1}^{n} \sum_{t=1}^{T} [p_{it} R_{i,t-L_i} - h_{it} I_{it}], \tag{16.83}$$

subject to

$$\sum_{ij \in S} a_{ij} R_{i,t-L_j} \leq \sum_{k \in S} C_{kt}, \quad \text{for all generated sets } S \text{ and periods } t, \tag{16.84}$$

$$\sum_{\tau=1}^{t} R_{i,t-L_i} - I_{it} + B_{it} = D_{it}, \quad \text{for all } i\text{; for all } t \leq T-1, \tag{16.85}$$

$$\sum_{\tau=1}^{t} R_{i,\tau-L_i} + B_{iT} = D_{iT}, \quad \text{for all } i, \tag{16.86}$$

$$\sum_{\tau=1}^{T} R_{i,\tau-L_i} + B_{iT} = D_{iT}, \quad \text{for all } i, \tag{16.87}$$

$$R_{it}, B_{it}, I_{it} \geq 0. \tag{16.88}$$

The authors analyze the number of decision variables and constraints in their different formulations and show that when alternative machine sets have a nested property, which occurs frequently in semiconductor manufacturing, this Direct Product Mix formulation provides a very compact model compared to alternative formulations. The nested property arises when a work center has machines of several technological generations, where each newer generation can perform all operations the previous generations could, as well as some additional new ones.

A drawback of the Direct Product Mix formulation developed above is its reliance on the assumption of uniform processing times across alternative machines as described above. Bermon and Hood (1999), in their study of production planning for IBM's semiconductor manufacturing operations, noted that this assumption was violated in their environment. Hung and Cheng (2002) extend the work of Leachman and Carmon (1992) by developing a formulation that does not require the uniformity assumption. In order to do this, they define a new set of partitioning variables that allocate the capacity of machines shared between machine sets to which they belong. Their computational experiments show that when the uniformity assumption on processing times holds, the Direct Product

Mix approach of Leachman and Carmon (1992) is preferable. However, when the uniformity assumption is violated, the Partition formulation developed by the authors remains valid.

Bermon and Hood (1999) present a slightly different model aimed at capacity planning that also addresses the problem of determining capacity in situations with alternative machine sets. Hung and Wang (1997) apply an approach similar to that of Leachman and Carmon (1992) to situations where alternative products can be used to meet a given demand, as by downgrading in electronics manufacturing.

16.7 Lot Sizing Models

As we have seen, the clearing function reflects the variability of the arrival and departure process. If a work center produces multiple products in lots, lot sizing determines the operation times of the orders and thus strongly influences the variability of the operation times. It also determines the total setup time during a period and thus influences the maximum output of the work center. Hence the lot sizes influence the average flow times and the WIP level at the work centers.

Considering the impact of lot sizes on average flow time and WIP means to anticipate consequences that become visible at the scheduling level. One way to achieve this is simultaneous lot sizing and scheduling. This economic lot scheduling problem (ELSP) has been studied extensively (for reviews, see Elmaghraby 1978; Graves 1981; Drexl and Kimms 1997; Pinedo and Chao 2005). Drum-Buffer-Rope-OPT, which has been described extensively in the 1980s, also performs simultaneous lot sizing and scheduling for the bottleneck work centers, distinguishing between *transfer* and (often larger) *process batches.* See Zäpfel and Missbauer (1993b) for related literature on OPT.

In accordance with the hierarchical structure of the manufacturing planning and control system that we assume throughout the chapter, detailed scheduling is performed locally within the production units. Thus we do not consider simultaneous lot sizing and scheduling or cyclic production. We assume that lot sizing passes the lots (production orders) to the order release function where they are released to the production units that perform sequencing. In this context, stochastic models of the manufacturing system are appropriate to anticipate the consequences of lot sizes on flow times and WIP.

Assuming a stationary state of the system, we can examine the relationship between the lot sizes and the long-term clearing function of a work center by means of an $M/G/1$ model producing products with identical data. We define m as the total demand rate (sum of the identical demand rates of the products), a the processing time per unit, r the setup time per lot, σ and v the standard deviation and coefficient of variation of the service times of the orders, x the lot size. a, r, and x are identical for all products. We assume that a setup is necessary for each lot. Considering setup time savings obtained by sequence optimization is described in Kekre (1984), Kekre (1987) and Missbauer (1997).

For simplicity we assume a given value for the coefficient of variation of the service times v that is independent of the common lot size x. This is sufficient for showing the structural properties. Missbauer, (2002b) and Karmarkar et al. (1985a, 1985b) discuss the relationship between lot sizes and the coefficients of variation of service times in the multi-product case. We also assume a Poisson arrival process, which is a rather strong assumption in the single-product case (see Kistner (1999) for a critique), but in the multi-product case with identical products this assumption is well justified (Missbauer 1999). Based on these assumptions we compute the arrival rate

$$\lambda = \frac{m}{x} \tag{16.89}$$

and the mean service rate

$$\mu = \frac{1}{r + ax}. \tag{16.90}$$

The mean waiting time $E[W_q]$ by the Pollaczek–Khinchine formula is then

$$E\left[W_q\right] = \frac{\rho^2 + \lambda^2\sigma^2}{2\lambda(1-\rho)} \tag{16.91}$$

with $\rho = \lambda/\mu = m(a + r/x)$, and the mean flow time

$$E[W_s] = E[W_q] + 1/\mu. \tag{16.92}$$

Substituting (16.89) for λ, (16.90) for μ and $\sigma = v(r + ax)$ into (16.91), we get the average waiting time as

$$E[W_q] = \frac{m(v^2+1)(ax+r)^2}{2[x(1-am) - mr]}. \tag{16.93}$$

Because of the assumption of Poisson input and the PASTA (Poisson arrival see time averages) property (Buzacott and Shanthikumar 1993, p. 54; Tijms 1994, p. 73 ff.) the average waiting time of the customers $E[W_q]$ must be identical to the average WIP at the server, measured in hours of work, given by the average remaining work $E[L_w]$). So we can write

$$E[L_w] = E[W_q]. \tag{16.94}$$

The average output in hours of work as a function of the average WIP $E(L_w)$ can be calculated from (16.93) and (16.94) as follows:

$$\text{Output} = m \cdot a = \frac{2axE[L_w]}{(ax+r)\,[2E[L_w] + (v^2+1)(ax+r)]} \tag{16.95}$$

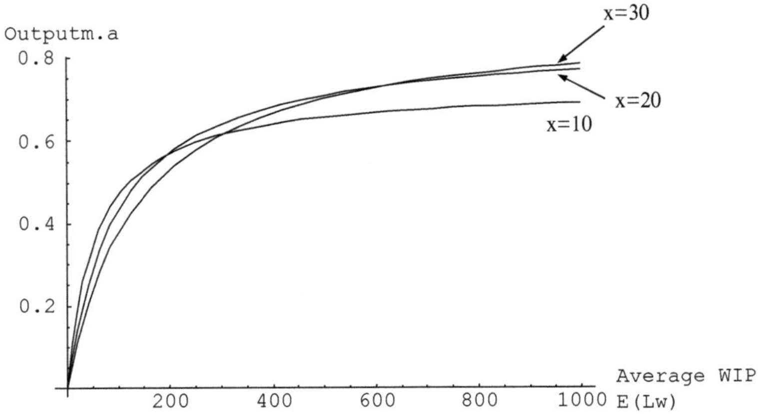

Fig. 16.16 Output as a function of WIP (16.95) for different values for the lot size $x(a = 5, r = 15)$

From (16.95) we see that a higher variability of the service times decreases the output for a given average WIP. The clearing function for different lot sizes is shown in Fig. 16.16.

This adds an important aspect to the lot sizing problem. Lot sizing should not only consider setup and inventory holding costs but also its impact on the clearing function, which can be regarded as capacitated lot sizing considering congestion effects. There are two ways to accomplish this

- Determination of standard lot sizes or lot sizing rules that take into account the impact of lot sizes on the clearing function (and hence on WIP and flow times that result from order release). The clearing function that results from the lot sizes is used for order release. In the extreme case lot sizes can be determined such that average flow time or WIP is minimized for the required output. (Note that usually this is in conflict with the traditional goal of minimizing the sum of setup and holding costs). Starting with (Karmarkar et al. 1985a, b; Karmarkar 1987) the majority of the literature on this topic is based on this idea (Missbauer 2002b).
- Defining a multi-dimensional clearing function with output as a function of WIP and lot sizes as independent variables:

$$X_{it} \leq f_i\left(W_{it}, x_{1t}, x_{2t}, \ldots, x_{Nt}\right). \tag{16.96}$$

  In this case the order release model optimizes the time-varying lot sizes and the amount of released work for all periods simultaneously. This recent research direction is presented in (Hwang and Uzsoy 2005). Note that the steady-state assumption implies that the changes in the characteristics of the arriving orders affect the clearing function within the same period.

Since lot sizes influence lead times, they also influence the lead time demand distribution, which in turn influences optimal safety stocks and reorder points. For this aspects, see Lambrecht et al. (1996) and Vaughan (2006). Little is known about multistage lot sizing considering congestion effects. For a model exploring this issue, see Missbauer (1999, 2002b).

16.8 Models Incorporating Uncertainty

Until this point we have treated all parameters of our various optimization formulations as deterministic, although in some cases they represent the expectation of a performance measure derived from an underlying stochastic system. However, in any industrial application all parameters of an optimization model, such as the cost estimates used in the objective function, the technological coefficients defining resource consumption by products, and especially the forecasts of future demand, are subject to significant uncertainties. The explicit treatment of uncertainty in optimization formulations of production planning problems is very much in its infancy, so we will focus on illustrating the issues in one particular area – that of uncertainty in demand forecasts, which requires the system to hold a certain amount of safety stock to maintain a given level of customer service.

It is well known that the amount of safety stock that needs to be held in a particular location to maintain a specified level of customer service is related to the distribution of the demand forecast over the replenishment lead time. To illustrate this relationship, consider a simple newsvendor model where the lead time is a random variable with mean μ_L and variance σ_L^2, and the demand rate has a mean of μ and a variance of σ^2. It is well known (e.g., Eppen and Martin 1988) that the optimal order level can be approximated by

$$\mu_L \mu + z_\alpha \sqrt{\mu_L \sigma^2 + \sigma_L^2 \mu^2}, \tag{16.97}$$

assuming that the lead time is normally distributed. Note that in the term under the square root, which denotes the standard deviation of the demand over the lead time, both the mean and the variability of the lead time interact with the mean and variability of the demand to influence the amount of safety stock required. However, the lead times, in turn, are determined by the utilization levels of the resources. While several approaches described in the previous section address this issue in terms of planning models, there have been few efforts to address this directly in the planning literature, although several stochastic models linking queues with inventory models have been proposed (e.g., Zipkin 1986; Liu et al. 2004).

In most current approaches in industry, many of which have their origin in the MRP literature, the amount of safety stock to be held in a particular location in a particular planning period is calculated outside of the planning model using actual or estimated parameters of the demand or demand forecast distribution. The planning model is then constrained to maintain this quantity of safety stock. Examples of such models are described at length by Wijngaard and Wortmann (1985) in the context of MRP systems, where the amount of safety stock to be maintained results in the release of additional orders to the system. Lambrecht et al. (1984a) show that the problem of determining the amount of safety stock in multistage production systems can be formulated as a Markov decision process or a dynamic program, but the size of the state space renders these approaches impractical from a computational point of view. They propose a number of heuristics, and conduct extensive computational experiments that examine the amount of safety lead time and safety stock required. Lambrecht et al. (1984b) perform computational experiments with the Markov decision process and examine the form of the optimal policies. Yano and Carlson (1988) propose a heuristic for setting safety stock in an assembly-type production system and examine its performance using simulation.

Graves (1986, 1988) develops an intriguing model that expressly links safety stock levels to the ability of the production system to react to changes, using a model that separates WIP and FGI and uses the proportional clearing function in Fig. 16.5 to model the production behavior of the facility based on planned lead times. However, his model assumes a stationary demand distribution, and the proportional clearing function may produce capacity-infeasible solutions at high WIP levels. He points out that distinguishing between WIP and FGI is significant since in many production systems WIP can serve some of the function of safety stock. This indicates the desirability of making safety stock decisions endogeneous in a production planning model that combines a realistic aggregate model of the behavior of congestion-prone capacitated production systems (such as that given by the clearing function formulations) with the explicit modeling of WIP and FGI as separate entities. We conjecture that such a model might well be capable of maintaining a given service level with significantly less safety stock in the form of finished goods between stages, since it would be able to recognize that WIP in the line would become available to meet demands, reducing the need for stocks of finished goods.

Leachman (1993) presents a large-scale linear programming framework for production planning in the semiconductor industry where safety stocks are addressed through the use of demand classes. Production required to replace safety stocks is modeled as a class of demand that has lower priority than firm customer orders, and capacity is allocated to these orders only if this will not compromise the ability of the system to meet firm customer orders. He suggests simulation of the production plan to obtain estimates of the variability of the resulting lead times this plan will impose on the system. These estimates of variability can then be used to determine safety stock levels to protect against variability in lead times. Hung and Chang (1999) elaborate further on this approach and give computational experiments examining its performance.

In the domain of production planning models, the chance-constrained (CC) formulations of Charnes and Cooper (1963) can be used to obtain deterministic equivalents to a number of production planning problems involving random variables. In the simplest version of this approach, consider demand to be the only source of uncertainty, and assume that demand in period t has a cumulative distribution function F_t. Note that I_t is a random variable due to the demand being random. Let us define U_t to be a target inventory level such that in each period t we produce $X_t = U_t - I_{t-1}$ units if $U_t > I_{t-1}$ and otherwise do not produce in that period.

Then we can write chance constraints of the form $P\{I_t \geq 0\} \geq 1 - \alpha$, where α denotes the acceptable probability of stockout in a period. The optimization problem is to determine the X_t. Defining $g_t = U_t - U_{t-1}$, implying $X_t = g_t + D_{t-1}$, yields

$$I_t = I_0 + \sum_{\tau=1}^{t} (X_\tau - D_\tau) = I_0 + \sum_{\tau=1}^{t} g_\tau - D_t. \tag{16.98}$$

This allows us to write the chance constraint as

$$P\left\{I_0 + \sum_{\tau=1}^{t} g_t \geq D_t\right\} \geq 1 - \alpha \tag{16.99}$$

implying

$$I_t = I_0 + \sum_{\tau=1}^{t} g_\tau \geq F^{-1}(1 - \alpha), \tag{16.100}$$

where the right-hand side is now a constant. Most existing CC production planning models are uncapacitated, and do not model WIP in any form. The chance constraints are thus developed to ensure that the finished goods inventory at the end of each period is positive with a given probability, corresponding to the service level desired. The most common approach to developing an objective function is to assume that the probabilities of constraint violation are sufficiently small that backorder costs can be neglected, as suggested by Bookbinder and Tan (1988) and Johnson and Montgomery (1974). Similar CC formulations of chance-constrained problems are given by Bookbinder and H'ng (1986), Gupta and Sengupta (1977), Sengupta (1972), Sengupta and Portillo-Campbell (1973), and Rakes et al. (1984), among others.

This approach does not appear to have been used much in recent years; stochastic programming (see, e.g., Birge and Louveaux 1997) has been preferred, for strong reasons related to the difficulties of the chance-constrained approach in modeling recourse actions (Blau 1974; Hogan et al. 1981; Charnes and Cooper 1983). The stochastic programming formulation is mathematically more complete in terms of its ability to model multiple stage decision problems with recourse actions possible at each stage. However, the large number of decision stages, corresponding to the time periods in planning problems encountered in industry, renders the use of stochastic programming computationally challenging, as suggested by Peters et al. (1977). The CC formulation has a number of difficulties – the desired probabilities of constraint violation need to be specified *a priori*, and the degree to which the constraints are violated is not accounted for in the objective function. From a practical perspective, the models are infeasible if it is not possible to satisfy all chance constraints with the desired probabilities, without providing the user any means of trading off service levels between products. In order to obtain tractable constraint sets, distributional assumptions must be made about the random variables on the right hand sides of the constraints. Extensive discussions of these issues can be found in, for example, Prekopa (1993) and Lejeune and Prekopa (2005).

However, the CC formulations also offer a number of advantages for practical implementation relative to stochastic programming, The first of these is that with varying degrees of approximation, depending on the degree to which the distributional assumptions on the random variables are violated, these models can be implemented using extensions of the LP formulations with which both practitioners and researchers are familiar. While pre-specifying the probabilities of constraint violation may be problematic in many application domains, in the context of production planning that we consider, the probability of constraint violation has a natural interpretation as the probability of a stockout. The need to pre-specify stockout probabilities may actually be an advantage in practice, as it forces users to think in terms of service levels, perhaps based on aggregating products into product families or customers into priority classes.

A specific kind of demand uncertainty can emerge in production planning models that decide on *aggregate* sales, production and inventory, usually over the seasonal cycle (sales and operations planning; see Vollmann, Berry et al. 2005, p. 60 ff.). Typically the products are aggregated into groups or families of products with similar demand pattern and resource requirements, and the decision variables are defined at this aggregate level. Since stockouts are defined at the level of individual products, nonnegativity of the aggregate inventory is not a sufficient condition for a feasible production plan. If the aggregate demand is considered as deterministic and the demand of individual products is uncertain with lower and upper bounds, the task is to find a *robust* aggregate production plan that allows a feasible solution at the level of individual products (disaggregation to obtain a feasible master production schedule). For this topic, see Lasserre and Mercé (1990) and Gfrerer and Zäpfel (1995).

16.9 Conclusions and Future Directions

The problems discussed in this chapter, of managing the release of work into a production system and allocating resource capacity among different products, constitutes one of the earliest applications of operations research to industrial planning and control problems, with literature dating back more than five decades. When viewed as part of a production planning and control hierarchy, the workload control approaches developed over the years focus on maintaining predictable lead time and throughput behavior in a stable demand environment, where the system can be expected to produce at a relatively constant rate. Traditional order release mechanisms, complemented by suitable methods for order acceptance and due date setting, are also recommended for make-to-order production where demand forecasts are difficult to obtain. This implies that the objective of these methods is to maintain a desirable pattern of aggregate material flow through the facility, which these techniques try to accomplish mainly by heuristic means with relatively little unifying theoretical support. The lack of a strong theoretical understanding of this area is evidenced by the fact that most of the existing knowledge from this type of research is in the form of results from simulation studies, which are sometimes contradictory, and often hard to generalize beyond the specific production system topology in which they were tested.

Focusing on optimization of aggregate material flows through the production system motivates the discussion of the second, higher level of the planning hierarchy, where a more aggregate plan for work release and capacity allocation take place. These models generally take a more aggregate perspective, with time being divided into discrete planning periods and material flows being viewed as a continuous medium as opposed to discrete jobs that must be handled as an integral unit. We have focused in particular on the rich literature on mathematical programming models of these problems, almost all of which can trace their ancestry to the work of Holt, Modigliani and their collaborators in the 1950s (Holt et al. 1955, 1956, 1960; Modigliani and Hohn 1955). It is interesting to note that until very recently, there has been a hiatus in research on these models; between the late 1970s and the late 1990s there are relatively few papers on formulation and modeling aspects of these problems, with the work of Leachman and his coworkers (Hackman and Leachman 1989; Leachman and Carmon 1992; Hung and Leachman 1996; Dessouky and Leachman 1997) being a significant exception. It is also interesting to note that the 1974 book by Johnson and Montgomery is still one of the best available references for most of the classical work in this area. One is left with the feeling that for many years this area was perceived as a "solved" problem with no further interesting research issues.

We hope that the discussion in this chapter will stimulate wider interest in both industry and academia in this area. While widely taught in academia and used in industry, the classical linear programming models have a number of limitations arising from their very aggregate, static approach to modeling production capacity. It is heartening that in recent years a growing number of researchers have begun to explore these problems anew (Pahl et al. 2005). The new approaches differ substantially from the classical approaches in their efforts to achieve solutions that are consistent with the queuing behavior of production systems, which is well studied (Buzacott and Shanthikumar 1993; Hopp and Spearman 2001), and thus tend to have nonlinear structure which, in many cases, can be addressed effectively in computational procedures.

A number of important research directions have been outlined in the chapter, but are worth summarizing again. The new nonlinear approaches, among which the clearing function approach appears to be the most studied, show considerable promise but need to be better understood both empirically and theoretically. The issues of how to derive clearing functions analytically in multistage systems when decisions at one stage affect the variability of arrivals, and hence the shape of the clearing function, at downstream stages needs to be examined. There also needs to be a better understanding of the implications of using steady-state queuing results to develop clearing functions for use in a dynamic, nonstationary environment, where the purpose of the planning process is to change release rates over time. A closely related and not well-understood issue is that of how changes in decision variable values at the boundaries between planning periods affect the implementability and execution of the plans obtained. To what extent insights on the transient behavior of queuing systems should be integrated into the models is not known today. In terms of empirical estimation of clearing functions, we have experimental results demonstrating that a simple least-squares fit to empirical data may result in very poor planning models, but there is no theoretically justified alternative approach available as yet.

The main alternative to clearing function models are lead time-oriented models. Fixed lead times do not recognize the load-dependence of the lead times in the case of time-varying capacity load. A fixed relationship between lead time distribution and capacity load in a period can lead to substantial modeling errors since it does not capture the dynamic characteristics of lead times. The iterative approaches of Hung and Leachman (1996) and Riaño et al. (2006) are very interesting, but their computational performance, especially their convergence characteristics, have not been tested extensively.

Similar concerns hold for most of the other approaches that have been suggested as alternatives. While both stochastic programming and chance constraint formulations have been proposed for addressing the issue of uncertainty inherent in most industrial applications, effective computational procedures are not available, and the implications of the formulations are not well understood. Most research on lot-sizing has focused on developing effective solution procedures for the resulting fixed-charge integer programming models, but the majority of these models use the same model of capacity as the classical linear programming models, and there is clearly much work to be done here.

Finally, from the point of view of industrial applications, it is notable that many of the proposed new approaches are significantly more complex in both their data and their computational requirements, and especially load-dependent lead times may complicate coordination between manufacturing departments. It is by no means obvious that the proposed new models are always superior to the classical models in all industrial environments. This requires developing a better, theory-based understanding of the conditions under which the additional complexity of the new models is justified over the well-understood classical models that have been the mainstay of industrial practice for several decades.

Acknowledgments The research of Reha Uzsoy was supported by the National Science Foundation under Grant DMI-0556136, by the Intel Research Council, by a software grant from Dash Optimization and an equipment grant from Intel Corporation.

References

Agnew C (1976) Dynamic modeling and control of some congestion prone systems. Oper Res 24(3):400–419

Andersson H, Axsater S et al. (1981) Hierarchical material requirements planning. Int J Prod Res 19(1):45–57

Anli OM, Caramanis M et al. (2007). Tractable supply chain production planning modeling non-linear lead time and quality of service constraints. J Manuf Syst 26(2):116–134

Anthony RN (1966) Planning and control systems: a framework for analysis. Harvard University Press, Cambridge

Asmundsson JM, Rardin RL et al. (2006) Tractable nonlinear production planning models for semiconductor wafer fabrication facilities. IEEE Trans Semicond Manuf 19:95–111

Asmundsson JM, Rardin RL et al. (2009) Production planning models with resources subject to congestion. Naval Res Log 56:142–157

Baker KR (1993) Requirements planning. In: Graves SC, Rinnooy Kan AHG, Zipkin PH. Logistics of production and inventory. Handbooks in operations research and management science, vol 3. Elsevier Science, Amsterdam, pp 571–627

Bergamaschi D, Cigolini R et al. (1997) Order review and release strategies in a job shop environment: a review and a classification. Int J Prod Res 35:399–420

Bermon S, Hood SJ (1999) Capacity optimization planning system (CAPS). Interfaces 29(5):31–50

Bertrand JWM, Wortmann JC (1981) Production control and information systems for component-manufacturing shops. Elsevier, Amsterdam

Bertrand JWM, Wortmann JC et al. (1990) Production control: a structural and design oriented approach. Elsevier, Amsterdam

Bertsimas D, Gamarnik D et al. (2003) From fluid relaxations to practical algorithms for high-multiplicity job shop scheduling: the holding cost objective. Oper Res 51(5):798–813

Bertsimas D, Sethuraman J (2002) From fluid relaxations to practical algorithms for job shop scheduling: the makespan objective. Math Program Series A 92:61–102

Birge JR, Louveaux F (1997) Introduction to stochastic programming. Springer, New York

Bitran GR, Haas EA et al. (1981) Hierarchical production planning: a single stage system. Oper Res 29(4):717–743

Bitran GR, Haas EA et al. (1982) Hierarchical production planning: a two-stage system. Oper Res 30(2):232–251

Bitran GR, Tirupati D (1993) Hierarchical production planning. Graves SC, Rinnooy Kan AHG, Zipkin PH Logistics of production and inventory. Handbooks in operations research and management science, vol. 4. Elsevier Science, Amsterdam, pp 523–568

Blau RA (1974) Stochastic programming and decision analysis: an apparent dilemma. Manage Sci 21(3):271–276

Bookbinder JH, H'ng BT (1986) Rolling horizon production planning for probabilistic time-varying demands. Int J Prod Res 24(6):1439–1458

Bookbinder JH, Tan JY (1988) Strategies for the probabilistic lot sizing problem with service level constraints. Manage Sci 34(9):1096–1108

Bowman EB (1956) Production scheduling by the transportation method of linear programming. Oper Res 4(1):100–103

Buzacott JA, Shanthikumar JG (1993) Stochastic models of manufacturing systems. Prentice-Hall, Englewood Cliffs

Byrne MD, Bakir MA (1999) Production planning using a hybrid simulation-analytical approach. Int J Prod Econ 59:305–311

Byrne MD, Hossain MM (2005) Production planning: an improved hybrid approach. Int J Prod Econ 93–94:225–229

Caramanis M, Pan H et al. (2001) A closed-loop approach to efficient and stable supply chain coordination in complex stochastic manufacturing. American Control Conference, Arlington, VA, 1381–1388

Carey M (1987) Optimal time-varying flows on congested networks. Oper Res 35(1):58–69

Carey M, Subrahmanian E (2000) An approach to modelling time-varying flows on congested networks. Transp Res B 34:157–183

Cassidy M (2003) Traffic flow and capacity. In: Hall RW (ed) Handbook of transportation science. Kluwer Academic, Dordrecht, pp 155–191

Charnes A, Cooper WW (1963) Deterministic equivalents for optimizing and satisficing under chance constraints. Oper Res 11:18–39

Charnes A, Cooper WW (1983) Response to "decision problems under risk and chance constrained programming: dilemmas in the transition". Manage Sci 29(6):750–753
Charnes A, Cooper WW et al. (1955) A model for optimizing production by reference to cost surrogates. Econometrica 23(3):307–323
Chen HB, Mandelbaum A (1991) Hierarchical modelling of stochastic networks part I: fluid models. In: Yao DD (ed) Stochastic modeling and analysis of manufacturing systems. Springer, New York
Cohen JW (1969) The Single server queue. North-Holland, Amsterdam
Cohen O (1988) The drum-buffer-rope (DBR) approach to logistics. In: Rolstadas A (ed) Computer-aided production management. Springer, New York
Davidson R, MacKinnon JG (1993) Estimation and inference in econometrics. Oxford University Press, New York
de Kok AG, Fransoo JC (2003) Planning supply chain operations: definition and comparison of planning concepts. In: de Kok AG, Graves SC (eds) OR Handbook on supply chain management. Elsevier, Amsterdam, pp 597–675
Dessouky MM, Leachman RC (1997) Dynamic models of production with multiple operations and general processing times. J Oper Res Soc 48(6):647–654
Drexl A, Kimms A (1997) Lot sizing and scheduling – survey and extensions. Eur J Oper Res 99:221–235
Elmaghraby SE (1978) The economic lot scheduling problem (ELSP): review and extensions. Manage Sci 24:587–598
Eppen G, Martin RK (1988) Determining safety stock in the presence of stochastic lead times. Manage Sci 34:1380–1390
Fine CH, Graves SC (1989) A tactical planning model for manufacturing subcomponents of mainframe computers. J Manuf Oper Manage 2:4–34
Forrester JW (1962) Industrial dynamics. MIT Press, Cambridge
Fredendall LD, Ojha D, Patterson W (2010) Concerning the theory of workload control. Eur J Oper Res 201:99–111
Gfrerer H, Zäpfel G (1995) Hierarchical model for production planning in the case of uncertain demand. Eur J Oper Res 86:142–161
Graves SC (1981) A review of production scheduling. Oper Res 29(4):646–675
Graves SC (1986) A tactical planning model for a job shop. Oper Res 34:552–533
Graves SC (1988) Safety stocks in manufacturing systems. J Manuf Oper Manage 1:67–101
Gunther HO, Van Beek P (2003) Advanced planning and scheduling solutions in process industry. Springer, Heidelberg
Gupta M (2005) Constraints management – recent advances and practices. Int J Prod Res 41(4):647–659
Gupta SK, Sengupta JK (1977) Decision rules in production planning under chance-constrained sales. Decision Sci 8:521–533
Hackman S (2008) Production economics. Springer, Berlin
Hackman ST, Leachman RC (1989) A general framework for modeling production. Manage Sci 35:478–495
Hanssmann F, Hess SW (1960) A linear programming approach to production and employment scheduling. Manage Technol 1(1):46–51
Harris FW (1915) Operations and cost. Factory management series. Shaw, Chicago
Hax AC, Candea D (1984) Production and inventory management. Prentice-Hall, Englewood Cliffs
Haxholdt C, Larsen ER et al. (2003) Mode locking and chaos in a deterministic queueing model with feedback. Manage Sci 49(6):816–830
Hendry LC, Kingsman BG (1991) A decision support system for job release in make to order companies. Int J Oper Prod Manage 11:6–16
Hogan AJ, Morris JG et al. (1981) Decision problems under risk and chance constrained programming: dilemmas in the transition. Manage Sci 27(6):698–716
Holt CC, Modigliani F et al. (1955) A linear decision rule for production and employment scheduling. Manage Sci 2(1):1–30

Holt CC, Modigliani F et al. (1956) Derivation of a linear rule for production and employment. Manage Sci 2(2):159–177

Holt CC, Modigliani F et al. (1960) Planning production, inventories and work force. Prentice Hall, Englewood Cliffs

Hopp WJ, Spearman ML (2001) Factory physics: foundations of manufacturing management. Irwin/McGraw-Hill, Boston

Hung YF, Chang CB (1999) Determining safety stocks for production planning in uncertain manufacturing. Int J Prod Econ 58:199–208

Hung YF, Cheng GJ (2002) Hybrid capacity modelling for alternative machine types in linear programming production planning. IIE Trans 34:157–165

Hung YF, Hou MC (2001) A production planning approach based on iterations of linear programming optimization and flow time prediction. J Chinese Inst Ind Engrs 18(3):55–67

Hung YF, Leachman RC (1996) A production planning methodology for semiconductor manufacturing based on iterative simulation and linear programming calculations. IEEE Trans Semicond Manufac 9(2):257–269

Hung YF, Wang QZ (1997) A new formulation technique for alternative material planning – an approach for semiconductor bin allocation. Comput Ind Eng 32(2):281–297

Hwang S, Uzsoy R (2005) A single stage multi-product dynamic lot sizing model with work in process and congestion. Research report, Laboratory for Extended Enterprises at Purdue, School of Industrial Engineering, Purdue University, West Lafayette

Irastorza JC, Deane RH (1974) A loading and balancing methodology for job shop control. AIIE Trans 6(4):302–307

Irdem DF, Kacar NB et al. (2008) An experimental study of an iterative simulation-optimization algorithm for production planning. In: Mason SJ, Hill R, Moench L, Rose O (eds) 2008 Winter simulation conference, Miami FL

Jackson JR (1955) Scheduling a production line to minimize maximum tardiness. University of California, Los Angeles

Jackson JR (1957) Networks of waiting lines. Opeartions Research 10(4):518–521

Johnson LA, Montgomery DC (1974) Operations research in production planning, scheduling and inventory control. Wiley, New York

Kanet JJ (1988) Load-limited order release in job shop scheduling systems. J Oper Manage 7:413–422

Karmarkar US (1987) Lot sizes, lead times and in-process inventories. Manage Sci 33(3):409–418

Karmarkar US (1989) Capacity loading and release planning with work-in-progress (WIP) and lead-times. J Manufac Oper Manage 2:105–123

Karmarkar US (1993) Manufacturing lead-times, order release and capacity loading. In: Graves SC, Rinnooy Kan AHG, Zipkin PH (eds) Logistics of production and inventory. Handbooks in operations research & management science, vol. 4. North-Holland, Amsterdam, pp 287–329

Karmarkar US, Kekre S et al. (1985a) Lotsizing in multimachine job shops. IIE Trans 13(3): 290–298

Karmarkar US, Kekre S et al. (1985b) Lot sizing and lead time performance in a manufacturing cell. Interfaces 15(2):1–9

Kekre S (1984) The effect of number of items processed at a facility on manufacturing lead time. Working paper series. University of Rochester, Rochester

Kekre S (1987) Performance of a manufacturing cell with increased product mix. IIE Trans 19(3):329–339

Kim B, Kim S (2001) Extended model for a hybrid production planning approach. International J Prod Econ 73:165–173

Kim JS, Leachman RC (1994) Decomposition method application to a large scale linear programming WIP projection model. Eur J Oper Res 74:152–160

Kim JS, Leachman RC et al. (1996) Dynamic release control policy for the semiconductor wafer fabrication lines. J Oper Res Soc 47(12):1516–1525

Kistner KP (1999) Lot sizing and queueing models: some remarks on Karmarkar's model. In: Leopold-Wildburger U, Feichtinger G, Kistner HP (eds) Modelling and Decisions in Economics: Essays in Honor of Franz Ferschl. Physica, Heidelberg, pp 173–188

Kleinrock L (1976) Queueing systems volume II: computer system applications. Wiley, New York

Koopmans T (ed) (1951) Activity analysis of production and allocation. Wiley, New York

Krämer W, Langenbach-Belz M (1976) Approximate formulae for the delay in queueing system GI/G/1. 8th International telegraphic congress. Melbourne, pp 235/1–235/8

Lambrecht MR, Chen S et al. (1996) A Lot sizing model with queueing delays: the issue of safety time. Eur J Oper Res 89:269–276

Lambrecht MR, Luyten R et al. (1984a) Protective inventories and bottlenecks in production systems. Eur J Oper Res 22:319–328

Lambrecht MR, Muckstadt JA et al. (1984b) Protective stocks in multi-stage production systems. Int J Prod Res 22:1001–1025

Land M (2004) Workload control in job shops, grasping the tap. Labyrinth, Ridderkerk

Lasserre JB, Mercé C (1990) Robust hierarchical production planning under uncertainty. Ann Oper Res 26(4):73–87

Lautenschläger M (1999) Mittelfristige Produktionsprogrammplanung mit auslastungsabhängigen Vorlaufzeiten. Peter Lang, Frankfurt am Main

Lautenschläger M, Stadtler H (1998) Modelling lead times depending on capacity utilization. Research report, Technische Universitat Darmstadt

Leachman RC (1993) Modeling techniques for automated production planning in the semiconductor industry. In: Ciriani TA, Leachman RC (eds) Optimization in industry: mathematical programming and modelling techniques in practice. Wiley, New York, pp 1–30

Leachman RC, Benson RF et al. (1996) IMPReSS: an automated production planning and delivery quotation system at Harris corporation – semiconductor sector. Interfaces 26:6–37

Leachman RC, Carmon TF (1992) On capacity modeling for production planning with alternative machine types. IIE Trans 24(4):62–72

Lejeune MA, Prekopa A (2005) Approximations for and convexity of probabilistically constrained problems with random right hand sides. RUTCOR research report. Rutgers University, New Jersey

Liu L, Liu X et al. (2004) Analysis and optimization of multi-stage inventory queues. Manage Sci 50:365–380

Lu S, Ramaswamy D et al. (1994) Efficient scheduling policies to reduce mean and variance of cycle time in semiconductor plants. IEEE Trans Semicond Manufac 7:374–388

Luss H (1982) Operations research and capacity expansion problems: a survey. Oper Res 30(5):907–947

Manne AS (1957) A note on the Modigliani-Hohn production smoothing model. Manage Sci 3(4):371–379

Manne AS (1960) On the job-shop scheduling problem. Oper Res 8(2):219–223

Medhi J (1991) Stochastic models in queuing theory. Academic, Amsterdam

Merchant DK, Nemhauser GL (1978a) A model and an algorithm for the dynamic traffic assignment problems. Transp Sci 12(3):183–199

Merchant DK, Nemhauser GL (1978b) Optimality conditions for a dynamic traffic assignment model. Transp Sci 12(3):200–207

Missbauer H (1997) Order release and sequence-dependent setup times. Int J Prod Econ 49:131–143

Missbauer H (1998) Bestandsregelung als Basis für eine Neugestaltung von PPS-Systemen. Physica, Heidelberg

Missbauer H (1999) Die Implikationen durchlauforientierter Losgrößenbildung für die Komplexität der Produktionsplanung und –steuerung. Zeitschrift für Betriebswirtschaft 69(2): 245–265

Missbauer H (2002a) Aggregate order release planning for time-varying demand. Int J Prod Res 40:688–718

Missbauer H (2002b) Lot sizing in workload control systems. Prod Plan Control 13:649–664

Missbauer H (2009) Models of the transient behaviour of production units to optimize the aggregate material flow. Int J Prod Econ 118(2):387–397
Missbauer H (forthcoming) Order release planning with clearing functions: a queueing-theoretical analysis of the clearing function concept. Int J Prod Econ
Missbauer H, Hauber W et al. (forthcoming). Developing a computerized scheduling system for the steelmaking - continuous casting process. In: Kempf KG, Keskinocak P, Uzsoy R (eds) Planning in the extended enterprise: a state of the art handbook. Springer, New York
Modigliani F, Hohn FE (1955) Production planning over time and the nature of the expectation and planning horizon. Econometrica 23(1):46–66
Neuts MF (1981) Matrix-geometric solutions in stochastic models. Johns Hopkins University Press, Baltimore
Nyhuis P, Wiendahl HP (2003) Logistische Kennlinien. Springer, Berlin
Orcun S, Uzsoy R et al. (2006) Using system dynamics simulations to compare capacity models for production planning. Winter simulation conference. Monterey, CA
Orlicky J (1975) Material requirements planning: the new way of life in production and inventory management. McGraw-Hill, New York
Pahl J, Voss S et al. (2005) Production planning with load dependent lead times. 4OR 3:257–302
Parker RG (1995) Deterministic scheduling theory. Chapman and Hall, London
Parrish SH (1987) Extensions to a model for tactical planning in a job shop environment. Operations Research Center. Massachusetts Institute of Technology, Cambridge, MA
Peeta S, Ziliaskopoulos AK (2001) Foundations of dynamic traffic assignment: the past, the present and the future. Network Spatial Econ 1(3–4):233–265
Perona M, Portioli A (1998) The impact of parameters setting in load oriented manufacturing control. Int J Prod Econ 55(133–142)
Peters RJ, Boskma K et al. (1977) Stochastic programming in production planning: a case with non-simple recourse. Statistica Neerlandica 31:113–126
Philipoom RR, Fry TD (1992) Capacity based order review/release strategies to improve manufacturing performance. Int J Prod Res 30:2559–2572
Pinedo M (1995) Scheduling theory, algorithms, and systems. Prentice-Hall, New Jersey
Pinedo M, Chao X (2005) Planning and scheduling in manufacturing and services. Springer, New York
Powell SG, Schultz KL (2004) Throughput in serial lines with state-dependent behaviour. Manage Sci 50(8):1095–1105
Prekopa A (1993) Programming under probabilistic constraint and maximizing a probability under constraints. Center for operations Research, Rutgers University, New Brunswick
Rakes TR, Franz LS et al. (1984) Aggregate production planning using chance-constrained goal programming. Int J Prod Res 22(4):673–684
Riaño G (2003) Transient behavior of stochastic networks: application to production planning with load-dependent lead times. School of Industrial and Systems Engineering. Georgia Institute of Technology, Atlanta
Riaño G, Hackman S et al. (2006) Transient behavior of queueing networks. School of Industrial and Systems Engineering, Georgia Institute of Technology, Atlanta
Riaño G, Serfozo R et al. (2003) Benchmarking of a stochastic production planning model in a simulation testbed. Winter simulation conference
Schneeweiß C (2003) Distributed decision making. Springer, Berlin
Selçuk B (2007) Dynamic performance of hierarchical planning systems: modeling and evaluation with dynamic planned lead times. Technische Universiteit Eindhoven, Eindhoven
Selçuk B, Fransoo JC et al. (2007) Work in process clearing in supply chain operations planning. IIE Trans 40:206–220
Sengupta JK (1972) Decision rules in stochastic programming under dynamic economic models. Swed J Econ 74:370–389
Sengupta JK, Portillo-Campbell JH (1973) A reliability programming approach to production planning. Int Stat Rev 41:115–127

Singhal J, Singhal K (2007) Holt, Modigliani, Muth and Simon's work and its role in the renaissance and and evolution of operations management. J Oper Manage 25:300–309
Smith SF (1993) Knowledge-based production management: approaches, results and prospects. Prod Plan Control 3(4):350–380
Spearman ML (1991) An analytic congestion model for closed production systems with IFR processing times. Manage Sci 37(8):1015–1029
Spearman ML, Woodruff DL et al. (1990) CONWIP: a pull alternative to Kanban. Int J Prod Res 28(5):879–894
Spitter JM, de Kok AG et al. (2005a) Timing production in LP models in a rolling schedule. Int J Prod Econ 93–94:319–329
Spitter JM, Hurkens CAJ et al. (2005b) Linear programming models with planned lead times for supply chain operations planning. Eur J Oper Res 163:706–720
Srinivasan A, Carey M et al. (1988) Resource pricing and aggregate scheduling in manufacturing systems. Graduate School of Industrial Administration, Carnegie-Mellon University, Pittsburgh
Stadtler H (1996) Hierarchische Produktionsplanung. Handwörterbuch der Produktionswirtschaft. Schäffer-Poeschel, Stuttgart, pp 631–641
Stadtler H, Kilger C (eds) (2008) Supply chain management and advanced planning: concepts, models, software and case studies. Springer-Verlag, Berlin
Stange K (1964) Die Anlauflösung für den einfachen exponentiellen Bedienungskanal (mit beliebig vielen Warteplätzen), der für t=0 leer ist. Unternehmenforschung 8:1–24
Stevenson M, Hendry LC (2006) Aggregate load-oriented workload control: a review and a reclassification of a key approach. Int J Prod Econ 104(2):676–693
Tang L, Liu J et al. (2001) A review of planning and scheduling systems and methods for integrated steel production. Eur J Oper Res 133:1–20
Tardif V, Spearman ML (1997) Diagnostic scheduling in finite-capacity production environments. Comput Ind Eng 32:867–878
Tatsiopoulos IP, Kingsman BP (1983) Lead time management. Eur J Oper Res 14:351–358
Tijms HC (1994) Stochastic models: an algorithmic approach. Wiley, New York
Uzsoy R, Lee CY et al. (1994) A review of production planning and scheduling models in the semiconductor industry part II: shop-floor control. IIE Trans Scheduling Logistics 26:44–55
Van Ooijen HPG (1996) Load-based work-order release and its effectiveness on delivery performance improvement. Eindhoven University of Technology, Eindhoven
Van Ooijen HPG, Bertrand JWM (2003) The effects of a simple arrival rate control policy on throughput and work-in-process in production systems with workload dependent processing rates. Int J Prod Econ 85(1):61–68
Vaughan TS (2006) Lot size effects on process lead time, lead time demand, and safety stock. Int J Prod Econ 100:1–9
Vepsalainen AP, Morton TE (1987) Priority rules for job shops with weighted tardiness costs. Manage Sci 33(8):1035–1047
Vepsalainen AP, Morton TE (1988) Improving local priority rules with global lead-time estimates: a simulation study. J Manufac Oper Manage 1:102–118
Vollmann TE, Berry WL et al. (1988) Manufacturing planning and control systems. Richard D. Irwin, Boston
Vollmann TE, Berry WL et al. (2005) Manufacturing planning and control for supply chain management. McGraw-Hill, New York
Voss S, Woodruff DL (2003) Introduction to computational optimization models for production planning in a supply chain. Springer, Berlin
Wagner HM, Whitin TM (1958) Dynamic version of the economic lot size model dynamic version of the economic lot size model. Manage Sci 5:89–96
Wiendahl HP (1995) Load oriented manufacturing control. Springer, Heidelberg
Wight O (1983) MRPII: unlocking America's productivity potential. Oliver Wight, Williston
Wijngaard J, Wortmann JC (1985) MRP and inventories. Eur J Oper Res 20:281–293
Yano CA, Carlson RC (1988) Safety stocks for assembly systems with fixed production intervals. J Manufac Oper Manage 1:182–201

Zäpfel G, Missbauer H (1993a) Production planning and control (PPC) systems including load-oriented order release – problems and research perspectives. Int J Prod Econ 30:107–122

Zäpfel G, Missbauer H (1993b) New concepts for production planning and control. Eur J Oper Res 67:297–320

Zäpfel G, Missbauer H et al. (1992) PPS-Systeme mit belastungs orientierter Auftragsfreigabe – Operationscharakteristika und Möglichkeiten zur Weiterentwicklung. Zeitschrift für Betriebswirtschaft 62:897–919

Zipkin PH (1986) Models for design and control of stochastic, multi-item batch production systems. Oper Res 34(1):91–104

Zipkin PH (1997) Foundations of inventory management. Irwin, Burr Ridge

Zweben M, Fox M (eds) (1994) Intelligent scheduling systems. Morgan Kaufman, San Francisco

Chapter 17
Aggregate Modeling of Manufacturing Systems

Erjen Lefeber and Dieter Armbruster

17.1 Introduction

Manufacturing systems can be modeled in several ways. In particular, during the design of a manufacturing system, discrete event modeling is an often used approach, cf. Banks (1998) and Cassandras and Lafortune (1999). Discrete event models often include a high level of detail. This high level of detail can be used to investigate the effect of all kinds of variables on the possible performance of the manufacturing system. However, when a manufacturing system is in operation, this model usually contains too much detail to keep all parameters up-to-date with the evolving current system. In addition, certain parameters cannot even be measured. Furthermore, running one scenario using a discrete event model takes several hours. Usually, discrete event models are only tailer made for answering specific problems. These models only contain part of the manufacturing system.

Another option might be to derive a less detailed model, in particular, for manufacturing planning and control or supply chain control. In this chapter, we discuss three classes of models, each at a different level of aggregation. We start with less detailed discrete event models based of effective process times (EPTs), where each workstation is modeled as a node in a queuing network. Next, in particular for the purpose of planning and control, we abstract from events and replace all discrete event queues with discrete time fluid queues. In addition, the throughput of each workstation is limited by a nonlinear function of the queue length, the clearing function, see also Chap. 16 of this book. Finally, we abstract from workstations and model manufacturing flow as a real fluid using continuum models. These models are scalable and suitable for supply chain control.

E. Lefeber (✉)
Department of Mechanical Engineering, Eindhoven University of Technology,
PO Box 513, Eindhoven, The Netherlands
e-mail: A.A.J.Lefeber@tue.nl

K.G. Kempf et al. (eds.), *Planning Production and Inventories in the Extended Enterprise*, 509
International Series in Operations Research & Management Science 151,
DOI 10.1007/978-1-4419-6485-4_17, © Springer Science+Business Media, LLC 2011

17.2 Effective Process Times

Building a discrete event model of an existing manufacturing system can be cumbersome, as manufacturing systems are prone to disturbances. Even though many disturbances can be modeled explicitly in highly detailed discrete event models, it is impossible to measure all sources of variability that might occur in a manufacturing system. In addition, highly detailed discrete event models are unsuitable for decision making due to their time-consuming simulation runs.

Instead of measuring detailed information, like raw process times, setup times, times to failures, times between repairs, operator behavior, etc., one can also try to measure the clean process time *including* other sources of additional waiting. This is the so-called *effective process time* (EPT), which has been introduced in Hopp and Spearman [2000] as the time seen by lots from a logistical point of view. In order to determine the EPT, they assume that the contribution of the individual sources of variability is known.

A similar description is given in Sattler [1996] where the EPT has been defined as all flow time except waiting for another lot. It includes waiting due to machine down time and operator availability and a variety of other activities. In Sattler [1996], it was also noticed that this definition of EPT is difficult to measure.

Instead of taking the bottom-up view of Hopp and Spearman [2000], a top-down approach can also be taken, as shown in Jacobs et al. (2001) and Jacobs et al. (2003), where algorithms have been introduced that enable determination of EPT realizations from a list of events. That is, instead of measuring each source of disturbances individually and derive an aggregate EPT distribution, one can also derive this EPT distribution from manufacturing data directly. In the remainder of this section, we illustrate for several situations how these EPTs can be measured from manufacturing data.

17.2.1 *A Single Lot Machine, No Buffer Constraints*

Consider a workstation consisting of one machine, which processes single lots (i.e., no batching) and assume that the Gantt chart of Fig. 17.1 describes a given time period.

- At $t = 0$, the first lot arrives at the workstation. After a setup, the processing of the lot starts at $t = 2$ and is completed at $t = 6$.
- At $t = 4$, the second lot arrives at the workstation. At $t = 6$ this lot could have been started, but apparently there was no operator available, so only at $t = 7$ the setup for this lot starts. Eventually, at $t = 8$, the processing of the lot starts and is completed at $t = 12$.
- The fifth lot arrives at the workstation at $t = 22$, processing starts at $t = 24$, but at $t = 26$ the machine breaks down. It takes until $t = 28$ before the machine has been repaired and the processing of the fifth lot continues. The processing of the fifth lot is completed at $t = 30$.

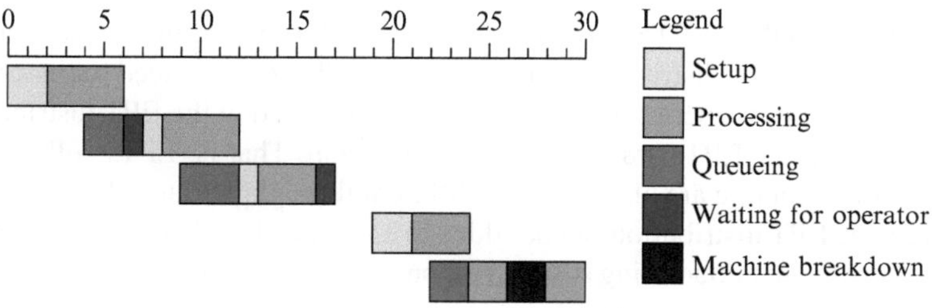

Fig. 17.1 Gantt chart of five lots at a single machine workstation

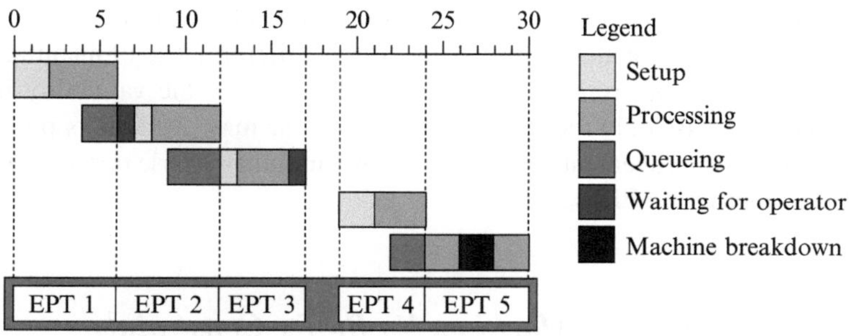

Fig. 17.2 EPT realizations of five lots at a single machine workstation

From a lot's point of view we observe:

- The first lot arrives at an empty system at $t = 0$ and departs from this system at $t = 6$. Its processing took 6 units of time.
- The second lot arrives at a nonempty system at $t = 4$ and needs to wait. At $t = 6$, the system becomes available and hence from $t = 6$ on there is no need for the second lot to wait. At $t = 12$, the second lot leaves the system, so from the point of view of this lot, its processing took from $t = 6$ till $t = 12$; the lot does not know whether waiting for an operator and a setup is part of its processing.
- The third lot sees no need for waiting after $t = 12$ and leaves the system at $t = 17$, so it assumes to have been processed from $t = 12$ till $t = 17$.

Following this reasoning, the resulting FPTs for lots are depicted in Fig. 17.2. Notice that only arrival and departure events of lots to a workstation are needed for determining the EPTs. Furthermore, none of the contributing disturbances needs to be measured.

In highly automated manufacturing systems, arrival and departure events of lots are being registered, so for these manufacturing systems, EPT realizations can be determined rather easily. These EPT realizations can be used in a relatively simple discrete event model of the manufacturing system, in this case a simple infinite FIFO queue. Such a discrete event model only contains the architecture of the manufacturing system, buffers, and machines. The process times of these machines are samples from their EPT-distribution as measured from real manufacturing data, or most often from the distribution fitted to that data. There is no need for incorporating machine failures, operators, etc., as this is all included in the EPT-distributions.

Furthermore, the EPTs are utilization independent. That is, EPTs collected at a certain throughput rate are also valid for different throughput rates. Also, machines with the same EPT-distribution can be added to a workstation. This makes it possible to study how the manufacturing system responds in case a new machine is added, or all kinds of other what-if-scenario's.

Finally, since EPT realizations characterize operational time variability, they can be used for performance measuring as explained in Ron and Rooda [2005]. Note that overall equipment effectiveness (OEE), which is widely used to quantify capacity losses in manufacturing equipment, directly relates to utilization, i.e., the fraction of time a workstation is busy. However, the performance of manufacturing systems is not only determined by utilization, but also by the variability in production processes. By only focusing on utilization, one may overlook opportunities for performance improvement by a reduction of variability. These opportunities are provided by measuring EPTs.

17.2.2 A Single Batch Machine, No Buffer Constraints

EPTs for equipment that serves batches of jobs have first been studied in Jacobs [2004] and Jacobs et al. [2006]. Consider a workstation consisting of one machine, which processes batches of jobs and assume that the Gantt chart of Fig. 17.3 describes a given time period. As we know from the previous section, only arrivals

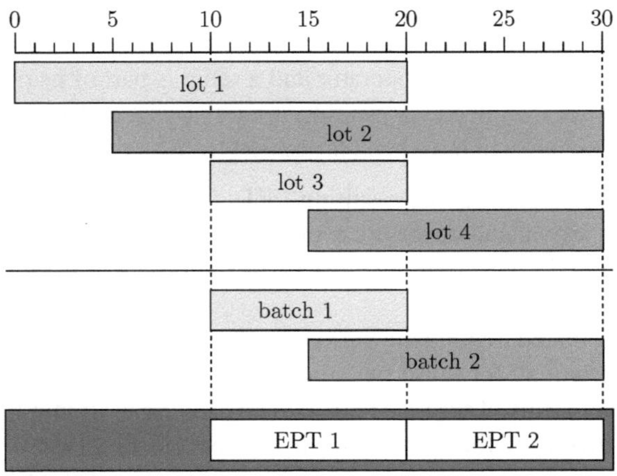

Fig. 17.3 Gantt chart of four lots (two batches) at a batch machine

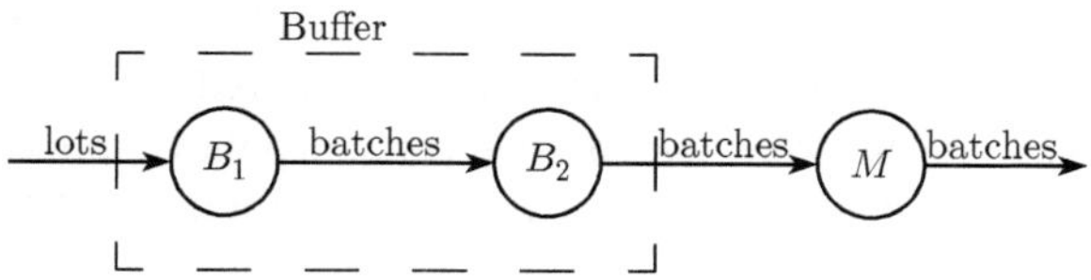

Fig. 17.4 Model of batch formation and queuing in front of a batch machine

and departures from jobs matter for determining EPTs, so in Fig. 17.3 we already abstracted from most disturbances. The only remaining issue is how to deal with the batching. For that purpose, we make a distinction between the policy for batch formation and the EPT of a batch. An other way of putting this is to assume that the buffer consists of two parts. A first part, B_1, in which lots are waiting to become batches, and a second part, B_2, where batches are queuing in front of the workstation, as depicted in Fig. 17.4. Taking this point of view, we can interpret Fig. 17.3 in the following way. At $t = 0$, the first lot arrives at the workstation in buffer B_1, waiting to become a batch with the third lot. At $t = 5$, the second lot arrives at the workstation in buffer B_1, waiting to become a batch with the fourth lot. At $t = 10$, the third lot arrives at the workstation in buffer B_1, resulting in the first batch to be formed. So at $t = 10$, the first batch moves from buffer B_1 to buffer B_2. At $t = 15$, the fourth lot arrives at the workstation in buffer B_1, resulting in the second batch to be formed. So at $t = 20$, the second batch moves from buffer B_1 to buffer B_2.

When we now look at the system consisting of buffer B_2 and the batch machine M, we have a system as we studied in the previous example. A system to which batches arrive, and which processes batches. The first batch arrives to this system at $t = 10$ and leaves the system at $t = 20$, the second batch arrives to this system at $t = 15$ and leaves the system at $t = 30$. Therefore, the first EPT runs from $t = 10$ till $t = 20$, the second EPT runs from $t = 20$ till $t = 30$.

Notice that using this approach, EPTs for batches only start as soon as a batch has been formed, or to be more precise: the batch that is processed finally. The period from $t = 0$ till $t = 10$, lot 1 was in the system and could have been processed as a batch of size 1. Therefore, one could argue that from the point of view of this lot, its EPT starts at $t = 0$. Also, one might say that as soon as lot 2 has arrived, a batch consisting of lots 1 and 2 could have been started, so the first EPT should have started at $t = 5$. This is not what we do, since we view batch formation as part of the way the system is controlled, not as a disturbance. As a result, we not only need to determine EPTs for batches, we also need to characterize the policy for batch formation. One way to deal with this is to include the policy for batch formation in the discrete event model which is actually being used in the manufacturing system under consideration. Another way to deal with this is to try to characterize the policy for batch formation in one way or the other, i.e., derive some "effective batch formation policy." The latter is still subject of current research.

As mentioned above, EPTs can also be used as performance measure. Notice that in case of batching, EPTs do not characterize capacity loss completely. Only the capacity loss given by the batches is characterized, including variability. Capacity loss due to a bad policy for batch formation is not captured in the EPT. This should be derived by analyzing the (effective) batch formation policy. Notice that again only arrival and departure event of lots are needed for determining the EPTs of batches.

17.2.3 A Multimachine Workstation, No Buffer Constraints

So far, we only considered workstations consisting of a single machine. However, workstations consisting of several machines in parallel can also be dealt with, see, e.g., Jacobs et al. [2003], Jacobs [2004] and Jacobs et al. [2006]. We do this in a similar way as we handled batching. That is, we view the decision of which lot is served by which machine again as part of the control system of the manufacturing system.

Consider a workstation consisting of two machines in parallel which both process single lots (i.e., no batching) and assume that the Gantt chart of Fig. 17.5 describes a given time period. Note that we abstracted from most disturbances like we did when we considered batching.

- At $t = 0$, the first lot arrives at the workstation. This lot is processed by Machine 1 and leaves this workstation at $t = 15$.
- At $t = 5$, the second lot arrives at the workstation. Even though Machine 2 is available, or at least not serving any job, this job is also processed by Machine 1 and leaves the workstation at $t = 25$.
- At $t = 10$, the third lot arrives at the workstation. This lot is processed by Machine 2 and leaves the workstation of $t = 30$.

The way we view this system, and is depicted in Fig. 17.6. We assume that the buffer consists of a dispatcher D which decides to which machine each lot will go. We assume that lots do not wait in this dispatcher, but immediately move on to a buffer in front of the machine at which they will finally be processed.

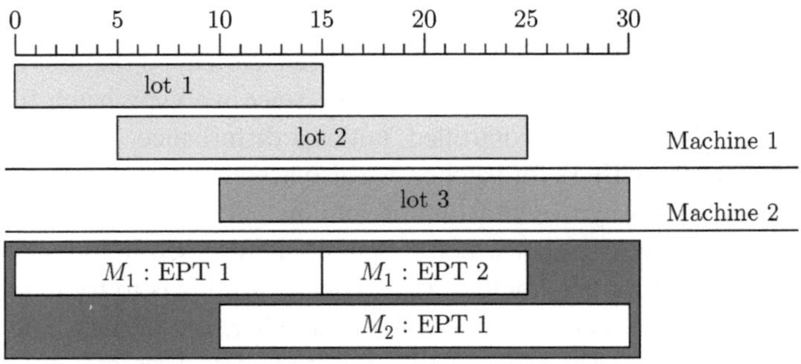

Fig. 17.5 Gantt chart of three lots at a workstation with two machines in parallel

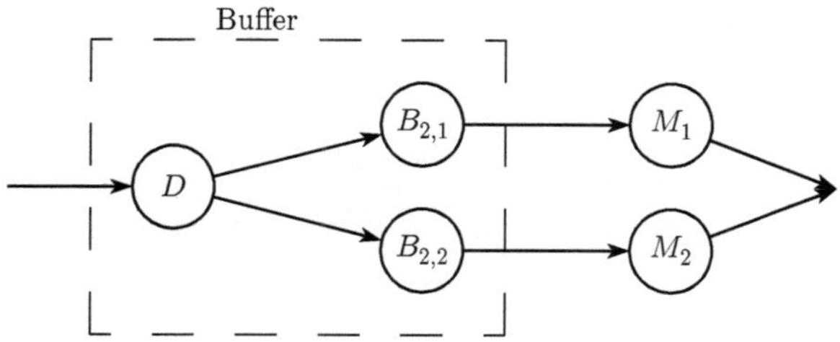

Fig. 17.6 Model of dispatching and queuing at a multimachine station

Using this abstraction, the EPTs as depicted in Fig. 17.5 follow straightforwardly for each separate machine. Notice again that the only data we need for determining the EPTs are arrival and departure event of lots. Also, we do not only need to determine the EPTs, but we also need to know the dispatching strategy. Either this policy is known from reality and can be implemented in the discrete event model, or an "effective dispatching policy" needs to be derived from manufacturing data. The latter is still a subject of current research. Furthermore, multimachine workstations with equipment that serves batches can easily be dealt with combining the results presented so far.

17.2.4 Finite buffers

In the preceding sections, we assumed infinite buffers or at least buffers that are large enough. This enabled us to analyze workstations in isolation. If buffer sizes are small and cannot be neglected, as for example in automotive industry, buffer sizes will explicitly be taken into account in the aggregate discrete event model. Therefore, the effect of blocking, will be explicitly taken into account by means of the discrete event model. Therefore, this disturbance should *not* be included in the EPT. To take into account the effect of blocking, a third event is needed. So far, we only needed arrival and departure events from lots. Or to be more precise: we needed *actual arrival* (AA) and *actual departure* (AD) events. For properly dealing with blocking we also need *possible departure* (PD) events, see also Kock et al. [2005, 2006a,c].

Consider a line of two machines in series, machine M_{j-1} and machine M_j, and assume that there is no buffer between these two machines. Let the Gantt chart of Fig. 17.7 describes a given time period, where we again abstracted from most disturbances.

- At $t = 0$, the first lot arrives at Machine M_{j-1}. At $t = 9$, this lot has been completed and moves to Machine M_j. Both the possible and actual departure at Machine M_{j-1} are at $t = 9$. Processing of the first lot at Machine M_j completes at $t = 22$.

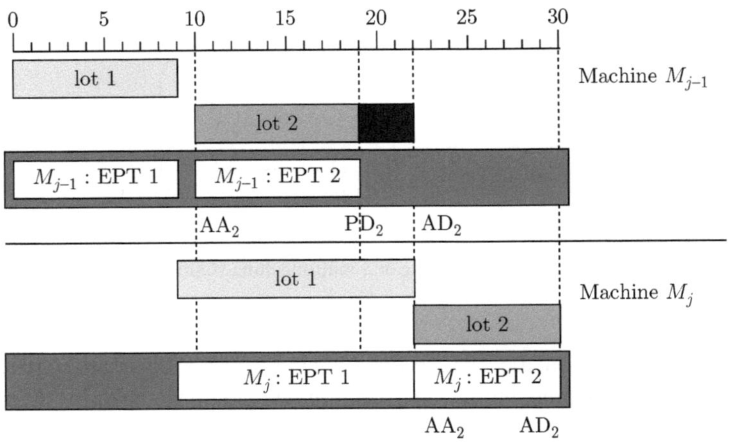

Fig. 17.7 Gantt chart of 2 lots at two sequential, unbuffered machines

- At $t = 10$, the second lot arrives at Machine M_{j-1}. At $t = 19$, this lot has been completed, but cannot yet move to Machine M_j. The possible departure for this lot is at $t = 19$. As Machine M_j only becomes available at $t = 22$, the actual departure at Machine M_{j-1} is at $t = 22$. The actual arrival at Machine M_j is at $t = 22$ for the second lot, and the actual departure at Machine M_j is at $t = 30$.

From the measured events, the EPTs follow readily. Since Machine M_{j-1} cannot help it to become blocked, the EPT for the second lot stops at $t = 19$, i.e., at the *possible departure* event. If we denote the jth EPT realization at Machine i as $\mathrm{EPT}_{i,j}$ we obtain

$$\mathrm{EPT}_{i,j} = \mathrm{PD}_{i,j} - \max\left(\mathrm{AA}_{i,j}, \mathrm{AD}_{i-1,j}\right), \tag{17.1}$$

where $\mathrm{AA}_{i,j} < \mathrm{PD}_{i,j} \leq \mathrm{AD}_{i,j}$ denote, respectively, the actual arrival, possible departure, and actual departure event at Machine i for lot j. By measuring only these three events at each machine, one is able to derive EPTs for each single job workstation in the manufacturing system.

17.2.5 Multilot Machines

By means of the results presented above, one is able to deal with both finite and infinite buffered multimachine workstations serving batches of jobs. In particular, multi might be one and batch sizes can also be one, so any kind of equipment can be dealt with which processes a single job at the time.

However, certain machines can start serving the next job before the previous one has left the machine. Typically these machines are some minifactories themselves. For these machines, we cannot use a simple queuing model. Therefore, for those machines, we cannot use the relation (17.1) to derive EPTs. A different aggregate model is needed for those kind of machines. First attempts for an aggregate model for multiple lot machines have been made in Eerden et al. [2006] and Kock et al. [2006b]. In particular, these models can also be used for aggregating parts of a manufacturing system. For the most recent results in this area, the interested reader may refer to Veeger et al. (2009) and http://se.wtb.tue.nl/~sereports.

17.3 Clearing Function Models

In the previous section, we derived how less detailed discrete event models can be build by abstracting from all kinds of disturbances like machine failure, setups, operator behavior, etc. By aggregating all disturbances into one EPT, a complex manufacturing system can be modeled as a relatively simple queueing network. Furthermore, the data required for this model can easily be measured from manufacturing data.

Even though this approach considerably reduces the complexity of discrete event models for manufacturing systems, this aggregate model is still unsuitable for manufacturing planning and control. Therefore, in this section, we introduce a next level of aggregation, by abstracting from events. Using the abstraction presented in the previous section, we can view a workstation as a node in a queueing network. In this section, we assume that such a node processes a deterministic continuous stream of fluid. That is, we consider this queue as a so-called fluid queue. In order not to loose the steady-state queueing relation between throughput and queue length, we impose this relation as a system constraint, the clearing function as introduced in Graves [1986], see also Chap. 20 of this book.

As an example, consider a manufacturing system consisting of two infinitely buffered workstations. Assume that Machine i has a mean EPT $t_{e,i}$ with a coefficient of variation $c_{e,i}$, i.e., a standard deviation of $c_{e,i} \cdot t_{e,i}$ for $i \in \{1, 2\}$. Let $u_0(k)$ denote the number of jobs started during the kth time period. Let $u_1(k)$ and $u_2(k)$ denote the utilization of Machines 1 and 2, respectively, during the kth time period. Furthermore, let $x_1(k)$ and $x_2(k)$ denote the buffer contents in workstations 1 and 2, respectively, at the beginning of the kth time period (i.e., the jobs in both buffer and machine), and let $x_3(k)$ denote the stored completed jobs or backlog at the beginning of the kth time period. Finally, let $d(k)$ denote the demand during the kth time period. Then we can write down the following discrete time fluid queue dynamics for this system

$$
\begin{aligned}
x_1(k+1) &= x_1(k) + u_0(k) - \frac{1}{t_{e,1}} u_1(k), \\
x_2(k+1) &= x_2(k) + \frac{1}{t_{e,1}} u_1(k) - \frac{1}{t_{e,2}} u_2(k), \\
x_3(k+1) &= x_3(k) + \frac{1}{t_{e,2}} u_2(k) - d(k).
\end{aligned}
\tag{17.2}
$$

Consider a workstation that consists of m identical servers in parallel which all have a mean effective processing time t_e and coefficient of variation c_e. Furthermore, assume that the coefficient of variation of the interarrival times is c_a and that the utilization of this workstation is $u < 1$. Then we know from the queuing theory of Takahasi and Sakasegawa [1977] that in steady state the mean number of jobs in this workstation is approximately given by

$$x = \frac{c_a^2 + c_e^2}{2} \cdot \frac{u^{\sqrt{2(m+1)}}}{m(1-u)} + u. \tag{17.3}$$

In Fig. 17.8, this relation has been depicted graphically. In the left-hand side of this figure, one can clearly see that for an increasing utilization, the number of jobs in this workstation increases nonlinearly. By swapping axes, this relation can be understood differently. Depending on the number of jobs in the workstation, a certain utilization can be achieved, or a certain throughput. This has been depicted in the right-hand side of Fig. 17.8. For the purpose of production planning, this effective clearing function provides an upper bound for the utilization of the workstation depending on the number of jobs in this workstation. Therefore, in addition to the model (17.2) we also have the constraints

$$\begin{aligned} \frac{c_{a,1}^2 + c_{e,1}^2}{2} \cdot \frac{u_1(k)^2}{1-u_1(k)} + u_1(k) &\le x_1(k), \\ \frac{c_{a,2}^2 + c_{e,2}^2}{2} \cdot \frac{u_2(k)^2}{1-u_2(k)} + u_2(k) &\le x_2(k). \end{aligned} \tag{17.4}$$

The clearing function model for production planning consists of the model (17.2) together with the constraints (17.4). When we want to use this clearing function model for production planning, we need the parameters c_e and c_a. In the previous section, we explained how EPTs can be determined for each workstation, which provides the parameter c_e for each workstation. In addition, for each workstation,

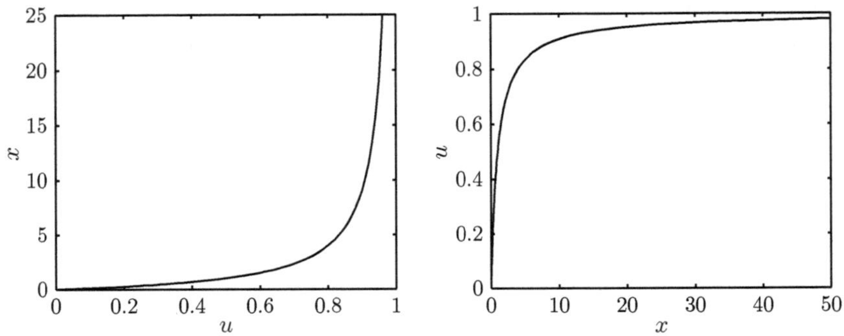

Fig. 17.8 Effective clearing function of (17.3) with $c_a = c_e = m = 1$

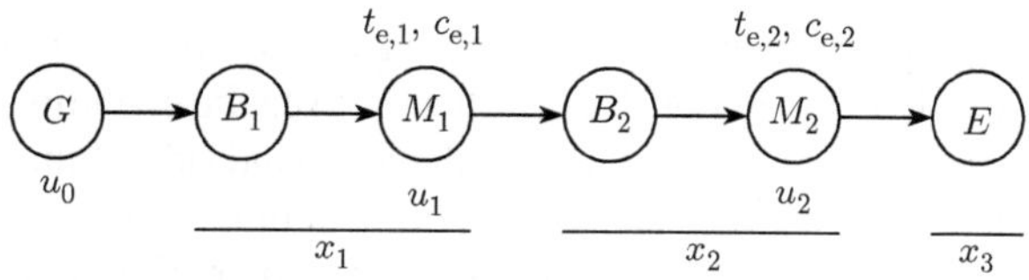

Fig. 17.9 Manufacturing system consisting of two workstations

the interarrival times of jobs can also be determined from arrival events, which provides the parameter c_a for each workstation. Therefore, both parameters can easily be determined from manufacturing data.

However, when applying this approach for production planning, one should carefully derive the EPTs. In particular, if the manufacturing execution system authorizes jobs for processing. In that case, the EPT of a lot cannot start before it has been authorized.

To illustrate this, consider the case depicted in Fig. 17.9. Assume that processing times of the workstations are exponentially distributed with means of, respectively, 0.21 and 0.23 h. Let an MPC production planning scheme be applied with time steps of 1 day (24 h) and a prediction horizon of 5 days. That is, consider a production planning scheme where each day a planning for the next 5 days is generated of which only the desired production levels for the first day are provided as targets (since the planning will be adjusted for the modified circumstances the next day). For this planning, the model (17.2) is used together with the constraints (17.4) and the obvious constraints that buffer contents and utilizations have to be nonnegative for each time period. We do allow for backlog, so x_3 is allowed to become negative. Assume that the goal is to minimize a linear cost function of the jobs in the system where the following customer demand is given:

$$d(k) = 90 + 10 \sin \frac{k\pi}{25}.$$

That is, a periodic demand with a period of 50 days (1,200 h) where demand varies between 80 and 100 jobs per day. This means that the bottleneck requires a utilization between 77 and 96%. Finally, assume that the shop floor implementation of meeting the required targets is by authorizing jobs equally distributed over time. So, if for a certain day a target of 96 jobs is set, every 15 min a new job is authorized.

Next, we consider two ways of determining EPTs. For the first (incorrect) method, we use (17.1) where the actual arrival event AA is the event of the arrival of a lot in the buffer. For the second (correct) method, we also use (17.1) for determining the EPTs, but in this case we use for the actual arrival event AA the latest of the following two events: the arrival of a lot in the buffer or the authorization of that lot for processing. In the latter case, we say that even when a lot has completed the service at the previous workstation, if it has not yet been authorized for processing, it cannot join the queue for processing and therefore actually has not yet arrived to that queue.

The difference in performance between these two ways of determining the actual arrival event AA is depicted in Fig. 17.10, where we see the evolution of the amount of jobs in the buffers and of the backlog. At the left-hand side of this figure, we see that every now and then wip levels explode. For example, around $t = 40{,}000$, we first see a backlog of about 400 lots and a little later the buffer contents in the first workstation reaches almost 5,000 lots. However, at the right-hand side of this figure, we see that the wip in the first workstation remains between 1 and 3 lots, the wip in the second workstation stays even between 2 and 3 lots, and no backlog occurs.

An explanation for this large difference in behavior can be understood if one looks at the EPT realizations. For the first method, the derived EPTs are presented in Fig. 17.11. Since we did not include any disturbances in our model, we know that the (mean) EPTs of the workstations should be 0.21 and 0.23, respectively. However, this is not what we see in Fig. 17.11. In the left-hand side of this figure, we see large EPT realizations every now and then. Also, we see periodic fluctuations in the EPT, implying that the realizations are utilization dependent, which they should not be. Recall that EPTs should be utilization independent. This periodic behavior becomes

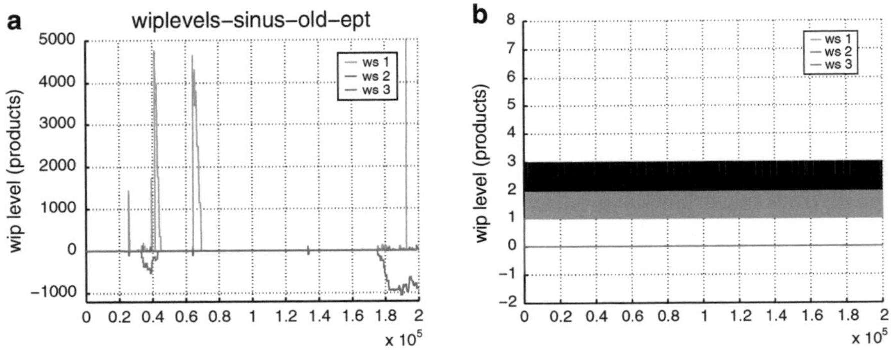

Fig. 17.10 (**a**) Resulting wip levels using incorrect EPT measurements. (**b**) Resulting wip levels using correct EPT measurements

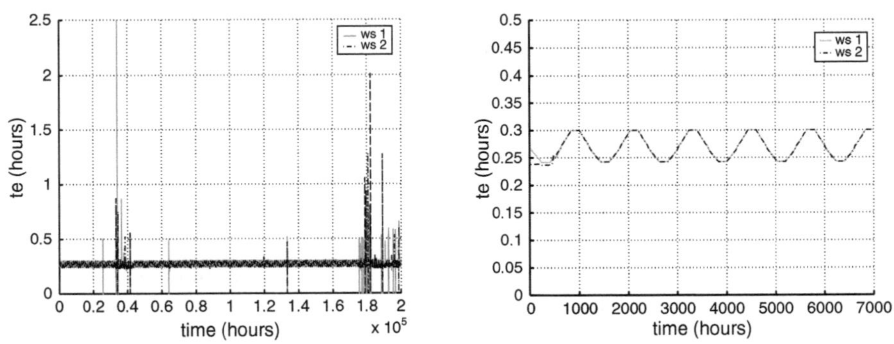

Fig. 17.11 (**a**) Incorrect EPT measurements (complete time horizon). (**b**) Incorrect EPT measurements (zoomed area)

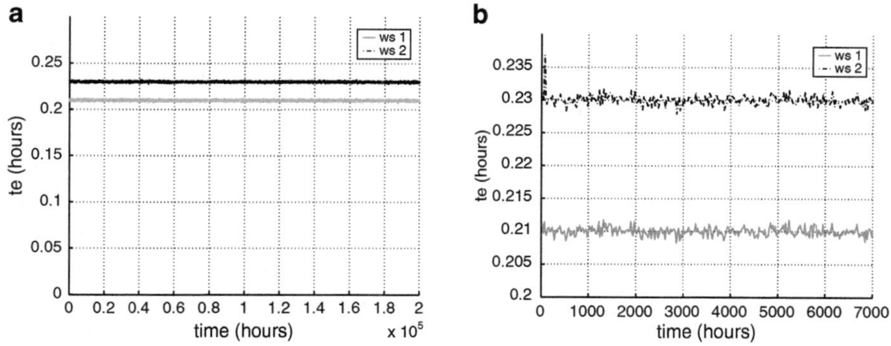

Fig. 17.12 (**a**) Correct EPT measurements (complete time horizon). (**b**) Correct EPT measurements (zoomed area)

even more clear when we zoom in on the first 7,000 time units, as depicted in the right-hand side. Furthermore, we see that the EPT realizations are also a little bit too large.

The explanation of these results is in the way EPTs are determined and the effect that this has on the production planning system. Assume that lots are waiting in the buffer and have not yet been authorized for production. Then they have to wait, even when the machine is idle. As a result, the EPT realization becomes larger. But larger EPT realizations imply that apparently less capacity is available at this machine. Therefore, for the next period, less jobs can be authorized for production. In this way, the planning system enters a viscous circle resulting in large excursions.

Indeed, if one uses as AA-event the moment when the lot has both arrived in the buffer and been authorized for production, better results are obtained, as seen in Fig. 17.12, In this figure, we see correct estimation of the EPT, where small fluctuations are only due to stochasticity. Also when we zoom in on the first 7,000 time units, no utilization dependency of EPT realizations can be found anymore.

17.4 Continuum Models

17.4.1 A Continuum of Production Stages

EPT and clearing function models can be developed for any arbitrary part of the production line. In particular, they can also be used to describe the aggregate behavior of a whole factory, replacing all the details of its production by, e.g., a clearing function relation that determines the outflux as a function of the current work in progress (WIP) in the factory. This will work well, if the associated cycle times through the factory are small and hence the change in WIP during a cycle time is also small (see Chap. 16 for a discussion for the proper timing of a clearing function). However, if the changes in influx are on a shorter timescale than the cycle time, we need to keep track of the time already spent in the factory by a given lot at a particular place in the production line. This can be done by adding delays into ordinary differential equation models or by modeling the flow of WIP through a factory explicitly via a transport equation.

Specifically, the fluid models that use EPT and clearing functions' approaches discussed in the previous sections are really a misnomer. While individual lots are aggregated into a continuum of products, we still consider individual machines or individual machine groups where a true fluid is characterized by two continuous independent variables, a time variable and a space variable. The appropriate spatial variable for a production flow characterizes the production stages or the degree of completion. We denote this variable with x and arbitrarily restrict it to the interval $[0, 1]$. Hence the fundamental variable that we consider is the product density (lot density) $\rho(x,t)$. Note that $\mathrm{d}W(0,t) = \rho(0,t)\mathrm{d}x$ is the WIP at the beginning of the factory, while $\mathrm{d}W(1,t) = \rho(1,t)\mathrm{d}x$ is the WIP at the end of the production line. For almost all manufacturing processes, especially for semiconductor fabs where lots leaving the factory have yet to be tested for their functionality, the fundamental equation describing the transport of a continuum of product through a continuum of production stages is given by a conservation equation for the product ρ.

$$\frac{\partial \rho(x,t)}{\partial t} + \frac{\partial F(\rho(x,t),x,t)}{\partial x} = 0 \tag{17.5}$$

where $F(\rho(x,t),x,t)$ is the flux at position x and time t which depends in a functional manner on ρ and possibly on the exact location x and time t. The influx is then given by

$$F(\rho(0,t),0,t) = \lambda(t) \tag{17.6}$$

the outflux is given by

$$F(\rho(1,t),1,t) = \mu(t) \tag{17.7}$$

and an initial WIP distribution is characterized as

$$\rho(x,0) = \rho_0(x) \tag{17.8}$$

Note that (17.5), (17.6) and (17.8) form an initial boundary value problem for a partial differential equation. If we are defining the flux as $F(x,t) = \rho(x,t)v(x,t)$ with v the fluid velocity then (17.6) is Little's law (Little 1961) averaged on timescales t and lengthscales x where $\lambda(t)$ is the average influx rate, $\rho(x,t)$ is the average WIP, and $v(x,t)$ is the inverse of the average cycle time.

Equations (17.5), (17.6) and (17.8) are a deterministic description of the flow of products through a factory. The resulting PDE is typically nonlinear and possibly nonlocal, however it is defined just on one spatial dimension. The computational effort to solve such a PDE is minimal. Hence this description is a candidate for a real-time decision tool simulating, e.g., the network of factories that make up a complicated supply chain or that describe the possible production options for a large company. The PDE models allow a user to explore different scenarios by varying the parameters that define the network of PDEs in real time. In addition, the PDE models are inherently time dependent allowing the study of non-equilibrium or transient behavior. The price paid for the convenience of fast time-dependent simulations is that the PDE solutions describe the average behavior of a certain factory under the conditions that define the simulation. Many production scenarios are highly volatile and the variances of output of WIP are as big or bigger than the means of the processes. In that case, a tool that predicts the mean behavior is not very useful but one can argue that such production processes are inherently unpredictable and that individual sample paths generated by a discrete event simulation are just as meaningless as the time evolution of the mean behavior. However, any process where the time dependence of the mean by itself provides useful information is a candidate for a successful description by partial differential equations. In the following, we will present short descriptions of the basic model and its refinements to capture more and more of the stochasticity of the process and of the detailed decision issues in production systems. References to more in-depth discussions are given. In Sect. 17.5, we will present open problems and directions for further improvements.

The fundamental reference for the idea of modeling production flow as a fluid is in Armbruster et al. [2006a]. Daganzo [2003] uses the idea of discrete kinematic waves to describe the inventory replenishment process in a supply chain. A recent paper (Göttlich et al. 2005) extends the idea to supply chain networks.

17.4.2 Flux Models

The fundamental modeling effort has been to find the right flux function F as a function of the WIP $\rho(x)$. Several first principle, heuristic and experimental attempts to find a good flux model have been discussed. Almost all of them are quasistatic or adiabatic models in the sense that the flux is not evolving in time but has a fixed functional relation to the WIP in the factory (a state equation) usually describing the functional dependence of outflux as a function of WIP in steady state. Hence any disturbance away from the state equation through, e.g., an increase in WIP caused by an increase in influx will lead to an instantaneous relaxation to the new throughput given by the state equation. The flux is written as $F = \rho v_{\text{eq}}$, $v_{\text{eq}} = v_{\text{eq}}(\rho) = 1/\tau(\rho)$ with v_{eq} the steady-state velocity and τ the average cycle time in steady state. Typical models are

- A traffic flow model (Greenshields 1935) with the equilibrium velocity

$$v_{\text{eq}}^{\text{LW}} = v_0 \left(1 - \frac{\rho}{\rho_{\max}}\right).$$

  Here v_0 is the "raw" velocity describing the flow through an empty factory, $\rho_{\max}$ is the density at which nothing moves any more in steady state and hence the density will increase without bounds (cf. a traffic jam). Note that the velocity at stage x depends only on the WIP at stage x. Such a property is valid for traffic models and for a-cyclic production systems where every production step is performed on a single dedicated machine set.

- A model describing the whole factory as an equivalent M/M/1 queue. In that case, we have the PASTA property and the cycle time becomes $\tau = 1/v_0(1+W)$ with W the length of the queue which here is $W = \int_0^1 \rho(x)\mathrm{d}x$, i.e., total WIP. The equilibrium velocity therefore becomes

$$v_{\mathrm{eq}}^{Q1} = \frac{v_0}{1+W}A.$$

  Notice that the M/M/1 model describes a re-entrant factory: since the equilibrium velocity is the same for all parts in the queue, any change in the length of the queue will affect all WIP in the factory uniformly. This is a crude model of a highly re-entrant factory where any increase in starts will lead to a slowdown everywhere inside the factory.
- A more sophisticated re-entrant factory model is given through the use of integration kernels $w(x,\xi)$

$$v_{\mathrm{eq}}^{Q2}(x,t) = \frac{v_0}{1+\int_0^1 w(x,\xi)\rho(\xi,t)\mathrm{d}\xi}$$

  The kernels $w(x,\xi)$ describe the influence of the competition for capacity from the product located at stage ξ on the product located at position x. E.g., assuming a re-entrant production with two passes through the same machines, then for $x \in [0, 0.5]$

$$w(x,\xi) = 0.5\delta(\xi - x) + 0.5\delta(\xi - (x+0.5)) \text{ and}$$
$$v_{\mathrm{eq}}^{Q2}(x,t) = \frac{v_0}{1 + 0.5\rho(x,t) + 0.5\rho(x+0.5,t)},$$

  with $v_{\mathrm{eq}}^{Q2}(x,t) = v_{\mathrm{eq}}^{Q2}(x+0.5,t)$.
- Detailed discrete event simulations can be used to determine the state equation through simulation. Given a DES model, we can determine average WIP in steady state for different throughputs. Assuming a clearing function model or a queuing model, we can then use least squares fits to parameterize the equilibrium throughput or the equilibrium velocity as $v_{\mathrm{eq}} = \Phi(\mathrm{WIP})$.

Figure 17.13 shows three different clearing functions for a line of 100 identical machines and an arrival process that is identical to the first machine process. The difference between the three different curves is due to different levels of variances. Notice that the capacity of the line, i.e., the horizontal asymptote for the clearing function as well as its curvature depends crucially on the stochasticity of the line. The interpolation is a least squares fit to an exponential model for the throughput μ as a function of the WIP W, $\mu = \mu_\infty(1+\exp(-kW))$ (Asmundsson et al. 2002).

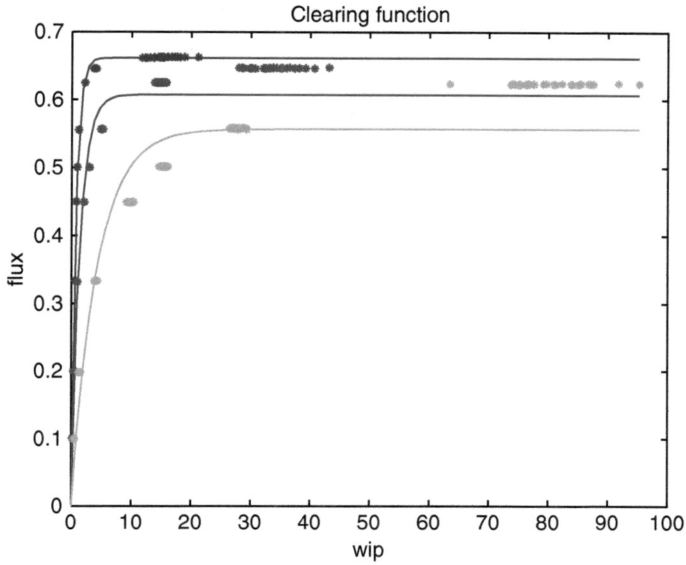

Fig. 17.13 Throughput as a function of WIP in steady state. From top to bottom, the three datasets represent coefficients of variations $c^2 = 0.1, 1$, and 6. Least squares interpolations are made for an exponential clearing function

It is obvious that the exponential decay is not a very good fit for moderate and high variances, suggesting that a low-order polynomial fit or a Pade approximation might work better. Nevertheless, only a few sets of discrete event simulations are necessary to get a general outline of the graph of the clearing function, allowing us to predict WIP and throughput times for arbitrary influxes. However, it is worth noting here that a clearing function characterizes the full state of a system—any change of the system may lead to a different clearing function. While this is obvious for the addition or removal of machines in the factory, the state is also characterized by the variances of the machines and the policies in the factory, in particular, by dispatch policies.

The major advantage of partial differential equation models is the fact that they are able to model time-dependent processes, e.g., transients. Figure 17.14a shows the average throughput for a seasonally varying input (sinusoidal) with a period of about 1 year. The noisy line comes from averaging 1,000 discrete event simulations of a model of a semiconductor factory (Perdaen et al. 2006). The continuous line shows the PDE simulation for the same experiment, where the PDE simulation is generated through a quasistatic model. The PDE simulation is quite good due to the fact that the influx varies slowly. Figure 17.14b shows the same experiment for a sinusoidal input that varies ten times faster. Now the PDE simulation seems to lag a bit relative to the discrete event simulation.

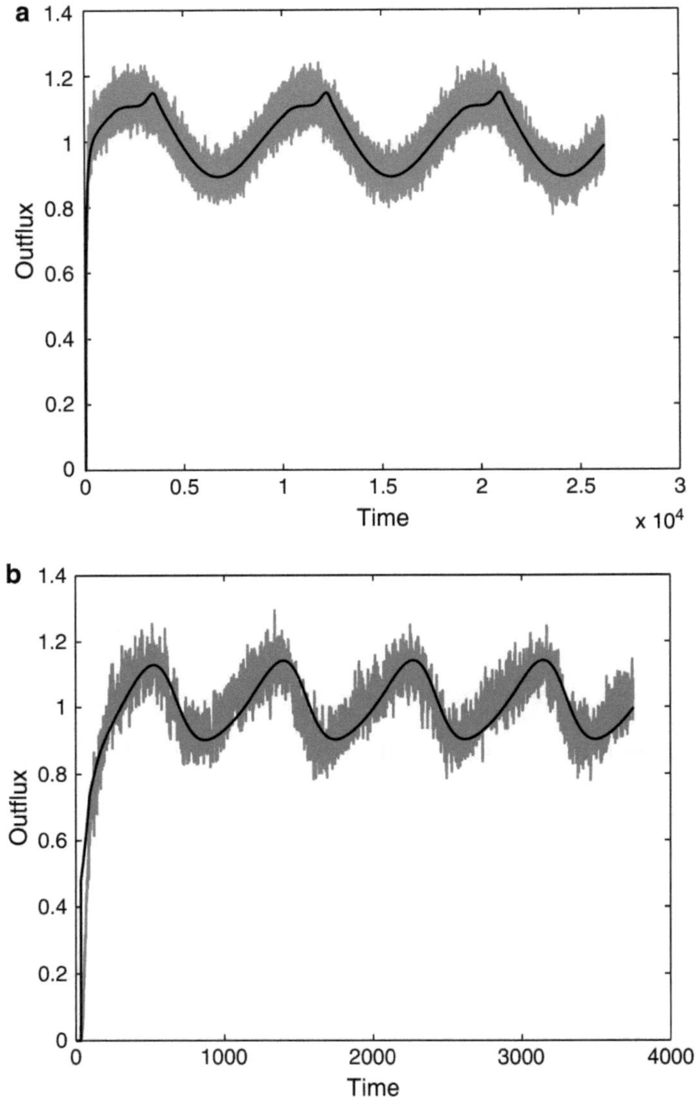

Fig. 17.14 Throughput as a function of time for a sinusoidally varying input (**a**) period of about one year, (**b**) period of about 1/10 of a year

17.4.3 Higher-Order Models and Extensions

Moment expansions. The quasistatic or adiabatic model is the zero order equation of a hierarchy of moment expansion models (Armbruster et al. 2004a). Moment expansions follow the approach of turbulence modeling or gas-dynamic modeling of transport processes (Cercignani 1988). Here, the fundamental quantity is a probability density distribution $f(x, v, t)$ where

$$f(x, v, t)\mathrm{d}x\,\mathrm{d}v\,\mathrm{d}t = \Pr\{\xi \in [x, x + \mathrm{d}x], \eta \in [v, v + \mathrm{d}v], \tau \in [t, t + \mathrm{d}t]\}$$

describes the probability to find a particle in an x-interval with a speed in a particular v-interval in a certain time interval. The time evolution of this probability density leads to a Boltzmann equation. That Boltzmann equation is equivalent to an infinite set of equations for the time evolution of the moments of the probability distribution with respect to the velocity v. As usual a heuristic cutoff is used to reduce the infinite set to a finite set. A two moment expansion is given as

$$\frac{\partial \rho}{\partial t} + \frac{\partial \rho v}{\partial x} = 0,$$

$$\frac{\partial v}{\partial t} + v\frac{\partial v}{\partial x} = 0.$$

Boundary conditions

$$\lambda(t) = \rho(0, t)v(0, t),$$

$$v(0, t) = \frac{v_0}{1 + W(t)},$$

reflect the idea that a lot that arrives at the end of the queue has an initial expectation of a cycle time given by the length of the queue in front of it. Assuming that the velocity is constant over the whole space interval we get that $\partial v/\partial t = 0$ and hence $v = v_{\mathrm{eq}}(\rho) = v_0/1 + W$, i.e., we have the explicit closure that leads to the quasistatic approach.

Diffusion. The quasistatic approach incorporates the influence of the stochasticity, in particular, the variances of the stochastic processes, only through a shift of the means (e.g., mean capacity, mean cycle time, etc.). A typical model that includes the variances explicitly is given through an advection diffusion equation

$$\frac{\partial \rho}{\partial t} + \frac{\partial F}{\partial x} = 0, \tag{17.9}$$

$$F = v_{\mathrm{eq}}\rho - D\frac{\partial \rho}{\partial x}, \tag{17.10}$$

where the advection process describes the deterministic evolution of the means and the diffusion process, parameterized through the diffusion coefficient D, models the behavior of a Brownian motion superimposed on these means. Armbruster and Ringhofer [2005] derive such an equation from first principles, based on a transport process that randomly updates the transport velocity from a density-dependent probability distribution. To model the re-entrant influence, the velocity is random

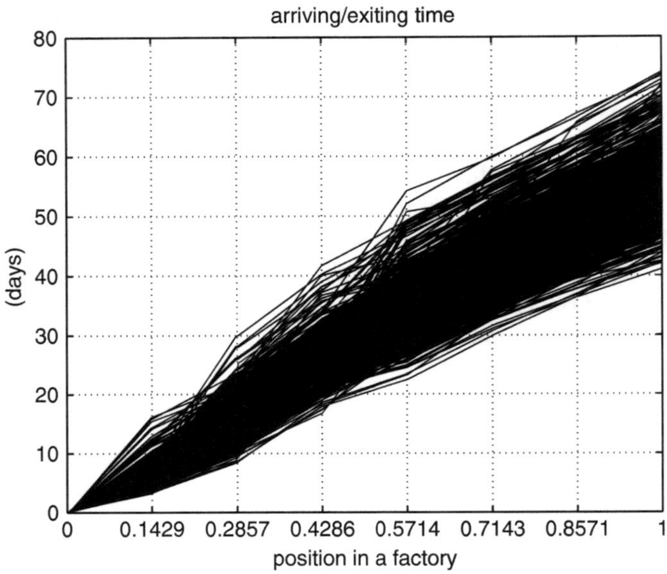

Fig. 17.15 Paths of 920 lots through an INTEL factory

in time but constant over all stages. A expansion based on an infinite number of machines and an infinite number of velocity updates of the associated Boltzmann equation leads to (17.10).

It is easy to show the presence of diffusion in real factory data as well as in discrete event simulations. Any state of the art production facility will be able to determine the exact location of any lot that goes through the factory at any given time. Figure 17.15 shows a crude approximation to the paths of 920 lots through a real INTEL factory. By starting all lots at the same place and time, the resulting fan in Fig. 17.15 is an indication of the diffusion process. Slicing the data in Fig. 17.15 at fixed times, we can generate histograms of the number of lots as a function of position in the factory. Figure 17.16 shows that, as expected from the central limit theorem, the distribution of WIP toward the end of the factory is reasonably well approximated by a normal distribution. Standard fitting procedures will allow us to determine the state equation v_{eq} and the diffusion coefficient D in (17.10) (Armbruster et al. 2004b).

17.4.4 Control of Production Lines

Having a differential equation model for a production line opens up the field of continuous control (see also Lefeber 2004; Göttlich et al. 2006). While there are still many open questions, two initial attempts have been successful.

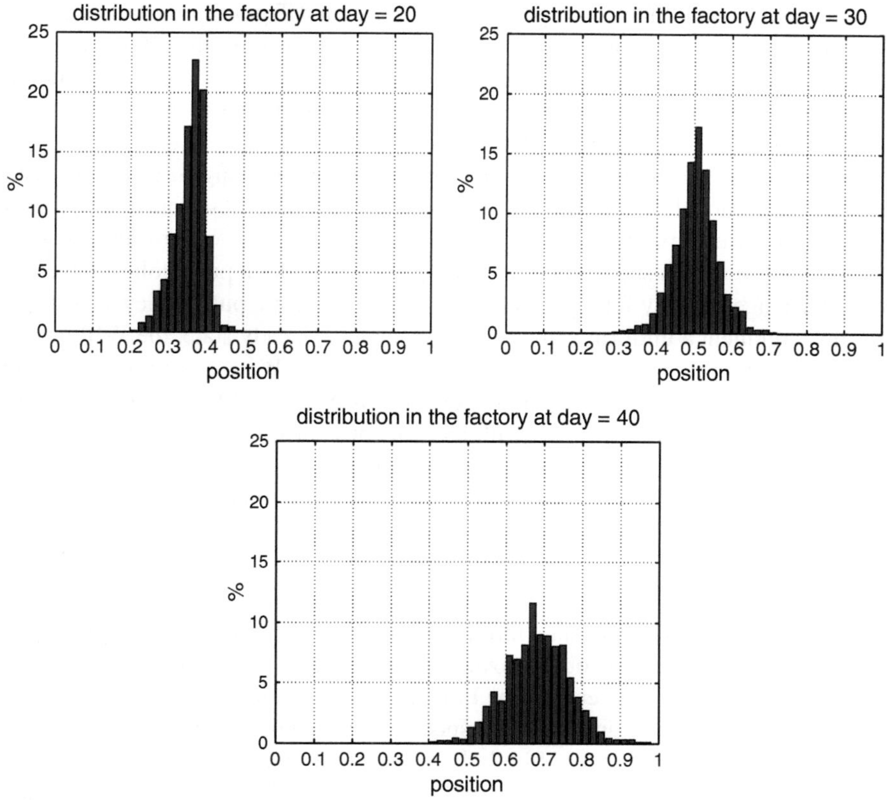

Fig. 17.16 Histograms of positions of the lots in the factory at time $t = 20$, $t = 30$, and $t = 40$

17.4.4.1 Control Via the Push-Pull Point

The cycle time through a semiconductor fab is several weeks. Hence, typically the starts into the fab are done "to plan" while the delivery out of the fab is "to orders." This reflects itself in the dispatch policies at the re-entrant machines. At the beginning of the factory, we have a push policy, favoring lots requiring early production stages over lots waiting for high production stages, whereas at the end of the factory we have a pull policy which tries to affect output by favoring the final steps over earlier steps. Somewhere in the middle of the factory there is a production stage where the push policy changes into a pull policy. That stage is called the push-pull point and it is one of the few possible control actuators inside the factory that might influence the output of the factory. In Perdaen et al. [2006], we have studied the use of changing the push-pull point to affect the tracking of a demand signal in a discrete event simulation of a semiconductor fab. We assume that we have a demand curve as a function of time, and a time interval in which the demand is of the order of magnitude of half of the total WIP of the factory. We then place the push-pull point in such a way that the demand over that time interval matches the total WIP downstream from the push-pull point.

The final result is that a push-pull control algorithm will not significantly improve the factory output for an open system where the WIP is uncontrolled. If we are using the push-pull algorithm together with a CONWIP policy, then the demand-outflux mismatch over a fixed time interval is reduced by a factor of 5–6, for a demand signal with a coefficient of variation $c = \sigma/\mu = 0.4$.

This control algorithm and its implementation have nothing to do per se with a continuum model of the factory. However, a continuum description provides a framework to understand the DES result: since the average cycle time for a lot under a pull policy is shorter than for a lot produced under a push policy, the associated average velocity for a pull policy is higher than for a push policy. Assuming for this argument a uniform velocity in the factory in steady state, the WIP profile $\rho(x) = \lambda/v$ will be constant, independent of x and t. We consider the upstream part of the production line as a homogeneous push line and the downstream part as a homogeneous pull line, each with its own constant velocity with $v_{\text{push}} < v_{\text{pull}}$. Since the throughput is the same everywhere and since $\rho v = \lambda$ has to hold, we get a jump in the WIP profile at the push-pull point by the amount

$$\frac{\rho_{\text{push}}}{\rho_{\text{pull}}} = \frac{v_{\text{pull}}}{v_{\text{push}}}. \tag{17.11}$$

Figure 17.17a shows the constant throughput and the discontinuous WIP profile.

When we now instantaneously move the PPP upstream by an amount Δx then the queues that were just upstream of the PPP and hence had the lowest priority on the line move up in priority and therefore speed up. Hence the product of $\rho_{\text{push}} v_{\text{pull}} > \lambda$, i.e., we create a flux bump. Similarly we create a flux dip by moving the PPP downstream. Keeping the PPP at its new location, the flux bump is downstream from the PPP and hence moves downstream with the constant speed v_{pull} pulling a WIP bump with it until they both exit the factory. During the time they exit, they will increase the outflux. Figure 17.17b, c shows this time evolution. After the WIP/flux bump has exited, the total WIP in the factory is lower and hence in order to satisfy the same demand, the push-pull point will have to move yet further upstream driving it toward the beginning of the factory.

In contrast, the time evolution of the flux bump for the PPP-CONWIP policy is illustrated in Fig. 17.18.

As the CONWIP policy is implemented by matching the starts to the outflux, once the WIP bump moves out of the factory, the starts will be increased to create a new WIP bump. In that way, the total throughput will stay high until the PPP point is moved downstream again. That will happen when the backlog has moved to zero and the sum of actual backlog and actual demand has decreased. In that way we have a policy that reverts all the time to a match between demand and outflux.

17.4.4.2 Creating an Arbitrary WIP Profile

One problem that represents a step to the practically more interesting problems (see Sect. 17.5) is the following: given a WIP profile $\rho_1(x), 0 \leq x \leq 1$ and a quasistatic model of a production system determined by $v_{\text{eq}} = \Phi(\text{WIP})$, what is the influx $\lambda(t)$

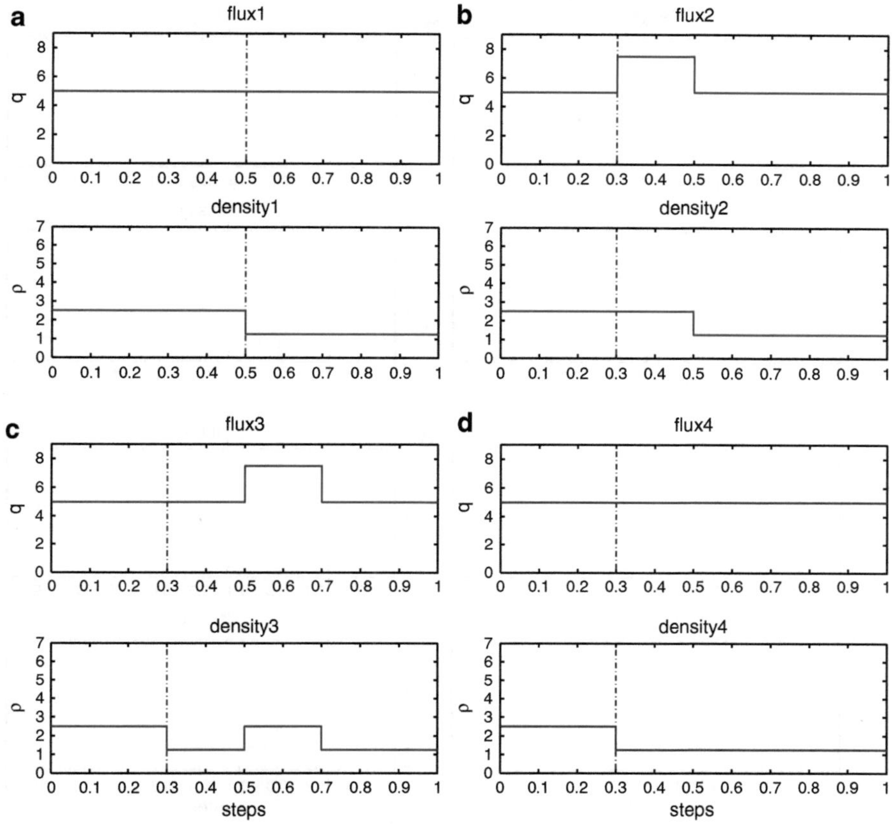

Fig. 17.17 Stages of creating a flux bump (**a**)–(**d**) show subsequent snapshots of the flux and WIP profile. For details and interpretation, see text

to generate a desired new WIP profile $\rho_2(x)$, subject to a time evolution determined by the PDE

$$\begin{aligned} &\rho_t + v_{\text{eq}}\rho_x = 0, && x \in (0,1)\,,\ t > 0. \\ &\lambda(t) = v(t)\rho(0,t), && t > 0. \end{aligned}$$

An implicit analytical solution involves the simple idea of letting the initial profile travel out through the right boundary while the new profile travels in through the left boundary.

$$\rho(x,t) = \begin{cases} \rho_1(x - \int_0^t v(s)\mathrm{d}s) & \text{if } \int_0^t v(s)\mathrm{d}s \le x \le 1 \\ \rho_2(1 + x - \int_0^t v(s)\mathrm{d}s) & \text{if } 0 \le x < \int_0^t v(s)\mathrm{d}s \le 1. \end{cases} \tag{17.12}$$

From (17.12), we can determine the influx $\lambda(t) = v(t)\rho_2(1-\int_0^t v(s)\mathrm{d}s)$. The transit time T for the initial profile $\rho_1(x)$ is defined by $1 = \int_0^T v(s)\mathrm{d}s$. Note that (17.12) is

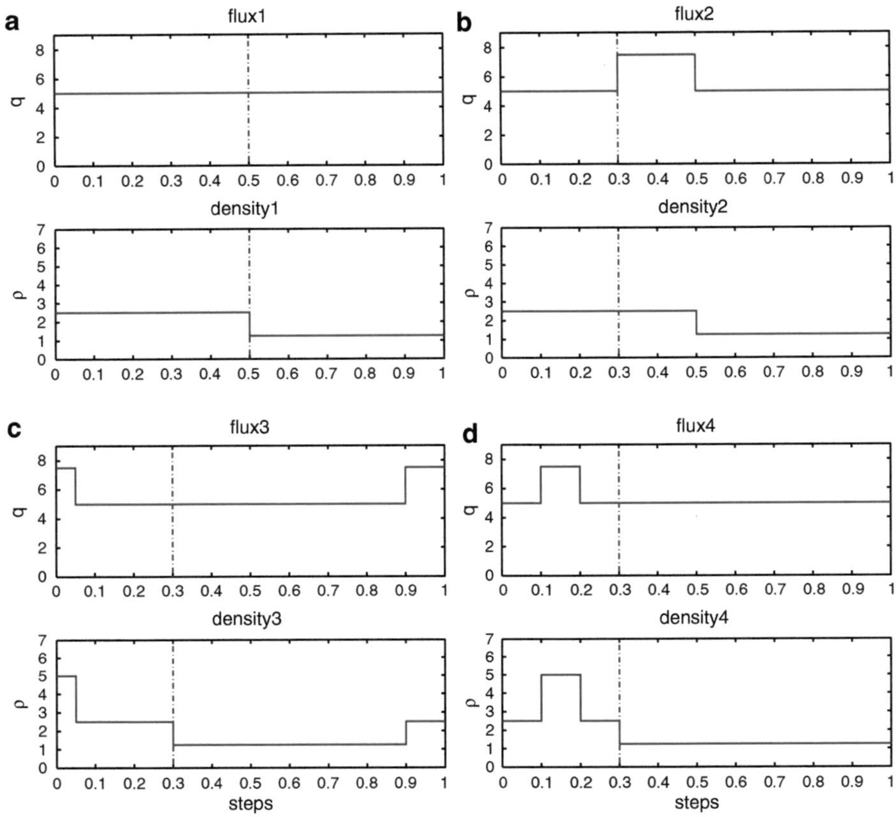

Fig. 17.18 Stages of creating a flux bump for a PPP-CONWIP policy (**a**)–(**d**) show subsequent snapshots of the flux and WIP profile. For details and interpretation, see text

a general solution for all time-dependent functions of velocity, especially including those based on the load $\int_0^1 \rho(x,t)\mathrm{d}x$. Furthermore, it is an implicit solution as the density $\rho(x,t)$ and hence the influx $\lambda(t)$ depend on the velocity $v(\rho(x,t),x,t)$ and its history.

A feasible numerical method to find an explicit solution for $\rho(x,t)$ and $\lambda(t)$ consists of the following steps:

1. Discretize in space and initialize $\rho(x_j,0)$ to $\rho_1(x_j)$ for all space points $j = 1\dots N$.
2. Determine $\rho(x_j,t_n+\delta t)$ by using a hyperbolic PDE solver and evaluate $v = v(t_n+\delta t)$. Integrate $\int_0^{t_n+\delta t} v(s)\mathrm{d}s$ and set $\rho(0,t_n+\delta t) = \rho_2(1-\int_0^{t_n+\delta t} v(s)\mathrm{d}s)$. Set $\lambda(t_n+\delta t) = v(t_n+\delta t)\rho(0,t_n+\delta t)$. Repeat until $\int_0^{t_n+\delta t} v(s)\mathrm{d}s = 1$

Figure 17.19a shows a starting profile $\rho_1(x)$ and an end profile $\rho_2(x)$. Figure 17.19b shows the influx $\lambda(t)$ that generates the new WIP profile for the state equation $v(t) = v_0/(1+\int_0^1 \rho(x,t)\mathrm{d}x)$.

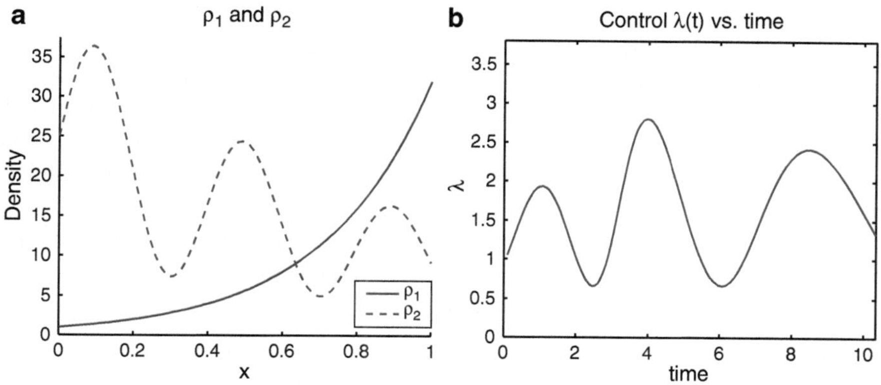

Fig. 17.19 (**a**) Two WIP profiles $\rho_1(x)$ and $\rho_2(x)$ and (**b**) the influx $\lambda(t)$ that transforms ρ_1 into ρ_2

17.5 Conclusions and Open Problems

We have presented three approaches to aggregate modeling of production lines: EPTs, clearing functions, and continuum models (PDEs). EPT is a tool to separate waiting for the availability of a machine from all other sources of variability that extend the processing time. EPTs are easy to measure and allow the development of discrete event simulations that aggregate many different and hard to characterize stochastic processes into one processing time. Alternatively, we can use EPTs to develop relatively simple queueing networks. We have shown that EPTs are utilization independent and that they can be defined for machines that work in parallel, for production lines with finite buffers, and for batch processes.

The next level of aggregation treats the products as a continuum and in that way loses the concept of an event. The resulting model consists of ordinary differential equations that reflect the queues in front of machines and their dynamics driven by the balance of influx and outflux. Together with the loss of the event, clearing function models also lose the stochastic behavior – a clearing function is a input-output relation that reflects the *average* behavior of the system that it is modeling. Simple queues allow an exact determination of the clearing function relationship but most networks require either off-line simulations or queueing approximations to determine the shape of the clearing function numerically.

Continuum models treat the whole production process as a continuum in products and a continuum in production steps. The resulting partial differential equations are typically hyperbolic and describe the movement of products through a factory as a WIP-wave. Different levels of scale and accuracy have been presented. The lowest level of accuracy is represented by a quasistatic approach that connects the PDE models to the clearing function models by using the clearing function as a state equation. The major advantage of continuum models is that they are scale independent, i.e., their simulation does not depend on the number of lots produced nor the number of stages that the lot is going through. A second advantage is that they allow the study of nonequilibrium and transient effects, something that can rarely be done in queueing models. Like the clearing function approach they are deterministic and typically represent the mean transport behavior, although the time evolution of higher-order moments can in principle be studied. PDE models can be extended to networks of factories (supply chains) (Armbruster et al. 2006a; Göttlich et al. 2005) and they can be set up to include policies (dispatch or global) (Armbruster et al. 2006b).

An interesting study for further research would be to compare the computational efforts as well as the performance of the four modeling approaches.

A major open problem for the continuum model approach is the following:

- In Armbruster and Ringhofer [2005] we have derived an advection diffusion equation from first principles that describe the mean time evolution of a certain stochastic production process. However, the process we used involved stochastically varying spatially homogeneous velocities which are not easily related to the usual characterization of the stochasticity of production. The latter is typically described through stochastically varying capacity reflecting the tool manufacturer's characterization of a machine through its time distribution for failure and its time distribution for repair. We are working on developing PDEs whose parameters are determined by a priori given distributions for those times.

Other open problems involve control and optimization of production:

- What is the influx $\lambda(t)$ that moves a production line from an equilibrium state with throughput d_1 to a new equilibrium state with throughput d_2 in shortest possible time.
- Given an initial WIP profile $\rho_0(x, t_0)$ and a demand signal $d(t)$ for $t_0 \leq t \leq t_0 + T$ for some time interval T. What is the input $\lambda(t)$ that minimizes the difference between the output and the demand over that time interval.

We are currently exploring variational methods analogous to optimal control problems for parabolic equations (Göttlich et al. 2006) to solve these optimal control problems.

Acknowledgments This work was supported in parts by NSF grant DMS 0604986. We thank Pascal Etman, Ad Kock, and Christian Ringhofer for many insightful discussions and Bas Coenen, Ton Geubbels, Michael Lamarca, Martijn Peeters, Dominique Perdaen, and William van den Bremer for computational assistance.

References

Armbruster D, Ringhofer C (2005) Thermalized kinetic and fluid models for reentrant supply chains. SIAM J Multiscale model Simul, 3(4):782–800

Asmundsson J, Uzsoy R, Rardin RL (2002) Compact nonlinear capacitymodelsfor supply chains: methodology. Technical report, Purdue University. preprint

Armbruster D, Marthaler D, Ringhofer C (2004a) Kinetic and fluid model hierarchies for supply chains. SIAM J Multiscale model and Simul 2(1):43–61

Armbruster D, Ringhofer C, Jo T-J (2004b) Continuous models for production flows. In: Proceedings of the 2004 American control conference, Boston, MA, pp 4589 – 4594

Armbruster D, Marthaler D, Ringhofer C, Kempf K, Jo T-J (2006a). A continuum model for a re-entrant factory. Oper Res 54(5):933–950

Armbruster D, Degond P, Ringhofer C (2006b) Kinetic and fluid models for supply chains supporting policy attributes. Technical report, Arizona State University, Department of Mathematics and Statistics. Accepted for publication in Transp Theory Stat Phys

Banks J (1998) Handbook of simulation: principles, methodology, advances, applications, and practice. John Wiley & Sons, Inc., USA

Cassandras CG, Lafortune S (1999) Introduction to discrete event systems. Kluwer Academic Publishers, Norwell, MA

Cercignani C (1988) The Boltzmann Equation and its applications. Springer Verlag, New York, NY

Daganzo CF (2003) A theory of supply chains. Springer Verlag, New York, NY

van der Eerden J, Saenger T, Walbrick W, Niesing H, Schuurhuis R (2006) Litho area cycle time reduction in an advanced semiconductor manufacturing line. In: Proceedings of the 2006 IEEE/SEMI advanced semiconductor manufacturing conference, Boston, MA, pp 114–119

Göttlich S, Herty M, Klar A (2005) Network models for supply chains. *Commun Math Sci* 3(4):545–559

Göttlich S, Herty M, Klar A (2006) Modelling and optimization of supply chains on complex networks. *Commun Math Sci* 4(2):315–330

Graves SC (1986) A tactical planning model for a job shop. *Oper Res* 34(4):522–533

Greenshields BD, Bibbins JR, Channing WS, Miller HH (1935) A study of highway capacity. *Highway research board proceedings*, vol 14, Part 1, Washington DC, pp 448–477

Hopp WJ, Spearman ML (2000) *Factory Physics*. 2nd edn. Irwin/McGraw-Hill, New York, NY

Jacobs JH, Etman LFP, van Campen EJJ, Rooda JE (2001) Quantifying operational time variability: the missing parameter for cycle time reduction. In: Proceedings of the 2001 IEEE/SEMI advanced semiconductor manufacturing conference, pp 1–10

Jacobs JH, Etman LFP, van Campen EJJ, Rooda JE (2003) Characterization of the operational time variability using effective processing times. *IEEE Trans Semicond Manuf*, 16(3):511–520

Jacobs JH (2004) Performance quantification and simulation optimization of manufacturing flow lines. Phd thesis, Eindhoven University of Technology, Eindhoven, The Netherlands

Jacobs JH, van Bakel PP, Etman LFP, Rooda JE (2006) Quantifying variability of batching equipment using effective process times. *IEEE Trans Semicond Manuf*, 19(2):269–275

Kock AAA, Wullems FJJ, Etman LFP, Adan IJBF, Rooda JE (2005) Performance evaluation and lumped parameter modeling of single server flowlines subject to blocking: an effective process time approach. In: Proceedings of the 5th international conference on analysis of manufacturing systems and production management, Zakynthos Island, Greece, pp 137–144

Kock AAA, Etman LFP, Rooda JE (2006a) Effective process time for multi-server tandem queues with finite buffers. SE Report 2006–08, Eindhoven University of Technology, Systems Engineering Group, Department of Mechanical Engineering, Eindhoven, The Netherlands. Submitted for publication. http://se.wtb.tue.nl/sereports/

Kock AAA, Etman LFP, Rooda, JE (2006b) Lumped parameter modelling of the litho cell. In: Proceedings of the INCOM06, vol 2. St. Etienne, France, pp 709–714

Kock AAA, Wullems FJJ, Etman LFP, Adan IJBF, Nijsse F, Rooda JE (2006c) Performance evaluation and lumped parameter modeling of single server flowlines subject to blocking: an effective process time approach. SE Report 2006–09, Eindhoven University of Technology, Systems Engineering Group, Department of Mechanical Engineering, Eindhoven, The Netherlands. Submitted for publication. Available via http://se.wtb.tue.nl/sereports/

Lefeber E (2004) Nonlinear models for control of manufacturing systems. In: Radons G, Neugebauer R (eds) Nonlinear dynamics of production systems, Weinheim, Germany, pp 69–81

Little JDC (1961) A proof for the queuing formula $l = \lambda w$. *Oper Res*, 9:383–387

Perdaen D, Armbruster D, Kempf K, Lefeber E (2007) Controlling a re-entrant manufacturing line via the push-pull point. Technical report, Arizona State University. preprint, in revision for the Int J Prod Res

de Ron AJ, Rooda JE (2005) Fab performance. *IEEE Tran Semicond Manuf*, 18(3):399–405

Takahasi K, Sakasegawa H (1977) A randomized response technique without making use of a randomizing device *Ann Ins Stat Math*, 29a:1–8

Sattler L (1996) Using queueing curve approximations in a fab to determine productivity improvements. In: Proceedings of the 1996 IEEE/SEMI advanced semiconductor manufacturing conference, Cambridge, MA, pp 140–145

Veeger CPL, Etman LFP, Lefeber E, Adan IJBF, van Herk J (2009) Predicting cycle time distributions for integrated processing workstations: an aggregate modeling approach. EURANDOM Report 2009-034, Eindhoven University of Technology, EURANDOM, Eindhoven, The Netherlands. Submitted for publication. Available via http://www.eurandom.tue.nl/reports/

Chapter 18
Robust Stability Analysis of Decentralized Supply Chains

Yanfeng Ouyang and Carlos Daganzo

In the supply chain literature, the term "bullwhip effect" refers to a phenomenon where the fluctuations in order sequence are usually greater upstream than downstream of the chain. Empirical observations have found that the orders placed by a supplier are often more variable than the actual quantities sold, and in multiechelon chains, even very steady customer demand can generate wildly fluctuating supplier orders several stages upstream (Lee et al. 1997a,b).

The bullwhip effect is of much practical importance. The term was originally coined by the Procter & Gamble Corporation to describe their empirical observations. In business schools, "beer games" are widely used to demonstrate its existence and pernicious effects (Goodwin and Franklin 1994, Kaminsky and Simchi-Levi 1998, Sterman 1989). The bullwhip effect is important because it results in huge operating costs for upstream suppliers because of operating inefficiencies (high costs) or lack of responsiveness (poor customer service and loss of customer goodwill), or both. Empirically, the bullwhip effect is estimated to inflate supply chain operating costs by 12.5–25% (Lee et al. 1997a,b). If the bullwhip effect is eliminated, the US grocery industry alone could save on the order of 30 billion dollars each year (Cooke 1993, Lee et al. 1997b).

The bullwhip effect phenomenon is first recognized in the 1950s (Forrester 1958, 1961, Magee 1956, Magee and Boodman 1967), and is also evident in macroeconomic data (Blinder 1986, Holt et al. 1960, Kahn 1987, Naish 1994, Ramey 1991). Later, simulations and games (Goodwin and Franklin 1994, Sterman 1989) reveal that it arises persistently, even if the games are unstructured. Recent research reveals that it is the suppliers, rationality that causes the bullwhip effect. To this end, a common and intuitive practice has been to analyze the bullwhip effect parametrically for one supply chain stage, by comparing the variances of the orders placed by the supplier and the customer. Lee et al. (1997a, 1997b) identified four operational causes of the bullwhip effect (demand forecast updating, order batching, rational shortage

Y. Ouyang (✉)
Department of Civil and Environmental Engineering, University of Illinois
at Urbana-Champaign, Urbana, IL 61801, USA
e-mail: yfouyang@illinois.edu

K.G. Kempf et al. (eds.), *Planning Production and Inventories in the Extended Enterprise*, International Series in Operations Research & Management Science 151,
DOI 10.1007/978-1-4419-6485-4_18,

gaming, and price fluctuation) and quantified their impacts for one retailer with an AR(1) customer demand process. Similar efforts, e.g., Baganha and Cohen [1998], Chen et al. [2000a,b], Graves [1999], So and Zheng [2003], were later made to study variants of the problem for other families of stationary and nonstationary demand processes (e.g., ARMA with or without a time trend). It has also been shown that in multiechelon chains, if the customer demand is ARMA or ARIMA, the upstream orders will also be ARMA or ARIMA (with different parameters) when suppliers follow certain ordering policies (Gaur et al. 2005, Gilbert 2005, Graves 1999, Zhang 2004). Aviv (2003) derives further results for order-up-to policies and general correlated demand processes. All these studies provide useful insights regarding the impact of the assumed demand processes and ordering policies on the bullwhip effect.

18.1 Need for Robust Analysis

In summary, most analyses in the literature have focused on (i) families of stationary and nonstationary customer demand processes (e.g., AR, ARMA, and ARIMA); (ii) specific inventory policies; and dominantly (iii) a single stage of a supply chain (Baganha and Cohen 1998, Chen et al. 2000a,b, Gaur et al. 2005, Gilbert 2005, Graves 1999, So and Zheng 2003). In reality, however, customer demand is hard to specify. Very often we do not know, and therefore cannot control, the customer demand. What's more, supply chains often contain multiple echelons, and the operating environment (e.g., supplier behavior and logistics systems) may be uncertain. We find it important to answer the following questions:

1. Can the conclusions for single supply chain stages and specific scenarios be generalized to long and complicated supply chains?
2. Can we separate the influences of the policies from those of the demand? If yes, are there robust policies that will avoid the bullwhip effect for arbitrary demand?
3. Very often supply chain operations are uncertain; suppliers may alter ordering policies based on economic growth, or lead times may be random. How do these additional uncertainties influence the bullwhip effect?

The answers to these questions are not trivial. The specialized knowledge we have accumulated for single-stage chains and preassumed demand processes does not generalize well to complicated multiechelon chains with arbitrary demand. The following example, taken from Ouyang and Daganzo [2006a], explains the reason.

Suppose the suppliers in a homogeneous multiechelon chain (e.g., retailer, wholesaler, manufacturer, and raw material provider) use the kanban-type "general replenishment rule" proposed in Dejonckheere et al. [2003],[1] which reduces the bullwhip effect for several AR demand processes. Now assume that the customer

[1] This linear policy utilizes current inventory position, current in-stock inventory, and past orders received. It is similar to the generalized kanban policy in Zipkin [2000]. For more details see Dejonckheere et al. [2003], pages 582–584.

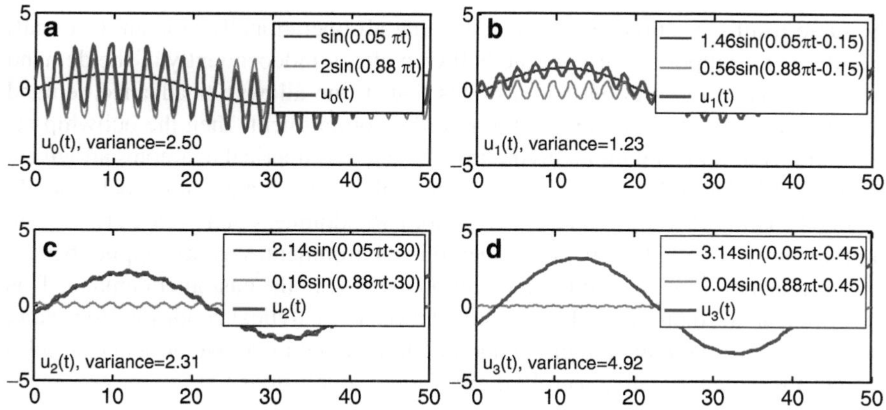

Fig. 18.1 Fluctuations of orders placed by members of a supply chain: (**a**) customer; (**b**) retailer; (**c**) wholesaler; and (**d**) manufacturer (Source: Ouyang and Daganzo, 2006a)

demand, $u_0(t)$, exhibits both seasonal fluctuations (with amplitude 1.0 and angular frequency 0.05π) and short-term variations (with amplitude 2.0 and angular frequency 0.88π); see Fig. 18.1(a). Applying the parameter values in Dejonckheere et al. [2003], we find that the amplitudes of the two wave components change by factors of 1.464 and 0.282, respectively, each time they pass through a supplier. Application of the policy at the first echelon results in the retailer orders $u_1(t)$ of Fig. 18.1(b). We see that the fluctuations indeed decrease for the first echelon, i.e., the retailer places orders more smoothly. Does this imply that the bullwhip effect does not exist in the chain?

A simple simulation at other echelons yields the wholesaler and manufacturer orders, $u_2(t)$ and $u_3(t)$, shown in Fig. 18.1(c)–(d), respectively. Note how the variance of the order sequences decreases for a few suppliers, and then increases. The reason is that the policy dampens the initially larger short-term fluctuations but amplifies the initially smaller seasonal fluctuations. This rather extreme example, however, leads to rather broadly applicable conclusions for practical customer demand processes. It clearly illustrates that variance amplification predictions obtained for single-echelon chains cannot be trivially extrapolated to multiechelon chains. It also suggests that in multiechelon chains, the bullwhip effect can be avoided only if fluctuations of all types (seasonal, short-term, etc.) are dampened. Hence, it is important to address our questions in the context of multiechelon chains, and emphasize robust results that would hold for all types of customer demand.

This chapter summarizes recent results on robust supply chain stability. Section 18.2 addresses questions 1 and 2, following Daganzo [2001, 2003, 2004], Ouyang [2005], Ouyang and Daganzo [2006a]. It first introduces basic notation for deterministically operated chains, and then presents two "frequency domain" approaches for the study of stability: harmonic analysis (Daganzo 2001, 2003, 2004) and transfer function analysis (Ouyang 2005, Ouyang and Daganzo 2006a). We show that these two approaches are based on the same concept and produce the same results. Section 18.2 also shows that all operationally efficient (rational) inventory control policies trigger the bullwhip effect, independently of the demand process. Finally, the section demonstrates that if one allows for advance demand information (ADI) by introducing future order commitments then the bullwhip effect can be eliminated without giving up efficiency. Numerical examples are given. Section 18.3 then develops a formulation in the time domain for stochastically operated chains (Ouyang and Daganzo 2006c). Randomness in this case arises from unpredictably varying factors in the operating environment, such as supplier behavior and transportation lead times, in addition to those in the customer demand. This addresses the third question. The section introduces analytical conditions to predict the existence of the bullwhip effect and bound its magnitude. Numerical examples are also presented. Finally, Sect. 18.4 gives some conclusions.

18.2 Deterministically Operated Chains

18.2.1 System Dynamics

Consider a multiechelon chain with $i = 1, 2, \ldots, I + 1$ suppliers and one final customer (treated as supplier $i = 0$), as shown in Fig. 18.2. Every supplier $(i = 0, 1, 2, \ldots, I)$ orders $u_i(t)$ items from its upstream neighbor at discrete times $t = \ldots, -2, -1, 0, 1, 2, \ldots$, and receives the items after a constant lead time $l_i = 0, 1, 2, \ldots$.[2] Physical shipments arrive at the beginning of every time period; suppliers inspect their inventories during the period; replenishment orders are then placed at the end of the period and received by the upstream neighbors immediately.

The conservation equations for the supplier's *inventory position* at time t, $x_i(t)$ and for the *in-stock inventory* at time t, $y_i(t)$ are:

$$x_i(t + 1) = x_i(t) + u_i(t) - u_{i-1}(t), \forall i = 1, 2, \ldots, \tag{18.1}$$

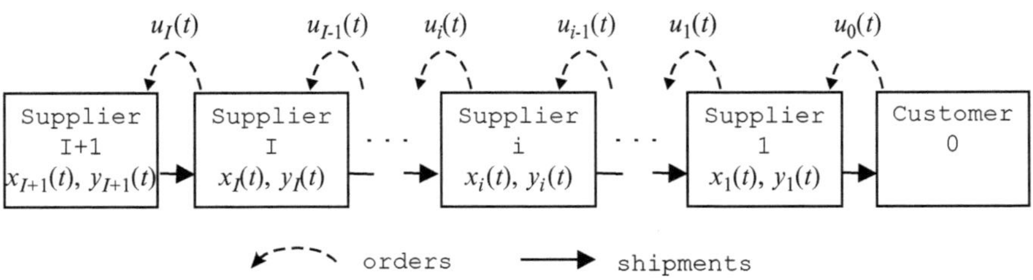

Fig. 18.2 A supply chain representation (Source: Ouyang and Daganzo 2006b)

[2] This constant lead time assumption is also used in the literature, e.g., Chen et al. [2000a,b], Gavirneni et al. [1999], Lee et al. [2000]. It is relaxed in Sect. 18.3.

and

$$y_i(t+1) = y_i(t) + u_i(t-l_i) - u_{i-1}(t), \forall i = 1,2,\ldots. \tag{18.2}$$

We focus on decentralized supply chain where suppliers act independently, placing orders based on private information; i.e., on all the inventory records and the histories of orders received and placed. Ouyang and Daganzo (2006a) shows that all the information available to supplier i at time t can be encapsulated in the following information set:

$$\begin{aligned}\mathcal{I}_i(t) := \{\, & x_i(t), x_i(t-1), \cdots, x_i(-\infty); y_i(t), y_i(t-1), \ldots, y_i(-\infty); \\ & u_{i-1}(t-1), u_{i-1}(t-2), \ldots, u_{i-1}(-\infty)\}.\end{aligned}$$

The most general linear and time-invariant (LTI) ordering policy can be written as follows:

$$u_i(t) = \gamma_i + A_i(P)x_i(t) + B_i(P)y_i(t) + C_i(P)u_{i-1}(t-1), i = 1,2,\ldots, \tag{18.3}$$

where parameter γ_i is a real number, and $A_i(\cdot), B_i(\cdot), C_i(\cdot)$ are polynomials with real coefficients. The symbol P is a backward lag operator; i.e., $P^k x(t) = x(t-k)$, $\forall k = 0,1,2,\ldots$. The polynomials $A_i(P)$ and $B_i(P)$ indicate the influence of inventory history, and $C_i(P)$ the history of orders received. Various definitions of polynomials $A_i(P)$, $B_i(P)$, and $C_i(P)$ then represent all LTI policies. Here, we give several examples.

Example 18.1 (Order-up-to with moving average demand forecasting). An order-up-to policy, where the "up-to level" is forecasted by a moving-average of orders received over r_i periods can be generally written as

$$u_i(t) = -x_i(t) + \frac{l_i}{r_i}\left[u_{i-1}(t-1) + \cdots + u_{i-1}(t-r_i)\right].$$

This policy can be denoted by:

$$A_i(P) = -1, B_i(P) = 0, C_i(P) = l_i \frac{1}{r_i}\left(1 + P + \cdots + P^{r_i - 1}\right). \tag{18.4}$$

Example 18.2 (Generalized kanban). The general replenishment rule in Dejonckheere et al. [2003], or generalized kanban policy in Zipkin [2000], can be generally written as

$$u_i(t) = ax_i(t) + by_i(t) + c\sum_{k=0}^{\infty}\alpha^k u_{i-1}(t-k-1),$$

where $-1 \le a < 0$, $-1 \le b \le 1, c > 0$, and $0 < \alpha < 1$. Note that in-stock inventory partly influences ordering decisions. This policy can be represented by

$$A_i(P) = -a, B_i(P) = b, C_i(P) = c(1 + \alpha P + \alpha^2 P^2 + \cdots). \tag{18.5}$$

Example 18.3 (Order-based). A family of "order-based" policies is defined in Daganzo [2001]:

$$\bar{u}_i(t) = \alpha \bar{u}_i(t-1) + \sum_{k=1}^{\infty} \beta_k \bar{u}_{i-1}(t-k), \forall i, t. \tag{18.6}$$

It can be shown that this policy is represented by

$$A(P) = \alpha - 1, B(P) = 0, C(P) = \sum_{k=0}^{\infty} c_k P^k, \text{ and } c_k = \alpha - 1 + \sum_{k'=1}^{k+1} \beta_{k'}.$$

18.2.2 Steady-State Properties

Equations (18.1)–(18.3) are combined to define the system dynamics. We assume that suppliers use "proper" policies; i.e., such that if the customer places orders of a constant size u^{∞}, then a steady state (or equilibrium) arises from the system dynamics where all the suppliers of the chain also place orders of the same size, u^{∞}, and maintain steady inventory positions x_i^{∞} and in-stock inventories y_i^{∞}. We also assume that the system is in this equilibrium for all $t \le 0$.

These steady-state variables reflect the supplier's inventory management philosophy. For every u^{∞}, a set of inventories x_i^{∞} and y_i^{∞} should be uniquely defined; i.e., the supplier should know exactly the amount of inventories it wants to keep when the demand is u^{∞}. Then any sustained small change in the equilibrium demand, du^{∞}, should be reflected by a change in the equilibrium inventory position, dx_i^{∞}.

Obviously, the equilibrium variables $(u^{\infty}, x_i^{\infty}, y_i^{\infty})$ must satisfy the system dynamics. Inserting these variables into (18.1)–(18.3), and after a few manipulations, we find

$$\frac{dx_i^{\infty}}{du^{\infty}} = \frac{1 + B_i(1)l_i - C_i(1)}{A_i(1) + B_i(1)}. \tag{18.7}$$

This quantity, called the "gain" in Daganzo [2001] and the "inventory gain" in Ouyang and Daganzo [2006a], is the marginal change in the equilibrium inventory position for a unit change in the equilibrium demand. Positive inventory gain means that the supplier will increase its inventory level if it perceives an increase in demand, and vice versa. This behavior is economically rational in most application contexts. (See Daganzo [2003, 2004] for more details.) In Sect. 18.2.5.1, it is shown that this rational behavior significantly influences the existence of the bullwhip effect.

18.2.3 A Robust Metric

The system dynamics can be equivalently expressed, following Ouyang and Daganzo [2006a], as homogeneous equations in terms of deviations (errors) from an equilibrium. Let $\bar{x}_i(t) := x_i(t) - x_i^\infty$, $\bar{y}_i(t) := y_i(t) - y_i^\infty$, and $\bar{u}_i(t) := u_i(t) - u^\infty$. Note that $\bar{x}_i(t) = \bar{y}_i(t) = \bar{u}_i(t) = 0$ for all $t = -\infty, \ldots, 0$ since the system is assumed to start from equilibrium. The system dynamics become:

$$\bar{u}_i(t) = A_i(P)\bar{x}_i(t) + B_i(P)\bar{y}_i(t) + C_i(P)\bar{u}_{i-1}(t-1),\ i = 1, 2, \ldots, \tag{18.8}$$

$$\bar{x}_i(t+1) = \bar{x}_i(t) + \bar{u}_i(t) - \bar{u}_{i-1}(t),\ i = 0, 1, \ldots, \tag{18.9}$$

$$\bar{y}_i(t+1) = \bar{y}_i(t) + \bar{u}_i(t - l_i) - \bar{u}_{i-1}(t),\ i = 0, 1, \ldots. \tag{18.10}$$

Note the similarity and equivalence between (18.8)–(18.10) and (18.1)–(18.3), except for the absence of the intercept term. Note from (18.8)–(18.10) that any realization of customer demand $\{\bar{u}_0(t)\}_{t=0}^\infty$ defines a unique upstream order sequence $\{\bar{u}_I(t)\}_{t=0}^\infty$.

A straightforward bullwhip effect metric is the ratio of the root mean square errors (RMSE) of (i) the order sequence received by the most upstream supplier, $\{\bar{u}_I(t)\}_{t=0}^\infty$, and (ii) the customer demand, $\{\bar{u}_0(t)\}_{t=0}^\infty$. This metric is directly analog to the conventional variance amplification measures. It, however, depends on the character of the input sequence.

Recall that it is desirable to investigate the bullwhip effect for customer demand with all possible types of fluctuations. When the customer demand is not well known, we propose using the *worst-case RMSE amplification* factor, W_I, across all possible customer demand sequences to certify (with the condition $W_I \leq 1$) that the RMSE is not amplified under any circumstances whatsoever; i.e., that the bullwhip effect does not arise. This is what we mean by robust analysis. We formalize this idea as follows:

Definition 18.1. Supplier $I + 1$ $(I > 0)$ in a supply chain described by (18.8)–(18.10) is said to experience no bullwhip effect if

$$W_I := \sup_{\forall \{\bar{u}_0(t)\} \neq 0} \left[\frac{\left(\sum_{t=0}^\infty \bar{u}_I^2(t)\right)^{\frac{1}{2}}}{\left(\sum_{t=0}^\infty \bar{u}_0^2(t)\right)^{\frac{1}{2}}} \right] \leq 1. \tag{18.11}$$

The quantity W_I is also called the L_2 *gain* in the control literature.[3] Obviously, if W_I satisfies (18.11), the supply chain is robust with respect to the bullwhip effect. This robustness idea has proven valuable in the design of complex machinery that has to operate reliably in unpredictable environments. By focusing on the worst-case scenario, engineers can separate the influence of the design (on which they focus their attention) from that of the environment. In our case, we can separate the

[3] This is unrelated to the "gain" or "inventory gain" defined earlier.

influence of the policy from that of the customer demand. This approach will help us design a robust supply chain, and thus address our second question. A robust design, i.e., one satisfying (18.11), will avoid the bullwhip effect and its pernicious economic consequences no matter what the customer does.

18.2.4 Frequency Domain Analysis

It is challenging to directly exhaust all arbitrary input-order sequences to find the worst RMSE amplification, as required by (18.11), especially when the supply chain is long and complex. Fortunately, we can simplify the task for LTI chains, by working in the "frequency domain" by means of transforms. The general idea is to decompose the customer orders into a set of sinusoidal waves (with specific amplitude for each frequency). When each of these waves is fed into the linear time-invariant supply chain, the output would also be sinusoidal with the same frequency. Then by superposing the sinusoidal output components, the upstream supplier order sequences can be obtained and analyzed. The goal is to find policies that do not increase the amplitude of any wave because, as should be intuitive, the RMSE of the output can exceed that of the input only if the amplitude of (at least) one wave is amplified by the policies – if the amplification property (18.11) is missing. The analysis in the frequency domain is much easier than checking (18.11) in the time domain because we only have to test the amplification factors over a range of frequencies. The second advantage is that the complexity of analyzing a multiple-stage chain is not significantly larger than that for a single supplier stage. We briefly show below two approaches that have been used in the literature to do this: harmonic analysis (Daganzo 2001, 2003, 2004) and transfer function analysis (Ouyang 2005, Ouyang and Daganzo 2006a). The two approaches are essentially the same, except for the specific transform used.

18.2.4.1 Harmonic Analysis

Daganzo [2001, 2003, 2004] first apply harmonic analysis to homogeneous supply chains, exploiting the fact that the system dynamics is a set of linear and homogeneous difference equations. The references examine an important subset of our LTI policies, present a simple test for properness and then examine the bullwhip effect with von Neumann's stability test.

The sequence $\{\bar{u}_0(t)\}_{t=0}^{\infty}$ is decomposed by a discrete Fourier transform (DFT) into a set of pure harmonic components, $\mathcal{A}_0(w)e^{-jwt}$ (where $j = \sqrt{-1}$), each with an angular frequency $w \in [0, 2\pi)$ and a complex amplitude $\mathcal{A}_0(w) \in \mathcal{C}$; i.e., $\bar{u}_0(t) = \frac{1}{2\pi}\int_0^{2\pi} \mathcal{A}_0(w)e^{jwt}dw$. The formula for the amplitudes of the sequence $\{\bar{u}_0(t)\}$ is $\mathcal{A}_0(w) = \sum_{t=-\infty}^{\infty} \bar{u}_0(t)e^{-jwt} = \sum_{t=0}^{\infty} \bar{u}_0(t)e^{-jwt}$.

Because (18.8)–(18.10) are linear and time-invariant, the following two things hold : (i) For any harmonic component, the output from each supplier stage of the linear chain is also harmonic with the same frequency but a different amplitude, $\mathcal{A}_i(w)\mathrm{e}^{-jwt}, i = 1, 2, \ldots, I$; (ii) the combined output, $\{\bar{u}_I(t)\}_{t=0}^{\infty}$, is the superposition of the harmonic outputs at the final stage, $\mathcal{A}_I(w)\mathrm{e}^{-jwt}$.

When the supply chain is homogeneous, all suppliers are alike. Every supplier stage amplifies the amplitude of the frequency component by the same factor; i.e., there exists $\xi(w) \in \mathcal{C}$ such that $\frac{\mathcal{A}_i(w)}{\mathcal{A}_0(w)} = [\xi(w)]^i, \forall i$. The system dynamics equations (18.8) – (18.10) are known to have solutions (called modes or waves) of the form

$$\bar{x}_i(t) = \frac{\xi(w) - 1}{\mathrm{e}^{-jw} - 1}[\xi(w)]^{i-1}\mathrm{e}^{-jwt}, \; \bar{y}_i(t) = \frac{\xi(w)\mathrm{e}^{jwl} - 1}{\mathrm{e}^{-jw} - 1}[\xi(w)]^{i-1}\mathrm{e}^{-jwt},$$
$$\bar{u}_i(t) = [\xi(w)]^i\mathrm{e}^{-jwt}, \forall w, i. \tag{18.12}$$

Each w now represents a mode; if

$$|\xi^*| := \sup_{w \in [0, 2\pi)} |\xi(w)| \leq 1, \tag{18.13}$$

there is no bullwhip effect. This is the von Neumann's stability test. As an example, insert the modes, $\bar{u}_i(t) = [\xi(w)]^i\mathrm{e}^{-jwt}$, into (18.6), and solve for $\xi(w)$; this yields:

$$\xi(w) = \frac{\sum_{k=1}^{\infty}\beta_k\mathrm{e}^{jwk}}{1 - \alpha\mathrm{e}^{jw}}. \tag{18.14}$$

Thus, for policy (18.6), the bullwhip effect does not arise if:

$$|\xi^*| := \sup_{w \in [0, 2\pi)} \left| \frac{\sum_{k=1}^{\infty}\beta_k\mathrm{e}^{-jwk}}{1 - \alpha\mathrm{e}^{-jw}} \right| \leq 1. \tag{18.15}$$

This condition can be easily checked for any given α and $\{\beta_k\}$. Daganzo (2004) and Ouyang and Daganzo [2006a] also show that for practical customer demand sequences,

$$\lim_{I \to \infty} \left[\frac{\left(\sum_{t=0}^{\infty}\bar{u}_I^2(t)\right)^{\frac{1}{2}}}{\left(\sum_{t=0}^{\infty}\bar{u}_0^2(t)\right)^{\frac{1}{2}}} \right]^{1/I} = |\xi^*|; \tag{18.16}$$

i.e., that $|\xi^*|$ describes the average amplification across stages for sufficiently long homogeneous chains.

18.2.4.2 Transfer Function Analysis

More recent work (Ouyang and Daganzo 2006a) applies conventional control theory to the complete set of LTI policies (18.3) and inhomogeneous supply chains. It uses the z-transform instead of the Fourier transform.[4] The z-transform of the order error sequence $\bar{u}_i(t)$ will be denoted as $U_i(z) := Z\{\bar{u}_i(t)\}, \forall i$. In general, $U_i(z) \in \mathcal{C}$ while $z \in \mathcal{C}$. Note that the z-transform of a given discrete sequence $\{f(t)\}_{-\infty}^{\infty}$ is defined by $Z\{f(t)\} := \sum_{-\infty}^{\infty} f(t)z^{-t}$. It is essentially the DFT after the substitution $z = e^{jw}$. Note that $U_i(z)$ and $\mathcal{A}_i(w)$ are related by $U_i(e^{jw}) = \mathcal{A}_i(w), \forall i$. We define the "stage-$i$ transfer function," $T_{i-1,i}(\cdot)$, as the counterpart to the amplification factor $\xi_{i-1,i}(\cdot)$[5] to describe the way supplier i transforms its input into an output; i.e.,

$$T_{i-1,i}(e^{jw}) := \frac{U_i(e^{jw})}{U_{i-1}(e^{jw})} = \frac{\mathcal{A}_i(w)}{\mathcal{A}_{i-1}(w)} = \xi_{i-1,i}(w). \tag{18.17}$$

Clearly, $U_i(z) = T_{i-1,i}(z)U_{i-1}(z)$. Then, the "transfer function" from customer demand input to supplier I order output, $T_I(\cdot)$, satisfies

$$\begin{aligned}\xi_I(w) &:= T_I(e^{jw}) := \frac{\mathcal{A}_I(w)}{\mathcal{A}_0(w)} = \frac{\mathcal{A}_I(w)}{\mathcal{A}_{I-1}(w)} \cdots \frac{\mathcal{A}_i(w)}{\mathcal{A}_{i-1}(w)} \cdots \frac{\mathcal{A}_1(w)}{\mathcal{A}_0(w)} \\ &= \prod_{i=1}^{I} T_{i-1,i}(e^{jw}),\end{aligned} \tag{18.18}$$

where

$$U_I(z) = T_I(z)U_0(z). \tag{18.19}$$

We can additionally define the "average transfer function", $\hat{T}_I(\cdot)$ representing the "average" amplification per stage, as:

$$\hat{T}_I(e^{jw}) := [T_I(e^{jw})]^{\frac{1}{I}}, \ \forall w \in [0, 2\pi).$$

In the homogeneous case, $\hat{T}_I = T_1 = \xi$.

Ouyang and Daganzo [2006a] show that for our general LTI policy (18.3):

$$T_{i-1,i}(z) = \frac{z^{-1}C_i(z^{-1}) - (z-1)^{-1}[A_i(z^{-1}) + B_i(z^{-1})]}{1 - (z-1)^{-1}[A_i(z^{-1}) + z^{-l_i} B_i(z^{-1})]}, \ \forall i. \tag{18.20}$$

The policy of supplier i is proper (stable in time) if all the poles of $T_{i-1,i}(z)$ are within the unit circle of the complex plane, $\{z : |z| < 1, z \in \mathcal{C}\}$. It will avoid the bullwhip effect if (see (18.11)):

$$\sup_{\forall w \in [0,2\pi)} |T_I(e^{jw})| = |\xi_I^*| \le 1. \tag{18.21}$$

[4] This is largely a matter of taste. The same results are obtained with both approaches. See Appendix A in Daganzo [2003] for more details.

[5] Subscripts are added to capture inhomogeneity.

Compared with the equivalent expression (18.11), formulae (18.15) and (18.21) have the advantage of minimizing functions over a real interval, whereas (18.11) involves a minimization over a set of infinite-dimensional realizations. Note again that in homogeneous chains, $W_I = W_1^I = \xi_I^* = (\xi^*)^I$. Thus,

$$\hat{T}_I(\mathrm{e}^{jw}) = \xi(w) = T_{0,1}(\mathrm{e}^{jw}).$$

Therefore, if a policy avoids the bullwhip effect with the robust metric for one stage, it avoids it for many. In other words, the robust metric avoids the problem illustrated by Fig. 18.1.

18.2.5 Analytical Formulas and Tests

This subsection discusses some additional results and managerial implications.

18.2.5.1 The Role of Inventory Gain

It was shown in Daganzo [2001, 2003, 2004] using concepts from the field of "conservation laws", that the bullwhip effect arises in a homogeneous chain if the inventory gain (common to all suppliers) is positive; i.e., if

$$\frac{dx^\infty}{du^\infty} = \frac{1 + B(1)l - C(1)}{A(1) + B(1)} > 0, \tag{18.22}$$

where the subscript i has been omitted. The result is interesting because it shows that rational supplier behavior *must* lead to the bullwhip effect in long chains. The result has been extended for inhomogeneous chains as follows:

Theorem 18.1. *(Ouyang and Daganzo 2006a) Supplier $I+1$ in an LTI supply chain described by (18.8)–(18.10) experiences the bullwhip effect if the average inventory gain of all the suppliers is positive; i.e., if*

$$\sum_{i=1}^{I} \frac{dx_i^\infty}{du^\infty} = \sum_{i=1}^{I} \frac{1 + B_i(1)l_i - C_i(1)}{A_i(1) + B_i(1)} > 0. \tag{18.23}$$

18.2.5.2 Quantification of the Bullwhip Effect

For ergodic processes, the time average of a statistical measure equals the ensemble average over the state space. If the customer demand is ergodic, the z-transform $U_i(z)$ for the orders placed by the ith supplier, obtained from (18.18)–(18.20), fully characterizes all the statistics of that order stream. For example, $U_i(z)$ yields a formula for the variance: since variance equals mean square error for an ergodic sequence, by Parseval's Theorem we have

Theorem 18.2. *(Ouyang and Daganzo 2006a) If the customer demand is ergodic, the mean square error (variance) of orders placed by supplier i, V_i, is*

$$V_i := \text{Var}\,(\bar{u}_i) = \frac{1}{2\pi}\int_{-\pi}^{\pi} U_i(\mathrm{e}^{jw}) \cdot U_i(\mathrm{e}^{-jw})\, dw, \forall i. \tag{18.24}$$

This formula is exact, but depends on the character of the input process, U_0, as can be seen from (18.18)–(18.20). This information, U_0, is obtained by taking the z-transform of the process $\{u_0(t)\}$.

Example 18.4 (Order-up-to with moving average demand forecasting). Ouyang and Daganzo (2006a) shows an example of the order-up-to policy, (18.4), where the "order-up-to level" is adjusted based on a two-period moving-average of orders received. In terms of errors, the policy is:

$$\bar{u}_i(t) = -\bar{x}_i(t) + \bar{u}_{i-1}(t-1) + \bar{u}_{i-1}(t-2), \forall i, t. \tag{18.25}$$

Its polynomials are: $A(P) = -1$, $B(P) = 0$, and $C(P) = 1 + P$. If the customer demand follows a standard i.i.d. Gaussian process, i.e., where $U_0(z) = 1, \forall |z| = 1$, then $U_i(z) = \left(\frac{(2z^2-1)}{z^3}\right)^i$, and

$$\begin{aligned} V_i &= \frac{1}{2\pi}\int_{-\pi}^{\pi} U_i(\mathrm{e}^{jw}) \cdot U_i(\mathrm{e}^{-jw})\, dw \\ &= \frac{1}{2\pi}\int_{-\pi}^{\pi} [5 - 4\cos(2w)]^i\, dw, \ \forall i. \end{aligned}$$

This gives the exact variance of supplier order sequences at various stages.

18.2.5.3 Effect of Advance Demand Information

Theorem 18.1 established a relationship between the bullwhip effect and positive inventory gain. To eliminate the former, we must restrict the latter. As explained in Daganzo [2001], efficient suppliers like to operate with positive gain, and this could be why the bullwhip effect is so prevalent. Thus, it is important to design policies that allow for arbitrary inventory gains without introducing the bullwhip effect.

Advance demand information (ADI) is found to effectively change the gain-bullwhip effect relationship. It has been first shown in Daganzo [2001] that order-up-to policies with ADI can avoid the bullwhip effect for any desired inventory gain.[6]

[6] Just-in-time chains can operate in this mode.

Then, Ouyang and Daganzo (2006a) found similar encouraging results for other LTI policies. Now, following Ouyang and Daganzo [2006a], we discuss the possibility of developing robust ordering policies that will allow suppliers to operate with any inventory gain while avoiding the bullwhip effect for any customer demand.

To provide ADI, suppliers inform their immediate upstream neighbors of the orders they will place in some future periods and commit to these quantities with a contract. Consider a generic supplier i. We assume that when it generates its commitment with supplier $i+1$ for the order $u_i(t)$ will be placed at time t (and delivered at $t + l_i$), it has also received commitments from supplier $i - 1$ up to time $t + h_i$, where h_i is a positive integer. Thus, when supplier i commits to order quantity $u_i(t)$ it has access to the following information:

$$\begin{aligned}\mathcal{I}_i(t) := \{\, & x_i(t), x_i(t-1), \ldots, x_i(-\infty); y_i(t), y_i(t-1), \ldots, y_i(-\infty); \\ & u_{i-1}(t+h_i-1), \ldots, u_{i-1}(t), \ldots, u_{i-1}(-\infty)\}.\end{aligned}$$

Suppose that these contracts are always fulfilled, and that the commitments received by suppliers are integrated into their policies to generate commitments for orders placed with their upstream neighbors, then the most general policy based on $\mathcal{I}_i(t)$ now is:

$$u_i(t) = \gamma_i + A_i(P)x_i(t) + B_i(P)y_i(t) + C_i(P)u_{i-1}(t+h_i-1), \ \forall i, t. \quad (18.26)$$

Note this is the same as (18.3) except that committed future orders have been incorporated into the policy.

It is easy to see that the incorporation of ADI into a policy characterized by polynomials $\{A_i, B_i, C_i\}$ does not change the relationship between x_i and u_i in the steady state. Thus, the inventory gain of the ordering policy remains the same. It is shown in Ouyang and Daganzo [2006a] that properness does not change either, and that the transfer function (18.20) becomes

$$\tilde{T}_{i-1,i}(z, h_i) = \frac{z^{-1}C_i(z^{-1}) - z^{-h_i}(z-1)^{-1}[A_i(z^{-1}) + B_i(z^{-1})]}{1 - (z-1)^{-1}[A_i(z^{-1}) + z^{-l_i}B_i(z^{-1})]}. \quad (18.27)$$

It is also shown that Theorem 18.1 becomes:

Theorem 18.3. *Supplier $I + 1$ in an LTI supply chain with ADI policy (18.26) experiences the bullwhip effect if*

$$\sum_{i=1}^{I} \frac{1 + B_i(1)l_i - C_i(1)}{A_i(1) + B_i(1)} > \sum_{i=1}^{I} h_i. \quad (18.28)$$

Note that the left-hand side of (18.28) is the same summation of inventory gains as that in (18.23), while the right-hand side changes from zero to the summation of ADI levels. Thus, when $\sum_{i=1}^{I} h_i > 0$, the supply chain may possibly be operated with the same inventory gain as before while avoiding the bullwhip effect.

And indeed, numerical examples suggest that introducing certain amount of ADI into a policy (with (18.26)) avoids the bullwhip effect for many practical policies. This opens the door for developing robust and economically appealing policies that are guaranteed to avoid the bullwhip effect for any possible customer demand. More details on the structure of such policies and practical implementation mechanisms have been studied in Ouyang and Daganzo [2006b].

18.2.6 Numerical Examples

Still following Ouyang and Daganzo [2006a], we now apply the above analytical results to three types of policies: (i) "order-up-to," (18.25); (ii) "generalized kanban," (18.5); and (iii) "order-based" (18.6). We assume a multiechelon homogeneous chain with lead time $l = 2$ at every stage. Since the chain is homogeneous, we can invoke Theorem 18.1 and use W_1 (for one-echelon) to test for the bullwhip effect. The results are summarized in Table 18.1.

The table shows that for both the order-up-to policy and generalized kanban policies, inventory gains are positive. This immediately says without the need for further analysis that the bullwhip effect exists for these two policies. Indeed, the amplification versus frequency plots in Figs. 18.3(a) and 18.3(b) (i.e., the relation $|T_{0,1}(e^{jw})|$ given by the transfer functions) have maxima $W_1 = 3.00$ and 1.68, respectively, confirming that the bullwhip effect arises. The last column in Table 18.1 gives one-stage transfer function when h periods of ADI are available. It is found by plotting the amplification factor (18.27) of the order-up-to policy that introducing ADI with $h = 2$ eliminates the bullwhip effect; while for the generalized kanban policy, we require $h = 3$.

For the order-based policy, the inventory gain is negative. Therefore, we cannot judge from this inequality whether the bullwhip effect arises. The definitive answer is found from the transfer function for one stage and the plot shown in Fig. 18.3(e). Note that $W_1 = 1.00$. Thus, the bullwhip effect does not arise with this policy for *any* realization of demand and for chains with *any* number of stages.

Figures 18.3(a)–18.3(f), taken from Ouyang and Daganzo [2006a], are simulations that illustrate these results. We use four customer demand processes:

- a family of three stationary AR(1) processes with i.i.d. Gaussian error terms $\varepsilon(t)$ of mean 0 and variance 1:

$$u_0(t+1) = \rho u_0(t) + (1-\rho)\varepsilon(t), \forall t;$$

  with $\rho = 0, 0.4, 0.8$; and
- a time-dependent process obtained by superposing our AR(1) process with $\rho = 0.4$ and a sinusoidal wave:

$$u_0(t+1) = 0.4u_0(t) + 0.6\varepsilon(t) + 7\sin(0.95\pi t), \forall t.$$

Table 18.1 Policy examples

Policy	Formula	Polynomials	Inventory gain $\frac{1+B(1)l-C(1)}{A(1)+B(1)}$	Transfer function $T_{0,1}(z)$	Worst-case amplification W_1	Transfer function with ADI $\tilde{T}_{0,1}(z,h)$
Order-up-to	$\bar{u}_i(t) = -\bar{x}_i(t) + \bar{u}_{i-1}(t-1)$ $+\bar{u}_{i-1}(t-2), \forall i,t$	$A(P) = -1$ $B(P) = 0$ $C(P) = 1 + P$	$\frac{1+0-2}{-1+0} = 1$	$\frac{2z^2-1}{z^3}$	3.00	$\frac{z^2+z^{2-h}-1}{z^3}$
Generalized kanban	$\bar{u}_i(t) = -\bar{x}_i(t)/8 - \bar{y}_i(t)/8$ $+\bar{u}_{i-1}(t-1)/2$ $+\bar{u}_{i-1}(t-2)/2, \forall i,t$	$A(P) = -1/8$ $B(P) = -1/8$ $C(P) = (1+P)/2$	$\frac{1-1/4-1}{-1/8-1/8} = 1$	$\frac{6z^2-4}{8z^3-7z^2+1}$	1.68	$\frac{4z^2+2z^{2-h}-4}{8z^3-7z^2+1}$
Order-based	$\bar{u}_i(t) = 0.5\bar{u}_i(t-1)$ $+0.3\bar{u}_{i-1}(t-1)$ $+0.2\bar{u}_{i-1}(t-2), \forall i,t$	$A(P) = -1/2$ $B(P) = 0$ $C(P) = -1/5$	$\frac{1+0+1/5}{-1/2+0} = -\frac{12}{5}$	$\frac{3z+2}{10z^2-5z}$	1.00	—

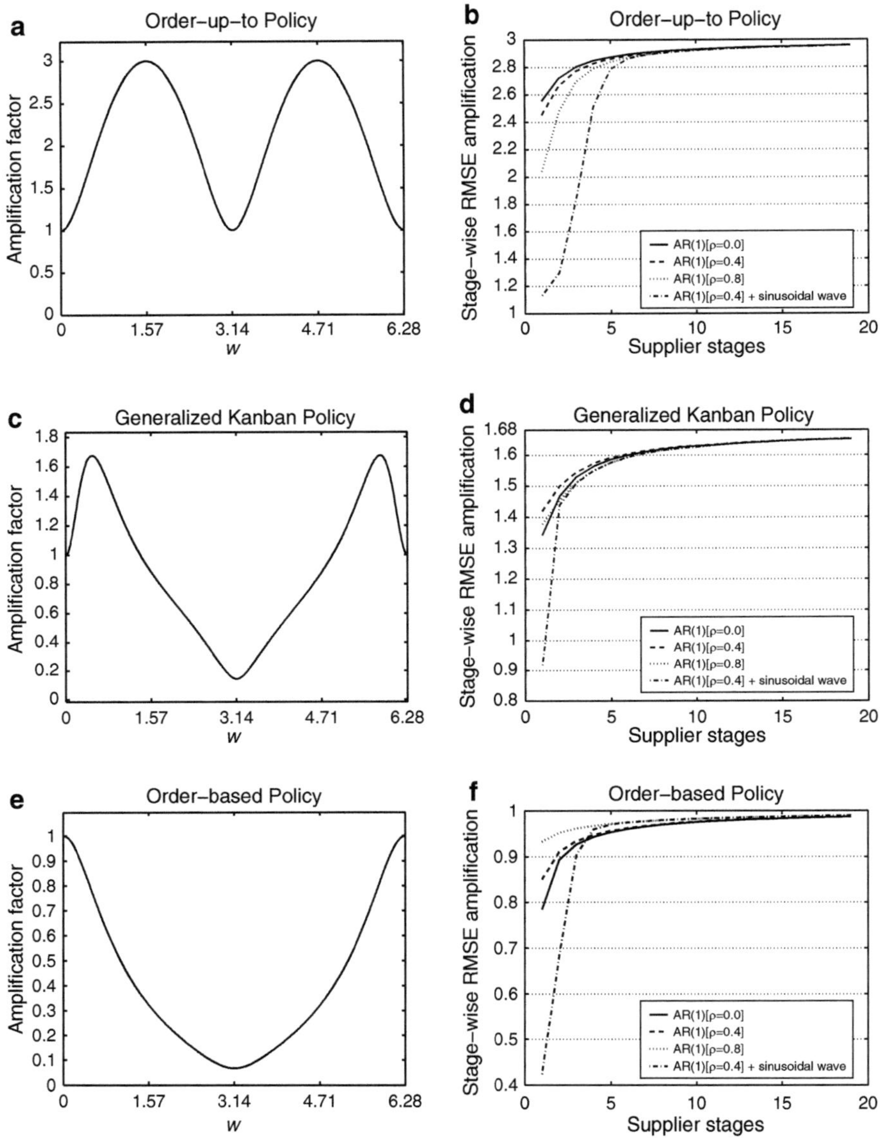

Fig. 18.3 Numerical results: (**a**) amplification factor for order-up-to policy; (**b**) simulated RMSE amplification at each stage for order-up-to policy; (**c**) amplification factor for kanban policy; (**d**) simulated RMSE amplification at each stage for kanban policy; (**e**) amplification factor for order-based policy; (**f**) simulated RMSE amplification at each stage for order-based policy. (Source: Ouyang and Daganzo, 2006a)

Of note:

1. In all cases (demand processes and policy), and as predicted by (18.24), the RMSE amplification factor at stage i, $[V_i/V_{i-1}]^{\frac{1}{2}}$, and thus the average RMSE amplification factor per stage, $[V_i/V_0]^{\frac{1}{2i}}$, both converge from below to W_1 as $i \to \infty$. This happens because the frequency component corresponding to W_1 becomes dominant upstream of the chain. This also justifies why in multiechelon chains we should be concerned with the worst-case.
2. For every policy, the curves are far apart for the first several stages ($i \leq 5$). But, as i increases, the curves converge. Thus, customer demand influences the RMSE amplification factor, but mostly in the stages just upstream of the customer. Farther upstream, where the bullwhip effect is most significant, the critical contributing factor is the policy.

18.3 Stochastically Operated Chains

In reality, supply chain operations are often uncertain. For example, the transportation network may suffer from unexpected congestion and hence yield random lead times. Price discounts may be offered from time to time, and as a result suppliers may alternate between aggressive and conservative policies. Earlier research, e.g., (Lee et al. 1997a,b), finds that suppliers' stochastic gaming behavior and price fluctuations exaggerate the bullwhip effect for specific ordering policies and customer demand processes.

This section addresses our third question; i.e., how would operational uncertainties affect the bullwhip effect for general supplier policies and customer demand? In order to do so, we need to cast the system dynamics in a way that will allow us to treat the system parameters (i.e., lead time l_i and the coefficients of A_i, B_i, C_i) as random variables. We assume that the stochasticity in these parameters can be captured by a Markovian jump linear system (MJLS). For background on this topic, see for example (Mariton 1990, Swaroop and Hedrick 1996, Seiler 2001). More specifically, we assume (i) that the parameters $\{A_i, B_i, C_i, l_i\}$ jump randomly among a finite set of modes, $\mathcal{M} = \{1, 2, \ldots, M\}$; and (ii) that the modes evolve as a Markov chain $\{\theta(t)\}$. Figure 18.4 illustrates the behavior of such supply chains. When the system is in mode $\theta(t)$, we can write the system parameters as $\{A_{i,\theta(t)}, B_{i,\theta(t)}, C_{i,\theta(t)}, l_{i,\theta(t)}\}$. Obviously, if there is only one mode (i.e., $M = 1$), then the supply chain reduces to the case studied in Sect. 18.2.

Following Ouyang and Daganzo [2006c], we present below the robust analytical conditions to diagnose the bullwhip effect. The impact of stochastic operating environment will be demonstrated using the MJLS framework.

We first analyze single-echelon chains in Sects. 18.3.1 and 18.3.2, and then generalize the results to multiechelon chains in Sect. 18.3.3.

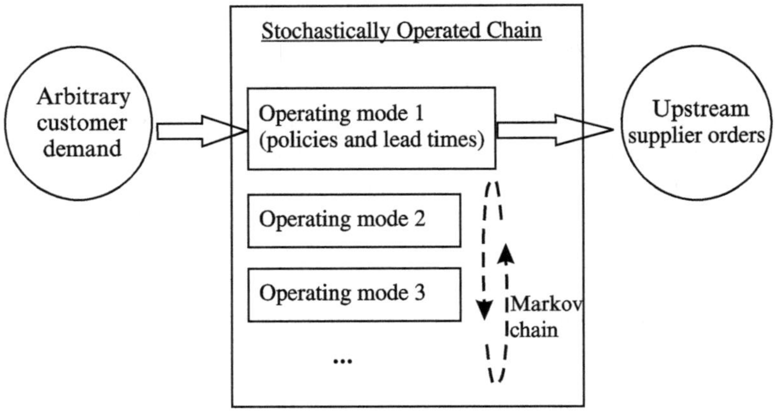

Fig. 18.4 Stochastic supply chain, represented as a Markovian jump linear system

18.3.1 Preliminaries: Choice of Metric

We assume for now that there is a single retailer ($I = 1$); thus the subscript i is momentarily dropped. Note that any realization of customer demand $\{\bar{u}_0(t)\}_{t=0}^{\infty}$ yields a random retailer order sequence $\{\bar{u}_1(t)\}_{t=0}^{\infty}$ that is dependent on the Markov chain $\{\theta(t)\}_{t=0}^{\infty}$. Thus, a robust metric for the bullwhip effect could be the maximum ratio of the RMSE of $\{\bar{u}_1(t)\}_{t=0}^{\infty}$ and $\{\bar{u}_0(t)\}_{t=0}^{\infty}$, for all possible realizations of customer demand *and* the Markov chain; i.e.,

$$\sup_{\forall\{\bar{u}_0(t)\}\neq 0, \forall\{\theta(t)\}} \left[\frac{\sum_{t=0}^{\infty} \bar{u}_1^2(t)}{\sum_{t=0}^{\infty} \bar{u}_0^2(t)}\right]^{\frac{1}{2}}, \tag{18.29}$$

where $\{\bar{u}_0(t)\}$ is square summable. This metric bounds the RMSE measure (L_2) of order sequences. Another possible metric would be the worst-case ratio between the maximum absolute order magnitudes; i.e.,

$$\sup_{\forall\{\bar{u}_0(t)\}\neq 0, \forall\{\theta(t)\}} \frac{\max_{t=0}^{\infty} |\bar{u}_1(t)|}{\max_{t=0}^{\infty} |\bar{u}_0(t)|}. \tag{18.30}$$

These two metrics have a nice scalability property similar to that of W_I. Specifically, if the bullwhip effect is avoided for one echelon of a homogeneous multiechelon chain, then it is avoided for all. These metrics are very robust, since they assume nothing about the random process underlying the chain operations, but if such process is known (and Markovian) it may be reasonable to evaluate the supremum of the *expected* RMSE amplification factor across the possible demand sequences. We propose the following:

$$*-W := \sup_{\forall\{\bar{u}_0(t)\}\neq 0} \left[\frac{E\left(\sum_{t=0}^{\infty} \bar{u}_1^2(t)\right)}{\sum_{t=0}^{\infty} \bar{u}_0^2(t)} \right]^{\frac{1}{2}}, \tag{18.31}$$

where $\{\bar{u}_0(t)\}$ is square summable. The expectation in the numerator of (18.31) (and all expectations without subscripts in this section) are taken across realizations of the Markov chain $\{\theta(t)\}$.

Clearly, W allows us to certify (with the condition $W \leq 1$) that the RMSE is not amplified under any customer demand whatsoever, subject to the scenarios represented in the Markovian operating modes. The choice among (18.29)–(18.31) is largely a practical matter. An advantage of (18.29) and (18.30) is that the supreme operation is commutative. Therefore, we can write (18.30), for example, as

$$\sup_{\forall\{\theta(t)\}} \left\{ \sup_{\forall\{\bar{u}_0(t)\}\neq 0} \left[\frac{\max_{t=0}^{\infty} |\bar{u}_1(t)|}{\max_{t=0}^{\infty} |\bar{u}_0(t)|} \right] \right\}. \tag{18.32}$$

Therefore, if for a specific policy one can prove that the inner supremum is less than one, for all realizations of $\{\theta(t)\}$ one is guaranteed that the bullwhip effect will not arise. Note that the analysis of the inner supremum is purely deterministic and that this method of proof can be applied even if $\{\theta(t)\}$ is non-Markovian, and the policy is nonlinear.

The advantage of (18.31) is that the expectation of RMSE is more closely related to economic costs. Furthermore, some systematic results for linear time-invariant chains with a known Markovian structure have already been derived for linear time-invariant chains with a Markovian structure (Ouyang and Daganzo 2006c). The rest of this section describes these results, following Ouyang and Daganzo [2006c]. But the job is far from done. Further research on metrics (18.29) and (18.30) may help expand the scope of problems that can be analyzed.

18.3.2 System Dynamics and Major Results

With our assumed operational uncertainty, mapping the problem into the frequency domain is no longer advantageous. Thus, we work in the time domain.

The dynamics of each mode of the chain, for a fixed value of $\theta(t)$, have a form similar to (18.8)–(18.10), characterized by $\{A_{\theta(t)}, B_{\theta(t)}, C_{\theta(t)}, l_{\theta(t)}\}$. Ouyang and Daganzo (2006c) show that (18.8)–(18.10) can be expressed in a form that directly relates order sequences $\{\bar{u}_1(t)\}$ and $\{\bar{u}_0(t)\}$, by eliminating the inventory variables. The result is:

$$\bar{u}_1(t+1) = \Phi_{\theta(t)}(P)\bar{u}_1(t) + \Psi_{\theta(t)}(P)\bar{u}_0(t), \tag{18.33}$$

where $\Phi_{\theta(t)}$ and $\Psi_{\theta(t)}$ are polynomials related to $A_{\theta(t)}, B_{\theta(t)}, C_{\theta(t)}, l_{\theta(t)}$ by

$$\begin{aligned}\Phi_{\theta(t)}(P) &:= [1 + A_{\theta(t)}(P) + P^{l_{\theta(t)}} B_{\theta(t)}(P)] \text{ and } \Psi_{\theta(t)}(P) \\ &:= [(1-P)C_{\theta(t)}(P) - B_{\theta(t)}(P) - A_{\theta(t)}(P)].\end{aligned}$$

We assume that both $\Phi_{\theta(t)}(P)$ and $\Psi_{\theta(t)}(P)$ have finite degrees across all modes; i.e., that

$$K := \max\left\{\deg \Phi_{\theta(t)}(\cdot), \deg \Psi_{\theta(t)}(\cdot) : \forall \theta(t)\right\} \leq \infty.$$

The coefficients of P^k in $\Phi_{\theta(t)}$ and $\Psi_{\theta(t)}$, denoted by $\alpha_{k,\theta(t)}$ and $\beta_{k,\theta(t)}$, respectively, $k = 0, 1, \ldots, K$, are obviously functions of the system parameters; i.e., of $l_{\theta(t)}$ and the coefficients of $A_{\theta(t)}$, $B_{\theta(t)}$, and $C_{\theta(t)}$.

We now eliminate the dependence of (18.33) on history prior to t, by augmenting the state into a $(K+1) \times 1$ column vector:

$$\mathbf{u}_i(t) := [\bar{u}_i(t), \bar{u}_i(t-1), \cdots, \bar{u}_i(t-K)]^{\mathrm{T}}, \ i = 0, 1. \tag{18.34}$$

The stochastic system dynamics then become:

$$\begin{aligned}\mathbf{u}_1(t+1) &= R_{\theta(t)} \cdot \mathbf{u}_1(t) + S_{\theta(t)} \cdot \mathbf{u}_0(t), \forall t > 0, \text{ and} \\ \mathbf{u}_i(t) &= \mathbf{0}, \forall t \leq 0, i = 0, 1.\end{aligned} \tag{18.35}$$

Here, $R_{\theta(t)}$ and $S_{\theta(t)}$ are $(K+1) \times (K+1)$ matrices of known constants that fully represent the dynamics of mode $\theta(t)$:

$$R_{\theta(t)} := \begin{bmatrix} \alpha_{0,\theta(t)} & \alpha_{1,\theta(t)} & \cdots & \alpha_{K-1,\theta(t)} & \alpha_{K,\theta(t)} \\ 1 & 0 & \cdots & 0 & 0 \\ 0 & 1 & \cdots & 0 & 0 \\ \vdots & \vdots & \ddots & \vdots & \vdots \\ 0 & 0 & \cdots & 1 & 0 \end{bmatrix},$$

$$S_{\theta(t)} := \begin{bmatrix} \beta_{0,\theta(t)} & \beta_{1,\theta(t)} & \cdots & \beta_{K,\theta(t)} \\ 0 & 0 & \cdots & 0 \\ \vdots & \vdots & \vdots & \vdots \\ 0 & 0 & \cdots & 0 \end{bmatrix}, \tag{18.36}$$

When $\theta(t) = m \in \mathcal{M}$, the system is in mode m, and we shall use $R_m := R_{\theta(t)}$ and $S_m := S_{\theta(t)}$ to denote the matrices of that specific mode. Note that the stochasticity of our supply chain can be represented by a matrix pair $\{R_{\theta(t)}, S_{\theta(t)}\}$ that changes randomly over time, as per the rules of our Markov chain. The transition probability matrix of the chain is defined as $\mathcal{P} = [p_{mn}]_{M \times M}$; i.e., $p_{mn} = \Pr\{\theta(t+1) = n | \theta(t) = m\}, \forall m, n \in \mathcal{M}$.

We define properness of a system governed by (18.35) as follows:

Definition 18.2. (Properness) For the system described by (18.35) with $\mathbf{u}_0(t) \equiv \mathbf{0}$, $\forall t$, the equilibrium state at $\mathbf{u}_1(t) = \mathbf{0}$ is *proper* if for every possible value of the initial state $\{\mathbf{u}_1(0), \theta(0)\}$,

$$\lim_{t\to\infty} E\{\|\mathbf{u}_1(t)\|^2 \,|\mathbf{u}_1(0), \theta(0)\} = 0,$$

where $\|\cdot\|$ denotes the Euclidean norm and the conditional expectation is taken across possible realizations of the Markov chain.

For deterministic chains, Definition 18.2 matches our definition of properness in Sect. 18.2. The following result, proven in Ouyang and Daganzo [2006c], can be used to certify properness and estimate the bullwhip effect:

Theorem 18.4. *A supply chain is proper and has bounded amplification $W < \gamma$ for some positive value γ if and only if: (i) system (18.35) is weakly controllable; i.e., for some $\theta(0) \in \mathcal{M}$, there exists a feasible transition sequence $\theta(0), \theta(1), \cdots, \theta(\tau-1)$ with $\tau < \infty$, such that*

$$\operatorname{rank}[S_{\theta(\tau-1)}, R_{\theta(\tau-1)}S_{\theta(\tau-2)}, \cdots, R_{\theta(\tau-1)}\cdots R_{\theta(1)}S_{\theta(0)}] = K+1;$$

and (ii) there exist $(K+1)\times(K+1)$ positive definite matrices, $G_n > 0, n \in \mathcal{M}$ and $H > 0$, that satisfy either of the following linear matrix inequality sets:

(a)

$$\begin{bmatrix} G_n & 0 \\ 0 & \gamma^2 H \end{bmatrix} - \sum_{m=1}^{M} p_{nm} \begin{bmatrix} R_m & S_m \\ E & 0 \end{bmatrix}^{\mathrm{T}} \begin{bmatrix} G_m & 0 \\ 0 & H \end{bmatrix} \begin{bmatrix} R_m & S_m \\ E & 0 \end{bmatrix} > 0, \forall n \in \mathcal{M}, \tag{18.37}$$

(b)

$$\begin{bmatrix} G_n & 0 \\ 0 & \gamma^2 H \end{bmatrix} - \begin{bmatrix} R_n & S_n \\ E & 0 \end{bmatrix}^{\mathrm{T}} \begin{bmatrix} \bar{G}_n & 0 \\ 0 & H \end{bmatrix} \begin{bmatrix} R_n & S_n \\ E & 0 \end{bmatrix} > 0,$$
$$\bar{G}_n := \sum_{m=1}^{M} p_{nm} G_m, \forall n \in \mathcal{M}, \tag{18.38}$$

where E is the identity matrix.

A restricted but very useful version of the theorem also applies without the controllability requirement. In this case, either (18.37) or (18.38) suffice to bound the amplification from above, but they are not necessary conditions. Weak controllability plays a role in the necessary part of the theorem because if the system is not weakly controllable (i.e., u_1 does not strongly depend on u_0) one can construct examples where u_1 cannot be made to vary enough in relation to u_0 (i.e., so that the amplification is bounded by γ) even if (18.37)–(18.38) are violated.

To implement this result, we need to verify weak controllability and solve a feasibility problem with matrix variables subject to linear matrix inequalities; i.e., (18.37) or (18.38). The search for feasible matrices G_n ($\forall n$) and H can be conducted by convex optimization. For example, for any γ (e.g., $\gamma = 1$) we can solve the following optimization problem:

$$\begin{aligned} \min\ & 1 \\ \text{s.t. } & (18.37) \text{ or } (18.38) \\ & \text{Tr}(H) = 1 \ \ (\text{scaling}) \\ & G_n > 0, \ \forall n \in \mathcal{M} \\ & H > 0. \end{aligned} \tag{18.39}$$

If this problem is feasible, then $W < \gamma$. For any given γ, problem (18.39) has $\frac{(M+1)}{2}(K+1)(K+2)$ scalar variables and is solvable in polynomial time with ellipsoid and interior-point algorithms. To determine the tightest bound, we can conduct a binary search for the minimum feasible γ. This approach is applied to solve the following problems in Ouyang and Daganzo [2006c].

Example 18.5 (Varying policies).

Suppose the business environment exhibits alternating seasons of recession (mode $m = 1$) and growth (mode $m = 2$), then according to an exogenous Markov chain with transition probability matrix we have

$$\mathcal{P} = \begin{bmatrix} 0.9 & 0.1 \\ 0.2 & 0.8 \end{bmatrix}.$$

The long-run probabilities for the two modes are $p_1^\infty = 2/3$, $p_2^\infty = 1/3$. Suppose, in recession, the retailer uses a conservative order-based policy (18.6), i.e.,

$$\bar{u}_1(t+1) = 0.5\bar{u}_1(t) + 0.3\bar{u}_0(t) + 0.2\bar{u}_0(t-1).$$

We know from Sect. 18.2.6 that this policy does not incur the bullwhip effect in LTI chains; i.e., $W = 1$ for this policy. During economic growth, the retailer uses an order-up-to policy (18.25) that yields:

$$\bar{u}_1(t+1) = 2\bar{u}_0(t) - \bar{u}_0(t-2).$$

Section 18.2.6 has shown that this policy incurs the bullwhip effect in LTI chains with $W = 3$. We now apply Theorem 18.4 to the full MJLS.

We define $\mathbf{u}_1(t) = [u_1(t), u_1(t-1), u_1(t-2)]^{\mathrm{T}}$ and $\mathbf{u}_0(t) = \left[u_0(t), u_0(t-1), u_0(t-2)\right]^{\mathrm{T}}$. Then

$$R_1 = \begin{bmatrix} 0.5 & 0 & 0 \\ 1 & 0 & 0 \\ 0 & 1 & 0 \end{bmatrix}, S_1 = \begin{bmatrix} 0.3 & 0.2 & 0 \\ 0 & 0 & 0 \\ 0 & 0 & 0 \end{bmatrix}, R_2 = \begin{bmatrix} 0 & 0 & 0 \\ 1 & 0 & 0 \\ 0 & 1 & 0 \end{bmatrix}, S_2 := \begin{bmatrix} 2 & 0 & -1 \\ 0 & 0 & 0 \\ 0 & 0 & 0 \end{bmatrix}.$$

It is trivial to verify that this system is weakly controllable. Given the data, we cannot find matrix variables to satisfy the conditions in Theorem 18.4 with $\gamma = 1$; the bullwhip effect will exist. A search over the solutions of (18.39) with different γ quickly reveals that the worst-case bound for RMSE amplification is $\gamma = 2.7054$. Interestingly, this exceeds the long-term average of the bounds of the deterministic modes, $(1 \cdot p_1^\infty + 3^2 \cdot p_2^\infty)^{1/2} = 1.91$.

Example 18.6 (Stochastic delivery delay).

We consider a simple case where transportation congestion randomly causes the lead time to vary from $l = 2$ (on-time) to $l = 3$ (delayed). The delay influences physical flow of items and hence the performance of policies based on in-stock inventory; e.g., the example of the generalized kanban policy:

$$\bar{u}_1(t) = -\bar{x}(t)/8 - \bar{y}(t)/8 + \bar{u}_0(t-1)/2 + \bar{u}_0(t-2)/2, \forall t.$$

There are four possible cases that would affect the in-stock inventory, depending on whether the orders scheduled to arrive in the previous and the current periods arrive on time or not. In each of these cases, the in-stock inventory is governed by a different dynamic equation:

Case 1. Previous and current arrivals both on-time. The in-stock inventory changes according to (18.2) with $l = 2$:

$$y_1(t+1) = y_1(t) + u_1(t-2) - u_0(t);$$

Case 2. Previous arrival on-time and current arrival delayed. No shipment arrives from upstream of the chain:

$$y_1(t+1) = y_1(t) - u_0(t);$$

Case 3. Previous arrival delayed and current arrival on-time. Two shipments arrive at the same time:

$$y_1(t+1) = y_1(t) + u_1(t-2) + u_1(t-3) - u_0(t);$$

Case 4. Previous and current arrivals both delayed. The in-stock inventory changes according to (18.2) with $l = 3$:

$$y_1(t+1) = y_1(t) + u_1(t-3) - u_0(t).$$

Suppose each shipment delay occurs independently with probability $p = 0.1$. Then, Cases 1–4 vary over time as a Markov chain with the following transition probability matrix:

$$\mathcal{P} = \begin{bmatrix} 0.9 & 0.1 & 0 & 0 \\ 0 & 0 & 0.9 & 0.1 \\ 0.9 & 0.1 & 0 & 0 \\ 0 & 0 & 0.9 & 0.1 \end{bmatrix}.$$

The long-term probability for the system to be in these fours modes are $0.81, 0.09, 0.09$, and 0.01, respectively.

One can think of these cases as the modes of an MJLS. Then, its R and S matrices, defined by (18.36), are:

$$R_1 := \begin{bmatrix} 7/8 & 0 & -1/8 & 0 \\ 1 & 0 & 0 & 0 \\ 0 & 1 & 0 & 0 \\ 0 & 0 & 1 & 0 \end{bmatrix}, R_2 := \begin{bmatrix} 7/8 & 0 & 0 & 0 \\ 1 & 0 & 0 & 0 \\ 0 & 1 & 0 & 0 \\ 0 & 0 & 1 & 0 \end{bmatrix}, R_3 := \begin{bmatrix} 7/8 & 0 & -1/8 & -1/8 \\ 1 & 0 & 0 & 0 \\ 0 & 1 & 0 & 0 \\ 0 & 0 & 1 & 0 \end{bmatrix},$$

$$R_4 := \begin{bmatrix} 7/8 & 0 & 0 & -1/8 \\ 1 & 0 & 0 & 0 \\ 0 & 1 & 0 & 0 \\ 0 & 0 & 1 & 0 \end{bmatrix}, S_1 = S_2 = S_3 = S_4 := \begin{bmatrix} 3/4 & 0 & -1/2 & 0 \\ 0 & 0 & 0 & 0 \\ 0 & 0 & 0 & 0 \\ 0 & 0 & 0 & 0 \end{bmatrix}.$$

We know that in LTI chains with modes 1, $W = 1.6763$; see Sect. 18.2.6. In LTI chains with modes 2–4, $W = 2.0000, 2.2.4395, 1.8716$, respectively.

It is trivial to verify that the MJLS system is again stochastically stable and weakly controllable. Given the data, we cannot find matrix variables to satisfy the conditions in Theorem 18.4 with $\gamma = 1$; the bullwhip effect will exist. A search quickly reveals that the bound for RMSE amplification is $\gamma = 1.6970$, which in this case is smaller than the long-term average of the bounds of the deterministic modes, $(1.6763^2 \cdot 0.81 + 2.0000^2 \cdot 0.09 + 2.4395^2 \cdot 0.09 + 1.8716^2 \cdot 0.01)^{1/2} = 1.7909$. In fact, when p varies between 0 and 1, the MJLS bound γ never exceeds the long-term average; see Fig. 18.5.

The above numerical examples show that the magnitude of the bullwhip effect in a stochastic chain is not trivially related to its magnitude in any of the modes, nor to its simple long-term average. Furthermore, operational uncertainties often degrade system performance. Theorem 18.4 can be used to quantify all these effects.

18.3.3 Multiechelon Stochastic Chains

The framework of Sect. 18.3.2 can be extended to multiechelon chains with $I + 1$ suppliers and a customer. We define the bullwhip effect metric for the most upstream order sequence $\{\bar{u}_I(t)\}$ as

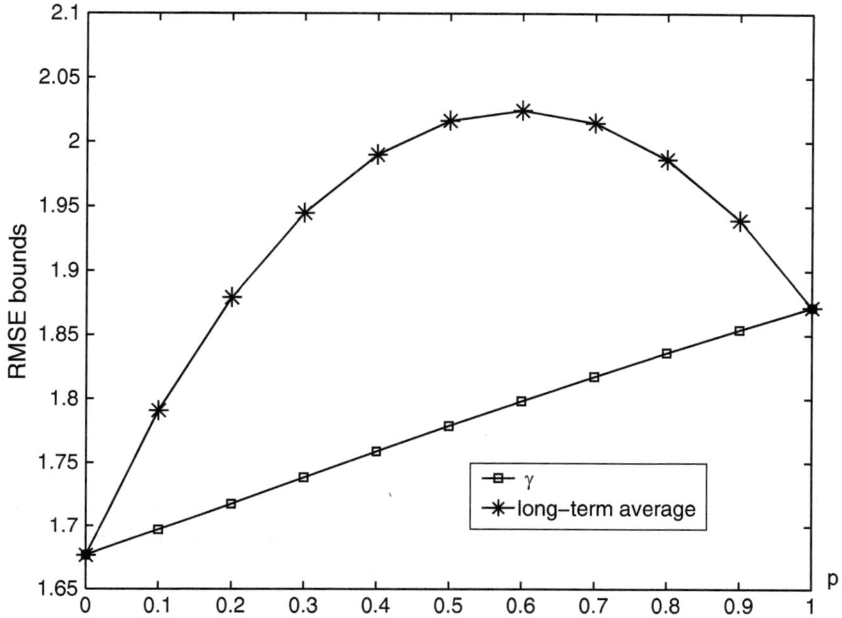

Fig. 18.5 RMSE bounds for supply chain with varying lead time

$$W_I := \sup_{\forall\{\bar{u}_0(t)\}\neq 0} \left[\frac{E\left(\sum_{t=0}^{\infty} \bar{u}_I^2(t)\right)}{\left(\sum_{t=0}^{\infty} \bar{u}_0^2(t)\right)} \right]^{\frac{1}{2}}.$$

We also define based on (18.34) the state vector of the entire chain:

$$\mathbf{U}_I(t) := \left[\mathbf{u}_I(t)^{\mathrm{T}}, \mathbf{u}_{I-1}(t)^{\mathrm{T}}, \ldots, \mathbf{u}_i(t)^{\mathrm{T}}, \ldots, \mathbf{u}_1(t)^{\mathrm{T}}\right]^{\mathrm{T}}$$

and the demand vector:

$$\mathbf{U}_0(t) := \left[\mathbf{u}_0(t)^{\mathrm{T}}, \mathbf{0}^{\mathrm{T}}, \ldots, \mathbf{0}^{\mathrm{T}}\right]^{\mathrm{T}}.$$

The system dynamics at every stage should be similar to (18.35); i.e.,

$$\mathbf{u}_i(t+1) = R_{i,\theta(t)} \cdot \mathbf{u}_1(t) + S_{i,\theta(t)} \cdot \mathbf{u}_0(t), \ \forall i = 1, 2, \ldots, I, \tag{18.40}$$

where the state space of Markov chain $\{\theta(t)\}$ now has multiple dimensions that capture the stochasticities at all supplier stages; i.e., as in matrix pairs $\{(R_{I,\theta(t)}, S_{I,\theta(t)}), ..., (R_{1,\theta(t)}, S_{i,\theta(t)})\}$. Then it is trivial to show that the system dynamics of the entire chain, with uncertainty, can be represented by

$$\mathbf{U}_I(t+1) = \mathcal{R}_{\theta(t)} \cdot \mathbf{U}_I(t) + \mathcal{S}_{\theta(t)} \cdot \mathbf{U}_0(t), \tag{18.41}$$

where

$$\mathcal{R}_{\theta(t)} := \begin{bmatrix} R_{I,\theta(t)} & S_{I,\theta(t)} & & & & \\ & R_{I-1,\theta(t)} & S_{I-1,\theta(t)} & & & \\ & & \ddots & \ddots & & \\ & & & & R_{2,\theta(t)} & S_{2,\theta(t)} \\ & & & & & R_{1,\theta(t)} \end{bmatrix},$$

$$\mathcal{S}_{\theta(t)} := \begin{bmatrix} \mathbf{0} & \mathbf{0} & \cdots & \mathbf{0} \\ \mathbf{0} & \mathbf{0} & \cdots & \mathbf{0} \\ \vdots & \vdots & & \vdots \\ \mathbf{0} & \mathbf{0} & \cdots & \mathbf{0} \\ S_{1,\theta(t)} & \mathbf{0} & \cdots & \mathbf{0} \end{bmatrix}.$$

This is the type of history-less system to which Theorem 18.4 applies. Therefore, all the results for single-echelon chains in previous sections continue to hold, if only $R_{(\cdot)}, S_{(\cdot)}$, and $H_{(\cdot)}$ are replaced by $\mathcal{R}_{(\cdot)}, \mathcal{S}_{(\cdot)}$, and a diagonal block matrix

$$\mathcal{H}_{(\cdot)} := \begin{bmatrix} H_{(\cdot)} & & & \\ & \mathbf{0} & & \\ & & \ddots & \\ & & & \mathbf{0} \end{bmatrix}.$$

18.4 Conclusions

In this chapter, we have demonstrated the importance of robust analysis in multiechelon chains where customer demand is uncertain. We have shown how to determine if policies used in a multistage chain prevent the bullwhip effect under all circumstances. We have also shown policies that will avoid the bullwhip effect independent of the customer demand.

We first presented analytical tests for deterministically operated chains. These chains were modeled and analyzed in the frequency domain, and analytical results were obtained for scenarios with or without the knowledge of customer demand. It was shown that different frequency-domain approaches in the literature are equivalent. The chapter also described the effect of ADI on the bullwhip effect. Finally, we allowed additional randomness to arise from unpredictably varying factors in the operating environment, such as supplier behavior and transportation lead times. Linear matrix inequality formulae enabled us to predict the bullwhip effect and bound its magnitude. Several numerical examples are shown.

The work presented in this chapter on robust analysis can and should be extended in several directions. First, the results in the chapter pertained to decentralized serial chains producing a single good, but general multicommodity networks with some level of coordination and information sharing among suppliers can be treated with similar approaches. Interesting effects can arise when the supplier-nodes can procure the same parts from different nodes. Second, randomness in the operating environment was assumed to be Markovian, stationary, and exogenous, but it should be possible to relax some of these assumptions. Finally, and very importantly for complex networks, attention should be given to the development and further study of robust market mechanisms to internalize the costs of order sequences; e.g., by buying and selling "bits" of order variability and ADI as mentioned at the end of Sect. 18.2.5.3. To ensure robustness, the metrics discussed in Sect. 18.3.1 should probably be used to measure these bits. Some exploratory research has recently been underway along these directions (Ouyang 2007, Ouyang and Li 2010).

Some recent research, however, also shows some evidence to suggest that the bullwhip effect is not as empirically prevalent as thought (Cachon 2007), or that its existence does not necessarily lead to a higher cost under certain conditions (Ridder et al. 1998). Obviously, further study is required to identify and systematically understand the mechanisms that are allowing some industries to avoid the disbenefits of the bullwhip effect.

References

Aviv Y (2003) A time-series framework for supply chain inventory management. Oper Res 51(2):210–227

Baganha MP, Cohen MA (1998) The stabilizing effect of inventory in supply chains. Oper Res 46(3):572–583

Blinder AS (1986) Can the production smoothing model of inventory behavior be saved? Q J Econ 101:431–454

Chen F, Drezner Z, Ryan J, Simchi-Levi D (2000a) Quantifying the bullwhip effect in a simple supply chain: The impact of forecasting, lead times, and information. Manage Sci 46(3): 436–443

Chen F, Ryan J, Simchi-Levi D (2000b) The impact of exponential smoothing forecasts on the bullwhip effect. Nav Res Logistics 47(4):271–286

Cooke JA (1993) The $30 Billion Promise. Traffic Manag 32:57–59

Cachon G (2007) In search of the bullwhip effect. Manuf Serv Oper Manage 9(4):457–479

Daganzo CF (2001) A theory of supply chains. Institute of Transportation Studies Research Report, UCB-ITS-RR-2001-7. University of California, Berkeley, CA, USA

Daganzo CF (2003) A Theory of Supply Chains. Springer, Heidelberg, Germany

Daganzo CF (2004) On the stability of supply chains. Oper Res 52(6):909–921

Dejonckheere J, Disney SM, Lambrecht MR, Towill DR (2003) Measuring and avoiding the bullwhip effect: A control theoretic approach. Eur J Oper Res 147:567–590

Forrester J (1958) Industrial dynamics, a major breakthrough for decision makers. Harv Bus Rev 36:37–66

Forrester J (1961) Industrial Dynamics. MIT Press, Cambridge, MA

Gaur V, Giloni A, Seshadri S (2005) Information sharing in a supply chain under ARMA demand. Manage Sci 51(6):961–969

Gavirneni S, Kapuscinski R, Tayur S (1999) Value of information in capacitated supply chains. Manage Sci 45(1):16–24

Gilbert K (2005) An ARIMA supply chain model. Manage Sci 51(2):305–310

Goodwin J, Franklin S (1994) The beer distribution game: using simulation to teach systems thinking. J Manage Dev 13(8):7–15
Graves S (1999) A single item inventory model for a nonstationary demand process. Manuf Serv Oper Manage 1:50–61
Holt CC, Modigliani F, Muth J, Simon HA (1960) Planning Production, Inventories and Work Force. Prentice Hall, Englewood Cliffs, NY
Kahn J (1987) Inventories and the volatility of production. Am Econ Rev 77:667–679
Kaminsky P, Simchi-Levi D (1998) The Computerized Beer Game: Teaching the Value of Integrated Supply Chain Management. In: Hau Lee, Shu Ming Ng (eds) Supply Chain and Technology Management, The Production and Operations Management Society.
Lee HL, Padmanabhan V, Whang S (1997a) The Bullwhip effect in supply chains. Sloan Manage Rev 38(3):93–102
Lee HL, Padmanabhan V, Whang S (1997b) Information Distortion in a Supply Chain: The Bullwhip Effect. Manage Sci 43(4):546–558
Lee HL, So KC, Tang CS (2000) The value of information sharing in a two level supply chain. Manage Sci 46(5):628–643
Magee JF (1956) Guides to inventory control (Part II). Harv Bus Rev 1956:106–116
Magee JF, Boodman D (1967) Production Planning and Inventory Control, 2nd edn. McGraw-Hill, NY
Mariton M (1990) Jump Linear Systems in Automatic Control. Marcel Dekker Inc., New York, USA
Naish HF (1994) Production smoothing in the linear quadratic inventory model. Econ J 104(425): 864–875
Ouyang Y (2005) System-level Stability and Optimality of Decentralized Supply Chains. Ph.D. Dissertation, University of California, Berkeley
Ouyang Y (2007) The effect of information sharing on supply chain stability and the bullwhip effect. Eur J Oper Res 182(3):1107–1121
Ouyang Y, Daganzo CF (2006) Characterization of the bullwhip effect in linear, time-invariant supply chains: some formulae and tests. Manage Sci 52(10):1544–1556
Ouyang Y, Daganzo CF (2006) Counteracting the bullwhip effect with decentralized negotiations and advance demand information. Physica A 363(1):14–23
Ouyang Y, Daganzo CF (2006) Robust tests for the bullwhip effect in supply chains with stochastic dynamics. Eur J Oper Res 185(1):340–353
Ouyang Y, Li X (2010) The bullwhip effect in supply chain networks. Eur J Oper Res 201:799–810
Ramey VA (1991) Nonconvex costs and the behavior of inventories J Polit Econ 99:306–334
Ridder A, van der Laan E, Salomon M (1998) How Larger Demand Variability May Lead to Lower Costs in the Newsvendor Problem. Oper Res 46(6):934–946
Seiler PJ (2001) Coordinated Control of Unmanned Aerial Vehicles. Ph.D. Dissertation, University of California, Berkeley
So KC, Zheng X (2003) Impact of supplier's lead time and forecast demand updating on retailer's order quantity variability in a two-level supply chain. Int J Prod Econ 86:169–179
Sterman J (1989) Modelling managerial behaviour: Misperceptions of feedback in a dynamic decision making experiment. Manage Sci 35(3):321–339
Swaroop D, Hedrick JK (1996) String stability of interconnected systems. IEEE Trans Automat Contr 41(3):349–357
Zhang X (2004) Evolution of ARMA demand in supply chains. Manuf Serv Oper Manage 6: 195–198
Zipkin PH (2000) Foundations of Inventory Management McGraw-Hill/Irwin, New York, USA

Chapter 19
Simulation in Production Planning: An Overview with Emphasis on Recent Developments in Cycle Time Estimation

Bruce E. Ankenman, Jennifer M. Bekki, John Fowler, Gerald T. Mackulak, Barry L. Nelson, and Feng Yang

19.1 Introduction

Based on the time horizon under consideration, production planning can be classified into strategic, tactical, and operational planning. Virtually, all manufacturing managers want on-time delivery, minimal work in process, short customer lead times (cycle times), and maximum utilization of resources. The goal of production planning is to strike a profitable balance among these conflicting objectives.

The challenges and benefits of using discrete-event simulation (DES) to model modern manufacturing systems were discussed in the earlier chapter by Fischbein et al. As a tool for production planning in these types of manufacturing systems, DES has been used in three primary ways: First, and most naively, it is used as a primary planning tool by simulating various possible production plans and picking the one that works best. In the second approach, DES is used in production planning to determine whether a production plan developed by a spreadsheet or a mathematical programming model is actually executable or is likely to provide acceptable performance, particularly in terms of cycle times. The third approach iterates between a mathematical programming model and the DES model until the variables describing the system (i.e., cycle time) estimates converge. The limitations of these approaches will be discussed in detail in the next section on traditional roles for simulation in production planning.

Rather than trying to cover the entire topic of simulation in production planning, the remainder of the chapter considers the single issue of using DES for cycle time planning. More specifically, we focus on the efficient generation of cycle time as a function of throughput in single-product environments (we briefly discuss extensions for multiple product environments). The chapter starts by briefly discussing why cycle time is important to manufacturers today and why it is important to be able to accurately estimate cycle times. Next, we describe the tools and techniques

B.E. Ankenman (✉)
Department of Industrial Engineering and Management Sciences, Northwestern University, 2145 Sheridan Road, Evanston, IL 60208, USA
e-mail: ankenman@northwestern.edu

K.G. Kempf et al. (eds.), *Planning Production and Inventories in the Extended Enterprise*, International Series in Operations Research & Management Science 151,
DOI 10.1007/978-1-4419-6485-4_19,

that can be used to estimate cycle times, specifically historical averages, queueing models, and DES models, with the focus primarily on the advantages that simulation can provide.

The usefulness of cycle time estimates can be greatly enhanced through the development of cycle time–throughput (CT–TH) curves that describe the cycle time of the factory as a function of the factory throughput. The latter part of the chapter presents state-of-the-art techniques for efficiently and accurately generating these curves for a single product and a static product mix (PM) and then briefly describes ongoing research for creating cycle time surfaces for multiple products with a changing PM.

19.2 Importance of Cycle Time Estimates

In today's business environment, cycle time is a critical measure of performance, and accurate cycle time estimation is a key component of successful production planning. Cycle time is important regardless of whether a company is in a *make to order or make to stock* production environment. The importance is rather obvious in the *make to order* environment because the customer must wait for the product to be manufactured and will change to another supplier if the quoted lead time is too long. In a *make to stock* environment, cycle time is important because long cycle times lead to an increased risk of product quality due to the fact that it will generally take long to detect manufacturing problems. Additionally, in the context of the entire supply chain, the mean and variability of product cycle times are of great importance to the production facility, as they impact the amount of safety stock that must be held. It is critical in all production environments to be able to not only shorten average cycle times but also accurately estimate cycle times and reduce cycle time variability.

Accurate cycle time estimation results in a more stable production environment (Chung and Lai 2006), and reducing cycle time variability makes planning easier and reduces the need for safety stocks in the supply chain. Moreover, shorter and less variable cycle times result in the production of a higher quality product, manufactured with improved responsiveness to customer needs and greater flexibility (Hopp and Spearman 2000). For service-driven manufacturing industries with intricate supply chains, accurate estimation of cycle time and timely delivery of product to both end customers and customers within the supply chain is as crucial as cost and product quality. A specific example of the importance of cycle time reduction to a complex manufacturing industry is given in the 2006 update of the International Technology Roadmap for Semiconductors (ITRS 2006). It states that the improvement of cycle time targets must be met in order to prevent slowing of the industry's growth. Cycle time reduction is listed in the roadmap as a difficult challenge for both the near term and the long term, and further evidence of the importance of cycle time reduction specifically to the semiconductor manufacturing industry is given in Nemoto et al. (2000), who demonstrate that significant financial benefits come from cycle time reduction in the ramp-up phases of manufacturing.

Within the published literature focused on cycle time and production planning, a wide variety of topics exist. Many papers present methods for reducing cycle times or cycle time variability of manufacturing systems based on lot release. For example, Ehteshami et al. (1992), e.g., illustrate the impact of hot lots on the cycle time distribution and Sivakumar and Chong (2001) perform a simulation-based study examining how selected input variables affect the 0.98 quantile of the cycle time distribution in a semiconductor backend manufacturing facility. They show that lot release to the first operation has the greatest impact on the cycle time distribution and found that reducing the variation in this factor yields a less variable cycle time distribution. Ko et al. (2004) followed up this work and proved that for M/G/1 and M/M/s systems, the cycle time is the lowest when the service rates do not vary.

Another area of the literature focuses specifically on manufacturing resources and their impact on cycle time. Recent examples of this are shown in Delp et al. (2005, 2006). In these papers, the availability adjusted x-factor and complete x-factor, both normalized system cycle time measures, are used to identify capacity-constraining machines in an effort to reduce mean cycle time. Schultz (2004), on the other hand, builds models for how spare parts inventory affects machine up-time and cycle time, while Sze et al. (2001) develop integer programming models to minimize the cycle time of printed circuit board assembly by determining the best assignment of components to machines.

In addition to papers that focus specifically on methods for reducing cycle time, a number of papers develop methodologies to more accurately predict cycle times. Liao and Wang (2004), e.g., present a neural network approach specifically for estimating cycle times for 300-mm semiconductor manufacturing systems with priority lots and material handling operations. Agrawal et al. (2000) highlight the downside of poor cycle time estimation, pointing out that underestimation results in overworked resources and overestimation in excessively long lead times and excess Work in Process (WIP) costs. To address the problem, they introduce a procedure based on accurate estimation of product cycle times using a backward scheduling approach, with an objective of minimizing cycle time. Cochran and Chen (2002), on the other hand, use a genetic algorithm to develop a daily production plan that simultaneously optimizes on-time delivery, bottleneck utilization, and effective Work in Process (WIP) for complex manufacturing facilities.

Recent research has also included the development of tools that incorporate accurate cycle time estimation in the job release portion of production planning. An example of this can be found in Asmundsson et al. (2006), who present a mathematical programming model for production planning based on functions, estimated via simulation, that estimate the expected output of a given resource over time. Their approach is able to capture the impact of slight increases in congestion on cycle time and results in a predicted plan, which closely matches the realized plan of the manufacturing system. Approaches combining DES with analytical approaches, such as those suggested by Asmundsson et al. (2006), have proven particularly effective for production-planning problems based on the accurate estimation of cycle time and cycle time parameters. The chapter by Missbauer and Uzsoy on optimization models in production systems provides a detailed review of related literature.

The wide body of published work on the importance of cycle time in production planning clearly illustrates the significance of the topic. Attention by researchers has been paid not only to directly reducing cycle times but also to accurately estimating cycle times for input into production planning systems. The most common approaches for estimating cycle times include the use of historical averages, queueing models, and simulation models. Each of these approaches has advantages and disadvantages in terms of the speed with which the cycle time estimates are obtained and hence the fidelity of the results.

When using historical averages, companies typically quote a customer delivery date as the historical average plus a safety margin of some kind. The safety margin is intended to take into account variability in the cycle time. Estimates based on historical averages are very quick to obtain, but are often inaccurate. For example, they do not consider increases in cycle time estimates that must be made based on increases in system utilization. Moreover, as product life cycle length decreases and product variability increases, historical averages are less useful, as they will not be accurate enough to predict the cycle time for the changes in PM that will result from the introduction of a new product (Herrmann and Chincholkar 2001).

For simple systems, queueing models provide a significant step up in sophistication from historical averages. They allow the incorporation of distributional considerations to account for the random nature of manufacturing and also take into account the utilization level of the system. For steady-state behavior, queueing models of simple systems allow analytical cycle time results to be obtained very quickly. Furthermore, since these results are generally point estimates, the comparisons between systems are very straightforward. Largely because of these advantages, a wide variety of literature exists on the application of queueing theory to manufacturing systems. Three examples include Papadopoulos and Heavey (1996) and Shantikumar et al. (2007), who give reviews of the application of queueing models to the manufacturing industry, and Connors et al. (1996) give a queueing network, designed to provide fast results for the analysis of semiconductor manufacturing facilities. By their nature, queueing models generate approximations of long-run (steady-state) cycle times that can be updated as plant conditions change. In the chapter of this book by Fischbein et al., the authors indicate that "there is never a time frame (or period) where a leading edge factory is ever in a 'steady state'." However, these approximations can still be useful in factory decision making.

If it is possible to develop a queueing model that is a reasonable representation of the original system, then it is generally a superior approach to simulation. However, there are many cases in which the system is too complex for accurate cycle time estimates using a queueing model. Simulation models allow the modeling of features such as complex routings, transient behavior, and dispatching policies, which are difficult (at best) to adequately capture in queueing analysis, making them the most commonly used tool for obtaining estimates of cycle time and other performance measures in complicated industries such as the semiconductor industry (Connors et al. 1996).

Simulation models have the ability to capture any desired level of detail about a production system, potentially yielding very accurate estimates of cycle time.

However, with increasing detail comes increased time for model development, maintenance, and execution. Fischbein et al. point out in their chapter of this handbook that the level of detail at which to build and validate a simulation model of a manufacturing system is a critical and often difficult decision. Too little detail results in a model that cannot answer the question it was built to answer; too much detail often results in a model that requires an unrealistic amount of data from the factory to provide meaningful results. Models with too much detail may take hours to run, limiting the types of "what-if" questions that can be answered in a reasonable time period. Additionally, highly detailed models require that a great deal of effort is spent for data collection, analysis, and maintenance. Fischbein et al. give suggestions for the level of detail that may be appropriate for modeling manufacturing systems (and estimating cycle times from those models), depending on the purpose and scope of the model.

Even with the difficulties of simulation modeling, it is still the most appropriate modeling choice for estimating cycle times in many complex manufacturing situations. However, given the potential difficulties of determining the appropriate level of model detail, long-run times, statistical significance of results, etc., attention must be paid to using simulation intelligently and efficiently.

19.3 Traditional Role of Simulation in Production Planning

Different methodologies and tools are available to support the planning of production, and these approaches can be broadly classified into two groups: analytic methodologies and simulation-based methodologies. Modeling techniques such as queueing networks, Markov chains, and mathematical programming are included in the first group. It is widely acknowledged that analytical techniques typically require high levels of abstraction and may not be able to represent some real-world situations. DES techniques address these issues and, as a result, provide a promising alternative for production planning. The major advantage of simulation is that it can provide accurate and highly detailed information on system performance. However, with this added detail and accuracy comes the disadvantage that detailed model or models must be constructed and maintained, making the process of performing a simulation study potentially time consuming. Yücesan and Fowler (2000) list the steps involved in simulating a manufacturing system as follows:

1. *Model Design*: (a) Identify the issues to be addressed; (b) plan the project; and (c) develop a conceptual model.
2. *Model Development*: (a) Choose a modeling approach; (b) build and test the model; and (c) verify and validate the model.
3. *Model Deployment*: (a) Experiment with the model; (b) analyze the results; and (c) implement the results for decision making.

Using these steps, the best decision/solution is selected through comparison of a moderate number of simulation scenarios defined from several combinations of input variables. This process may be very time consuming, as models of complex manufacturing systems may take several hours to complete a single run (Fowler and Rose 2004), limiting our ability to uncover the best decisions in critical practical settings. This limitation arises because a simulation "is not inherently optimizing; rather it is descriptive of the performance of a given configuration of the system" (Conway et al. 1959). To make simulation an effective tool for production planning, hybrid approaches that integrate analytical methods and computer simulation have been proposed in an effort to achieve the advantages of both while avoiding their disadvantages. We distinguish three major themes in the literature: simulation optimization, metamodeling, and the integration of mathematical programming and simulation. We review the second of these themes below. Both simulation optimization and the integration of mathematical programming and simulation are covered in the chapters of this handbook by Missbauer and Uzsoy and Fischbein et al., respectively.

19.4 Simulation and Metamodeling

A metamodel is a mathematical approximation of the input–output relationship that is implied by the simulation model, and it can be represented by $\mathbf{y} = g(\mathbf{x}, \boldsymbol{\beta})$. Here, $\mathbf{x}$ and $\mathbf{y}$ are vector-valued input and output, respectively, $\boldsymbol{\beta}$ is the vector of unknown parameters, which will be estimated based on simulation data, and g could be, for instance, the expected-value surface. When the metamodeling is successful, the resulting fitted metamodel provides a functional relationship between decision variables (input) and performance of interest (output) while possessing the high fidelity of DES.

Four general goals of metamodeling are identified by Kleijnen and Sargent (2000) for simulation and metamodeling: (a) Understanding the problem entity; (b) predicting values of the output or response variable; (c) performing optimization; and (d) aiding verification and validation. Here, we primarily address goals (b) and (c). Once a metamodel is in hand, optimization can be carried out using deterministic optimization procedures (Fu et al. 2005). However, as pointed out by Nelson (2004), metamodels provide a stronger support for planning than simulation optimization does in a broad range of decision-making contexts. In many situations, an objective function representing the optimization goal is very difficult to formulate, which hinders the use of optimization methods, whereas a metamodel provides a comprehensive response surface over the feasible region of decision variables and allows for tradeoff analysis and evaluations of system performance.

The major issues in metamodeling include: (a) the choice of a functional form for the metamodel; (b) the design of experiments (DOEs) to collect data via simulation; and (c) the assessment of the adequacy of the fitted metamodel. See Barton (1998) and Kleijnen et al. (2005) for details.

The traditional techniques for constructing a metamodel have been based on parametric polynomial response surface (PRS) approximations. Kleijnen (1993) applied a polynomial metamodel on a case study concerning a decision support system for production planning in metal tube manufacturing. In Shang and Tadikamalla (1993), PRS models are used to optimize the yield with respect to various input factors including lot size, input buffer capacity, etc. The application of metamodels to flexible manufacturing system (FMS) design has been demonstrated by Kleijnen (1988), and Lim and Cochran (1990) used metamodels in the context of shop floor control.

Metamodeling approaches include polynomial regression, splines, radial basis functions, neural networks, spatial correlation models, and frequency–domain approximations (Barton 1998). Here, we briefly review the use of neural networks as a metamodeling tool in production planning in light of its wide application and demonstrate effectiveness in terms of providing response predictions in diverse areas (Vellido et al. 1999). Sabuncuoglu and Touhami (2002) conducted an investigation of the potential use of neural networks as simulation metamodels in manufacturing. Chyssolouris et al. (1990) proposed a neural network approach for the design of manufacturing systems: The neural network was developed to map from desired performance measures to suitable design parameters such as the number of resources for each work center of a job shop. Mollaghasemi et al. (1998) applied a neural network metamodel to a real-world application involving the test operations of a major semiconductor manufacturing plant and suggested a suitable design in terms of scheduling rules and the number of testers to achieve a set of performance goals.

19.5 CT–TH Evaluation via Simulation on Demand

Many man-hours are invested in developing and exercising simulation models of manufacturing systems – models that include critical details that are difficult or impossible to incorporate into simple load calculations or queueing approximations. Unfortunately, simulation models can be clumsy tools for tactical decision making. Models of complex manufacturing systems often take several hours for a single run; however, even if a simulation model could be executed in just a few minutes of elapsed time, it would often require hours to perform the necessary experimentation to produce an accurate and precise solution. Optimization via simulation (where some combination of simulation outputs is maximized or minimized) is even more problematic as a tactical decision-making tool. The analyst must develop an objective function incorporating tradeoffs that are not easily quantified.

The difficulties discussed above have motivated researchers to pursue methods for increasing the efficiency of model building and model execution (Fowler et al. 2001, 2005; Mackulak et al. 2005; Park et al. 2002). Incorporating simulation into both a tactical and an operational role within manufacturing operations requires significant reductions in both run time and experimentation effort. Unfortunately, most of these studies have not obtained an order of magnitude improvement to what remains the most serious impediment to tactical simulation usage in manufacturing:

It still takes too long to get useful simulation results from a full-sized factory model, particularly when one is interested in obtaining cycle time estimates for a variety of start-rate scenarios.

The *simulation on demand* concept directly addresses this fundamental weakness of simulation within manufacturing. "Simulation on demand" focuses on the *efficiency of obtaining useful simulation results when needed* rather than on the efficiency of a single simulation replication. The premise is to exploit the availability of large quantities of idle computer resources by running a simulation model as a background process on lightly used computer(s) or overnight (or over a weekend) when computers and simulation software licenses are often idle. The simulation will execute a series of carefully selected design points that enable an accurate response surface map of the output to be created. After a sufficient number of design points have been created, the model of the output can be used to produce an interpolated estimate of a design point that has never been simulated. This approach builds a complete response surface map (cRSM) over the area of interest and represents a bridge between the flexibility of simulation and the insight provided by analytical queueing models.

While the cRSM approach to simulation on demand clearly encounters some of the challenges (i.e., poor data systems) raised in the earlier chapter by Fischbein et al., the concepts underlying the cRSM approach are predicated on a large body of research conducted on the efficient and accurate generation of CT–TH curves. The following sections will provide the fundamental concepts behind the usefulness and application of CT–TH curves.

19.5.1 The Implied CT–TH Curve

CT–TH curves are often employed as decision-making tools in manufacturing settings (Atherton and Dayhoff 1986; Spence and Welter 1987; Bitran and Tirupati 1989; Brown et al. 1997). A CT–TH curve displays the projected average cycle time plotted against throughput rate, or start rate, with cycle time defined as the time from entering to leaving the system. These curves are useful for planning at both the strategic and tactical levels.

Decisions regarding the impact on cycle time of a 2% increase in start rate can be widely different depending on the shape of the curve and the distance from the knee. For example, if a semiconductor wafer fab has a curve as illustrated in Fig. 19.1 and is operating at the level of 22,000 wafer starts per month, it will experience only a minor change in average cycle time by ramping up an additional 440 wafer starts. Alternatively, if the factory is operating on the same curve but is at 22,500 wafer starts per month, a 440 wafer start increase drastically alters cycle time. In both cases, we called for a 440 wafer start increase, but drastically different outcomes resulted from what seemed to be the same action. Management therefore needs to develop CT–TH curves, if they are interested in predicting the impact of start rate changes on average cycle time.

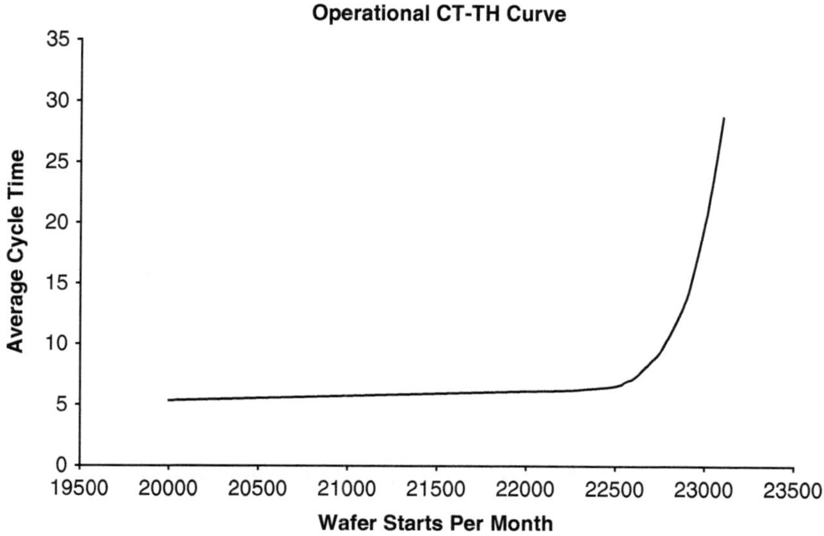

Fig. 19.1 A sample cycle time–throughput (CT–TH) curve

Although all factories have a CT–TH curve, the operation of a factory occurs at a specific *point* on the curve. While it is common to have a predicted cycle time associated with a specific start rate, it is uncommon for a manager to understand whether changes in cycle time result from moving along the curve or from shifting from one curve to another. A factory operates on only one curve at a time, but shifts from one curve to another occur when changes to the factory are made such that the average cycle time for a given number of wafer starts changes. Examples of such changes include changes to scheduling or dispatching policies or changes in the capacity of critical resources. For example, in the curve shown in Fig. 19.1, the average cycle time at 23,000 wafer starts per month is approximately 15. If additional capacity was added to the bottleneck, the average cycle time at the same 23,000 wafer starts would be reduced, causing the factory operations to shift to a new curve. The ability to review the curves associated with factory operations and the impact of jumping from one curve to another has motivated a deeper investigation into the creation and use of these curves as more than strategic planning tools.

Unfortunately, the simple collection and analysis of past throughput history is insufficient for curve generation. It is unlikely that an operating factory has experienced a sufficient number of changes *along the same curve* to allow creation of the curve. For example, a factory seldom operates on the flat portion of the curve where equipment utilizations are in the less than 50% range. It is also unlikely that the factory has carefully ramped up production start rates over the most rapidly changing portion of the curve, so the estimation of the shape in this region becomes problematic. In fact, as described previously, every time the factory changes dispatch policy or adds more equipment, it may not just be moving along the curve *but may in fact be*

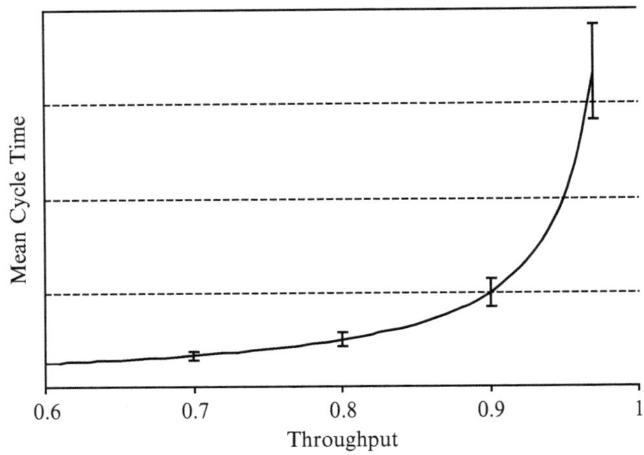

Fig. 19.2 Precision along a CT–TH curve using equal allocation of simulation effort

shifting to an entirely new curve. The technique of CT–TH curve generation requires the collection of large amounts of representative data; often, even when data have been collected, they have not been maintained in an appropriate manner to permit such an analysis. As a result, other than for the simplest of systems, simulation is the preferred method of data generation.

Several different design points must be simulated to generate a CT–TH curve. A careful selection of the design points can lead to minimal simulation expense. Various authors have discussed methods for generating a CT–TH curve and how to select these design points (see Park et al. 2001; Fowler et al. 2001; Yang et al. 2007). Other authors have presented methods for determining an appropriate allocation of simulation effort to the design points of the CT–TH curve being simulated, so as to obtain nearly equal absolute or relative precision (Leach et al. 2005).

The method commonly used by practitioners to generate a CT–TH curve via simulation is to allocate an equal amount of simulation effort to each throughput rate being simulated (Fig. 19.2). As throughput rate approaches capacity, the cycle time and the variance of the cycle time increase. Figure 19.2 illustrates that by equally allocating simulation effort to all design points, we will have a CT–TH curve that is less precise as we approach capacity, a clearly undesirable characteristic.

19.6 Building Single-Product CT–TH Curves via Simulation

The earlier chapter by Fischbein et al. points out that in modern manufacturing industries, single-PM factories are no longer common. However, even with a rapidly changing PM, decisions still need to be made, and often the best option is to make those decisions based on the current PM. In this section, we present issues that arise when estimating CT–TH curves via simulation and some methods for resolving those issues. To carefully define the problems and solutions requires a mathematical treatment.

We begin by considering single-product CT–TH curves in detail. Note that when we say "single product," we could be considering a CT–TH curve of a facility that produces only one product or focusing on one product out of many provided that the relative "mix" of the various products (as defined below) remains the same at all levels of system throughput (i.e., the facility is producing products to fulfill a projected mix of customer demands). Here, PM refers to the relative throughput of different existing products, not the introduction of new products or the phasing out of older products. The section concludes with discussions of emerging areas, including estimating CT–TH–PM (product mix) surfaces, modeling capacity expansion/reallocation, and accounting for batching.

Throughout this section, we represent a manufacturing system as a (possibly complex) network of queues (work centers, material handling equipment, etc.), the products as customers flowing through that network, and cycle time as the end-to-end time from product release to completion.

Define the following:

$$\boldsymbol{\lambda} = (\lambda_1, \ldots, \lambda_K) \text{vector of start rates for } K \text{ products (products/time),}$$

$$x = \text{utilization of the bottleneck station(s) in the facility, } 0 < x < 1, \text{ and}$$

$$\boldsymbol{\alpha} = (\alpha_1, \ldots, \alpha_K) \text{ product mix vector where } \alpha_k = \lambda_k / \sum_{h=1}^{K} \lambda_h.$$

Without loss of generality, we will only consider the cycle time of product 1, and denote its steady-state cycle time as $C(\boldsymbol{\lambda}) = C(x, \boldsymbol{\alpha})$, a random variable with unknown distribution that depends on the start rates. Notice that if we know the processing capacity of the bottleneck station, then specifying $(x, \boldsymbol{\alpha})$ is equivalent to specifying $\boldsymbol{\lambda}$. We will drop the dependence on $\boldsymbol{\alpha}$ in the single-product case. For the CT random variable to have a limiting distribution ("steady state"), among other things, the system logic and driving inputs must not be changing over time. As with queueing models of cycle time, "steady state" is only a useful approximation to aid decision making since real manufacturing systems change frequently, and in some wafer fabs may not actually be reached due to long cycle times and short product life cycles. Often, these approximations can be helpful in factory decision making. When they are not particularly helpful and one is interested in reasonably short-term behavior, clearing functions can be very useful. These are discussed in detail in the chapter by Missbauer and Uzsoy.

Generically, let $c_r(\boldsymbol{\lambda}) = c_r(x, \boldsymbol{\alpha}) = \mathrm{E}[C^r(x, \boldsymbol{\alpha})], r = 1, 2, \ldots$ be noncentral moments of the steady-state CT; we drop the subscript r when we refer to the mean (first moment). For the marginal variance of CT, let $\sigma^2(x, \boldsymbol{\alpha}) = \mathrm{Var}[C(x, \boldsymbol{\alpha})] = c_2(x, \boldsymbol{\alpha}) - c(x, \boldsymbol{\alpha})^2$, and let the asymptotic variance (defined below) of the rth moment be $\tau_r^2(x, \boldsymbol{\alpha})$. Denote the pth percentile of CT by $c_{p\%}(x, \boldsymbol{\alpha})$. To obtain percentiles, we will argue later that it is better to estimate them indirectly via moment-based approximations, rather than directly via order statistics, which is why we need to define $c_r(x, \boldsymbol{\alpha})$.

To estimate moments of CT, we will make one or more replications of a (typically large) number of individual product CTs. Let $C_{ij}(x, \alpha)$ be the jth observed CT from the ith replication, for $i = 1, 2, \ldots, m(x, \alpha)$ and $j = 1, 2, \ldots, l(x, \alpha)$. Our "steady state" assumption corresponds to requiring that $C_{ij}(x, \alpha)$ converges in distribution to $C(x, \alpha)$ as $j \to \infty$ for any i. Examples of conditions in which the assumptions are met are relatively easy to find, but in cases where this convergence assumption is not met, the results may not be accurate.

19.7 Point-by-Point Estimation of Single-Product CT–TH Curves

Perhaps the most straightforward way to generate a CT–TH curve via simulation is to select a fine grid of throughputs, say $0 < x_1 < x_2 < \cdots < x_d < 1$, and run simulation experiments at each one to estimate $c_r(x)$. We could, equivalently, select a grid of release rates λ (which correspond to throughput in steady state). However, when we fit CT–TH curves to data, there are a number of advantages to standardizing TH, so that system capacity always corresponds to a TH of 1. We shall adopt this convention from this point.

Unfortunately, even this approach has pitfalls, which are perhaps easiest to understand by examining a simple queueing model. Suppose that a manufacturing facility could be represented as an M/M/1 queue with service rate $\mu = 1$. Then it is well known that

$$c(x) = \frac{1}{1-x}$$

implies that the mean CT increases dramatically as the TH approaches system capacity. Similarly,

$$\sigma^2(x) = \frac{1}{(1-x)^2}$$

shows that the variance of the CT also explodes as $x \to 1$. Higher moments have similar behavior.

More importantly for estimating CT moments, however, we have the following: Consider making one or more long replications and averaging the observed CTs from each one to estimate $c(x)$. For the ith replication, let $\overline{C}_i(x) = \sum_{j=1}^{l(x)} C_{ij}(x)/l(x)$ be the sample mean CT, where $l(x)$ is the total number of observed cycle times from replication i. Then it can be shown that the asymptotic variance of the sample mean for the M/M/1 queue for x near 1 is

$$\tau^2(x) = \lim_{l(x)\to\infty} l(x)\mathrm{Var}[\overline{C}_i(x)] \propto \frac{1}{(1-x)^4}.$$

The asymptotic variance is useful because we can argue that for long enough run length $l(x)$

$$\mathrm{Var}[\overline{C}_i(x)] \approx \frac{\tau^2(x)}{l(x)},$$

and therefore that

$$\mathrm{Var}[\overline{C}(x)] \approx \frac{\tau^2(x)}{l(x)m(x)},$$

where $\overline{C}(x)$ is the average of the $m(x)$ replication averages. Thus, to maintain a constant relative error across the entire grid of x points, the quantity

$$\frac{\sqrt{\dfrac{\tau^2(x)}{l(x)m(x)}}}{c(x)}$$

must remain constant over x. This implies that the run length or the number of replications must grow proportional to $1/(1-x)^2$ as we approach system capacity. Naively making replications of equal length, or an equal number of replications, at each design point x will lead to mean CT estimates of widely different quality, and the problem only gets worse for higher moments. In Sect. 19.8, we present experimental design tools that address this issue. In the next few subsections, we assume that $(l(x), m(x))$ are somehow given and focus on how we use the data.

19.7.1 Point-by-Point Moment-Based Percentile Estimates

DES models have traditionally been used to generate estimates of average cycle time, and much work has been done on reducing the simulation run time for large-scale production systems to obtain these estimates more quickly. However, even with decreasing run times, there are still no efficient and easily implemented methods for obtaining accurate estimates of cycle time percentiles. Much of the reason for this is that percentiles are more difficult to compute than simple averages and can require excessive data storage.

A direct percentile estimate is one in which the estimate is a function of the raw data itself. Order statistics are traditionally used for this purpose. To obtain a cycle time percentile estimate from a discrete event simulation model using order statistics, the cycle time values are simply collected and ordered from low to high. The desired percentile is then directly selected from the sorted data. For example, to estimate the 95th percentile of cycle time from 100,000 observations, select the 95,000th smallest data point. Clearly, a drawback of this solution technique is that all the observations must be sorted and then stored to obtain the estimate. Even with rapidly increasing computing power, sorting and storing hundreds of millions of samples required to estimate some percentiles is still unreasonable (Chen and Kelton 2001).

An indirect percentile estimate is one in which the estimate is a function of data parameters (i.e., sample mean, sample variance, etc.) rather than the raw data itself. Indirect estimation techniques have the advantage of requiring less data storage, but may also be less accurate or have higher variance than the direct estimation techniques. One method for indirectly estimating cycle time percentiles was presented in McNeill et al. (2003). Their technique is a moment-based approach for estimating cycle time percentiles at a given throughput level and PM and is based on the Cornish–Fisher expansion (CFE) (Cornish and Fisher 1937). It utilizes the first four terms of the CFE and takes into account the first four moments of the cycle time distribution, allowing accurate percentile estimates to be generated for a variety of cycle time distributions found in manufacturing systems.

The CFE is an asymptotic series used to approximate normalized percentiles from any distribution, given a percentile from the standard normal distribution and the distribution's moments. The principle behind the CFE is that if a set of moments of a true and fitted distribution agree, the percentiles of the fitted distribution can be regarded as an approximation to the percentiles from the true distribution. The percentiles from the fitted distribution are expressed as an asymptotic series. The terms of the expansion are polynomial functions of the corresponding standardized percentile of the normal distribution, and the coefficients are functions of the standardized moments. In essence, the CFE is a moment-based correction for a non-normal random variable, and when sample moments are used in place of theoretical moments, the approximation becomes an estimator. In such cases, it is important to have accurate, low-variance estimators of the sample moments. Equations for consistent moment estimates can be found in Kenney and Keeping (1954) and are also given in Bekki et al. (2006).

Since the CFE is an asymptotic series, it is important to determine the appropriate number of terms to include. Adding additional terms to the CFE requires estimating higher and higher moments, which are, in turn, more and more difficult to estimate. Bekki et al. (2010) found that using the first four terms of the expansion clearly yields the best results for the widest variety of percentiles. Equation (19.1), given below, gives the first four terms of the CFE, where $\hat{\mu}$ is the sample mean, $\hat{\sigma}$ is the sample standard deviation, $\hat{\gamma}_1$ is the sample standardized central skewness, $\hat{\gamma}_2$ is the sample standardized central excess kurtosis, and z_α is the α percentile from the standard normal distribution. Here, y^α represents the desired percentile estimate from the original sample distribution, while w_α represents the corresponding percentile estimate from the standardized distribution:

$$y^\alpha = \hat{\mu} + \hat{\sigma} w_\alpha,$$
$$w_\alpha = z_\alpha + 1/6\left(z_\alpha^2 - 1\right)\hat{\gamma}_1 + 1/24\left(z_\alpha^3 - 3z_\alpha\right)\hat{\gamma}_2 - 1/36\left(2z_\alpha^3 - 5z_\alpha\right)\hat{\gamma}_1^2. \quad (19.1)$$

To obtain a percentile estimate of the cycle time distribution from a DES model using the CFE, running totals of the appropriate sums of squares, sums of cubes, etc., must be kept at each simulated throughput x to calculate the moment estimates during each simulation run. At the conclusion of each simulation run, (19.1) is used in conjunction with the moment estimates to obtain a percentile estimate $\hat{y}^{\alpha}$. Detailed descriptions of the percentile estimation procedure, including an approach for building confidence intervals around percentile estimates, can be found in McNeill et al. (2003) and Bekki et al. (2010).

Bekki et al. (2010) used this approach successfully in extensive experiments specialized to the kind of data we expect to find in manufacturing. Experimental systems ranged in complexity from a single M/M/1 queue to a model representing a full semiconductor manufacturing factory. First-in-first-out dispatching was used at all workstations. Their results show that the CFE technique provides accurate and precise results, is easy to implement, and requires storage of only five values used to calculate sample moments, regardless of the number of cycle time observations collected. Additionally, if there is a need to collect additional data, adding these new data to moment estimates for indirect estimation is much easier than adding the data to a sorted list of cycle time observations used for direct estimation. Finally, the approach has the advantage of being able to generate multiple percentile estimates from a single set of simulation runs by simply changing the z_{α} value in (19.1).

Figure 19.3 illustrates one of the benefits to the decision maker of generating multiple percentile estimates from a single set of simulation runs. This plot was generated from the output of a single set of runs and gives the complete range of cycle time percentiles for a model representing a full semiconductor manufacturing factory at 90% loading. The figure gives an estimate of the inverse of the cycle time distribution's cumulative distribution function at this traffic intensity and provides a valuable tool for quoting lead times. For example, it illustrates that if the factory was being run at 90% loading, a lead time of 1,400 hours could be met approximately 90% of the time, while a quoted lead time of 1,300 hours could only be met approximately 50% of the time.

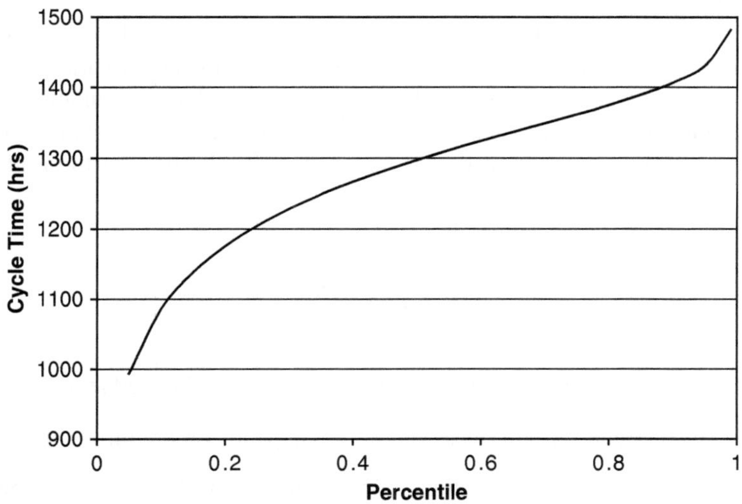

Fig. 19.3 Complete range of percentile estimates for a model of a semiconductor manufacturing facility at 90% loading

Finally, additional publications on moment-based cycle time percentile estimators indicate that the CFE-based approach in McNeill et al. (2003) and Bekki et al. (2010) requires modification in settings where non-FIFO dispatching rules are employed. Approaches combining data transformations with the CFE are proposed in McNeill et al. (2005), Bekki et al. (2006), and Bekki et al. (2009). The approach in Bekki et al. (2009) combines a power transformation with the CFE and was shown to provide a viable method for cycle time percentile estimation from DES models of manufacturing systems employing non-FIFO dispatching rules. Moreover, results showed that as models become larger and closer to today's complex manufacturing environments, the approach yielded even better accuracy and greater robustness to changes in the cycle time distribution. The transformation-based approach harnesses the benefits of the original CFE approach, making it a very attractive cycle time percentile estimation procedure for models of manufacturing systems.

19.7.2 Interpolation Between Points

A repercussion of point-by-point CT–TH curve estimation is that some sort of interpolation is needed to estimate CT properties at throughputs x that were not simulated. If the grid points are packed closely enough, then a simple linear interpolation may be adequate. However, as the M/M/1 example illustrated, exceptionally long runs may be required at the higher levels of TH, which argues against running simulations at a very fine grid. In Sect. 19.8, we describe methods for fitting models motivated by queueing theory to a very small set of carefully selected design points. In this section, we briefly discuss an intermediate strategy: using a sophisticated interpolation scheme.

In a series of papers, Kleijnen and van Beers (Kleijnen and van Beers 2004, 2005; van Beers and Kleijnen 2003, 2004) describe how the interpolation method of kriging can be adapted to the output of discrete-event, stochastic simulations in general, and queueing simulations in particular. In its simplest form, kriging estimates $c(x)$ by a weighted average of the estimated mean cycle times at the grid points $\mathbf{x} = (x_1, x_2, \ldots, x_d)$; i.e.,

$$\hat{c}(x) = \sum_{h=1}^{d} \hat{\beta}_h(x, \mathbf{x})\overline{C}(x_h),$$

where $\overline{C}(x_h)$ is the average of all the observations of CT obtained at utilization x_h, $\sum_{h=1}^{d} \hat{\beta}_h(x, \mathbf{x}) = 1$ and $\hat{\beta}_h(x, \mathbf{x})$ is a weight that depends on the throughput x to be interpolated, the grid $\mathbf{x}$ and the simulation outputs themselves. Loosely speaking, the kriging estimator gives more weight to CT estimates at grid points x_h that are closer to the point x to be interpolated.

The power of kriging comes from treating the unknown function – $c(x)$ in this case – as a realization of a random function, say $c(x) = Z(x) + \varepsilon(x)$, where $Z(x)$ is the "signal" function we are interested in deducing, and $\varepsilon(x)$ is a noise process representing the inherent output variability of the simulation at any throughput x. The random function $Z(x)$ is assumed to exhibit spatial covariance $\mathrm{Cov}(Z(x), Z(x')) = R(x, x')$, which is typically treated as only depending on the distance $|x - x'|$. By specifying a parametric form for this covariance function, its parameters can be estimated from the output data, and, given the spatial covariance function, the optimal (in a precise statistical sense) weights $\hat{\beta}_h(x, \mathbf{x})$ can be derived. Thus, kriging is a data-driven interpolation.

The difficulty introduced by stochastic simulation is the addition of output noise $\varepsilon(x)$, which is in a sense confounded with the uncertainty represented by $Z(x)$ and may itself depend on the throughput x (in fact, we know from the above M/M/1 example that the variance of CT differs dramatically across the range of x). See Kleijnen and van Beers (2005) for an analysis of this issue.

Because kriging is an interpolation method, it favors a finer grid (more design points x) than the queueing-motivated models we describe below. There is also no guarantee that the kriging estimator will exhibit known properties of the response function (that $c(x)$ is nondecreasing in x, for instance). However, kriging has the advantages that it is general purpose, it will not be subject to the lack of fit inherent in a poorly chosen metamodel, and it works largely without change for interpolating higher moments than the mean. Further, kriging extends naturally to a multidimensional independent variable, like the PM $\boldsymbol{\alpha}$. Good additional references include Barton and Meckesheimer (2006) and Santner et al. (2003).

19.7.3 Efficient CT–TH Curve Generation

Fowler et al. (2001) investigate the use of two common variance reduction techniques (common random numbers and antithetic variates) in efficiently generating CT–TH curves that linearly interpolate a set of (TH, CT) points. In their paper, the term "efficient" reflects the capability to provide a simulation-based CT–TH curve with an acceptable precision and accuracy using limited available resources. The goal was to generate CT–TH curves more economically, so that the cost of analysis could be reduced, allowing companies to make better manufacturing capacity management decisions. The experimentation in this paper included simulating an M/M/1 queueing system and a system with five stations in series (a special case of a Jackson queueing network). The results showed that common random numbers were effective when there was an adequate computing budget, but introduced too much bias when the computing budget was not large enough. On the other hand, the results showed that antithetic variates were effective for small or large computing budgets.

19.8 Building CT–TH Curves by Design

19.8.1 Model-Based Designs

One of the primary approaches to designing an experiment is to select the location of design points to optimize some criterion function; this is often called optimal design of experiments (see Pukelsheim 2006). Typically, this criterion is related to the variance/covariance matrix of the parameters in the model. The most popular optimization criterion is D-optimality, which seeks to minimize the volume of the joint confidence region of all the parameters in the model. Park et al. (2002) use the D-optimality criterion to choose design points for building the CT–TH curve. Since it concerns the confidence interval of the parameters of a model, D-optimality assumes that a model is specified for the response curve. Park et al. (2002) suggest two nonlinear regression models. The first accounts for the presence of batching in the factory, which is common in industries such as semiconductor manufacturing. The presence of batching changes the shape of the CT–TH curve from the one similar to that shown in Fig. 19.1 to a bowl-shaped curve, with dramatic increases in cycle time both at very low frequencies of wafer starts (in these cases, lots wait a long time for batches to be formed) as well as at very high frequencies of wafer starts. The first nonlinear regression model, suggested by Park et al., is appropriate when using batching policies, such as the full batch policy or the minimum batch size policy with a minimum greater than 1, that increase the cycle time at low levels of throughput. The second model suggested by Park et al., which has cycle time as a monotonically increasing function of the throughput, is used when there is no batching or a greedy batch policy is employed. The two models are shown below. Notice that both models have the cycle time exploding as the throughput, x, nears the capacity (β_2). If the throughput is normalized to the capacity, then $\beta_2 = 1$.

$$c(x) = \frac{\beta_1 x}{\beta_2 - x} - \beta_3, \tag{19.2}$$

$$c(x) = \frac{\beta_3}{x} + \frac{\beta_1 x}{\beta_2 - x} - \beta_4. \tag{19.3}$$

Both these models are generalizations of the CT–TH curve of a G/G/1 queue.

The experimental design for fitting these models is a selection of throughput values at which exhaustive simulations are conducted and the average steady-state cycle time is recorded. Nonlinear regression is used to estimate the parameters, and thus the linear approximation to the variance/covariance matrix is used to approximate the D-criterion. The candidate design points are placed at regular intervals from zero throughput to one, where one represents full capacity. The experimental design procedure recommended is a sequential procedure that starts with the minimum number of design points, which are required to support the model [three in the case of model (19.2) and four in the case of model (19.3)]. The D-criterion can be expressed as a function of the location of the design points. The initial points are selected as the set of three [or four in the case of model (19.2)] candidate points that maximize the D-criterion. All the other candidate points are then ranked according to the D-criterion for entry into the design if needed. After simulations are conducted at the initial points, the model parameters are estimated. Each additional candidate point is added sequentially in the predefined order until the parameter estimates no longer change by an appreciable amount (1% was used as a stopping criterion). This method was validated through construction of a CT–TH curve for a semiconductor wafer fabrication facility.

Another approach to building CT–TH curves was proposed by Cheng and Kleijnen (1999), hereafter CK, where they generalized the CT–TH curve of the M/M/1 curve as shown below:

$$c(x) = f(x) \sum_{l=0}^{t} \beta_l x^l + \varepsilon(x) = \frac{\sum_{l=0}^{t} \beta_l x^l}{1-x} + \varepsilon(x), \tag{19.4}$$

where $f(x) = 1/(1-x)$ and it is assumed that throughput, x, is scaled from zero to one (again $x = 1$ represents full capacity). CK only deal with the case of no batching or a greedy batch policy and thus the model in (19.4) is a more general form of the Park et al. (2002) model in (19.2). To fit (19.4) to the simulation data, CK develop a linear regression model since the only nonlinear part of the equation, $f(x)$, is known and can be dealt with through a transformation. The variance of the error term in (19.4) depends on x as

$$\mathrm{Var}[\varepsilon(x)] = [h(x)\sigma]^2, \tag{19.5}$$

where $h(x)$ is assumed to be known from asymptotic theory or other considerations.

The design of the experiment consists of the location of the design points $\mathbf{x} = (x_1, x_2, \ldots, x_m)$ and the fraction of a total of N replications assigned to those points $\boldsymbol{\pi} = (\pi_1, \pi_2, \ldots, \pi_m)$. The design is constructed to minimize a criterion called PM, which is a scaled version of the weighted-average variance of the estimated expected response over the throughput range of interest.

The CK procedure for fitting the model (19.4) can be summarized as follows. Given $f(x)$, $h(x)$, a maximum value of t and a fixed budget of N replications, find the optimal design $(\mathbf{x}, \boldsymbol{\pi})$ by minimizing PM. With the design points $\mathbf{x}$ fixed, carry out simulation experiments sequentially and adjust the allocation $\mathbf{x}$. Once the total number of runs has been exhausted, use backward selection to decide the appropriate polynomial order of model (19.4) and obtain the fitted curve.

The CK method leaves open the question of how to specify $f(x)$ and $h(x)$, which affect the design of the experiment and, more importantly, the adequacy of model (19.4) to represent the true CT–TH curve. When these two functions are known, CK is highly effective and efficient, and works within a fixed budget. However, for complicated manufacturing systems, there is not likely to be sufficient information to infer such characteristics. In other words, obtaining good choices for $f(x)$ or $h(x)$, although not impossible, is difficult in practice. Further, we have strong empirical evidence (Allen 2003; Johnson et al. 2004) that the $f(x)$ and $h(x)$ used by CK can be far from correct in realistic manufacturing simulations. Since model (19.2) used in Park et al. (2002) is a specific instance of (19.4), the same weakness can be attributed to their method as well.

In summary, the procedures by Park et al. (2002) and CK are both interesting and useful methods of experimental design for fitting a model such as that given in (19.4), but cases may arise in practice where these models are not sufficiently accurate to produce useful CT–TH curves.

19.8.2 Fixed Budget Variance-Based Designs

Practitioners commonly allocate an equal amount of simulation effort to each of several throughput rates when generating a CT–TH curve; this is referred to as *naive sampling*. In the case of a CT–TH curve, where cycle time variance is known to increase rapidly as throughput approaches capacity, naive sampling is likely to lead to widely varying precision at the throughput rates simulated. Since CT–TH curves support start-rate decisions, it is reasonable to assume that widely varying precision along the curve is undesirable.

In Leach et al. (2005) and Fowler et al. (2008), several different traffic intensities, i.e., design points are simulated in order to generate a CT–TH curve. Let D be the set of design points. The objective of these papers is to determine the allocation of a fixed budget of simulation effort to the design points being simulated that achieves nearly equal absolute or relative precision along the curve. The papers differ in the method of estimating the variance of the mean response estimates. In Leach et al. (2005), an approximation to the asymptotic variance is used and in Fowler et al. (2008) pilot runs are used.

The percentage of replications allocated to design point j for the absolute precision case using asymptotic variance $\tau^2(j)$ is

$$\pi(j) \doteq \frac{\left(t_{1-\alpha/2,m(j)-1}\right)^2 \tau^2(j)}{\sum_{k\in D}\left(\left(t_{1-\alpha/2,m(k)-1}\right)^2 \tau^2(k)\right)}. \tag{19.6}$$

The percentage of runs allocated to design point j for the absolute precision case using the sample variance from pilot runs ($S^2_{\overline{X}}(j)$) is

$$\pi(j) \doteq \frac{\left(t_{1-\alpha/2,m(j)-1}\right)^2 S^2_{\overline{X}}(j)}{\sum_{k\in D}\left(\left(t_{1-\alpha/2,m(k)-1}\right)^2 S^2_{\overline{X}}(k)\right)}. \tag{19.7}$$

The asymptotic variance approach has the advantage of requiring only one stage of simulation runs, but it requires approximating the asymptotic variance of cycle time at each design point, which can be difficult for complex systems. The pilot run approach does not have this limitation, but it requires two stages: the pilot runs and the production runs. In both papers, the proposed method is shown to outperform naive sampling.

19.8.3 Precision-Based Design

A precision-driven DOE strategy was proposed in Yang et al. (2007) to sequentially build up simulation experiments for the efficient generation of CT–TH curves. It allows the user to specify a precision level and is able to provide a fitted curve with desired precision by running simulation. We summarize the method in this section.

The estimation of the CT–TH curve is based on two statistical regression models (19.8) and (19.9), the forms of which are both motivated by heavy traffic queueing analysis and supported by extensive investigation of realistic manufacturing systems. One is called the expected cycle time (ECT) model:

$$c(x) = E[\overline{C}_i(x)] = \frac{\sum_{l=0}^{t} \beta_l x^l}{(1-x)^p}, i = 1, 2, \ldots, m(x) \tag{19.8}$$

which characterizes the relationship between the expected cycle time and normalized throughput x over a range of interest $[x_L, x_U]$. Unknown parameters are the polynomial coefficients $\boldsymbol{\beta}$, polynomial order t, and the exponent p. As explained earlier, the sample mean CT, $\overline{C}_i(x)$, obtained from the ith simulation replication performed at x will be used as the data points to which the CT–TH models are fit. The variance of $\overline{C}_i(x)$ depends on x and is represented by the following variance model:

$$\text{Var}[\overline{C}_i(x)] = \frac{\sigma^2}{(1-x)^{2q}}. \tag{19.9}$$

Both σ^2 and q are unknown parameters. With the sample mean CT data $\{\overline{C}_i(x), i = 1, 2, \ldots, m(x)\}$ at different values of x, the sample variance of $\overline{C}_i(x)$ can also be estimated over x, from which the variance model (19.9) can be fitted. With the estimated parameter $\hat{q}$, transforming the response $\overline{C}_i(x)$ by multiplying with $(1-x)^q$ will yield a constant variance and results in a standard nonlinear regression model:

$$c(x) \times (1-x)^q = E[\overline{C}_i(x) \times (1-x)^q] = (1-x)^{q-p} \sum_{l=0}^{t} \beta_l x^l = (1-x)^r \sum_{l=0}^{t} \beta_l x^l, \tag{19.10}$$

where $\boldsymbol{\beta}$, t, and the exponent r are unknown parameters. Thus, given a $\{\overline{C}_i(x), i = 1, 2, \ldots, m(x)\}$ dataset, model fitting is performed in two steps: (a) Fit the variance model (19.9) and obtain the q estimate and (b) use estimated parameter $\hat{q}$ to stabilize variance for the original observations $\overline{C}_i(x)$ and then fit model (19.10). The estimators of the ECT model (19.8) are obtained indirectly by noting that the coefficients $\boldsymbol{\beta}$ in model (19.8) coincide with those in (19.10), and p is estimated by the difference between the q and r estimates.

The goal is to obtain a precisely estimated CT–TH curve that helps manufacturers decide at what throughput they should run the system. Thus, Yang et al. (2007) evaluate the goodness of the fit by the relative error achieved on the ECT response estimators. Since the curve fitting is based on the nonlinear regression performed on models (19.9) and (19.10), variance estimates can be obtained on the estimated parameters in (19.9) and (19.10). Yang et al. (2007) let the user specify a target precision, say $\gamma\%$, which is defined as the relative error on the ECT estimator:

$$\gamma\% = \frac{\sqrt{\text{Var}[\hat{c}(x)]}}{\hat{c}(x)} \times 100.$$

Once fitted curves have been obtained, the relative error of the ECT estimate $\hat{c}(x)$ can be approximated for any throughput x over $[x_L, x_U]$. The user can choose to check the precision achieved at a throughput level of particular interest, or at a number of points in $[x_L, x_U]$ before they declare that a fitted curve with desired precision has been generated.

For efficient estimation of the CT–TH models presented above, DOE methodologies are developed to collect simulation data sequentially. The experiment design consists of the location of design points, the throughput levels at which simulations will be executed, the allocation of computational effort, and the number of simulation replications assigned to each design point. The best choice of experiment design depends on the true ECT and variance curves, which are unknown at the stage of designing experiments. In light of this, Yang et al. (2007) developed the YAN procedure to approach the DOE problem in a sequential manner. The model curves are estimated ever more precisely as more simulation data are obtained, and further experimentation is guided by the current best estimate of the models. This design and modeling process is continued until the prespecified precision $\gamma\%$ is achieved on the ECT response estimator.

To demonstrate the effectiveness of the YAN procedure, Yang et al. (2007) applied it to a number of systems to generate their corresponding CT–TH curves. The systems explored included analytically tractable queueing models and realistic semiconductor manufacturing systems. For simple queueing models such as M/M/1/FIFO, M/M/1/SPT (nonpreemptive shortest processing time first), and M/M/1/LPT (nonpreemptive longest processing time first), the true CT–TH curves can be derived analytically, and hence the quality of the simulation-based model estimation can be evaluated easily. The real wafer fab considered is provided by the Modeling and Analysis for Semiconductor Manufacturing Lab at Arizona State University (www.eas.asu.edu/~masmlab/). Since the true underlying curve is unknown in this case, "nearly true" ECT estimates were obtained by intensive simulation (running simulation until the standard error of the expected cycle time estimate was essentially zero). All the numerical experiments show that YAN is able to generate high-quality CT–TH curves with desired precision. Comparisons were also performed, which show that YAN can be more efficient than the procedure proposed by Cheng and Kleijnen (1999).

19.9 Multiproduct CT–TH Curves

The focus of this chapter has been on CT as a function of TH. However, especially in today's complex manufacturing environments, PM can also affect CTs, even if the overall system throughput is unchanged. However, fitting CT–TH–PM surfaces via simulation is a much more challenging problem and is a focus of current research by the chapter authors.

To see why the problem is difficult, consider perhaps the simplest nontrivial multiproduct queueing model, a multiproduct M/G/1 queue. For this model, the overall product release rate is λ, while the release rate for product k is $\lambda_k = \alpha_k \lambda$. If the service time for product type k has mean $1/\mu_k$ and variance ς_k^2, then we can show that for, say, product 1

$$c(x) = \beta_0 + \frac{\sum_{k=1}^{K} \beta_k \alpha_k}{1 - x}, \tag{19.11}$$

where $x = \lambda \sum_{k=1}^{K} \frac{\alpha_k}{\mu_k} = \sum_{k=1}^{K} \frac{\lambda_k}{\mu_k}$, $\beta_0 = \frac{1}{\mu_1}$ and $\beta_k = \frac{\lambda}{2}\left(\frac{1}{\mu_k^2} + \varsigma_k^2\right)$. Unfortunately, a realistic manufacturing system consisting of multiple work stations will exhibit behavior that is more like what one would expect from a sum of functions of the form (19.11), implying a model with lots of parameters that will be difficult to fit by observing only the overall product cycle times. In addition, the bottleneck stations that are actually operating at utilization x will change if x is held fixed but the PM changes, introducing sharp ridges in the response surface.

Hung et al. (2003) attacked this problem by attempting to fit polynomials in the release rates λ_k. Realizing that such models could not provide a good global fit over the entire release-rate space, they used classification and regression trees to produce a data-driven partition of this space into subregions that are well represented by low-order polynomials in λ_k.

If a capacity model exists, then a system-driven partitioning of the TH-PM space into regions of constant bottleneck may be attained. Within these regions, it may be possible to interpolate between CT–TH curves fit for a collection of fixed PMs, using either a general-purpose interpolator such as kriging or a model-based interpolation suggested by (19.11).

Clearly, precisely modeling a high-dimensional CT–TH–PM surface will require many more runs of (perhaps) very complex simulation models, which will not be feasible if runs are made one at a time with a human analyst's intervention. Fortunately, modern, distributed computing environments are making it easier to distribute simulations of distinct cases to idle computers and collect the results.

This may eventually force a change in traditional software licensing arrangements that charge "per CPU" used, which would be a substantial impediment to the paradigm described here.

19.10 Future Research

Production planning was one of the early areas of application for DES. However, until recently, DES has been used primarily to evaluate proposed production plans, often in an ad hoc manner and generally to estimate cycle times. In this chapter, we have described some recent efforts to use simulation more directly in production planning and focused on the estimation of CT–TH curves via simulation. A significant limitation of the work we discussed was that it has focused on characterizing the steady-state behavior of manufacturing systems, and the resulting CT–TH curves are particularly suitable for supporting long-term decision making in manufacturing (e.g., capacity expansion). For medium-term production planning over a time horizon of weeks or months, it has long been recognized that stationary behavior may not be indicative of the system performance (Uzsoy et al. 1992; Papadopolous et al. 1993): due to factors such as short product life cycles and frequent change of technologies, manufacturing systems may never be operated in its steady state (see the chapter by Missbauer and Uzsoy in this handbook and the paper by Shanthikumar et al. 2007). Nevertheless, time-dependent behavior is rarely considered in the context of production planning due to the fact that transient analysis of real systems is notoriously difficult.

As with steady-state performance, the existing literature has used both computer simulation and analytical models to address the transient behavior of queueing systems. The former frequently becomes too computationally demanding as mentioned earlier. The analytical methods for transient analysis, on the other hand, have focused on developing numerical solutions to a set of time-dependent ordinary differential equations (ODEs) describing the system's dynamic behavior (see the review of Ingolfsson et al. 2007); the construction of these ODEs relies heavily on restrictive assumptions such as the Markov property, and solving the typically large set of ODEs is also computationally challenging.

We believe that to accurately capture the transient performance, a metamodeling approach is also able to overcome the limitations of simulation and analytical methods. As opposed to mapping a static regression model representing the system's steady-state behavior, here, a number of transfer function models (TFMs) can be estimated through simulation, thus quantifying the time-dependent performance. The resulting TFMs are difference equations, like the discrete approximations of the ODEs provided by an analytical approach. Thus, they embody the high fidelity of simulation and allow for prompt "what-if" analysis.

In the near future, we expect to see the development of methods that allow the capacity of the factory to be changed and methods for nongreedy batching policies (such as full batch policies) that lead to a complete response surface mapping that is not monotonically increasing. Finally, we anticipate the development of methods that determine revenue maximal production plans with cycle time constraints.

Acknowledgment This work was partially supported by Semiconductor Research Corporation Grant Numbers 2004-OJ-1224 and 2004-OJ-1225 and by grants DMII 0140441 and DMII 0140385 from the National Science Foundation.

References

Agrawal A, Minis I, Nagi R (2000) Cycle time reduction by improved MRP-based production planning. Int J Prod Res 28:4823–4841.

Asmundsson J, Rardin RL, Uzsoy R (2006) Tractable nonlinear production planning models for semiconductor wafer fabrication facilities. IEEE Trans Semicond Manufact 19:95–111.

Atherton RW, Dayhoff JE (1986) Signature analysis: simulation of inventory, cycle time and throughput trade-offs in wafer fabrication. IEEE Trans Components Hybrids Manufact Technol 9(4):498–507.

Barton RR (1998) Simulation metamodels. In: Medeiros DJ, Watson EF, Carson JS, Manivannan MS (eds) Proceedings of the 1998 winter simulation conference. IEEE, Piscataway, New Jersey, pp 167–174.

Barton RR, Meckesheimer M (2006) Metamodel-based simulation optimization. In: Henderson SG, Nelson BL (eds) Handbook in OR & MS 13. Elsevier, North Holland, pp 535–574.

Bekki JM, Fwoler JW, Mackulak GT, Nelson BL (2010) Indirect cycle time quantile estimation using the Cornish-Fisher expansion. IIE Trans 42:31–44.

Bekki JM, Fowler JW, Mackulak GT, Kulahci M (2009) Estimation of cycle time quantiles in manufacturing environments employing non-FIFO dispatching policies. J Simul 3:69–83.

Bekki J, Mackulak GT, Fowler JW (2006) Indirect cycle-time percentile Estimation for non-FIFO dispatching policies. In: Perrone LF, Wieland FP, Liu J, Lawson BG (eds) Proceedings of the 2006 winter simulation conference, pp 1825–1835.

Bitran GR, Tirupati D (1989) Tradeoff curves, targeting and balancing in manufacturing queueing networks. Oper Res 37(4):547–564.

Brown S, Chance F, Fowler JW et al. (1997) A centralized approach to factory simulation. Future Fab Int 1(3):83–86.

Chen EJ, Kelton WD (2001) Quantile and histogram estimation. In: Peters BA, Smith JS, Medeiros DJ, Rohrer MW (eds) Proceedings of the 2001 winter simulation conference, pp 451–459.

Cheng RCH, Kleijnen JPC (1999) Improved design of queueing simulation experiments with highly heteroscedastic responses. Oper Res 47:762–777.

Chryssolouris G, Lee M, Pierce J et al. (1990) Use of neural networks for the design of manufacturing systems. Manufact Rev 3:187–194.

Chung S-H, Lai C-M (2006) Job releasing and throughput planning for wafer fabrication under demand fluctuating make-to-stock environment. Int J Adv Manufact Technol 31:316–327.

Cochran JK, Chen H-N (2002) Generating daily production plans for complex manufacturing facilities using multi-objective genetic algorithms. Int J Prod Res 40:4147–4167.

Connors DP, Feigin GE, Yao DD (1996) Queueing network model for semiconductor manufacturing. IEEE Trans Semicond Manufact 9:412–427.

Conway RW, Johnson BM, Maxwell WL (1959) Some problems of digital systems simulation. ManageSci 6:92–110.

Cornish EA, Fisher RA (1937) Moments and cumulants in the specification of distributions. Revue de l'Institut International de Statistique 5:307–320.

Delp D, Si J, Fowler J (2006) The development of the complete x-factor contribution measurement for improving cycle time and cycle time variability. IEEE Trans Semicond Manufact 19:352–362.

Delp D, Si J, Hwang Y, Pei B et al. (2005) Availability adjusted x-factor. Int J Prod Res 43: 3933–3953.

Ehteshami B, Pétrakian RG, Shabe PM (1992) Trade-offs in cycle time management: hot lots. IEEE Trans Semicond Manufact 3: 101–106.

Fowler J, Rose O (2004) Grand challenges in modeling and simulation of complex manufacturing systems. Simul Trans Soc Comput Simulat Int 80(9):469–476.

Fowler JW, Leach SE, Mackluak GT, Nelson BL (2008) Variance-based sampling for simulating cycle time-throughput curves using simulation-based estimates. J Simul 2:69–80.

Fowler JW, Mackulak GT, Ankenman BE et al. (2005) Procedures for efficient cycle time-throughput curve generation. In: Proceedings of the NSF 2005 DMII Grantees Conference, pp 1–8.

Fowler JW, Park S, Mackulak GT, Shunk DL (2001) Efficient cycle time-throughput curve generation using fixed sample size procedure. Int J Prod Res 39(12):2595–2613.

Fu MC, Glover F, April J (2005) Simulation optimization: a review, new developments, and applications. In: Kuhl ME, Steiger NM, Armstrong FB, Joines JA (eds) Proceedings of the 2005 Winter Simulation Conference. IEEE, Piscataway, New Jersey, pp 83–95.

Herrmann JW, Chincholkar MM (2001) Reducing throughput time during product design. J Manufact Syst 20:416–428.

Hopp WJ, Spearman ML (2000) Factory physics, 2nd edn. Irwin McGraw-Hill, Boston.

Hung YC, Michailidis G, Bingham DR (2003) Developing efficient simulation methodology for complex queueing networks. In: Chick S, Sanchez PJ, Ferrin D, Morrice DJ (eds) Proceedings of the 2003 Winter Simulation Conference. IEEE, Piscataway, New Jersey, pp 512–519.

Ingolfsson A, Akhmetshina E, Budge S, Li Y et al. (2007) A survey and experimental comparison of service level approximation methods for non-stationary M/M/s queueing systems. INFORMS J Comput 19(2): 201–214.

International Technology Roadmap for Semiconductors 2006 update. Available via http://www.itrs.net/Links/2006Update/FinalToPost/10_Factory_2006Update.pdf. Accessed 1 2007.

Kenney JF, Keeping ES (1954) Mathematics of statistics, part 1. D. Van Nostrand Company, Inc., Princeton, New Jersey.

Kleijnen JPC (1988) Experimental design and regression analysis in simulation: an FMS case study, Eur J Oper Res 33:257–261.

Kleijnen JPC (1993) Simulation and optimization in production planning: a case study, Decis Supp Syst 9:269–280.

Kleijnen JPC, Sanchez SM, Lucas TW et al. (2005) State-of-the-art review: a user's guide to the brave new world of designing simulation experiments. INFORMS J Comput 17(3):263–289.

Kleijnen JPC, Sargent JPC (2000) A methodology for fitting and validating metamodels in simulation. Eur J Oper Res 120:14–29.

Kleijnen JPC, van Beers WCM (2004) Application-driven sequential designs for simulation experiments: Kriging metamodeling. J Oper Res Soc 55:876–883.

Kleijnen JPC, van Beers WCM (2005) Robustness of Kriging when interpolating in random simulation with heterogeneous variances: some experiments. Eur J Oper Res 165:826–834.

Ko SS, Serfozo R, Sivakumar AI (2004) Reducing cycle times in manufacturing and supply chains by input and service rate smoothing. IIE Trans 36:145–153.

Leach SE, Fowler JW, Mackulak GT et al. (2005) Asymptotic variance-based sampling for simulating cycle time-throughput curves. Working paper ASUIE-ORPS-2005–003, Arizona State University.

Liao DY, Wang CN (2004) Neural-network based delivery time estimates for prioritized 300-mm automatic material handling operations. IEEE Trans Semicond Manufact 17:324–332.

Mackulak GT, Fowler JW, Park S et al. (2005) A three phase simulation methodology for generating accurate and precise cycle time-throughput curves. Int J Simul Process Model 1:36–47.

McNeill J, Mackulak G, Fowler J (2003) Indirect estimation of cycle time percentiles from discrete event simulation models using the Cornish-Fisher expansion. In: Chick S, Sánchez PJ, Ferrin D, Morrice DJ (eds) Proceedings of the 2003 winter simulation conference. IEEE, Piscataway, New Jersey, pp 1377–1382.

McNeill J, Nelson BL, Fowler JW et al. (2005) Cycle-time percentile estimation in systems employing dispatching rules. In: Kuhl ME, Steiger NM, Armstrong FB, Joines JA (eds) Proceedings of the 2005 winter simulation. IEEE, Piscataway, New Jersey, pp 751–755.

Mollaghasemi M, LeCroy K, Georgiopoulos M (1998) Application of neural networks and simulation modeling in manufacturing system design. Interfaces 28: 100–114.

Nelson BL (2004) 50th anniversary article: stochastic simulation research in management science. Manage Sci 50:855–868.

Nemoto K, Akcali E, Uzsoy R (2000) Quantifying the benefits of cycle time reduction in semiconductor wafer fabrication. IEEE Trans Electron Pack Manufact 23:39–47.

Papadopoulos HT, Heavey C (1996) Queueing theory in manufacturing systems analysis and design: a classification of models for production and transfer lines. Eur J Oper Res 92:1–27.

Papadopolous HT, Heavey C Browne J (1993) Queueing theory in manufacturing systems analysis and design, 1st edn. Springer, New York.

Park S, Fowler JW, Mackulak GT et al. (2002) D-optimal sequential experiments for generating a simulation-based cycle time-throughput curve. Oper Res 50(6):981–990.

Pukelsheim F (2006) Optimal design of experiments. SIAM, Philadelphia.

Sabuncuoğlu I, Touhami S (2002) Simulation metamodeling with neural networks: an experimental investigation. Int J Prod Res 40(11): 2483–2505.

Santner TJ, Williams BJ, Notz WI (2003) The design and analysis of computer experiments. Springer-Verlag, NY.

Schultz C (2004) Spare parts inventory and cycle time reduction. Int J Prod Res 42:759–776.

Shang JS, Tadikamalla PR (1993) Output maximization of a CIM system: simulation and statistical approach. Int J Prod Res 31(1):19–41.

Shantikumar JG, Ding S, Zhang MT (2007) Queueing theory for semiconductor manufacturing systems: a survey and open problems. IEEE Trans Automat Sci Eng 4(4):513–522.

Sivakumar AI, Chong CS (2001) A simulation based analysis of cycle time distribution, and throughput in semiconductor backend manufacturing. Comput Ind 45:59–78.

Spence AM, Welter DJ (1987) Capacity planning of a photolithography work cell in a wafer manufacturing line. In Proceedings of the IEEE international conference on robotics and automation, Raleigh, NC, Piscataway, NJ, pp 702–708.

Sze MT, Fi P, Lee WB (2001) Modeling the component assignment problem in PCB assembly. Assembly Autom 21:55–60.

Uzsoy R, Lee CY, Martin Vega L (1992)A review of production planning and scheduling in the semiconductor industry, part I: system characteristics, performance evaluation, and production planning. IIE Trans 24(4):47–61.

van Beers WCM, Kleijnen JPC (2003) Kriging for interpolation in random simulation. J Oper Res Soc 54:255–262.

van Beers WCM, Kleijnen JPC (2004) Kriging in simulation: a survey. In: Ingalls RG, Rossetti MD, Smith JS, Peters BA (eds) Proceedings of the 2004 winter simulation conference. IEEE, Piscataway, New Jersey, pp 113–121.

Vellido A, Lisboa PJG, Vaughan J (1999) Neural networks in business: a survey of applications (1992–1998). Expert Syst Appl 17:51–70.

Yang F, Ankenman BE, Nelson BL (2007) Efficient generation of cycle time-throughput curves through simulation and metamodeling. Naval Res Logist 54:78–93.

Yücesan E, Fowler JW (2000) Simulation analysis of manufacturing and logistics systems. Kluwer Encyclopedia Prod Manufact Manage 687–697, ISBN: 978-0-7923-8630-8.

Chapter 20
Simulation-Optimization in Support of Tactical and Strategic Enterprise Decisions

Juan Camilo Zapata, Joesph Pekny, and Gintaras V. Reklaitis

20.1 Introduction

The modern enterprise has developed highly complex supply chains in order to efficiently satisfy demand while remaining competitive. Supply chains have become distributed global networks that encompass not only the manufacture and delivery of goods but also the activities associated with their development. Moreover, local "here and now" decisions must be made in the presence of future uncertainty while also considering their global and long-term implications. This coupling of wide problem scope with multiple sources of internal and external uncertainties, such as production line breakdowns, raw material availability, market demand, exchange rate fluctuations, developmental failures, etc., has resulted in supply chain decision-making processes that are of high complexity and a very large scale (Zapata et al. 2008).

The need for techniques capable of determining the optimal set of decisions for this kind of systems has motivated the development of stochastic programming, stochastic dynamic programming, and simulation optimization. Stochastic programming and stochastic dynamic programming rely on the ability to articulate a tractable mathematical formulation of the system, which can be very difficult for complex supply chain applications. Furthermore, owing to the large size of problem spaces, nonlinearity of objective functions and constraint, and the discrete nature of many decisions, the resulting stochastic program may not always be solvable using state-of-the-art stochastic programming methods. Hence, the focus of this chapter is on simulation optimization, which couples the flexibility of discrete event simulation to accommodate arbitrary stochastic elements and model the dynamics and complexities of real-world systems without the need to develop formal mathematical models, and the ability of optimization schemes to systematically search the decision space. However, similar to stochastic programming and related techniques, simulation optimization can easily become computationally very demanding, and

J.F. Pekny (✉)
College of Engineering, Purdue University, West Lafayette, IN 47907, USA
e-mail: pekny@purdue.edu

K.G. Kempf et al. (eds.), *Planning Production and Inventories in the Extended Enterprise*, International Series in Operations Research & Management Science 151,
DOI 10.1007/978-1-4419-6485-4_20,

thus requires that a range of sometimes rather subtle issues be addressed effectively to obtain a viable trade-off between solution time, modeling effort, and solution quality.

The chapter is organized as follows. Section 20.2 provides a summary of the different simulation-optimization methods that are available, aimed at guiding the reader in the selection of the most adequate technique for his/her particular problem. Section 20.3 presents two industrial case studies in which simulation optimization was used to support the decision-making process. Finally, concluding remarks are presented in Section 20.4.

20.2 Simulation-Optimization Solution Strategies

This section reviews the existing simulation-optimization methods, including their strengths and weaknesses. The aim of the review is to explain at a conceptual level the underlying algorithms and provide relevant references. To facilitate the presentation of the different methods we start by formalizing the problem in a mathematical sense. The problem to be solved can be expressed as

$$\min_{\theta \in \Theta}(\max)\ J\ (\theta), \qquad (20.1)$$

where θ is the decision vector of p parameters, the feasible region $\Theta \subset \Re^p$ is the set of possible values of the parameter θ, and $J(\theta) = E\left[L(\theta, \omega)\right]$ represents the expected value of a performance measure $L(\theta, \omega)$. Notice that $L(\theta, \omega)$ is a random variable that can take different values depending on the specific realizations of the stochastic effects of the system, ω. Therefore, the problem exhibits not only the typical challenges of finding an optimal solution but also those of estimating the performance measure.

In general, simulation-optimization methods are classified based on the continuous or discrete nature of the decision space (Fu 1994). In addition, methods for discrete variables are further cataloged according to the number of feasible solutions (small or large (including infinite)), and the ordered (i.e., represents different levels or degrees of the underlying characteristic (e.g., safety inventory level)) or unordered (i.e., represents categories that cannot be quantified (e.g., queue discipline)) nature of the variables. Figure 20.1 shows the classification scheme and the methods that fall in each class. It is important to highlight though that different classes of methods are often used in combination within a single computational scheme. On the continuous side, methods that mix response surface methodologies (RSM) and stochastic approximation (SA) have been developed (Ho 1992). In the case of discrete variables, hybrid approaches that combine different methods within a class, as well as methods in different classes, have been proposed. For example, Hall and Bowden (1996) combine metaheuristics with pattern search; Nozari and Morris (1984) combine ranking and selection (R&S) and pattern search, and Pichitlamken and Nelson (2003) combine R&S and metaheuristics.

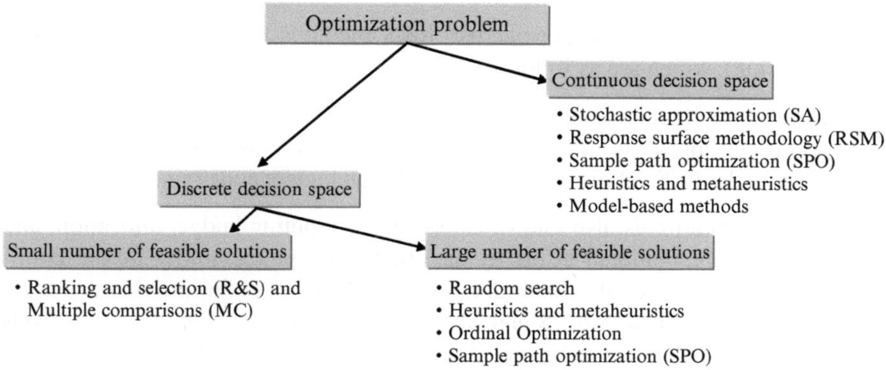

Fig. 20.1 Classification of simulation-optimization techniques

20.2.1 Small Number of Discrete Feasible Solutions: Ranking and Selection (R&S) and Multiple Comparisons (MCs)

The techniques available for problems with a small number of feasible solutions focus on the exhaustive comparison of all feasible solutions rather than on the search algorithms (Fu 2002). The presence of uncertainties transforms the comparison process into an inference exercise that uses the statistical machinery developed for the calculation of confidence intervals.

The basic concept behind Multiple Comparisons (MC) is very simple. The differences in performance measure, $\hat{J}(\theta_i) - \hat{J}(\theta_j)$, for some kind of pairwise comparison of the possible solutions are estimated from simulations. Then, the corresponding confidence intervals are examined in search of an absolute winner (i.e., in the case of an all-pairwise comparison, the θ_i whose confidence intervals in regard all other possible solutions are strictly negative (strictly positive)). However, it is not possible to guarantee a solution a priori since the confidence intervals may not be tight enough. Therefore, all the techniques in this class are aimed at exploiting the opportunities presented by simulation to reduce variance (e.g., common random numbers) and hence tighten the confidence intervals using the minimum possible number of simulations (Fu 1994).

Ranking and selection also uses confidence intervals but within the context of the correct selection concept. These methods measure in some way how far the chosen solution is from the optimal one. In general, two approaches have been proposed to measure that "distance." The first is known as the indifference zone. In this case, the objective is to obtain a solution that is within a certain range (indifference zone), δ, of the optimal solution, θ_*, with a specified probability of correct selection (PCS), P^*, (i.e., $P\{J(\theta_i) - J(\theta_*) < \delta\} \geq P^*$). The second approach, referred to as subset selection, guarantees that with a certain probability, a particular group of solutions chosen from the original set will contain at least one solution, θ_s, that is within a specified indifference zone (i.e., $P\{J(\theta_s) - J(\theta_*) < \delta\} \geq P^*$).

From an implementation perspective, R&S methods follow two formulations (Fu et al. 2005), which are as follows:

1. Minimize the number of simulations subject to the PCS exceeding a given level (a traditional approach that offers little control over computational requirements).
2. Maximize the PCS subject to a given simulation budget constraint.

The latter formulation is also known as optimal computing budget allocation, and manages the computational effort by sacrificing the predictability of the confidence levels. Swisher et al. (2004) and Kim and Nelson (2006) provide extensive lists of references for both R&S and MC methods. These two classes of methods were originally considered to be two different strategies (Fu 1994), but Nelson and Matejcik (1995) established the connection between the two by showing that most indifference zone procedures can also provide confidence intervals for a certain type of multiple comparison method.

20.2.2 Large Number of Discrete Feasible Solutions

20.2.2.1 Random Search

Random search methods move successively from one feasible solution to a neighboring one based on probabilistic arguments. All methods in this class (see the review by Banks (1998)) follow the same algorithmic structure as follows:

1. Initialization with a feasible solution
2. Probabilistic generation of a new decision vector, obtained from a set of neighboring feasible solutions
3. Estimation of performance measures and comparison with the values from the previous iteration
4. Evaluation of stopping criteria and return to Step 2 if not satisfied

The methods in this class are characterized by the definition of the neighborhood (the set in which the algorithm can move from one solution to another in a single iteration), the selection strategy of the next decision vector, and the manner in which the optimum is chosen. A representative example of this class of methods is simulated annealing, which attempts to achieve a global optimum by allowing moves leading to nonimproving solutions with a certain probability that depends on the stage of the procedure. Nonimproving moves leading to a poorer solution are more likely to be accepted early in the process; as the search progresses towards a global optimum, the probability of accepting non-improving moves tends to zero. A step-by-step description of a version of the method for a minimization problem is as follows (Alrefaei and Andradottir 1999):

Step 1. Initialize the decision variables, θ_0, the number of iterations, $n = 0$, the optimal solution, $\theta_0^* = \theta_0$, and $A_0(\theta) = C_0(\theta) = 0$ for each θ, where $A_i(\theta)$ is the sum of all the estimates of the performance measure $J(\theta)$,

$\hat{J}(\theta_n)$, obtained from simulations in the $_i$ first iterations and $C_i(\theta)$ is the number of replicates in the $_i$ first iterations.

Step 2. Generate a neighbor solution, θ'_n, of the current point θ_n based on the chosen transition probability matrix, $R(\cdot,\cdot)$. This means that for all $\theta \in N(\theta_n)$, where $N(\theta_n)$ is the neighborhood of θ_n the probability of being selected in the next iteration is given by $P(\theta'_n = \theta) = R(\theta_n, \theta)$.

Step 3. Estimate $\hat{J}(\theta_n)$, and $\hat{J}(\theta'_n)$, using simulation. If $\hat{J}(\theta'_n) \leq \hat{J}(\theta_n)$, then let $\theta_{n+1} = \theta'_n$. Otherwise, sample a uniform distribution $U_n \sim U[0,1]$ and an exponential distribution $e_n \sim \exp\left[\hat{J}(\theta_n) \leq \hat{J}(\theta'_n) / T\right]$, and if $U_n \leq \dot{e}_n$ then let $\theta_{n+1} = \theta'_n$. Otherwise let $\theta_{n+1} = \theta_n$. Notice that T(known as the temperature) is the iteration-dependent parameter used to decrease the probability of accepting nonimproving moves as the number of iterations increase.

Step 4. Let $n = n + 1$, $A_n(\theta) = A_{n-1}(\theta) + \hat{J}(\theta)$, and $C_n(\theta) = C_{n-1}(\theta) + 1$, for $\theta = \theta_n$ or θ'_n, and $A_n(\theta) = A_{n-1}(\theta)$ and $C_n(\theta) = C_{n-1}(\theta)$ for all θ that have been explored but are different from θ_n and θ'_n $\left(\theta \in \Theta_E \backslash \{\theta_n \text{ or } \theta'_n\}\right)$. Finally, select the θ associated with the smallest average value of the performance measure, $A_n(\theta) / C_n(\theta)$, from the set of decisions explored, Θ_E, $\left(\min_{\theta \in \Theta_E} A_n(\theta) / C_n(\theta)\right)$.

Ideally, the method returns to Step 2 and the algorithm is repeated until convergence is reached. However, resource and time limitations may require the use of a user-defined stopping criteria, such as number of iterations, a threshold value for the performance measure, etc.

The advantages of random search methods are their model independence (i.e., no explicit mathematical model of the system needs to be developed) and the existence of theoretical convergence proofs under certain conditions. However, in practice, convergence can be slow and dependent on the selection of the neighborhood structure (Banks 2005) and does not scale well with the number of variables in the decision space.

20.2.2.2 Ordinal Optimization

These methods are based on the observation that in most cases it is much easier, in terms of computation, to directly find the ordering among candidate solutions than it is to estimate the performance measure of each candidate solution and rank the solutions based on this measure. This idea can be explained with the following simple example (Fu et al. 2005). Assume that there are only two possible decision vectors θ_1 and θ_2, and the decision maker wants to know which of the two results in the smallest expected value of the performance measure ($J(\theta_1) < J(\theta_2)$ or $J(\theta_1) > J(\theta_2)$). One approach can be to estimate each of the expected performance measures independently, $\hat{J}(\theta_1)$ and $\hat{J}(\theta_2)$, until the standard error for each estimate is less than the indifference amount, ε, and compare the resulting values. On the other hand, an ordinal optimization method would define a variable $X = L(\theta_1) - L(\theta_2)$ and determine whether $E[X]$ is positive or negative. The latter strategy is more efficient because the estimation of $E[X]$ requires lesser number of simulations when compared with the estimation of $\hat{J}(\theta_1)$ and $\hat{J}(\theta_2)$. Swisher et al. (2004) provide an extensive list of references for this technique.

20.2.3 Continuous Decision Variables

20.2.3.1 Stochastic Approximation (SA)

Stochastic approximation refers to a group of methods that attempt to mimic the gradient search method traditionally used in deterministic optimization. As in its deterministic counterpart, SA searches for a local optimum to the problem given by (20.1), that satisfies the first-order condition

$$\hat{\nabla} J(\theta) = 0, \tag{20.2}$$

where $\hat{\nabla} J(\theta)$ represents the estimated gradient of the performance function. The analogy to the deterministic case also applies to the general structure of SA algorithms which are based on the following iterative form:

$$\theta_{n+1} = \prod\nolimits_{\Theta} \left(\theta_n - a_n \hat{\nabla} J(\theta_n)\right). \tag{20.3}$$

Here θ_n is the solution vector at the beginning of iteration n, $\{a_n\}$ is a positive sequence of step sizes, and $\prod_{\Theta}$ represents some projection back into the feasible set Θ when the iteration leads to a solution outside the set. Similar to that in the deterministic case, the algorithm determines at each iteration the value of the decision vector based on the gradient and the step size values calculated for that iteration and the value of the decision vector in the previous iteration. Algorithms referred to as Robust SA algorithms differ slightly in that they use the iterative process based on (20.3), but instead of returning the final value of the decision vector as the optimum, they return an average (e.g., moving horizon or exponentially weighted moving average) of a certain number of iterates to reduce the variance in the estimation (Fu 2002). The set of constraints that determine the feasible region also exhibit some differences when compared with the deterministic case. In general, the feasible region is determined by a mix of deterministic and probabilistic constraints. Probabilistic constraints limit the probability of constraint violations but not the magnitude of the violations. They are expressed as follows:

$$P\left(f(\theta,\omega) \geq 0\right) > 1 - \alpha, \tag{20.4}$$

where f is the random vector representing the left hand side of a set of constraints whose realizations depend on the set of decisions θ, and the presence of uncertainties ω. P is the vector of probabilities of violating the constraints, and α ($\alpha \in (0,1)$) are the tolerance levels of the decision maker to these violations. Notice, however, that only a few algorithms can handle this kind of constraint (see Andradottir (1998) and Kushner and Yin (2003) and references hereafter).

For all SA algorithms to properly converge it is required that the step size goes to zero at a rate that is not too fast (to avoid premature convergence to a suboptimal solution), and not too slow (to ensure eventual convergence). Mathematically, these conditions are commonly represented as $\sum_{n=1}^{\infty} a_n = \infty$ and $\sum_{n=1}^{\infty} a_n^2 < \infty$. In addition, it is required that the bias of the objective function gradient estimate, $\hat{\nabla} J(\theta)$ in (20.3), goes to zero (Fu 1994). In theory, the appropriate step size rate can be achieved using a simple harmonic series ($a_n = a/n$), but in practice this choice results in slow convergence rates. Since the performance of any SA algorithm is quite sensitive to this sequence, researchers have developed different strategies aimed to speed up convergence. Heuristic decrements in step size have been proposed (e.g., Chapter 9 in Banks (1998) and references therein), as well as the use of a constant step size in the early stages of the iterative process followed by heuristic decrements (Fu 2002).

The need to obtain unbiased estimates of the objective function gradient in an efficient manner has motivated most of the different techniques used in SA. The remainder of this section provides an introduction to each of these developments and a summary of their main strengths and limitations.

Finite Differences (FD)

Similar to that in numerical differentiation the idea is to use a secant as an approximation to the gradient (a tangent). Therefore, the value of $\hat{\nabla} J(\theta)$ at iteration n is given by

$$\hat{\nabla} J_n = \left[\hat{\nabla}_1 J_n \ldots \hat{\nabla}_p J_n\right], \tag{20.5}$$

where $\hat{\nabla}_i J_n$ can be calculated using forward differences as follows:

$$\hat{\nabla}_i J_n = \frac{\hat{J}\left(\theta_n + c_n e_i\right) - \hat{J}\left(\theta_n\right)}{c_n} \tag{20.6}$$

or central differences as follows:

$$\hat{\nabla}_i J_n = \frac{\hat{J}\left(\theta_n + c_n e_i\right) - \hat{J}\left(\theta_n - c_n e_i\right)}{2c_n}. \tag{20.7}$$

Here e_j denotes the ith unit vector (e.g., $e_i = (0,\ldots,0,1,0,\ldots 0)$) and c_n a small positive number that can take a different value at each iteration. The use of forward or central differences is driven by the trade-off between estimation bias and computational burden. The calculation based on central differences requires the simulation of $2p$ sets with $\theta_n \pm c_n e_i$ values, while the one based on forward differences requires only p+1 simulations. However, the estimators obtained using central differences usually have smaller bias than those obtained using forward differences, which often leads to a smaller number of iterations, n.

When finite differences are used to obtain the gradients for (20.3), the SA technique is called the KieferWolfowitz algorithm (Kiefer and Wolfowitz 1952). This algorithm has the following two important advantages with respect to other SA techniques: (1) implementation is straightforward due to its simplicity, and (2) it is not model-dependent (i.e., no explicit mathematical model of the system needs to be developed, which means that this technique can be applied to systems with any level of complexity). However, the KieferWolfowitz algorithm converges to the true local optimum only when very small c_n values (i.e., $c_n \to 0$) are used. The problem of using small c_n values is that the estimated gradients, $\hat{\nabla} J_n$, exhibit large variances that often slow down the convergence rate. This limitation has been addressed in some situations by using common random numbers (Fu 1994).

Simultaneous Perturbation (SP)

This technique as well as the other gradient estimation techniques in the remainder of this section was developed in response to the significant computational requirements of methods based on finite differences. SP uses the same conceptual framework of finite differences, but reduces the number of simulations required by perturbing all components of the decision vector simultaneously. Specifically, the value for any $\hat{\nabla}_i J_n$ can be obtained from the results of the simulations for just two sets of θ values, $(\theta_n + c_n \Delta_n)$ and $(\theta_n - c_n \Delta_n)$, with the following expression:

$$\hat{\nabla} J_n = \frac{\hat{J}\,(\theta_n + c_n \Delta_n) - \hat{J}\,(\theta_n - c_n \Delta_n)}{2 c_n \Delta_n}, \tag{20.8}$$

where $\Delta_n = \left(\Delta_{n1}, \ldots, \Delta_{np}\right)$ represents a vector of independent identically distributed (i.i.d.) random perturbations with zero mean. Though the elements of Δ_n may be assigned different kinds of distributions according to the specific characteristics of the problem at hand (Spall 1999). Sadegh and Spall (1998) showed that the optimal distribution for these elements, based on asymptotic distribution results, is a symmetric Bernoulli (i.e., the probability of success is 0.5). In a more recent development, deterministic perturbation sequences have been proposed to enhance the convergence rate of the stochastic approximation method based on simultaneous perturbation (SPSA) (Bhatnagar 2003).

Spall (1992) found that the SPSA method was superior (i.e., the difference between the actual minimum value and the estimated one was smaller for the same amount of computational effort) to the KieferWolfowitz algorithm for a fairly complicated numerical study. He also proved that both approaches have the same asymptotic convergence rate in spite of SPSA's significantly lower computational requirements at each iteration. In theory, the superiority of the SPSA method grows with the dimension p of the decision vector as the computational burden of SPSA is independent of its dimension. However, this potential for higher efficiency can only be realized if the number of iterations required to converge to the global optimum does not increase to a level that exceeds the savings obtained by reducing the number of simulations in each iteration. To realize as much of that potential as possible, not only must the selection of a_n in (20.3) be carefully made (as in any other SA technique), but also that of c_n and Δ_n in (20.8). Though the selection is problem-dependent and there are no universal rules, Spall (1998, 1999) provides some recommendations as a starting point.

Perturbation Analysis (PA)

Though the name may lead one to think that there is some kind of connection between PA and SP, these two techniques use completely different conceptual frameworks. PA does not explore through simulation the region around the decision vector θ_n to determine in which direction to move for the next iteration; instead, it determines such a direction by using only the output of the simulations with the current value of θ_n. Hence PA infers the behavior of the system with $\theta_n + \Delta\theta_n$, where $\Delta\theta_n$ is a small perturbation, from the information obtained with θ_n, and uses it to estimate the gradient for the next iteration. This may seem a little too "magical;" in the words of the developers of the technique (Ho and Cao 1991): "At first this may sound counterintuitive, stemming from the philosophical belief that one cannot get something for nothing. A more concrete and equally intuitive objection is that sample paths of the simulation under θ_n and $\theta_n + \Delta\theta_n$ will in general sooner or later become totally different as they evolve in time even for very small $\Delta\theta_n$". However, all the techniques belonging to the PA class accomplish this seemingly impossible objective by using some kind of sample path analysis. The pioneering technique in the field is known as infinitesimal perturbation analysis (IPA). The basic idea behind IPA is that it is possible to reconstruct a perturbed path from a nominal one by keeping track of the changes in the timing of events, if the sequence of events in the simulation do not change (the critical timing path stays constant). The relevance of IPA is that it is capable of simultaneously implementing such an accounting exercise for a multitude of perturbations, and when applicable is highly efficient (i.e., exhibits fast convergence) (Ho and Cao 1991). However, it is only suitable for continuous performance measures and requires complete knowledge of the underlying model, that is, an explicit model that relates inputs and outputs has to be available. In broad terms, the method can be described in three steps. The first step consists in developing a recursive (e.g., indexed by the number of customers arriving) explicit mathematical model that relates the outputs of the simulation that are part of the performance measure (e.g., inventory levels and total time of a customer in the system) with those random variables (e.g., quantity produced and service time at the teller) which depend directly on the decision variables (e.g., base stock and mean value of service time). Next, the model is differentiated with respect to the decision variables and the results are substituted into the function that represents the expected value of the performance measure gradient. Finally, the differentials of the random variables with respect to the decision variables are substituted based on the perturbation generation rule and the gradient is calculated using simulation outputs The perturbation generation rule allows one to obtain the changes on the random variable, ω, caused by changes in the decision variables in terms of the information collected from simulation runs with θ_n. The perturbation generation rule is given by

$$\frac{d\omega}{d\theta} = \frac{dF^{-1}(\theta,\xi)}{d\theta}, \tag{20.9}$$

where $F(\theta,\omega)$ is the cumulative distribution function of ξ with parameter θ, and random variable ω; and ξ is a random variable independent of θ (e.g., if ω is exponentially distributed, with mean θ, $\omega = F^{-1}(\theta,\xi) = -\ln(1-\xi)\cdot\theta$ and $\xi \sim U[0,1)$). However, from an implementation perspective, it is more convenient to use an equivalent formula that does not require the form of the inverse function:

$$\frac{d\omega}{d\theta} = \frac{dF(\theta,\omega)\,d\theta}{dF(\theta,\omega)\,d\omega}. \tag{20.10}$$

To clarify the method let us consider a very simple problem (Fu 1994) (For a more complex case in the context of inventory management refer to Tayur et al. (1999)). Find the mean service time θ of a first come first serve single server M/M/1 queue, which minimizes the sum of expected mean time in the system over a given number of customers served, L:

$$\min_{\theta\in\Theta} E\left[\frac{1}{N}\sum_{i=1}^{N} T_i\right], \tag{20.11}$$

where T_i is the time in the system for the ith customer and N is the number of customers served. Note that T_i is the only output of the simulation that is part of the performance measure, $L = \left[1/N\sum_{i=1}^{N} T_i\right]$. Under the conditions described, T_i satisfies the recursive Lindley equation

$$T_{i+1} = \omega_{i+1} + \begin{cases} T_i - A_{i+1} & \text{if } T_i \geq A_{i+1} \\ 0 & \text{if } T_i < A_{i+1} \end{cases}, \tag{20.12}$$

where ω_i is the service time for the ith customer and A_i is the interarrival time between the $(i-1)$ th and the ith customers. Notice that ω_i is the random variable that depends on the decision variable, which implies that (20.12) is the explicit model referred to in the first step of the method. Continuing with the method, (20.12) is differentiated to obtain

$$\frac{dT_{i+1}}{d\theta} = \frac{d\omega_{i+1}}{d\theta} + \begin{cases} dT_i/d\theta & \text{if } T_i \geq A_{i+1} \\ 0 & \text{if } T_i < A_{i+1} \end{cases}. \tag{20.13}$$

and expression (20.13) is substituted recursively into itself for each customer, and the resulting expressions in terms of $\mathrm{d}\omega_i \,/\, \mathrm{d}\theta$ are substituted into the expression for the expectation of the gradient of L, $E\left[\mathrm{d}L \,/\, \mathrm{d}\theta\right] = E\left[1\,/N\,\sum_{i=1}^{N}\mathrm{d}T_i \,/\, \mathrm{d}\theta\right]$, to obtain

$$E\left[\frac{\mathrm{d}L}{\mathrm{d}\theta}\right] = \sum_{s=1}^{S}\left[\frac{1}{n_s}\sum_{i=1}^{n_s}\sum_{j=1}^{i}\frac{\mathrm{d}\omega_{(j,s)}}{\mathrm{d}\theta}\right] \tag{20.14}$$

where S is the number of simulations, n_s the number of customers served in simulation s, and the (j,s) subscript denotes the jth customer in the sth simulation. Alternatively, the expectation can be estimated with the output from a single simulation. This is possible because the system is regenerative, which means that at random times $0 = t_0 < t_1 < t_2 < \cdots$ the future of the stochastic process becomes a probabilistic replica of itself. Therefore, instead of using results from multiple simulations, it is enough to extend the duration of only one run and split it into i.i.d. periods (regenerative cycles). In this case, the expression for the expectation is given by

$$E\left[\frac{\mathrm{d}L}{d\theta}\right] = \frac{1}{N}\sum_{m=1}^{M}\sum_{i=1}^{n_m}\sum_{j=1}^{i}\frac{\mathrm{d}\omega_{(j,m)}}{\mathrm{d}\theta}, \tag{20.15}$$

where M is the number of regenerative cycles, n_m the number of customers served in the mth regenerative cycle, and the (j,m) subscript denotes the jth customer in the mth busy period, i.e., $(j,m) = j + \sum_{i=1}^{m-1} n_i$.

The final step of the method requires the substitution of the random variable differentials. As ω_i is exponentially distributed, by using (20.10), it can be shown that $\mathrm{d}\omega_{(j,s)} \,/\, \mathrm{d}\theta \;=\; \omega_{(j,s)} \,/\, \theta_n$. Therefore, the final expression for the expectation of the gradient is given by

$$E\left[\frac{\mathrm{d}L}{\mathrm{d}\theta}\right] = \sum_{s=1}^{S}\left[\frac{1}{n_s}\sum_{i=1}^{n_s}\sum_{j=1}^{i}\frac{\omega_{(j,s)}}{\theta_n}\right]. \tag{20.16}$$

The use of IPA to estimate the performance measure gradients in the SA approach provides a framework that converges faster than that based on finite differences. However, it has the following two important drawbacks: (1) it is model-dependent and (2) it estimates $E\left[\mathrm{d}L \,/\, \mathrm{d}\theta\right]$ instead of the desired $\mathrm{d}E[L] \,/\, \mathrm{d}\theta$. Clearly, the second aspect is not an issue when the expectation (integration) and differentiation operators can be interchanged. However, in most cases this is only possible when L is almost surely continuous with respect to θ (Ho and Cao 1991). From an implementation perspective this means that it has to be possible to develop a "transformation" that allows one to represent the system in terms of random variables whose distributions do not depend on decision variables, and the performance function based on the transformation has to be continuous in θ for almost every ω.

In addition, as mentioned above, if the CPT changes or the performance measure is discontinuous, IPA is not valid. Though this problem has been addressed for some conditions using the same conceptual framework (Fu and Hu 1997), such extensions are not as straightforward as IPA.

Likelihood Ratio (LR)

As in the case of PA, LR methods, which are also known as score function (SF) methods, use only the output of the simulations with the current value of the decision vector θ_n to estimate the gradient of the expected value of the performance measure. The methods in this class require milder continuity requirements for the performance measure $L(\theta, \omega)$, than those stipulated by PA. This is possible as the gradient is calculated by differentiating the probability distribution function of the performance measure instead of the performance measure itself. However, applicability of this idea is limited to problems whose decision variables θ are parameters of the distributions that represent the uncertainty in the system. This means that decision variables that are not part of the characterization of the uncertainties, such as reordering points and safety inventory levels, cannot be part of θ. To overcome this limitation, Rubinstein and coworkers (Kleijnen and Rubinstein 1996; Rubinstein and Shapiro 1993) have proposed transformations for some types of problems that move the parameters lying outside the characterization of the probability distributions into them.

LR methods are strongly connected to the importance sampling concept, commonly used to derive estimators with a reduced variance. The basic idea behind LR methods can be illustrated by deriving the gradient of the performance measure for static systems (i.e., a system that does not evolve in time, such as reliability problems) with probability density functions that only depend on a single parameter. In this case, the expected value of the performance measure has the form

$$E[L] = \int L(\omega)\mathrm{d}F(\theta, \omega) = \int L(\omega) f(\theta, \omega)\mathrm{d}\omega, \tag{20.17}$$

where $F(\theta, \omega)$ represents the cumulative distribution of ω and $f(\theta, \omega)$ the density function. Differentiating (20.17) with respect to θ, interchanging integration and differentiation, and multiplying and dividing by $f(\theta, \omega)$ we obtain

$$\begin{aligned} \frac{\partial E[L]}{\partial \theta} &= \frac{\partial}{\partial \theta} \int L(\omega) f(\theta, \omega)\mathrm{d}\omega = \int L(\omega) \frac{\partial f(\theta, \omega)}{\partial \theta} \mathrm{d}\omega \\ &= \int L(\omega) \frac{\partial f(\theta, \omega)}{\partial \theta} \frac{f(\theta, \omega)}{f(\theta, \omega)} \mathrm{d}\omega = \int L(\omega) \frac{\partial \ln f(\theta, \omega)}{\partial \theta} f(\theta, \omega)\mathrm{d}\omega \\ &= E\left[L(\omega) \frac{\partial \ln f(\theta, \omega)}{\partial \theta}\right] = E\left[L(\omega) S^{(1)}(\theta, \omega)\right] \end{aligned} \tag{20.18}$$

where $S^{(1)}$ is called the efficient score function. Notice that multiplication and division by $f(\theta, \omega)$ are necessary to obtain an expression that has the mathematical form of an expectation. The expectation form is convenient because it allows the estimation of the desired quantity from simulated data by averaging it over a given set of realizations. Therefore, (20.18) can be expressed as

$$\frac{\mathrm{d}E[L]}{\mathrm{d}\theta} = \frac{1}{N}\sum_{i=1}^{N} L(\omega_i) S^{(1)}(\theta, \omega_i) \tag{20.19}$$

where N is the total number of simulations, and $L(\omega_i)$ and $S^{(1)}(\theta, \omega_i)$ are particular realizations of $L(\omega)$ and $S^{(1)}(\theta, \omega)$, respectively. Though (20.19) is readily implementable, it usually does not result in the fastest possible convergence. Equation (20.18) can be improved from a variance reduction perspective by exploiting the ideas behind importance sampling. Specifically, by multiplying and dividing the integrand in (20.17) by $g(\omega)$, where $g(\omega)$ is a probability distribution whose support (set of values of ω for which $g(\omega)$ is strictly greater than zero) is included in the support of $f(\theta, \omega)$ for every θ, the gradient of $E[L]$ can be expressed as

$$\frac{\mathrm{d}E[L]}{\mathrm{d}\theta} = \int L(\omega)\frac{\partial W(\theta, \omega)}{\partial \theta}\mathrm{d}G(\omega) = E\left[L(Z)\frac{\partial W(\theta, Z)}{\partial \theta}\right] \tag{20.20}$$

where $G(\omega)$ is the cumulative probability distribution of $g(\omega)$, $W(\theta, \omega) = f(\theta, \omega) \,/\, g(\omega)$, Z is a random variable with density $g(\omega)$, and $\partial W(\theta, Z) \,/\, \partial \theta = W(\theta, Z) S^{(1)}(\theta, Z)$. Notice that the change in the random variable is not more that a change in notation to emphasize that the expectation is with respect to $g(\omega)$, instead of $f(\theta, \omega)$. The estimator of the gradient in this case is given by

$$\frac{\mathrm{d}E[L]}{\mathrm{d}\theta} = \frac{1}{N}\sum_{i=1}^{N} L(Z_i) W(\theta, Z_i) S^{(1)}(\theta, Z_i). \tag{20.21}$$

Though it is a common strategy to select $g(\omega) = f(\theta_0, \omega)$ for some fixed value θ_0, the accuracy of the estimator is determined by its variance, which depends on $g(\omega)$. Therefore, the selection of $g(\omega)$ and the calculation of the estimator's variance are integral parts of this technique. Rubinstein and Shapiro (1993) provide a complete presentation of this methodology in the context of static as well as for dynamic systems (e.g., queuing networks).

In terms of strengths and weaknesses, the use of gradients estimated with LR for SA can result in very rapidly converging algorithms because LR exploits the structure of the performance measure. However, it requires complete knowledge of the density function of the uncertainties, careful selection of $g(\omega)$, and the satisfaction of certain regularity conditions which guarantee the interchangeability of the differentiation and integration operators in (20.18). Specifically, a function $h(\omega)$ with finite Lebesgue integral that satisfies

$$|L(\omega)\partial f(\theta,\omega) \,/\, \partial(\omega)| \leq h(\omega) \tag{20.22}$$

has to exist.

Finally, it is important to note that LR often works in systems where IPA fails and can be more easily extended to higher derivative estimates for higher order Newton-like methods than (20.3). However, when IPA works, the gradients usually have much lower variances than those obtained with LR methods (Fu 1994).

Frequency Domain Analysis

Frequency domain analysis (FDA) estimates the gradient of the expected value of the performance measure by using harmonic analysis. The method is based on the idea that the change in the magnitude of the performance measure, caused by perturbing the vector of decision variables θ, with sinusoidal functions allows the determination of the sensitivity of the system to each of those variables in a single simulation. In theory, the use of distinct frequencies for θ_i makes possible the estimation of each variable's contribution to the performance measure. In this method, θ is perturbed according to

$$\theta(t) = \overline{\theta} + \alpha \sin(wt), \tag{20.23}$$

where $\overline{\theta}$ is a vector of nominal values for the decision variables, α is the vector of oscillation amplitudes, w is the vector of oscillation frequencies, called the driving frequencies, and $t = 1, 2, \ldots, T$ is a "time" index. Notice that t is rarely the simulation time, instead it is a problem specific discrete label for the transient entities processed through the simulation (e.g., number of customers).

Conceptually, the method exploits the orthogonality (i.e., if g and f are two functions, they are orthogonal if $\int_a^b f(x)g(x)\mathrm{d}x = 0$) of the harmonic basis (i.e., sine and cosine), to isolate the impact of each decision variable on the performance measure gradient. Specifically, the method assumes that the performance measure can be approximated by a polynomial meta-model that can be transformed into a trigonometric (harmonic) one using (20.23). The polynomial meta-model is obtained by assuming that the relationship between L and $\theta(t)$ can be locally approximated around $\overline{\theta}$ by a second-order Taylor expansion:

$$\begin{aligned} L\left(t|\theta(t)\right) &= L(\overline{\theta}) + \sum_{j=1}^{p} \sum_{\tau=-\infty}^{\infty} g_j(\tau)(\theta_j(t-\tau) - \overline{\theta}_j) \\ &\quad + \sum_{j=1}^{p} \sum_{\tau=-\infty}^{\infty} g_{jj}(\tau)(\theta_j(t-\tau) - \overline{\theta}_j)^2 \end{aligned}$$

$$+\sum_{j=1}^{p-1}\sum_{m=j+1}^{p}\sum_{\tau=-\infty}^{\infty}\sum_{\nu=-\infty}^{\infty} g_{jm}(\tau,\nu)\left(\theta_j(t-\tau)-\overline{\theta}_j\right)$$
$$\times\left(\theta_m(t-\nu)-\overline{\theta}_m\right)+O\left(\left\|\theta(t)-\overline{\theta}\right\|_{\infty}^{3}\right)+\varepsilon\left(t|\theta(t)\right) \quad (20.24)$$

where p is the number of decision variables; $\|\cdot\|_{\infty}$ the infinity norm, $O(\cdot)$, the order of magnitude of the error generated by the truncation of the Taylor series, $\varepsilon\left(t|\theta\left(t\right)\right)$ represents the stochastic part of the model, and g are the so called memory filters that weight past values of $\theta(t)$. The Taylor expansion is advantageous as the summations of filters in each term of (20.24) can be obtained through regression analysis from simulation results, and can be associated with the gradient and higher order differentials of the expected value of the performance measure, $J(\theta)$. The relationship can be derived by setting $\theta(t)=\overline{\theta}$ in (20.24) and differentiating it with respect to the decision variables:

$$\sum_{\tau=-\infty}^{\infty} g_j(\tau)=\frac{\partial J(\theta)}{\partial\theta_j},$$
$$\sum_{\tau=-\infty}^{\infty} g_{jj}(\tau)=\frac{\partial^2 J(\theta)}{2\partial\theta_j^2},$$
$$\sum_{\tau=-\infty}^{\infty}\sum_{\nu=-\infty}^{\infty} g_{jm}(\tau,\nu)=\frac{\partial^2 J(\theta)}{\partial\theta_j\partial\theta_m} \quad (20.25)$$

It is important to note that the following three assumptions are behind this derivation: (1) $\varepsilon(t|\theta\left(t\right))$ has a stationary covariance (i.e., it is fixed for all t) with mean 0, 2. The summation of covariances from all the time periods is bounded, and (3) $J(\theta)$ is relatively smooth (i.e., twice continuously differentiable) (Ho and Cao 1991).

By substituting (20.23) into (20.24), the following form of the meta-model is obtained:

$$L\left(t\,|\theta\left(t\right)\right) = L(\overline{\theta})+\sum_{j=1}^{p}\sum_{\tau=-\infty}^{\infty} g_j\left(\tau\right)a_j\sin\left(w_j(t-\tau)\right)$$
$$+\sum_{j=1}^{p}\sum_{\tau=-\infty}^{\infty} g_{jj}(\tau)a_j^2\sin^2\left(w_j(t-\tau)\right)$$
$$+\sum_{j=1}^{p-1}\sum_{m=j+1}^{p}\sum_{\tau=-\infty}^{\infty}\sum_{\nu=-\infty}^{\infty} g_{jm}(\tau,\nu)\,a_j a_m\sin\left(w_j\left(t-\tau\right)\right)$$
$$\times\sin\left(w_m\left(t-\nu\right)\right)+O\left(\left\|\tilde{\theta}(t)-\overline{\theta}\right\|_{\infty}^{3}\right)+\varepsilon\left(t\,|\,\theta(t)\right) \quad (20.26)$$

where $\tilde{\theta}(t) - \overline{\theta} = \left(a_1 \sin(w_1 t), a_2 \sin(w_2 t), \ldots, a_p \sin(w_p t)\right)$. Equation (20.26) can be further manipulated using trigonometric identities and some algebra to derive the following more convenient form:

$$\begin{aligned} L\left(t|\tilde{\theta}(t)\right) &= B(0) + \sum_{j=1}^{p}\left[A(w_j)\sin(w_j t) + B(w_j)\cos(w_j t)\right] \\ &\quad + \sum_{j=1}^{p}\left[A(2w_j)\sin(2w_j t) + B(2w_j)\cos(2w_j t)\right] \\ &\quad + \sum_{j=1}^{p-1}\sum_{m=j+1}^{p}\left[A(w_j \pm w_m)\sin((w_j \pm w_m)t)\right] + B(w_j \pm w_m) \\ &\quad \times \cos((w_j \pm w_m)t) + O\left(\left\|\tilde{\theta}(t) - \overline{\theta}\right\|_{\infty}^{3}\right) + \varepsilon(t \mid \tilde{\theta}(t)) \end{aligned} \quad (20.27)$$

where the coefficients $A(w_j)$, $A(2w_j)$, $A(w_j \pm w_m)$, $B(0)$, $B(w_j)$, $B(2w_j)$, and $B(w_j \pm w_m)$ are in terms of the memory filters and sinusoidal functions; for instance, $A(w_j) = a_j \sum_{\tau=-\infty}^{\infty} g_j(\tau)\cos(w_j \tau)$. Therefore, by taking the limit as $w_j \to 0$ and using the results in (20.25) it can be proved that the gradient and higher order derivatives of the performance measure can be obtained from the estimates of these coefficients (Jacobson and Schruben 1999). Specifically, the estimate of the ith component of the gradient has the following form:

$$\frac{\partial E[L]}{\partial \theta_i} = \frac{2}{a_i T}\sum_{t=1}^{T} L\left(\theta\left(t\,|\tilde{\theta}(t)\right)\right)\sin(w_i t), \quad (20.28)$$

where $i = 1, \ldots, p$, and $t \mid \tilde{\theta}(t)$ denotes that $L(\theta)$ is sampled at each "time" t from a simulation in which the input for the decision vector is given by (20.23). From a theoretical perspective, the main advantage of the FDA method is the combination of model independence (excluding the indexing issue) and minimum simulation requirements. However, the determination of the specific values of the frequencies w_i is not a trivial task as they have to be selected in such a way that aliasing (i.e., the effect that causes different signals to become indistinguishable) is prevented, and the need to make them tend to 0 ($w \to 0$) translates into very long simulation horizons. In addition, the method is limited to systems in steady-state and exhibits an unavoidable trade-off between the variance of the gradient estimator (the larger α the better) and its bias (the smaller α the better) (Jacobson and Schruben 1999).

Finally, regarding the indexing issue, it is important to note that although simple indices based on the concept of transient entities processed are limited to very simple systems (Fu 1994), Hazra et al. (1997) have suggested a strategy to discretize the global simulation clock of the simulation and used it as a "time" index that can fit any kind of system. However, this approach can be difficult, if not impossible, to implement in some commercially available discrete event simulation software.

20.2.3.2 Response Surface Methodology (RSM)

RSM encompasses two types of strategies. The first consists of the use of regression techniques to construct an approximate functional relationship (meta-model) between the decision variables and the performance measure that fits the entire decision space, Θ, or a subset of Θ, and the subsequent use of optimization methods on the meta-model to analytically estimate an optimum (Wan and coworkers (2006) provide an example of this technique in the context of the pharmaceutical industry). The second strategy, known as sequential RSM, follows a philosophy similar to SA, consisting of three steps that are repeated iteratively until a convergence criterion is satisfied. First, a meta-model in the region surrounding the decision vector obtained in the previous iteration is constructed. Next, the meta-model is differentiated to obtain a functional form of the gradient, and substituted into

$$\theta_{n+1} = \theta_n - a_n \hat{\nabla} J(\theta_n). \tag{20.29}$$

Finally, the next iterate for the decision vector is computed from (20.29) through a line search. In spite of the similarities between sequential RSM and the SA methods discussed above, RSM differs due to its inability to mathematically show an asymptotical convergence, and the use of functional forms for the gradient instead of numerical values.

In the literature, the most commonly found RSM algorithm is a mix of the two philosophies described above, which uses a two-phase design of experiments based polynomial regression strategy (Fu 1994). In Phase I, first-order experimental designs (i.e., consider only linear (main) effects) are used iteratively until the linear response surface becomes inadequate (i.e., the interaction effects become larger than the main effects), while in Phase II, a quadratic response surface (fitted using second-order experimental designs) of the area identified in Phase I is used to analytically determine the optimum.

The most important considerations in the implementation of any RSM method are the inclusion of variance reduction techniques (e.g., common random variables, control random variables, etc.) and the selection of the experimental designs. Every type of design provides a different trade-off between variance (due to sample variation) and bias (due to poor model fit), which results in a particular performance of the algorithm. Jacobson and Schruben (1989), Safizadeh (1990) and Kleijnen in Banks (1998) provide an exhaustive set of references for RSM strategies, including algorithms that allow the inclusion of deterministic constraints.

The attractiveness of conventional RSM methods is rooted in their applicability to any kind of system. However, its "black box" nature that does not allow rigorous convergence analysis, its typical blind search of the solution space (which usually leads to the excessive use of simulation runs in unimportant areas), and the limited ability of low-degree polynomials to fit complex functions (which can provide poor results when the performance measure is represented by functions with sharp ridges and flat valleys (Azadivar 1999)) has limited its use to simple problems. Though the last two shortcomings have been addressed with the use of better model fitting methods (e.g., Wan et al. (2005) show that RSM may perform considerably better than SPA when support vector machines are used for model regression), these techniques are considerably more involved from a statistical perspective than traditional regression techniques.

20.2.3.3 Sample Path Optimization (SPO)

Conceptually, the methods in this class use an approach similar to the first type of RSM strategy described above. Specifically, the system is sampled multiple times, and the information collected is used to generate a functional approximation of the performance measure that is optimized using deterministic optimization tools. The main difference between SPO and RSM is that the latter uses regression techniques to obtain the functional approximation, whereas the first uses an explicit model obtained from first principles (like IPA and LR), or completely avoids the need for an explicit model by exploiting the structure of the problem. Though there are no specific rules to derive the explicit model (the key step in SPO), the basic idea is to be able to generate expressions in which the expected value of the performance measure is explicitly represented in terms of decision variables and random variables independent of the decision variables:

$$\hat{J}(\theta) = \frac{1}{N} \sum_{i=1}^{N} h(\theta, \xi_i), \tag{20.30}$$

where N is the number of simulations and ξ_i the ith realization of the θ independent random variables. In some cases, a model with this kind of structure can be directly derived, but in most of real-world problems that is not possible. Therefore, similar to that in IPA, transformations have to be implemented to obtain objective functions with underlying random variables independent of θ. This means that any model suitable for IPA can be solved with SPO. Alternatively, for some problems in which the effect of the decision variables only enters the problem through the distribution of the underlying random variables, approximations based on likelihood ratios and importance sampling have been developed to obtain the required functional form of the performance measure (e.g., Banks (1998), Rubinstein and Shapiro (1993), and Shapiro (1996) and references therein). Finally, in some cases it is possible to develop routines that do not require the derivation of an explicit model. A very simple example of this subgroup of problems is the allocation of a fixed amount of buffer space among a group of servers such that the time to overflow the system is maximized. The SPO strategy is to run the simulation multiple times with the buffers between the servers unconstrained until the total availability of buffer space is exhausted and then to chose the most frequent allocation. Healy and Fu (1992, 1997) and Healy and Schruben (1991) provide a complete presentation of this example and more involved problems, including cases with discrete decision spaces.

In general, SPO has several important advantages, which are as follows: (1) the strategy that uses explicit models can deal with problems in which the decision variables are subject to constraints of the type $E\,[k(\theta)] < 0$, where $k(\theta)$ can be derived in the same way as the performance measure, (2) it can be easily implemented in commercial simulators, due to its modularity (i.e., first simulation and second optimization), and (3) it can be applied to some problems with discrete decision spaces. However, it also has considerable limitations such as the following: (1) it is restricted to systems that have reached a steady state, (2) it usually requires a lot more evaluations than SA (Azadivar 1999; Fu and Healy 1992), (3) similar to IPA and LR, it is problem-specific (due to the need explicit models), (4) its effectiveness is highly dependent on the ability to develop explicit models that allow the calculation of first- and second-order derivatives (usually required by deterministic nonlinear optimization techniques), and (5) The solutions provided by SPO methods may not be optimal as this technique solves the problem: $E[\min_{\theta\in\Theta} L(\theta,\omega)]$ instead of the desired problem: $\min_{\theta\in\Theta} E\,[L(\theta,\omega)]$.

20.2.4 Metaheuristics

A metaheuristic is a general framework consisting of black-box procedures that can be applied to different kinds of problems without significant changes. The applicability of these techniques to problems with continuous or discrete decision spaces is dictated by the particular structure of the method and the way in which it is adapted to the problem at hand. Metaheuristics are the dominant strategies used in commercial optimization software for simulation optimization (Fu 2002), as well as in the solution of large-scale problems. This is due to the fact that many of the methods mentioned above are model-dependent and/or require a high level of expertise for their implementation. In this section, we provide a short description of the metaheuristics available for simulation optimization and a set of relevant references. Special attention is given to the most commonly used methods, genetic algorithms (GA), tabu search (TS), and scatter search (SS).

20.2.4.1 Pattern Search

Pattern search methods are sequential algorithms that move from iteration to iteration based on some characteristic or pattern in the observations, instead of relying on gradients or randomization. Conceptually, these techniques try to use some form of memory but at a very basic level and are mentioned here mostly for historical purposes. The most important techniques in this class are:

1. The Hooke and Jeeves method (1961), which is based on the idea that if a direction has produced an improvement in the estimated performance measure, then one should continue moving in that direction,
2. The simplex method (Jacobson and Schruben 1989) and references therein), not to be confused with the classical algorithm for linear programming, which compares the estimated performance measures from an initial set of possible solutions, eliminates the worst performer, and replaces it by a new one determined by the centroid of the remaining solutions, and
3. The complex method (Azadivar (1999) and references therein), which is the simplex method modified to handle constrained problems.

20.2.4.2 Genetic Algorithms (GA)

The set of GA is one of a class of algorithms inspired by the biological principles of evolution known as evolutionary algorithms. This technique searches for the optimal decision vector, θ, based on a performance measure (fitness function, in GA terminology), by iteratively updating a population of good decision vectors. The decision vector associated with each member of the population is encoded as a string of symbols (genes) that form a chromosome, and is generated from the members of the population in the previous iteration through random genetic operators (Sect. 20.3.2 provides a specific example). In general, the algorithm can be described as follows:

Step 1. Initialize the population with a set of members generated using previous knowledge of the problem and/or a random process, and estimate their performance according to the chosen fitness function. The number of simulations required for the estimation of the performance measure is determined by a stopping criterion such as confidence intervals, convergence efficiency, or computational budget.

Step 2. Create new chromosomes (reproduction) by using genetic operators. The best known operators are crossover and mutation. Crossover consists in the random exchange between two members (parents) of part of their chromosomes, and mutation is a random alteration of some of the genes in a given member.

Step 3. Estimate the fitness function of the newly created chromosomes and select from this group and the population in the previous iteration (generation) the members of the next generation based on the superiority of their estimated performance measures.

Step 4. Check for "convergence": stop or go to Step 2. Genetic algorithms are not guaranteed to converge; therefore, the definite termination condition is usually specified as a maximal number of generations or an acceptable fitness level for the best individual.

It is important to note that though the general structure of the algorithm always follows these steps, specific procedures for encoding, initialization, reproduction, and selection can be chosen based on the problem at hand to enhance the performance of the algorithm. Reeves and Rowe (2003) provide an excellent guide to the GA technique.

20.2.4.3 Scatter Search (SS)

Similar to GA, scatter search is a population-based evolutionary algorithm. However, it uses a completely different approach for the generation of new population members (decision vectors). Specifically, the members of the population (called the reference set) are combined in a systematic way, instead of randomly. The combination strategies are generalized forms of linear combinations that consider at least all pairs of members in the reference set. SS also differs from GA in the size of the population; reference sets tend to be small compared with the populations used in GA. In general, SS algorithms can be described as follows (Laguna 2003):

Step 1. Generate a starting set of decision vectors as diverse as possible and apply heuristics to these vectors in an attempt to improve their performance. From the resulting population, choose the vectors with the best estimated performance to be part of the initial reference set. Notice that the notion of "best" is not only limited to the value of the performance measure; a solution may be added to the reference set if it improves the diversity of the set, regardless of the performance measure.

Step 2. Create new members consisting of systematic generalized linear combinations of two or more members of the current reference set.

Step 3. Apply the heuristic process used in Step 1 to improve the members created in Step 2.

Step 4. Extract a collection of the "best" improved solutions from Step 3 and use them to replace the worst performing members in the reference set. If the reference set does not change, stop. Otherwise go to step 2.

Laguna (2003) provides a complete presentation of this methodology, including the different member combination strategies available and their suitability according to the type of problem at hand.

20.2.4.4 Tabu Search (TS)

As in random search (Sect. 20.2.2.1), TS explores the solution space by moving successively from one feasible solution to a neighboring one. However, instead of using probabilistic arguments to guide the search, it uses a strategy based on the ideas of adaptive memory and responsive exploration. This means that TS redefines the solution neighborhood at each iteration based on the information previously collected to avoid visiting already explored areas or areas characterize by poor performance.

The method accomplishes this by selecting certain attributes or combination of attributes that cannot be part of the new solutions (are tabu). The memory structure used in TS uses two types of information, namely, explicit and attributive. The explicit part is captured by recording good solutions or highly attractive but unexplored neighborhoods of those good solutions; while the attributive part records information about solution attributes that change in going from one solution to another (e.g., increase in the risk level of a portfolio of projects). Glover and Laguna (1997) provide an exhaustive presentation of the concepts and applications of TS.

20.2.5 Other Methods

In addition to the methods just described, there are simulation-optimization techniques which, in spite of not being widely used at present, could be viable options for specific problems or could become so as they are further developed. This group of methods includes neural networks (Glover et al. 1999), branch and bound for discrete systems (Norkin et al. 1998), nested partitions (Shi and Olafsson 2000), and the collection of algorithms known as model-based methods. Model-based methods, instead of generating actual solutions, construct probability distributions for the solution space that can be used to estimate where the best solutions are located. The following techniques belong to this group: swarm intelligence, estimation of distribution algorithms (EDAs), the cross-entropy (CE) method and model reference adaptive search (Fu et al. 2005).

20.3 Two Industrial Problems

In this section two case studies based on actual industrial problems are presented to illustrate the potential of simulation optimization as a decision support tool. The presentation of each case study includes a short description of the problem, a discussion supporting the selection of a specific simulation-optimization method, and a summary of the implementation of the method and the results obtained.

20.3.1 Inventory Management

Any enterprise that manufactures products faces uncertainties in a range of factors such as demand, prices and availability of raw materials, production lead times, currency exchange variability, etc. Some of these factors directly affect the profitability of the enterprise by limiting the operating margins, while others have an indirect impact such as inability to meet customer needs or the accumulation of excess inventory. The inability to meet customer needs results in both loss of "here and now" and long-term profit as poorly served customers may not come back. Therefore, in any industrial setting, customer satisfaction level (CSL; the expected value of the ability to meet customer demand) is recognized as an important performance measure. A high level of customer satisfaction can be achieved by maintaining high inventories to hedge against uncertainty (e.g., fluctuations in demand or availability of raw material). However, additional inventory entails increased holding cost (including opportunity cost of invested capital and warehouse space). Decision makers attempt to minimize the impact of this trade-off between customer satisfaction and inventory holding cost on the profitability of the enterprise by specifying different safety stock levels for each product across the supply chain.

A great deal of work has been done to develop analytical strategies that allow the determination of the optimal allocation of safety stocks (Jung et al. 2004). However, those strategies fall short when the enterprise manufactures multiple products that share production facilities with limited capacity and scheduling constraints, experience significant queue effects and lead times, and faces uncertain demand from several customers. This kind of environment is common to many industrial and pharmaceutical manufacturers, including the particular case we were confronted with.

We looked into the operation of the supply chain of a major US polyethylene producer whose main source of uncertainty is demand and who wanted to reach specific levels of customer satisfaction. The company uses a decision-making strategy in which CSLs are specified by top management according to certain strategic considerations and aggregated data, while minimization of the cost of delivering the products is left to planners and the people in operations. Thus, the problem to be addressed is the determination of how much, where and when to produce, and the safety stock levels for each product. The company has two production sites, which have different layouts and capacities that directly supply the seven sales regions into which USA is divided. It produces five types of polyethylene (A, B, C, D, and E) in ten different grades (0–9), in two types of packaging (box or bag) for a total of 100 (5 types $\times$ 10 grades $\times$ 2 packages) stock keeping units (SKUs). The demand for each SKU is characterized as a normal distribution, whose mean value changes on a weekly basis according to internal forecasting models.

The first step in developing a simulation-optimization strategy for a problem is to determine which group of techniques (continuous or discrete, and small or large number of feasible solutions) is appropriate according to the characteristics of the solution space and the limitations of each method. In this case, it is clear that the inventory levels can take any integer value, which due to the combinatorial nature of the problem rules out any of the algorithms that fall under the "small number of discrete feasible solutions" class. The discrete character of the decision space could be also used to disregard the methods for systems with continuous decision spaces, but the levels of inventories required by an operation like the one here considered are high enough that the use of such techniques in conjunction with rounding needs to be considered as they may provide near optimal solutions. In the remaining class, "large number of discrete feasible solutions," ordinal optimization and SPO can be ignored. The first method is disregarded due to the size of the decision space, and the second due to our inability to develop an explicit model that characterizes the performance measure (customer service level). The lack of an explicit model is also the reason to disregard the methods for continuous decision spaces SA+PA and SA+LR. The SA+FDA also has to be disregarded as it requires the system to reach steady state, a condition that is not achievable in this problem due to the seasonal demand fluctuations. This leaves us with the following set of potential solution methods: random search, SA+FD, SA+RSM, metaheuristics, and any of these four methods in combination with one of the methods under the "small number of discrete feasible solutions" class.

Once the options have been narrowed down based on the solution space and the limitations of the methods, the selection process has to be driven by the strengths of the remaining options with regard to the problem at hand. For the problem considered here it is important to understand the connection between CSL, defined as service level, the production strategy, and the safety stock level of a product under uncertain demand. Over a given range of demand variance there are three possible operational regimes. In regime I, production facilities have sufficient spare capacity to cope with any change in demand. Therefore, in this regime, a relatively low or even zero safety stock level may be sufficient to achieve the desired customer satisfaction. In regime II, the production capacity maybe quite strained when the demand for different products spike at some point in time. In this regime, if there is not enough safety stock, the CSL for some products sharing production facilities may fail to reach their target values. Finally, in regime III, the capacity available cannot satisfy the combined expected demands of the different products. In this regime, the safety stock and production resources must be assigned strategically to meet the demands of some customers in preference to the others. For the problem at hand, the sites owned by the company have enough capacity to operate in regimes I and II. This means that no customer priority has to be used to allocate production capacity and therefore any desired level of inventory for each product is realizable. Notice that this condition and the hierarchical decision-making strategy used by the company (i.e., tactical decisions such as service level dictate operational goals) allows for the use of a decomposition strategy. The idea is to use a multilevel optimization approach instead of an integrated approach in which production quantities along the time horizon and safety inventory levels are considered together in a massive stochastic program. The multilevel strategy is composed of a simulation-optimization strategy that determines the optimal stock levels based on long-term customer satisfaction, and deterministic (expected values) rolling horizon planning and scheduling optimizations, embedded in the simulation, which allocate production resources by minimizing cost. Figure 20.2a illustrates the "outer" optimization on the safety stock levels, and Fig. 20.2b the inner problem in which the simulation of the system constantly interacts with the planning and scheduling models in a rolling horizon fashion. The planning model is formulated as an LP for a 3-month horizon that takes into account production, transportation, inventory holding, and shortage costs; whereas the schedule is generated for 40 days using the VirtECS scheduling software (Advanced Process Combinatorics Inc., 2004).

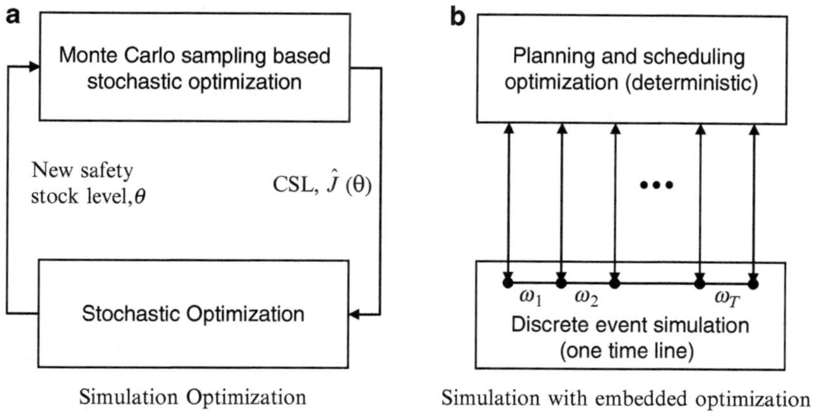

Fig. 20.2 Configuration of simulation and optimization strategies

The outer optimization problem can be mathematically represented as follows:

$$\min_{\theta} J(\theta) = \sum_{i=1}^{100} \mu_i \left| L_i^{\Delta}(\theta) \right| \tag{20.31}$$

subject to

$$L_i(\theta) + L_i^{\Delta}(\theta) \geq L_i^{\text{target}} \quad \forall i \tag{20.32}$$

where μ_i is the penalty for missing the target CSL for product i, $\theta = (\theta_{11}, \ldots, \theta_{is})$ is the decision vector including the safety stock levels of each product i in each production facility s, $L_i(\theta)$ is the CSL (expected value of the probability of fully meeting every demand for product i), and $L_i^{\Delta}(\theta)$ is the deviation with respect to the target CSL, L_i^{target}. Notice that the CSLs are the only variables in the objective function (20.31). This condition combined with the fact that the level of customer satisfaction is a monotonic increasing function of θ (the larger the safety stock the higher the customer satisfaction), implies that the best local adjustment to each decision variable has to be inversely proportional to the magnitude of the penalty resulting from deviating from the target CSL, $\mu_i \left| L_i^{\Delta}(\theta) \right|$. Though the adjustment is local in the sense that it does not consider the effects and constraints associated with the embedded planning and scheduling problems, the monotonic nature of CSL(θ) guarantees convergence to a global optimal solution if the estimator of $L_i(\theta)$ is unbiased. Therefore, an efficient simulation-optimization strategy for this problem should be capable of exploiting the fact that if the performance measure improves in a particular direction, then one should continue moving in that direction. The only method in the shortlisted group capable of doing that is the pattern search metaheuristic. This metaheuristic was then selected and implemented in the recursive algorithm below:

Step 1. Initialize safety stock levels, θ_{is}^{n}, where $n = 0$ for all i and s
Step 2. Estimate $J_i\ (\theta^n)$ and $L_i^{\Delta}\ (\theta^n)$ using simulation
Step 3. Check for convergence of the estimated performance measure - if $\left|\hat{J}\ (\theta^n) - \hat{J}\ (\theta^{n-1})\right| \leq \varepsilon$ stop. Otherwise, continue
Step 4. Calculate the new safety stock level

$$\theta_{is}^{n+1} = \theta_{is}^{n} + \alpha\beta_{is}\left(\mu_i\ \hat{L}_i^{\Delta}\ (\theta)\right)$$

where α is a step size factor that can be adjusted by trial and error, and β_{is} is the distribution factor, which represents the ratio of product supply from each production site in the previous iteration
Step 5. Check for convergence, if $\left|\hat{J}\ (\theta^n) - \hat{J}\,(\theta^{n-1})\right| \leq \varepsilon$ stop. Otherwise, go to Step 2.

The algorithm was used to solve a case in which the coefficient of variation of the different demands was assumed to be 30%. Figure 20.3, where $Ai - x$ denotes the

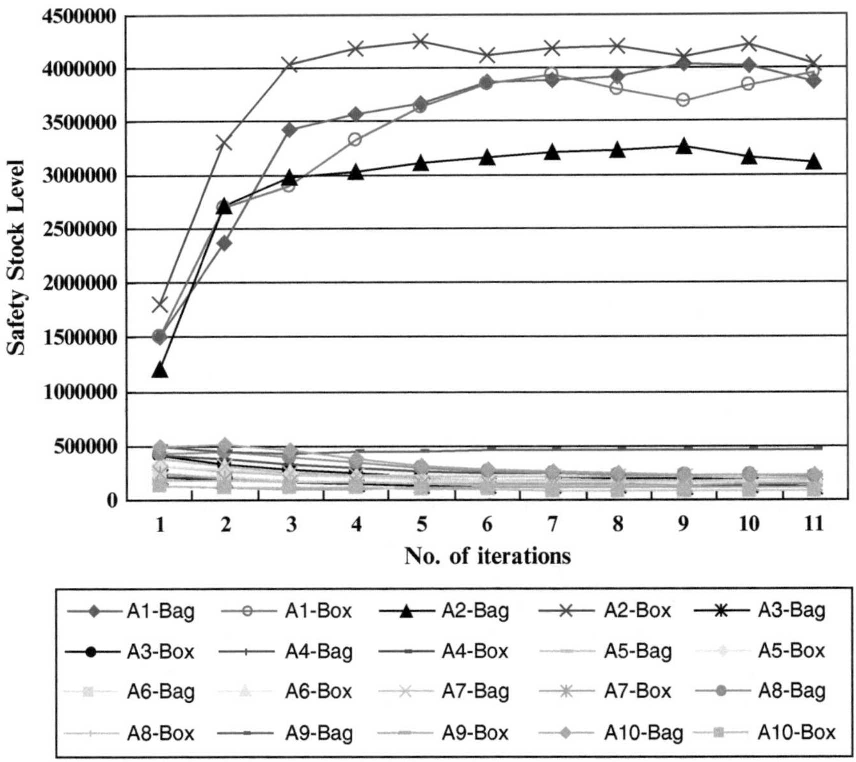

Fig. 20.3 Trajectory of safety stock levels of A type products at plant 1 as the optimization proceeds

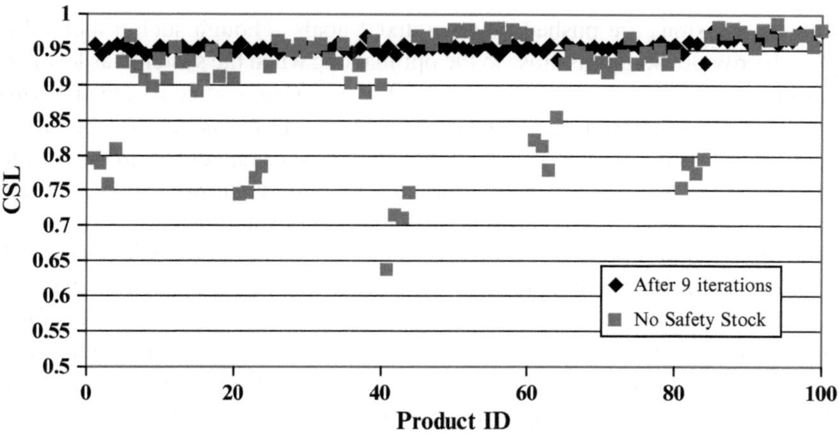

Fig. 20.4 Comparison of the Customer Satisfaction Levels (CSL) with and without safety stock

final product of type A and grade i packaged in facility x, shows the iterative process for the safety stocks of the type A products in one of the production facilities when the starting values are zero. It is important to note that four products, A0-bag, A0-box, A1-box, and A1-bag, make up 80% of the demand for type A and the rest, from A2-bag to A9-box, make up the remaining 20% (the same is true for the rest of the polyethylene types). As expected, the products with a larger demand need higher safety stocks in order to cope with the 30% variability. Figure 20.4 summarizes the estimated CSLs without safety stock, $\hat{L}_i\ (\theta^0)$, and after nine iterations of the algorithm $\hat{L}_i\ (\theta^9)$, showing the efficiency of the computational framework in solving the outer optimization problem. Notice that the change is more pronounced in the group of major products (the first four type-grade-package triplets that take 80% of the demand) which go from the 0.6–0.8 range to levels very close to the 0.95 target, and in some of the minor products that show lower CSLs in the presence of safety stock. The latter counterintuitive result can be attributed to the additional strain imposed on production by the increase in the safety stock levels of the major products.

20.3.2 Portfolio Selection of New Compounds to be Developed in the Pharmaceutical Industry

The hierarchical decision-making strategy mentioned in the previous problem is not only used when dealing with tactical (e.g., set service level) and operational decisions (e.g., set safety stock levels), but also when strategic decisions need to be made. This means that strategic decisions are usually made based on aggregated data, representing the capacity of the organization at lower levels, and those decisions are pushed down as fixed goals. Though such a hierarchical approach provides solutions close to the optimal one when the system has low levels of uncertainty; that is rarely the case in highly uncertain and constrained environments. A good solution at the tactical and operational levels can be obtained for the specific goals dictated by the strategic decision makers, but the quality of these goals with respect to the attainable optimum remains unknown. Such a situation does arise in the context of pharmaceutical products development.

The selection of a portfolio of drugs to be developed is a strategic decision that has uncertain financial implications on the order of billions of dollars which are only realized over the long term (decades). This decision-making process is further complicated by the low probability of success of new compounds (high attrition rates), unpredictable changes in regulations, technologies and health trends, dependencies between projects (drugs) from a variety of perspectives, uncertainties in terms of duration and cost in each stage of the development process, and limited human and capital resources. In addition, as in any other kind of portfolio there are solutions that have the same exposure to risk, but a different level of rewards. Therefore, the problem to be addressed consists in choosing a prioritized portfolio on the reward-risk-efficient frontier (i.e., the portfolios with the maximum level of rewards for a given level of risk) for the level of risk considered acceptable by the enterprise. Notice that such a selection, in addition to being influenced by all the uncertainties mentioned above, is constrained by the limited amount of renewable (e.g., equipment) and nonrenewable resources (e.g., budget for clinical trials), and the strategies used by decision makers at the tactical and operational levels to allocate them. Therefore, the optimization strategy has to be able to capture the impact of these constraints on the behavior of the system.

There are three major stages in the lifecycle of a new drug, which are: discovery, development and commercialization. The discovery stage tends to be highly unpredictable and case specific, while the other two follow a well-defined path. This situation, coupled with the limited availability of the renewable and nonrenewable resources necessary to simultaneously develop all the compounds rated as promising by discovery (lead molecules), has directed all the attention, from a modeling and optimization perspective, to the development and commercialization stages. Once a molecule is promoted to the status of a lead molecule, it goes through a network of tasks similar to that shown in Fig. 20.5. Though small variations in the drug development lifecycle occur from company to company, Fig. 20.5 depicts a fairly realistic model of what happens in this kind of industry. In the figure, tasks are represented by rectangles, while decision points are presented as diamonds. In general, these tasks can be classified into two groups, evaluation and commercialization, and manufacturing. The purpose of the tasks in the first group (upper row in Fig. 20.5) is to determine the safety and efficacy of the drug and satisfy all requirements to make it commercially available if these two aspects are favorable. The second group (lower row in Fig. 20.5) encompasses all the tasks necessary to scale up the laboratory procedures into commercial size manufacturing facilities. A complete explanation of the activities covered by each task can be found in Blau et al. (2004). We examined

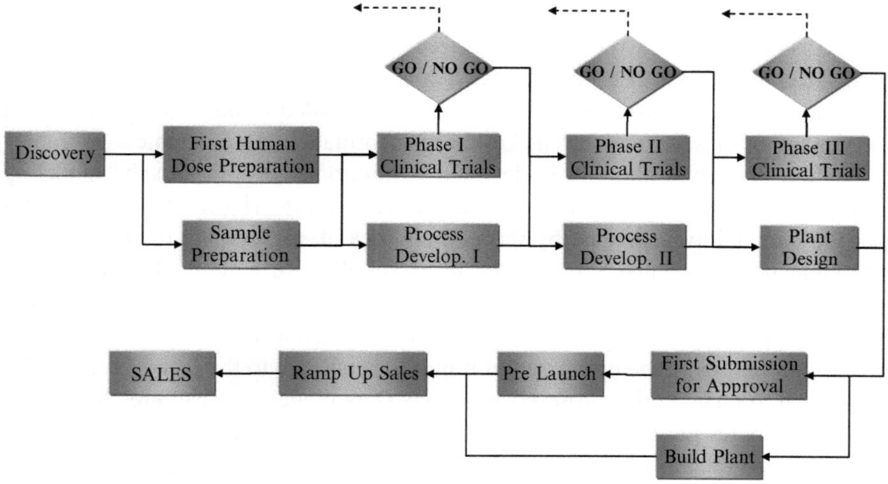

Fig. 20.5 Schematic of a pharmaceutical R&D pipeline model

the portfolio of a US-based pharmaceutical company that had a total of nine lead compounds with a 20-year patent protection whose development process can be approximated by the model in Fig. 20.5.

As in the previous case study, the first step in determining a suitable simulation-optimization method for the problem is to narrow down the options based on the characteristics of the solution space. The fact that a group of compounds and their corresponding priorities need to be selected from a finite set eliminates all techniques under the "continuous decision space" class. The number of potential strategies can be further reduced by taking into account the combinatorial nature of the problem. The nine compounds and their priorities can be mixed and matched into almost one million different permutations, ruling out any strategy in the "small number of feasible solutions" class, and ordinal optimization. SPO is also disregarded due to our inability to develop a model that characterizes the performance measure in terms of the decision variables. This leave us with the following set of potential strategies: random search, metaheuristics, and any of these two in combination with one of the methods under "small number of discrete feasible solutions." The final choice of a method is driven by the strengths of the remaining options relative to the problem at hand. In portfolio problems, the desired outcome is not just a single optimal point but a characterization of the efficient reward-risk frontier. Hence, the use of a random search method, though feasible, would be highly inefficient as it would be necessary to run it multiple times to construct the efficient frontier. With this point of departure, a trial and error process was implemented to find a metaheuristic capable of solving the problem. Tabu search was examined, but was discarded as it was not possible to stop the method from getting stuck in certain areas of the solution space. In the second iteration, a GA was tested with excellent results. This method was selected not only because it provided the desired output (i.e., an efficient frontier), but also because it allowed a natural representation for the decision variable, a vector of prioritized projects. There is currently no formal structured way to select a metaheuristic; it is more an art than a science. Though some directions are provided in the references provided in this chapter, the black box nature of these approaches makes their performance unpredictable.

Before describing the GA in detail, it is important to point out the modeling assumptions and simplifications used in the case study. The model only considers the uncertainty generated by the probabilities of success/failure at the end of the clinical trials, which are modeled by Bernoulli distributions. The rest of the potentially uncertain variables (costs, sales per year, and task durations) are approximated with their mean values. These model simplifications were necessary not due to limitations in the optimization framework but due to the lack of reliable information to characterize those uncertainties. The model also captures four types of dependencies between projects, which are as follows: (1) resource dependencies, (2) manufacturing cost dependencies, (3) financial return dependencies, and (4) technical success dependencies. Learning curve effects frequently lead to resource dependencies. A common example occurs when the development times are reduced for the trailing candidate of two functionally similar drug types. Cost dependencies occur when the combined cost of a development activity for two drug candidates is less than the sum of their individual costs because of resource sharing. For example, it may be possible to use the same production facilities for two chemically or biologically similar drug candidates. Financial return dependencies occur when there is synergism or competition in the marketplace. For example, cannibalization can occur when two drug candidates are aimed at developing products that compete with each other in the marketplace. Technical dependencies occur when the technical success or failure of a drug candidate affects the probability of technical success of an as-yet-untested trailing drug candidate. For example, two drug candidates might be developed to release an active ingredient in a controlled fashion. If the precedent candidate is successful, the probability of success of the as-yet-untested second candidate will be increased. The specific realizations of the dependencies considered in this problem are described by Blau et al. (2004).

The final consideration for the model is the representation of the strategy used to allocate and reallocate resources after a project failure and at the end of each year. The resource allocation policies were obtained following the framework conceived by Varma (2005), which uses a simulation of the task network in Fig. 20.5 and an observer. The simulation includes an integer program (IP) for short-term resource allocation that can assign three different levels of resources (associated with specific durations) to each task, namely, most likely (ML) value, and a certain percentage below and above of the most likely value. The observer tallies each of the outputs from the IP and determines the allocation policies by relating the most frequent decisions observed to the corresponding realization of the pipeline state space. This minimizes the size of the state space (composition of the portfolio and development stage of each compound) while keeping as much information as possible by breaking it into drug states $S_i = \{\mathrm{DS}_i, \mathrm{NLEV}_i, \mathrm{NHEV}_i\}$, where DSi is the development stage of drug i, NLVEi, the number of drugs having lower expected value than drug i in the same development stage, and NHEVi is the number of drugs having a higher expected value than drug i in the same development stage.

The optimization problem to be solved by the GA can be mathematically expressed for the specific case in which rewards are measured by the expected positive net present value (EPNPV) and risk by the probability of losing money ($P\,(\mathrm{NPV}(\theta) < 0)$) as:

$$\min_{\theta} J\,(\theta) = \mathrm{EPNPV}\,(\theta) \tag{20.33}$$

subject to

$$P(\mathrm{NPV}(\theta) < 0) < \beta \tag{20.34}$$

where NPV is the net present value, θ is the prioritized portfolio of drugs, and β is the upper bound for the probability of losing money that needs to be varied in the $(0, 1)$ interval to obtain the efficient frontier.

The GA is encoded such that each gene contains the number of a drug candidate (with 0 indicating that a project was not selected), and its position in the chromosome represents the priority given to the compound. For example, the chromosome 203000400 corresponds to a portfolio that consists of three compounds: 2, 3 and 4, of which compound 2 has the highest priority. A fitness function Zk of the following form is used:

$$Z_k = \alpha\left(\frac{\mathrm{EPNPV}_k - \mathrm{EPNPV}_{\mathrm{min}}}{\mathrm{EPNPV}_{\mathrm{max}} - \mathrm{EPNPV}_{\mathrm{min}} + \gamma}\right) + (1-\alpha)\left(\frac{\mathrm{Risk}_{\mathrm{max}} - \mathrm{Risk}_k}{\mathrm{Risk}_{\mathrm{max}} - \mathrm{Risk}_{\mathrm{min}} + \gamma}\right) \tag{20.35}$$

where EPNPVmin and EPNPVmax are the minimum and maximum expected positive net present values, respectively, in the current population; $\mathrm{Risk}_{\mathrm{min}}$ and $\mathrm{Risk}_{\mathrm{max}}$ are the maximum and minimum risk levels in the current population, measured as the probability of losing money, γ is a small positive number that prevents division by zero, and α weights the present value vs. the level of risk in a convex linear combination. The GA proceeds to find chromosomes that improve the fitness function by generating new chromosomes through the use of some genetic operators and estimating the fitness function values using simulations of the model in Fig. 20.5 (Zapata et al. 2008). Notice that the NPV and PNPV for each simulation can be calculated from the discounted development costs accumulated as the compounds move through the pipeline and the returns realized when the drug hits the market.

The GA was run for different percentages of the amount of resources that can be allocated above or below the ML value, including a base case in which reallocation of resources was implemented based on the original priorities given by the GA sequence (i.e., no information about realized uncertainties and the state of the pipeline is used) and no flexibility in the quantity of resources was considered. Figure 20.6 presents the results for the base case. All the points corresponding to the maximum EPNPV for a given level of risk are linked to form an approximate reward-risk-efficient frontier. At first sight, it looks like its shape reflects the

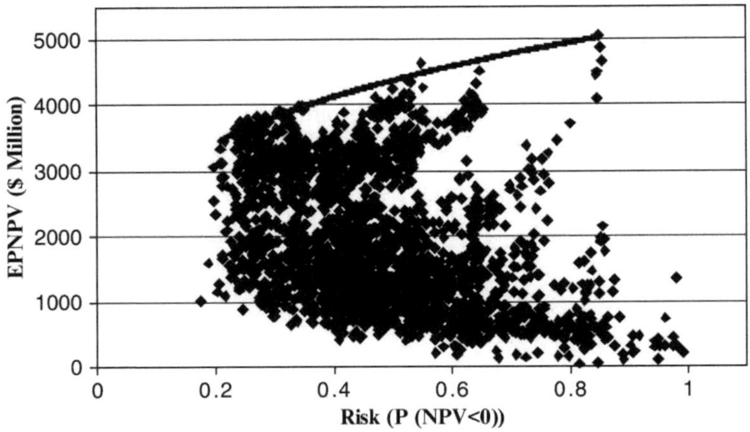

Fig. 20.6 Efficient frontier base case

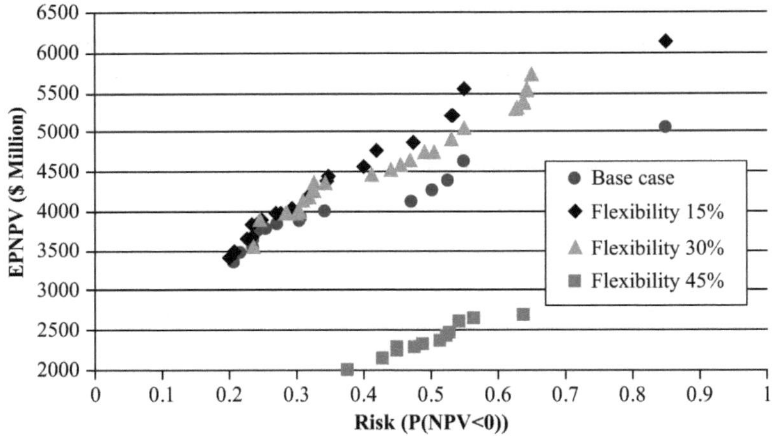

Fig. 20.7 Efficient frontiers for with and without dynamic resource allocation

general form found by Markowitz in financial portfolios (Luenberger 1998), but a closer look reveals that the direct correlation between rewards and risk is violated in the middle section; the depression in the efficient frontier implies that there are efficient portfolios which bear more risk but result in lower rewards). This counterintuitive result was not observed when flexibility in allocating resources was considered. Figure 20.7 shows the dominating portfolios for the three different dynamic resource allocation cases considered. The compositions of the portfolios on the efficient frontier in the base case and those with dynamic resource allocation are remarkably different in the region where the depression is found. These results are significant as they reveal that it is not possible to decouple the strategic and tactical decision-making processes without becoming substantially suboptimal. Therefore simulation-optimization strategies like the one here presented are essential to be able to accurately model the system and optimize it to improve the quality of the decisions made.

However, it is important to highlight that the computational burden required to solve the problem was very high. It took between 3 and 5 days on a 64 bit Sun-Sparc Ultra-Enterprise with 25, 400 MHz processors, and 8 M CPU cache per processor to run each case. This burden is bearable if we consider that these kinds of decisions are commonly made every 6 months, but would be unacceptable in a decision-making process that has to be repeated with a much greater frequency.

20.4 Conclusions

A summary of the simulation-optimization methods currently available was provided. Our discussion was organized by classifying methods into those intended for small discrete, large discrete, and continuous decision spaces. In the first category, the number of feasible solutions is small and therefore the focus of the methods is on the exhaustive comparison of possible solutions through statistical inference. The size of large discrete and continuous decision spaces shifts the focus to methods based on search algorithms, with the exception of ordinal optimization that uses statistical inference to exhaustively compare possible solutions. The majority of the methods in these two categories are model-independent and therefore can be applied to any problem. However, this very advantage is responsible for slow convergence rates (random search), unpredictable convergences (RSM and metaheuristics), and high computational burden (SA with FD and RSM). Though in principle two methods, SA with SP (for any system) and SA with FDA (for systems in steady state), are immune to these issues, the difficulty in parameterizing them results for the most part in slow convergence rates during execution. By contrast, the three model-dependent methods, SA with PA and LR and some types of SPO, tend to exhibit a faster convergence but are applicable to a limited number of very simple problems. At the end, the selection of a method for most problems is more an art than a science and requires a significant amount of trial and error. This situation has led practitioners to mainly use metaheuristics (especially GA and SS) and RSM due to their flexibility to accommodate any type of problem and their relative simplicity.

The chapter also presented two industrial case studies, in which simulation-optimization methods were successfully used. The case studies served to illustrate not only the implementation of a few methods, but also to highlight some of the considerations that are relevant in the selection of a method. From these case studies and the initial discussion in this chapter is evident that simulation optimization is the right tool to support several complex industrial decision-making processes.

However, in general, simulation optimization requires a significant level of technical sophistication from the user, especially in the area of statistics, as well as large amounts of computational resources.

References

Alrefaei MH, Andradottir S (1999) Simulated annealing algorithm with constant temperature for discrete stochastic optimization. Manage Sci 45:748–764.

Andradottir S (1998) Review of simulation optimization techniques. Presented at 1998 Winter Simulation Conference, Washington, DC, USA.

Azadivar F (1999) Simulation optimization methodologies. Presented at 1999 Winter Simulation Conference, Phoenix, AZ, USA.

Banks J (1998) Handbook of simulation:principles, methodology, advances, applications, and aractice. Wiley, New York.

Banks J (2005) Discrete-event system simulation. 4th edn. Pearson Prentice Hall, Upper Saddle River, NJ.

Bhatnagar S et al (2003) Two-timescale simultaneous perturbation stochastic approximation using deterministic perturbation sequences. ACM Trans Model Comput Simul 13:180–209.

Blau GE et al (2004) Managing a portfolio of interdependent new product candidates in the pharmaceutical industry. J Prod Innov Manage 21:227–245.

Fu MC (1994) Optimization via simulation: a review. Ann Oper Res 53:199–247.

Fu MC (2002) Optimization for simulation: theory vs. practice. INFORMS J Comput 14:192–215.

Fu MC, Healy KJ (1992) Simulation optimization of inventory systems. Presented at 1992 Winter simulation conference, Arlington, VA, USA.

Fu MC, Healy KJ (1997) Techniques for optimization via simulation: an experimental study on an (s, S) inventory system. IIE Trans (Institute of Industrial Engineers) 29:191–199.

Fu M, Hu J-Q (1997) Conditional Monte Carlo: gradient estimation and optimization, applications. Kluwer Academic, Boston.

Fu MC, Glover FW, April J (2005) Simulation optimization: a review, new developments, and applications. Presented at 2005 winter simulation conference, Orlando, FL, USA.

Glover F, Laguna M (1997) Tabu Search. Boston, MA: Kluwer Academic.

Glover F, Kelly JP, Laguna M (1999) New advances for wedding optimization and simulation. Presented at 1999 winter simulation conference, Phoenix, AZ, USA.

Hall JD, Bowden RO (1996) Simulation optimization for a manufacturing problem. Presented at Southeastern simulation conference, Huntsville, AL, USA. Society for Computer Simulation.

Hazra MM, Morrice DJ, Park SK (1997) Simulation clock-based solution to the frequency domain experiment indexing problem. IIE Trans (Institute of Industrial Engineers), 29, 769–782.

Healy, K. and Schruben, L.W (1991) Retrospective simulation response optimization. Presented at 1991 winter simulation conference, Phoenix, AZ, USA.

Henderson, S.G.and Nelson, B.L (2006) Handbooks in operations research and management science: simulation. Elsevier, Amsterdam.

Ho Y-C, Cao X-R (1991) Perturbation analysis of discrete event dynamic Systems. Kluwer Academic, Boston, MA.

Ho YC et al (1992) Optimizing discrete event dynamic systems via the gradient surface method. Presented at 30th IEEE conference on decision and control part 1 (of 3), Brighton, England.

Hooke R, Jeeves TA (1961) Direct search solution of numerical and statistical problems. J ACM 8:212.

Jacobson SH, Schruben LW (1989) Techniques for simulation response optimization. Oper Res Lett 8:1–9.

Jacobson SH, Schruben L (1999) Harmonic analysis approach to simulation sensitivity analysis. IIE Trans (Institute of Industrial Engineers) 31:231–243.

Jacobson SH, Buss AH, Schruben LW (1991) Driving frequency selection for frequency domain simulation experiments. Oper Res 39:917.

Jung JY et al (2004) A simulation based optimization approach to supply chain management under demand uncertainty. Comput Chem Eng 28:2087–2106.

Kiefer JC, Wolfowitz, J (1952) Stochastic estimation of the maximum of a regression function. Bull Am Math Soc 58:465–465

Kleijnen JPC, Rubinstein RY (1996) Optimization and sensitivity analysis of computer simulation models by the score function method. Eur J Oper Res 88:413–427.

Kushner HJ, Yin G (2003) Stochastic approximation and recursive algorithms and applications. 2nd edn. Applications of mathematics, vol. 35. Springer, New York xxii, p. 474.

Laguna M, Martâi R (2003) Scatter search: methodology and implementations in C. Kluwer Academic, Boston, MA.

Luenberger DG (1998) Investment science. Oxford University Press, New York.

Nelson BL, Matejcik FJ (1995) Using common random numbers for indifference-zone selection and multiple comparisons in simulation. Manage Sci 41:1935.

Norkin VI, Pflug GC, Ruszczynski A (1998) A branch and bound method for stochastic global optimization. Math Program 83:425–450.

Nozari A, Morris JS (1984) Application of an optimization procedure to steady-state simulation. Presented at 1984 winter simulation conference, Dallas, TX, USA.

Pichitlamken J, Nelson BL (2003) A combined procedure for optimization via simulation. ACM Trans Model Comput Simul 13:155–179.

Reeves CR, Rowe JE (2003) Genetic algorithms: principles and perspectives: a guide to GA theory. Kluwer Academic, Boston, MA.

Rubinstein RY, Shapiro A (1993) Discrete event systems: sensitivity analysis and stochastic optimization by the score function method. Wiley Chichester, NY.

Sadegh P, Spall JC (1998) Optimal random perturbations for stochastic approximation using a simultaneous perturbation gradient approximation. IEEE Trans Autom Control 43:1480–1484.

Safizadeh MH (1990) Optimization in simulation. Current issues and the future outlook. Naval Res Logist 37:807–825.

Shapiro A (1996) Simulation based optimization. Presented at 1996 winter simulation conference, Coronado, CA, USA.

Shi L, Olafsson S (2000) Nested partitions method for global optimization. Oper Res 48:390–407.

Spall JC (1992) Multivariate stochastic approximation using a simultaneous perturbation gradient approximation. IEEE Trans on Autom Control 37:332–341.

Spall JC (1998) Implementation of the simultaneous perturbation algorithm for stochastic optimization. IEEE Trans Aerospace Electron Syst 34:817–823.

Spall JC (1999) Stochastic optimization and the simultaneous perturbation method. Presented at 1999 winter simulation conference, Phoenix, AZ, USA.

Swisher JR et al (2004) A survey of recent advances in discrete input parameter discrete-event simulation optimization. IIE Trans 36:591–600.

Tayur S, Ganeshan R, Magazine M (1999) Quantitative models for supply chain management. Kluwer Academic, Boston, MA.

Varma VA (2005) Development of computational models for strategic and tactical management of pharmaceutical R&D pipelines. PhD Thesis, Purdue University.

Varma VA et al (2007) Enterprise-wide modeling and optimization an overview of emerging research challenges and opportunities. Comput Chem Eng 31:692–711.

Wan X, Pekny JF, Reklaitis GV (2005) Simulation-based optimization with surrogate models application to supply chain management Comput Chem Eng 29:1317–1328.

Wan X, Pekny JF Reklaitis GV (2006) Simulation based optimization for risk management in multi-stage capacity expansion. Presented at computer-aided chemical engineering, 21: 16th European symposium on computer aided process engineering and 9th International symposium on process systems engineering.

Zapata JC, Varma VA, Reklaitis GV (2008) Impact of tactical and operational policies in the selection of a new product portfolio. Comput Chem Eng 32:307–319.

Author Index

C

I

J

K

M

Subject Index

T

U

MIX
Papier aus verantwortungsvollen Quellen
Paper from responsible sources
FSC® C105338

If you have any concerns about our products,
you can contact us on
ProductSafety@springernature.com

In case Publisher is established outside the EU,
the EU authorized representative is:
Springer Nature Customer Service Center GmbH
Europaplatz 3, 69115 Heidelberg, Germany

Printed by Libri Plureos GmbH
in Hamburg, Germany